# Advanced Mathematics for Engineers and Scientists with Worked Examples

*Advanced Mathematics for Engineers and Scientists with Worked Examples* covers the core to advanced topics in mathematics required for science and engineering disciplines. It is primarily designed to provide a comprehensive, straightforward, and step-by-step presentation of mathematical concepts to engineers, scientists, and general readers. It moves from simple to challenging areas, with carefully tailored worked examples of different degrees of difficulty. Mathematical concepts are deliberately linked with appropriate engineering applications to reinforce their value and are aligned with topics taught in major overseas curriculums.

This book is written primarily for students at Levels 3 and 4 (typically in the early stages of a degree in engineering or a related discipline) or for those undertaking foundation degree, Higher National Diploma (HND), International Foundation Year (IFY), and International Year One (IYO) courses with maths modules. It is organised into four main parts:

- Trigonometry
- Advanced Mathematics
- Matrices and Vectors
- Calculus

Each of the above four parts is divided into two or more chapters, and each chapter can be used as a stand-alone guide with no prior knowledge assumed. Additional exercises and resources for each chapter can be found online. To access this supplementary content, please go to www.dszak.com.

**Shefiu Zakariyah** is a Chartered Engineer (CEng) and a senior member of IEEE (SMIEEE), with over a decade of experience in teaching, content development and assessment, and technical training and consulting.

# Advanced Mathematics for Engineers and Scientists with Worked Examples

Shefiu Zakariyah

Routledge
Taylor & Francis Group
LONDON AND NEW YORK

Cover image: © Shutterstock

First published 2025
by Routledge
4 Park Square, Milton Park, Abingdon, Oxon OX14 4RN

and by Routledge
605 Third Avenue, New York, NY 10158

*Routledge is an imprint of the Taylor & Francis Group, an informa business*

© 2025 Shefiu Zakariyah

The right of Shefiu Zakariyah to be identified as author of this work has been asserted in accordance with sections 77 and 78 of the Copyright, Designs and Patents Act 1988.

All rights reserved. No part of this book may be reprinted or reproduced or utilised in any form or by any electronic, mechanical, or other means, now known or hereafter invented, including photocopying and recording, or in any information storage or retrieval system, without permission in writing from the publishers.

*Trademark notice*: Product or corporate names may be trademarks or registered trademarks, and are used only for identification and explanation without intent to infringe.

*British Library Cataloguing-in-Publication Data*
A catalogue record for this book is available from the British Library

*Library of Congress Cataloging-in-Publication Data*
Names: Zakariyah, Shefiu, author.
Title: Advanced mathematics for engineers and scientists with worked examples / Shefiu Zakariyah.
Description: Abingdon, Oxon ; New York, NY : Routledge, 2024. | Includes bibliographical references.
Identifiers: LCCN 2023044241 | ISBN 9781032665108 (hardback) | ISBN 9781032663272 (paperback) | ISBN 9781032665122 (ebook)
Subjects: LCSH: Engineering mathematics. | Mathematical physics.
Classification: LCC TA330 .Z185 2024 | DDC 620.001/51–dc23/eng/20240226
LC record available at https://lccn.loc.gov/2023044241

ISBN: 978-1-032-66510-8 (hbk)
ISBN: 978-1-032-66327-2 (pbk)
ISBN: 978-1-032-66512-2 (ebk)

DOI: 10.1201/9781032665122

Typeset in Times
by Deanta Global Publishing Services, Chennai, India

Access the Support Material: www.routledge.com/9781032665108

*To the stimuli*
*Parents (late), sister (late), brother (late), and grandparents (late)*
*and*
*Wife, children, siblings, and friends*

# Contents

About the Author .................................................................. xiv
How to Use This Book ............................................................. xv
Acknowledgements ............................................................... xvi
Abbreviations ...................................................................... xvii
Greek Alphabet .................................................................... xviii
Mathematical Operators and Symbols ............................................ xix
Physical Quantities ................................................................ xxi
Prefixes Denoting Powers of Ten .................................................. xxii

**Chapter 1**    Trigonometric Functions I ............................................ 1

     1.1    Introduction ............................................................... 1
     1.2    Trigonometric Ratios ..................................................... 1
     1.3    Ratios of Special Angles ................................................. 7
     1.4    Inverse Trigonometric Ratios ............................................ 10
     1.5    Quadrants ................................................................. 17
     1.6    Graphs of Trigonometric Functions .................................... 35
           1.6.1    Sine Function .................................................. 35
           1.6.2    Cosine Function ................................................ 37
           1.6.3    Tangent Function .............................................. 40
     1.7    Chapter Summary ....................................................... 44
     1.8    Further Practice ......................................................... 46

**Chapter 2**    Trigonometric Functions II ........................................... 47

     2.1    Introduction ............................................................. 47
     2.2    Reciprocal of Trigonometric Functions ................................ 47
           2.2.1    Sec Function .................................................... 47
           2.2.2    Cosec Function ................................................. 49
           2.2.3    Cot Function .................................................... 49
     2.3    Inverse Trigonometric Functions ....................................... 56
           2.3.1    Arcsin Function ................................................ 56
           2.3.2    Arccos Function ................................................ 58
           2.3.3    Arctan Function ................................................ 58
     2.4    Transformation of Trigonometric Functions ........................... 62
           2.4.1    Translation ..................................................... 62
           2.4.2    Stretch .......................................................... 67
     2.5    Chapter Summary ....................................................... 78
     2.6    Further Practice ......................................................... 79

**Chapter 3**    Trigonometric Identities I ............................................ 80

     3.1    Introduction ............................................................. 80
     3.2    Fundamentals ........................................................... 80
     3.3    The Addition Formulas ................................................. 86
     3.4    Multiple Angles ......................................................... 96

|  |  | 3.4.1 | The Double-Angle Formulas .................................... | 96 |
|  |  | 3.4.2 | The Triple-Angle Formulas ..................................... | 101 |
|  |  | 3.4.3 | The Half-Angle Formulas ...................................... | 103 |
|  | 3.5 | Chapter Summary ........................................................ | | 115 |
|  | 3.6 | Further Practice ........................................................... | | 116 |

**Chapter 4** Trigonometric Identities II ................................................. 117

|  | 4.1 | Introduction ............................................................... | 117 |
|  | 4.2 | The Factor Formulas .................................................... | 117 |
|  | 4.3 | The $R$ Addition Formulas ............................................. | 128 |
|  | 4.4 | Chapter Summary ........................................................ | 147 |
|  | 4.5 | Further Practice ........................................................... | 148 |

**Chapter 5** Solving Trigonometric Equations ........................................ 149

|  | 5.1 | Introduction ............................................................... | | 149 |
|  | 5.2 | Methods of Solving Trigonometric Equations ........................... | | 149 |
|  |  | 5.2.1 | Sketching a Graph ............................................. | 150 |
|  |  | 5.2.2 | CAST Diagram ................................................. | 152 |
|  |  | 5.2.3 | Quadrant Formula .............................................. | 153 |
|  | 5.3 | Changing the Interval of Transformed Functions ........................ | | 159 |
|  |  | 5.3.1 | Type 1: $y = \sin ax$, $y = \cos ax$, $y = \tan ax$ ..................... | 159 |
|  |  | 5.3.2 | Type 2: $y = \sin(x \pm b)$, $y = \cos(x \pm b)$, $y = \tan(x \pm b)$ ........ | 165 |
|  |  | 5.3.3 | Type 3: $y = \sin(ax \pm b)$, $y = \cos(ax \pm b)$, $y = \tan(ax \pm b)$ ..... | 167 |
|  | 5.4 | Trigonometric Equations Involving Identities ........................... | | 173 |
|  |  | 5.4.1 | The Fundamentals .............................................. | 173 |
|  |  | 5.4.2 | Addition Formulas ............................................. | 178 |
|  |  | 5.4.3 | Multiple Angle ................................................. | 179 |
|  |  | 5.4.4 | The Factor Formulas ........................................... | 191 |
|  |  | 5.4.5 | $R$ Formulas ..................................................... | 196 |
|  | 5.5 | Chapter Summary ........................................................ | | 201 |
|  | 5.6 | Further Practice ........................................................... | | 202 |

**Chapter 6** The Binomial Expansion .................................................. 203

|  | 6.1 | Introduction ............................................................... | 203 |
|  | 6.2 | What Is a Binomial Expression? ........................................ | 203 |
|  | 6.3 | Pascal's Triangle ......................................................... | 204 |
|  | 6.4 | Factorials .................................................................. | 212 |
|  | 6.5 | Notation for Combination ............................................... | 214 |
|  | 6.6 | Relationship between Pascal's Triangle and Combination ............. | 215 |
|  | 6.7 | Binomial Expansions for $(1 \pm ax)^n$ .................................. | 216 |
|  |  | 6.7.1 | When $n$ Is a Positive Integer ................................. | 216 |
|  |  | 6.7.2 | When $n$ Is Negative ........................................... | 221 |
|  |  | 6.7.3 | When $n$ Is a Fraction .......................................... | 222 |
|  |  | 6.7.4 | Validity of Binomial Expression ............................... | 223 |
|  | 6.8 | Binomial Expansion: $(p + q)^n$ ........................................ | 227 |
|  |  | 6.8.1 | Method 1 ....................................................... | 228 |
|  |  | 6.8.2 | Method 2 ....................................................... | 228 |

Contents

|     | 6.9  | Approximation | 234 |
|     | 6.10 | Chapter Summary | 236 |
|     | 6.11 | Further Practice | 238 |

**Chapter 7** Partial Fractions .................................................. 239

|     | 7.1 | Introduction | 239 |
|     | 7.2 | What Is a Partial Fraction | 239 |
|     | 7.3 | Resolving into Partial Fractions | 241 |
|     | 7.4 | Types of Partial Fractions | 242 |
|     |     | 7.4.1 Non-Repeated Linear Factors | 242 |
|     |     | 7.4.2 Repeated Linear Factors | 249 |
|     |     | 7.4.3 Irreducible Non-Repeated Non-Linear Factors | 256 |
|     |     | 7.4.4 Irreducible Repeated Non-Linear Factors | 269 |
|     |     | 7.4.5 Improper Fractions | 272 |
|     | 7.5 | Chapter Summary | 279 |
|     | 7.6 | Further Practice | 279 |

**Chapter 8** Complex Numbers I ................................................. 280

|     | 8.1 | Introduction | 280 |
|     | 8.2 | Complex Numbers | 280 |
|     |     | 8.2.1 $j$ Operator | 280 |
|     |     | 8.2.2 Imaginary Number | 287 |
|     | 8.3 | Fundamental Operations of Complex Numbers | 288 |
|     |     | 8.3.1 Addition and Subtraction | 288 |
|     |     | 8.3.2 Multiplication | 289 |
|     |     | 8.3.3 Division | 291 |
|     |     |     8.3.3.1 Division by Real Numbers | 291 |
|     |     |     8.3.3.2 Division by Complex Numbers | 292 |
|     | 8.4 | Forms of Complex Numbers | 297 |
|     |     | 8.4.1 Cartesian Form | 297 |
|     |     | 8.4.2 Polar Form | 298 |
|     |     | 8.4.3 Exponential Form | 303 |
|     | 8.5 | Conversion between Various Forms | 304 |
|     | 8.6 | Chapter Summary | 308 |
|     | 8.7 | Further Practice | 310 |

**Chapter 9** Complex Numbers II ................................................ 311

|     | 9.1 | Introduction | 311 |
|     | 9.2 | Argand Diagrams | 311 |
|     | 9.3 | Equations Involving Complex Numbers | 313 |
|     | 9.4 | Phasor and $j$ Operator | 318 |
|     | 9.5 | De Moivre's Theorem | 329 |
|     |     | 9.5.1 Roots of Complex Numbers | 331 |
|     |     | 9.5.2 Powers of Trigonometric Functions and Multiple Angles | 336 |
|     | 9.6 | Logarithm of a Complex Number | 340 |
|     | 9.7 | Applications | 341 |
|     | 9.8 | Chapter Summary | 347 |
|     | 9.9 | Further Practice | 350 |

| | | | |
|---|---|---|---|
| **Chapter 10** | Matrices | | 351 |
| | 10.1 | Introduction | 351 |
| | 10.2 | Matrix Notation | 351 |
| | 10.3 | Types of Matrices | 352 |
| | 10.4 | Matrix Operation | 357 |
| | | 10.4.1 Addition | 357 |
| | | 10.4.2 Subtraction | 359 |
| | | 10.4.3 Scalar-Matrix Multiplication | 360 |
| | | 10.4.4 Matrix-Matrix Multiplication | 364 |
| | 10.5 | Determinant and Its Properties | 372 |
| | | 10.5.1 Single Matrix (1 by 1) | 372 |
| | | 10.5.2 Two-by-Two Matrix | 373 |
| | | 10.5.3 Three-by-Three Matrix | 375 |
| | | 10.5.4 Determinants of a Higher-Order Matrix | 378 |
| | | 10.5.5 Properties of Determinants | 379 |
| | 10.6 | Inverse of a Square Matrix and Its Properties | 381 |
| | | 10.6.1 Inverse of a $2 \times 2$ Matrix | 381 |
| | | 10.6.2 Inverse of Any Square Matrix | 384 |
| | 10.7 | Solving a System of Linear Equations | 389 |
| | | 10.7.1 Cramer's Rule | 390 |
| | |     10.7.1.1 Simultaneous Equations in Two Unknowns | 390 |
| | |     10.7.1.2 Simultaneous Equations in Three Unknowns | 393 |
| | | 10.7.2 Matrix Form | 398 |
| | |     10.7.2.1 Simultaneous Equations in Two Unknowns | 399 |
| | |     10.7.2.2 Simultaneous Equations in Three Unknowns | 401 |
| | 10.8 | Chapter Summary | 405 |
| | 10.9 | Further Practice | 407 |
| **Chapter 11** | Vectors | | 408 |
| | 11.1 | Introduction | 408 |
| | 11.2 | Scalar and Vector | 408 |
| | 11.3 | Free Vector | 409 |
| | 11.4 | Vector Notation | 410 |
| | 11.5 | Collinearity | 410 |
| | 11.6 | Zero Vector | 410 |
| | 11.7 | Negative Vector | 410 |
| | 11.8 | Equality of Vectors | 411 |
| | 11.9 | Unit Vector | 413 |
| | 11.10 | Cartesian Representation of Vectors | 413 |
| | 11.11 | Magnitude of a Vector | 414 |
| | 11.12 | Position Vectors | 418 |
| | 11.13 | Addition and Subtraction of Vectors | 421 |
| | | 11.13.1 Addition of Vectors | 422 |
| | | 11.13.2 Subtraction of Vectors | 423 |
| | 11.14 | Multiplication | 429 |
| | | 11.14.1 Multiplication by a Scalar | 429 |
| | | 11.14.2 Scalar (or Dot) Product | 430 |

Contents

|  |  |  |
|---|---|---|
| 11.15 | Vector Equation of a Line | 443 |
| 11.16 | Intersection of Two Lines | 449 |
| 11.17 | Angle between Two Lines | 454 |
| 11.18 | Chapter Summary | 457 |
| 11.19 | Further Practice | 460 |

**Chapter 12** Fundamentals of Differentiation .................. 461

- 12.1 Introduction ........................ 461
- 12.2 What Is Differentiation? ............ 461
- 12.3 Limits ............................. 462
- 12.4 Gradient ........................... 463
  - 12.4.1 Straight Line ................ 463
  - 12.4.2 Non-Linear Graph or Curve .... 465
    - 12.4.2.1 Drawing a Tangent ....... 465
    - 12.4.2.2 Numerical Method ........ 467
    - 12.4.2.3 Analytical Method ....... 470
    - 12.4.2.4 Derivative of a Function . 471
    - 12.4.2.5 Notations for a Derivative of a Function ... 472
- 12.5 Standard derivatives ............... 478
  - 12.5.1 Power (or Polynomial) Functions ... 478
  - 12.5.2 Exponential Functions ........ 493
  - 12.5.3 Logarithmic Functions ........ 496
  - 12.5.4 Trigonometric Functions ...... 498
- 12.6 Higher-Order Derivatives ........... 503
- 12.7 Chapter Summary ................... 507
- 12.8 Further Practice ................... 510

**Chapter 13** Advanced Differentiation .................. 511

- 13.1 Introduction ....................... 511
- 13.2 Rules of Differentiation ........... 511
  - 13.2.1 Chain Rule ................... 512
  - 13.2.2 Product Rule ................. 524
  - 13.2.3 Quotient Rule ................ 533
  - 13.2.4 Combined Rule ................ 541
- 13.3 Implicit Differentiation ........... 543
- 13.4 Logarithmic Differentiation ........ 548
- 13.5 Parametric Differentiation ......... 554
- 13.6 Chapter Summary ................... 563
- 13.7 Further Practice ................... 564

**Chapter 14** Applications of Differentiation .................. 565

- 14.1 Introduction ....................... 565
- 14.2 Rate of Change ..................... 565
  - 14.2.1 Velocity ..................... 565
  - 14.2.2 Acceleration ................. 566
  - 14.2.3 Jerk ......................... 566
- 14.3 Tangent and Normal Line Equation ... 566

|   |   |   |
|---|---|---|
| | 14.4 | Increasing and Decreasing Functions . . . . . . . . . . . . . . . . . . . . . . . . . . . . . . . . . . . 579 |
| | 14.5 | Stationary Points . . . . . . . . . . . . . . . . . . . . . . . . . . . . . . . . . . . . . . . . . . . . . . . . . . . . 583 |
| | | 14.5.1 Minimum and Maximum Points . . . . . . . . . . . . . . . . . . . . . . . . . . . . 584 |
| | | 14.5.2 Inflexion . . . . . . . . . . . . . . . . . . . . . . . . . . . . . . . . . . . . . . . . . . . . . . . . 597 |
| | 14.6 | Curve Sketching . . . . . . . . . . . . . . . . . . . . . . . . . . . . . . . . . . . . . . . . . . . . . . . . . . . 602 |
| | 14.7 | Chapter Summary . . . . . . . . . . . . . . . . . . . . . . . . . . . . . . . . . . . . . . . . . . . . . . . . . . 611 |
| | 14.8 | Further Practice . . . . . . . . . . . . . . . . . . . . . . . . . . . . . . . . . . . . . . . . . . . . . . . . . . . 612 |

**Chapter 15** Indefinite Integration . . . . . . . . . . . . . . . . . . . . . . . . . . . . . . . . . . . . . . . . . . . . . . . . . . . . 613

|   |   |   |
|---|---|---|
| | 15.1 | Introduction . . . . . . . . . . . . . . . . . . . . . . . . . . . . . . . . . . . . . . . . . . . . . . . . . . . . . . 613 |
| | 15.2 | Indefinite Integral . . . . . . . . . . . . . . . . . . . . . . . . . . . . . . . . . . . . . . . . . . . . . . . . . 613 |
| | | 15.2.1 Constant of Integration . . . . . . . . . . . . . . . . . . . . . . . . . . . . . . . . . . . . 614 |
| | | 15.2.2 Integrating to Determine the Equation of a Curve . . . . . . . . . . . . . . 616 |
| | | 15.2.3 Standard Integral Forms . . . . . . . . . . . . . . . . . . . . . . . . . . . . . . . . . . . 616 |
| | |     15.2.3.1 Integrating Power of $x$ or $x^n$ ($n \neq -1$) . . . . . . . . . . . . . . 616 |
| | |     15.2.3.2 Integrating Rational $\frac{1}{x}$ and $\frac{1}{ax+b}$ . . . . . . . . . . . . . . . . 632 |
| | |     15.2.3.3 Integrating Exponential Functions $e^x$ . . . . . . . . . . . . . . 634 |
| | |     15.2.3.4 Integrating Trigonometric Functions . . . . . . . . . . . . . . . . 636 |
| | 15.3 | Chapter Summary . . . . . . . . . . . . . . . . . . . . . . . . . . . . . . . . . . . . . . . . . . . . . . . . . . 644 |
| | 15.4 | Further Practice . . . . . . . . . . . . . . . . . . . . . . . . . . . . . . . . . . . . . . . . . . . . . . . . . . . 646 |

**Chapter 16** Definite Integration . . . . . . . . . . . . . . . . . . . . . . . . . . . . . . . . . . . . . . . . . . . . . . . . . . . . . 647

|   |   |   |
|---|---|---|
| | 16.1 | Introduction . . . . . . . . . . . . . . . . . . . . . . . . . . . . . . . . . . . . . . . . . . . . . . . . . . . . . . 647 |
| | 16.2 | Definite Integral . . . . . . . . . . . . . . . . . . . . . . . . . . . . . . . . . . . . . . . . . . . . . . . . . . . 647 |
| | | 16.2.1 Area Under a Curve . . . . . . . . . . . . . . . . . . . . . . . . . . . . . . . . . . . . . . . 652 |
| | | 16.2.2 Area of a Curve Above and Below $x$-Axis . . . . . . . . . . . . . . . . . . . . 665 |
| | | 16.2.3 Area Bounded by a Curve and a Line . . . . . . . . . . . . . . . . . . . . . . . . 670 |
| | | 16.2.4 Area Bounded by Two Curves . . . . . . . . . . . . . . . . . . . . . . . . . . . . . . 674 |
| | | 16.2.5 Integrating to Infinity . . . . . . . . . . . . . . . . . . . . . . . . . . . . . . . . . . . . . . 677 |
| | 16.3 | Chapter Summary . . . . . . . . . . . . . . . . . . . . . . . . . . . . . . . . . . . . . . . . . . . . . . . . . . 678 |
| | 16.4 | Further Practice . . . . . . . . . . . . . . . . . . . . . . . . . . . . . . . . . . . . . . . . . . . . . . . . . . . 680 |

**Chapter 17** Advanced Integration I . . . . . . . . . . . . . . . . . . . . . . . . . . . . . . . . . . . . . . . . . . . . . . . . . . 681

|   |   |   |
|---|---|---|
| | 17.1 | Introduction . . . . . . . . . . . . . . . . . . . . . . . . . . . . . . . . . . . . . . . . . . . . . . . . . . . . . . 681 |
| | 17.2 | Integration of Complicated Functions . . . . . . . . . . . . . . . . . . . . . . . . . . . . . . . . 681 |
| | | 17.2.1 Binomial of the Form $(ax+b)^n$ . . . . . . . . . . . . . . . . . . . . . . . . . . . . 682 |
| | | 17.2.2 Linear Functions of the Form $f'(ax+b)$ . . . . . . . . . . . . . . . . . . . . . 685 |
| | | 17.2.3 Integration of the Form $\int \frac{f'(x)}{f(x)}\, dx$ . . . . . . . . . . . . . . . . . . . . . . . 690 |
| | | 17.2.4 Integration of the Form $\int [f'(x)].[f(x)]\, dx$ |
| | |         and $\int [f'(x)].[f(x)]^n\, dx$ . . . . . . . . . . . . . . . . . . . . . . . . . . . . . . . . 694 |
| | | 17.2.5 Integration of the Form $\int \left[\frac{du}{dx}\right].[f'(u)].dx$ . . . . . . . . . . . . . . . . 698 |
| | 17.3 | Integration by Partial Fraction . . . . . . . . . . . . . . . . . . . . . . . . . . . . . . . . . . . . . . . 701 |
| | 17.4 | Integration by Substitution . . . . . . . . . . . . . . . . . . . . . . . . . . . . . . . . . . . . . . . . . . 704 |
| | 17.5 | Chapter Summary . . . . . . . . . . . . . . . . . . . . . . . . . . . . . . . . . . . . . . . . . . . . . . . . . . 715 |
| | 17.6 | Further Practice . . . . . . . . . . . . . . . . . . . . . . . . . . . . . . . . . . . . . . . . . . . . . . . . . . . 716 |

Contents                                                                                    xiii

**Chapter 18**   Advanced Integration II ............................................... 717
        18.1   Introduction ..................................................... 717
        18.2   Parametric Integration .......................................... 717
        18.3   Integration by Parts (Integration of Products) ................... 721
        18.4   Integration of Trigonometric Functions .......................... 738
              18.4.1   Power of Sine and Cosine Functions .................... 738
              18.4.2   Product of Sine and Cosine Functions .................. 741
              18.4.3   Other Trigonometric Functions ........................ 742
        18.5   Chapter Summary ............................................... 750
        18.6   Further Practice ................................................. 750

**Chapter 19**   Application of Integration ............................................ 751
        19.1   Introduction ..................................................... 751
        19.2   Numerical Integration ........................................... 751
              19.2.1   Trapezium Rule ....................................... 752
              19.2.2   Simpson's Rule ....................................... 759
              19.2.3   Calculating Error ..................................... 766
        19.3   Applications of Integration ...................................... 773
              19.3.1   Areas and Volumes ................................... 774
                      19.3.1.1   Area ......................................... 774
                      19.3.1.2   Volumes of Solids of Revolution ................ 775
              19.3.2   Mean Values .......................................... 785
              19.3.3   Root Mean Square (RMS) Value ....................... 786
        19.4   Chapter Summary ............................................... 792
        19.5   Further Practice ................................................. 793

Appendix ..................................................................... 795
Glossary ..................................................................... 809
Index ........................................................................ 818

# About the Author

**Shefiu Zakariyah** is a Chartered Engineer (Engineering Council UK) with more than 15 years of experience in research, and teaching engineering and mathematics courses within the higher education sectors. In this period, he has worked for Loughborough University, the University of Greenwich, the University of Cambridge, the University of Malaya, and the University of Namibia. Currently, Shefiu is an Associate Lecturer with the University of Derby, an Ofqual subject matter specialist, a Materials Developer for NCUK, and a technical consultant. He also serves as a Professional Registration Advisor (PRA) and Professional Registration Interviewer (PRI) for the Institution of Engineering and Technology (IET) supporting engineers and technicians to gain professional recognition. Shefiu served as an executive member of the IET Manufacturing Network and played a significant role in the Laser Institute of America, serving on the technical subcommittee for Control Measures and Testing (TSC-4). Furthermore, Shefiu was a member of the standard subcommittee responsible for the Safe Use of Lasers in Research, Development, and Testing (ANSI SSC-8-Z136-8). Shefiu has worked with learners from pre-university to postgraduate levels and is committed to making complex concepts accessible.

# How to Use This Book

## Users

- A detailed explanation is provided for each concept using simple and digestible language so that you can relax and enjoy studying this book. This is essential in order to enrich a self-study style of learning and to meet any need for remote study. If you decide to skip any part, you can head to the summary section where you will find a succinct yet rich overview of the chapter.
- You are advised to attempt the worked examples before checking the solution. This is to measure your understanding and check if there is any gap. Additional practice for each chapter is available at www.dszak.com.
- Where applicable, alternative method(s) to solve a problem are provided in a box following the first method introduced.
- Key formulas are numbered as (**X.Y**), where **X** is the chapter number and **Y** represents the ordinal position of the equation in the chapter. In other words, an equation that is numbered as (**2.10**) is the tenth equation in **Chapter 2**. A similar style is used for figures and tables.
- Important facts, terms, and formulas are clearly given in **bold type** (and placed in a box as *Formula OR Equation*) for emphasis and ease of reference.
- If you encounter a term or symbol, which might have been introduced but not explained in the section you are reading, the author has provided a range of useful resources at the beginning and end of this book, including a glossary and mathematical symbols, that you can easily refer to. Also, the index is useful to cross-reference other applications and usage of the same term or symbol.
- Whilst chapters are designed to be stand-alone, it will however be helpful in a few cases that you read related chapter(s) or section(s). Also to mention that this book does not follow the order of any syllabus, as it is intended for a wider audience.
- The first three chapters are designed to lay a foundation in arithmetic and algebra, equipping the reader with the essential principles and building a robust base from which learners can expand their mathematical understanding and skills. For those already familiar with the subject matter, a quick review of these sections may suffice.

## Instructors

- A set of PowerPoint slides for each chapter is provided online at www.dszak.com. Instructors may modify these materials for educational purposes, provided that proper credit is given.
- Also, a bank of questions is available for adaptation with due acknowledgement.

# Acknowledgements

This book has come to fruition due to unparalleled support and contributions from many people that it becomes impossible to mention them all, I'm sincerely grateful for their help and hope that they will accept this general expression of gratitude. I want to start by thanking the staff at Routledge, Taylor & Francis, for their confidence and support throughout this process, especially: Tony Moore (Senior Editor), Lillian Woodall (Project Manager at Deanta), Gabriella Williams (then Editorial Assistant), Frazer Merritt (then Editorial Assistant), and Aimee Wragg (Editorial Assistant). I can't repay your hard work, dedication, and patience.

My friends and colleagues in the teaching profession and industries have been very helpful and generous with their time in carefully reviewing this book and giving me constructive feedback. In this regard, I am very grateful to (listed in alphabetical order): Dr Abdelhalim Azbaid El Ouahabi (*PhD*), Dr Aliyu Ahmad (*PhD*), Bouchra Mohamed Lemine, Burket Ali (Head of Engineering and Computing, DMUIC), Dr Mayowa Kassim Aregbesola (*PhD*, previously with George Mason University), Dr Paul Richard Hammond (*MSc, PhD*), Dr Richard Welford (*DPhil*), Dr Sarbari Mukherjee (*MSc, PhD*), Salma Okhai (*BSc* Combined Science, PGCE (Secondary, DMUIC), Dr Shamsudeen Hassan (*PhD, CEng, MEI*), and Tim Wilmshurst (*MSc, FIET, CEng*).

Finally but most importantly, I owe a lot to my family who have been my source of inspiration and motivation. I would like to thank my wife: Khadijah Olaniyan (*B.A.*, Loughborough University), and my daughters and sons (fondly called 'Prince(ss)' and 'Doctor') for everything, including proofing and helpful tips.

# Abbreviations

| Term | Explanation |
|---|---|
| AC | Alternating Current |
| ACW | Anti-clockwise |
| CAH | Cosine–Adjacent–Hypotenuse |
| CAST | Cosine, All, Sine, and Tangent |
| CCW | Counter Clockwise |
| CW | Clockwise |
| d.p. | Decimal Place(s) |
| DC | Direct Current |
| HCF | Highest Common Factor |
| LCM | Lowest Common Multiple |
| LHS | Left-Hand Side |
| LIATE | Logarithmic–Inverse Trigonometric–Algebraic–Trigonometric–Exponential (function) |
| RHS | Right-Hand Side |
| s.f. | Significant Figure(s) |
| SI | Système International d'unités (International System of Units) |
| SOH | Sine–Opposite–Hypotenuse |
| TOA | Tangent–Opposite–Adjacent |
| wrt | With Respect To |

# Greek Alphabet

| Letter | | Name |
|---|---|---|
| **Upper Case** | **Lower Case** | |
| A | $\alpha$ | Alpha |
| B | $\beta$ | Beta |
| Γ | $\gamma$ | Gamma |
| Δ | $\delta$ | Delta |
| E | $\epsilon$ or $\varepsilon$ | Epsilon |
| Z | $\zeta$ | Zeta |
| H | $\eta$ | Eta |
| Θ | $\theta$ or $\vartheta$ | Theta |
| I | $\iota$ | Iota |
| K | $\kappa$ | Kappa |
| Λ | $\lambda$ | Lambda |
| M | $\mu$ | Mu |
| N | $\nu$ | Nu |
| Ξ | $\xi$ | Xi |
| O | $o$ | Omicron |
| Π | $\pi$ | Pi |
| P | $\rho$ or $\varrho$ | Rho |
| Σ | $\sigma$ or $\varsigma$ | Sigma |
| T | $\tau$ | Tau |
| Y | $\upsilon$ | Upsilon |
| Φ | $\phi$ or $\varphi$ | Phi |
| X | $\chi$ | Chi |
| Ψ | $\psi$ | Psi |
| Ω | $\omega$ | Omega |

# Mathematical Operators and Symbols

| Sign | Name | Sign | Name |
|---|---|---|---|
| $=$ | equal to | $\mathbb{Q}$ | set of rational numbers |
| $\neq$ | not equal to | $\mathcal{R}$ | real part |
| $+$ | addition (or plus) | $\mathfrak{J}$ | imaginary part |
| $-$ | subtraction | $\%$ | percentage |
| $\pm$ | plus or minus | $°$ | degree |
| $\mp$ | minus or plus | $°F$ | degrees Fahrenheit |
| $\times$ | multiplication (or times) | $°C$ | degrees Celsius |
| $\div$ | division | $|k|$ | modulus of $k$ (or absolute value of $k$). Also determinant of $k$ |
| $/$ | division slash | | |
| $\infty$ | infinity | $*$ | asterisk operator (or multiply by) |
| $\equiv$ | identical to | $\bullet$ | bullet operator or dot operator (or multiply by) |
| $\not\equiv$ | not identical to | | |
| $\sqrt{\phantom{x}}$ | radical sign (or square root) | $\cdots$ | horizontal ellipsis |
| $\sqrt[3]{\phantom{x}}$ | cube root | $\vdots$ | vertical ellipsis |
| $\sqrt[4]{\phantom{x}}$ | fourth root | $\therefore$ | therefore |
| $\sqrt[n]{x}$ | $n$th root of $x$ | $\because$ | because |
| $!$ | factorial | $\sum$ | summation (or sigma) |
| $\propto$ | proportional to | $::$ | proportion |
| $\sim$ | similar to/approximately | $:$ | ratio (or such that) |
| $\approx$ | almost equal to | log | logarithm to base 10 |
| $\simeq$ | asymptotically equal to | ln | natural logarithm (or logarithm to base $e$) |
| $\not\simeq$ | not asymptotically equal to | | |
| $\not\approx$ | not almost equal to | $\rightarrow$ | tend to |
| $<$ | less than | $\emptyset$ or $\{\}$ | empty set |
| $\not<$ | not less than | $\in$ | element of (or belongs to) |
| $>$ | greater than | $\notin$ | not an element of |
| $\not>$ | not greater than | $\ni$ | contains as member |
| $\leq$ | less than or equal to | $\not\ni$ | does not contain as member |
| $\nleq$ | neither less than nor equal to | $\ngeq$ | neither greater than nor equal to |
| $\geq$ | greater than or equal to | $\lneq$ | less than but not equal to |
| $\mathbb{N}$ | natural number | $\gneq$ | greater than but not equal to |
| $\mathbb{Z}$ | set of integers | $\ll$ | much less than/far less than |
| $\mathbb{R}$ | set of real numbers | $\lll$ | very much less than/far far less than |

| Sign | Name | Sign | Name |
|---|---|---|---|
| $\gg$ | much greater than/far greater than | ⊾ | right angle with arc |
| $\ggg$ | very much greater than/far far greater than | △ | right triangle |
| | | $\perp$ | perpendicular to (or orthogonal to) |
| $\Delta$ | increment/change | $\parallel$ | parallel to |
| $\delta$ | delta | $\nparallel$ | not parallel to |
| $\lim_{x \to \infty} f(x)$ | limiting value of $f(x)$ as $x \to \infty$ | # | equal and parallel to |
| $f'(x)$ | $f$ dash of $x$ (or $f$ prime of $x$) | $f(x)$ | $f$ of $x$ (or $f$ function $x$) |
| $f'''(x)$ | $f$ triple dash of $x$ | $f''(x)$ | $f$ double dash of $x$ |
| $f^{-1}(x)$ | inverse $f$ of $x$ (or inverse $f$ function $x$) | $f^n(x)$ | $n$th derivative of $f(x)$ with respect to $x$ |
| $\frac{\delta y}{\delta x}$ | delta $y$ over delta $x$ | $y'$ | $y$ dash (or $y$ prime) |
| $\frac{dy}{dx}$ | derivative of $y$ with respect to $x$ | $\frac{\partial y}{\partial x}$ | partial derivative of $y$ with respect to $x$ |
| $\frac{d^n y}{dx^n}$ | $n$th derivative of $y$ with respect to $x$ | $\frac{d^2 y}{dx^2}$ | dee 2 $y$ dee $x$ squared |
| $\int$ | integral | $\frac{\delta y}{\delta x}$ | delta $y$ over delta $x$ |
| $\int y\,dx$ | the indefinite integral of $y$ with respect $x$ | $\int f(x)\,dx$ | the indefinite integral of $f(x)$ with respect $x$ |
| $\iint$ | double integral | | |
| $z^*$ | complex conjugate of $z$ | $\int_b^a y\,dx$ | the definite integral of $y$ with respect $x$ between the limits $a$ and $b$ |
| $\underline{a}$ | vector $a$ | | |
| $\bar{a}$ | vector $a$, $a$ bar | $\iiint$ | triple integral |
| $\angle$ | angle | $\mathbf{a}$ | vector $a$ |
| $\measuredangle$ | measured angle | $OA$, $\overrightarrow{OA}$ | vector $OA$ |
| $\llcorner$ | right angle | $\hat{a}$ | $a$ hat |

# Physical Quantities

## TABLE 1
### Fundamental (or Base) Quantities

| Quantity Name | Unit Name | Unit Symbol |
|---|---|---|
| Length ($l$) | metre | m |
| Mass ($m$) | kilogram | kg |
| Time ($t$) | second | s |
| Electric Current ($I$) | ampere | A |
| Temperature ($\theta$) | kelvin | K |
| Amount ($n$) | mole | mol |
| Luminosity (L) | candela | cd |

## TABLE 2
### Some Common Derived Quantities

| Quantity Name | Unit Name | Unit Symbol |
|---|---|---|
| Area ($A$) | metre squared | m$^2$ |
| Volume ($V$) | metre cubed | m$^3$ |
| Density ($\rho$) | kilogram per metre cubed | kgm$^{-3}$ |
| Linear velocity ($u$ or $v$) | metre per second | ms$^{-1}$ |
| Linear acceleration ($a$) | metre per second squared | ms$^{-2}$ |
| Force ($F$) | newton | N |
| Pressure ($p$) | newton per metre squared or Pascal | Nm$^{-2}$ or Pa |
| Frequency ($f$) | hertz | Hz |
| Time period ($T$) | second | s |
| Energy ($E$) | joule | J |
| Work ($W$) | joule | J |
| Power ($P$) | watt | W |
| Voltage ($V$) | volt | V |
| Electric charge ($Q$ or $q$) | coulomb | C |
| Resistance ($R$) | ohm | $\Omega$ |
| Conductance ($G$) | siemens or per ohm or mho | S or $\Omega^{-1}$ |
| Reactance ($X$) | ohm | $\Omega$ |
| Impedance ($Z$) | ohm | $\Omega$ |
| Admittance ($Y$) | siemens or per ohm or mho | S or $\Omega^{-1}$ |
| Capacitance ($C$) | farad | F |
| Inductance ($L$) | henry | H |

# Prefixes Denoting Powers of Ten

| Prefix | Symbol | Exponential Form | Equivalent | Value |
|---|---|---|---|---|
| peta- | P | $10^{15}$ | 1 000 000 000 000 000 | one quadrillion |
| tera- | T | $10^{12}$ | 1 000 000 000 000 | one trillion |
| giga- | G | $10^{9}$ | 1 000 000 000 | one billion |
| mega- | M | $10^{6}$ | 1 000 000 | one million |
| kilo- | k | $10^{3}$ | 1 000 | one thousand |
| hecto- | h | $10^{2}$ | 100 | one hundred |
| deca- | da | $10^{1}$ | 10 | one ten |
| deci- | d | $10^{-1}$ | 0.1 | one-tenth |
| centi- | c | $10^{-2}$ | 0.01 | one-hundredth |
| milli- | m | $10^{-3}$ | 0.001 | one-thousandth |
| micro- | μ | $10^{-6}$ | 0.000 001 | one-millionth |
| nano- | n | $10^{-9}$ | 0.000 000 001 | one-billionth |
| pico- | p | $10^{-12}$ | 0.000 000 000 001 | one-trillionth |
| femto- | f | $10^{-15}$ | 0.000 000 000 000 001 | one-quadrillionth |

**NOTE**

- From power −3 down and +3 up, the powers are all in multiple of three.
- Positive powers have capital letter symbols except for deca, hecto, and kilo, while negative powers have small letter symbols.
- 'kilo' is the only small letter symbol in the positive multiples of three.

# 1 Trigonometric Functions I

## Learning Outcomes

Once you have studied the content of this chapter, you should be able to:

- State trigonometric ratios
- Calculate the sine, cosine, and tangent of any given angle
- Discuss quadrants and their applications
- Determine trigonometric ratios of special angles (0°, 30°, 45°, 60°, and 90°)
- Evaluate inverse trigonometric ratios
- Use the exact sine, cosine, and tangent values of special angles
- Draw the graphs of sine, cosine, and tangent functions

## 1.1 INTRODUCTION

Trigonometry is an essential branch of mathematics that deals with specific functions and their behaviours in relation to angles. Any process that involves a fixed repeated motion can be represented or modelled by what is termed as a sinusoidal waveform. This waveform has the general form $A \sin(x + \theta)$, whether it is a pendulum bob, a weight on a spring, or a swing. In fact, any forward and backward or up and down movement that is maintained at a constant speed between two fixed points can be included in this topic. We will start in this chapter by looking at the trigonometric ratios of sine, cosine, and tangent functions, and carry on our discussion on trigonometry in the subsequent four chapters.

## 1.2 TRIGONOMETRIC RATIOS

By trigonometric ratios, we mean the sine, cosine, and tangent of an angle. They are usually shortened as $\sin \theta$, $\cos \theta$, and $\tan \theta$, respectively, where $\theta$ represents the amount of turn or the angle. Note that we can use any other letter or symbol instead of $\theta$.

We will use a right-angled triangle as the basis of our definition. Let us consider triangle $ABC$ shown in Figure 1.1.

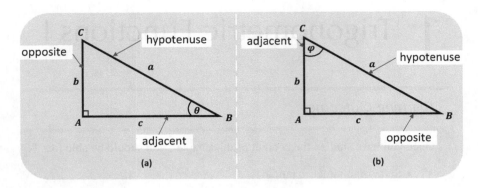

**FIGURE 1.1** Naming the three sides of a right-angled triangle relative to: (a) angle $\theta$, and (b) angle $\varphi$.

The three sides are $|BC|$, $|AC|$, and $|AB|$ corresponding to sides $a$, $b$, and $c$, respectively. For our case, we need to label the sides as:

- **Hypotenuse**

This is the longest side of the triangle or the side opposite to the 90-degree angle. Note that the biggest angle in a right-angled triangle is $90°$ and the side opposite to it must be the longest. In this case, it is $|BC|$ or $a$. We should be familiar with this name from our study of Pythagoras' theorem.

- **Opposite**

Unlike hypotenuse, the opposite side is not static but depends on the referenced angle. Either $|AB|$ or $|AC|$ can be the opposite. As we are given angle $\theta$ here in Figure 1.1(a), corresponding to $\angle B$ or $\widehat{B}$, the opposite is therefore $|AC|$ or $b$.

- **Adjacent**

Linguistically, it means 'next to', 'nearer to', or 'close to'. But to what? Well, it is the side next to the given angle. Since the given angle here is $\theta$, the adjacent side is therefore $|AB|$ or $c$.

In summary, in a right-angled triangle, the hypotenuse is the longest side and both the opposite and adjacent sides can flip between the remaining two sides, subject to the referenced angle. To illustrate this, the reference angle in Figure 1.1(b) is $\varphi$ (corresponding to $\angle C$ or $\widehat{C}$), therefore the opposite and adjacent are now $|AB|$ and $|AC|$, respectively.

Now that we know the conventional names for the three sides of a right-angled triangle, what is the definition of the sine, cosine, and tangent of an angle? They are ratios of a pair of sides. Using Figure 1.1(a), they are given as:

$$\boxed{\sin\theta = \frac{|AC|}{|BC|}} \quad \boxed{\cos\theta = \frac{|AB|}{|BC|}} \quad \boxed{\tan\theta = \frac{|AC|}{|AB|}} \tag{1.1}$$

The reference to the vertices of a triangle and the sides (or their lengths) can change but the names of each side in relation to a given angle will not. Hence, our definitions of the three trigonometric

# Trigonometric Functions I

ratios will be based on these names:

$$\boxed{\sin \theta = \frac{\text{Opposite}}{\text{Hypotenuse}} = \frac{O}{H}} \quad \boxed{\cos \theta = \frac{\text{Adjacent}}{\text{Hypotenuse}} = \frac{A}{H}} \quad \boxed{\tan \theta = \frac{\text{Opposite}}{\text{Adjacent}} = \frac{O}{A}} \quad (1.2)$$

To remember the ratios, we have a very popular mnemonic **SOH–CAH–TOA** (pronounced as 'sow-car-tour'), which implies:

- **SOH**: **S**ine–**O**pposite–**H**ypotenuse
- **CAH**: **C**osine–**A**djacent–**H**ypotenuse
- **TOA**: **T**angent–**O**pposite–**A**djacent

Note that tangent is a ratio of sine to cosine. In other words:

$$\boxed{\tan \theta = \frac{\sin \theta}{\cos \theta}} \quad (1.3)$$

The above identity will be proved shortly in the worked examples. Note that the Adjacent, Opposite, and Hypotenuse sides are shortened to Adj., Opp., and Hyp., respectively. Another point to note is that since the hypotenuse is the longest side, the sine and cosine of any angle will always be less than one.

Now it is time to try some examples.

## Example 1

Using $\triangle ABC$ and $\triangle XYZ$ shown in Figure 1.2, determine the exact value of the following.

**a)** $\sin \theta$  **b)** $\sin \varphi$  **c)** $\cos \theta$  **d)** $\cos \varphi$  **e)** $\tan \theta$  **f)** $\tan \varphi$

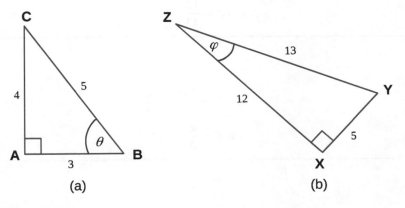

**FIGURE 1.2** Example 1.

What did you get? Find the solution below to double-check your answer.

### Solution to Example 1

**a)** $\sin \theta$
**Solution**

$$\sin \theta = \frac{\text{Opp}}{\text{Hyp}} = \frac{4}{5}$$
$$\therefore \sin \theta = 0.8$$

---

**b)** $\sin \varphi$
**Solution**

$$\sin \varphi = \frac{\text{Opp}}{\text{Hyp}} = \frac{5}{13}$$
$$\therefore \sin \varphi = \frac{5}{13}$$

---

**c)** $\cos \theta$
**Solution**

$$\cos \theta = \frac{\text{Adj}}{\text{Hyp}} = \frac{3}{5}$$
$$\therefore \cos \theta = 0.6$$

---

**d)** $\cos \varphi$
**Solution**

$$\cos \varphi = \frac{\text{Adj}}{\text{Hyp}} = \frac{12}{13}$$
$$\therefore \cos \varphi = \frac{12}{13}$$

---

**e)** $\tan \theta$
**Solution**

$$\tan \theta = \frac{\text{Opp}}{\text{Adj}} = \frac{4}{3}$$
$$\therefore \tan \theta = \frac{4}{3}$$

---

**f)** $\tan \varphi$
**Solution**

$$\tan \varphi = \frac{\text{Opp}}{\text{Adj}} = \frac{5}{12}$$
$$\therefore \tan \varphi = \frac{5}{12}$$

# Trigonometric Functions I

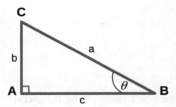

**FIGURE 1.3** Example 2.

Let us try another example.

## Example 2

Using the triangle $ABC$ in Figure 1.3, prove that $\tan\theta = \frac{\sin\theta}{\cos\theta}$.

What did you get? Find the solution below to double-check your answer.

## Solution to Example 2

$$\sin\theta = \frac{\text{Opp}}{\text{Hyp}} = \frac{b}{a}$$

$$\cos\theta = \frac{\text{Adj}}{\text{Hyp}} = \frac{c}{a}$$

$$\tan\theta = \frac{\text{Opp}}{\text{Adj}} = \frac{b}{c}$$

We have two ways to show (or prove) this identity:

**Option 1** Using the RHS

$$\frac{\sin\theta}{\cos\theta} = \frac{\left(\frac{b}{a}\right)}{\left(\frac{c}{a}\right)}$$

$$= \frac{b}{a} \div \frac{c}{a}$$

$$= \frac{b}{a} \times \frac{a}{c} = \frac{b}{c}$$

$$\therefore \frac{\sin\theta}{\cos\theta} = \frac{b}{c} = \tan\theta$$

**Option 2** Using the LHS

$$\tan\theta = \frac{b}{c}$$

$$= \frac{b}{c} \times \frac{a}{a} = \frac{ba}{ca} = \frac{ab}{ca}$$

$$= \frac{a}{c} \times \frac{b}{a} = \frac{1}{\cos\theta} \times \sin\theta$$

$$\therefore \tan\theta = \frac{\sin\theta}{\cos\theta}$$

Final example to try.

### Example 3

Given that $\sin \varphi = \frac{2}{3}$, determine the exact value of the following:

a) $\cos \varphi$                                         b) $\tan \varphi$

What did you get? Find the solution below to double-check your answer.

### Solution to Example 3

We need to sketch the triangle first as shown in Figure 1.4. Now given that

$$\sin \varphi = \frac{\text{Opp}}{\text{Hyp}} = \frac{2}{3}$$

Using Pythagoras' theorem, we have

$$3^2 = 2^2 + x^2$$
$$x^2 = 3^2 - 2^2 = 5$$
$$\therefore x = \sqrt{5}$$

**a) $\cos \varphi$**
**Solution**

$$\cos \varphi = \frac{\text{Adj}}{\text{Hyp}} = \frac{\sqrt{5}}{3}$$
$$\therefore \cos \varphi = \frac{1}{3}\sqrt{5}$$

**b) $\tan \varphi$**
**Solution**

$$\tan \varphi = \frac{\text{Opp}}{\text{Adj}} = \frac{2}{\sqrt{5}}$$
$$\therefore \tan \varphi = \frac{2}{5}\sqrt{5}$$

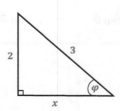

**FIGURE 1.4** Solution to Example 3.

# Trigonometric Functions I

## 1.3 RATIOS OF SPECIAL ANGLES

By special angles, it usually implies the naturally occurring angles such as **30°**, **45°**, and **60°**. Angles **0°** and **90°** are commonly included in this group. Their exact trigonometric ratios (or values) can easily be determined and used in solving problems. Let's look at them one after another.

- **Angle 45°**

For this case, we have chosen an isosceles triangle shown in Figure 1.5. The length of the equal sides is chosen to be 1 unit. Using Pythagoras' theorem, the hypotenuse is $\sqrt{1^2 + 1^2} = \sqrt{2}$. Because this is a right-angled isosceles triangle, the three angles are 90°, 45°, and 45°.

Referring to Figure 1.5, we have:

$$\sin 45° = \frac{\text{Opp}}{\text{Hyp}} = \frac{1}{\sqrt{2}} = \frac{\sqrt{2}}{2} \qquad \cos 45° = \frac{\text{Adj}}{\text{Hyp}} = \frac{1}{\sqrt{2}} = \frac{\sqrt{2}}{2} \qquad \tan 45° = \frac{\text{Opp}}{\text{Adj}} = \frac{1}{1} = 1$$

Note that the choice of 1 unit for the two sides is arbitrary and only for simplicity, although the results will not be affected if a different value is chosen.

- **Angles 30° and 60°**

For this case, we will use an equilateral triangle $ABC$ (Figure 1.6), with each side having 2 units. Draw a perpendicular bisector $AA'$ from vertex $A$ to side $BC$. A perpendicular bisector is a line that meets another line at a right angle (i.e., perpendicular) and divides it equally. The two minor triangles are right-angled and symmetrical.

Using Pythagoras' theorem, we have $|AA'| = \sqrt{2^2 - 1^2} = \sqrt{3}$. We can use either of the two small triangles $AA'B$ or $AA'C$ to define these special angles, but we will choose $\triangle AA'C$ for this case.

With reference to Figure 1.6, we can define 30° as:

$$\sin 30° = \frac{\text{Opp}}{\text{Hyp}} = \frac{1}{2} \qquad \cos 30° = \frac{\text{Adj}}{\text{Hyp}} = \frac{\sqrt{3}}{2} \qquad \tan 30° = \frac{\text{Opp}}{\text{Adj}} = \frac{1}{\sqrt{3}} = \frac{\sqrt{3}}{3}$$

Similarly, we define 60° as:

$$\sin 60° = \frac{\text{Opp}}{\text{Hyp}} = \frac{\sqrt{3}}{2} \qquad \cos 60° = \frac{\text{Adj}}{\text{Hyp}} = \frac{1}{2} \qquad \tan 60° = \frac{\text{Opp}}{\text{Adj}} = \frac{\sqrt{3}}{1} = \sqrt{3}$$

- **Angles 0° and 90°**

We need to show this differently from others. In the right-angled triangle (Figure 1.7), $\alpha$ and $\beta$ are complimentary because $\alpha + \beta = 90°$. Any increase in one angle results in a corresponding decrease in the other. If $\alpha$ approaches 0°, then $\beta$ tends to 90°.

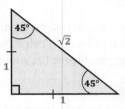

**FIGURE 1.5** Determining angle 45°.

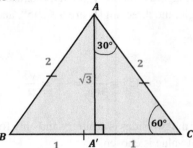

**FIGURE 1.6** Determining angles 30° and 60°.

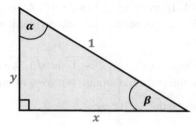

**FIGURE 1.7** Determining angles 0° and 90°.

At $\alpha = 90°$, we have $\beta = 0°$, $x = 1$, and $y = 0$, thus

$$\sin \beta = \sin 0° = \frac{\text{Opp}}{\text{Hyp}} = \frac{0}{1} = 0 \quad \Big| \quad \cos \beta = \cos 0° = \frac{\text{Adj}}{\text{Hyp}} = \frac{1}{1} = 1 \quad \Big| \quad \tan \beta = \tan 0° = \frac{\text{Opp}}{\text{Adj}} = \frac{0}{1} = 0$$

At $\alpha = 0°$, we have $\beta = 90°$, $x = 0$, and $y = 1$, thus

$$\sin \beta = \sin 90° = \frac{\text{Opp}}{\text{Hyp}} = \frac{1}{1} = 1 \quad \Big| \quad \cos \beta = \cos 90° = \frac{\text{Adj}}{\text{Hyp}} = \frac{0}{1} = 0 \quad \Big| \quad \tan \beta = \tan 0° = \frac{\text{Opp}}{\text{Adj}} = \frac{1}{0} = \infty$$

The special angles are summarised in Table 1.1.

# Trigonometric Functions I

**TABLE 1.1**
Special Angles (0°, 30°, 45°, 60°, and 90°)

|              | 0 | 30°                  | 45°                  | 60°                  | 90° |
|--------------|---|----------------------|----------------------|----------------------|-----|
| $\sin\theta$ | 0 | $\frac{1}{2}$        | $\frac{\sqrt{2}}{2}$ | $\frac{\sqrt{3}}{2}$ | 1   |
| $\cos\theta$ | 1 | $\frac{\sqrt{3}}{2}$ | $\frac{\sqrt{2}}{2}$ | $\frac{1}{2}$        | 0   |
| $\tan\theta$ | 0 | $\frac{\sqrt{3}}{3}$ | 1                    | $\sqrt{3}$           | ∞   |

Great stuff! Let's try one example or two.

### Example 4

Using trigonometric ratios, prove the following identities.

**a)** $\sin\theta = \cos(90° - \theta)$ 　　　　　　**b)** $\cos\varphi = \sin(90° - \varphi)$

What did you get? Find the solution below to double-check your answer.

### Solution to Example 4

**a)** $\sin\theta = \cos(90° - \theta)$
**Solution**
To prove this, we will consider $\triangle ABC$ where $\angle B = \theta$. Since this is a right-angled triangle, angles $B$ and $C$ are complimentary (i.e., $\widehat{B} + \widehat{C} = 90°$ or $\widehat{C} = 90° - \theta$) as shown in the diagram (Figure 1.8). Now we have:

$$\sin\theta = \frac{\text{Opp}}{\text{Hyp}} = \frac{|AC|}{|BC|} \qquad\qquad \cos(90° - \theta) = \frac{\text{Adj}}{\text{Hyp}} = \frac{|AC|}{|BC|}$$

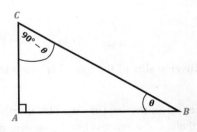

**FIGURE 1.8** Solution to Example 4(a).

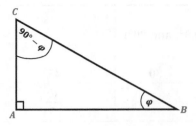

**FIGURE 1.9** Solution to Example 4(b).

As the RHS of both are equal, therefore:

$$\sin\theta = \cos(90° - \theta)$$

**NOTE**
This is known as the ratio of complimentary angle. For example:

- $\sin 15° = \cos 75°$
- $\sin\frac{2}{5}\pi = \cos\frac{\pi}{10}$

---

**b)** $\cos\varphi = \sin(90° - \varphi)$
**Solution**
Similarly
Now we have (Figure 1.9):

$$\cos\varphi = \frac{\text{Adj}}{\text{Hyp}} = \frac{|AB|}{|BC|} \qquad \sin(90° - \varphi) = \frac{\text{Opp}}{\text{Hyp}} = \frac{|AB|}{|BC|}$$

As the RHS of both are equal, therefore:

$$\cos\varphi = \sin(90° - \varphi)$$

---

## 1.4 INVERSE TRIGONOMETRIC RATIOS

We have defined the trigonometric ratio of sine, cosine, and tangent of angle. Now we need to work backward, i.e., finding the angle whose trigonometric ratio is given or known. This is called **inverse trigonometry** or **inverse trigonometric ratios**.

For example, given an angle $\theta$ such that its sine is known to be 0.5, we will write this as:

$$\sin\theta = 0.5$$

The inverse of sine is written as:

$$\theta = \sin^{-1}(0.5)$$

which is read as '$\theta$ **is equal to inverse sine of 0.5**', or '$\theta$ **is equal to the sine inverse of 0.5**', or '$\theta$ **is equal to arc sine 0.5**'.

Notice the power of negative one (−1) after sine (or sin). This is only a notation of the inversion and does not indicate power. Nevertheless, the process can be shown, though technically incorrect, as laid out below.

# Trigonometric Functions I

| | |
|---|---|
| $\sin \theta = 0.5$ | Divide both sides by sine |
| $\theta = \dfrac{0.5}{\sin}$ | Note that $\dfrac{1}{x}$ can be written as $x^{-1}$ |
| $\theta = \sin^{-1} 0.5$ | Note that the power is just to illustrate inversion |

Back to our focus, what does $\sin^{-1}(0.5)$ represent? It simply means an angle whose sine is 0.5. Similarly, $\cos^{-1}(0.6)$, is read as **'cos inverse of 0.6'**, or **'arc cosine 0.6'**, is an angle whose cosine has a value of 0.6. If $\alpha$ is a natural number, then $\tan^{-1} \alpha$ (read as **'tan inverse of $\alpha$'** or **'arc tan $\alpha$'**) is an angle whose tangent has a value of $\alpha$.

In a table of trigonometric functions, now obsolete, we can easily check inverse functions. There are now special keys for arcsin, arccos, and arctan (or $\sin^{-1}$, $\cos^{-1}$, and $\tan^{-1}$, respectively) on most scientific calculators. In MATLAB and Excel, the commands used for inverse trigonometric functions are **asin**, **acos**, and **atan**.

Now it is time to try some examples.

## Example 5

Using your calculator (or any suitable calculating device), determine the following angles in degrees. Present your answers correct to 2 s.f.

a) $\sin^{-1} 0.5$  
b) $\cos^{-1} \dfrac{1}{\sqrt{2}}$  
c) $\tan^{-1} \dfrac{1}{\sqrt{3}}$  
d) $\sin^{-1}(-0.345)$  
e) $\cos^{-1}(-0.156)$  
f) $\tan^{-1}(-2.5)$  

What did you get? Find the solution below to double-check your answer.

## Solution to Example 5

### HINT

In this case, we will use angle $\theta$ to denote the required angle.

**a) $\sin^{-1} 0.5$**
**Solution**

$$\theta = \sin^{-1} 0.5$$
$$\therefore \theta = 30° \text{ (2 s.f.)}$$

**b)** $\cos^{-1} \dfrac{1}{\sqrt{2}}$

**Solution**

$$\theta = \cos^{-1} \dfrac{1}{\sqrt{2}}$$
$$\therefore \theta = \mathbf{45°(2\ s.f.)}$$

---

**c)** $\tan^{-1} \dfrac{1}{\sqrt{3}}$

**Solution**

$$\theta = \tan^{-1} \dfrac{1}{\sqrt{3}}$$
$$\therefore \theta = \mathbf{30°\ (2\ s.f.)}$$

---

**d)** $\sin^{-1}(-0.345)$

**Solution**

$$\theta = \sin^{-1}(-0.345) = -20.1818°$$
$$\therefore \theta = \mathbf{-20°\ (2\ s.f.)}$$

---

**e)** $\cos^{-1}(-0.156)$

**Solution**

$$\theta = \cos^{-1}(-0.156) = 98.9748°$$
$$\therefore \theta = \mathbf{99°\ (2\ s.f.)}$$

---

**f)** $\tan^{-1}(-2.5)$

**Solution**

$$\theta = \tan^{-1}(-2.5) = -68.1986°$$
$$\therefore \theta = \mathbf{-68°\ (2\ s.f.)}$$

---

Let us try another set of examples.

## Example 6

Using your calculator (or any suitable calculating device), determine the following angles in radians. Present your answers correct to 2 d.p. where necessary.

**a)** $\sin^{-1} \dfrac{1}{\sqrt{2}}$      **b)** $\cos^{-1} 0.2501$      **c)** $\tan^{-1} \sqrt{3}$

**d)** $\sin^{-1}(-0.8792)$      **e)** $\cos^{-1}(-0.1089)$      **f)** $\tan^{-1}(-18)$

# Trigonometric Functions I

What did you get? Find the solution below to double-check your answer.

### Solution to Example 6

**HINT**

In this case, we will use angle $\theta$. Also, remember to change your calculator from degree to radian.

**a)** $\sin^{-1} \dfrac{1}{\sqrt{2}}$

**Solution**

$$\theta = \sin^{-1} \dfrac{1}{\sqrt{2}}$$

$$\therefore \theta = \dfrac{1}{4}\pi$$

**b)** $\cos^{-1} 0.2501$

**Solution**

$$\theta = \cos^{-1} 0.2501 = 1.3180^c$$

$$\therefore \theta = \mathbf{1.32^c} \text{ (2 d.p.)}$$

**c)** $\tan^{-1} \sqrt{3}$

**Solution**

$$\theta = \tan^{-1} \sqrt{3}$$

$$\therefore \theta = \dfrac{1}{3}\pi$$

**d)** $\sin^{-1}(-0.8792)$

**Solution**

$$\theta = \sin^{-1}(-0.8792) = -1.0742^c$$

$$\therefore \theta = \mathbf{-1.07^c} \text{ (2 d.p.)}$$

**e)** $\cos^{-1}(-0.1089)$

**Solution**

$$\theta = \cos^{-1}(-0.1089) = 1.6799^c$$

$$\therefore \theta = \mathbf{1.68^c} \text{ (2 d.p.)}$$

**f)** $\tan^{-1}(-18)$

**Solution**

$$\theta = \tan^{-1}(-18) = -1.5153^c$$

$$\therefore \theta = \mathbf{-1.52^c} \text{ (2 d.p.)}$$

Another set of examples to try.

### Example 7

Without using a calculator, determine the value of the following. Present your answers in exact form.

a) $\sin\left(\arccos\frac{1}{2}\right)$
b) $\tan\left(\arcsin\frac{\sqrt{2}}{2}\right)$

What did you get? Find the solution below to double-check your answer.

### Solution to Example 7

a) $\sin\left(\arccos\frac{1}{2}\right)$

**Solution**
This is a double function. We will attempt the inner function and then move to the outer. Thus

$$\theta = \arccos\frac{1}{2} = 60°$$

Hence,

$$\sin\left(\arccos\frac{1}{2}\right) = \sin 60° = \frac{\sqrt{3}}{2}$$

$$\therefore \sin\left(\arccos\frac{1}{2}\right) = \frac{\sqrt{3}}{2}$$

b) $\tan\left(\arcsin\frac{\sqrt{2}}{2}\right)$

**Solution**
Again, this is a double function. We will attempt the inner function and then move to the outer. Thus

$$\theta = \arcsin\frac{\sqrt{2}}{2} = 45°$$

Hence,

$$\tan\left(\arcsin\frac{\sqrt{2}}{2}\right) = \tan 45° = 1$$

$$\therefore \tan\left(\arcsin\frac{\sqrt{2}}{2}\right) = 1$$

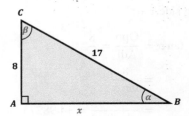

**FIGURE 1.10** Example 8.

One final example to try.

### Example 8

Find the missing angles and side in $\triangle ABC$ (Figure 1.10).

---

What did you get? Find the solution below to double-check your answer.

### Solution to Example 8

$$\sin \alpha = \frac{\text{Opp}}{\text{Hyp}} = \frac{8}{17}$$

$$\alpha = \sin^{-1}\left(\frac{8}{17}\right) = 28.0725°$$

$$\therefore \alpha = \mathbf{28.1°}$$

We know that the total angle in a triangle is 180°. Hence

$$\alpha + \beta = 90°$$

$$\beta = 90° - 28.1°$$

$$\therefore \beta = \mathbf{61.9°}$$

We can equally determine $\beta$ using a trigonometric ratio as:

$$\cos \beta = \frac{\text{Adj}}{\text{Hyp}} = \frac{8}{17}$$

$$\beta = \cos^{-1}\left(\frac{8}{17}\right) = 61.9275°$$

$$\therefore \beta = \mathbf{61.9°}$$

Now we need to find the missing side $x$ as:

$$\tan \alpha = \frac{\text{Opp}}{\text{Adj}} = \frac{8}{x}$$

$$x = \frac{8}{\tan \alpha}$$

$$= \frac{8}{\tan 28.1°} = 14.9827 \text{ unit}$$

$$\therefore x = \mathbf{15 \text{ unit}}$$

$x$ can also be found by using

$$\tan \beta = \frac{\text{Opp}}{\text{Adj}} = \frac{x}{8}$$

$$x = 8 \tan \beta$$

$$= 8 \tan 61.9° = 14.9827 \text{ unit}$$

$$\therefore x = \mathbf{15 \text{ unit}}$$

Or

$$\sin \beta = \frac{\text{Opp}}{\text{Hyp}} = \frac{x}{17}$$

$$x = 17 \sin \beta$$

$$= 17 \sin 61.9° = 14.9962 \text{ unit}$$

$$\therefore x = \mathbf{15 \text{ unit}}$$

**ALTERNATIVE METHOD**

This is a right-angled triangle so we can use Pythagoras' theorem, where

$$a^2 = b^2 + c^2$$

In this case, $a = 17$, $b = 8$, and $c = x$.

$$c^2 = a^2 - b^2$$

$$c = \sqrt{a^2 - b^2} = \sqrt{17^2 - 8^2}$$

$$c = \sqrt{225} = 15 \text{ unit}$$

$$\therefore x = \mathbf{15 \text{ unit}}$$

# Trigonometric Functions I

Now since we have all the three sides, we can find the missing angles as:

$$\sin \alpha = \frac{\text{Opp}}{\text{Hyp}} = \frac{8}{17}$$

$$\alpha = \sin^{-1}\left(\frac{8}{17}\right) = 28.0725°$$

$$\therefore \alpha = \mathbf{28.1°}$$

$$\cos \beta = \frac{\text{Adj}}{\text{Hyp}} = \frac{8}{17}$$

$$\beta = \cos^{-1}\left(\frac{8}{17}\right) = 61.9275°$$

$$\therefore \beta = \mathbf{61.9°}$$

## 1.5 QUADRANTS

Our discussion of trigonometric ratios in the previous section only covers angles in the range $0 \leq \theta \leq 90°$. To be able to fully define our trigonometric ratios, we need to cover the full range of one revolution, which is $0 \leq \theta \leq 360°$. This can be achieved by the introduction of quadrants.

Consider an $x$–$y$ Cartesian plane shown in Figure 1.11(a). We can also view this as a circle or an orange slice, divided into four equal parts, with each part known as a **quadrant**.

Conventionally, we start our measurement from the positive $x$-axis (or the top right portion) and move counter-clockwise, as shown by the direction of the arrow in Figure 1.11(a). Each quadrant has a size

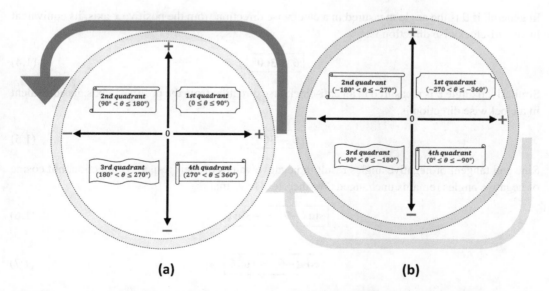

**FIGURE 1.11** Four quadrants and their respective angle intervals illustrated (when the angle is measured from the positive $x$-axis: (a) in an anti-clockwise direction, and (b) in a clockwise direction).

## TABLE 1.2
**Clockwise Angles and Their Anti-Clockwise Equivalents Illustrated for the Four Quadrants**

| Quadrant | Anti-Clockwise Measurement Range | Clockwise Measurement Range |
|---|---|---|
| First | $0 \leq \theta \leq 90°$ | $-270° < \theta \leq -360°$ |
| Second | $90° < \theta \leq 180°$ | $-180° < \theta \leq -270°$ |
| Third | $180° < \theta \leq 270°$ | $-90° < \theta \leq -180°$ |
| Fourth | $270° < \theta \leq 360°$ | $0 \leq \theta \leq -90°$ |

of 90 degrees. Thus, the first quadrant refers to angle $\theta$ such that $0 \leq \theta \leq 90°$. The second, third, and fourth quadrants are $90° < \theta \leq 180°$, $180° < \theta \leq 270°$, and $270° < \theta \leq 360°$, respectively.

What happens if the angles are measured in a clockwise direction? Since this is contrary to convention, the quadrant numbering remains the same (i.e., the first is the top right section, the second is the top left section, and so on), however, the angle will be negative, as shown in Figure 1.11(b). This is very similar to what happens to a vector quantity.

Notice from Figure 1.11 and from our intuitive understanding of positions around a circle that we can find the equivalent of an angle measured in both clockwise and anti-clockwise directions. With the aid of Table 1.2, we can say that $\theta = -360°$ in the clockwise direction is the same as $\theta = 0°$ in the anti-clockwise measurement. Also, $\theta = -270°$ is the same as $\theta = 90°$, and similarly, $\theta = -90°$ is the the same as when $\theta = 270°$.

In general, if $\theta$ is the angle measured in a clockwise direction from the positive $x$-axis, its equivalent in an anti-clockwise direction is:

$$\boxed{\theta + 360°} \tag{1.4}$$

Similarly, if angle $\theta$ is measured in an anti-clockwise direction from the positive $x$-axis, its equivalent in a clockwise direction is:

$$\boxed{\theta - 360°} \tag{1.5}$$

Sine and tangent of negative angles result in negative values of their respective function, but cosine of negative angles remains unchanged. We therefore have that

$$\boxed{\sin(-\theta) = -\sin\theta} \tag{1.6}$$

$$\boxed{\cos(-\theta) = \cos\theta} \tag{1.7}$$

$$\boxed{\tan(-\theta) = -\tan\theta} \tag{1.8}$$

# Trigonometric Functions I

Let's try some examples.

### Example 9

Determine the quadrant of an angle $\theta$ if:

a) $\theta = 39°$  
b) $\theta = 148°$  
c) $\theta = 313°$  
d) $\theta = -300°$  
e) $\theta = -215°$  
f) $\theta = -76°$  

What did you get? Find the solution below to double-check your answer.

### Solution to Example 9

**a)** $\theta = 39°$  
**Solution**

$$\theta \text{ is in the first quadrant}$$
$$\because 0 \leq \theta \leq 90°$$

---

**b)** $\theta = 148°$  
**Solution**

$$\theta \text{ is in the second quadrant}$$
$$\because 90° < \theta \leq 180°$$

---

**c)** $\theta = 313°$  
**Solution**

$$\theta \text{ is in the fourth quadrant}$$
$$\because 270° < \theta \leq 360°$$

---

**d)** $\theta = -300°$  
**Solution**

If $\theta = -300°$ it implies a clockwise measurement, therefore:

$$\theta = -300° + 360° = 60°$$

$$\therefore \theta \text{ is in the first quadrant}$$
$$\because 0 \leq \theta \leq 90°$$

### ALTERNATIVE METHOD

Alternatively, if $\theta$ is negative, then we can use the clockwise measurement range (Table 1.2), therefore:

$$\theta \text{ is in the first quadrant}$$
$$\because -270° < \theta \leq -360°$$

---

**e)** $\theta = -215°$
**Solution**
If $\theta = -215°$, it implies a clockwise measurement, therefore:

$$\theta = -215° + 360° = 145°$$

$$\therefore \theta \text{ is in the second quadrant}$$
$$\because 90° < \theta \leq 180°$$

---

**f)** $\theta = -76°$
**Solution**
If $\theta = -76°$, it implies a clockwise measurement, therefore:

$$\theta = -76° + 360° = 284°$$

$$\therefore \theta \text{ is in the fourth quadrant}$$
$$\because 270° < \theta \leq 360°$$

---

Now let us look at the trigonometric ratios in the four quadrants one after the other.

**First Quadrant** $0 \leq \theta \leq 90°$

We will be looking at the trigonometric ratios for angles between 0 and 90 degrees. In Figure 1.12, the unit arm $OA$, which is always positive, swings to have a projection of $OB$ on the y-axis and $OC$ on the x-axis. Notice that $|OC|$ is equal to $|BA|$ just like $|OB| = |CA|$.

Using $\triangle OAC$ and noting the signs of x-axis and y-axis in this quadrant, which represent the projection of our $OA$, we can state that:

$$\sin \theta = \frac{\text{Opp}}{\text{Hyp}} = \frac{|AC|}{|OA|} \qquad \cos \theta = \frac{\text{Adj}}{\text{Hyp}} = \frac{|OC|}{|OA|} \qquad \tan \theta = \frac{\text{Opp}}{\text{Adj}} = \frac{|AC|}{|OC|}$$

In summary, sine, cosine, and tangent are all positive in the first quadrant.

Before we proceed to other quadrants, it is essential to make it clear that the adjacent is always the x-axis and $\sin \theta$ (as hypotenuse is a unit arm $OA$) and the opposite is always the y-axis and $\cos \theta$ (as hypotenuse is a unit arm $OA$). The hypotenuse remains the same as a positive 1 unit $OA$.

# Trigonometric Functions I

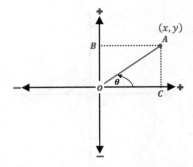

**FIGURE 1.12** Determining trigonometric ratios using phasor diagram (first quadrant).

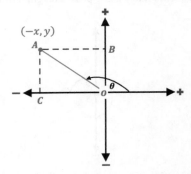

**FIGURE 1.13** Determining trigonometric ratios using phasor diagram (second quadrant).

**Second Quadrant** $90 < \theta \leq 180°$

The second quadrant covers angles between 90 and 180 degrees. Again, we use the arm $OA$ as a swing, which is positive (Figure 1.13). Using $\triangle OAC$, we can state that:

$$\sin \theta = \frac{\text{Opp}}{\text{Hyp}} = \frac{|AC|}{|OA|} \qquad \cos \theta = \frac{\text{Adj}}{\text{Hyp}} = \frac{-|OC|}{|OA|} = -\frac{|OC|}{|OA|} \qquad \tan \theta = \frac{\text{Opp}}{\text{Adj}} = \frac{|AC|}{-|OC|} = -\frac{|AC|}{|OC|}$$

In summary, sine is positive, but cosine and tangent are negative in the second quadrant.

**Third Quadrant** $180° < \theta \leq 270°$

The third quadrant covers angles between 180 and 270 degrees (Figure 1.14). Using $\triangle OAC$, we can state that:

$$\sin \theta = \frac{\text{Opp}}{\text{Hyp}} = \frac{-|AC|}{|OA|} = -\frac{|AC|}{|OA|} \qquad \cos \theta = \frac{\text{Adj}}{\text{Hyp}} = \frac{-|OC|}{|OA|} = -\frac{|OC|}{|OA|} \qquad \tan \theta = \frac{\text{Opp}}{\text{Adj}} = \frac{-|AC|}{-|OC|} = \frac{|AC|}{|OC|}$$

In summary, tangent is positive, but sine and cosine are negative in the third quadrant.

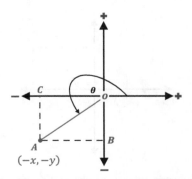

**FIGURE 1.14** Determining trigonometric ratios using phasor diagram (third quadrant).

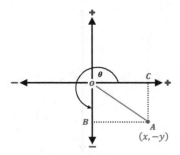

**FIGURE 1.15** Determining trigonometric ratios using phasor diagram (fourth quadrant).

**Fourth Quadrant** $270° < \theta \leq 360°$

The fourth quadrant covers angles between 270 and 360 degrees. Using $\triangle OAC$, we can state that (Figure 1.15):

$$\sin \theta = \frac{\text{Opp}}{\text{Hyp}} = \frac{-|AC|}{|OA|} = -\frac{|AC|}{|OA|} \quad \bigg| \quad \cos \theta = \frac{\text{Adj}}{\text{Hyp}} = \frac{|OC|}{|OA|} = \frac{|OC|}{|OA|} \quad \bigg| \quad \tan \theta = \frac{\text{Opp}}{\text{Adj}} = \frac{-|AC|}{|OC|} = -\frac{|AC|}{|OC|}$$

In summary, cosine is positive, but sine and tangent are negative in the fourth quadrant.

That's easy, right? Let's quickly put together all the above 'summaries' in a table for easy access.

We can see from Table 1.3 that each of the trigonometric ratios is positive in two separate quadrants and negative in the remaining two quadrants. Figure 1.16 attempts to illustrate this.

Let's quickly provide some notes about Figure 1.16.

- **A** stands for **a**ll, which means that all three trigonometric ratios are positive here. Obviously, this confirms our earlier definition of the ratios using a triangle for angles up to and including 90°.
- **S** stands for **S**ine, which means that ONLY sine is positive in the second quadrant.
- **T** stands for **T**angent, which means that ONLY tangent is positive in the third quadrant.
- **C** stands for **C**osine, which means that ONLY cosine is positive in the fourth quadrant.

# Trigonometric Functions I

**TABLE 1.3**

**Signs (Positive and Negative) of the Three Trigonometric Ratios in the Four Quadrants Illustrated**

|  | 1st Quadrant | 2nd Quadrant | 3rd Quadrant | 4th Quadrant |
|---|---|---|---|---|
| **sin $\theta$** | Positive | Positive | Negative | Negative |
| **cos $\theta$** | Positive | Negative | Negative | Positive |
| **tan $\theta$** | Positive | Negative | Positive | Negative |

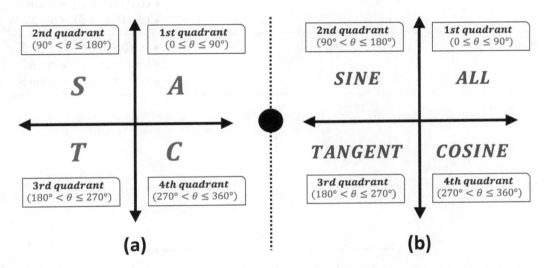

**FIGURE 1.16** Signs (positive and negative) of the three trigonometric ratios in the four quadrants illustrated.

Choose any mnemonic to remember this, but one of the commonly used ones is **ACTS**, going clockwise from the first quadrant or **CAST** going anticlockwise from the fourth quadrant. We can also consider **ASTAC**, obtained **from All–Sine–TAngent–Cosine** going anti-clockwise or **A**ll **S**ilver **T**oy **C**ars (**ASTC**).

In general, we are always given angles in the first quadrant but how do we find other angles within one revolution (i.e., 360) that is equal to the given one in the first quadrant? Table 1.4 provides the equations to be used for each of the three trigonometric functions, given that the angle in the first quadrant is known to be $\theta$.

We have derived the trigonometric ratios of angles 0 and 90° earlier with a right-angled triangle. We will repeat it here but include angles 180°, 270°, and 360° in our derivation.

Consider Figure 1.17 where arm $OA$ of 1 unit swings in such a way that its angle with respect to the positive $x$-axis is $\theta$. We are interested in special positions where point $A$ is at $B$ ($\theta = 90°$), $C$ ($\theta = 180°$), $D$ ($\theta = 270°$), and $E$ ($\theta = 0 = 360°$). We will also consider that $|OB| = -|OD|$ and $|OC| = -|OE|$.

## TABLE 1.4
### Equations for Determining Equivalent Angles in the Four Quadrants

| Quadrant | Equation | Illustration |
|---|---|---|
| First | NA | NA |
| Second | $180° - \theta$ | • $\sin(180° - \theta) = \sin\theta$<br>• $\cos(180° - \theta) = -\cos\theta$<br>• $\tan(180° - \theta) = -\tan\theta$ |
| Third | $180° + \theta$ | • $\sin(180° + \theta) = -\sin\theta$<br>• $\cos(180° + \theta) = -\cos\theta$<br>• $\tan(180° + \theta) = \tan\theta$ |
| Fourth | $360° - \theta$ | • $\sin(360° - \theta) = -\sin\theta$<br>• $\cos(360° - \theta) = \cos\theta$<br>• $\tan(360° - \theta) = -\tan\theta$ |

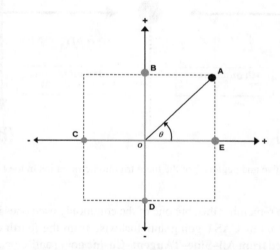

**FIGURE 1.17** Determining trigonometric ratios using phasor diagram (for angles **0°, 90°, 180°,** and **360°**).

**(1) $\theta = 0$**

At $\theta = 0$ is the same as at $\theta = 360°$, which is when the projection arm *OA* is entirely on the positive *x*-axis and there is no vertical component (or zero vertical *y*-axis component). Hence, the trigonometric ratios are:

$$\sin 0° = \frac{\text{Opp}}{\text{Hyp}} = \frac{0}{1} = 0 \qquad \cos 0° = \frac{\text{Adj}}{\text{Hyp}} = \frac{1}{1} = 1 \qquad \tan 0° = \frac{\text{Opp}}{\text{Adj}} = \frac{0}{1} = 0$$

# Trigonometric Functions I

## (2) $\theta = 90°$

When arm $OA$ swings to point $B$, the angle swept will be $90°$. At this point, the projection arm $OA$ is entirely on the positive $y$-axis and there is no horizontal component (or zero $x$-axis component). Hence, the trigonometric ratios for this case are:

$$\sin 90° = \frac{\text{Opp}}{\text{Hyp}} = \frac{1}{1} = \mathbf{1} \qquad \cos 90° = \frac{\text{Adj}}{\text{Hyp}} = \frac{0}{1} = \mathbf{0} \qquad \tan 90° = \frac{\text{Opp}}{\text{Adj}} = \frac{1}{0} = \infty$$

## (3) $\theta = 180°$

When arm $OA$ swings to point $C$, the angle swept will be $\theta = 180°$. At this point, the projection arm $OA$ is entirely on the negative $x$-axis and there is no vertical component (or zero $y$-axis component). Hence, the trigonometric ratios for this case are:

$$\sin 180° = \frac{\text{Opp}}{\text{Hyp}} = \frac{0}{1} = \mathbf{0} \qquad \cos 180° = \frac{\text{Adj}}{\text{Hyp}} = \frac{-1}{1} = \mathbf{-1} \qquad \tan 180° = \frac{\text{Opp}}{\text{Adj}} = \frac{0}{1} = \mathbf{0}$$

Alternatively, using the second-quadrant identities where $\theta = 180°$, we have

$$\sin 180° = \sin(180° - 180°) = \sin 0 = 0$$
$$\cos 180° = -\cos(180° - 180°) = -\cos 0 = -1$$
$$\tan 180° = -\tan(180° - 180°) = -\tan 0 = 0$$

## (4) $\theta = 270°$

When arm $OA$ swings to point $D$, the angle swept will be $\theta = 270°$. At this point, the projection arm $OA$ is entirely on the negative $y$-axis and there is no horizontal component (or zero $x$-axis component). Hence, the trigonometric ratios for this case are:

$$\sin 270° = \frac{\text{Opp}}{\text{Hyp}} = \frac{-1}{1} = \mathbf{-1} \qquad \cos 270° = \frac{\text{Adj}}{\text{Hyp}} = \frac{0}{1} = \mathbf{0} \qquad \tan 270° = \frac{\text{Opp}}{\text{Adj}} = \frac{-1}{0} = -\infty$$

Alternatively, using the third quadrant identities where $\theta = 270°$, we have

$$\sin 270° = -\sin(180° + 90°) = -\sin 90° = -1$$
$$\cos 270° = -\cos(180° + 90°°) = -\cos 90° = 0$$
$$\tan 270° = \tan(180° + 90°) = \tan 90° = \infty$$

## (5) $\theta = 360°$

Same as at $\theta = 0$.

Alternatively, using the fourth quadrant identities where $\theta = 360°$, we have

$$\sin 360° = \sin(360° - 360°) = -\sin 0 = 0$$
$$\cos 360° = \cos(360° - 360°) = \cos 0 = 1$$
$$\tan 360° = -\tan(360° - 360°) = -\tan 0 = 0$$

## TABLE 1.5
**Special Angles (Angles between 90° and 360°)**

|            | 0 | 90° | 180° | 270°        | 360° |
|------------|---|-----|------|-------------|------|
| sin θ      | 0 | 1   | 0    | −1          | 0    |
| cos θ      | 1 | 0   | −1   | 0           | 1    |
| tan θ      | 0 | ∞   | 0    | −∞/∞        | 0    |

This (second set of) special angles is summarised in Table 1.5.

Now it is time to look at a few examples.

### Example 10

Without using a calculator, determine the exact values of the following:

**a)** $\sin 120°$  **b)** $\sin 210°$  **c)** $\sin 315°$  **d)** $\sin(-120°)$

What did you get? Find the solution below to double-check your answer.

### Solution to Example 10

**a)** $\sin 120°$
**Solution**
$\theta = 120°$ is in the second quadrant, so sine will be **positive**. Therefore:

$$\sin 120° = \sin(180° - 120°)$$
$$= \sin 60°$$
$$\therefore \sin 120° = \frac{\sqrt{3}}{2}$$

**b)** $\sin 210°$
**Solution**
$\theta = 210°$ is in the third quadrant, so sine will be **negative**. Therefore:

$$\sin 210° = -\sin(180° + 30°)$$
$$= -\sin 30°$$
$$\therefore \sin 210° = -\frac{1}{2}$$

# Trigonometric Functions I

**c)** $\sin 315°$

**Solution**

$\theta = 315°$ is in the fourth quadrant, so sine will be **negative**. Therefore:

$$\sin 315° = -\sin(360° - 315°)$$
$$= -\sin 45°$$
$$\therefore \sin 315° = -\frac{\sqrt{2}}{2}$$

---

**d)** $\sin(-120°)$

**Solution**

If $\theta = -120°$, it implies a clockwise measurement, therefore:

$$\theta = -120° + 360° = 240°$$

$\theta = 240°$ is in the third quadrant, so it will be **negative**. Therefore:

$$\sin(-120°) = \sin 240°$$
$$= -\sin(180° + 60°)$$
$$= -\sin 60°$$
$$\therefore \sin(-120°) = -\frac{\sqrt{3}}{2}$$

---

More examples to try.

## Example 11

Without using a calculator, determine the exact values of the following:

**a)** $\cos 150°$   **b)** $\cos 240°$   **c)** $\cos 330°$   **d)** $\cos(-135°)$

What did you get? Find the solution below to double-check your answer.

## Solution to Example 11

**a)** $\cos 150°$

**Solution**

$\theta = 150°$ is in the second quadrant, so cosine will be **negative**. Therefore:

$$\cos 150° = -\cos(180° - 150°)$$
$$= -\cos 30°$$
$$\therefore \cos 150° = -\frac{\sqrt{3}}{2}$$

**b)** cos 240°
**Solution**
$\theta = 240°$ is in the third quadrant, so cosine will be **negative**. Therefore:

$$\cos 240° = -\cos(180° + 60°)$$
$$= -\cos 60°$$
$$\therefore \cos 240° = -\frac{1}{2}$$

---

**c)** cos 330°
**Solution**
$\theta = 330°$ is in the fourth quadrant, so cosine will be **positive**. Therefore:

$$\cos 330° = \cos(360° - 330°)$$
$$= \cos 30°$$
$$\therefore \cos 330° = \frac{\sqrt{3}}{2}$$

---

**d)** cos (−135°)
**Solution**
If $\theta = -135°$, it implies a clockwise measurement, therefore:

$$\theta = -135° + 360° = 225°$$

$\theta = 225°$ is in the third quadrant, so it will be **negative**. Therefore:

$$\cos -135° = \cos 225°$$
$$= -\cos(180° + 45°)$$
$$= -\cos 45°$$
$$\therefore \cos(-135°) = -\frac{\sqrt{2}}{2}$$

---

Another set of examples to try.

**Example 12**

Without using a calculator, determine the exact values of the following:

**a)** tan 120°　　　**b)** tan 225°　　　**c)** tan 330°　　　**d)** tan (−90°)

Trigonometric Functions I

What did you get? Find the solution below to double-check your answer.

## Solution to Example 12

**a)** tan 120°
**Solution**
θ = 120° is in the second quadrant, so tan will be **negative**. Therefore:

$$\tan 120° = -\tan(180° - 120°)$$
$$= -\tan 60°$$
$$\therefore \tan 120° = -\sqrt{3}$$

**b)** tan 225°
**Solution**
θ = 225° is in the third quadrant, so tan will be **positive**. Therefore:

$$\tan 225° = \tan(180° + 45°)$$
$$= \tan 45°$$
$$\therefore \tan 225° = 1$$

**c)** tan 330°
**Solution**
θ = 330° is in the fourth quadrant, so tan will be **negative**. Therefore:

$$\tan 330° = -\tan(360° - 330°)$$
$$= -\tan 30°$$
$$\therefore \tan 330° = -\frac{\sqrt{3}}{3}$$

**d)** tan(−90°)
**Solution**
If θ = −90°, it implies a clockwise measurement, therefore:

$$\theta = -90° + 360° = 270°$$

θ = 270° is in the third quadrant, so tan will be **positive**. Therefore:

$$\tan -90° = \tan 270°$$
$$= \tan(180° + 90°)$$
$$= \tan 90°$$
$$\therefore \tan(-90°) = \infty$$

Before we finish this section, it is essential to know whether there is a way to solve the trigonometric ratios of angles greater than 360°. Yes, we can do this! The clue to this is when we said that θ = 0 is the same as θ = 360°.

We can similarly say this for any other angle above $\theta = 360°$. In a clearer tone, angle $1°$ is the same as $361°$, because (taking 0 as our reference point) rotating by $1°$ and $361°$ from the origin will take you to the same point. You can try and confirm the sine, cosine, and tangent of the following pairs:

$$30° \text{ and } 390°, \quad 45° \text{ and } 405°, \quad \textbf{AND} \quad 213° \text{ and } 573°$$

What did you get? Check these too:

$$30° \text{ and } -330°, \quad 45° \text{ and } -315°, \quad \textbf{AND} \quad 213° \text{ and } -147°.$$

The above exercise shows that when multiples of $360°$ are added to or subtracted from an angle $\theta$, the trigonometric ratios of the angle remain unchanged. We can write this in standard format as:

$$\boxed{\sin\theta = \sin(\theta + 360°n) = \sin(\theta - 360°n)} \quad (1.9)$$

$$\boxed{\cos\theta = \cos(\theta + 360°n) = \cos(\theta - 360°n)} \quad (1.10)$$

$$\boxed{\tan\theta = \tan(\theta + 360°n) = \tan(\theta - 360°n)} \quad (1.11)$$

where $n$ is any positive integer.

Let's bid the section a farewell with some examples.

## Example 13

Determine the equivalents of the following trigonometric functions in the range $0 \leq \theta \leq 360°$.

a) $\sin 470°$      b) $\sin 915°$      c) $\sin(-265°)$      d) $\sin(-945°)$

What did you get? Find the solution below to double-check your answer.

## Solution to Example 13

**a)** $\sin 470°$
**Solution**
Since $\theta > 360°$, we need to subtract $360°n$, where $n = 1$ in this case (Figure 1.18). Therefore:

$$\sin 470° = \sin(470° - 360°)$$
$$= \sin 110°$$

# Trigonometric Functions I

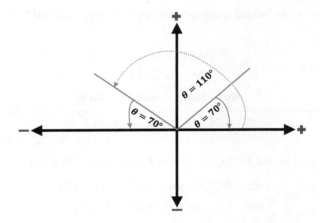

**FIGURE 1.18** Solution to Example 13(a).

This is the first value in the second quadrant, the other positive value will be in the first quadrant and is

$$\sin 110° = \sin(180° - 110°)$$
$$= \sin 70°$$
$$\therefore \sin 470° = \sin 110° = \sin 70°$$

**NOTE**
We've included a graph for this example for illustration only, though this is not required.

---

**b)** $\sin 915°$

**Solution**

$\theta > 360°$, so we need to subtract $360°n$, where $n = 2$ in this case. Therefore:

$$\sin 915° = \sin(915° - 360° \times 2)$$
$$= \sin(915° - 720°)$$
$$= \sin 195°$$

This is the first value in the third quadrant, the other positive value will be in the fourth quadrant and is

$$\sin 195° = \sin(360° - 15°)$$
$$= \sin 345°$$
$$\therefore \sin 915° = \sin 195° = \sin 345°$$

---

**c)** $\sin(-265°)$

**Solution**

$\theta < 360°$, so we need to add $360°n$, where $n = 1$ in this case. Therefore:

$$\sin(-265°) = \sin(-265° + 360°)$$
$$= \sin 95°$$

This is the first value in the second quadrant, the other positive value will be in the first quadrant and is

$$\sin 95° = \sin(180° - 95°)$$
$$= \sin 85°$$
$$\therefore \sin(-265°) = \sin 95° = \sin 85°$$

---

**d)** $\sin(-945°)$
**Solution**
Since $\theta < 360°$, we need to add $360°n$, where $n = 3$ in this case. Therefore:

$$\sin(-945°) = \sin(-945° + 360° \times 3)$$
$$= \sin(-945° + 1080°)$$
$$= \sin 135°$$

This is the first value in the second quadrant, the other positive value will be in the first quadrant and is

$$\sin 135° = \sin(180° - 135°)$$
$$= \sin 45°$$
$$\therefore \sin(-945°) = \sin 135° = \sin 45°$$

---

Another set of examples to try.

### Example 14

Determine the equivalents of the following trigonometric functions in the range $0 \le \theta \le 360°$.

**a)** $\cos 395°$  **b)** $\cos 900°$  **c)** $\cos(-1360°)$  **d)** $\cos(-270°)$

What did you get? Find the solution below to double-check your answer.

### Solution to Example 14

**a)** $\cos 395°$
**Solution**
$\theta > 360°$, so we need to subtract $360°n$, where $n = 1$ in this case. Therefore:

$$\cos 395° = \cos(395° - 360°)$$
$$= \cos 35°$$

This is the first value in the first quadrant, the other positive value will be in the fourth quadrant and is

$$\cos 35° = \cos(360° - 35°)$$
$$= \cos 325°$$
$$\therefore \cos 395° = \cos 35° = \cos 325°$$

**b)** cos 900°
**Solution**
θ > 360°, so we need to subtract 360°n, where n = 2 in this case. Therefore:

$$\cos 900° = \cos(900° - 360° \times 2)$$
$$= \cos 180°$$

This is the first value in the second quadrant, the other positive value will be in the third quadrant and is

$$\cos 180° = \cos(180° + 0°)$$
$$= \cos 180°$$
$$\therefore \mathbf{\cos 900° = \cos 180° = \cos 180°}$$

**NOTE**
This is tricky because there is only one equivalent for the cosine function, or we can say that there exists a repeated equivalent. The latter would imply that 180° is a second quadrant as well as a third quadrant value.

---

**c)** cos(−1360°)
**Solution**
θ < 360°, so we need to add 360°n, where n = 4 in this case. Therefore:

$$\cos(-1360°) = \cos(-1360° + 360° \times 4)$$
$$= \cos 80°$$

This is the first value in the first quadrant, the other positive value will be in the fourth quadrant and is

$$\cos 80° = \cos(360° - 80°)$$
$$= \cos 280°$$
$$\therefore \mathbf{\cos(-1360°) = \cos 80° = \cos 280°}$$

---

**d)** cos(−270°)
**Solution**
θ < 360°, so we need to add 360°n, where n = 1 in this case. Therefore:

$$\cos(-270°) = \cos(-270° + 360°)$$
$$= \cos 90°$$

This is the first value in the first quadrant, the other positive value will be in the fourth quadrant and is

$$\cos 90° = \cos(360° - 90°)$$
$$= \sin 270°$$
$$\therefore \mathbf{\cos(-270°) = \cos 90° = \cos 270°}$$

One final set of examples to try.

## Example 15

Determine the equivalents of the following trigonometric functions in the range $0 \leq \theta \leq 360°$.

**a)** $\tan 978°$      **b)** $\tan 515°$      **c)** $\tan(-300°)$      **d)** $\tan(-45°)$

What did you get? Find the solution below to double-check your answer.

## Solution to Example 15

**a)** $\tan 978°$
**Solution**
$\theta > 360°$ so, we need to subtract $360°n$, where $n = 2$ in this case. Therefore:

$$\tan 978° = \tan(978° - 360° \times 2)$$
$$= \tan 258°$$

This is the first value in the third quadrant, the other will be in the first quadrant and is

$$\tan 258° = \tan(180° + 78°)$$
$$= \tan 78°$$
$$\therefore \tan 978° = \tan 258° = \tan 78°$$

---

**b)** $\tan 515°$
**Solution**
$\theta > 360°$, so we need to subtract $360°n$, where $n = 1$ in this case. Therefore:

$$\tan 515° = \tan(515° - 360°)$$
$$= \tan 155°$$

This is the first value in the second quadrant, the other positive value will be in the fourth quadrant and is

$$\tan 155° = \tan(360° - 25°)$$
$$= \tan 335°$$
$$\therefore \tan 515° = \tan 155° = \tan 335°$$

**NOTE**
For this example and the one above, we could have simply added or subtracted $180°$ from the first value. This is a property of tangent function, which will be discussed later in this chapter.

**c)** tan (−300°)
**Solution**
$\theta < 360°$, so we need to add $360°n$, where $n = 1$ in this case. Therefore:

$$\tan(-300°) = \tan(-300° + 360°)$$
$$= \tan 60°$$

This is the first value in the first quadrant, the other positive value will be in the third quadrant and is

$$\tan 60° = \tan(180° + 60°)$$
$$= \tan 240°$$
$$\therefore \tan(-300°) = \tan 60° = \tan 240°$$

**d)** tan (−45°)
**Solution**
$\theta < 360°$, so we need to add $360°n$, where $n = 1$ in this case. Therefore:

$$\tan(-45°) = \tan(-45° + 360°)$$
$$= \tan 315°$$

This is the first value in the fourth quadrant, the other positive value will be in the second quadrant and is

$$\tan 315° = \tan(180° - 45°)$$
$$= \tan 135°$$
$$\therefore \tan(-45°) = \tan 315° = \tan 135°$$

## 1.6 GRAPHS OF TRIGONOMETRIC FUNCTIONS

We have now looked, albeit numerically, at the 'outputs' of the three trigonometric functions (sine, cosine, and tangent) as angle $\theta$ varies. In this section and the ones to follow, our focus is to visually examine these functions as a plot of the function against the angle. For this, our independent variable (x-axis) is the angle $\theta$ and the dependent variable (y-axis) is $\sin \theta$, $\cos \theta$, or $\tan \theta$. Like any graphical representation or method, we will see how the graphs of these functions can be used to solve problems relating to trigonometric functions.

### 1.6.1 Sine Function

Figure 1.19 is a plot of $y = \sin \theta$ in the interval $0 \leq \theta \leq 360°$. This can be achieved by creating a table as shown in Table 1.6 or by using a graphical calculator or any suitable software. In any case, the output is the same.

We have now modified the interval $-720° \leq \theta \leq 720°$ and re-plotted the function to cover four revolutions or cycles (Figure 1.20).

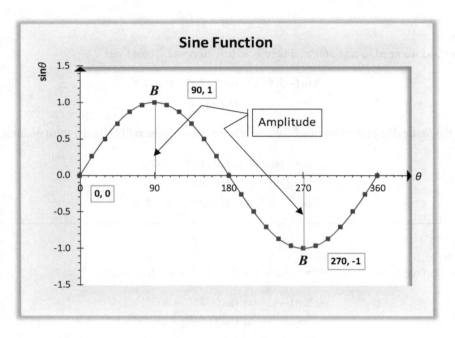

**FIGURE 1.19** Sine function in the interval $0 \leq \theta \leq 360°$ illustrated.

**TABLE 1.6**
**Sine Function in the Interval $0° \leq \theta \leq 360°$ Illustrated**

| $\theta$ | 0° | 30° | 60° | 90° | 120° | 150° | 180° | 210° | 240° | 270° | 300° | 330° | 360° |
|---|---|---|---|---|---|---|---|---|---|---|---|---|---|
| $\sin\theta$ | 0 | 0.5 | 0.866 | 1 | 0.866 | 0.5 | 0 | −0.5 | −0.866 | −1 | −0.866 | −0.5 | 0 |

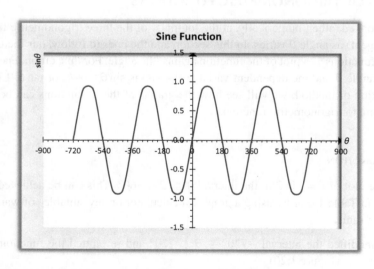

**FIGURE 1.20** Sine function in the interval $-720° \leq \theta \leq 720°$ illustrated.

# Trigonometric Functions I

From the above plots, we can summarise the behaviour of a sine function as:

**Note 1** The graph forms a wave-like pattern, and it is said to **oscillate**.

**Note 2** The maximum value of $\sin\theta$ is 1 and occurs at 90 degrees. It repeats itself every 360° to the left or right of the graph (i.e., $90° \pm 360°n$). These points are called the **peaks** or **crests** of the sine function.

**Note 3** The minimum value of $\sin\theta$ is $-1$ and occurs at 270 degrees. It repeats every 360° to the left or right of the graph (i.e., $270° \pm 360°n$). These points are called the **troughs**.

**Note 4** It has (0, 0) coordinates, which coincide with the origin and where the graph crosses the $x$-axis. It repeats every 180° to the left or to the right of the graph (i.e., $\pm 180°n$). These are the points where there are no vibrations of the medium particles.

**Note 5** The shape of the function in the interval $0 \leq \theta \leq 360°$ repeats indefinitely in both directions of the $x$-axis. We therefore say that the sine function is **periodic** with a **periodicity** of 360°. One complete cycle is called a **period**.

**Note 6** Since $\sin(-x) = -\sin x$, the sine function has a **rotational symmetry** about the origin (0, 0) such that it looks the same anytime it is rotated 180 degrees about the origin. It is therefore an odd function.

**Note 7** Sine has another symmetry about $\theta = 90°$.

**Note 8** The value of a vertical line drawn from the $x$-axis to the maximum point ($A$) or minimum point ($B$) is called the **Amplitude** of the function, and it is denoted with a letter **A**. It is also the same as half the distance between the crest and trough.

## 1.6.2 Cosine Function

In Figure 1.21, the cosine function $y = \cos\theta$ has been plotted in the same way as the sine function. First in the interval $0° \leq \theta \leq 360°$ (Figure 1.21) and then extended to $-720° \leq \theta \leq 720°$ (Figure 1.22).

We will notice that the two functions are identical in shape. Apart from the starting point, one may not be able to distinguish between them. Can you try to identify which one is sine and which one is cosine in the plot in Figure 1.23?

Did you say that $f(x)$ is the sine function and $g(x)$ is the cosine function? This is a good attempt, but the answer is the other way around, i.e. $g(x)$ is sine and $f(x)$ is cosine.

**TABLE 1.7**

**Cosine Function in the Interval $0° \leq \theta \leq 360°$ Illustrated**

| $\theta$ | 0° | 30° | 60° | 90° | 120° | 150° | 180° | 210° | 240° | 270° | 300° | 330° | 360° |
|---|---|---|---|---|---|---|---|---|---|---|---|---|---|
| $\cos\theta$ | 1 | 0.866 | 0.5 | 0 | $-0.5$ | $-0.866$ | $-1$ | $-0.866$ | $-0.5$ | 0 | 0.5 | 0.866 | 1 |

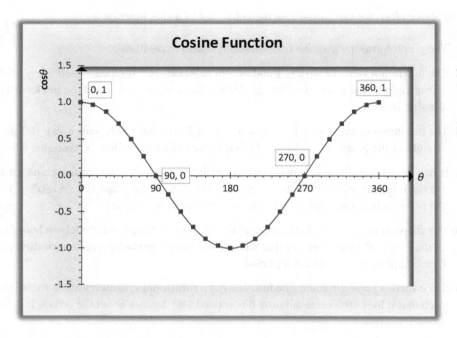

**FIGURE 1.21** Cosine function in the interval $0 \leq \theta \leq 360°$ illustrated.

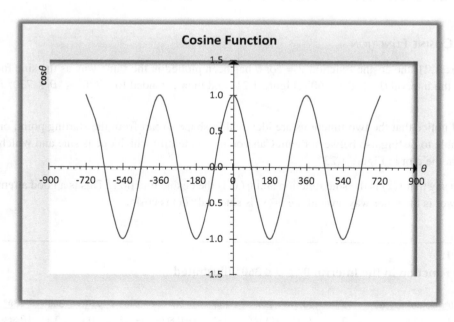

**FIGURE 1.22** Cosine function in the interval $-720° \leq \theta \leq 720°$ illustrated.

# Trigonometric Functions I

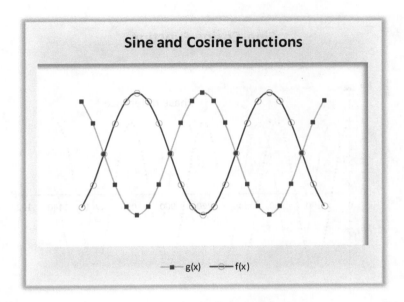

**FIGURE 1.23** Sine and cosine functions indistinguishably drawn.

All notes about the sine function are equally applicable to the cosine function, but there are additional points summarised as follows:

**Note 1** The maximum value of $\cos\theta$ is 1 and occurs at 0 degrees. It repeats itself every 360° to the left or right of the graph, i.e., $\pm 360°n$. For example, it is 1 at $\pm 360°$, $\pm 720°$, etc.

**Note 2** The minimum value of $\cos\theta$ is $-1$ and occurs at 180 degrees. It repeats every 360° to the left or right of the graph, i.e., $270° \pm 360°n$.

**Note 3** It has (0, 1) coordinates, which coincide with where the graph crosses the $y$-axis.

**Note 4** Since $\cos(-x) = \cos x$, the cosine function is symmetrical about the $y$-axis or $\theta = 0$. It is therefore an even function.

**Note 5** There is a phase difference of 90 degrees between the sine and cosine functions, with cosine leading the sine (Figure 1.24). Alternatively, we can say that sine lags the cosine function.

**Note 6** The 90° phase difference between the two functions is equivalent to a horizontal shift. We can therefore derive one from the other.

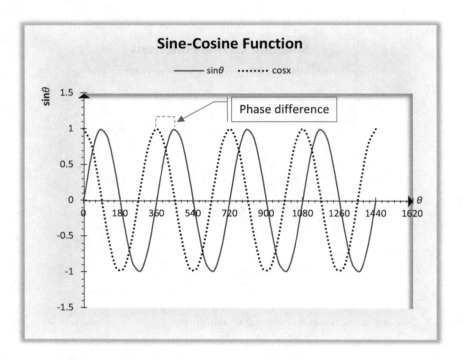

**FIGURE 1.24** Phase difference illustrated.

### 1.6.3 Tangent Function

Figure 1.25 is a plot of the function $y = \tan \theta$ in the interval $0 \leq \theta \leq 360°$ (Table 1.8). The interval is extended to $-360° \leq \theta \leq 360°$ (Figure 1.26).

It is obvious that the tangent function is remarkably different from the other two functions (Figure 1.26). Let us summarise the key features of this function between 0 and 360 degrees as:

**Note 1** The curve is not continuous but has three distinct parts within $0 \leq \theta \leq 360°$ or two distinct parts within $0 \leq \theta \leq 180°$.

**Note 2** The graph rises rapidly as the angle approaches 90° and 270°, but it never touches the vertical lines. These lines are called **vertical asymptotes** and are shown with dotted lines in Figure 1.25. We can simply say that this occurs at an odd multiple of 90°.

**Note 3** There are discontinuities at 90° and 270° as the tangent function is **undefined** at these points. In other words, the value at these angles is infinite. A calculator will return an error message when you try $\tan 90°$ or $\tan 270°$.

**Note 4** It has (0, 0) coordinates, which coincide with the origin and where the graph crosses the $x$-axis. It repeats at every 180° to the left or right of the graph, i.e., $\pm 180° n$.

**Note 5** Unlike sine and cosine functions, the tangent function does not have minimum and maximum values. Or more technically, the maximum and minimum values are $\infty$ and $-\infty$, respectively.

**Note 6** Tangent is also periodic, but it repeats itself every 180°. Thus, its **periodicity** is 180°.

Trigonometric Functions I

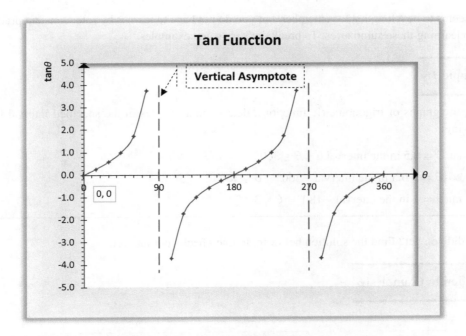

**FIGURE 1.25** Tangent function in the interval $0° \leq \theta \leq 360°$ illustrated.

**TABLE 1.8**
**Tangent Function in the Interval $0° \leq \theta \leq 360°$ Illustrated**

| $\theta$ | 0° | 30° | 60° | 90° | 120° | 150° | 180° | 210° | 240° | 270° | 300° | 330° | 360° |
|---|---|---|---|---|---|---|---|---|---|---|---|---|---|
| $\tan \theta$ | 0 | 0.5774 | 1.7321 | ∞ | −1.7321 | −0.5774 | 0 | 0.5774 | 1.7321 | ∞ | −1.7321 | −0.5774 | 0 |

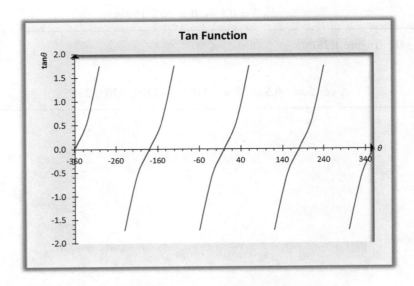

**FIGURE 1.26** Tangent function in the interval $-360° \leq \theta \leq 360°$ illustrated.

This seems to be a long list to remember, but you do not have to learn it by rote; constant practice is key to learning these summaries. To break, let's try some examples.

### Example 16

Using the graphs of trigonometric functions, determine all angles in the specified interval for the following:

a) $\sin \theta = 0.5$ in the interval $0 \leq \theta \leq 360°$

b) $\cos \theta = -0.5$ in the interval $-360° \leq \theta \leq 360°$

c) $\tan \theta = 1$ in the interval $-180° \leq \theta \leq 360°$

What did you get? Find the solution below to double-check your answer.

### Solution to Example 16

**HINT**

Apply the following:

- Sketch or draw a graph of the function $y = \sin \theta$, $y = \cos \theta$, or $y = \tan \theta$.
- Draw a horizontal line $l_1$ at the given value of the function and draw a vertical line $l_2$ where $l_1$ meets the plot of the function.
- Read the values off the $x$-axis. This is the angle you are looking for.

**a)** $\sin \theta = 0.5$ (Figure 1.27)
**Solution**

$$\therefore \sin \theta = 0.5 \rightarrow \theta = 30°, 150°$$

**b)** $\cos \theta = -0.5$ (Figure 1.28)
**Solution**

$$\therefore \cos \theta = -0.5 \rightarrow \theta = -240°, -120°, 120°, 240°$$

# Trigonometric Functions I

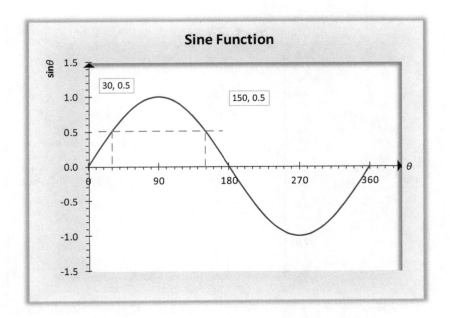

**FIGURE 1.27** Solution to Example 16(a).

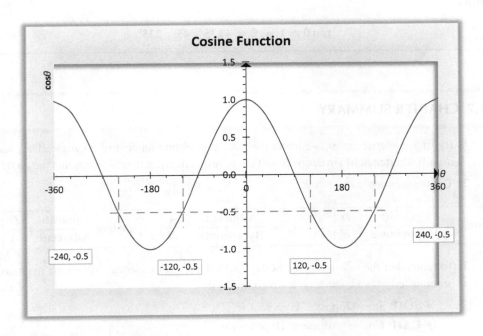

**FIGURE 1.28** Solution to Example 16(b).

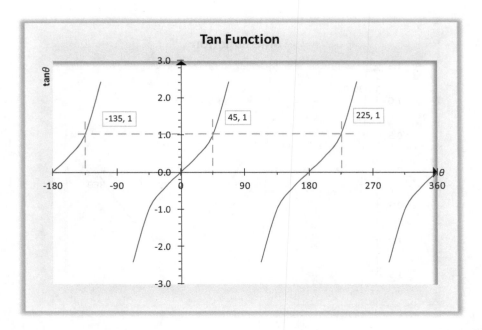

**FIGURE 1.29** Solution to Example 16(c).

**c)** $\tan \theta = 1$ (Figure 1.29)
**Solution**

$$\therefore \tan \theta = 1 \rightarrow \theta = -135°, \ 45°, \ 225°$$

---

## 1.7 CHAPTER SUMMARY

1) By trigonometric ratios, we mean the sine, cosine, and tangent of an angle. They are usually shortened as $\sin \theta$, $\cos \theta$, and $\tan \theta$, respectively, where $\theta$ represents the angle.

2) The trigonometric ratios are:

$$\sin \theta = \frac{\text{Opposite}}{\text{Hypotenuse}} = \frac{O}{H} \qquad \cos \theta = \frac{\text{Adjacent}}{\text{Hypotenuse}} = \frac{A}{H} \qquad \tan \theta = \frac{\text{Opposite}}{\text{Adjacent}} = \frac{O}{A}$$

3) To remember the ratios, we use **SOH–CAH–TOA** (as one word without the hyphens, pronounced as 'sow-car-tour'), which implies:
   - **SOH**: **S**ine–**O**pposite–**H**ypotenuse
   - **CAH**: **C**osine–**A**djacent–**H**ypotenuse
   - **TOA**: **T**angent–**O**pposite–**A**djacent

# Trigonometric Functions I

4) Tangent is a ratio of sine to cosine and is given by:

$$\boxed{\tan\theta = \frac{\sin\theta}{\cos\theta}}$$

5) By special angles, it usually implies the naturally occurring angles such as **30°**, **45°**, and **60°**. Angles **0°** and **90°** are commonly included in this group.

6) In general, if $\theta$ is the angle measured in a clockwise direction from the positive *x*-axis, its equivalent in an anti-clockwise direction is:

$$\boxed{\theta + 360°}$$

7) Similarly, if angle $\theta$ measured in an anti-clockwise direction from the positive *x*-axis, its equivalent in a clockwise direction is:

$$\boxed{\theta - 360°}$$

8) Sine and tangent of negative angles result in negative values of their respective functions, but cosine of negative angles remains unchanged.

$$\boxed{\sin(-\theta) = -\sin\theta} \text{ OR } \boxed{\cos(-\theta) = \cos\theta} \text{ OR } \boxed{\tan(-\theta) = -\tan\theta}$$

9) Quadrants
   - Sine, cosine, and tangent are all positive in the first quadrant.
   - Sine is positive, but cosine and tangent are negative in the second quadrant, and we use $180° - \theta$.
   - Tangent is positive, but sine and cosine are negative in the third quadrant, and we use $180° + \theta$.
   - Cosine is positive, but sine and tangent are negative in the fourth quadrant, and we use $360° - \theta$.

10) When multiples of 360° are added to or subtracted from an angle $\theta$, the trigonometric ratios of the angle remain unchanged. We can write this in standard format as:

$$\boxed{\sin\theta = \sin(\theta + 360°n) = \sin(\theta - 360°n)}$$

$$\boxed{\cos\theta = \cos(\theta + 360°n) = \cos(\theta - 360°n)}$$

$$\boxed{\tan\theta = \tan(\theta + 360°n) = \tan(\theta - 360°n)}$$

where *n* is any positive integer.

11) Sine and cosine functions
    - The graph forms a wave-like pattern, and it is said to **oscillate**.
    - The maximum value of $\sin\theta$ is 1 and occurs at 90 degrees, and the maximum value of $\cos\theta$ is 1 and occurs at 0 degree. These points are called the **peaks** or **crests** of the sine/cosine function.

- The minimum value of $\sin\theta$ is $-1$ and occurs at 270 degree, and the minimum value of $\cos\theta$ is $-1$ and occurs at 180 degrees. These points are called the **troughs** of the sine/cosine function.
- Sine and cosine functions are **periodic** with a **periodicity** of 360°.
- Sine has (0, 0) coordinates, which coincide with the origin and where the graph crosses the $x$-axis. Cosine has (0, 1) coordinates, which is where the graph crosses the $y$-axis.
- Since $\sin(-x) = -\sin x$, the sine function exhibits **rotational symmetry** about the origin ((0, 0)) such that it appears identical when rotated 180 degrees around the origin.
- Since $\cos(-x) = \cos x$, the cosine function is symmetrical about the $y$-axis.
- The value of a vertical line drawn from the $x$-axis to the maximum point, or minimum point is called the **Amplitude** of the function and it is denoted with a letter **A**.
- There is a phase difference of 90 degrees between sine and cosine functions, with cosine leading the sine. Alternatively, we can say that sine lags the cosine function.

12) Tangent function

- The curve is not continuous but has three distinct parts within $0 \leq \theta \leq 360°$ or two distinct parts within $0 \leq \theta \leq 180°$.
- The graph rises rapidly as the angle approaches 90° and 270°, but it never touches the vertical lines. These lines are called **vertical asymptotes**.
- There are discontinuities at 90° and 270° as the tangent function is **undefined** at these points.
- Tangent has (0, 0) coordinates, which coincide with the origin and where the graph crosses the $x$-axis. It repeats at every 180° to the left or right of the graph, i.e., $\pm 180°n$.
- Unlike sine and cosine function, tangent function does not have minimum and maximum values. Or more technically, the maximum and minimum values are $\infty$ and $-\infty$ respectively.
- Tangent is also periodic, but it repeats itself every 180°. Thus, its **periodicity** is 180°.

****

## 1.8 FURTHER PRACTICE

To access complementary contents, including additional exercises, please go to www.dszak.com.

# 2 Trigonometric Functions II

## Learning Outcomes

Once you have studied the content of this chapter, you should be able to:

- Discuss the graphs of inverse trigonometric functions
- Discuss the graphs of reciprocal functions
- Evaluate the transformations of sine, cosine, and tangent functions

## 2.1 INTRODUCTION

This chapter will, in continuation of our discussion on trigonometric functions, look at both the reciprocals and the inverse of the three trigonometric functions that were covered in the previous chapter. It will also discuss the transformation of these functions.

## 2.2 RECIPROCAL OF TRIGONOMETRIC FUNCTIONS

The reciprocal of sine, cosine, and tangent are called cosecant, secant, and cotangent, respectively; they are shortened to cosec, sec, and cot. Notice that the third letter of each name shows the reciprocal we are referring to. cosec is further abbreviated to csc as a three-letter convention.

In general, a reciprocal of a number, variable, or function is one over it. Hence, the ratios of the reciprocals of trigonometric functions are given as:

$$\csc\theta = \frac{1}{\sin\theta} = \frac{\text{Hyp}}{\text{Opp}} \qquad \sec\theta = \frac{1}{\cos\theta} = \frac{\text{Hyp}}{\text{Adj}} \qquad \cot\theta = \frac{1}{\tan\theta} = \frac{\text{Adj}}{\text{Opp}} \qquad (2.1)$$

It is often easier to find the main trigonometric ratio and flip the answer to obtain its reciprocal. For example, to find $\cot x$, find $\tan x$ and take the reciprocal of the answer.

What follows is a description of the behaviour of each of the three reciprocals.

### 2.2.1 Sec Function

Figure 2.1 is a plot of $\sec x$ and $\cos x$ on the same chart in the range $-2\pi < x < 2\pi$.

Most of the features of the cosine function hold for sec with a few observations as noted in Table 2.1. In fact, secant is to cosecant as cosine is to sine.

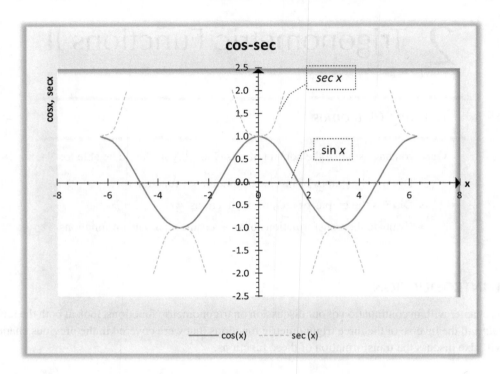

**FIGURE 2.1** Sec function in the interval $-2\pi < x < 2\pi$ illustrated.

**TABLE 2.1**
**Characteristics of Sec Function Explained**

|  | $\sec x$ |
|---|---|
| Note 1 | As $\sec x$ is a reciprocal of $\cos x$, it implies that it is undefined when $\cos x = 0$. Hence, $\sec x$ has vertical asymptotes at these points, i.e., $x = \frac{1}{2}\pi \pm n\pi$, where $n$ is any positive integer. |
| Note 2 | As angle $x$ approaches $\frac{1}{2}\pi$ from the left, $\cos x$ is positive but approaches zero whilst $\sec x$ is positive but approaches infinity (or the vertical asymptote). On the other hand, if angle $x$ approaches $\frac{1}{2}\pi$ from the right, $\cos x$ is negative but approaches zero whilst $\sec x$ is also negative but approaches negative infinity. |
| Note 3 | $\sec x$ is maximum when $\cos x$ is minimum and vice versa. |
| Note 4 | The maximum and minimum values for $\sec x$ are $-1$ and $1$, respectively; an exact reciprocal of $\cos x$. |
| Note 5 | Like $\cos x$, $\sec x$ has a periodicity of $2\pi$ and the $y$-axis acts as its line of symmetry. |

# Trigonometric Functions II

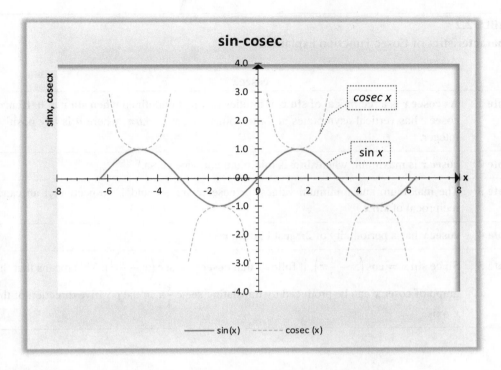

**FIGURE 2.2** Cosec function in the interval $-2\pi < x < 2\pi$ illustrated.

## 2.2.2 Cosec Function

Figure 2.2 is a plot of $\csc x$ and $\sin x$ on the same chart in the range $-2\pi < x < 2\pi$.

Most of the features of the sine function hold for cosec with a few observations as noted in Table 2.2.

## 2.2.3 Cot Function

Figure 2.3 is a plot of $\cot x$ and $\tan x$ on the same chart in the range $-2\pi < x < 2\pi$.

Most of the features of the tangent function hold for cot with a few observations as noted in Table 2.3.

Now let's look at examples.

### Example 1

Without using a calculator, determine the exact value of the following:

a) $\sec 330°$  b) $\csc 240°$  c) $\cot 135°$  d) $\cot(-135°)$  e) $\sec\left(-\dfrac{4\pi}{3}\right)$  f) $\csc\left(-\dfrac{3\pi}{2}\right)$

### TABLE 2.2
### Characteristics of Cosec Function Explained

| | cosec $x$ |
|---|---|
| Note 1 | As **cosec** $x$ is a reciprocal of **sin** $x$, it implies that it is undefined when **sin** $x = 0$. Hence, **cosec** $x$ has vertical asymptotes at these points, i.e., $x = \pm n\pi$, where $n$ is any positive integer. |
| Note 2 | **cosec** $x$ is maximum when **sin** $x$ is minimum and vice versa. |
| Note 3 | The maximum and minimum values for **cosec** $x$ are $-1$ and $1$, respectively; an exact reciprocal of **sin** $x$. |
| Note 4 | **cosec** $x$ has a periodicity of $2\pi$ just like **sin** $x$. |
| Note 5 | Since $\sin x = \cos\left(x - \frac{1}{2}\pi\right)$, it follows that $\operatorname{cosec} x = \sec\left(x - \frac{1}{2}\pi\right)$. This means that the graph of **cosec** $x$ can be produced by translating **sec** $x$ $\frac{1}{2}\pi$ in the positive direction of the $x$-axis. |

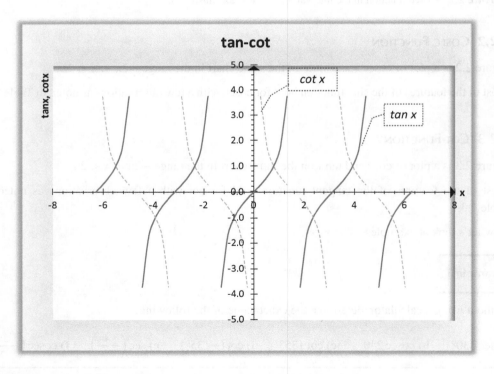

**FIGURE 2.3** Cot function in the interval $-2\pi < x < 2\pi$ illustrated.

Trigonometric Functions II

**TABLE 2.3**
**Characteristics of Cot Function Explained**

| | cot $x$ |
|---|---|
| Note 1 | As **cot $x$** is a reciprocal of **tan $x$**, it means that it is undefined when **tan $x$ = 0**, which is the same as when **sin $x$ = 0**. Hence, **cot $x$** has vertical asymptotes at these points, i.e., at $x = \pm n\pi$, where $n$ is any positive integer. |
| Note 2 | **cot $x$ = 0** when **tan $x$** is undefined, which implies that it crosses the $x$-axis at these points, i.e., $x = \frac{1}{2}\pi \pm n\pi$ where $n$ is any positive integer. The reverse is the case, i.e., **cot $x$** is undefined, **tan $x$** crosses the $x$-axis. |
| Note 3 | When **tan $x$** is positive and small, **cot $x$** will be positive but large and vice versa. The same trend is observed for the negative $y$-axis region. |
| Note 4 | **cot $x$** has a periodicity of $\pi$ just like **tan $x$**. |

What did you get? Find the solution below to double-check your answer.

**Solution to Example 1**

**a) sec 330°**
**Solution**

$$\sec\theta = \frac{1}{\cos\theta}$$

$\theta = 330°$ is in the fourth quadrant, so it will be **positive** for cosine. Therefore, we have

$$\sec 330° = \frac{1}{\cos 330°} = \frac{1}{\cos(360° - 330°)}$$
$$= \frac{1}{\cos 30°} = \frac{1}{\left(\frac{\sqrt{3}}{2}\right)} = \frac{2}{\sqrt{3}}$$

$$\therefore \sec 330° = \frac{2}{\sqrt{3}}$$

**NOTE**
We can rationalise $\frac{2}{\sqrt{3}}$ to obtain $\frac{2\sqrt{3}}{3}$, so **sec 330°** $= \frac{2\sqrt{3}}{3}$.

**b)** cosec 240°
**Solution**

$$\csc \theta = \frac{1}{\sin \theta}$$

$\theta = 240°$ is in the third quadrant, so it will be **negative** for sine. Therefore, we have

$$\csc 240° = \frac{1}{\sin 240°} = -\frac{1}{\sin(180° + 60°)}$$
$$= -\frac{1}{\sin 60°} = -\frac{1}{\left(\frac{\sqrt{3}}{2}\right)} = -\frac{2}{\sqrt{3}}$$

$$\therefore \csc 240° = -\frac{2}{\sqrt{3}} \text{ or } -\frac{2\sqrt{3}}{3}$$

---

**c)** cot 135°
**Solution**

$$\cot \theta = \frac{1}{\tan \theta}$$

$\theta = 135°$ is in the second quadrant, so it will be **negative** for tangent. Therefore, we have

$$\cot 135° = \frac{1}{\tan 135°} = -\frac{1}{\tan(180° - 135°)}$$
$$= -\frac{1}{\tan 45°} = -\frac{1}{1} = -1$$

$$\therefore \cot 135° = -1$$

---

**d)** cot(−135°)
**Solution**

$$\cot \theta = \frac{1}{\tan \theta}$$

If $\theta = -135°$, it implies a clockwise measurement, therefore:

$$\theta = 360° - 135° = 225°$$

$\theta = 225°$ is in the third quadrant, so it will be **positive** for tangent. Therefore, we have

$$\cot(-135°) = \cot 225° = \frac{1}{\tan 225°}$$
$$= \frac{1}{\tan(180° + 45°)}$$
$$= \frac{1}{\tan 45°} = \frac{1}{1} = 1$$

$$\therefore \cot(-135°) = 1$$

# Trigonometric Functions II

**e)** $\sec\left(-\frac{4\pi}{3}\right)$

**Solution**

$$\sec\theta = \frac{1}{\cos\theta}$$

If $\theta = -\frac{4\pi}{3}$, it implies a clockwise measurement, therefore:

$$\theta = 2\pi - \frac{4\pi}{3} = \frac{2\pi}{3}$$

$\theta = \frac{2\pi}{3}$ is in the second quadrant, so it will be **negative** for cosine. Therefore, we have

$$\sec\left(-\frac{4\pi}{3}\right) = \sec\frac{2\pi}{3} = \frac{1}{\cos\frac{2\pi}{3}}$$

$$= -\frac{1}{\cos\left(\pi - \frac{2\pi}{3}\right)}$$

$$= -\frac{1}{\cos\frac{\pi}{3}} = -\frac{1}{\left(\frac{1}{2}\right)} = -2$$

$$\therefore \sec\left(-\frac{4\pi}{3}\right) = -2$$

---

**f)** $\operatorname{cosec}\left(-\frac{3\pi}{2}\right)$

**Solution**

$$\operatorname{cosec}\theta = \frac{1}{\sin\theta}$$

If $\theta = -\frac{3\pi}{2}$, it implies a clockwise measurement, therefore:

$$\theta = 2\pi - \frac{3\pi}{2} = \frac{\pi}{2}$$

$\theta = \frac{\pi}{2}$ is in the first quadrant, so it will be **positive** for sine. Therefore, we have

$$\operatorname{cosec}\left(-\frac{3\pi}{2}\right) = \operatorname{cosec}\frac{\pi}{2} = \frac{1}{\sin\frac{\pi}{2}}$$

$$= \frac{1}{1} = 1$$

$$\therefore \operatorname{\mathbf{cosec}}\left(-\frac{3\pi}{2}\right) = \mathbf{1}$$

Another example to try.

### Example 2

Using trigonometric ratios, prove that $\tan\theta = \cot(90° - \theta)$.

---

What did you get? Find the solution below to double-check your answer.

### Solution to Example 2

To prove this, we will consider $\triangle ABC$ where $\angle B = \theta$ (Figure 2.4). Since this is a right-angled triangle, angles $B$ and $C$ are complimentary (i.e., $\angle B + \angle C = 90°$ or $\angle C = 90° - \theta$) as shown in Figure 2.4.

We have:

$$\tan\theta = \frac{\text{Opp}}{\text{Adj}} = \frac{|AC|}{|AB|} \qquad \cot(90° - \theta) = \frac{\text{Adj}}{\text{Opp}} = \frac{|AC|}{|AB|}$$

As the RHS of both are equal, therefore:

$$\tan\theta = \cot(90° - \theta)$$

#### NOTE

This is known as the ratio of complimentary angles. By the same method, we can prove that $\cot\theta = \tan(90° - \theta)$. What this means in practice is that, for example,

$$\tan 70° = \cot 20° \qquad \tan\frac{3}{7}\pi = \cot\frac{\pi}{14} \qquad \cot 71.5° = \tan 18.5° \qquad \cot\frac{3}{8}\pi = \tan\frac{\pi}{8}$$

---

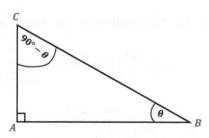

**FIGURE 2.4** Solution to Example 2.

Trigonometric Functions II

Another set of examples to try.

### Example 3

Given that $\cos \alpha = \frac{11}{14}$, such that angle $\alpha$ is acute, determine the exact value of the following:

a) $\sec \alpha$  b) $\operatorname{cosec} \alpha$  c) $\cot \alpha$

What did you get? Find the solution below to double-check your answer.

### Solution to Example 3

We need to sketch the triangle first as shown in Figure 2.5.
Given that

$$\cos \alpha = \frac{\text{Adj}}{\text{Hyp}} = \frac{11}{14}$$

Using Pythagoras' theorem, we have

$$14^2 = 11^2 + x^2$$
$$x^2 = 14^2 - 11^2 = 75$$
$$\therefore x = 5\sqrt{3}$$

**a) $\sec \alpha$**
**Solution**

$$\sec \alpha = \frac{\text{Hyp}}{\text{Adj}} = \frac{14}{11}$$
$$\therefore \sec \alpha = \frac{14}{11}$$

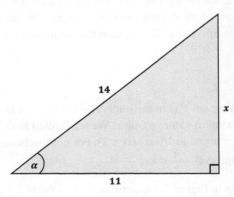

FIGURE 2.5  Solution to Example 3.

**ALTERNATIVE METHOD**

$$\sec\alpha = \frac{1}{\cos\alpha} = \frac{1}{\left(\frac{11}{14}\right)} = \frac{14}{11}$$

$$\therefore \sec\alpha = \frac{14}{11}$$

**b)** cosec $\alpha$
**Solution**

$$\text{cosec}\,\alpha = \frac{\text{Hyp}}{\text{Opp}} = \frac{14}{5\sqrt{3}}$$

$$\therefore \text{cosec}\,\alpha = \frac{14}{15}\sqrt{3}$$

**c)** cot $\alpha$
**Solution**

$$\cot\alpha = \frac{\text{Adj}}{\text{Opp}} = \frac{11}{5\sqrt{3}}$$

$$\therefore \cot\alpha = \frac{11}{15}\sqrt{3}$$

## 2.3 INVERSE TRIGONOMETRIC FUNCTIONS

In Chapter 1, we discussed the inverse trigonometric ratios. This will be explained here again but with a focus on their graphs. In general, trigonometric functions are many-to-one functions; this is to say that more than one value of independent variable $x$ will give the same value of dependent variable ($\sin x$, $\cos x$, or $\tan x$). For example, $\sin 0 = \sin\pi = \sin 2\pi = 0$, $\cos 0 = \cos\pi = \cos(-\pi) = -1$, and $\tan 0 = \tan\pi = \tan(-\pi) = 0$.

Therefore, to plot the inverse functions, we need to restrict the domain such that the trigonometric ratio behaves as though it is a one-to-one function. This will be clearer once we start looking at each function. One common feature of all the trigonometric inverses is that they act as mirrors of their respective functions along the line $y = x$. This is therefore indicated on each of the graphs.

### 2.3.1 ARCSIN FUNCTION

Figure 2.6 is a plot of **arcsin** (or $\sin^{-1} x$) in the interval $-1 \leq x \leq 1$, which corresponds to a range of values from minimum to maximum values of $\sin x$. We are limited to this range because an absolute value of $x$ must be less than or equal to 1 (i.e., $|x| \leq 1$). For comparison, we also plot $\sin x$ and $x = y$ on the same chart using the interval $-\frac{\pi}{2} \leq x \leq \frac{\pi}{2}$.

The main feature of the graph in Figure 2.6 is summarised in Table 2.4.

# Trigonometric Functions II

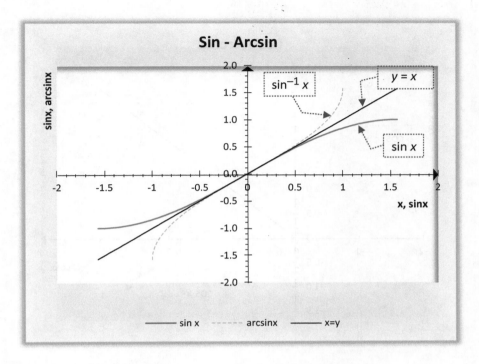

**FIGURE 2.6** Arcsin function in the interval $-1 \leq x \leq 1$ illustrated.

**TABLE 2.4**
**Arcsin Function in the Interval $-1 \leq x \leq 1$ and Sine Function in the Interval $-\frac{\pi}{2} \leq x \leq \frac{\pi}{2}$ Compared**

|  | $\sin x$ |  | $\sin^{-1} x$ |
|---|---|---|---|
| Note 1 | The domain is limited to $-\frac{\pi}{2} \leq x \leq \frac{\pi}{2}$. | Note 1 | The domain is $-1 \leq x \leq 1$. |
| Note 2 | The range is $-1 \leq \sin x \leq 1$. | Note 2 | The range is $-\frac{\pi}{2} \leq \sin^{-1} x \leq \frac{\pi}{2}$. |
| Note 3 | The graph passes through the origin $(0, 0)$ only for the range taken. | Note 3 | The graph passes through the origin $(0, 0)$. |
| Note 4 | The maximum and minimum points remain as $1$ and $-1$, respectively. This is irrespective of the portion taken as interval, which fulfils the condition of a one-to-one function. | Note 4 | The coordinates of the endpoints are $\left(-1, -\frac{\pi}{2}\right)$ and $\left(1, \frac{\pi}{2}\right)$. This will remain unchanged even if there is a transformation in $\sin x$ as the domain must be limited to $-1 \leq x \leq 1$. |

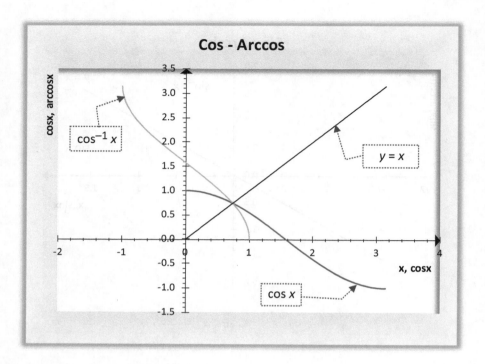

**FIGURE 2.7**  Arccos function in the interval $-1 \leq x \leq 1$ illustrated.

### 2.3.2 ARCCOS FUNCTION

Figure 2.7 is a plot of **arccos** (or $\cos^{-1} x$) in the interval $1 \leq x \leq 1$, which corresponds to a range of values from minimum to maximum values of $\cos x$. We are limited to this range for the same reason given for $\sin x$ above. For comparison, we also plot $\cos x$ and $x = y$ on the same chart using the interval $0 \leq x \leq \pi$.

The main feature of the graph in Figure 2.7 is summarised in Table 2.5.

### 2.3.3 ARCTAN FUNCTION

Figure 2.8 is a plot of **arctan** (or $\tan^{-1} x$). Inverse tangent is valid for any value of $x$ such that $-\infty \leq x \leq \infty$, but we've limited the interval to a range so that we have a one-to-one mapping. For comparison, we also plot $\tan x$ and $x = y$ on the same chart using the $-\frac{\pi}{2} \leq x \leq \frac{\pi}{2}$.

The main feature of the graph in Figure 2.8 is summarised in Table 2.6.

# Trigonometric Functions II

## TABLE 2.5
### Arccos Function in the Interval $-1 \leq x \leq 1$ and Cosine Function in the Interval $0 \leq x \leq \pi$ Compared

|        | $\cos x$ |        | $\cos^{-1} x$ |
|--------|----------|--------|---------------|
| Note 1 | The domain is limited to $0 \leq x \leq \pi$. | Note 1 | The domain is $-1 \leq x \leq 1$. |
| Note 2 | The range is $-1 \leq \cos x \leq 1$. | Note 2 | The range is $0 \leq \cos^{-1} x \leq \pi$. |
| Note 3 | The graph passes through the $x$-axis at $\left(\frac{\pi}{2}, 0\right)$ and the $y$-axis at $(0, 1)$. If we take other ranges, e.g., $\pi \leq x \leq 2\pi$, it will have different intercepts. | Note 3 | The graph passes through the $x$-axis at $(1, 0)$ and the $y$-axis at $\left(0, \frac{\pi}{2}\right)$. |
| Note 4 | The maximum and minimum points remain as $1$ and $-1$, respectively. This is irrespective of the portion taken as interval, which fulfils the condition of a one-to-one function. | Note 4 | The coordinates of the endpoints are $(-1, -\pi)$ and $(1, \pi)$. This will remain unchanged even if there is a transformation in $\cos x$ as the domain must be limited to $-1 \leq x \leq 1$. |

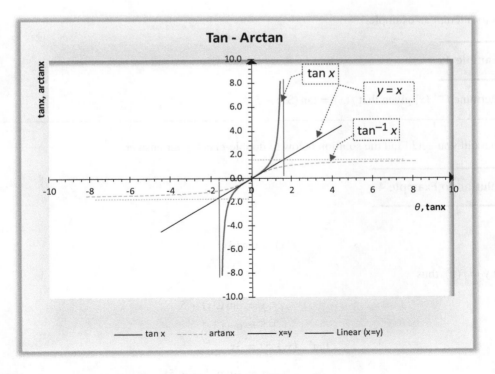

**FIGURE 2.8** Arctan function in the interval $-\infty \leq x \leq \infty$ illustrated.

## TABLE 2.6
**Arctan Function in the Interval $-\infty \leq x \leq \infty$ and Tangent Function in the Interval $-\frac{\pi}{2} \leq x \leq \frac{\pi}{2}$ Compared**

|  | $\tan x$ |  | $\tan^{-1} x$ |
|---|---|---|---|
| Note 1 | The domain is limited to $-\frac{\pi}{2} < x < \frac{\pi}{2}$. | Note 1 | The domain is $-\infty < x < \infty$. Note that the domain does not include the $\pm\infty$ as $\tan x$ is undefined at $\pm\pi$. We have vertical asymptotes at $y = -\frac{\pi}{2}$ and $y = \frac{\pi}{2}$. |
| Note 2 | The range is $-\infty < \tan x < \infty$. Note that the range does not include the $\pm\infty$ as $\tan x$ is undefined at $\pm\pi$. We have vertical asymptotes at $x = \frac{\pi}{2}$ and $y = \frac{\pi}{2}$. | Note 2 | The range is $-\frac{\pi}{2} < \tan^{-1} x < \frac{\pi}{2}$. This is the interval of the answer you will get when you compute the arctan function on a calculator for example |
| Note 3 | The graph passes through the origin $(0, 0)$ only for the range taken. | Note 3 | The graph passes through the origin $(0, 0)$. |

Let us attempt an example.

### Example 4

Determine $f^{-1}(x)$ given that $f(x) = \tan(3x) - 5$.

What did you get? Find the solution below to double-check your answer.

### Solution to Example 4

$$f(x) = \tan(3x) - 5$$

Let $y = f(x)$, thus

$$y = \tan(3x) - 5$$
$$y + 5 = \tan(3x)$$
$$\tan^{-1}(y + 5) = 3x$$
$$x = \frac{1}{3}\tan^{-1}(y + 5)$$

# Trigonometric Functions II

Now substitute $x$ for $f^{-1}(x)$ and $y$ for $x$.

$$\therefore f^{-1}(x) = \frac{1}{3}\tan^{-1}(x+5)$$

Another example for us to try.

### Example 5

Sketch the graph of $y = 2\arcsin x$ in the interval $-1 \leq x \leq 1$ and state the range.

What did you get? Find the solution below to double-check your answer.

### Solution to Example 5

**Solution**

The graph of $y = 2\arcsin(x)$ is the same as $y = \arcsin(x)$ except the former is stretched vertically by a factor of 2. Hence, the range is also double and is

$$-\pi \leq 2\arcsin(x) \leq \pi$$

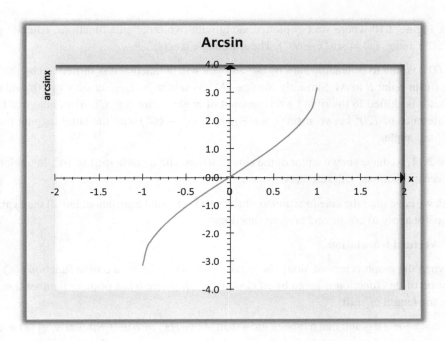

**FIGURE 2.9** Solution to Example 5.

## 2.4 TRANSFORMATION OF TRIGONOMETRIC FUNCTIONS

Transformation is the change in shape of a graph function, and this can occur in several ways. Here, we will be looking at two main categories of changes to trigonometric functions, namely translation and stretch.

### 2.4.1 TRANSLATION

This is the movement (without a change in shape) of the sine, cosine, and tangent graphs along the $x$–$y$ plane and we have two main types.

**Type 1   Horizontal translation**

This is when the graph is moved along the $x$-axis right or left. Given a sine function $f(x) = \sin x$, a translation of this function is given by $g(x) = \sin(x \pm c)$ where $c$ is a positive number (i.e., $c > 0$), denoting the angle of shift.

$g_1(x) = \sin(x + c)$ is a graph that has been moved to the left of $f(x)$ and is said to be **leading** $\sin x$ by an angle of $c$. Similarly, $g_2(x) = \sin(x - c)$ is a graph that has been moved to the right $f(x)$ and is said to be **lagging** $\sin x$ by an angle of $c$.

Note that the angle at which maximum and minimum occur will be shifted accordingly. In general, given that $\sin x$ is transformed into $\sin(x \pm c)$, it follows that:

**a)** maximum point in $\sin x$ occurs at $90°$, $450°$, $810°$, etc., which corresponds to $90° \mp c$, $450° \mp c$, $810° \mp c$, etc., in $\sin(x \pm c)$.

**b)** minimum point in $\sin x$ occurs at $270°$, $630°$, $990°$, etc., which implies to $270° \mp c$, $630° \mp c$, $990° \mp c$, etc., in $\sin(x \pm c)$.

Consider Figure 2.10 where $\sin x$ is plotted along with two other sine functions, namely: $g_1(x) = \sin(x + 60°)$ and $g_2(x) = \sin(x - 60°)$. The shift angle $c$ is $60°$.

$\sin(x + 60°)$ is said to be leading $\sin x$ by $60°$ and you will notice that it is shifted left by this amount of angle (from point $B$ to $A$). Similarly, $\sin(x - 60°)$ is said to be lagging $\sin x$ by $60°$ and you will notice that it is shifted to the right by this amount of angle (point $B$ to $C$). Also, note that there is a phase difference of $120°$ between $\sin(x + 60°)$ and $\sin(x - 60°)$ with the latter lagging the former by this phase angle.

In Figure 2.11, we have shown a plot of the sine function with a phase shift of $10°$. In total, there are nine different functions with reference to $\sin x$.

Although we have used the sine function to illustrate a horizontal translation, but all the explanations given equally apply to cosine and tangent functions.

**Type 2   Vertical translation**

This is when the graph is moved along the $y$-axis up or down. Given a cosine function $f(x) = \cos x$, a translation of this function is given by $g(x) = \cos x \pm b$ where $b$ is a positive number (i.e., $b > 0$), denoting the length of shift.

$g_1(x) = \cos x + b$ is a graph that has been moved up above $f(x)$ by $b$ unit. Similarly, $g_2(x) = \cos x - b$ is a graph that has been moved down below $f(x)$ by $b$ unit.

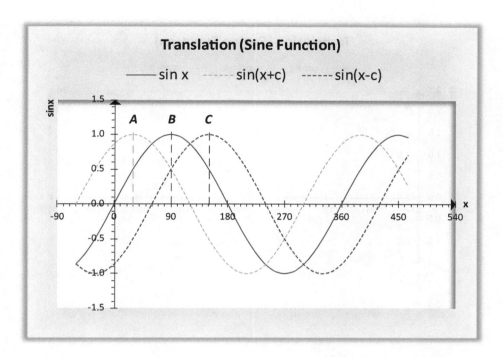

**FIGURE 2.10** Horizontal translation of sine function with a phase angle of **60°** illustrated.

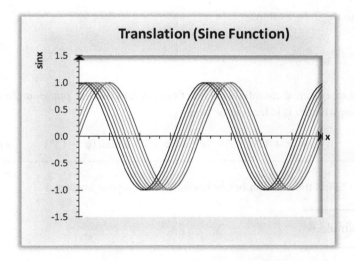

**FIGURE 2.11** Horizontal translation of sine function with a phase angle of **10°** illustrated.

Consider Figure 2.12 where $\cos x$ is plotted along with two other cosine functions, namely $g_1(x) = \cos x + 1$ and $g_2(x) = \cos x - 1$. The translation length $b$ is 1 unit. Notice that $\cos x + 1$ has been moved up by 1 unit (from point $B$ to $A$) and $\cos x - 1$ has been moved down by the same amount (from point $B$ to $C$). No change to the points of maximum and minimum, but the $y$-coordinate would have been shifted by $\pm b$. Consequently, the amplitude remains unchanged. The $x$-coordinate of the maximum and minimum points also remain the same, but the $y$-coordinate is shifted by $\pm b$.

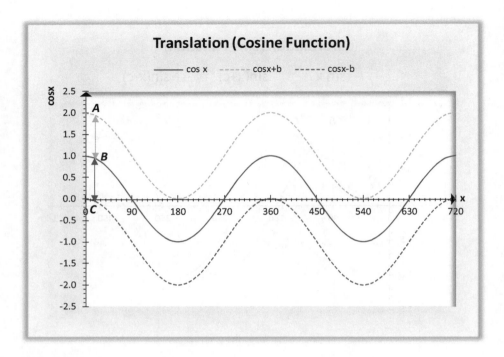

**FIGURE 2.12** Vertical translation of cosine function at 1 unit illustrated.

It's a good time to try some examples.

### Example 6

Using $\sin \varphi$, $\cos \varphi$, or $\tan \varphi$ as reference functions, determine the phase angle of the following functions and state whether it is leading or lagging.

**a)** $\sin(\varphi - \pi)$    **b)** $\cos(\varphi + 120°)$    **c)** $\cos \varphi$    **d)** $\tan(\varphi - 15°)$    **e)** $\tan(\varphi + 1.2\pi)$

What did you get? Find the solution below to double-check your answer.

### Solution to Example 6

#### HINT

We need to write the function in the form $\sin(x \pm c)$, where $c$ is the phase angle. If

- the sign between $x$ and $c$ is positive, it is a lead.
- the sign between $x$ and $c$ is negative it is a lag.

Trigonometric Functions II 65

**a)** $\sin(\varphi - \pi)$
**Solution**

$\therefore$ **Phase angle is $\pi$, it's lagging $\sin\varphi$**

**b)** $\cos(\varphi + 120°)$
**Solution**

$\therefore$ **Phase angle is 120°, it's leading $\cos\varphi$**

**c)** $\cos\varphi$
**Solution**

$\therefore$ **Phase angle is 0, it's in phase with $\cos\varphi$ (neither lagging or leading)**

**d)** $\tan(\varphi - 15°)$
**Solution**

$\therefore$ **Phase angle is 15°, it's lagging $\tan\varphi$**

**e)** $\tan(\varphi + 1.2\pi)$
**Solution**

$\therefore$ **Phase angle is $1.2\pi$, it's leading $\tan\varphi$**

Another set of examples to try.

## Example 7

Determine the coordinates of the maximum and minimum points of the following functions in the interval $0 \leq x \leq 360°$ or $0 \leq x \leq 2\pi$. Sketch the graph in each case.

**a)** $y_1 = 3 + \sin(x + 30°)$ 
**b)** $y_2 = \cos\left(x - \dfrac{\pi}{3}\right) - 5$

What did you get? Find the solution below to double-check your answer.

## Solution to Example 7

**HINT**

- We need to vary the range based on the phase angle.
- The $y$-coordinate is $A \pm b$, where $A$ is the amplitude and $b$ is the vertical shift. In this example, $A = 1$. Given that $y = A\sin(x + \theta) + b$ or $y = A\cos(x + \theta) + b$.

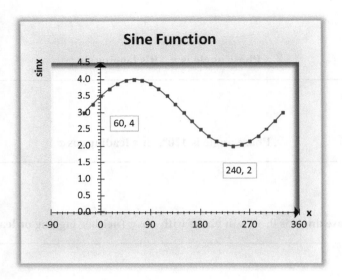

**FIGURE 2.13** Solution to Example 7(a).

**a)** $y_1 = 3 + \sin(x + 30°)$
**Solution**
The interval is now $0 \leq x + 30° \leq 360°$ or $-30° \leq x \leq 330°$.
The sine function is maximum at 90°, therefore the $x$-coordinate is:

$$x + 30° = 90°$$
$$x = 60°$$

The $y$-coordinate is:

$$A + b = 1 + 3 = 4$$

The sine function is minimum at 270°, therefore the $x$-coordinate is:

$$x + 30° = 270°$$
$$x = 240°$$

The $y$-coordinate is:

$$A + b = -1 + 3 = 2$$
$$\therefore (60°, 4), (240°, 2)$$

---

**b)** $\cos\left(x - \frac{\pi}{3}\right) - 5$
**Solution**
The interval is now $0 \leq x - \frac{\pi}{3} \leq 2\pi$ or $\frac{\pi}{3} \leq x \leq \frac{7}{3}\pi$.

# Trigonometric Functions II

The cosine function is maximum at 0 and $2\pi$, therefore the $x$-coordinate is:

$$x - \frac{\pi}{3} = 0 \quad \text{AND} \quad x - \frac{\pi}{3} = 2\pi$$
$$x = \frac{\pi}{3} \qquad\qquad\qquad x = \frac{7}{3}\pi$$

The $y$-coordinate is:

$$A + b = 1 - 5 = -4$$

The cosine function is minimum at $\pi$, therefore the $x$-coordinate is:

$$x - \frac{\pi}{3} = \pi$$
$$x = \frac{4}{3}\pi$$

The $y$-coordinate is

$$A + b = -1 - 5 = -6$$
$$\therefore \left(\frac{1}{3}\pi, -4\right), \left(\frac{7}{3}\pi, -4\right), \left(\frac{4}{3}\pi, -6\right)$$

The graph of $\cos\left(x - \frac{\pi}{3}\right) - 5$ is shown in Figure 2.14.

## 2.4.2 Stretch

This is when a trigonometric function is 'transformed' along either the $x$-axis or $y$-axis. This transformation can result in magnification and demagnification of the shape. By extension, it also produces

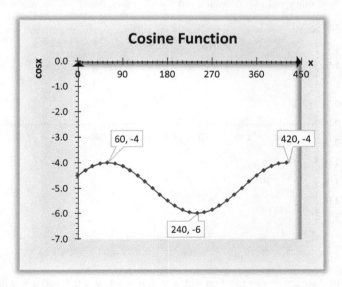

**FIGURE 2.14** Solution to Example 7(b).

a reflection of the shape. Again, we will consider two types here, namely vertical and horizontal stretch.

**Type 1 Vertical stretch**

This is when the graph is either stretched or squashed in the $y$-axis direction. The general form for this type is shown below for the three trigonometric functions.

$$\boxed{y = n \sin x} \quad \boxed{y = n \cos x} \quad \boxed{y = n \tan x} \tag{2.2}$$

where $n$ represents the amount of vertical stretch. If:

- $n < 0$ (i.e., $n$ is negative), then the shape is reflected (or mirrored) in the $x$-axis.
- $0 < n < 1$, the shape is reduced or squashed in the $y$-axis.
- $n > 1$, the shape is enlarged or stretched out in the $y$-axis.

The above three situations are shown in Figure 2.15, where $n = 1$ (default), $n = -1$ (reflection or mirror), $n = 0.5$ (squash or reduction), and $n = 2$ (stretch or enlargement).

**Type 2 Horizontal stretch**

This is when the graph is stretched (or squashed) in the $x$-axis direction. The general form for this is:

$$\boxed{y = \sin nx} \quad \boxed{y = \cos nx} \quad \boxed{y = \tan nx} \tag{2.3}$$

The above implies that the graph has been horizontally stretched by a factor of $\frac{1}{n}$. Have you noticed the difference? The bigger the value of $n$, the more the shape is squashed and vice versa. If:

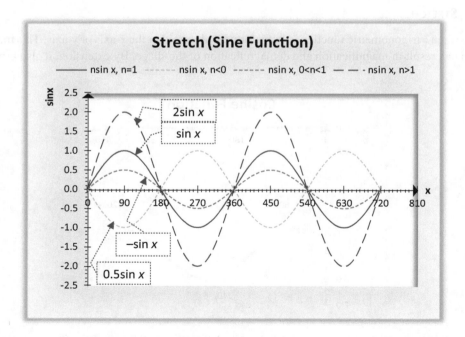

**FIGURE 2.15** Vertical stretch of sine function with $n = 1$, $n = -1$, $n = 0.5$, and $n = 2$ illustrated.

- $n < 0$ (i.e., $n$ is negative), then the shape is reflected (or mirrored) in the $x$-axis. As the cosine function is symmetrical about the origin, the shape remains unchanged for $\pm n$. For example, the shape for $\cos 3x$ is the same as $\cos(-3x)$, where $n = 3$ and $n = -3$, respectively.
- $0 < n < 1$, the shape is enlarged or stretched outwardly in the $x$-axis direction. This represents a low-frequency oscillating system.
- $n > 1$, the shape is reduced or squashed inwardly in the $x$-axis direction. This represents a high-frequency oscillating system.

The above three situations are shown in Figure 2.16, where $n = 1$ (default), $n = -2$ (reflection or mirror), $n = 0.5$ (stretch outwards or enlarged), and $n = 3$ (squashed inwardly or reduced).

Notice that there is no sign of reflection in $\cos(-2x)$ though $n$ is negative, contrary to expectation. This is because $\cos(-2x) = \cos 2x$. We have the same effect using sine here but use $n = -1$ instead, to clearly show the mirror effect.

Before we leave this topic, let us write the general expressions for the three trigonometric functions as:

$$\boxed{y = a\sin(bx \pm c)} \quad \boxed{y = a\cos(bx \pm c)} \quad \boxed{y = a\tan(bx \pm c)} \tag{2.4}$$

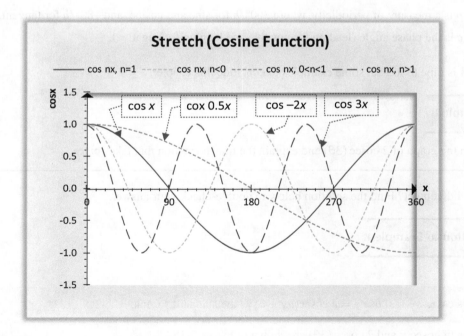

**FIGURE 2.16** Horizontal stretch of cosine function with $n = 1$, $n = -2$, $n = 0.5$, and $n = 3$ illustrated.

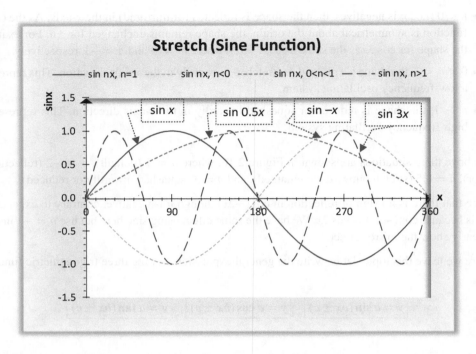

**FIGURE 2.17** Horizontal stretch of sine function with $n = 1$, $n = -1$, $n = 0.5$, and $n = 3$ illustrated.

where

- $a$ is the amplitude of the function,
- $b$ is a measure of periodicity, where $360°/b$ for sine and cosine, and $180°/b$ for tangent, and
- $c$ is the phase angle (leading for positive and lagging for negative).

Now it is time to look at some examples to conclude this chapter.

### Example 8

Sketch the graph of $2 \operatorname{cosec}(3\theta)$ and explain the transformation that takes place.

What did you get? Find the solution below to double-check your answer.

### Solution to Example 8

**HINT**

In this case, we will use $y = a \operatorname{cosec}(bx + c)$ as the standard format.

Graphs of $\operatorname{cosec} \theta$ and $2 \operatorname{cosec}(3\theta)$ are shown in Figure 2.18.

# Trigonometric Functions II

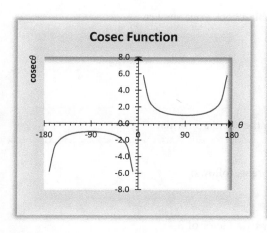
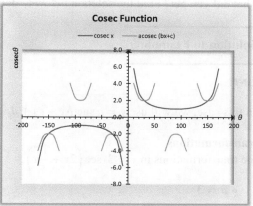

**FIGURE 2.18** Solution to Example 8.

The graph of $2 \cosec(3\theta)$ is like $\cosec \theta$ with the following transformations:

- $a = 2$
  - This means that $2 \cosec(3\theta)$ is stretched vertically by a factor of 2.
  - It also implies that its minimum and maximum points are doubled to 2 and $-2$, respectively.

- $b = 3$
  - This implies that $2 \cosec(3\theta)$ is stretched horizontally by a factor of $\frac{1}{3}$.
  - It also changes its periodicity to $120°$ (i.e., $\frac{360°}{3}$) or repeats itself at every $120°$ and the shape pattern has tripled. Asymptotes now occur at $= \pm\frac{1}{3}n\pi$.

- $c = 0$
  - This implies that $2 \cosec(3\theta)$ has not been translated horizontally (left or right).

## Example 9

Describe the transformations that take place in $y = 3 \sec\left(2x + \frac{1}{3}\pi\right) - 1$ and give the coordinates of the minimum points in the interval $0 \leq x \leq 2\pi$.

What did you get? Find the solution below to double-check your answer.

> **Solution to Example 9**

**HINT**

In this case, we will use $y = a\sec(bx + c) + d$ as the standard format.

**Transformations**
The transformations in $y = 3\sec\left(2x + \frac{1}{3}\pi\right) - 1$ are as follows:

- $a = 3$
  - This means that $y$ is stretched vertically by a factor of 3.
  - It also implies that its minimum and maximum points are multiplied by a factor of 3 and thus become 3 and $-3$, respectively.
- $b = 2$
  - This implies that $y$ is stretched horizontally by a factor of $\frac{1}{2}$.
  - It also changes its periodicity to 180° or repeats itself at every 180° (and the shape pattern has doubled in the graph provided).
- $c = \frac{1}{3}\pi$
  - This implies that $y$ is leading by this amount or it has been translated to the left of the horizontal $x$-axis by this amount.
- $d = -1$
  - This means that the graph is shifted down by 1 unit.

---

**Coordinates of minimum points**

- $\sec x$ is minimum at $(0, 1)$ and $(2\pi, 1)$.
- $a = 3$ means $(0, 3)$ and $(2\pi, 3)$.
- $b = 2$ means $(0, 3)$, $(\pi, 3)$, and $(2\pi, 3)$ as the periodicity is now $\pi$. Equally, there is now a shift of $\frac{1}{6}\pi$ to the left. The new coordinates of minimum points are $(0 - \frac{1}{6}\pi, 3)$, $(\pi - \frac{1}{6}\pi, 3)$, and $\left(2\pi - \frac{1}{6}\pi, 3\right)$.
- $(-\frac{1}{6}\pi, 3)$ is not in the stated range or intervals. We are therefore left with two coordinates $(\frac{5}{6}\pi, 3)$ and $\left(\frac{11}{6}\pi, 3\right)$.
- $-1$ indicates that the minimum point is one unit less than 3.
- Hence, the coordinates of $y = 3\sec\left(2x + \frac{1}{3}\pi\right) - 1$ are

$$\left(\frac{5}{6}\pi, 2\right), \left(\frac{11}{6}\pi, 2\right)$$

# Trigonometric Functions II

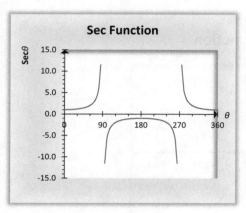 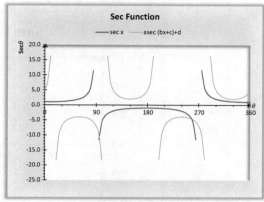

**FIGURE 2.19** Solution to Example 9.

Graphs of $\sec x$ and $3\sec\left(2x + \frac{1}{3}\pi\right) - 1$ are shown in Figure 2.19.

## Example 10

On the same axes, sketch the graph of $y = \sin x$ and $y = -3\sin 2x$ in the range $0 \leq x \leq 360°$. Comment on the shape of $y = 3\sin(-2x)$.

What did you get? Find the solution below to double-check your answer.

## Solution to Example 10

- Sketch or draw the graph of $y = \sin x$ in the stated range (Figure 2.20).
- Now the second function is $y = -3\sin 2x$. $-3$ means that we need to reflect the graph in the $x$-axis (because of the minus) and stretch it vertically by a factor of 3. The result of these two actions is shown in Figure 2.21.
- The 2 in $-3\sin 2x$ implies that the function is stretched by a factor of $\frac{1}{2}$ or the periodicity is $180°\left(=\frac{360°}{2}\right)$. Hence, we need to fit two of the shape of $\sin x$ within the same range (Figure 2.22).
- $3\sin(-2x)$ will give the same result as $-3\sin 2x$ as shown above. This is because the reflection made by the negative $-3$ is provided by $-2$. Alternatively, we can say that since $\sin(-x) = -\sin x$, then $-3\sin 2x = 3\sin(-2x)$.
- In addition, we can equally say that $3\sin 2x$ is the same as $-3\sin(-2x)$. This is because the double reflection in the latter function simply means no reflection.

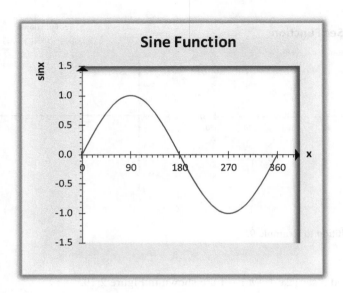

**FIGURE 2.20** Solution to Example 10 – Part I.

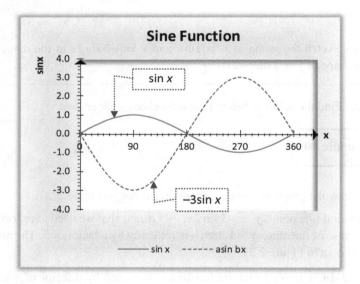

**FIGURE 2.21** Solution to Example 10 – Part II.

### Example 11

On the same axes, sketch the graph of $y = \cos x$ and $y = 2\cos\left(\frac{3}{2}x\right)$ in the range $-180° \leq x \leq 180°$. Use your graphs to solve $\cos x = 1$ and $2\cos\left(\frac{3}{2}x\right) = 1$.

Trigonometric Functions II

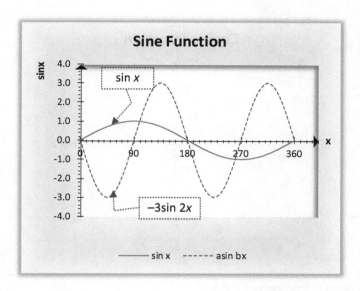

**FIGURE 2.22** Solution to Example 10 – Part III.

What did you get? Find the solution below to double-check your answer.

## Solution to Example 11

- Sketch or draw the graph of $y = \cos x$ in the stated range (Figure 2.23).
- The second function is $y = 2\cos\left(\frac{3}{2}x\right)$. 2 means that we need to stretch it vertically by a factor of 2 and horizontally by a factor of $\frac{2}{3}$.
- Draw a horizontal line at $y = 1$ to cross the two functions. Draw a vertical line at each of the intersecting points and read off the values (Figure 2.24).

For $\cos x = 1$, there is only one solution:

$$(0°, 1)$$

For $2\cos\left(\frac{3}{2}x\right) = 1$, there are four solutions:

$$(-160°, 1), (-80°, 1), (80°, 1), (160°, 1)$$

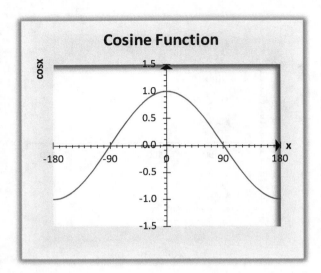

**FIGURE 2.23** Solution to Example 11 – Part I.

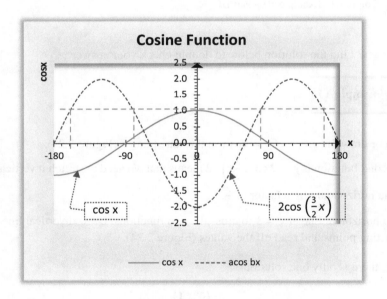

**FIGURE 2.24** Solution to Example 11 – Part II.

One final example to try.

### Example 12

On the same axes, sketch the graph of $\tan x$ and $1.5 \tan(2x + 30°)$ in the range $-180° \leq x \leq 180°$.

Trigonometric Functions II

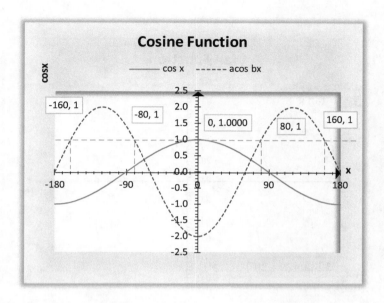

**FIGURE 2.25** Solution to Example 11 – Part III.

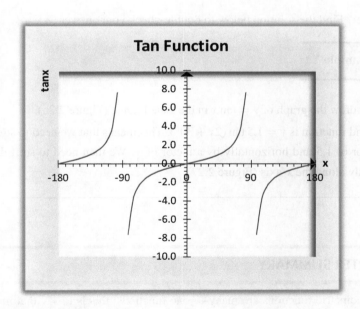

**FIGURE 2.26** Solution to Example 12 – Part I.

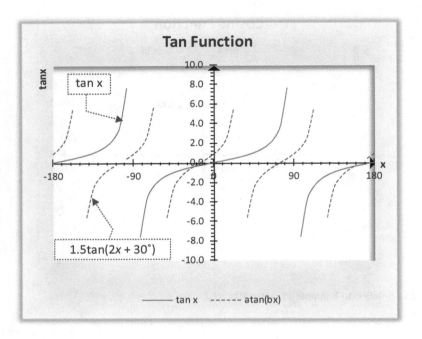

**FIGURE 2.27** Solution to Example 12 – Part II.

What did you get? Find the solution below to double-check your answer.

**Solution to Example 12**

- Sketch or draw the graph of $y = \tan x$ in the stated range (Figure 2.26).
- The second function is $y = 1.5 \tan(2x + 30°)$. This means that we need to stretch it vertically by a factor of 1.5 and horizontally by a factor of $\frac{1}{2}$. We then need to shift the graph by 30° backwardly along the $x$-axis (Figure 2.27).

## 2.5 CHAPTER SUMMARY

1) Trigonometric functions are many-to-one functions; this is to say that more than one value of independent variable $x$ will give the same value of dependent variable ($\sin x$, $\cos x$, or $\tan x$).
2) To plot the inverse functions, we need to restrict the domain such that the trigonometric ratio behaves as though it is a one-to-one function.
3) **arcsin** (or $\sin^{-1} x$) in the interval $1 \leq x \leq 1$

# Trigonometric Functions II

- The domain is $-1 \leq x \leq 1$.
- The range is $-\frac{\pi}{2} \leq \sin^{-1} x \leq \frac{\pi}{2}$.
- The graph passes through the origin $(0, 0)$.
- The coordinates of the endpoints are $\left(-1, -\frac{\pi}{2}\right)$ and $\left(1, \frac{\pi}{2}\right)$. This will remain unchanged even if there is a transformation in $\sin x$ as the domain must be limited to $-1 \leq x \leq 1$.

4) **arccos** (or $\cos^{-1} x$) in the interval $1 \leq x \leq 1$

   - The domain is $-1 \leq x \leq 1$.
   - The range is $0 \leq \cos^{-1} x \leq \pi$.
   - The graph passes through the $x$-axis at $(1, 0)$ and $y$-axis at $\left(0, \frac{\pi}{2}\right)$.
   - The coordinates of the endpoints are $(-1, -\pi)$ and $(1, \pi)$. This will remain unchanged even if there is a transformation in $\cos x$ as the domain must be limited to $-1 \leq x \leq 1$.

5) **arctan** (or $\tan^{-1} x$) in the interval $-\infty < x < \infty$

   - The domain is $-\infty < x < \infty$. Note that the domain does not include the $\pm \infty$ as $\tan x$ is undefined at $\pm \pi$. We have vertical asymptotes at $y = -\frac{\pi}{2}$ and $y = \frac{\pi}{2}$.
   - The range is $-\frac{\pi}{2} < \tan^{-1} x < \frac{\pi}{2}$. This is the interval of the answer you will get when you compute arctan function on a calculator for example
   - The graph passes through the origin $(0, 0)$.

6) In general, a reciprocal of a number, variable, or function is one over it. Hence, we define the reciprocals of trigonometric functions as:

$$\csc \theta = \frac{1}{\sin \theta} = \frac{\text{Hyp}}{\text{Opp}} \qquad \sec \theta = \frac{1}{\cos \theta} = \frac{\text{Hyp}}{\text{Adj}} \qquad \cot \theta = \frac{1}{\tan \theta} = \frac{\text{Adj}}{\text{Opp}}$$

7) Transformation is the change in shape of a graph function, and this can be translation, stretch, or reflection.

****

## 2.6 FURTHER PRACTICE

To access complementary contents, including additional exercises, please go to www.dszak.com.

# 3 Trigonometric Identities I

## Learning Outcomes

Once you have studied the content of this chapter, you should be able to:

- Use common trigonometric identities
- Use trigonometric identities to prove trigonometric relations
- Use addition formulas involving sine, cosine, and tangent functions
- Use double, triple, and half-angle formulas involving sine, cosine, and tangent functions

## 3.1 INTRODUCTION

In the current chapter, we will look at how identities play a major role in trigonometric functions and the relationships that exist between the six commonly used functions, namely sine (sin), cosine (cos), tangent (tan), cotangent (cot), secant (sec), and cosecant (cosec or csc).

## 3.2 FUNDAMENTALS

In this section, our discussion will centre around what is loosely known as the 'fundamentals' of trigonometric identities, their relevance, and wide usage in trigonometry. The first two are:

$$\tan \theta = \frac{\sin \theta}{\cos \theta} \quad \textbf{AND} \quad \sin^2 \theta + \cos^2 \theta = 1$$

Note that $\sin^2 \theta = (\sin \theta)^2$ and $\cos^2 \theta = (\cos \theta)^2$, which is a notation used to represent the positive powers of trigonometric functions. Thus, we do not say $\sin^{-2} \theta$ and $\frac{1}{\sin^2 \theta}$ are the same for example, i.e. $\sin^{-2} \theta \neq \frac{1}{\sin^2 \theta}$. Note that sine without mentioning the angle is meaningless just as the radical $\sqrt{\phantom{x}}$ without a number or variable cannot be operated.

There are two more common identities that are directly obtained from $\sin^2 \theta + \cos^2 \theta = 1$. How? Let's illustrate this very quickly.

- **Dividing $\sin^2 \theta + \cos^2 \theta = 1$ by $\cos^2 \theta$**

$$\frac{\sin^2 \theta}{\cos^2 \theta} + \frac{\cos^2 \theta}{\cos^2 \theta} = \frac{1}{\cos^2 \theta}$$

$$\therefore \tan^2 \theta + 1 = \sec^2 \theta$$

$$\because \sec \theta = \frac{1}{\cos \theta}$$

# Trigonometric Identities I

Note that the sign $\therefore$ implies 'because'.

- **Dividing $\sin^2 \theta + \cos^2 \theta = 1$ by $\sin^2\theta$**

$$\frac{\sin^2\theta}{\sin^2\theta} + \frac{\cos^2\theta}{\sin^2\theta} = \frac{1}{\sin^2\theta}$$

$$\therefore 1 + \cot^2\theta = \csc^2\theta$$

$$\because \csc\theta = \frac{1}{\sin\theta}$$

We can put the four 'fundamental' identities together as:

$$\boxed{\tan\theta = \frac{\sin\theta}{\cos\theta}} \quad \boxed{\sin^2\theta + \cos^2\theta = 1} \quad \boxed{\tan^2\theta + 1 = \sec^2\theta} \quad \boxed{\cot^2\theta + 1 = \csc^2\theta} \quad (3.1)$$

That's all for now. Let's try some examples.

## Example 1

Using appropriate identities, prove the following:

a) $\dfrac{\sin^2\varphi}{1+\cos\varphi} = 1 - \cos\varphi$

b) $\cot\theta - \cos\theta \csc\theta = 0$

c) $1 - \sin x \cos x \cot x = \sin^2 x$

d) $\dfrac{1-\sin^2\alpha}{\cos^2\alpha} = (\sec\alpha - \tan\alpha)(\sec\alpha + \tan\alpha)$

What did you get? Find the solution below to double-check your answer.

## Solution to Example 1

**HINT**

The proof is to show that the LHS expression is equal to the RHS expression. To demonstrate this, we need to take one side and simplify until we obtain the other side.

a) $\dfrac{\sin^2\varphi}{1+\cos\varphi} = 1 - \cos\varphi$

**Solution**

$$\frac{\sin^2\varphi}{1+\cos\varphi} = 1 - \cos\varphi$$

We will start with the LHS as:

$$\frac{\sin^2\varphi}{1+\cos\varphi} = \frac{1-\cos^2\varphi}{1+\cos\varphi}$$
$$= \frac{(1+\cos\varphi)(1-\cos\varphi)}{1+\cos\varphi}$$
$$\therefore \frac{\sin^2\varphi}{1+\cos\varphi} = 1-\cos\varphi$$

---

**b)** $\cot\theta - \cos\theta\csc\theta = 0$

**Solution**

$$\cot\theta - \cos\theta\csc\theta = 0$$

We will start with the LHS as:

$$\cot\theta - \cos\theta\csc\theta = \frac{\cos\theta}{\sin\theta} - \cos\theta\frac{1}{\sin\theta}$$
$$= \frac{\cos\theta}{\sin\theta} - \frac{\cos\theta}{\sin\theta} = 0$$
$$\therefore \; \boldsymbol{\cot\theta - \cos\theta\csc\theta = 0}$$

---

**c)** $1 - \sin x \cos x \cot x = \sin^2 x$

**Solution**

Taking the LHS expression, we have

$$1 - \sin x \cos x \cot x = 1 - \sin x \cos x \left(\frac{\cos x}{\sin x}\right)$$
$$= 1 - \cos^2 x$$
$$\therefore 1 - \sin x \cos x \cot x = \sin^2 x$$

---

**d)** $\frac{1-\sin^2\alpha}{\cos^2\alpha} = (\sec\alpha - \tan\alpha)(\sec\alpha + \tan\alpha)$

**Solution**

$$\frac{1-\sin^2\alpha}{\cos^2\alpha} = (\sec\alpha - \tan\alpha)(\sec\alpha + \tan\alpha)$$

We will start with the LHS as:

$$\frac{1-\sin^2\alpha}{\cos^2\alpha} = \frac{1}{\cos^2\alpha} - \frac{\sin^2\alpha}{\cos^2\alpha}$$
$$= \sec^2\alpha - \tan^2\alpha$$
$$= (\sec\alpha - \tan\alpha)(\sec\alpha + \tan\alpha)$$
$$\therefore \; \boldsymbol{\frac{1-\sin^2\alpha}{\cos^2\alpha} = (\sec\alpha - \tan\alpha)(\sec\alpha + \tan\alpha)}$$

# Trigonometric Identities I

Another example to try.

**Example 2**

Simplify $\csc^2\beta - 2\cot\beta - 4$ such that the final answer is in a factored form containing only the cotangent function.

What did you get? Find the solution below to double-check your answer.

**Solution to Example 2**

$$\csc^2\beta - 2\cot\beta - 4 = (\cot^2\beta + 1) - 2\cot\beta - 4$$
$$= \cot^2\beta - 2\cot\beta - 3$$
$$= (\cot\beta - 3)(\cot\beta + 1)$$
$$\therefore \csc^2\beta - 2\cot\beta - 4 = (\cot\beta - 3)(\cot\beta + 1)$$

Another example to try.

**Example 3**

Show that

$$2\sin^2 x - \cos^2 x - \frac{2}{\csc x} = (3\sin x + 1)(\sin x - 1)$$

What did you get? Find the solution below to double-check your answer.

**Solution to Example 3**

$$2\sin^2 x - \cos^2 x - \frac{2}{\csc x} = (3\sin x + 1)(\sin x - 1)$$

We will start with the LHS as:

$$2\sin^2 x - \cos^2 x - \frac{2}{\csc x} = 2\sin^2 x - (1 - \sin^2 x) - 2(\sin x)$$
$$= 2\sin^2 x - 1 + \sin^2 x - 2\sin x$$
$$= 3\sin^2 x - 2\sin x - 1$$
$$= (3\sin x + 1)(\sin x - 1)$$
$$\therefore 2\sin^2 x - \cos^2 x - \frac{2}{\csc x} = (3\sin x + 1)(\sin x - 1)$$

Another set of examples.

## Example 4

Angles $x$ and $y$ are acute such that $\sec x = \frac{17}{8}$ and $\tan y = \frac{5}{12}$. Using appropriate identities, determine the exact value of the following:

a) $\sin x$      b) $\tan x$      c) $\cot y$      d) $\csc y$      e) $\cos y$

What did you get? Find the solution below to double-check your answer.

## Solution to Example 4

### HINT

- In this case, we are not required to sketch as we did in the previous chapter, instead, we are limited to using identities.
- Obviously, there will be more than one way to deal with each case. We will try to show an alternative approach where it is possible.
- As $x$ and $y$ are acute angles, only the positive root is taken and valid.
- Remember to express the answer as a fraction, as the exact value is required.

**a)** $\sin x$
**Solution**
We know that
$$\sin^2 x + \cos^2 x = 1$$
This implies that
$$\sin^2 x = 1 - \cos^2 x$$
But
$$\cos x = \frac{1}{\sec x} = \frac{1}{\left(\frac{17}{8}\right)} = \frac{8}{17}$$
Therefore,
$$\sin^2 x = 1 - \left(\frac{8}{17}\right)^2$$
$$\sin x = \sqrt{\frac{225}{289}}$$
$$\therefore \sin x = \frac{15}{17}$$

# Trigonometric Identities I

**b)** $\tan x$

**Solution**

$$\tan x = \frac{\sin x}{\cos x} = \sin x \sec x$$

$$= \left(\frac{15}{17}\right)\left(\frac{17}{8}\right) = \frac{15}{8}$$

$$\therefore \tan x = \frac{15}{8}$$

**ALTERNATIVE METHOD**

$$\tan^2 x = \sec^2 x - 1$$

$$= \left(\frac{17}{8}\right)^2 - 1 = \frac{225}{64}$$

$$\tan x = \sqrt{\frac{225}{64}}$$

$$\therefore \tan x = \frac{15}{8}$$

**c)** $\cot y$

**Solution**

$$\cot y = \frac{1}{\tan y}$$

$$= \frac{1}{\frac{5}{12}} = \frac{12}{5}$$

$$\therefore \cot y = \frac{12}{5}$$

**d)** $\operatorname{cosec} y$

**Solution**

$$\cot^2 y + 1 = \operatorname{cosec}^2 y$$

$$\operatorname{cosec}^2 y = \frac{1}{\tan^2 y} + 1$$

$$= \frac{1}{\left(\frac{5}{12}\right)^2} + 1 = \frac{169}{25}$$

$$\operatorname{cosec} y = \sqrt{\frac{169}{25}}$$

$$\therefore \operatorname{cosec} y = \frac{13}{5}$$

**e)** cos y
**Solution**

$$\sin^2 y + \cos^2 y = 1$$

This implies that

$$\cos^2 y = 1 - \sin^2 y$$

But

$$\sin y = \frac{1}{\operatorname{cosec} y}$$

Therefore,

$$\cos^2 y = 1 - \frac{1}{\operatorname{cosec}^2 y}$$

$$\cos^2 y = 1 - \frac{1}{\left(\frac{13}{5}\right)^2} = 1 - \left(\frac{5}{13}\right)^2$$

$$\cos y = \sqrt{\frac{144}{169}}$$

$$\therefore \cos y = \frac{12}{13}$$

## 3.3 THE ADDITION FORMULAS

Given any two angles $A$ and $B$, the trigonometric ratios of their addition (sum) or subtraction (difference) are achieved using addition formulas, which are summarised in Table 3.1 for sine, cosine, and tangent.

**TABLE 3.1**
**Addition Formulas for the Three Trigonometric Ratios**

|  | Addition | Subtraction |
|---|---|---|
| Sine | $\sin(A+B) = \sin A \cos B + \sin B \cos A$ | $\sin(A-B) = \sin A \cos B - \sin B \cos A$ |
| Cosine | $\cos(A+B) = \cos A \cos B - \sin A \sin B$ | $\cos(A-B) = \cos A \cos B + \sin A \sin B$ |
| Tangent | $\tan(A+B) = \dfrac{\tan A + \tan B}{1 - \tan A \tan B}$ | $\tan(A-B) = \dfrac{\tan A - \tan B}{1 + \tan A \tan B}$ |

(3.2)

# Trigonometric Identities I

The following should be noted about the above formulas.

**Note 1** It is collectively called **Addition Formulas**, though it consists of addition (or sum) and subtraction (or difference).

**Note 2** Angles $A$ and $B$ represent any acute or obtuse angle.

**Note 3** The formulas can be derived. In fact, once the sine and cosine of addition of $A$ and $B$ are derived, it is much easier to find the other trigonometric functions, i.e., $\tan(A+B)$, $\sin(A-B)$, $\cos(A-B)$, and $\tan(A-B)$.

**Note 4** The formulas for subtraction can be easily obtained if $B$ is replaced with $-B$ in the corresponding addition formula and then simplified. We will prove this in the worked examples.

**Note 5** Since the tangent of an angle is the ratio of sine to cosine of the same angle, it is easy to find the addition formula for tangent from the corresponding addition formulas of sine and cosine.

**Note 6** We've shown only one variant of the tangent formula. There are others that are also valid and will be demonstrated in the worked examples.

Now it is time to look at some examples.

### Example 5

Use addition formulas to prove that $\tan(A+B) = \dfrac{\tan A + \tan B}{1 - \tan A \tan B}$.

What did you get? Find the solution below to double-check your answer.

### Solution to Example 5

**HINT**

We will need the following:

- $\sin(A+B) = \sin A \cos B + \sin B \cos A$
- $\cos(A+B) = \cos A \cos B - \sin A \sin B$

$$\tan(A+B) = \frac{\sin(A+B)}{\cos(A+B)}$$

Substitute the addition formulas as:

$$= \frac{\sin A \cos B + \sin B \cos A}{\cos A \cos B - \sin A \sin B}$$

Divide both the numerator and denominator by $\cos A \cos B$

$$= \frac{\left(\dfrac{\sin A \cos B}{\cos A \cos B}\right) + \left(\dfrac{\sin B \cos A}{\cos A \cos B}\right)}{\left(\dfrac{\cos A \cos B}{\cos A \cos B}\right) - \left(\dfrac{\sin A \sin B}{\cos A \cos B}\right)}$$

Now simplify

$$= \frac{\left(\frac{\sin A}{\cos A}\right) + \left(\frac{\sin B}{\cos B}\right)}{(1) - \left[\left(\frac{\sin A}{\cos A}\right)\left(\frac{\sin B}{\cos B}\right)\right]}$$

$$= \frac{(\tan A) + (\tan B)}{(1) - [(\tan A)(\tan B)]}$$

$$= \frac{\tan A + \tan B}{1 - \tan A \tan B}$$

$$\therefore \tan(A + B) = \frac{\tan A + \tan B}{1 - \tan A \tan B}$$

Another example to try.

### Example 6

Use addition formulas to prove the following identities.

a) $\sin(A - B) = \sin A \cos B - \sin B \cos A$

b) $\cos(A - B) = \cos A \cos B + \sin A \sin B$

c) $\tan(A - B) = \frac{\tan A - \tan B}{1 + \tan A \tan B}$

What did you get? Find the solution below to double-check your answer.

### Solution to Example 6

**HINT**

Apply the following:

- Start with the addition formula, replace $B$ with $-B$ and simplify.
- Apply the fact that $\sin(-B) = -\sin B$, $\cos(-B) = \cos B$, and $\tan(-B) = -\tan B$.

a) $\sin(A - B) = \sin A \cos B - \sin B \cos A$
**Solution**

$$\sin(A + B) = \sin A \cos B + \sin B \cos A$$

Now replace $B$ with $-B$ and simplify, we have

$$\sin(A + (-B)) = \sin A \cos(-B) + \sin(-B) \cos A$$
$$= \sin A \cos B - \sin B \cos A$$
$$\therefore \sin(A - B) = \sin A \cos B - \sin B \cos A$$

# Trigonometric Identities I

**b)** $\cos(A - B) = \cos A \cos B + \sin A \sin B$
**Solution**

$$\cos(A + B) = \cos A \cos B - \sin A \sin B$$

Now replace $B$ with $-B$ and simplify, we have

$$\cos(A + (-B)) = \cos A \cos(-B) - \sin A \sin(-B)$$
$$= \cos A \cos B - \sin A (-\sin B) = \cos A \cos B + \sin A \sin B$$
$$\therefore \cos(A - B) = \cos A \cos B + \sin A \sin B$$

---

**c)** $\tan(A - B) = \frac{\tan A - \tan B}{1 + \tan A \tan B}$
**Solution**

$$\tan(A + B) = \frac{\tan A + \tan B}{1 - \tan A \tan B}$$

Now replace $B$ with $-B$ and simplify, we have

$$\tan(A + (-B)) = \frac{\tan A + \tan(-B)}{1 - \tan A \tan(-B)}$$
$$= \frac{\tan A - \tan B}{1 - \tan A (-\tan B)}$$
$$\therefore \tan(A - B) = \frac{\tan A - \tan B}{1 + \tan A \tan B}$$

---

Another set of examples to try.

## Example 7

Prove the following identities.

**a)** $\tan(A + B) = \frac{1 + \cot A \tan B}{\cot A - \tan B}$  **b)** $\tan(A + B) = \frac{\tan A \cot B + 1}{\cot B - \tan A}$

**c)** $\tan(A + B) = \frac{\cot A + \cot B}{\cot A \cot B - 1}$

---

What did you get? Find the solution below to double-check your answer.

## Solution to Example 7

**a)** $\tan(A + B) = \frac{1 + \cot A \tan B}{\cot A - \tan B}$
**Solution**

$$\tan(A + B) = \frac{\sin(A + B)}{\cos(A + B)}$$

Substitute the addition formulas for both sine and cosine

$$= \frac{\sin A \cos B + \sin B \cos A}{\cos A \cos B - \sin A \sin B}$$

Divide both the numerator and denominator by $\sin A \cos B$

$$= \frac{\left\{\dfrac{\sin A \cos B}{\sin A \cos B}\right\} + \left\{\dfrac{\sin B \cos A}{\sin A \cos B}\right\}}{\left\{\dfrac{\cos A \cos B}{\sin A \cos B}\right\} - \left\{\dfrac{\sin A \sin B}{\sin A \cos B}\right\}}$$

Now simplify

$$= \frac{1 + \left(\dfrac{\sin B \cos A}{\sin A \cos B}\right)}{\left(\dfrac{\cos A}{\sin A}\right) - \left(\dfrac{\sin B}{\cos B}\right)}$$

$$= \frac{1 + \left(\dfrac{\cos A}{\sin A}\right)\left(\dfrac{\sin B}{\cos B}\right)}{\left(\dfrac{\cos A}{\sin A}\right) - \left(\dfrac{\sin B}{\cos B}\right)}$$

$$= \frac{1 + \cot A \tan B}{\cot A - \tan B}$$

$$\therefore \tan(A + B) = \frac{1 + \cot A \tan B}{\cot A - \tan B}$$

---

**b)** $\tan(A + B) = \dfrac{\tan A \cot B + 1}{\cot B - \tan A}$

**Solution**

$$\tan(A + B) = \frac{\sin(A + B)}{\cos(A + B)}$$

Substitute the addition formulas for both sine and cosine

$$= \frac{\sin A \cos B + \sin B \cos A}{\cos A \cos B - \sin A \sin B}$$

Divide both the numerator and denominator by $\cos A \sin B$

$$= \frac{\left\{\dfrac{\sin A \cos B}{\cos A \sin B}\right\} + \left\{\dfrac{\sin B \cos A}{\cos A \sin B}\right\}}{\left\{\dfrac{\cos A \cos B}{\cos A \sin B}\right\} - \left\{\dfrac{\sin A \sin B}{\cos A \sin B}\right\}}$$

# Trigonometric Identities I

Now simplify

$$= \frac{\left(\frac{\sin A \cos B}{\cos A \sin B}\right) + 1}{\left(\frac{\cos B}{\sin B}\right) - \left(\frac{\sin A}{\cos A}\right)}$$

$$= \frac{\left(\frac{\sin A}{\cos A}\right)\left(\frac{\cos B}{\sin B}\right) + 1}{\left(\frac{\cos B}{\sin B}\right) - \left(\frac{\sin A}{\cos A}\right)}$$

$$= \frac{\tan A \cot B + 1}{\cot B - \tan A}$$

$$\therefore \tan(A+B) = \frac{\tan A \cot B + 1}{\cot B - \tan A}$$

---

c) $\tan(A+B) = \frac{\cot A + \cot B}{\cot A \cot B - 1}$

**Solution**

$$\tan(A+B) = \frac{\sin(A+B)}{\cos(A+B)}$$

Substitute the addition formulas for both sine and cosine

$$= \frac{\sin A \cos B + \sin B \cos A}{\cos A \cos B - \sin A \sin B}$$

Divide both the numerator and denominator by $\sin A \sin B$

$$= \frac{\left\{\frac{\sin A \cos B}{\sin A \sin B}\right\} + \left\{\frac{\sin B \cos A}{\sin A \sin B}\right\}}{\left\{\frac{\cos A \cos B}{\sin A \sin B}\right\} - \left\{\frac{\sin A \sin B}{\sin A \sin B}\right\}}$$

Now simplify

$$= \frac{\left(\frac{\cos B}{\sin B}\right) + \left(\frac{\cos A}{\sin A}\right)}{\left[\left(\frac{\cos A}{\sin A}\right)\left(\frac{\cos B}{\sin B}\right)\right] - (1)}$$

$$= \frac{(\cot B) + (\cot A)}{[(\cot A)(\cot B)] - (1)}$$

$$\therefore \tan(A+B) = \frac{\cot A + \cot B}{\cot A \cot B - 1}$$

More examples to try.

### Example 8

Use addition formulas to show the following trigonometric identities.

a) $\sin(\pi - \theta) = \sin\theta$    b) $\cos(\pi + \theta) = -\cos\theta$    c) $\tan(2\pi - \theta) = -\tan\theta$

What did you get? Find the solution below to double-check your answer.

### Solution to Example 8

**HINT**

Note that:
- $\sin\pi = \sin 0 = 0$
- $\cos\pi = \cos 0 = -1$

a) $\sin(\pi - \theta) = \sin\theta$
**Solution**

$$\sin(\pi - \theta) = \sin\theta$$

Let's take the LHS and say $A = \pi$ and $B = \theta$. Substitute and simplify as:

$$\sin(\pi - \theta) = \sin\pi\cos\theta - \sin\theta\cos\pi$$
$$= \sin 0\cos\theta - \sin\theta\cos\pi$$
$$= -\sin\theta \times (-1) = \sin\theta$$
$$\therefore \sin(\pi - \theta) = \sin\theta$$

This agrees with the second quadrant formula for sine.

b) $\cos(\pi + \theta) = -\cos\theta$
**Solution**

$$\cos(\pi + \theta) = -\cos\theta$$

Let's take the LHS and say $A = \pi$ and $B = \theta$. Substitute and simplify as:

$$\cos(\pi + \theta) = \cos A\cos B - \sin A\sin B$$
$$= \cos\pi\cos\theta - \sin\pi\sin\theta$$
$$= (-1)\cos\theta - (0)\sin\theta$$
$$= (-1) \times \cos\theta = -\cos\theta$$
$$\therefore \cos(\pi + \theta) = -\cos\theta$$

This agrees with the third quadrant formula for cosine.

# Trigonometric Identities I

**c)** $\tan(2\pi - \theta) = -\tan\theta$

**Solution**

$$\tan(2\pi - \theta) = -\tan\theta$$

Let's take the LHS and say $A = 2\pi$ and $B = \theta$. Substitute and simplify as:

$$\tan(2\pi - \theta) = \frac{\tan A - \tan B}{1 + \tan A \tan B}$$
$$= \frac{\tan 2\pi - \tan\theta}{1 + \tan 2\pi \tan\theta}$$
$$= \frac{(0) - \tan\theta}{1 + (0)\tan\theta}$$
$$= \frac{0 - \tan\theta}{1 + 0} = \frac{-\tan\theta}{1}$$
$$= -\tan\theta$$

Because $\tan 2\pi = \tan 0 = 0$

$$\therefore \tan(2\pi - \theta) = -\tan\theta$$

**ALTERNATIVE METHOD**

$$\tan(2\pi - \theta) = \frac{\sin(2\pi - \theta)}{\cos(2\pi - \theta)}$$

Let's take the LHS and say $A = 2\pi$ and $B = \theta$. Substitute and simplify as:

$$= \frac{\sin A \cos B - \sin B \cos A}{\cos A \cos B - \sin A \sin B}$$
$$= \frac{\sin 2\pi \cos\theta - \sin\theta \cos 2\pi}{\cos 2\pi \cos\theta - \sin 2\pi \sin\theta}$$
$$= \frac{(0)\cos\theta - \sin\theta(1)}{(1)\cos\theta - (0)\sin\theta}$$
$$= \frac{0 - \sin\theta}{\cos\theta + 0} = \frac{-\sin\theta}{\cos\theta}$$
$$= -\tan\theta$$

$$\therefore \tan(2\pi - \theta) = -\tan\theta$$

Let's conclude this part with this set of examples.

## Example 9

Using appropriate addition formulas, determine the exact value of the following:

**a)** $\sin 135°$  **b)** $\tan 15°$  **c)** $\cos 75°$  **d)** $\sin 225°$  **e)** $\tan 105°$

What did you get? Find the solution below to double-check your answer.

### Solution to Example 9

**HINT**

Apply the following:

- Decide on how to combine the special angles (**30°**, **45°**, **60°**, and **90°**) to make up the given angle using addition or subtraction.
- Replace the chosen angles with $A$ and $B$ and apply the relevant identities.
- Apply the exact values of the special angles. You may need to refer to Table 1.1 in Chapter 1.

**a)** $\sin 135°$
**Solution**

$$\sin 135° = \sin(A + B)$$
$$= \sin(90° + 45°)$$

Let's make $A = 90°$ and $B = 45°$. Now substitute and simplify as:

$$\sin(90° + 45°) = \sin A \cos B + \sin B \cos A$$
$$= \sin 90° \cos 45° + \sin 45° \cos 90°$$
$$= (1)\left(\frac{\sqrt{2}}{2}\right) + \left(\frac{\sqrt{2}}{2}\right)(0)$$
$$= \frac{\sqrt{2}}{2} + 0 = \frac{\sqrt{2}}{2}$$
$$\therefore \sin 135° = \sin(90° + 45°) = \frac{\sqrt{2}}{2}$$

**b)** $\tan 15°$
**Solution**

$$\tan 15° = \tan(A - B)$$
$$= \tan(60° - 45°)$$

Let's make $A = 60°$ and $B = 45°$. Now substitute and simplify as:

$$\tan(60° - 45°) = \frac{\tan A - \tan B}{1 + \tan A \tan B}$$
$$= \frac{\tan 60° - \tan 45°}{1 + \tan 60° \tan 45°}$$
$$= \frac{\sqrt{3} - 1}{1 + \left(\sqrt{3}\right)(1)}$$
$$= \frac{\sqrt{3} - 1}{1 + \sqrt{3}}$$

Trigonometric Identities I

Rationalising the denominator, we have

$$= \frac{\sqrt{3}-1}{1+\sqrt{3}} \times \frac{1-\sqrt{3}}{1-\sqrt{3}} = \frac{(\sqrt{3}-1)(1-\sqrt{3})}{(1+\sqrt{3})(1-\sqrt{3})}$$

$$= \frac{\sqrt{3}-3-1+\sqrt{3}}{1-3} = \frac{2\sqrt{3}-4}{-2} = \frac{2(\sqrt{3}-2)}{-2} = -(\sqrt{3}-2)$$

$$\therefore \tan 15° = \tan(60°-45°) = 2-\sqrt{3}$$

---

**c) cos 75°**
**Solution**

$$\cos 75° = \cos(A+B)$$
$$= \cos(45°+30°)$$

Let's make $A = 45°$ and $B = 30°$. Now substitute and simplify as:

$$\cos(45°+30°) = \cos A \cos B - \sin A \sin B$$
$$= \cos 45° \cos 30° - \sin 45° \sin 30°$$
$$= \left(\frac{\sqrt{2}}{2}\right)\left(\frac{\sqrt{3}}{2}\right) - \left(\frac{\sqrt{2}}{2}\right)\left(\frac{1}{2}\right)$$
$$= \frac{\sqrt{6}}{4} - \frac{\sqrt{2}}{4} = \frac{\sqrt{6}-\sqrt{2}}{4}$$

$$\therefore \cos 75° = \cos(45°+30°) = \frac{1}{4}\left(\sqrt{6}-\sqrt{2}\right)$$

---

**d) sin 225°**
**Solution**

$$\sin 225° = \sin(A+B)$$
$$= \sin(180°+45°)$$

Let's make $A = 180°$ and $B = 45°$. Now substitute and simplify as:

$$\sin(180°+45°) = \sin A \cos B + \sin B \cos A$$
$$= \sin 180° \cos 45° + \sin 45° \cos 180°$$
$$= (0)\left(\frac{\sqrt{2}}{2}\right) + \left(\frac{\sqrt{2}}{2}\right)(-1)$$
$$= 0 - \frac{\sqrt{2}}{2}$$

$$\therefore \sin 225° = \sin(180°+45°) = -\frac{\sqrt{2}}{2}$$

**e)** tan 105°
**Solution**

$$\tan 105° = \tan(A + B)$$
$$= \tan(60° + 45°)$$

Let's make $A = 60°$ and and $B = 45°$. Now substitute and simplify as:

$$\tan(60° + 45°) = \frac{\tan A + \tan B}{1 - \tan A \tan B}$$
$$= \frac{\tan 60° + \tan 45°}{1 - \tan 60° \tan 45°}$$
$$= \frac{\sqrt{3} + 1}{1 - (\sqrt{3})(1)}$$
$$= \frac{\sqrt{3} + 1}{1 - \sqrt{3}}$$

Rationalising the denominator, we have

$$= \frac{\sqrt{3} + 1}{1 - \sqrt{3}} \times \frac{1 + \sqrt{3}}{1 + \sqrt{3}} = \frac{(\sqrt{3} + 1)(1 + \sqrt{3})}{(1 - \sqrt{3})(1 + \sqrt{3})}$$
$$= \frac{\sqrt{3} + 3 + 1 + \sqrt{3}}{1 - 3} = \frac{2\sqrt{3} + 4}{-2} = \frac{2(\sqrt{3} + 2)}{-2} = -(\sqrt{3} + 2)$$

$$\therefore \tan 105° = \tan(60° + 45°) = -2 - \sqrt{3}$$

## 3.4 MULTIPLE ANGLES

In this section, we are concerned with the identities $n$ (and $\frac{1}{n}$) multiples of a given angle, such that $n$ is a positive integer. Each member of this family is similar, as they are obtained through the addition formulas. For brevity, we will only show three different values of $n$, namely double ($n = 2$), triple ($n = 3$), and half ($n = \frac{1}{2}$). The formulas resulting from these three different values of $n$ are, respectively, called double-angle formulas, triple-angle formulas, and half-angle formulas.

### 3.4.1 THE DOUBLE-ANGLE FORMULAS

This is when $n = 2$, and thus it is called double-angle formulas. The identities are summarised in Table 3.2, but their derivation will be covered in the worked examples.

From the information above, it is evident that double-angle formulas represent a specific type of addition formulas where $A = B$. Notice that the cosine has three variants for a double-angle formula. The first is obtained directly from the addition formulas and the last two are derived from the first one using $\sin^2 A + \cos^2 A = 1$.

# Trigonometric Identities I

## TABLE 3.2
### Double-Angle Formulas for the Three Trigonometric Ratios

| Function | Identities |
|---|---|
| Sine | $\sin 2A = 2\sin A \cos A$ |
| Cosine | $\cos 2A = \cos^2 A - \sin^2 A$ \quad $\cos 2A = 2\cos^2 A - 1$ \quad $\cos 2A = 1 - 2\sin^2 A$ |
| Tangent | $\tan 2A = \dfrac{2\tan A}{1 - \tan^2 A}$ |

(3.3)

Let's go for some examples to illustrate the above identities.

### Example 10

Using addition formulas, prove the following double-angle identities.

a) $\sin 2A = 2\sin A \cos A$
b) $\cos 2A = \cos^2 A - \sin^2 A$, $\cos 2A = 2\cos^2 A - 1$, and $\cos 2A = 1 - 2\sin^2 A$
c) $\tan 2A = \dfrac{2\tan A}{1 - \tan^2 A}$

What did you get? Find the solution below to double-check your answer.

### Solution to Example 10

**HINT**

Apply the following:

- Start with addition formulas, replace $B$ with $A$, and simplify.
- We may need to apply $\sin^2 A + \cos^2 A = 1$.

a) $\sin 2A = 2\sin A \cos A$
**Solution**

$$\sin(A + B) = \sin A \cos B + \sin B \cos A$$

Now replace $B$ with $A$, we have

$$\sin(A + A) = \sin A \cos A + \sin A \cos A$$
$$= 2\sin A \cos A$$
$$\therefore \sin 2A = 2\sin A \cos A$$

**b)** $\cos 2A = \cos^2 A - \sin^2 A = 2\cos^2 A - 1 = 1 - 2\sin^2 A$

**Solution**

$$\cos(A + B) = \cos A \cos B - \sin A \sin B$$

Now replace $B$ with $A$, we have:

$$\cos(A + A) = \cos A \cos A - \sin A \sin A$$
$$= \cos^2 A - \sin^2 A$$
$$\therefore \cos 2A = \cos^2 A - \sin^2 A$$

From $\sin^2 A + \cos^2 A = 1$, we have $\sin^2 A = 1 - \cos^2 A$. Substitute this into the above double-angle formula as:

$$\cos 2A = \cos^2 A - \sin^2 A$$
$$= \cos^2 A - (1 - \cos^2 A)$$
$$= \cos^2 A + \cos^2 A - 1$$
$$\therefore \cos 2A = 2\cos^2 A - 1$$

Similarly, $\cos^2 A = 1 - \sin^2 A$. We can now substitute this into either of the two identities above. Using the first identity, we have:

$$\cos 2A = \cos^2 A - \sin^2 A$$
$$= (1 - \sin^2 A) - \sin^2 A$$
$$= 1 - \sin^2 A - \sin^2 A$$
$$\therefore \cos 2A = 1 - 2\sin^2 A$$

Or using the second (derived) identity, we have:

$$\cos 2A = 2\cos^2 A - 1$$
$$= 2(1 - \sin^2 A) - 1$$
$$= 2 - 2\sin^2 A - 1$$
$$\therefore \cos 2A = 1 - 2\sin^2 A$$

---

**c)** $\tan 2A = \dfrac{2\tan A}{1 - \tan^2 A}$

**Solution**

$$\tan(A + B) = \frac{\tan A + \tan B}{1 - \tan A \tan B}$$

Now replace $B$ with $A$, we have:

$$\tan(A + A) = \frac{\tan A + \tan A}{1 - \tan A \tan A}$$
$$= \frac{\tan A + \tan A}{1 - \tan A \tan A}$$
$$\therefore \tan 2A = \frac{2\tan A}{1 - \tan^2 A}$$

# Trigonometric Identities I

Another example to try.

### Example 11

Using the double-angle formulas, determine the exact value of the following trigonometric expressions.

**a)** $\sin 75° \cos 75°$  **b)** $\cos^2 \frac{5}{3}\pi - \sin^2 \frac{5}{3}\pi$

What did you get? Find the solution below to double-check your answer.

### Solution to Example 11

**HINT**

Apply the following:

- Pay attention to the question format and select the appropriate double-angle identities.
- We will need our knowledge of special angles, since we are looking for exact values.

**a)** $\sin 75° \cos 75°$
**Solution**
For this, the appropriate identity is

$$\sin 2A = 2 \sin A \cos A$$

Divide both sides by 2, we have

$$\frac{1}{2} \sin 2A = \sin A \cos A$$

In comparison, we have

$$\sin 75° \cos 75° = \frac{1}{2} \sin 2(75°)$$
$$= \frac{1}{2} \sin 150° = \frac{1}{2} \sin (180° - 150°)$$
$$= \frac{1}{2} \sin 30° = \frac{1}{2}\left(\frac{1}{2}\right) = \frac{1}{4}$$
$$\therefore \sin 75° \cos 75° = \frac{1}{4}$$

**b)** $\cos^2 \frac{5}{3}\pi - \sin^2 \frac{5}{3}\pi$
**Solution**
For this, the appropriate identity is

$$\cos 2A = \cos^2 A - \sin^2 A$$

In comparison, we have
$$A = \frac{5}{3}\pi$$
$$2A = \frac{10}{3}\pi$$

Therefore
$$\cos^2\frac{5}{3}\pi - \sin^2\frac{5}{3}\pi = \cos 2A = \cos 2\left(\frac{5}{3}\pi\right)$$
$$= \cos\frac{10}{3}\pi$$

$\frac{10}{3}\pi$ is more than one revolution, so we need to take $n$ revolution, where $n = 1$ for this case. Thus
$$= \cos\left(\frac{10}{3}\pi - 2\pi\right) = \cos\frac{4}{3}\pi$$

$\frac{4}{3}\pi$ is in the third quadrant, so we have
$$\cos\frac{4}{3}\pi = -\cos\left(\pi + \frac{1}{3}\pi\right)$$
$$= -\cos\frac{1}{3}\pi = -\frac{1}{2}$$
$$\therefore \cos^2\frac{5}{3}\pi - \sin^2\frac{5}{3}\pi = -\frac{1}{2}$$

---

One final example to try.

## Example 12

Prove that $\cos^4 x - \sin^4 x \equiv \cos 2x$.

---

What did you get? Find the solution below to double-check your answer.

## Solution to Example 12

Let's start with the LHS, we have
$$\cos^4 x - \sin^4 x = \left(\cos^2 x\right)^2 - \left(\sin^2 x\right)^2$$

This is a case of a difference of two squares. Thus,
$$\left(\cos^2 x\right)^2 - \left(\sin^2 x\right)^2 = \left(\cos^2 x - \sin^2 x\right)\left(\cos^2 x + \sin^2 x\right)$$
$$= \left(\cos^2 x - \sin^2 x\right)(1) = \cos^2 x - \sin^2 x$$
$$= \cos 2x$$
$$\therefore \cos^4 x - \sin^4 x \equiv \cos 2x$$

# Trigonometric Identities I

## 3.4.2 THE TRIPLE-ANGLE FORMULAS

This is when $n = 3$. The identities are summarised in Table 3.3, but they will be derived in the worked examples. Note that the identities shown represent a type for each, which ensures that they are homogeneous in the function being defined. For example, the triple identity for sine consists of the sine function only and the same for the other two functions.

It is time to look at some examples.

### Example 13

Using addition formulas, prove the following triple-angle identities.

**a)** $\sin 3A$          **b)** $\cos 3A$          **c)** $\tan 3A$

What did you get? Find the solution below to double-check your answer.

### Solution to Example 13

**HINT**

Apply the following:

- Start with addition formula and simplify.
- Apply the double-angle formulas and simplify.
- We will need to apply $\sin^2 A + \cos^2 A = 1$.
- There are other possibilities depending on how we carry out the substitution process.

**a)** $\sin 3A$
**Solution**

$$\sin(A + 2A) = \sin 2A \cos A + \sin A \cos 2A$$

Using double angle formula, we have:

$$= (2 \sin A \cos A) \cos A + \sin A \left(1 - 2 \sin^2 A\right)$$
$$= 2 \sin A \cos^2 A + \sin A - 2\sin^3 A$$

From $\sin^2 A + \cos^2 A = 1$, we have $\cos^2 A = 1 - \sin^2 A$. Substitute this into the above as:

$$= 2 \sin A \left(1 - \sin^2 A\right) + \sin A - 2\sin^3 A$$
$$= 2 \sin A - 2 \sin^3 A + \sin A - 2\sin^3 A$$
$$\therefore \sin 3A = 3 \sin A - 4 \sin^3 A$$

## TABLE 3.3
**Triple-Angle Formulas for the Three Trigonometric Ratios**

| Function | Identities |
|---|---|
| Sine | $\sin 3A = 3\sin A - 4\sin^3 A$ |
| Cosine | $\cos 3A = 4\cos^3 A - 3\cos A$ |
| Tangent | $\tan 3A = \dfrac{2\tan A - \tan^3 A}{1 - 3\tan^2 A}$ |

(3.4)

**ALTERNATIVE METHOD**

If we instead substitute $\cos 2A = 2\cos^2 A - 1$, we have

$$\sin(A + 2A) = \sin 2A \cos A + \sin A \cos 2A$$

Using double angle formula, we have:

$$= (2\sin A \cos A)\cos A + \sin A (2\cos^2 A - 1)$$
$$= 2\sin A \cos^2 A + 2\sin A \cos^2 A - \sin A$$

From $\sin^2 A + \cos^2 A = 1$, we have $\cos^2 A = 1 - \sin^2 A$. Substitute this into the above as:

$$= 2\sin A (1 - \sin^2 A) + 2\sin A (1 - \sin^2 A) - \sin A$$
$$= 2\sin A - 2\sin^3 A + 2\sin A - 2\sin^3 A - \sin A$$
$$\therefore \sin 3A = 3\sin A - 4\sin^3 A$$

**b)** $\cos 3A$
**Solution**

$$\cos(A + 2A) = \cos 2A \cos A - \sin 2A \sin A$$

Using double angle formula, we have:

$$= (\cos 2A)\cos A - (\sin 2A)\sin A$$
$$= (2\cos^2 A - 1)\cos A - (2\sin A \cos A)\sin A$$
$$= 2\cos^3 A - \cos A - 2\sin^2 A \cos A$$

From $\sin^2 A + \cos^2 A = 1$, we have $\sin^2 A = 1 - \cos^2 A$. Substitute this into the above as:

$$= 2\cos^3 A - \cos A - 2(1 - \cos^2 A)\cos A$$
$$= 2\cos^3 A - \cos A - 2\cos A + 2\cos^3 A$$
$$= 2\cos^3 A + 2\cos^3 A - \cos A - 2\cos A$$
$$\therefore \cos 3A = 4\cos^3 A - 3\cos A$$

Trigonometric Identities I

**c)** $\tan 3A$
**Solution**

$$\tan(3A) = \tan(2A + A) = \frac{\tan 2A + \tan A}{1 - \tan 2A \tan A}$$

Using double angle formula, we have:

$$= \frac{\left[\frac{2\tan A}{1-\tan^2 A}\right] + \tan A}{1 - \left[\frac{2\tan A}{1-\tan^2 A}\right]\tan A}$$

$$= \frac{\left[\frac{2\tan A}{1-\tan^2 A} + \tan A\right]}{1 - \left[\frac{2\tan^2 A}{1-\tan^2 A}\right]} = \frac{\left\{\frac{2\tan A}{1-\tan^2 A} + \tan A\right\}}{\left\{\frac{1-\tan^2 A}{1-\tan^2 A} - \frac{2\tan^2 A}{1-\tan^2 A}\right\}}$$

$$= \frac{\left\{\frac{2\tan A}{1-\tan^2 A} + \frac{\tan A(1-\tan^2 A)}{1-\tan^2 A}\right\}}{\left\{\frac{1-\tan^2 A}{1-\tan^2 A} - \frac{2\tan^2 A}{1-\tan^2 A}\right\}}$$

$$= \frac{2\tan A + \tan A(1-\tan^2 A)}{1-\tan^2 A} \div \frac{1-\tan^2 A - 2\tan^2 A}{1-\tan^2 A}$$

$$= \frac{2\tan A + \tan A(1-\tan^2 A)}{1-\tan^2 A} \times \frac{1-\tan^2 A}{1-\tan^2 A - 2\tan^2 A}$$

$$= \frac{2\tan A + \tan A(1-\tan^2 A)}{1-\tan^2 A - 2\tan^2 A} = \frac{2\tan A + \tan A - \tan^3 A}{1 - 3\tan^2 A}$$

$$\therefore \tan 3A = \frac{3\tan A - \tan^3 A}{1 - 3\tan^2 A}$$

---

### 3.4.3 THE HALF-ANGLE FORMULAS

This is when $n = \frac{1}{2}$. The identities are summarised in Table 3.4, and their derivation will be covered in the worked examples.

You may have noticed that the above are very similar to their double-angle equivalents. Yes, this is true; we've only changed $A$ to $\frac{1}{2}A$.

**TABLE 3.4**
**Half-Angle Formulas for the Three Trigonometric Ratios**

| Function | Identities |
|---|---|
| Sine | $\sin A = 2\sin\frac{1}{2}A \cos\frac{1}{2}A$ |
| Cosine | $\cos A = \cos^2\frac{1}{2}A - \sin^2\frac{1}{2}A \quad\bigg|\quad \cos A = 2\cos^2\frac{1}{2}A - 1 \quad\bigg|\quad \cos A = 1 - 2\sin^2\frac{1}{2}A$ |
| Tangent | $\tan A = \dfrac{2\tan\frac{1}{2}A}{1 - \tan^2\frac{1}{2}A}$ |

(3.5)

Now it is time to look at some examples.

### Example 14

Using addition formulas, prove the following half-angle identities.

a) $\sin A = 2\sin\frac{1}{2}A \cos\frac{1}{2}A$

b) $\cos A = \cos^2\frac{1}{2}A - \sin^2\frac{1}{2}A$, $\cos A = 2\cos^2\frac{1}{2}A - 1$, **and** $\cos A = 1 - 2\sin^2\frac{1}{2}A$

c) $\tan A = \dfrac{2\tan\frac{1}{2}A}{1-\tan^2\frac{1}{2}A}$

# Trigonometric Identities I

What did you get? Find the solution below to double-check your answer.

## Solution to Example 14

**HINT**

Apply the following:

- Start with addition formulas, replace $B$ with $A$, and simplify.
- Replace $2A$ with $A$, because $\sin 2A = \sin(A+A)$ implies $\sin A = \sin\left(\frac{1}{2}A + \frac{1}{2}A\right)$.
- We may need to apply another identity ($\sin^2 \frac{1}{2}A + \cos^2 \frac{1}{2}A = 1$), which is a variant of $\sin^2 A + \cos^2 A = 1$.

**a)** $\sin A = 2\sin\frac{1}{2}A \cos\frac{1}{2}A$

**Solution**

$$\sin(A+B) = \sin A \cos B + \sin B \cos A$$

Now replace $B$ with $A$, we have

$$\sin 2A = \sin(A+A) = \sin A \cos A + \sin A \cos A$$
$$= 2\sin A \cos A$$

Thus

$$\sin A = \sin\left(\frac{1}{2}A + \frac{1}{2}A\right) = 2\sin\frac{1}{2}A \cos\frac{1}{2}A$$
$$\therefore \sin A = 2\sin\frac{1}{2}A \cos\frac{1}{2}A$$

---

**b)** $\cos A = \cos^2\frac{1}{2}A - \sin^2\frac{1}{2}A = 2\cos^2\frac{1}{2}A - 1 = 1 - 2\sin^2\frac{1}{2}A$

**Solution**

$$\cos(A+B) = \cos A \cos B - \sin A \sin B$$

Now replace $B$ with $A$, we have

$$\cos 2A = \cos(A+A) = \cos A \cos A - \sin A \sin A$$
$$= \cos^2 A - \sin^2 A$$

Thus

$$\cos A = \cos\left(\frac{1}{2}A + \frac{1}{2}A\right) = \cos\frac{1}{2}A \cos\frac{1}{2}A - \sin\frac{1}{2}A \sin\frac{1}{2}A$$

$$\therefore \cos A = \cos^2\frac{1}{2}A - \sin^2\frac{1}{2}A$$

For the same argument used in double-angle formulas and by applying $\sin^2 \frac{1}{2}A + \cos^2 \frac{1}{2}A = 1$, we can show that the other variants are:

$$\cos A = \cos^2 \frac{1}{2}A - \sin^2 \frac{1}{2}A$$
$$= \cos^2 \frac{1}{2}A - \left(1 - \cos^2 \frac{1}{2}A\right)$$
$$= \cos^2 \frac{1}{2}A + \cos^2 \frac{1}{2}A - 1$$
$$\therefore \cos A = 2\cos^2 \frac{1}{2}A - 1$$

Similarly,

$$\cos A = \cos^2 \frac{1}{2}A - \sin^2 \frac{1}{2}A$$
$$= \left(1 - \sin^2 \frac{1}{2}A\right) - \sin^2 \frac{1}{2}A$$
$$= 1 - \sin^2 \frac{1}{2}A - \sin^2 \frac{1}{2}A$$
$$\therefore \cos A = 1 - 2\sin^2 \frac{1}{2}A$$

Or using the second derived identity, we have

$$\cos A = 2\cos^2 \frac{1}{2}A - 1$$
$$= 2\left(1 - \sin^2 \frac{1}{2}A\right) - 1$$
$$= 2 - 2\sin^2 \frac{1}{2}A - 1$$
$$\therefore \cos A = 1 - 2\sin^2 \frac{1}{2}A$$

---

**c)** $\tan A = \dfrac{2\tan \frac{1}{2}A}{1 - \tan^2 \frac{1}{2}A}$

**Solution**

$$\tan(A + B) = \frac{\tan A + \tan B}{1 - \tan A \tan B}$$

Now replace $B$ with $A$, we have

$$\tan(A + A) = \frac{\tan A + \tan A}{1 - \tan A \tan A}$$
$$= \frac{\tan A + \tan A}{1 - \tan A \tan A}$$

# Trigonometric Identities I

Thus

$$\tan A = \tan\left(\frac{1}{2}A + \frac{1}{2}A\right) = \frac{2\tan\frac{1}{2}A}{1 - \tan^2\frac{1}{2}A}$$

$$\therefore \tan A = \frac{2\tan\frac{1}{2}A}{1 - \tan^2\frac{1}{2}A}$$

Another set of examples to try.

## Example 15

Given that $t = \tan\frac{1}{2}A$, show that:

a) $\sin A = \frac{2t}{1+t^2}$  b) $\cos A = \frac{1-t^2}{1+t^2}$

What did you get? Find the solution below to double-check your answer.

## Solution to Example 15

**HINT**

Apply the following:

- Start with the half-angle addition formulas.
- We need to apply $\sin^2\frac{1}{2}A + \cos^2\frac{1}{2}A = 1$, which is a variant of $\sin^2 A + \cos^2 A = 1$.

a) $\sin A = \frac{2t}{1+t^2}$
**Solution**

$$\sin A = 2\sin\frac{1}{2}A \cos\frac{1}{2}A$$

Divide the RHS by 1

$$= \frac{2\sin\frac{1}{2}A \cos\frac{1}{2}A}{1}$$

Use $\sin^2\frac{1}{2}A + \cos^2\frac{1}{2}A$ to replace 1 on the RHS.

$$= \frac{2\sin\frac{1}{2}A \cos\frac{1}{2}A}{\sin^2\frac{1}{2}A + \cos^2\frac{1}{2}A}$$

Divide both the numerator and denominator by $\cos^2 \frac{1}{2}A$.

$$= \frac{\left(\frac{2\sin\frac{1}{2}A\cos\frac{1}{2}A}{\cos^2\frac{1}{2}A}\right)}{\left(\frac{\sin^2\frac{1}{2}A+\cos^2\frac{1}{2}A}{\cos^2\frac{1}{2}A}\right)} = \frac{\frac{2\sin\frac{1}{2}A\cos\frac{1}{2}A}{\left(\cos\frac{1}{2}A\right)\left(\cos\frac{1}{2}A\right)}}{\left(\frac{\sin^2\frac{1}{2}A}{\cos^2\frac{1}{2}A} + \frac{\cos^2\frac{1}{2}A}{\cos^2\frac{1}{2}A}\right)}$$

Now simplify

$$= \frac{2\left(\frac{\sin\frac{1}{2}A}{\cos\frac{1}{2}A}\right)}{\left(\frac{\sin^2\frac{1}{2}A}{\cos^2\frac{1}{2}A}+1\right)} = \frac{2\left(\tan\frac{1}{2}A\right)}{\left(\tan^2\frac{1}{2}A\right)+1}$$

$$\therefore \sin A = \frac{2t}{1+t^2}$$

---

**b)** $\cos A = \frac{1-t^2}{1+t^2}$

**Solution**

$$\cos A = \cos^2\frac{1}{2}A - \sin^2\frac{1}{2}A$$

$$= \frac{\cos^2\frac{1}{2}A - \sin^2\frac{1}{2}A}{1}$$

$$= \frac{\cos^2\frac{1}{2}A - \sin^2\frac{1}{2}A}{\sin^2\frac{1}{2}A + \cos^2\frac{1}{2}A}$$

Divide both the numerator and denominator by $\cos^2\frac{1}{2}A$

$$= \frac{\left(\frac{\cos^2\frac{1}{2}A-\sin^2\frac{1}{2}A}{\cos^2\frac{1}{2}A}\right)}{\left(\frac{\sin^2\frac{1}{2}A+\cos^2\frac{1}{2}A}{\cos^2\frac{1}{2}A}\right)} = \frac{\frac{\cos^2\frac{1}{2}A}{\cos^2\frac{1}{2}A} - \frac{\sin^2\frac{1}{2}A}{\cos^2\frac{1}{2}A}}{\left(\frac{\sin^2\frac{1}{2}A}{\cos^2\frac{1}{2}A} + \frac{\cos^2\frac{1}{2}A}{\cos^2\frac{1}{2}A}\right)}$$

# Trigonometric Identities I

Now simplify

$$\frac{\left(1 - \frac{\sin^2 \frac{1}{2}A}{\cos^2 \frac{1}{2}A}\right)}{\left(\frac{\sin^2 \frac{1}{2}A}{\cos^2 \frac{1}{2}A} + 1\right)} = \frac{1 - \tan^2 \frac{1}{2}A}{\tan^2 \frac{1}{2}A + 1}$$

$$\therefore \cos A = \frac{1 - t^2}{1 + t^2}$$

Another example to try.

## Example 16

$\alpha$ and $\beta$ are two acute angles such that $\tan \alpha = \frac{4}{3}$ and $\sin \beta = \frac{12}{13}$. Using an appropriate identity or identities, determine the exact value of the following:

a) $\sin(\alpha - \beta)$  b) $\cos(\alpha + \beta)$  c) $\tan 2\alpha$  d) $\sin 3\beta$

What did you get? Find the solution below to double-check your answer.

## Solution to Example 16

**HINT**

- In this case, it will be helpful to find all the relevant trigonometric ratios ($\sin \alpha$, $\cos \alpha$, $\cos \beta$, and $\tan \beta$) before proceeding to the questions.
- You can use the fact that $\alpha = 53.13°$ and $\beta = 67.38°$, correct to 2 d.p., to quickly check your answers.

Let's start by determining the following:

| • $\sin \alpha$ | • $\cos \alpha$ | • $\cos \beta$ | • $\tan \beta$ |
|---|---|---|---|
| $\cot^2 \alpha + 1 = \operatorname{cosec}^2 \alpha$ $\operatorname{cosec}^2 \alpha = \dfrac{1}{\tan^2 \alpha} + 1$ $= \dfrac{1}{\left(\tfrac{4}{3}\right)^2} + 1$ $= \dfrac{25}{16}$ $\operatorname{cosec} \alpha = \sqrt{\dfrac{25}{16}}$ $= \dfrac{5}{4}$ $\therefore \sin \alpha = \dfrac{4}{5}$ | $\cos^2 \alpha = 1 - \sin^2 \alpha$ $\cos \alpha = \sqrt{1 - \sin^2 \alpha}$ $= \sqrt{1 - \left(\tfrac{4}{5}\right)^2}$ $\therefore \cos \alpha = \dfrac{3}{5}$ | $\cos \beta = \sqrt{1 - \sin^2 \beta}$ $= \sqrt{1 - \left(\tfrac{12}{13}\right)^2}$ $\therefore \cos \beta = \dfrac{5}{13}$ | $\tan \beta = \dfrac{\sin \beta}{\cos \beta}$ $= \dfrac{\left(\tfrac{12}{13}\right)}{\left(\tfrac{5}{13}\right)} =$ $\therefore \tan \beta = \dfrac{12}{5}$ |

We can now use the above results in our calculations below.

**a)** $\sin(\alpha - \beta)$
**Solution**

$$\sin(\alpha - \beta) = \sin \alpha \cos \beta - \sin \beta \cos \alpha$$
$$= \left(\tfrac{4}{5}\right)\left(\tfrac{5}{13}\right) - \left(\tfrac{12}{13}\right)\left(\tfrac{3}{5}\right)$$
$$= \dfrac{20}{65} - \dfrac{36}{65}$$
$$= \dfrac{20 - 36}{65} = -\dfrac{16}{65}$$
$$\therefore \sin(\alpha - \beta) = -\dfrac{16}{65}$$

**b)** $\cos(\alpha + \beta)$
**Solution**

$$\cos(\alpha + \beta) = \cos \alpha \cos \beta - \sin \alpha \sin \beta$$
$$= \left(\tfrac{3}{5}\right)\left(\tfrac{5}{13}\right) - \left(\tfrac{4}{5}\right)\left(\tfrac{12}{13}\right)$$
$$= \dfrac{15}{65} - \dfrac{48}{65} = \dfrac{15 - 48}{65} = -\dfrac{33}{65}$$
$$\therefore \cos(\alpha + \beta) = -\dfrac{33}{65}$$

# Trigonometric Identities I

**c)** $\tan 2\alpha$
**Solution**

$$\tan 2\alpha = \frac{2\tan\alpha}{1-\tan^2\alpha}$$

$$= \frac{2\left(\frac{4}{3}\right)}{1-\left(\frac{4}{3}\right)^2} = \frac{\left(\frac{8}{3}\right)}{\left(1-\frac{16}{9}\right)} = \frac{\left(\frac{8}{3}\right)}{\left(-\frac{7}{9}\right)} = \left(\frac{8}{3}\right) \times \left(-\frac{9}{7}\right)$$

$$\therefore \tan 2\alpha = -\frac{24}{7}$$

**d)** $\sin 3\beta$
**Solution**

$$\sin 3\beta = 3\sin\beta - 4\sin^3\beta$$

$$= 3\left(\frac{12}{13}\right) - 4\left(\frac{12}{13}\right)^3$$

$$= \left(\frac{36}{13}\right) - \left(\frac{6912}{2197}\right)$$

$$\therefore \sin 3\beta = -\frac{828}{2197}$$

A final set of examples for us to try.

## Example 17

Using appropriate identities, determine the exact value of the following:

**a)** $\sin^2 \frac{1}{8}\pi$    **b)** $\cos^2 \frac{1}{8}\pi$    **c)** $\tan^2 \frac{1}{8}\pi$

What did you get? Find the solution below to double-check your answer.

## Solution to Example 17

**HINT**

Apply the following:

- We will be using the double-angle (or half-angle) formula.
- We need to apply $\sin^2 \frac{1}{2}A + \cos^2 \frac{1}{2}A = 1$ or $\sin^2 A + \cos^2 A = 1$.

**a)** $\sin^2 \frac{1}{8}\pi$
**Solution**

$$\cos 2A = 1 - 2\sin^2 A$$

Let $A = \frac{1}{8}\pi$, thus

$$\cos 2\left(\frac{1}{8}\pi\right) = 1 - 2\sin^2\left(\frac{1}{8}\pi\right)$$
$$\cos \frac{1}{4}\pi = 1 - 2\sin^2\left(\frac{1}{8}\pi\right)$$

But $\cos \frac{1}{4}\pi = \frac{1}{\sqrt{2}}$, thus

$$\frac{1}{\sqrt{2}} = 1 - 2\sin^2\left(\frac{1}{8}\pi\right)$$
$$2\sin^2\left(\frac{1}{8}\pi\right) = 1 - \frac{1}{\sqrt{2}} = \frac{\sqrt{2}-1}{\sqrt{2}}$$
$$\therefore \sin^2\left(\frac{1}{8}\pi\right) = \frac{\sqrt{2}-1}{2\sqrt{2}} = \frac{2-\sqrt{2}}{4}$$

**ALTERNATIVE METHOD**

$$\sin A = 2\sin \frac{1}{2}A \cos \frac{1}{2}A$$

Let $A = \frac{1}{4}\pi$, thus

$$\sin A = 2\sin \frac{1}{2}A \cos \frac{1}{2}A$$
$$\sin \frac{1}{4}\pi = 2\sin \frac{1}{2}\left(\frac{1}{4}\pi\right) \cos \frac{1}{2}\left(\frac{1}{4}\pi\right)$$
$$= 2\sin \frac{1}{8}\pi \cos \frac{1}{8}\pi$$

But $\sin \frac{1}{4}\pi = \sqrt{\frac{1}{2}}$, thus

$$\sqrt{\frac{1}{2}} = 2\sin \frac{1}{8}\pi \cos \frac{1}{8}\pi$$

Square both sides

$$\left(\sqrt{\frac{1}{2}}\right)^2 = \left(2\sin \frac{1}{8}\pi \cos \frac{1}{8}\pi\right)^2$$
$$\frac{1}{2} = 4\sin^2 \frac{1}{8}\pi \cos^2 \frac{1}{8}\pi$$

# Trigonometric Identities I

Note that $\sin^2 \frac{1}{8}\pi + \cos^2 \frac{1}{8}\pi = 1$, which implies that $\cos^2 \frac{1}{8}\pi = 1 - \sin^2 \frac{1}{8}\pi$. Therefore:

$$\frac{1}{2} = 4\sin^2 \frac{1}{8}\pi \cos^2 \frac{1}{8}\pi$$

$$\frac{1}{2} = 4\sin^2 \frac{1}{8}\pi \left(1 - \sin^2 \frac{1}{8}\pi\right)$$

$$\frac{1}{2} = 4\sin^2 \frac{1}{8}\pi - 4\sin^4 \frac{1}{8}\pi$$

$$8\sin^4 \frac{1}{8}\pi - 8\sin^2 \frac{1}{8}\pi + 1 = 0$$

Let $\sin^2 \frac{1}{8}\pi = x$, therefore

$$8x^2 - 8x + 1 = 0$$

Using quadratic formula, we have $a = 8$, $b = -8$, and $c = 1$. Hence

$$x = \frac{-b \pm \sqrt{b^2 - 4ac}}{2a}$$

$$= \frac{-(-8) \pm \sqrt{(-8)^2 - 4(8)(1)}}{2 \times 8} = \frac{8 \pm 4\sqrt{2}}{16}$$

$$= \frac{2 \pm \sqrt{2}}{4}$$

Thus

$$\sin^2 \frac{1}{8}\pi = \frac{2 \pm \sqrt{2}}{4}$$

---

**b)** $\cos^2 \frac{1}{8}\pi$

**Solution**

$$\cos 2A = 2\cos^2 A - 1$$

Let $A = \frac{1}{8}\pi$, thus

$$\cos 2\left(\frac{1}{8}\pi\right) = 2\cos^2\left(\frac{1}{8}\pi\right) - 1$$

$$\cos \frac{1}{4}\pi = 2\cos^2\left(\frac{1}{8}\pi\right) - 1$$

But $\cos\frac{1}{4}\pi = \frac{\sqrt{2}}{2}$, thus

$$\frac{\sqrt{2}}{2} = 2\cos^2\left(\frac{1}{8}\pi\right) - 1$$

$$2\cos^2\left(\frac{1}{8}\pi\right) = \frac{\sqrt{2}}{2} + 1 = \frac{\sqrt{2}+2}{2}$$

$$\cos^2\left(\frac{1}{8}\pi\right) = \frac{\sqrt{2}+2}{4}$$

---

**c)** $\tan^2\frac{1}{8}\pi$

**Solution**

$$\tan 2A = \frac{2\tan A}{1 - \tan^2 A}$$

Let $A = \frac{1}{8}\pi$, thus

$$\tan 2\left(\frac{1}{8}\pi\right) = \frac{2\tan\left(\frac{1}{8}\pi\right)}{1 - \tan^2\left(\frac{1}{8}\pi\right)}$$

$$\tan\frac{1}{4}\pi = \frac{2\tan\left(\frac{1}{8}\pi\right)}{1 - \tan^2\left(\frac{1}{8}\pi\right)}$$

But $\tan\frac{1}{4}\pi = 1$, thus

$$1 = \frac{2\tan\left(\frac{1}{8}\pi\right)}{1 - \tan^2\left(\frac{1}{8}\pi\right)}$$

or

$$1 - \tan^2\left(\frac{1}{8}\pi\right) = 2\tan\left(\frac{1}{8}\pi\right)$$

$$\tan^2\left(\frac{1}{8}\pi\right) + 2\tan\left(\frac{1}{8}\pi\right) - 1 = 0$$

Let $\tan\frac{1}{8}\pi = x$, therefore,

$$x^2 + 2x - 1 = 0$$

Using quadratic formula, we have $a = 1$, $b = 2$, and $c = -1$. Hence

$$x = \frac{-b \pm \sqrt{b^2 - 4ac}}{2a}$$

$$= \frac{-2 \pm \sqrt{2^2 - 4(1)(-1)}}{2 \times 1} = \frac{-2 \pm 2\sqrt{2}}{2}$$

$$= -1 \pm \sqrt{2}$$

Thus

$$\tan \frac{1}{8}\pi = -1 + \sqrt{2} \ \ or \ \ \tan \frac{1}{8}\pi = -1 - \sqrt{2}$$

Because $\tan \frac{1}{8}\pi$ is positive and acute, the only valid answer is

$$\tan \frac{1}{8}\pi = \sqrt{2} - 1$$

## 3.5 CHAPTER SUMMARY

1) Identities play a major role in trigonometric functions and the relationships that exist between the six commonly used functions, namely sine (sin), cosine (cos), tangent (tan), cotangent (cot), secant (sec), and cosecant (cosec or csc).

2) The four 'fundamental' identities are:

$$\boxed{\tan\theta = \frac{\sin\theta}{\cos\theta}} \quad \boxed{\sin^2\theta + \cos^2\theta = 1} \quad \boxed{\tan^2\theta + 1 = \sec^2\theta} \quad \boxed{\cot^2\theta + 1 = \csc^2\theta}$$

3) Addition formulas for the three trigonometric ratios:

- **Sine**

$$\boxed{\sin(A+B) = \sin A \cos B + \sin B \cos A}$$
$$\boxed{\sin(A-B) = \sin A \cos B - \sin B \cos A}$$

- **Cosine**

$$\boxed{\cos(A+B) = \cos A \cos B - \sin A \sin B}$$
$$\boxed{\cos(A-B) = \cos A \cos B + \sin A \sin B}$$

- **Tangent**

$$\boxed{\tan(A+B) = \frac{\tan A + \tan B}{1 - \tan A \tan B}} \ \text{AND} \ \boxed{\tan(A-B) = \frac{\tan A - \tan B}{1 + \tan A \tan B}}$$

4) Double-angle formulas for the three trigonometric ratios:

- **Sine**

$$\boxed{\sin 2A = 2 \sin A \cos A}$$

- **Cosine**

$$\boxed{\cos 2A = \cos^2 A - \sin^2 A}$$

$$\boxed{\cos 2A = 2\cos^2 A - 1}$$

$$\boxed{\cos 2A = 1 - 2\sin^2 A}$$

- Tangent

$$\tan 2A = \frac{2\tan A}{1 - \tan^2 A}$$

5) Triple-angle formulas for the three trigonometric ratios:

- Sine

$$\sin 3A = 3\sin A - 4\sin^3 A$$

- Cosine

$$\cos 3A = 4\cos^3 A - 3\cos A$$

- Tangent

$$\tan 3A = \frac{2\tan A - \tan^3 A}{1 - 3\tan^2 A}$$

6) Half-angle formulas for the three trigonometric ratios:

- Sine

$$\sin A = 2\sin \frac{1}{2}A \cos \frac{1}{2}A$$

- Cosine

$$\cos A = \cos^2 \frac{1}{2}A - \sin^2 \frac{1}{2}A$$

$$\cos A = 2\cos^2 \frac{1}{2}A - 1$$

$$\cos A = 1 - 2\sin^2 \frac{1}{2}A$$

- Tangent

$$\tan A = \frac{2\tan \frac{1}{2}A}{1 - \tan^2 \frac{1}{2}A}$$

\*\*\*\*

## 3.6 FURTHER PRACTICE

To access complementary contents, including additional exercises, please go to www.dszak.com.

# 4 Trigonometric Identities II

## Learning Outcomes

Once you have studied the content of this chapter, you should be able to:

- Use factor formulas involving sine, cosine, and tangent functions
- Use $R$ addition formulas to solve problems involving trigonometric functions
- Determine the maximum and minimum values of a sum of two trigonometric functions

## 4.1 INTRODUCTION

This chapter will, in continuation of our discussion on trigonometric identities, look at two more essential identities, namely factor formulas and $R$ addition formulas.

## 4.2 THE FACTOR FORMULAS

Given any two angles $A$ and $B$, the trigonometric ratios of the factor formulas, also called the product formulas, are summarised in Table 4.1.

You will notice here that we added and/or subtracted trigonometric ratios of two different angles $A$ and $B$. This is in contrast to when we compute the trigonometric ratios of addition or subtraction of two angles on the LHS.

Another variant of the last identity is $\cos A - \cos B = 2 \sin \frac{1}{2}(A+B) \sin \frac{1}{2}(B-A)$. This is because $\sin \frac{1}{2}(B-A) = -\sin \frac{1}{2}(A-B)$. Also note that since $\sin A + \sin B = \sin A - \sin(-B)$, it is possible to obtain **sin A − sin B** by replacing **B** with **−B** in the RHS formula of **sin A + sin B**, just like in the addition formula for the sine function.

The above identities may not seem to be very relevant here but are key to solving some integral functions in calculus that we shall cover later in this book.

## TABLE 4.1
**Factor Formulas (Sum to Product) for Trigonometric Ratios**

|        | Identities |
|--------|------------|
| Sine   | $\sin A + \sin B = 2 \sin \frac{1}{2}(A+B) \cos \frac{1}{2}(A-B)$ |
|        | $\sin A - \sin B = 2 \sin \frac{1}{2}(A-B) \cos \frac{1}{2}(A+B)$ |
| Cosine | $\cos A + \cos B = 2 \cos \frac{1}{2}(A+B) \cos \frac{1}{2}(A-B)$ |
|        | $\cos A - \cos B = -2 \sin \frac{1}{2}(A+B) \sin \frac{1}{2}(A-B)$ |

$$\text{(4.1)}$$

Let's try some examples.

### Example 1

Using appropriate methods, prove the following factor formulas or identities:

a) $\sin x + \sin y = 2 \sin \frac{1}{2}(x+y) \cos \frac{1}{2}(x-y)$   b) $\sin x - \sin y = 2 \sin \frac{1}{2}(x-y) \cos \frac{1}{2}(x+y)$

c) $\cos x + \cos y = 2 \cos \frac{1}{2}(x+y) \cos \frac{1}{2}(x-y)$   d) $\cos x - \cos y = -2 \sin \frac{1}{2}(x+y) \sin \frac{1}{2}(x-y)$

What did you get? Find the solution below to double-check your answer.

### Solution to Example 1

**HINT**

- For this proof, we need to use addition formulas (see Section 3.3) and they are reproduced below.

$$\sin(A+B) = \sin A \cos B + \sin B \cos A \quad ---\text{(i)}$$
$$\sin(A-B) = \sin A \cos B - \sin B \cos A \quad ---\text{(ii)}$$

# Trigonometric Identities II

$$\cos(A+B) = \cos A \cos B - \sin A \sin B \quad ---\text{(iii)}$$
$$\cos(A-B) = \cos A \cos B + \sin A \sin B \quad ---\text{(iv)}$$

- We will need to solve pairs of simultaneous linear equations, but will not show full workings.

**a)** $\sin x + \sin y = 2\sin\frac{1}{2}(x+y)\cos\frac{1}{2}(x-y)$

**Solution**
Add (i) with (ii)

$$\sin(A+B) + \sin(A-B) = 2\sin A \cos B \quad --\text{(v)}$$

Let

$$x = A + B \quad ---\text{(vi)}$$
$$y = A - B \quad ---\text{(vii)}$$

Solving (vi) and (vii) simultaneously, we have

$$A = \frac{1}{2}(x+y) \quad ---\text{(viii)}$$
$$B = \frac{1}{2}(x-y) \quad ---\text{(ix)}$$

Substituting vi, vii, viii, and ix in equation v, we have

$$\therefore \sin x + \sin y = 2\sin\frac{1}{2}(x+y)\cos\frac{1}{2}(x-y)$$

---

**b)** $\sin x - \sin y = 2\sin\frac{1}{2}(x-y)\cos\frac{1}{2}(x+y)$

**Solution**
(i) minus (ii)

$$\sin(A+B) - \sin(A-B) = 2\sin B \cos A \quad (\text{x})$$

Substituting vi, vii, viii, and ix in equation x, we have

$$\therefore \sin x - \sin y = 2\sin\frac{1}{2}(x-y)\cos\frac{1}{2}(x+y)$$

---

**c)** $\cos x + \cos y = 2\cos\frac{1}{2}(x+y)\cos\frac{1}{2}(x-y)$

**Solution**
(iii) plus (iv)

$$\cos(A+B) + \cos(A-B) = 2\cos A \cos B \quad --\text{(xi)}$$

Substituting vi, vii, viii, and ix in equation xi, we have

$$\therefore \cos x + \cos y = 2\cos\frac{1}{2}(x+y)\cos\frac{1}{2}(x-y)$$

**d)** $\cos x - \cos y = -2\sin\frac{1}{2}(x+y)\sin\frac{1}{2}(x-y)$
**Solution**
(iii) minus (iv)

$$\cos(A+B) - \cos(A-B) = -2\sin A \sin B \quad --\text{(xii)}$$

Substituting vi, vii, viii, and ix in equation xii, we have

$$\therefore \cos x - \cos y = -2\sin\frac{1}{2}(x+y)\sin\frac{1}{2}(x-y)$$

Another set of examples for us to try.

### Example 2

Using factor formulas or any suitable method, express the following as a product of two trigonometric ratios:

**a)** $\sin 3t + \sin 3t$  **b)** $\sin 3\alpha - \sin 4\beta$  **c)** $\cos 7\theta + \cos 5\theta$

What did you get? Find the solution below to double-check your answer.

### Solution to Example 2

**a)** $\sin 3t + \sin 3t$
**Solution**
**Step 1:** Let's choose a formula

$$\sin x + \sin y = 2\sin\frac{1}{2}(x+y)\cos\frac{1}{2}(x-y)$$

**Step 2:** Let's substitute such that $x = 3t$ and $y = 3t$

$$\sin 3t + \sin 3t = 2\sin\frac{1}{2}(3t+3t)\cos\frac{1}{2}(3t-3t)$$

$$= 2\sin\frac{1}{2}(6t)\cos\frac{1}{2}(0)$$

$$= 2\sin 3t \cos 0$$

$$\therefore \sin 3t + \sin 3t = 2\sin 3t$$

# Trigonometric Identities II

### ALTERNATIVE METHOD

This is a simple algebraic addition. If we let $3x = t$, then $\sin 3x + \sin 3x$ is the same as $a + a = 2a$. Hence, $\sin 3x + \sin 3x = 2 \sin 3x$. This demonstrates the validity of the factor formula.

### NOTE

As a general rule, the factor formulas are not intended for when $A = B$ as we've seen in this example.

**b)** $\sin 3\alpha - \sin 4\beta$

**Solution**

**Step 1:** Let's choose a formula

$$\sin x - \sin y = 2 \sin \frac{1}{2}(x - y) \cos \frac{1}{2}(x + y)$$

**Step 2:** Let's substitute such that $x = 3\alpha$ and $y = 4\beta$

$$\sin 3\alpha - \sin 4\beta = 2 \sin \frac{1}{2}(3\alpha - 4\beta) \cos \frac{1}{2}(3\alpha + 4\beta)$$

$$\therefore \sin 3\alpha - \sin 4\beta = 2 \sin \frac{1}{2}(3\alpha - 4\beta) \cos \frac{1}{2}(3\alpha + 4\beta)$$

**c)** $\cos 7\theta + \cos 5\theta$

**Solution**

**Step 1:** Let's choose a formula

$$\cos x + \cos y = 2 \cos \frac{1}{2}(x + y) \cos \frac{1}{2}(x - y)$$

**Step 2:** Let's substitute such that $x = 7\theta$ and $y = 5\theta$

$$\cos 7\theta + \cos 5\theta = 2 \cos \frac{1}{2}(7\theta + 5\theta) \cos \frac{1}{2}(7\theta - 5\theta)$$

$$= 2 \cos \frac{1}{2}(12\theta) \cos \frac{1}{2}(2\theta) = 2 \cos 6\theta \cos \theta$$

$$\therefore \cos 7\theta + \cos 5\theta = 2 \cos 6\theta \cos \theta$$

Another set of examples to try.

### Example 3

Without a calculator, use factor formulas to simplify the following:

**a)** $\cos \frac{5}{6}\pi + \cos \frac{1}{6}\pi$      **b)** $\sin 40° - \sin 50°$      **c)** $\cos 60° + \cos 30°$

What did you get? Find the solution below to double-check your answer.

**Solution to Example 3**

**HINT**

- Addition formulas will be appropriate in most cases.
- We also need special angles, but remember that this requires using factor formulas.

**a)** $\cos \frac{5}{6}\pi + \cos \frac{1}{6}\pi$
**Solution**

$$\cos A + \cos B = 2\cos \frac{1}{2}(A+B)\cos \frac{1}{2}(A-B)$$

Let's substitute such that $A = \frac{5}{6}\pi$ and $B = \frac{1}{6}\pi$. Therefore,

$$\cos \frac{5}{6}\pi + \cos \frac{1}{6}\pi = 2\cos \frac{1}{2}\left(\frac{5}{6}\pi + \frac{1}{6}\pi\right)\cos \frac{1}{2}\left(\frac{5}{6}\pi - \frac{1}{6}\pi\right)$$

$$= 2\cos \frac{1}{2}(\pi)\cos \frac{1}{2}\left(\frac{2}{3}\pi\right)$$

$$= 2\cos \frac{1}{2}\pi \cos \frac{1}{3}\pi$$

Because $\cos \frac{1}{2}\pi = 0$, $\cos \frac{1}{3}\pi = \frac{1}{2}$, we have

$$= 2(0)\left(\frac{1}{2}\right)$$

$$\therefore \cos \frac{5}{6}\pi + \cos \frac{1}{6}\pi = 0$$

---

**b)** $\sin 40° - \sin 50°$
**Solution**

$$\sin x - \sin y = 2\sin \frac{1}{2}(x-y)\cos \frac{1}{2}(x+y)$$

Let's substitute such that $x = 40°$ and $y = 50°$. Therefore,

$$\sin 40° - \sin 50° = 2\sin \frac{1}{2}(40° - 50°)\cos \frac{1}{2}(40° + 50°)$$

$$= 2\sin \frac{1}{2}(-10°)\cos \frac{1}{2}(90°)$$

$$= 2\sin(-5°)\cos 45°$$

$$= 2\cos 45° \sin(-5°)$$

# Trigonometric Identities II

Because $\cos 45° = \frac{\sqrt{2}}{2}$ and $\sin(-5°) = -\sin 5°$, we have

$$= 2\left(\frac{\sqrt{2}}{2}\right)(-\sin 5°) = -\sqrt{2}\sin 5°$$

$$\therefore \sin 40° - \sin 50° = -\sqrt{2}\sin 5°$$

No further simplification is possible.

---

**c)** $\cos 60° + \cos 30°$

**Solution**

$$\cos x + \cos y = 2\cos\frac{1}{2}(x+y)\cos\frac{1}{2}(x-y)$$

Let's substitute such that $x = 60°$ and $y = 30°$. Therefore,

$$\cos 60° + \cos 30° = 2\cos\frac{1}{2}(60° + 30°)\cos\frac{1}{2}(60° - 30°)$$

$$= 2\cos\frac{1}{2}(90°)\cos\frac{1}{2}(30°)$$

$$= 2\cos 45° \cos 15°$$

Because $\cos 45° = \frac{\sqrt{2}}{2}$, we have

$$= 2\left(\frac{\sqrt{2}}{2}\right)(\cos 15°) = \sqrt{2}\cos 15°$$

We can use an addition formula to simplify $\cos 15°$ as

$$\cos 15° = \cos(60° - 45°)$$

$$= \cos 60° \cos 45° + \sin 60° \sin 45°$$

$$= \frac{1}{2} \times \frac{\sqrt{2}}{2} + \frac{\sqrt{3}}{2} \times \frac{\sqrt{2}}{2}$$

$$= \frac{\sqrt{2}}{2}\left(\frac{1}{2} + \frac{\sqrt{3}}{2}\right)$$

$$\therefore \cos 15° = \frac{\sqrt{2}}{2}\left(\frac{1+\sqrt{3}}{2}\right)$$

Note that we could have used $\cos 15° = \cos(45° - 30°)$, the result will still be the same. Hence, we can say that

$$\sqrt{2}\cos 15° = \sqrt{2} \times \frac{\sqrt{2}}{2}\left(\frac{1+\sqrt{3}}{2}\right)$$

$$= \frac{\sqrt{2} \times \sqrt{2}}{2}\left(\frac{1+\sqrt{3}}{2}\right)$$

$$= \frac{2}{2}\left(\frac{1+\sqrt{3}}{2}\right)$$

$$\therefore \cos 60° + \cos 30° = \frac{1+\sqrt{3}}{2}$$

## ALTERNATIVE METHOD

Without using the factor formulas (and subsequently addition formulas), we could have simplified this quicker and better as:

$$\cos 60° + \cos 30° = \frac{1}{2} + \frac{\sqrt{3}}{2}$$
$$= \frac{1+\sqrt{3}}{2}$$

This also shows that $\cos 60° + \cos 30° \neq \cos(60° + 30°)$, otherwise the answer would have been 0 because $\cos 90° = 0$.

Also, we can expand $\cos 15°$ using $45°$ and $30°$ as

$$\cos 15° = \cos(45° - 30°)$$
$$= \cos 45° \cos 30° + \sin 45° \sin 30°$$
$$= \frac{\sqrt{2}}{2} \times \frac{\sqrt{3}}{2} + \frac{\sqrt{2}}{2} \times \frac{1}{2}$$
$$= \frac{\sqrt{2}}{2}\left(\frac{\sqrt{3}}{2} + \frac{1}{2}\right)$$
$$\therefore \cos 15° = \frac{\sqrt{2}}{2}\left(\frac{1+\sqrt{3}}{2}\right)$$

Another example to try.

## Example 4

Using appropriate factor formulas, express the following as a sum of two trigonometric ratios:

**a)** $\sin 4\theta \cos \theta$ **b)** $\cos 5\varphi \cos 7\varphi$ **c)** $2 \sin 60° \sin 20°$ **d)** $2 \sin 25° \cos 35°$

What did you get? Find the solution below to double-check your answer.

## Solution to Example 4

### HINT

We need to look critically at the product. If it is a:

- product of two cosines then it is a sum of cosine
- product of two sines then it is a difference of cosine

# Trigonometric Identities II

- product of a sine and a cosine, where the coefficient of the sine angle is greater than that of the cosine, then it is a sum of sine
- product of a sine and a cosine, where the coefficient of the sine angle is less than that of the cosine, then it is a difference of sine

**a)** $\sin 4\theta \cos \theta$

**Solution**

It is obvious that the factor formula for this is

$$\sin x + \sin y = 2 \sin \tfrac{1}{2}(x+y) \cos \tfrac{1}{2}(x-y)$$

$$\tfrac{1}{2}\{\sin x + \sin y\} = \sin \tfrac{1}{2}(x+y) \cos \tfrac{1}{2}(x-y)$$

Now, we have that

$$\sin 4\theta \cos \theta = \sin \tfrac{1}{2}(x+y) \cos \tfrac{1}{2}(x-y)$$

Comparing both sides, we have

$$\tfrac{1}{2}(x+y) = 4\theta$$

$$x + y = 8\theta$$

Also

$$\tfrac{1}{2}(x-y) = \theta$$

$$x - y = 2\theta$$

Solving the above two equations, we have

$$x = 5\theta,\ y = 3\theta$$

Hence

$$\mathbf{\sin 4\theta \cos \theta = \tfrac{1}{2}\{\sin 5\theta + \sin 3\theta\}}$$

---

**b)** $\cos 5\varphi \cos 7\varphi$

**Solution**

It is obvious that the factor formula for this is

$$\cos x + \cos y = 2 \cos \tfrac{1}{2}(x+y) \cos \tfrac{1}{2}(x-y)$$

$$\tfrac{1}{2}\{\cos x + \cos y\} = \cos \tfrac{1}{2}(x+y) \cos \tfrac{1}{2}(x-y)$$

Now, we have that

$$\cos 5\varphi \cos 7\varphi = \cos \tfrac{1}{2}(x+y) \cos \tfrac{1}{2}(x-y)$$

Comparing both sides, we have

$$\frac{1}{2}(x+y) = 5\varphi$$
$$x+y = 10\varphi$$

Also

$$\frac{1}{2}(x-y) = 7\varphi$$
$$x-y = 14\varphi$$

Solving the above two equations, we have

$$x = 12\varphi, \ y = -2\varphi$$

Hence

$$\cos 5\varphi \cos 7\varphi = \frac{1}{2}\{\cos 12\varphi + \cos(-2\varphi)\} = \frac{1}{2}\{\cos 12\varphi + \cos 2\varphi\}$$

---

c) $2 \sin 60° \sin 20°$
**Solution**
It is obvious that the factor formula for this is

$$\cos x - \cos y = -2 \sin \frac{1}{2}(x+y) \sin \frac{1}{2}(x-y)$$
$$-\{\cos x - \cos y\} = 2 \sin \frac{1}{2}(x+y) \sin \frac{1}{2}(x-y)$$

Then we have that

$$2 \sin 60° \sin 20° = 2 \sin \frac{1}{2}(x+y) \sin \frac{1}{2}(x-y)$$

Comparing both sides, we have

$$\frac{1}{2}(x+y) = 60°$$
$$x+y = 120°$$

Also

$$\frac{1}{2}(x-y) = 20°$$
$$x-y = 40°$$

Solving the above two equations, we have

$$x = 80°, \ y = 40°$$

Hence

$$2 \sin 60° \sin 20° = -\{\cos 80° - \cos 40°\} = \cos 40° - \cos 80°$$

# Trigonometric Identities II

**d)** $2 \sin 25° \cos 35°$
**Solution**
It is obvious that the factor formula for this is

$$\sin x - \sin y = 2 \sin \frac{1}{2}(x-y) \cos \frac{1}{2}(x+y)$$

Then we have that

$$2 \sin 25° \cos 35° = 2 \sin \frac{1}{2}(x-y) \cos \frac{1}{2}(x+y)$$

Comparing both sides, we have

$$\frac{1}{2}(x+y) = 35°$$
$$x + y = 70°$$

Also

$$\frac{1}{2}(x-y) = 25°$$
$$x - y = 50°$$

Solving the above two equations, we have

$$x = 60°, \ y = 10°$$

Hence

$$2 \sin 25° \cos 35° = \sin 60° - \sin 10°$$

---

In the above examples, we have used the factor formulas or addition–product formulas we presented on page 118. However, we have a standard set of formulas that are used to convert product to sum, and these are given in Table 4.2.

In the formulas in Table 4.2, $A > B$.

---

**TABLE 4.2**
**Factor (Product–Addition) Formulas for Trigonometric Ratios**

|  | Identities |
|---|---|
| **Sine–cosine** | $2 \sin A \cos B = \sin(A+B) + \sin(A-B)$ |
|  | $2 \cos A \sin B = \sin(A+B) - \sin(A-B)$ |
| **Cosine** | $2 \cos A \cos B = \cos(A+B) + \cos(A-B)$ |
| **Sine** | $2 \sin A \sin B = -\cos(A-B) - \cos(A+B)$ |

(4.2)

Now, let's try Example 4(a) using the above identity. We are given $\sin 4\theta \cos \theta$. This is a mix of sine and cosine functions with an angle in sine greater than that of cosine, so we will need the first identity.

$$2 \sin A \cos B = \sin(A+B) + \sin(A-B)$$

which implies that

$$\sin A \cos B = \frac{1}{2}\sin(A+B) + \frac{1}{2}\sin(A-B)$$

Looking at $\sin 4\theta \cos \theta$, we have $A = 4\theta$ and $B = \theta$. Now substitute as:

$$\sin 4\theta \cos \theta = \frac{1}{2}\sin(4\theta + \theta) + \frac{1}{2}\sin(4\theta - \theta)$$
$$= \frac{1}{2}\sin(5\theta) + \frac{1}{2}\sin(3\theta)$$

Hence

$$\sin 4\theta \cos \theta = \frac{1}{2}\{\sin 5\theta + \sin 3\theta\}$$

This solution seems quicker using the newly presented formula, but we can also use addition–product formulas for product–addition problems. You can now try more questions using the formulas in Table 4.2.

## 4.3 THE R ADDITION FORMULAS

We have covered both sine and cosine functions and their properties in Chapter 1 and understand that they are identical except that they have a phase difference of half a pi rad ($\frac{\pi}{2}$) or 90° between them. The question now is what happens if two signals with sine and cosine functions are mixed. Mathematically, this means adding sine and cosine together, i.e.,

$$y = \sin x + \cos x$$

Let's look at Figure 4.1, which shows a plot of $y = \sin x$, $y = \cos x$, and $y = \sin x + \cos x$ in the interval $0 \le x \le 360°$. It is not surprising that the sum of sine and cosine functions exhibits a wave-like transformed pattern.

A close look at the plot (Figure 4.1) will provide us with the following summary about $\sin x + \cos x$.

**Note 1** It is vertically stretched. Its amplitude, and its maximum and minimum points are different from those of its components i.e., $\sin x$ and $\cos x$.

**Note 2** It looks like $y = \sin x$ but is shifted horizontally to the left. Similarly, it looks like $y = \cos x$ but shifted horizontally to the right. In AC circuits, these are the same as leading and lagging, respectively.

**Note 3** The shift angle is the same in both, though this is because they have the same amplitude of 1, i.e. the coefficient of $\sin x$ and $\cos x$ are both 1.

**Note 4** If the shift angle or phase difference is $\alpha$ and the vertical stretch factor is represented by $R$, then we can represent $y = \sin x + \cos x$ by either $y = R\sin(x + \alpha)$ or $y = R\cos(x - \alpha)$.

# Trigonometric Identities II

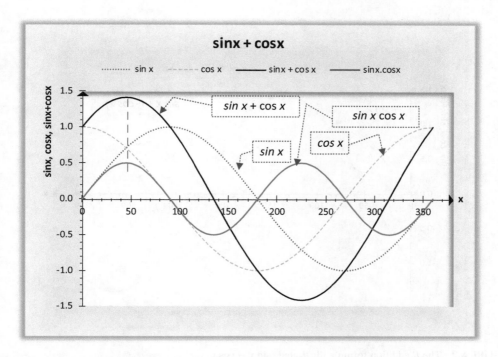

**FIGURE 4.1** The $R$ addition formula illustrated ($\sin x + \cos x$).

The above looks interesting, as it helps us to represent a sum of sine and cosine functions as a 'transformed' sine or 'transformed' cosine function. What if we want to find the difference of sine and cosine functions? The process is the same. We only need to understand that $y = \sin x - \cos x$ is the addition of $\sin x$ and negation of $\cos x$, i.e., $-\cos x$. This is shown in Figure 4.2 using the same interval of $0 \leq x \leq 360°$.

All that we've said about sum also holds for the difference of the two functions except for one aspect, which is that we can represent $y = \sin x - \cos x$ by either $y = R \sin(x - \alpha)$ or $y = R \cos(x + \alpha)$. Notice the sign change between $x$ and $\alpha$.

Does the above analysis hold if the coefficient of sine and cosine is not one or the same? Yes it does, as can be seen in Figure 4.3 for $y = 3 \sin x + 4 \cos x$.

Table 4.3 provides a summary of the $R$ addition formulas.

where $a$, $b$, and $R$ are positive real numbers and $\alpha$ is an acute angle, such that $0 < \alpha < 90°$.

We can further state that:

a) $R = \sqrt{a^2 + b^2}$ is valid for all cases and always positive, i.e., $R > 0$. Note that this is equal to the amplitude of the resulting sine or cosine function.

b) $\alpha = \tan^{-1}\left(\dfrac{b}{a}\right)$ or $\alpha = \tan^{-1}\left(-\dfrac{a}{b}\right)$ depending on the original expression, the sign of $a$ and $b$, and the $R$ equivalent we have chosen.

c) We need to be careful about the sign, especially when $a$ and $b$ are both negative, which can cancel each other out.

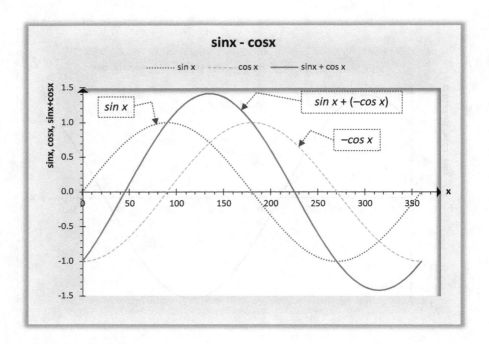

**FIGURE 4.2** The $R$ addition formula illustrated ($\sin x - \cos x$).

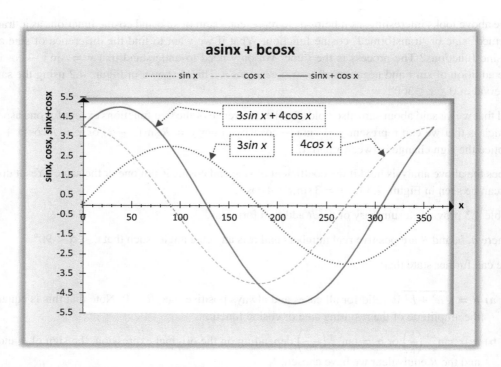

**FIGURE 4.3** The $R$ addition formula illustrated ($3\sin x + 4\cos x$).

# Trigonometric Identities II

**TABLE 4.3**
**The $R$ Addition Formulas Illustrated**

| Function | Sum | Difference |
|---|---|---|
| Sine | $a\sin x + b\cos x = R\sin(x+\alpha)$ | $a\sin x - b\cos x = R\sin(x-\alpha)$ |
| Cosine | $a\sin x + b\cos x = R\cos(x-\alpha)$ | $a\sin x - b\cos x = R\cos(x+\alpha)$ |

(4.3)

We will illustrate two cases to show how to find $R$ and $\alpha$, but once you know the trick it may be less relevant, as will be demonstrated shortly in this section.

**Case 1** $a\sin x + b\cos x \equiv R\sin(x+\alpha)$

Given that

$$a\sin x + b\cos x \equiv R\sin(x+\alpha)$$

Let's consider the RHS expression and apply the addition formula to it as:

$$R\sin(x+\alpha) \equiv R\sin x\cos\alpha + R\sin\alpha\cos x$$

Now we have that

$$a\sin x + b\cos x \equiv R\sin x\cos\alpha + R\sin\alpha\cos x$$
$$\equiv (R\cos\alpha)\sin x + (R\sin\alpha)\cos x$$

Comparing the like parts, we have

$$a\sin x = (R\cos\alpha)\sin x$$
$$\therefore a = R\cos\alpha \quad ----\text{(i)}$$

and

$$b\cos x = (R\sin\alpha)\cos x$$
$$\therefore b = R\sin\alpha \quad ----\text{(ii)}$$

(ii) divided by (i), we have

$$\frac{R\sin\alpha}{R\cos\alpha} = \frac{b}{a}$$

Simplifying this, we have

$$\frac{\sin\alpha}{\cos\alpha} = \frac{b}{a}$$
$$\tan\alpha = \frac{b}{a}$$
$$\therefore \alpha = \tan^{-1}\left(\frac{b}{a}\right)$$

Square (i) and (ii)

$$a^2 = R^2\cos^2\alpha \quad ----\text{(iii)}$$
$$b^2 = R^2\sin^2\alpha \quad ----\text{(iv)}$$

(iii) + (iv)

$$a^2 + b^2 = R^2\cos^2\alpha + R^2\sin^2\alpha$$
$$a^2 + b^2 = R^2\left(\cos^2\alpha + \sin^2\alpha\right)$$
$$\therefore R^2 = a^2 + b^2$$
$$\because \cos^2\alpha + \sin^2\alpha = 1$$

**Case 2** $a\sin x + b\cos x \equiv R\cos(x - \alpha)$

Given that

$$a\sin x + b\cos x \equiv R\cos(x - \alpha)$$

Let's consider the RHS expression and apply the addition formula to it as:

$$R\cos(x - \alpha) \equiv R\cos x\cos\alpha + R\sin x\sin\alpha$$

Now we have that

$$a\sin x + b\cos x \equiv R\cos x\cos\alpha + R\sin x\sin\alpha$$

Comparing the like parts, we have

$$a\sin x = R\sin x\sin\alpha$$
$$\therefore a = R\sin\alpha \quad ----\text{(i)}$$

and

$$b\cos x = R\cos x\cos\alpha$$
$$\therefore b = R\cos\alpha \quad ----\text{(ii)}$$

(i) divided by (ii), we have

$$\frac{R\sin\alpha}{R\cos\alpha} = \frac{a}{b}$$

Simplifying this, we have

$$\frac{\sin\alpha}{\cos\alpha} = \frac{a}{b}$$
$$\tan\alpha = \frac{a}{b}$$
$$\therefore \boldsymbol{\alpha = \tan^{-1}\left(\frac{a}{b}\right)}$$

# Trigonometric Identities II

Square (i) and (ii)

$$a^2 = R^2 \sin^2 \alpha \qquad ----\text{(iii)}$$
$$b^2 = R^2 \cos^2 \alpha \qquad ----\text{(iv)}$$

(iii) + (iv)

$$a^2 + b^2 = R^2 \cos^2 \alpha + R^2 \sin^2 \alpha$$
$$a^2 + b^2 = R^2 \left( \cos^2 \alpha + \sin^2 \alpha \right)$$
$$\therefore R^2 = a^2 + b^2$$
$$\because \cos^2 \alpha + \sin^2 \alpha = 1$$

Let's wrap up our discussion with examples.

### Example 5

$A$ is a positive real number and $\beta$ is an acute angle such that $-90° < \beta < 90°$. Determine the exact value of $A$ and the angle $\beta$ in the following. Present your answer correct to the nearest degree.

a) $4 \sin \theta + 8 \cos \theta \equiv A \sin (\theta + \beta)$
b) $4 \sin \theta + 8 \cos \theta \equiv A \cos (\theta + \beta)$
c) Using your results above, sketch the graph of $4 \sin \theta + 8 \cos \theta$ in the interval $-180° < \theta < 180°$ and state the coordinates of its maximum and minimum points.

What did you get? Find the solution below to double-check your answer.

### Solution to Example 5

#### HINT

> Although the approach shown above will be illustrated, but note that the amplitude ($A$) is easy to determine. Also, once we're used to this, there is an even easier way to arrive at an answer using the addition formulas.

a) $4 \sin \theta + 8 \cos \theta \equiv A \sin (\theta + \beta)$
**Solution**
Given that

$$4 \sin \theta + 8 \cos \theta \equiv A \sin (\theta + \beta)$$

Let's consider the RHS expression and apply the addition formula as:

$$A \sin (\theta + \beta) \equiv A \sin \theta \cos \beta + A \sin \beta \cos \theta$$

Now we have that

$$4 \sin \theta + 8 \cos \theta \equiv A \sin \theta \cos \beta + A \sin \beta \cos \theta$$

Comparing the like parts, we have

$$4 \sin \theta = A \sin \theta \cos \beta$$
$$\therefore 4 = A \cos \beta \qquad ----(i)$$

and

$$8 \cos \theta = A \sin \beta \cos \theta$$
$$\therefore 8 = A \sin \beta \qquad ----(ii)$$

(ii) divided by (i), we have

$$\frac{A \sin \beta}{A \cos \beta} = \frac{8}{4}$$

Simplifying this, we have

$$\frac{\sin \beta}{\cos \beta} = 2$$
$$\tan \beta = 2$$
$$\therefore \boldsymbol{\beta = \tan^{-1}(2) = 63°}$$

Square (i) and (ii)

$$4^2 = A^2 \cos^2 \beta \qquad ----(iii)$$
$$8^2 = A^2 \sin^2 \beta \qquad ----(iv)$$

(iii) + (iv)

$$4^2 + 8^2 = A^2 \cos^2 \beta + A^2 \sin^2 \beta$$
$$4^2 + 8^2 = A^2 \left( \cos^2 \beta + \sin^2 \beta \right)$$
$$\therefore A^2 = 4^2 + 8^2$$
$$A = \sqrt{4^2 + 8^2} = 4\sqrt{5}$$
$$\because \cos^2 \beta + \sin^2 \beta = 1$$

Therefore,

$$\boldsymbol{A = 4\sqrt{5}, \quad \beta = 63°}$$

---

**b)** $4 \sin \theta + 8 \cos \theta \equiv A \cos (\theta + \beta)$
**Solution**
Given that

$$4 \sin \theta + 8 \cos \theta \equiv A \cos (\theta + \beta)$$

Let's consider the RHS expression and apply the addition formulas as:

$$A \cos (\theta + \beta) \equiv A \cos \theta \cos \beta - A \sin \beta \sin \theta$$

# Trigonometric Identities II

Now we have that

$$4\sin\theta + 8\cos\theta \equiv A\cos\theta\cos\beta - A\sin\beta\sin\theta$$

Comparing the like part, we have

$$4\sin\theta = -A\sin\beta\sin\theta$$
$$4 = -A\sin\beta$$
$$\therefore -4 = A\sin\beta \quad ----(i)$$

and

$$8\cos\theta = A\cos\theta\cos\beta$$
$$\therefore 8 = A\cos\beta \quad ----(ii)$$

(i) divided by (ii), we have

$$\frac{A\sin\beta}{A\cos\beta} = -\frac{4}{8}$$

Simplifying this, we have

$$\frac{\sin\beta}{\cos\beta} = -0.5$$
$$\tan\beta = -0.5$$
$$\therefore \boldsymbol{\beta = \tan^{-1}(-0.5) = -27°}$$

Square (i) and (ii)

$$(-4)^2 = A^2\sin^2\beta \quad ----(iii)$$
$$8^2 = A^2\cos^2\beta \quad ----(iv)$$

(iii) + (iv)

$$4^2 + 8^2 = A^2\cos^2\beta + A^2\sin^2\beta$$
$$4^2 + 8^2 = A^2(\cos^2\beta + \sin^2\beta)$$
$$\therefore A^2 = 4^2 + 8^2$$
$$A = \sqrt{4^2 + 8^2} = 4\sqrt{5}$$
$$\because \cos^2\beta + \sin^2\beta = 1$$

Therefore,

$$\boldsymbol{A = 4\sqrt{5}, \quad \beta = -27°}$$

**c)** Sketch of $4\sin\theta + 8\cos\theta$
**Solution**
We can use either of the two results to transform the respective graph.

**Option 1** Using results in (a)

From (a) we have that $\mathbf{4\sin\theta + 8\cos\theta = 4\sqrt{5}\sin(\theta + 63°)}$

It is easier to sketch (Figure 4.4) the RHS from $\sin\theta$ by:

- stretching sine function vertically by a factor of $\mathbf{4\sqrt{5}}$, and
- translating it horizontally (shift to the left) by $\mathbf{63°}$.

We know that $\sin\theta$ has a maximum coordinate of $(90°, 1)$ and a minimum coordinates of $(270°, -1)$ in the given interval. Therefore, for $4\sqrt{5}\sin(\theta + 63°)$, we just need to multiply the $y$-coordinate by $4\sqrt{5}$ and subtract $63°$ from the $x$-coordinate. Hence

$$\text{Max} : \left(27°, 4\sqrt{5}\right)$$
$$\text{Min} : \left(207°, -4\sqrt{5}\right)$$

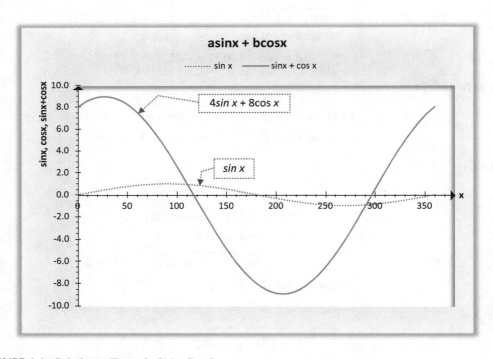

**FIGURE 4.4** Solution to Example 5(c) – Part I.

# Trigonometric Identities II

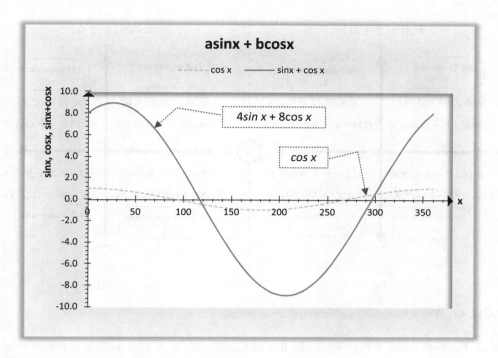

**FIGURE 4.5** Solution to Example 5(c) – Part II.

**Option 2** Using results in (b)

From (b), we have that $4\sin\theta + 8\cos\theta = 4\sqrt{5}\cos(\theta - 27°)$

It is easier to sketch (Figure 4.5) the LHS from the cosine function by:

- stretching cosine function vertically by a factor of $4\sqrt{5}$, and
- translating it horizontally (shift to the right) by $27°$.

We know that $\cos\theta$ has maximum coordinates of $(0°, 1)$ and $(360°, 1)$, and of $(180°, -1)$ in the given interval. Therefore, for $4\sqrt{5}\cos(\theta - 27°)$, we just need to multiply the $y$-coordinate by $4\sqrt{5}$ and add $27°$ to the $x$-coordinate. Hence

$$\text{Max}: \left(27°, \ 4\sqrt{5}\right)$$
$$\text{Min}: \left(207°, -4\sqrt{5}\right)$$

You may have noticed that there is a second maximum of $\left(387°, \ 4\sqrt{5}\right)$ if we follow what was mentioned above, but $387°$ is outside the given range so it is not a valid answer.

---

The above method seems long and can be boring. We therefore need a quicker way to deal with this, and the answer is a new but experience-based approach. Given the general expression $a\sin x + b\cos x$, we can use the following tips to solve any trigonometric expression involving the sum and difference of two functions:

**Note 1** You need either a sine equivalent $R\sin(x \pm \alpha)$ or a cosine equivalent $R\cos(x \pm \alpha)$. If this is not given or stated, decide which one you want.

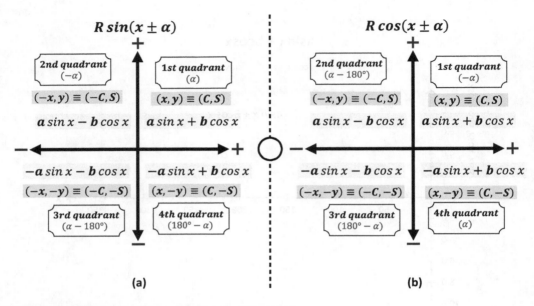

**FIGURE 4.6** The $R$ formulas – author's method illustrated.

**Note 2** For both cases, $R$ is positive and it is obtained using $R = \sqrt{a^2 + b^2}$. It is irrelevant if the coefficient of sine, cosine, or both is positive or negative.

**Note 3** For $R\sin(x \pm \alpha)$, $\alpha = \tan^{-1}\left(\frac{b}{a}\right)$ and for $R\cos(x \pm \alpha)$, $\alpha = \tan^{-1}\left(\frac{a}{b}\right)$. The trick here is that $\alpha$ is the *'tan inverse of the coefficient of the function you are not aiming for over the one you want'*. You can decide how you would remember this. This resembles the current divider rule of two resistors in parallel used when analysing electric circuits.

**Note 4** For this approach, the phase angle (leading or lagging) $\alpha$ is such that $-180° < \alpha < 180°$.

**Note 5** Remove the sign in the answer above, i.e., get the modulus of $\alpha$ or $|\alpha|$.

**Note 6** Finally, adjust $\alpha$ by getting the quadrant phase for the desired expression using Figure 4.6. Consider sine as though it is on the vertical axis ($y$) whilst cosine is on the horizontal axis ($x$). For example, the first quadrant is $(x, y)$, which is equivalent to $(C, S)$ where $C$ and $S$ represent cosine and sine, respectively. This is the same when $a$ and $b$ are positive, i.e., $(a, b)$.

Let's summarise the new method in four steps as:

**Step 1:** Choose (or decide) which of these two you want: $R\sin(x + \alpha)$ or $R\cos(x + \alpha)$.

**Step 2:** Find $R$ using $R = \sqrt{a^2 + b^2}$. This is a positive value and is the same for both cosine and sine formulas.

**Step 3:** Determine $|\alpha|$. For $R\sin(x + \alpha)$, $|\alpha| = \left|\tan^{-1}\left(\frac{b}{a}\right)\right|$; and for $R\cos(x + \alpha)$, $|\alpha| = \left|\tan^{-1}\left(\frac{a}{b}\right)\right|$.

**Step 4:** Adjust $\alpha$ based on the quadrant as shown in Table 4.4.

We will illustrate this using $\sin x + \cos x$, where $|a| = |b|$ for the four possible scenarios as shown in Tables 4.5 and 4.6, and Figure 4.7.

# Trigonometric Identities II

**TABLE 4.4**
**The R Formulas – Author's Method Illustrated**

| Quadrant | Adjusted $\alpha$ (for sine) | Adjusted $\alpha$ (for cosine) |
|---|---|---|
| First | $\alpha$ | $-\alpha$ |
| Second | $-\alpha$ | $\alpha - 180°$ |
| Third | $\alpha - 180°$ | $180° - \alpha$ |
| Fourth | $180° - \alpha$ | $\alpha$ |

**TABLE 4.5**
**Analysis for $R \sin(x \pm \alpha)$**

| Expression | Quadrant | $\|\alpha\|$ | Adjusted $\alpha$ | $R = \sqrt{a^2 + b^2}$ | $R\sin(x+\alpha)$ |
|---|---|---|---|---|---|
| $\sin x + \cos x$ | First | $\left\|\tan^{-1}\left(\frac{1}{1}\right)\right\| = 45°$ | $45°$ | $\sqrt{1^2+1^2} = \sqrt{2}$ | $\sqrt{2}\sin(x+45°)$ |
| $\sin x - \cos x$ | Second | $\left\|\tan^{-1}\left(\frac{-1}{1}\right)\right\| = 45°$ | $-45°$ | $\sqrt{1^2+(-1)^2} = \sqrt{2}$ | $\sqrt{2}\sin(x-45°)$ |
| $-\sin x - \cos x$ | Third | $\left\|\tan^{-1}\left(\frac{-1}{-1}\right)\right\| = 45°$ | $= (45° - 180°)$ $= -135°$ | $\sqrt{(-1)^2+(-1)^2} = \sqrt{2}$ | $\sqrt{2}\sin(x-135°)$ |
| $-\sin x + \cos x$ | Fourth | $\left\|\tan^{-1}\left(\frac{1}{-1}\right)\right\| = 45°$ | $= (180° - 45°)$ $= 135$ | $\sqrt{(-1)^2+1^2} = \sqrt{2}$ | $\sqrt{2}\sin(x+135°)$ |

Let's test our new method with some examples.

## Example 6

Express the following sum (or difference) of two trigonometric functions in $R\sin(x+\alpha)$ or $R\sin(x-\alpha)$, where $-180° < \alpha < 180°$. Give $R$ in exact form and $\alpha$ in degrees correct to 1 d.p. where applicable.

a) $3\cos x + 4\sin x$
b) $2\sqrt{3}\sin x - 2\cos x$
c) $-3\cos x - \sqrt{2}\sin x$
d) $7\cos x - 6\sin x$

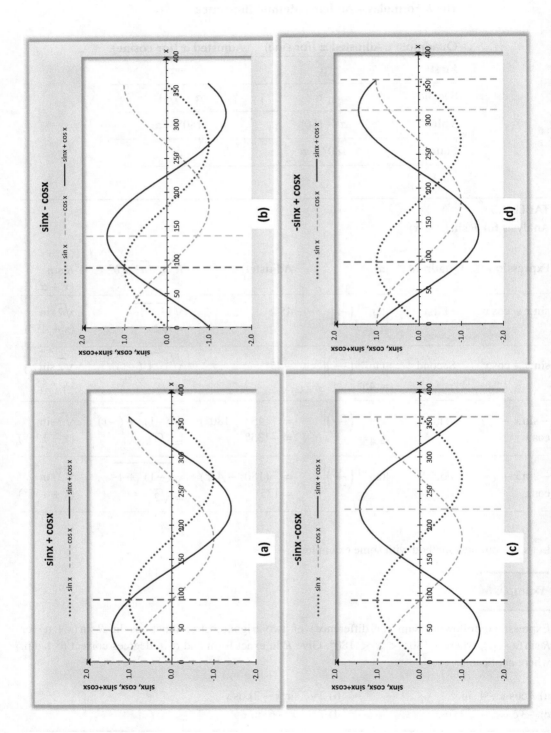

**FIGURE 4.7** The $R$ formulas: (a) first quadrant ($\sin x + \cos x$), (b) second quadrant $\sin x - \cos x$, (c) third quadrant $-\sin x - \cos x$, and (d) fourth quadrant $-\sin x + \cos x$.

# TABLE 4.6
## Analysis for $R\cos(x \pm \alpha)$

| Expression | Quadrant | $\|\alpha\|$ | Adjusted $\alpha$ | $R = \sqrt{a^2+b^2}$ | $R\cos(x+\alpha)$ |
|---|---|---|---|---|---|
| $\sin x + \cos x$ | First | $\left\|\tan^{-1}\left(\frac{1}{1}\right)\right\| = 45°$ | $-45°$ | $\sqrt{1^2+1^2} = \sqrt{2}$ | $\sqrt{2}\sin(x-45°)$ |
| $\sin x - \cos x$ | Second | $\left\|\tan^{-1}\left(\frac{1}{-1}\right)\right\| = 45°$ | $=(45°-180°) = -135°$ | $\sqrt{1^2+(-1)^2} = \sqrt{2}$ | $\sqrt{2}\sin(x-135°)$ |
| $-\sin x - \cos x$ | Third | $\left\|\tan^{-1}\left(\frac{-1}{-1}\right)\right\| = 45°$ | $=(180°-45°) = 135$ | $\sqrt{(-1)^2+(-1)^2} = \sqrt{2}$ | $\sqrt{2}\sin(x+135°)$ |
| $-\sin x + \cos x$ | Fourth | $\left\|\tan^{-1}\left(\frac{-1}{1}\right)\right\| = 45°$ | $45°$ | $\sqrt{(-1)^2+1^2} = \sqrt{2}$ | $\sqrt{2}\sin(x+45°)$ |

What did you get? Find the solution below to double-check your answer.

### Solution to Example 6

**a)** $3\cos x + 4\sin x$
**Solution**

$$R = \sqrt{a^2+b^2}$$
$$= \sqrt{3^2+4^2} = 5$$
$$|\alpha| = \left|\tan^{-1}\left(\frac{3}{4}\right)\right| = 36.9°$$

This is the first quadrant, the adjusted $\alpha = 36.9°$.

$$\therefore 3\cos x + 4\sin x = 5\sin(x+36.9°)$$

**b)** $2\sqrt{3}\sin x - 2\cos x$
**Solution**

$$R = \sqrt{a^2+b^2}$$
$$= \sqrt{\left(2\sqrt{3}\right)^2+(-2)^2} = 4$$
$$|\alpha| = \left|\tan^{-1}\left(\frac{-2}{2\sqrt{3}}\right)\right| = 30°$$

This is the second quadrant, the adjusted $\alpha = -30°$.

$$\therefore 2\sqrt{3}\sin x - 2\cos x = 4\sin(x - 30°)$$

---

**c)** $-3\cos x - \sqrt{2}\sin x$
**Solution**

$$R = \sqrt{a^2 + b^2}$$
$$= \sqrt{(-3)^2 + (-\sqrt{2})^2} = \sqrt{11}$$
$$|\alpha| = \left|\tan^{-1}\left(\frac{-3}{-\sqrt{2}}\right)\right| = 64.8°$$

This is the third quadrant, the adjusted $\alpha = 64.8° - 180° = -115.2°$.

$$\therefore -3\cos x - \sqrt{2}\sin x = \sqrt{11}\sin(x - 115.2°)$$

---

**d)** $7\cos x - 6\sin x$
**Solution**

$$R = \sqrt{a^2 + b^2}$$
$$= \sqrt{7^2 + (-6)^2} = \sqrt{85}$$
$$|\alpha| = \left|\tan^{-1}\left(\frac{7}{-6}\right)\right| = 49.4°$$

This is the fourth quadrant, the adjusted $\alpha = 180° - 49.4° = 130.6°$.

$$\therefore 7\cos x - 6\sin x = \sqrt{85}\sin(x + 130.6°)$$

---

Here is another set of examples.

### Example 7

Express the following sum (or difference) of two trigonometric functions in $R\cos(x + \alpha)$ or $R\cos(x - \alpha)$, where $-180° < \alpha < 180°$. Give $R$ in exact form and $\alpha$ in degrees correct to 1 d.p. where applicable.

a) $\sqrt{5}\sin x + \sqrt{3}\cos x$      b) $2\sin x - \sqrt{5}\cos x$

c) $-\sin x - 3\cos x$      d) $3\cos x - \sqrt{6}\sin x$

# Trigonometric Identities II

What did you get? Find the solution below to double-check your answer.

### Solution to Example 7

**a)** $\sqrt{5}\sin x + \sqrt{3}\cos x$
**Solution**

$$R = \sqrt{a^2 + b^2}$$
$$= \sqrt{\left(\sqrt{5}\right)^2 + \left(\sqrt{3}\right)^2} = 2\sqrt{2}$$
$$|\alpha| = \left|\tan^{-1}\left(\frac{\sqrt{5}}{\sqrt{3}}\right)\right| = 52.2°$$

This is the first quadrant, the adjusted $\alpha = -52.2°$.

$$\therefore \sqrt{5}\sin x + \sqrt{3}\cos x = 2\sqrt{2}\cos(x - 52.2°)$$

---

**b)** $2\sin x - \sqrt{5}\cos x$
**Solution**

$$R = \sqrt{a^2 + b^2}$$
$$= \sqrt{2^2 + \left(-\sqrt{5}\right)^2} = 3$$
$$|\alpha| = \left|\tan^{-1}\left(\frac{2}{-\sqrt{5}}\right)\right| = 41.8°$$

This is the second quadrant, the adjusted $\alpha = 41.8° - 180° = -138.2°$.

$$\therefore 2\sin x - \sqrt{5}\cos x = 3\cos(x - 138.2°)$$

---

**c)** $-\sin x - 3\cos x$
**Solution**

$$R = \sqrt{a^2 + b^2}$$
$$= \sqrt{(-1)^2 + (-3)^2} = \sqrt{10}$$
$$|\alpha| = \left|\tan^{-1}\left(\frac{-1}{-3}\right)\right| = 18.4°$$

This is the third quadrant, the adjusted $\alpha = 180° - 18.4° = 161.6°$.

$$\therefore -\sin x - 3\cos x = \sqrt{10}\cos(x + 161.6°)$$

**d)** $3\cos x - \sqrt{6}\sin x$
**Solution**

$$R = \sqrt{a^2 + b^2}$$
$$= \sqrt{3^2 + \left(-\sqrt{6}\right)^2} = \sqrt{15}$$
$$|\alpha| = \left|\tan^{-1}\left(\frac{-\sqrt{6}}{3}\right)\right| = 39.2°$$

This is the fourth quadrant, the adjusted $\alpha = 39.2°$.

$$\therefore 3\cos x - \sqrt{6}\sin x = \sqrt{15}\cos(x + 39.2°)$$

Here is a final set of examples for us to try.

### Example 8

State the minimum and maximum values of the following trigonometric expression.

a) $\sqrt{13}\sin\left(\theta - \frac{1}{3}\pi\right)$      b) $3\sqrt{2}\cos\theta + 2$
c) $\cos\theta - \sqrt{6}\sin\theta$      d) $\sqrt{5}\sin\theta - 2\cos\theta - \sqrt{3}$
e) $3 - 4\sin\theta - 5\cos\theta$      f) $1 - 2\sqrt{3}\cos\theta + \sqrt{13}\cos\left(\theta - \frac{1}{2}\pi\right)$

What did you get? Find the solution below to double-check your answer.

### Solution to Example 8

**HINT**

Where appropriate, first find the $R$ value using $R = \sqrt{a^2 + b^2}$.

**a)** $\sqrt{13}\sin\left(\theta - \frac{1}{3}\pi\right)$
**Solution**
Using the standard expression and comparing it with

$$A\sin(\theta + \alpha)$$

we therefore have
- Minimum

$$-\sqrt{13}$$

- Maximum

$$\sqrt{13}$$

**b)** $3\sqrt{2}\cos\theta + 2$
**Solution**
Using the standard expression

$$A\cos(\theta + \alpha)$$

and comparing, we therefore have the minimum and maximum values as $-3\sqrt{2}$ and $3\sqrt{2}$, respectively. But there is a vertical translation of 2, hence

- Minimum

$$2 - 3\sqrt{2}$$

- Maximum

$$2 + 3\sqrt{2}$$

---

**c)** $\cos\theta - \sqrt{6}\sin\theta$
**Solution**

$$R = \sqrt{a^2 + b^2}$$
$$= \sqrt{1^2 + \left(-\sqrt{6}\right)^2} = \sqrt{7}$$

Hence

- Minimum

$$-\sqrt{7}$$

- Maximum

$$\sqrt{7}$$

---

**d)** $\sqrt{5}\sin\theta - 2\cos\theta - \sqrt{3}$
**Solution**

$$R = \sqrt{a^2 + b^2}$$
$$= \sqrt{\left(\sqrt{5}\right)^2 + (-2)^2} = 3$$

Therefore, we have the minimum and maximum values as $-3$ and $3$, respectively. But there is a vertical translation of $-\sqrt{3}$, hence

- Minimum

$$-3 - \sqrt{3} = -\left(3 + \sqrt{3}\right)$$

- Maximum

$$3 - \sqrt{3}$$

e) $3 - 4\sin\theta - 5\cos\theta$
**Solution**

$$R = \sqrt{a^2 + b^2}$$
$$= \sqrt{(-4)^2 + (-5)^2} = \sqrt{41}$$

Therefore, we have the minimum and maximum values as $-\sqrt{41}$ and $\sqrt{41}$, respectively. But there is a vertical translation of 3, hence

- Minimum

$$3 - \sqrt{41}$$

- Maximum

$$3 + \sqrt{41}$$

---

f) $1 - 2\sqrt{3}\cos\theta + \sqrt{13}\cos\left(\theta - \frac{1}{2}\pi\right)$
**Solution**
We need to first apply an identity to simplify the expression, which means

$$\cos\left(\theta - \frac{1}{2}\pi\right) = \sin\theta$$

Hence

$$1 - 2\sqrt{3}\cos\theta + \sqrt{13}\cos\left(\theta - \frac{1}{2}\pi\right) = 1 - 2\sqrt{3}\cos\theta + \sqrt{13}\sin\theta$$

Now, we have the right format and then have

$$R = \sqrt{a^2 + b^2}$$
$$= \sqrt{\left(-2\sqrt{3}\right)^2 + \left(\sqrt{13}\right)^2} = 5$$

Therefore, we have the minimum and maximum values as $-5$ and $5$, respectively. But there is a vertical translation of 1, hence

- Minimum

$$1 - 5 = -4$$

- Maximum

$$1 + 5 = 6$$

# 4.4 CHAPTER SUMMARY

1) Factor formulas (addition–product) for trigonometric ratios:

   - **Sine**

   $$\sin A + \sin B = 2 \sin \frac{1}{2}(A+B) \cos \frac{1}{2}(A-B)$$

   $$\sin A - \sin B = 2 \sin \frac{1}{2}(A-B) \cos \frac{1}{2}(A+B)$$

   - **Cosine**

   $$\cos A + \cos B = 2 \cos \frac{1}{2}(A+B) \cos \frac{1}{2}(A-B)$$

   $$\cos A - \cos B = -2 \sin \frac{1}{2}(A+B) \sin \frac{1}{2}(A-B)$$

2) Product–addition formulas for trigonometric ratios:

   - **Sine–cosine**

   $$2 \sin A \cos B = \sin(A+B) + \sin(A-B)$$

   $$2 \cos A \sin B = \sin(A+B) - \sin(A-B)$$

   - **Cosine**

   $$2 \cos A \cos B = \cos(A+B) + \cos(A-B)$$

   - **Sine**

   $$2 \sin A \sin B = -\cos(A-B) - \cos(A+B)$$

   In the above formulas $A > B$.

3) R formulas for trigonometric ratios:

   - **Sine**

   $$a \sin x + b \cos x = R \sin(x + \alpha) \quad \text{AND} \quad a \sin x - b \cos x = R \sin(x - \alpha)$$

   - **Cosine**

   $$a \sin x + b \cos x = R \cos(x - \alpha) \quad \text{AND} \quad a \sin x - b \cos x = R \cos(x + \alpha)$$

   where $a$, $b$, and $R$ are positive real numbers and $\alpha$ is an acute angle such that $0 < \alpha < 90°$.

We can further state that:
- $R = \sqrt{a^2 + b^2}$ is valid for all cases and always positive.
- $\alpha = \tan^{-1}\left(\pm\frac{b}{a}\right)$ or $\alpha = \tan^{-1}\left(\pm\frac{a}{b}\right)$ depending on the original expression, the sign of $a$ and $b$, and the $R$ equivalent we have chosen.
- We need to be careful about the sign especially when $a$ and $b$ are both negative, which can cancel each other out.

****

## 4.5 FURTHER PRACTICE

To access complementary contents, including additional exercises, please go to www.dszak.com.

# 5 Solving Trigonometric Equations

## Learning Outcomes

Once you have studied the content of this chapter, you should be able to:

- Solve trigonometric equations
- Change the given interval based on the transformed functions
- Use trigonometric identities when solving trigonometric equations

## 5.1 INTRODUCTION

A trigonometric equation (linear and non-linear) is an equation involving trigonometric ratios. As such, solving this type of equation requires that one is equipped with skills relating to trigonometric functions and identities, and how to solve equations in general.

Recall that to solve an equation, one needs to find the value of the unknown quantity or variable, which will satisfy the equality of the expression on both sides of the equal sign (=). For example, to solve a linear equation $2x - 1 = 0$ means to find the value of $x$ that would make the LHS expression (i.e., $2x - 1$) equal to the RHS expression (i.e., 0). For this case, $x = \frac{1}{2}$. This is the only solution and nothing more. If the equation is quadratic, we expect two values (repeated or non-repeated).

The above is exactly what we will be doing here with trigonometric equations. However, there are a couple of particularities.

**Note 1** Due to their periodicity, there are unlimited solutions for any given trigonometric equation. Therefore, we need to know the range of values (angles) required. A calculator will only give us one answer, usually in the first quadrant for positive values.

**Note 2** Unlike polynomials, we cannot foretell the number of solutions in trigonometric equations. The bigger the interval the more the solutions, but two solutions are generally expected for every 360-degree interval.

## 5.2 METHODS OF SOLVING TRIGONOMETRIC EQUATIONS

It is important to be acquainted with methods of solving trigonometric equations. We will use a simple trigonometric equation to illustrate each of these methods. Our choice for this exercise is a sine function. Let's say, we have to solve the following trigonometric equation:

$$\sin x = -0.5, \quad -360° \leq x \leq 360°$$

Before proceeding to the methods, let's quickly comment on the above equation. $\sin x = -0.5$ in the interval $-360° \le x \le 360°$ means that we should provide all possible values of $x$ that satisfies $\sin x$ being equal to $-0.5$ between $-360°$ and $360°$. Consequently, the answer(s) must not be more than 360 degrees or less than $-360$ degrees. Now, let's go through the three methods one after the other.

## 5.2.1 Sketching a Graph

This method is very straightforward, provided that the trigonometric equation has been reduced to its simplest form, i.e., $\sin x = a$, $\cos x = a$, or $\tan x = a$, where $a$ is an irrational number. We will later consider cases where a trigonometric equation should be simplified first.

$\sin x = -0.5$ in $-360° \le x \le 360°$ can be solved graphically by following these steps:

**Step 1:** In case it is not one of the special angles (i.e., $0°$, $30°$, $45°$, $60°$, and $90°$), use your calculator to find the inverse function. In other words,

$$\sin x = -0.5$$
$$x = \sin^{-1}(-0.5)$$
$$= -30°$$

This will give you the value to be shown on the graph apart from the maximum/minimum value(s). You should use symmetry (and other relevant properties) to show other points.

**Step 2:** Use the key features of the sine function to sketch it. To be specific, use the maximum/minimum values of relevant special angles to sketch the function as accurately as possible, as shown in Figure 5.1.

**Step 3:** Draw a horizontal line, passing through $f(x) = -0.5$ and cutting through the entire function (Figure 5.2).

**Step 4:** Draw a vertical line at each point of intersection of the curve with the horizontal line drawn in Step 3 (Figure 5.3).

**Step 5:** Read off the values on the horizontal axis at which the vertical lines drawn in Step 4 meet the x-axis, which are $-150°$, $-30°$, $210°$, and $330°$. These are the solutions of the given trigonometric equation.

**NOTE**

Instead of drawing the sine function that would cover the given range $-360° \le x, \le 360°$, the following will produce the same results:

a) Draw a complete cycle of $0° \le x \le 360°$.

b) Find the values of the function in this range, usually two values. For the given sine function, they are where sine is negative, and these are $210°$ and $330°$.

c) Use the periodicity property to find other values. For the sine function use $\theta \pm 360°n$, where $\theta$ is the angle(s) obtained within the $360°$ range. For the current case, we need to simply subtract $360°$, because adding $360°$ would make us go above the upper limit of the range. Thus, $210° - 360° = -150°$ and $330° - 360° = -30°$ are the other solutions. The complete solutions are therefore $-150°$, $-30°$, $210°$, and $330°$. These are the same values as previously obtained.

Solving Trigonometric Equations 151

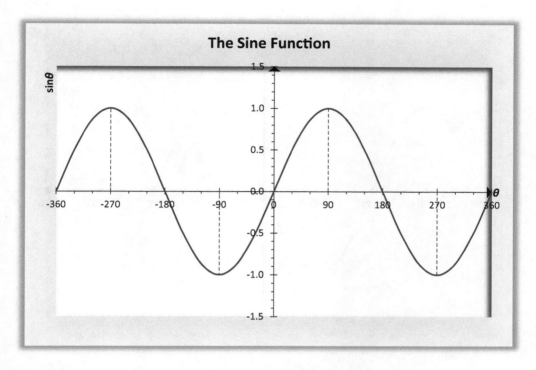

**FIGURE 5.1** Solving trigonometric equations graphically, step 2 illustrated.

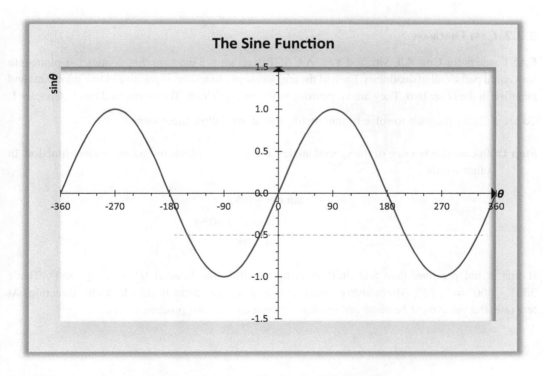

**FIGURE 5.2** Solving trigonometric equations graphically, step 3 illustrated.

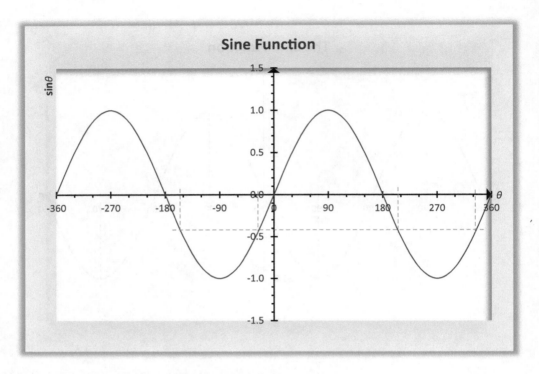

**FIGURE 5.3** Solving trigonometric equations graphically, step 4 illustrated.

### 5.2.2 CAST DIAGRAM

**CAST** stands for **C**os, **A**ll, **S**in, and **T**an. A CAST diagram is a diagram that divides $x$–$y$ plane into four equal parts, called quadrants. Each of the trigonometric functions is positive in two quadrants and negative in the other two. They are all positive in the first quadrant. We've covered this in Chapter 1.

To use a CAST diagram to solve trigonometric equations, follow these steps:

**Step 1:** In case it is not one of the special angles, use your calculator to find the inverse function. In other words

$$\sin x = -0.5$$
$$x = \sin^{-1}(-0.5)$$
$$= -30°$$

If this is not positive, then add 360° to bring it within the interval $0° \leq x \leq 360°$. That's $330° - 360° = -30°$. Alternatively, consider it as a measurement in the clockwise direction. At any rate, the value must be acute and positive, measured from the positive $x$-axis.

# Solving Trigonometric Equations

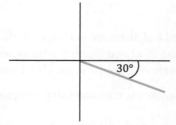

**FIGURE 5.4** Solving trigonometric equations using CAST diagram, step 2 illustrated.

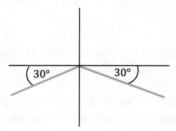

**FIGURE 5.5** Solving trigonometric equations using CAST diagram, step 3 illustrated.

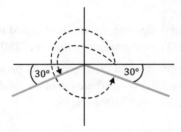

**FIGURE 5.6** Solving trigonometric equations using CAST diagram, step 4 illustrated.

**Step 2:** Locate your answer from step 1 above in CAST in clockwise (CW) or counter-clockwise (CCW) measurement. Figure 5.4 shows it in the CW format.

**Step 3:** This is a sine function, which is negative in the third and fourth quadrants. Therefore, mirror the angle in the fourth quadrant in step 2 above to the third quadrant as indicated in Figure 5.5.

**Step 4:** Conventionally, angles are measured counter-clockwise from the positive $x$-axis. Hence, the two angles are $180° + 30° = 210°$ and $360° - 30° = 330°$. These are indicated in Figure 5.6.

**Step 5:** Use the periodicity property to find other values as previously explained, by adding or subtracting multiples of $360°$ (or $2\pi$ if angle is in radians). Thus, $210° - 360° = -150°$ and $30° - 360° = -330°$ are the other solutions. The complete solutions are, therefore, $-150°$, $-30°$, $210°$, and $330°$ as before.

## 5.2.3 QUADRANT FORMULA

Quadrant formulas were introduced and discussed in Chapter 1. As a quick review, they are $180° - \theta$, $180° + \theta$, and $360° - \theta$ for the second, third, and fourth quadrants, respectively. No formula is required for the first quadrant. This third method is similar to the CAST method discussed above.

Follow these steps to solve a trigonometric equation using the quadrant formula:

**Step 1:** If not a special angle, use your calculator to find its inverse function. In other words

$$\sin x = -0.5$$
$$x = \sin^{-1}(-0.5)$$
$$= -30°$$

If this is not positive, then add 360° to bring it within the range $0° \leq x \leq 360°$, which is $-30° + 360° = 330°$.

**Step 2:** Decide which quadrant is the answer in step 1 above and determine its 'pair' quadrant. (Note that each trigonometric function is positive in two quadrants and negative in the other two). For this case, it is the fourth quadrant.

Since sine is negative in this quadrant, we must find where sine is also negative. This is the third quadrant.

**Step 3:** Use $360° - \theta$ to find $\theta$ and then use the result obtained to find the sine value in the third quadrant using $180° + \theta$. Thus, the second solution is $180° + \theta = 180° + 30° = 210°$.

**Step 4:** Use the periodicity property to find other values as previously explained. Thus, $210° - 360° = -150°$ and $330° - 360° = -30°$ are the other solutions. The complete solutions are therefore $-150°, -30°, 210°,$ and $330°$ as before.

We are now done with the three methods. Let's try some examples.

### Example 1

Solve the following trigonometric equations in the range $0 \leq x \leq 360°$. Present your answer correct to the nearest degree.

a) $2\cos x = 0.8$  
b) $\sqrt{3}\tan x = -1$  
c) $3\cos x + 1 = \sqrt{2}$  
d) $1 - \sin x = \sqrt{3}$  
e) $7\sin x - 5 = \sqrt{6}$  
f) $\sqrt{3}\csc x + 3 = 1$

# Solving Trigonometric Equations

What did you get? Find the solution below to double-check your answer.

## Solution to Example 1

**HINT**

- In the examples to follow, we will try to swap between the methods introduced above.
- The answer obtained from a calculator is usually the first quadrant unless it is negative.

**a)** $2 \cos x = 0.8$
**Solution**
**Method:** Formula

$$2 \cos x = 0.8$$
$$\cos x = 0.4$$
$$x = \cos^{-1}(0.4) = 66°$$

Since cosine is positive in the first and fourth quadrants, we now need to find the solution in the fourth quadrant using

$$360° - x = 360° - 66° = 294°$$
$$\therefore x = 66°, \, 294°$$

---

**b)** $\sqrt{3} \tan x = -1$
**Solution**
**Method:** CAST

$$\sqrt{3} \tan x = -1$$
$$\tan x = -\frac{1}{\sqrt{3}} \text{ or } -\frac{\sqrt{3}}{3}$$
$$x = \tan^{-1}\left(-\frac{\sqrt{3}}{3}\right) = -30°$$

A negative angle is measured from the positive x-axis in a clockwise direction. Since tangent is negative in the second and fourth quadrants, our measurements are taken in these quadrants. This is illustrated in Figure 5.7.

$$\therefore x = 150°, \, 330°$$

---

**c)** $3 \cos x + 1 = \sqrt{2}$
**Solution**
**Method:** Graph

$$3 \cos x + 1 = \sqrt{2}$$

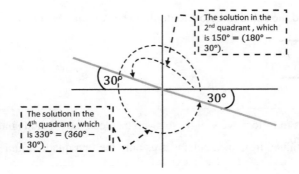

**FIGURE 5.7** Solution to Example 1(b).

**Step 1:** Re-arranging this, we have

$$\cos x = \frac{\sqrt{2}-1}{3}$$

**Step 2:** Draw the graph of cosine in the stated range (Figure 5.8).
**Step 3:** $\frac{\sqrt{2}-1}{3} = 0.1381$ (4 d.p.). Mark this point on the vertical axis and draw a horizontal line through it.
**Step 4:** Draw vertical lines at the point of intersections.

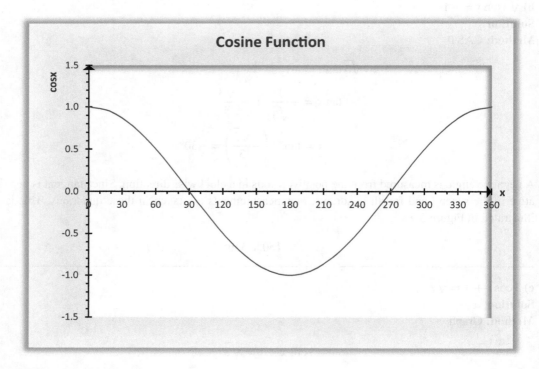

**FIGURE 5.8** Solution to Example 1(c) – Part I.

# Solving Trigonometric Equations

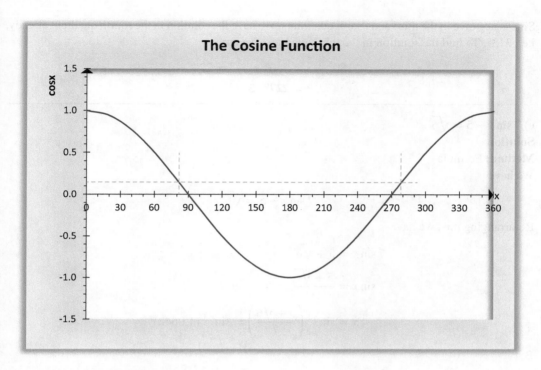

**FIGURE 5.9** Solution to Example 1(c) – Part II.

Note that

$$x = \cos^{-1}\left(\frac{\sqrt{2}-1}{3}\right) = 82°$$

This will help in indicating the relevant angles shown in Figure 5.9.

**Step 5:** Read the values on the horizontal axis.

$$\therefore x = 82°, \; 278°$$

---

**d)** $1 - \sin x = \sqrt{3}$
**Solution**
**Method:** Formula

$$1 - \sin x = \sqrt{3}$$

**Step 1:** Re-arranging this, we have

$$\sin x = 1 - \sqrt{3}$$
$$x = \sin^{-1}\left(1 - \sqrt{3}\right) = -47°$$

**Step 2:** Add 360° to bring the above into a positive equivalent within the stated range as −47° + 360° = 313°.

Sine is negative in the third and fourth quadrants. We have the solution in the fourth quadrant already, i.e., 313°. To find the solution in the third quadrant, we use

$$180° + x = 180° + 47° = 227°$$
$$\therefore x = \mathbf{227°, \ 313°}$$

---

e) $7 \sin x - 5 = \sqrt{6}$
**Solution**
**Method:** Formula
We have

$$7 \sin x - 5 = \sqrt{6}$$

Re-arranging this, we have

$$7 \sin x = 5 + \sqrt{6}$$
$$\sin x = \frac{5 + \sqrt{6}}{7}$$
$$x = \sin^{-1}\left(\frac{5 + \sqrt{6}}{7}\right) = \sin^{-1}(1.0642)$$

But

$$\frac{5 + \sqrt{6}}{7} = 1.064 \ (4 \ s.f.)$$

which implies that

$$\frac{5 + \sqrt{6}}{7} > 1$$

Hence, the above trigonometric equation has no solutions as the maximum/minimum value for the sine function is ±1.

---

f) $\sqrt{3} \operatorname{cosec} x + 3 = 1$
**Solution**
**Method:** Formula

$$\sqrt{3} \operatorname{cosec} x + 3 = 1$$
$$\sqrt{3} \operatorname{cosec} x = -2$$
$$\operatorname{cosec} x = -\frac{2}{\sqrt{3}}$$

We need to change $\operatorname{cosec} x$, so we have

$$\operatorname{cosec} x = \frac{1}{\sin x} = -\frac{2}{\sqrt{3}}$$
$$\sin x = -\frac{\sqrt{3}}{2}$$
$$x = \sin^{-1}\left(-\frac{\sqrt{3}}{2}\right) = -60°$$

Solving Trigonometric Equations

If $x = -60°$, it implies a clockwise measurement, therefore:

$$x = 360° - 60° = 300°$$

Since sine is negative in the third and fourth quadrants, we now need to find the solution in the third quadrant using

$$180° + x = 180° + 60° = 240°$$

$$\therefore x = 240°, \ 300°$$

**NOTE**
Any time we have equations with reciprocals of trigonometric functions, we will need to change sec, cosec, and cot to cos, sin, and tan, respectively, before proceeding to solve for the unknown.

## 5.3 CHANGING THE INTERVAL OF TRANSFORMED FUNCTIONS

In the preceding section and the examples that followed, we limited ourselves to what could be termed as a default form of trigonometric equations. In other words, we used $y = \sin x$, $y = \cos x$, and $y = \tan x$. When trigonometric functions are transformed, the interval of the solutions will change accordingly. We will now discuss three distinct transformations and how to solve problems involving each type.

### 5.3.1 TYPE 1: $y = \sin ax$, $y = \cos ax$, $y = \tan ax$

This is simply a horizontal stretch where $a$ is the stretching factor. Consequently, we have that:

a) the intervals will be $a$ times **the original interval**, and
b) the number of solutions will be $a$ times the **original number of solutions**.

As an illustration, to solve $y = \cos 2x$ in the range $-\pi \leq x \leq \pi$ would imply solving the same in the range $-2\pi \leq 2x \leq 2\pi$. Notice the multiplier 2 in the interval, which has extended the range. Consequently, the number of solutions will be doubled depending on the angles. Table 5.1 shows more examples of transformation along the horizontal axis.

We will not discuss the vertical stretch (or stretch along the y-axis) as this does not change the value of $x$. It is a good time to roll out some examples.

### Example 2

Solve the following trigonometric equations in the range $0 < \theta < \pi$. Present your answer correct to 3 s.f. where necessary.

a) $5 - 2\cos(-3\theta) = 6$   b) $\sqrt{3}\sin\left(\frac{2}{3}\theta\right) - 3\cos\left(\frac{2}{3}\theta\right) = 0$   c) $\tan 2\theta - 1 = \sqrt{3}$

## TABLE 5.1
### Changing the Interval of Transformed Functions (Vertical Stretch) Illustrated

|   | Equation | Original Range | New Range |
|---|---|---|---|
| 1) | $y = \sin 2\theta$ | $0 \leq \theta \leq 2\pi$ | $0 \leq 2\theta \leq 4\pi$ |
| 2) | $y = \tan 3\theta$ | $-180° < \theta < 180°$ | $-540° < 3\theta < 540°$ |
| 3) | $y = \cos\left(\frac{3}{2}\theta\right)$ | $0 \leq \theta < 180°$ | $0 \leq \frac{3}{2}\theta < 270°$ |
| 4) | $y = \tan 0.5\theta$ | $0 < \theta \leq 2\pi$ | $0 < 0.5\theta \leq \pi$ |

What did you get? Find the solution below to double-check your answer.

### Solution to Example 2

**a)** $5 - 2\cos(-3\theta) = 6$
**Solution**
The original interval is $0 < \theta < \pi$, the new range for this case is $0 < 3\theta < 3\pi$. Given that

$$5 - 2\cos(-3\theta) = 6$$

**Step 1:** Re-arranging this, we have

$$-2\cos(-3\theta) = 6 - 5$$
$$-2\cos(-3\theta) = 1$$
$$\cos(-3\theta) = -\frac{1}{2}$$

Or

$$\cos(-3\theta) = -0.5$$

**Step 2:** Recall that $\cos(-3\theta) = \cos 3\theta$, thus

$$\cos 3\theta = -0.5$$
$$3\theta = \cos^{-1}(-0.5) = -\frac{\pi}{3}$$

**Step 3:** To make it easy, it is customary to substitute as:

$$z = 3\theta$$

Hence

$$z = -\frac{\pi}{3}$$

# Solving Trigonometric Equations

**Step 4:** Cosine is symmetry about the origin (i.e., $\cos 3\theta = -\cos 3\theta$) so

$$z = \cos^{-1}(-0.5) = \cos^{-1}(0.5) = \frac{\pi}{3}$$

**Step 5:** Cosine is negative in the second and third quadrants, thus we have

$$z = \pi - \frac{\pi}{3} = \frac{2}{3}\pi \text{ and } z = \pi + \frac{\pi}{3} = \frac{4}{3}\pi$$

Thus, we can see that the two values within $2\pi$ range are:

$$z = \frac{2}{3}\pi, \frac{4}{3}\pi$$

**Step 6:** Since we need values up to $3\pi$, let's add $2\pi$ to have

$$z = \frac{2}{3}\pi + 2\pi = \frac{8}{3}\pi \text{ and } z = \frac{4}{3}\pi + 2\pi = \frac{10}{3}\pi$$

The second solution (i.e. $\frac{10}{3}\pi$) is clearly outside our range, hence the complete solutions are

$$z = \frac{2}{3}\pi, \frac{4}{3}\pi, \frac{8}{3}\pi$$

**Step 7:** Recall that $z = 3\theta$, thus

$$3\theta = z = \frac{2}{3}\pi, \frac{4}{3}\pi, \frac{8}{3}\pi$$

$$\theta = \frac{2}{9}\pi, \frac{4}{9}\pi, \frac{8}{9}\pi$$

$$\therefore \theta = \frac{2}{9}\pi, \frac{4}{9}\pi, \frac{8}{9}\pi$$

### ALTERNATIVE METHOD

The solutions are shown in Figure 5.10 by plotting $y = \cos 3\theta$ in the interval of $0 < \theta < \pi$. Alternatively, the solutions can be obtained by plotting $y = \cos z$ in the interval of $0 < \theta < 3\pi$. In this case, each solution (Figure 5.11) should be divided by 3 to obtain the values of $\theta$.
In both cases, use the steps outlined in the graphical method to obtain the solutions from the graphs. We've shown three points in each case; these are observed to be approximately equal to the analytical solutions.

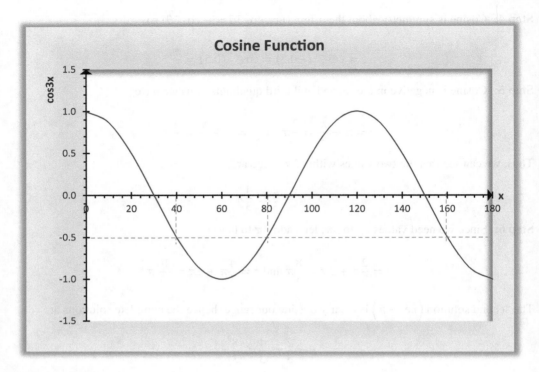

**FIGURE 5.10** Solution to Example 2(a) – Part I.

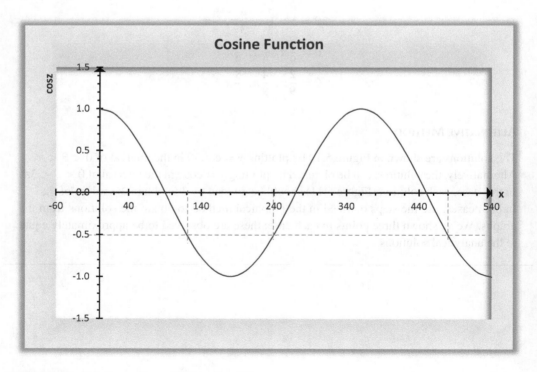

**FIGURE 5.11** Solution to Example 2(a) – Part II.

# Solving Trigonometric Equations

**b)** $\sqrt{3} \sin\left(\frac{2}{3}\theta\right) - 3\cos\left(\frac{2}{3}\theta\right) = 0$

**Solution**

Given that

$$\sqrt{3} \sin\left(\frac{2}{3}\theta\right) - 3\cos\left(\frac{2}{3}\theta\right) = 0$$

**Step 1:** Re-arranging this, we have

$$\sqrt{3} \sin\left(\frac{2}{3}\theta\right) = 3\cos\left(\frac{2}{3}\theta\right)$$

**Step 2:** Divide both sides by $\sqrt{3}$, we have

$$\frac{\sin\left(\frac{2}{3}\theta\right)}{\cos\left(\frac{2}{3}\theta\right)} = \frac{3}{\sqrt{3}}$$

We can simplify the RHS (or rationalise the denominator) to have

$$\frac{\sin\left(\frac{2}{3}\theta\right)}{\cos\left(\frac{2}{3}\theta\right)} = \sqrt{3}$$

**Step 3:** But $\tan\theta = \frac{\sin\theta}{\cos\theta}$, thus we have

$$\tan\left(\frac{2}{3}\theta\right) = \sqrt{3}$$

$$\frac{2}{3}\theta = \tan^{-1}\left(\sqrt{3}\right) = \frac{1}{3}\pi$$

**Step 4:** We can now say that if the original interval is $0 < \theta < \pi$, the new range is $0 < \frac{2}{3}\theta < \frac{2}{3}\pi$.

$$\frac{2}{3}\theta = \tan^{-1}\left(\sqrt{3}\right) = \frac{1}{3}\pi$$

**Step 5:** Tangent is positive in the first and third quadrants. We've already obtained the solution in the first quadrant (i.e., $\frac{1}{3}\pi$). As the upper limit of the range is less than $\pi$, there is no need to find the third quadrant value. We can, therefore, conclude that there is only one solution which is

$$\frac{2}{3}\theta = \frac{1}{3}\pi$$

**OR**

$$\theta = \frac{3}{2}\left(\frac{1}{3}\pi\right) = \frac{1}{2}\pi$$

$$\therefore \theta = \frac{1}{2}\pi$$

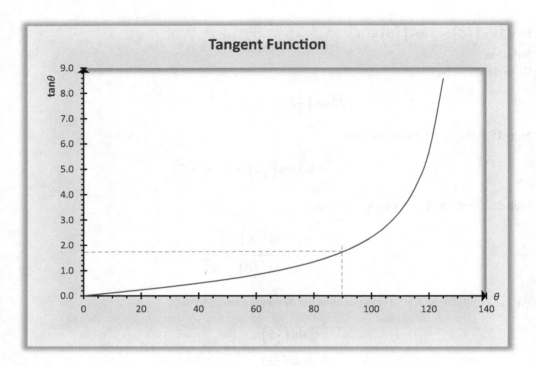

**FIGURE 5.12** Solution to Example 2(b).

---

**ALTERNATIVE METHOD**

The same solution is obtained by plotting (Figure 5.12).

---

**c)** $\tan 2\theta - 1 = \sqrt{3}$

**Solution**

The original interval is $0 < \theta < \pi$, and the new range for this is $0 < 2\theta < 2\pi$. Given that

$$\tan 2\theta - 1 = \sqrt{3}$$

**Step 1:** Re-arranging this, we have

$$\tan 2\theta = 1 + \sqrt{3}$$
$$2\theta = \tan^{-1}\left(1 + \sqrt{3}\right) = 1.2199$$

**Step 2:** Tangent is positive in the first and third quadrants. We've already obtained the solution in the first quadrant (i.e., 1.2199 rad). The solution in the third quadrant is

$$\pi + 1.2199 = 4.3615 \text{ rad}$$

Thus, we can see that the two values within $2\pi$ range are

$$2\theta = 1.2199, \ 4.3615$$

Solving Trigonometric Equations

**TABLE 5.2**
**Changing the Interval of Transformed Functions (Horizontal Translation) Illustrated**

|    | Equation | Original Range | New Range |
|----|----------|----------------|-----------|
| 1) | $y = \cos\left(\theta - \frac{\pi}{6}\right)$ | $0 < \theta < \pi$ | $-\frac{\pi}{6} < \left(\theta - \frac{\pi}{6}\right) < \frac{5}{6}\pi$ |
| 2) | $y = \sin(\theta + 35°)$ | $-180° \le \theta < 180°$ | $-145° \le (\theta + 35°) < 215°$ |
| 3) | $y = \tan(\theta - 0.5\pi)$ | $0 < \theta \le 4\pi$ | $-\frac{1}{2}\pi < (\theta - 0.5\pi) \le \frac{7\pi}{2}$ |

**Step 3:** These above solutions have already covered all possible solutions in the intended range, therefore

$$2\theta = 1.2199, \ 4.3615$$
$$\theta = 0.60995, \ 2.18075$$
$$\therefore \theta = \mathbf{0.610, \ 2.18}$$

### 5.3.2 Type 2: $y = \sin(x \pm b)$, $y = \cos(x \pm b)$, $y = \tan(x \pm b)$

This is a horizontal translation on the $x$-axis, where $b$ is a positive real number and represents the amount of translation left or right. Consequently, we have that:

a) if the given interval is $\theta_1 \le x \le \theta_2$, the new interval will be $\theta_1 \pm b \le (x \pm b) \le \theta_2 \pm b$, and

b) the number of solutions will, mostly, remain unchanged.

To illustrate, solving $y = \tan(x - 15°)$ in the range $-90° \le x \le 90°$ would imply solving the same in the range $-90° - 15° \le (x - 15°) \le 90° - 15°$, i.e., $-105° \le x - 15° \le 75°$. Table 5.2 shows more examples of horizontal translation on the $x$-axis.

---

Solving problems of this type is not any different from type 1 above. Let's try an example.

### Example 3

Solve the trigonometric equation $2\sin(\theta - 15°) - \sqrt{3} = 0$ in the range $0 < \theta < 360°$.

---

What did you get? Find the solution below to double-check your answer.

### Solution to Example 3

**Solution**
The original interval is $0 < \theta < 360°$, the new range for this is $-15° < (\theta - 15°) < 345°$.

$$2\sin(\theta - 15°) - \sqrt{3} = 0$$

**Step 1:** Re-arranging this equation, we have

$$\sin(\theta - 15°) = \frac{\sqrt{3}}{2}$$

Therefore

$$\theta - 15° = \sin^{-1}\left(\frac{\sqrt{3}}{2}\right) = 60°$$

**Step 2:** Sine is positive in the first and second quadrants. The second quadrant is obtained using

$$180° - 60° = 120°$$

Thus, we can see that the two values within $2\pi$ range are:

$$\theta - 15° = 60°,\ 120°$$

**Step 3:** Let's find $\theta$ by re-arranging the above as:

$$\theta = 60° + 15° = 75° \text{ and } \theta = 120° + 15° = 135°$$

We can say

$$\therefore \theta = 75°,\ 135°$$

### ALTERNATIVE METHOD

The solutions are shown by plotting $y = \sin(\theta - 15°)$ in the interval of $0 < \theta < 360°$. We can then draw a line (Figure 5.13) through $y = \frac{\sqrt{3}}{2}$.

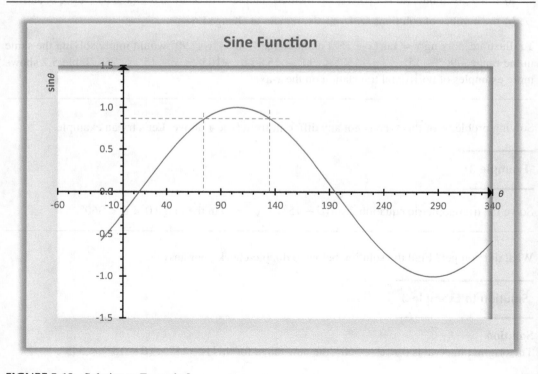

**FIGURE 5.13** Solution to Example 3.

# Solving Trigonometric Equations

## TABLE 5.3
Changing the Interval of Transformed Functions (Horizontal Translation and Stretch) Illustrated

|    | Equation | Original Range | New Range |
|----|----------|----------------|-----------|
| 1) | $y = \sin\left(\frac{3}{5}\theta - 20°\right)$ | $-180° \leq \theta \leq 360°$ | $-128° \leq \left(\frac{3}{5}\theta - 20°\right) \leq 196°$ |
| 2) | $y = \cos(2\theta + 60°)$ | $0° < \theta < 180°$ | $60° < (2\theta + 60°) < 420°$ |
| 3) | $y = \tan\left(3\theta + \frac{1}{4}\pi\right)$ | $\frac{1}{2}\pi \leq \theta < 2\pi$ | $\frac{7}{4}\pi \leq \left(3\theta + \frac{1}{4}\pi\right) < \frac{25}{4}\pi$ |

### 5.3.3 Type 3: $y = \sin(ax \pm b)$, $y = \cos(ax \pm b)$, $y = \tan(ax \pm b)$

This last type is a combination of type 1 and type 2 (i.e., a horizontal stretch and translation in a single trigonometric equation). We can say that if the given interval is $\theta_1 \leq x \leq \theta_2$, the new interval will be $a\theta_1 \pm b \leq (ax \pm b) \leq a\theta_2 \pm b$. Notice that, first we need to multiply the limits (lower and upper) by the coefficient of the angle and then add/subtract the phase angle. This order is very essential. To illustrate, solving $y = \cos(3x - 40°)$ in the range $0° < x < 180°$ would imply solving the same in the range $0° - 40° < 3x - 40° < (3 \times 180°) - 40°$, i.e., $-40° < 3x - 40° < 500°$.

Table 5.3 shows more examples of this type.

It is a good time to try a couple of examples.

### Example 4

Solve the following trigonometric equations in the range $-180° < x < 180°$.

a) $\tan\left(\frac{5}{4}x - 110°\right) = -5$

b) $\sqrt{2}\sin(3x + 40°) + 1 = 0$

What did you get? Find the solution below to double-check your answer.

### Solution to Example 4

a) $\tan\left(\frac{5}{4}x - 110°\right) = -5$

**Solution**

The original interval is $-180° < x < 180°$, the new range for this is $-335° < \left(\frac{5}{4}x - 110°\right) < 115°$.

Given that

$$\tan\left(\frac{5}{4}x - 110°\right) = -5$$

**Step 1:** Let $z = \frac{5}{4}x - 110°$, thus

$$\tan z = -5$$
$$z = \tan^{-1}(-5) = -78.7°$$

**Step 2:** Tangent is negative in the second and fourth quadrants and its value in the second quadrant is obtained using

$$180° - 78.7° = 101.3°$$

The fourth quadrant is not required since the upper limit of our new interval is 115°.

**Step 3:** Our $z$ values now are

$$z = -78.7°, \ 101.3°$$

**Step 4:** As we've reached the upper limit, we now need to subtract 180° as:

$$z = -78.7° - 180 = -258.7°$$

If we subtract 180° again, we will be out of the range. Hence, the values of $z$ in between $-335°$ and $115°$ are:

$$z = -258.7°, -78.7°, \ 101.3°$$

**Step 5:** Remember that $z = \frac{5}{4}x - 110°$, thus

$$\frac{5}{4}x - 110° = z = -258.7°, -78.7°, \ 101.3°$$

$x$ is obtained by adding 110° to each value and dividing each result by $\frac{5}{4}$ (or multiplying by $\frac{4}{5}$) as:

- **When $z = -258.7°$**

  $\frac{5}{4}x - 110° = -258.7°$

  $\frac{5}{4}x = -258.7°$
  $\phantom{\frac{5}{4}x =} +110°$

  $\frac{5}{4}x = -148.7°$

  $x = \frac{4}{5}(-148.7°)$

  Hence $x = -118.96°$

- **When $z = -78.7°$**

  $\frac{5}{4}x - 110° = -78.7°$

  $\frac{5}{4}x = -78.7°$
  $\phantom{\frac{5}{4}x =} +110°$

  $\frac{5}{4}x = 31.3°$

  $x = \frac{4}{5}(31.3°)$

  Hence $x = 25.04°$

- **When $z = 101.3°$**

  $\frac{5}{4}x - 110° = 101.3°$

  $\frac{5}{4}x = 101.3°$
  $\phantom{\frac{5}{4}x =} +110°$

  $\frac{5}{4}x = 211.3°$

  $x = \frac{4}{5}(211.3°)$

  Hence $x = 169.04°$

$$\therefore x = -119°, 25°, \ 169°$$

### ALTERNATIVE METHOD

The solutions are shown by plotting $\tan\left(\frac{5}{4}x - 110°\right) = -5$ in the interval of $-180° < x < 180°$. We then draw a line (Figure 5.14) through $y = -5$.

# Solving Trigonometric Equations

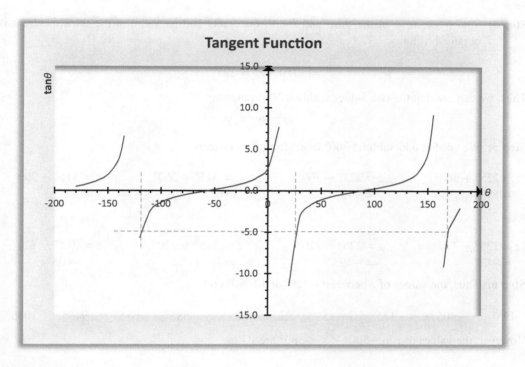

**FIGURE 5.14** Solution to Example 4(a).

**b)** $\sqrt{2}\sin(3x + 40°) + 1 = 0$
**Solution**
The original interval is $-180° < x < 180°$, the new range for this is $-500° < x < 580°$. Given that

$$\sqrt{2}\sin(3x + 40°) + 1 = 0$$

**Step 1:** Re-arranging this, we have

$$\sqrt{2}\sin(3x + 40°) = -1$$
$$\sin(3x + 40°) = -\frac{1}{\sqrt{2}}$$
$$\sin(3x + 40°) = -\frac{\sqrt{2}}{2}$$

**Step 2:** We can let $z = 3x + 40°$, thus

$$\sin z = -\frac{\sqrt{2}}{2}$$
$$z = \sin^{-1}\left(-\frac{\sqrt{2}}{2}\right) = -45°$$

**Step 3:** We have that

$$\sin(-45°) = \sin(360° - 45°) = \sin 315°$$

**Step 4:** Sine is negative in the third and fourth quadrants and its value in the third quadrant is obtained using

$$180° + 45° = 225°$$

Thus, we can see that the two values within a 360° range are

$$z = 225°, \ 315°$$

**Step 5:** We need to add/subtract 360° twice from the above as:

$$z = 225° + 360° \qquad z = 225° - 360° \qquad z = 315° + 360° \qquad z = 315° - 360°$$
$$= 585° \qquad\qquad = -135° \qquad\qquad = 675° \qquad\qquad = -45°$$

Also

$$z = 225° + 720° \qquad z = 225° - 720° \qquad z = 315° + 720° \qquad z = 315° - 720°$$
$$= 945° \qquad\qquad = -495° \qquad\qquad = 1035° \qquad\qquad = -405°$$

**Step 6:** Thus, the values of $z$ between $-720°$ and $1080°$ are:

$$-495° \quad -405° \quad -135° \quad -45° \quad 225° \quad 315° \quad 585° \quad 675° \quad 945° \quad 1035°$$

However, the values of $z$ in $-500° < x < 580°$ are six as:

$$-495° \quad\quad -405° \quad\quad -135° \quad\quad -45° \quad\quad 225° \quad\quad 315°$$

**Step 7:** Remember that $z = 3x + 40°$, thus

$$3x + 40° = z = -495°, -405°, -135°, -45°, 225°, 315°$$

$x$ is obtained by subtracting 40° from each value and then dividing each result by 3 as:

- When $z = -495°$

$$3x + 40° = -495°$$
$$3x = -495° - 40°$$
$$3x = -535°$$
$$x = \frac{1}{3}(-535°)$$

Hence

$$x = -178.33°$$

- When $z = -405°$

$$3x + 40° = -405°$$
$$3x = -405° - 40°$$
$$3x = -445°$$
$$x = \frac{1}{3}(-445°)$$

Hence

$$x = -148.33°$$

Solving Trigonometric Equations

- **When $z = -135°$**

$$3x + 40° = -135°$$
$$3x = -135° - 40°$$
$$3x = -175°$$
$$x = \frac{1}{3}(-175°)$$

Hence

$$x = -58.33°$$

- **When $z = -45°$**

$$3x + 40° = -45°$$
$$3x = -45° - 40°$$
$$3x = -85°$$
$$x = \frac{1}{3}(-85°)$$

Hence

$$x = -28.33°$$

- **When $z = 225°$**

$$3x + 40° = 225°$$
$$3x = 225° - 40°$$
$$3x = 185°$$
$$x = \frac{1}{3}(185°)$$

Hence

$$x = 61.67°$$

- **When $z = 315°$**

$$3x + 40° = 315°$$
$$3x = 315° - 40°$$
$$3x = 275°$$
$$x = \frac{1}{3}(275°)$$

Hence

$$x = 91.67°$$

$$\therefore x = -178.3°, -148.3°, -58.3°, -28.3°, 61.7°, 91.7°$$

### ALTERNATIVE METHOD

The solutions are shown by plotting $\sqrt{2}\sin(3x + 40°) + 1 = 0$ in the interval of $-180° < x < 180°$. We then draw a line through $y = -\frac{\sqrt{2}}{2}$.

The solutions (Figure 5.15) are shown by plotting $\sqrt{2\sin(z + 40°) + 1} = 0$ in the interval of $-500° < x < 580°$. We then draw a line through $y = -\frac{\sqrt{2}}{2}$.

In this case, $x$ is obtained by subtracting 40° from each value and then dividing each result by 3 (Figure 5.16).

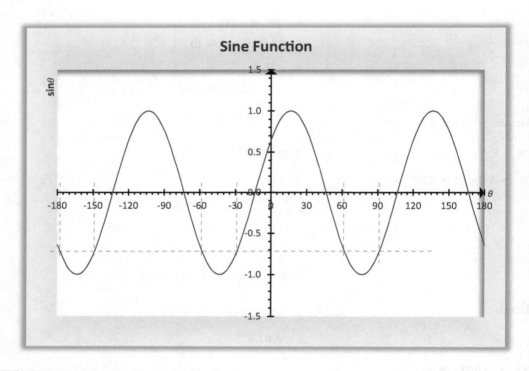

**FIGURE 5.15** Solution to Example 4(b) – Part I.

Solving Trigonometric Equations

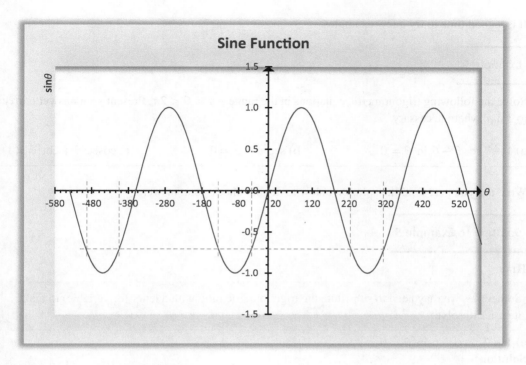

**FIGURE 5.16** Solution to Example 4(b) – Part II.

## 5.4 TRIGONOMETRIC EQUATIONS INVOLVING IDENTITIES

Our focus here is to highlight the relationship between trigonometric identities (discussed in Chapters 3 and 4) and equations. When a trigonometric equation involves secant, cosecant, and cotangent, it is sometimes easier to convert them at the beginning of our workings to cosine, sine, and tangent, respectively. Alternatively, this conversion can be delayed until the point at which we need to find the inverse function.

For clarity and ease of reference, we have grouped trigonometric equations under six categories.

### 5.4.1 THE FUNDAMENTALS

In this first group, we will show how the fundamental trigonometric identities can be employed in solving trigonometric equations. To be clear, by 'fundamental' we mean the following listed identities, which were introduced in Chapter 3.

$$\boxed{\tan\theta = \frac{\sin\theta}{\cos\theta}} \quad \boxed{\sin^2\theta + \cos^2\theta = 1} \quad \boxed{\tan^2\theta + 1 = \sec^2\theta} \quad \boxed{\cot^2\theta + 1 = \operatorname{cosec}^2\theta} \quad (5.1)$$

Note that the third and fourth identities follow from the second identity (by dividing by $\cos^2\theta$ to obtain the third and by $\sin^2\theta$ to obtain the fourth identity).

It is a good time to try a few examples.

### Example 5

Solve the following trigonometric equations in the range $-\pi < \theta < 2\pi$. Present your answer correct to 3 d.p. where necessary.

a) $1 - 7\cos\theta - 6\sin^2\theta = 0$     b) $\sec^2\theta - 5 = 0$     c) $\text{cosec}^2\theta + \cot^2\theta = 1$

What did you get? Find the solution below to double-check your answer.

### Solution to Example 5

**HINT**

Sometimes we may need to substitute the trigonometric ratio with a letter (or variable) to make it easier to solve.

a) $1 - 7\cos\theta - 6\sin^2\theta = 0$
**Solution**
We have that

$$1 - 7\cos\theta - 6\sin^2\theta = 0$$

**Step 1:** To solve this, we need to have the equation entirely in sine or cosine. For this case, it is a good option to express it in cosine. We can achieve this by using the identity:

$$\sin^2\theta = 1 - \cos^2\theta$$

**Step 2:** Let's re-write this as:

$$1 - 7\cos\theta - 6\left(1 - \cos^2\theta\right) = 0$$
$$1 - 7\cos\theta - 6 + 6\cos^2\theta = 0$$
$$6\cos^2\theta - 7\cos\theta - 5 = 0$$

**Step 3:** The above looks like a quadratic equation, it is in fact a quadratic (trigonometric) equation. We can either proceed to solve it in its current form or employ substitution to make it easier. We will go with the second option. Hence, let

$$x = \cos\theta$$

Thus, upon substitution we have

$$6x^2 - 7x - 5 = 0$$

**Step 4:** We are free to use any of the methods for solving a quadratic equation (see *Foundation Mathematics for Engineers and Scientists with Worked Examples* by the same author), we will go for factorisation here.

# Solving Trigonometric Equations

$$6x^2 - 7x - 5 = 0$$
$$(3x - 5)(2x + 1) = 0$$

Thus, we have:

$$3x - 5 = 0 \quad \textbf{OR} \quad 2x + 1 = 0$$
$$x = \frac{5}{3} \qquad\qquad\qquad x = -\frac{1}{2}$$

**Step 5:** Let's substitute back now. In other words,

$$\cos\theta = \frac{5}{3} > 1 \quad \textbf{OR} \quad \cos\theta = -\frac{1}{2}$$

$\cos\theta = \frac{5}{3}$ is not a valid solution, as the maximum value of $\cos\theta$ is 1. Hence, the only solution is

$$\cos\theta = -\frac{1}{2}$$

which implies that

$$\theta = \cos^{-1}\left(-\frac{1}{2}\right) = \frac{2}{3}\pi$$

**Step 6:** Cosine is negative in the second and third quadrants, so we have

$$\theta = \pi + \frac{1}{3}\pi = \frac{4}{3}\pi$$

Thus, we can see that the two values within $2\pi$ range are

$$\frac{2}{3}\pi, \ \frac{4}{3}\pi$$

**Step 7:** Cosine is symmetrical about the $y$-axis, so we can have

$$-\frac{2}{3}\pi, \ -\frac{4}{3}\pi$$

But $-\frac{4}{3}\pi$ is outside the stated range. Therefore, the solutions of the equation in the range are

$$\theta = -\frac{2}{3}\pi, \ \frac{2}{3}\pi, \frac{4}{3}\pi$$

---

**b)** $\sec^2\theta - 5 = 0$
**Solution**
Given that

$$\sec^2\theta - 5 = 0$$

**Step 1:** Let's use the identity $\tan^2\theta + 1 = \sec^2\theta$, thus

$$\left(\tan^2\theta + 1\right) - 5 = 0$$
$$\tan^2\theta - 4 = 0$$
$$(\tan\theta + 2)(\tan\theta - 2) = 0$$

**Step 2:** Let $y = \tan\theta$, we have

$$y + 2 = 0 \quad \text{OR} \quad y - 2 = 0$$
$$y = -2 \qquad\qquad\qquad y = 2$$

In other words,

$$\tan\theta = -2 \qquad\qquad \tan\theta = 2$$
$$\theta = \tan^{-1}(-2) \quad \text{OR} \quad \theta = \tan^{-1}(2)$$
$$= -1.107 \qquad\qquad = 1.107$$

**Step 3:** Add/subtract multiples of $\pi$, we have

$$-1.107 + \pi = 2.035 \quad -1.107 - \pi = -4.249 \quad 1.107 + \pi = 4.249 \quad 1.107 - \pi = -2.035$$

**Step 4:** Therefore, the solutions of the equation in the range $-\pi < \theta < 2\pi$ are:

$$\theta = -2.035, \ -1.107, \ 1.107, \ 2.035, \ 4.249, \ 5.176$$

---

### ALTERNATIVE METHOD

**Step 1:** Using difference of two squares, factorise

$$\sec^2\theta - 5 = 0$$
$$\left(\sec\theta - \sqrt{5}\right)\left(\sec\theta + \sqrt{5}\right) = 0$$

Thus, we have:

$$\sec\theta - \sqrt{5} = 0 \quad \text{OR} \quad \sec\theta + \sqrt{5} = 0$$
$$\sec\theta = \sqrt{5} \qquad\qquad\qquad \sec\theta = -\sqrt{5}$$

**Step 2:** When $\sec\theta = \sqrt{5}$, use $\sec\theta = \frac{1}{\cos\theta}$. Thus

$$\frac{1}{\cos\theta} = \sqrt{5}$$
$$\cos\theta = \frac{1}{\sqrt{5}}$$
$$\theta = \cos^{-1}\left(\frac{1}{\sqrt{5}}\right) = 1.107$$

**Step 3:** Cosine is positive in the first and fourth quadrants, so its value in the fourth quadrant is

$$2\pi - 1.107 = 5.176$$

**Step 4:** Using the symmetry of cosine about the $y$-axis, we have the last solution as $-1.107$.

Solving Trigonometric Equations

**Step 5:** When $\sec\theta = -\sqrt{5}$, use $\sec\theta = \frac{1}{\cos\theta}$. Thus

$$\frac{1}{\cos\theta} = -\sqrt{5}$$

$$\cos\theta = -\frac{1}{\sqrt{5}}$$

$$\theta = \cos^{-1}\left(\frac{1}{\sqrt{5}}\right) = 1.107$$

$$\theta = \cos^{-1}\left(-\frac{1}{\sqrt{5}}\right) = 2.034$$

**Step 6:** Cosine is negative in the second and third quadrants, so its value in the third quadrant is

$$2\pi - 2.034 = \mathbf{4.249}$$

**Step 7:** Using the symmetry of cosine about the $y$-axis, we have the last solution as

$$-2.034$$

**Step 8:** Therefore, the solutions of the equation in the range $-\pi < \theta < 2\pi$ are:

$$\therefore \theta = -2.034, -1.107, 1.107, 2.034, 4.249, 5.176$$

---

c) $\operatorname{cosec}^2\theta + \cot^2\theta = 1$
**Solution**
We have

$$\operatorname{cosec}^2\theta + \cot^2\theta = 1$$

**Step 1:** Let's use the identity $\cot^2\theta + 1 = \operatorname{cosec}^2\theta$, thus

$$(\cot^2\theta + 1) + \cot^2\theta = 1$$
$$\cot^2\theta + \cot^2\theta = 0$$
$$2\cot^2\theta = 0$$
$$\cot\theta = 0$$

But

$$\cot\theta = \frac{1}{\tan\theta}$$

Thus, we have

$$\frac{1}{\tan \theta} = 0$$

$$\tan \theta = \frac{1}{0} = \infty$$

$$\theta = \tan^{-1}(\infty) = \frac{\pi}{2}$$

**Step 2:** Tangent is positive in the first and third quadrants, so its value in the third quadrant is

$$\pi + \frac{\pi}{2} = \frac{3\pi}{2}$$

**Step 3:** The above are solutions in the range 0 to $2\pi$. To get the value in the range $-\pi$ to 0 subtract $\pi$ from the first quadrant value, we have

$$\frac{\pi}{2} - \pi = -\frac{\pi}{2}$$

**Step 4:** Therefore, the solutions of the equation in the range $-\pi < \theta < 2\pi$ are

$$\therefore \theta = -\frac{\pi}{2}, \frac{\pi}{2}, \frac{3\pi}{2}$$

### 5.4.2 Addition Formulas

In this case, we will solve equations which require the use of addition formulas. These have been covered previously, but reproduced in Table 5.4 for quick access and reference.

It is a good time to try an example to illustrate this.

### Example 6

Solve $\sin\left(x + \frac{\pi}{6}\right) = 2 \cos x$ in the interval $-\pi < x < \pi$. Present your answer in exact form.

**TABLE 5.4**
**Addition Formulas**

|  | Addition | Subtraction |
|---|---|---|
| Sine | $\sin(A+B) = \sin A \cos B + \sin B \cos A$ | $\sin(A-B) = \sin A \cos B - \sin B \cos A$ |
| Cosine | $\cos(A+B) = \cos A \cos B - \sin A \sin B$ | $\cos(A-B) = \cos A \cos B + \sin A \sin B$ |
| Tangent | $\tan(A+B) = \dfrac{\tan A + \tan B}{1 - \tan A \tan B}$ | $\tan(A-B) = \dfrac{\tan A - \tan B}{1 + \tan A \tan B}$ |

(5.2)

# Solving Trigonometric Equations

What did you get? Find the solution below to double-check your answer.

## Solution to Example 6

**Solution**
We have that

$$\sin\left(x + \frac{\pi}{6}\right) = 2\cos x$$

**Step 1:** Let's apply the addition formula to the LHS as follows:

$$\sin\left(x + \frac{\pi}{6}\right) = \sin x \cos\frac{\pi}{6} + \sin\frac{\pi}{6}\cos x$$

$$= \left(\frac{\sqrt{3}}{2}\right)\sin x + \left(\frac{1}{2}\right)\cos x$$

**Step 2:** Let's substitute the result in the original equation as:

$$\left(\frac{\sqrt{3}}{2}\right)\sin x + \left(\frac{1}{2}\right)\cos x = 2\cos x$$

$$\sqrt{3}\sin x + \cos x = 4\cos x$$

$$\sqrt{3}\sin x = 3\cos x$$

$$\frac{\sin x}{\cos x} = \frac{3}{\sqrt{3}}$$

$$\tan x = \sqrt{3}$$

$$x = \tan^{-1}\left(\sqrt{3}\right) = \frac{\pi}{3}$$

**Step 3:** This solution is in the first quadrant where tangent is positive. Tangent is also positive in the third quadrant, but this is outside the stated range. To get the value in the negative part, subtract $\pi$ as:

$$\frac{\pi}{3} - \pi = -\frac{2}{3}\pi$$

**Step 4:** The solutions of the equation in the stated range are therefore:

$$x = -\frac{2}{3}\pi, \quad x = \frac{1}{3}\pi$$

---

### 5.4.3 Multiple Angle

In this section, we'll explore examples where the formulas for multiple-angle identities play a crucial role in solving trigonometric equations. They are reproduced in Tables 5.5–5.7 for quick reference.

**TABLE 5.5**
**Double-Angle Formulas**

| Function | Identities | | |
|---|---|---|---|
| Sine | $\sin 2A = 2\sin A \cos A$ | | |
| Cosine | $\cos 2A = \cos^2 A - \sin^2 A$ | $\cos 2A = 2\cos^2 A - 1$ | $\cos 2A = 1 - 2\sin^2 A$ |
| Tangent | $\tan 2A = \dfrac{2\tan A}{1 - \tan^2 A}$ | | |

**TABLE 5.6**
**Triple-Angle Formulas**

| Function | Identities |
|---|---|
| Sine | $\sin 3A = 3\sin A - 4\sin^3 A$ |
| Cosine | $\cos 3A = 4\cos^3 A - 3\cos A$ |
| Tangent | $\tan 3A = \dfrac{2\tan A - \tan^3 A}{1 - 3\tan^2 A}$ |

**TABLE 5.7**
**Half-Angle Formulas**

| Function | Identities | | |
|---|---|---|---|
| Sine | $\sin A = 2\sin\dfrac{1}{2}A \cos\dfrac{1}{2}A$ | | |
| Cosine | $\cos A = \cos^2\dfrac{1}{2}A - \sin^2\dfrac{1}{2}A$ | $\cos A = 2\cos^2\dfrac{1}{2}A - 1$ | $\cos A = 1 - 2\sin^2\dfrac{1}{2}A$ |
| Tangent | $\tan A = \dfrac{2\tan\dfrac{1}{2}A}{1 - \tan^2\dfrac{1}{2}A}$ | | |

(5.3)

# Solving Trigonometric Equations

It's a good time to attempt a few examples.

## Example 7

Solve the following trigonometric equations in the range $-\pi < \theta < 2\pi$. Present your answer correct to 3 d.p. where necessary.

a) $3 \csc 2\theta - 5 = 0$

b) $\sin\left(\frac{1}{2}\theta\right) + \cos\theta = 0$

c) $\cot\theta = 3\sin 2\theta$

d) $\frac{1-\cos 2\theta}{\sin^2\theta - \cos^2\theta} = 2$

e) $\tan 3\theta = \tan\theta$

f) $2\cos 2\theta - \sin\theta + 1 = 0$

What did you get? Find the solution below to double-check your answer.

## Solution to Example 7

### HINT

Sometimes we may need to substitute the trigonometric ratio with a letter (or variable) to make it easier to solve.

**a)** $3 \csc 2\theta - 5 = 0$

**Solution**

Given that $3 \csc 2\theta - 5 = 0$

**Step 1:** Let's simplify

$$3 \csc 2\theta - 5 = 0$$
$$3 \csc 2\theta = 5$$
$$\csc 2\theta = \frac{5}{3}$$
$$\frac{1}{\sin 2\theta} = \frac{5}{3}$$
$$\sin 2\theta = \frac{3}{5}$$

**Step 2:** Interval

The original interval is $-\pi < \theta < 2\pi$, and the new range for this is $-2\pi < 2\theta < 4\pi$. Therefore

$$2\theta = \sin^{-1}\left(\frac{3}{5}\right) = 0.644$$

**Step 3:** Sine is positive in the first and second quadrants, so its value in the second quadrant is

$$\pi - 0.644 = 2.498$$

Thus, we can see that the two values between 0 and $2\pi$ are

$$2\theta = 0.644, \ 2.498$$

**Step 4:** Since we need values up to $4\pi$, let's add $2\pi$ to have

$$2\theta = 0.644 + 2\pi \quad \text{AND} \quad 2\theta = 2.498 + 2\pi$$
$$= 6.927 \qquad\qquad\qquad = 8.781$$

**Step 5:** Subtract $2\pi$ to obtain a solution in the range $-2\pi$ to $0$

$$2\theta = 0.644 - 2\pi \quad \text{AND} \quad 2\theta = 2.498 - 2\pi$$
$$= -5.639 \qquad\qquad\qquad = -3.785$$

The complete solutions are:

$$2\theta = -5.639, -3.785, 0.644, 2.498, 6.927, 8.781$$

**Step 6:** Divide each answer by 2 to have the values of $\theta$ in the stated range as:

$$\theta = -\frac{5.639}{2}, -\frac{3.785}{2}, \frac{0.644}{2}, \frac{2.498}{2}, \frac{6.927}{2}, \frac{8.781}{2}$$

**Step 7:** The solutions of the equation in the range $-\pi < \theta < 2\pi$ are

$$\therefore \theta = -2.820, -1.893, 0.322, 1.249, 3.464, 4.391$$

---

**b)** $\sin\left(\frac{1}{2}\theta\right) + \cos\theta = 0$

**Solution**

Given that

$$\sin\left(\frac{1}{2}\theta\right) + \cos\theta = 0$$

**Step 1:** Let's simplify using the identity $\cos\theta = 1 - 2\sin^2\left(\frac{1}{2}\theta\right)$ as

$$\sin\left(\frac{1}{2}\theta\right) + \cos\theta = 0$$
$$\sin\left(\frac{1}{2}\theta\right) + 1 - 2\sin^2\left(\frac{1}{2}\theta\right) = 0$$
$$2\sin^2\left(\frac{1}{2}\theta\right) - \sin\left(\frac{1}{2}\theta\right) - 1 = 0$$

**Step 2:** Let $\sin\left(\frac{1}{2}\theta\right) = y$, we have

$$2y^2 - y - 1 = 0$$
$$(2y + 1)(y - 1) = 0$$

**Step 3:** Solve the quadratic equation

$$2y + 1 = 0 \qquad\qquad y - 1 = 0$$
$$y = -\frac{1}{2} \quad \text{OR} \quad y = 1$$
$$\therefore \sin\left(\frac{1}{2}\theta\right) = -\frac{1}{2} \qquad \therefore \sin\left(\frac{1}{2}\theta\right) = 1$$

Solving Trigonometric Equations

**Step 4:** Remember that we need to reduce the range by half. Our working range is thus $-\frac{1}{2}\pi < \theta < \pi$.

**Step 5:** When $\sin\left(\frac{1}{2}\theta\right) = -\frac{1}{2}$

$$\frac{\theta}{2} = \sin^{-1}\left(-\frac{1}{2}\right) = -\frac{\pi}{6}$$

$-\frac{\pi}{6} \pm 2\pi$ will take us outside the required range, so the above is the only valid answer, thus

$$\theta = -\frac{\pi}{3}$$

**Step 6:** When $\sin\left(\frac{1}{2}\theta\right) = 1$

$$\frac{\theta}{2} = \sin^{-1}(1) = \frac{\pi}{2}$$

The next maximum value for sine is at $\theta > \frac{3}{2}\pi$, so the above is the only valid answer, thus

$$\theta = \pi$$

**Step 7:** Therefore, the solutions of the equation in the range $-\pi < \theta < 2\pi$ are

$$\therefore \theta = -\frac{\pi}{3}, \pi$$

A graphical solution is shown in Figure 5.17.

---

c) $\cot\theta = 3\sin 2\theta$
**Solution**
Given that

$$\cot\theta = 3\sin 2\theta$$

**Step 1:** Simplify using relevant identities

$$\frac{\cos\theta}{\sin\theta} = 3(2\sin\theta\cos\theta)$$

$$\frac{\cos\theta}{\sin\theta} = 6\sin\theta\cos\theta$$

$$\cos\theta = 6\sin^2\theta\cos\theta$$

$$6\sin^2\theta\cos\theta - \cos\theta = 0$$

$$\cos\theta(6\sin^2\theta - 1) = 0$$

**NOTE**
Do not divide both sides by $\cos\theta$, otherwise some solutions will be missed out.

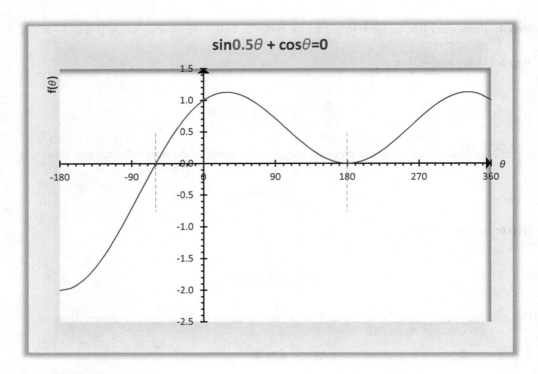

**FIGURE 5.17** Solution to Example 7(b).

**Step 2:** Therefore, we have

$$\cos\theta = 0$$
$$\theta = \cos^{-1}(0) = \frac{\pi}{2}$$

**OR**

$$6\sin^2\theta - 1 = 0$$
$$6\sin^2\theta = 1$$
$$\sin^2\theta = \frac{1}{6}$$
$$\sin\theta = \pm\sqrt{\frac{1}{6}} = \pm\frac{\sqrt{6}}{6}$$

**Step 3:** When $\cos\theta = 0$

$$\theta = \cos^{-1}(0) = \frac{\pi}{2}$$

The other minimum value of cosine within the 0 to $2\pi$ interval is $\frac{3\pi}{2}$. Using the symmetry of cosine about the y-axis, we have $-\frac{\pi}{2}$ and $-\frac{3\pi}{2}$. Unfortunately, the last value is outside the required range.

**Step 4:** When $\sin\theta = \frac{\sqrt{6}}{6}$

$$\theta = \sin^{-1}\frac{\sqrt{6}}{6} = \mathbf{0.421}$$

The other value in the second quadrant is

$$\theta = \pi - 0.421 = \mathbf{2.721}$$

Solving Trigonometric Equations

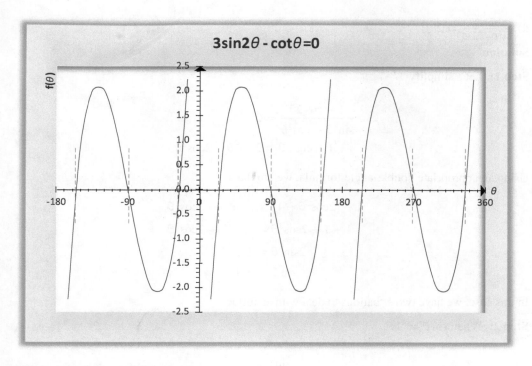

**FIGURE 5.18** Solution to Example 7(c).

Unfortunately using the periodicity of $-2\pi$ will take us out of the required range.

**Step 5:** When $\sin \theta = -\frac{\sqrt{6}}{6}$

$$\sin \theta = -\frac{\sqrt{6}}{6}$$

$$\theta = \sin^{-1}\left(-\frac{\sqrt{6}}{6}\right) = -\mathbf{0.421}$$

**Step 6:** The value of sine is negative in the third and fourth quadrants and are:

$$\theta = \pi + 0.421 \qquad \text{AND} \qquad \theta = 2\pi - 0.421$$
$$= \mathbf{3.563} \qquad\qquad\qquad = \mathbf{5.862}$$

**Step 7:** Using periodicity of $-2\pi$ will have

$$\theta = 3.561 - 2\pi \qquad \text{AND} \qquad \theta = 5.862 - 2\pi$$
$$= -\mathbf{2.722} \qquad\qquad\qquad = -\mathbf{0.421}$$

**Step 8:** The solutions of the equation in the range $-\pi < \theta < 2\pi$ are

$$\therefore \theta = -2.722, \ -\frac{\pi}{2}, \ -0.421, \ 0.421, \ \frac{\pi}{2}, \ 2.722, \ 3.563, \ \frac{3\pi}{2}, \ 5.862$$

d) $\frac{1-\cos 2\theta}{\sin^2\theta - \cos^2\theta} = 2$

**Solution**

**Step 1:** Let's simplify

$$\frac{1-\cos 2\theta}{\sin^2\theta - \cos^2\theta} = 2$$

$$1 - \cos 2\theta = 2\sin^2\theta - 2\cos^2\theta$$

Using an appropriate double-angle formula, we will have

$$1 - (2\cos^2\theta - 1) = 2\sin^2\theta - 2\cos^2\theta$$
$$1 + 1 - 2\cos^2\theta = 2\sin^2\theta - 2\cos^2\theta$$
$$2\sin^2\theta = 2$$
$$\sin\theta = \pm 1$$

In this case, we have two situations to deal with as follows.

**Step 2:** When $\sin\theta = 1$

$$\sin\theta = 1$$

which implies that

$$\theta = \sin^{-1}(1) = \frac{\pi}{2}$$

This is the only maximum value of sine within the 0 to $2\pi$ interval.

**Step 3:** When $\sin\theta = -1$

$$\sin\theta = -1$$

which implies that

$$\theta = \sin^{-1}(-1) = -\frac{\pi}{2}$$

The sine function is minimum only at $\frac{3\pi}{2}$ within the 0 to $2\pi$ interval.

**Step 4:** The solutions of the equation in the range $-\pi < \theta < 2\pi$ are

$$\theta = -\frac{\pi}{2}, \frac{\pi}{2}, \frac{3\pi}{2}$$

# Solving Trigonometric Equations

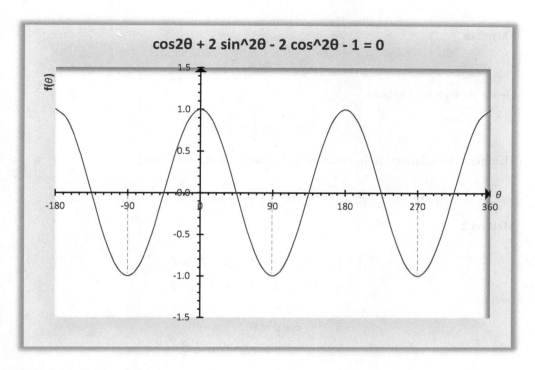

**FIGURE 5.19** Solution to Example 7(d).

## ATERNATIVE METHODS

**Method 1**

$$\frac{1 - \cos 2\theta}{\sin^2\theta - \cos^2\theta} = 2$$

$$\frac{1 - \cos 2\theta}{-(\cos^2\theta - \sin^2\theta)} = 2$$

$$\frac{1 - \cos 2\theta}{-(\cos 2\theta)} = 2$$

$$1 = \cos 2\theta - 2\cos 2\theta$$

$$\cos 2\theta = -1$$

$$2\theta = \cos^{-1}(-1) = \pi$$

This is the only minimum value of cosine within the 0 to $2\pi$ interval. For this case, we need to double the interval to $-2\pi < \theta < 4\pi$. Let's use $\pm 2\pi$ periodicity property to find other values in the new range as:

$$2\theta = \pi + 2\pi \qquad \text{AND} \qquad 2\theta = \pi - 2\pi$$
$$= 3\pi \qquad \qquad \qquad = -\pi$$

Therefore

$$2\theta = -\pi, \pi, 3\pi$$

Divide through by 2 to have

$$\theta = -\frac{1}{2}\pi, \frac{1}{2}\pi, \frac{3}{2}\pi$$

Therefore, the solutions of the equation in the range $-\pi < \theta < 2\pi$ are:

$$\theta = -\frac{\pi}{2}, \frac{\pi}{2}, \frac{3\pi}{2}$$

**Method 2**

$$1 - \left(1 - 2\sin^2\theta\right) = 2\sin^2\theta - 2\cos^2\theta$$
$$1 - 1 + 2\sin^2\theta = 2\sin^2\theta - 2\cos^2\theta$$
$$2\cos^2\theta = 0$$
$$\cos\theta = 0$$
$$\theta = \cos^{-1}(0) = \frac{\pi}{2}$$

The other minimum value of cosine within 0 to $2\pi$ interval is $\frac{3\pi}{2}$. Using the symmetry of cosine about the y-axis, we have $-\frac{\pi}{2}$ and $-\frac{3\pi}{2}$.

---

**e)** $\tan 3\theta = \tan \theta$
**Solution**
Given that

$$\tan 3\theta = \tan \theta$$

**Step 1:** Let's simplify using triple-angle formula

$$\tan 3\theta = \tan \theta$$
$$\frac{2\tan\theta - \tan^3\theta}{1 - 3\tan^2\theta} = \tan\theta$$
$$2\tan\theta - \tan^3\theta = \tan\theta\left(1 - 3\tan^2\theta\right)$$
$$2\tan\theta - \tan^3\theta = \tan\theta - 3\tan^3\theta$$
$$2\tan\theta - \tan^3\theta - \tan\theta + 3\tan^3\theta = 0$$
$$\tan\theta + 2\tan^3\theta = 0$$
$$\tan\theta\left(1 + 2\tan^2\theta\right) = 0$$

Solving Trigonometric Equations

**Step 2:** Solve the equation

$$\tan\theta = 0 \quad \text{OR} \quad 1 + 2\tan^2\theta = 0$$
$$2\tan^2\theta = -1$$
$$\tan\theta = \pm\sqrt{-\frac{1}{2}}$$

$\sqrt{-\frac{1}{2}}$ is not a real number, hence the only valid solution is

$$\tan\theta = 0$$

which implies that

$$\theta = \tan^{-1}(0) = 0$$

**Step 3:** Tangent is also positive in the third quadrant, so we have

$$\theta = \pi$$

**Step 4:** Let's use the $\pm\pi$ periodicity property as:

$$\theta = \pi + \pi = 2\pi \text{ AND } \theta = 0 - \pi = -\pi$$

**Step 5:** The solutions of the equation in the range $-\pi < \theta < 2\pi$ are:

$$\therefore \theta = -\pi,\ 0,\ \pi, 2\pi$$

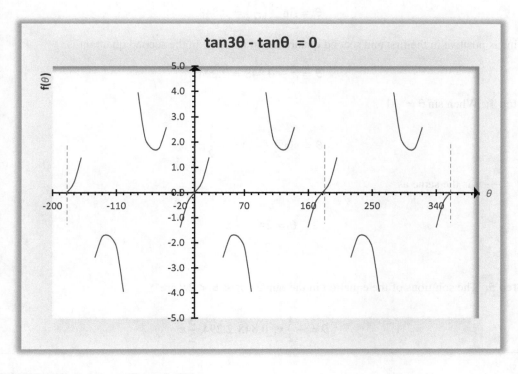

**FIGURE 5.20** Solution to Example 7(e).

**f)** $2\cos 2\theta - \sin\theta + 1 = 0$
**Solution**
Given that

$$2\cos 2\theta - \sin\theta + 1 = 0$$

**Step 1:** Let's simplify using double-angle formula, we have

$$2\left(1 - 2\sin^2\theta\right) - \sin\theta + 1 = 0$$
$$2 - 4\sin^2\theta - \sin\theta + 1 = 0$$
$$4\sin^2\theta + \sin\theta - 3 = 0$$
$$(4\sin\theta - 3)(\sin\theta + 1) = 0$$

**Step 2:** Solve the above quadratic equation

$$4\sin\theta - 3 = 0 \qquad \text{OR} \qquad \sin\theta + 1 = 0$$
$$\sin\theta = \tfrac{3}{4} \qquad\qquad\qquad \sin\theta = -1$$
$$\theta = \sin^{-1}\left(\tfrac{3}{4}\right) \qquad\qquad \theta = \sin^{-1}(-1)$$
$$= 0.848 \qquad\qquad\qquad = -\tfrac{1}{2}\pi$$

**Step 3:** When $\sin\theta = \tfrac{3}{4}$

$$\theta = \sin^{-1}\left(\tfrac{3}{4}\right) = 0.848$$

Sine is positive in the first and second quadrants, so its value in the second quadrant is

$$\theta = \pi - 0.848 = 2.294$$

**Step 4:** When $\sin\theta = -1$

$$\theta = \sin^{-1}(-1) = -\tfrac{1}{2}\pi$$

$\theta = -\tfrac{1}{2}\pi$ is the same as

$$\theta = 2\pi - \tfrac{1}{2}\pi$$
$$= \tfrac{3}{2}\pi$$

**Step 5:** The solutions of the equation in the range $-\pi < \theta < 2\pi$ are

$$\therefore \theta = -\tfrac{1}{2}\pi,\ 0.848,\ 2.294,\ \tfrac{3}{2}\pi$$

# Solving Trigonometric Equations

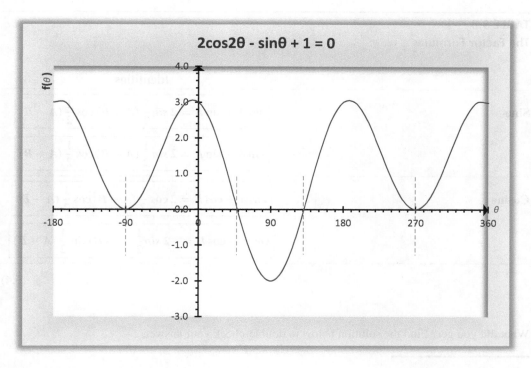

**FIGURE 5.21** Solution to Example 7(f).

## 5.4.4 The Factor Formulas

As with the previous categories, we will use the factor formulas here, and they are reproduced in Table 5.8.

It is a good time to try a couple of examples.

### Example 8

Solve the following trigonometric equations in the indicated intervals and sketch the graph to confirm your answer. Present your answer correct to 3 s.f. where necessary.

a) $\cos 4x + \sin x - \cos 2x = 0$ ; $0 < x < \pi$
b) $\cos(30° + x)\cos(30° - x) = \frac{1}{\sqrt{2}}$ ; $-90° < x < 180°$

**TABLE 5.8**
**The Factor Formulas**

|  | Identities |
|---|---|
| Sine | $\sin A + \sin B = 2 \sin \frac{1}{2}(A+B) \cos \frac{1}{2}(A-B)$ |
|  | $\sin A - \sin B = 2 \sin \frac{1}{2}(A-B) \cos \frac{1}{2}(A+B)$ |
| Cosine | $\cos A + \cos B = 2 \cos \frac{1}{2}(A+B) \cos \frac{1}{2}(A-B)$ |
|  | $\cos A - \cos B = -2 \sin \frac{1}{2}(A+B) \sin \frac{1}{2}(A-B)$ |

(5.4)

What did you get? Find the solution below to double-check your answer.

**Solution to Example 8**

a) $\cos 4x + \sin x - \cos 2x = 0; \ 0 < x < \pi$
**Solution**
Given that
$$\cos 4x + \sin x - \cos 2x = 0$$

**Step 1:** Let's simplify

$$\cos 4x + \sin x - \cos 2x = 0$$
$$\cos 4x - \cos 2x + \sin x = 0$$
$$[\cos 4x - \cos 2x] + \sin x = 0$$

**Step 2:** Applying the factor formula to the difference of cosine, we have

$$\left[-2 \sin \frac{1}{2}(4x+2x) \sin \frac{1}{2}(4x-2x)\right] + \sin x = 0$$
$$-2 \sin 3x \sin x + \sin x = 0$$
$$\sin x - 2 \sin 3x \sin x = 0$$
$$\sin x (1 - 2 \sin 3x) = 0$$

Solving Trigonometric Equations

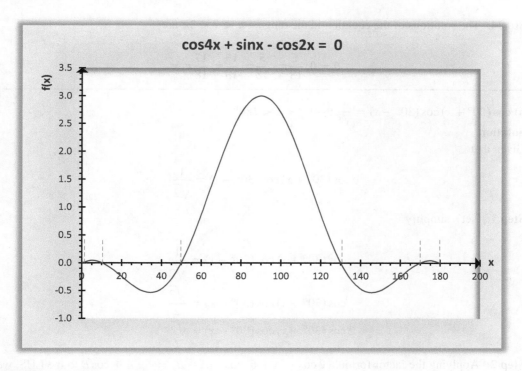

**FIGURE 5.22** Solution to Example 8(a).

**Step 3:** Solve the equation

$$\sin x = 0$$
$$x = \sin^{-1} 0 = \mathbf{0}, \boldsymbol{\pi}$$

**OR**

$$1 - 2\sin 3x = 0$$
$$\sin 3x = \frac{1}{2}$$
$$3x = \sin^{-1}(0.5) = \frac{1}{6}\pi, \frac{5}{6}\pi$$

For this case, the new interval is $0 < 3x < 3\pi$. Hence, we will add $2\pi$ to each of the above two solutions as:

$$2\pi + \frac{1}{6}\pi = \frac{13}{6}\pi \quad \text{AND} \quad 2\pi + \frac{5}{6}\pi = \frac{17}{6}\pi$$

The full solution for the second part is

$$3x = \sin^{-1}(0.5) = \frac{1}{6}\pi, \frac{5}{6}\pi, \frac{13}{6}\pi, \frac{17}{6}\pi$$

Divide through by 3:

$$x = \frac{1}{18}\pi, \frac{5}{18}\pi, \frac{13}{18}\pi, \frac{17}{18}\pi$$

**Step 4:** The solutions of the equation in the range $-\pi < \theta < 2\pi$ are

$$\therefore \theta = 0, \frac{1}{18}\pi, \frac{5}{18}\pi, \frac{13}{18}\pi, \frac{17}{18}\pi, \pi$$

---

**b)** $\cos(30° + x)\cos(30° - x) = \frac{1}{\sqrt{2}}$ ; $-90° < x < 180°$

**Solution**

Given that

$$\cos(30° + x)\cos(30° - x) = \frac{1}{\sqrt{2}}$$

**Step 1:** Let's simplify

$$\cos(30° + x)\cos(30° - x) = \frac{1}{\sqrt{2}}$$

$$\cos(30° + x)\cos(30° - x) = \frac{\sqrt{2}}{2}$$

$$2\cos(30° + x)\cos(30° - x) = \sqrt{2}$$

**Step 2:** Applying the factor formula $2\cos\frac{1}{2}(A+B)\cos\frac{1}{2}(A-B) = \cos A + \cos B$ to the LHS, we have

$$2\cos(30° + x)\cos(30° - x) = 2\cos\frac{1}{2}(A+B)\cos\frac{1}{2}(A-B)$$

Or

$$\cos(30° + x)\cos(30° - x) = \cos\frac{1}{2}(A+B)\cos\frac{1}{2}(A-B)$$

This implies that

$$\frac{1}{2}(A+B) = 30° + x \qquad ---\text{(i)}$$

$$\frac{1}{2}(A-B) = 30° - x \qquad ---\text{(ii)}$$

Solving (i) and (ii) simultaneously, we have

$$A = 60° \qquad ---\text{(iii)}$$

$$B = 2x \qquad ---\text{(iv)}$$

**Step 3:** Now substitute the value of $A$ and $B$ in the RHS of the identity given in step 2.

$$2\cos(30° + x)\cos(30° - x) = \cos 60° + \cos 2x$$

Solving Trigonometric Equations

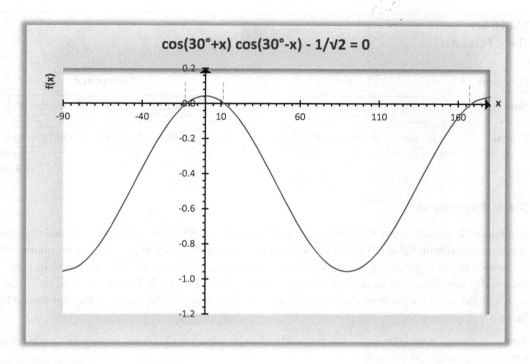

**FIGURE 5.23** Solution to Example 8(b).

**Step 4:** Let's now simplify

$$2\cos(30° + x)\cos(30° - x) = \sqrt{2}$$
$$\cos 60° + \cos 2x = \sqrt{2}$$
$$\cos 2x = \sqrt{2} - \cos 60° = \sqrt{2} - \frac{1}{2}$$
$$2x = \cos^{-1}\left(\frac{2\sqrt{2} - 1}{2}\right) = \mathbf{23.9°, \ 336.1°}$$

The new interval for this is $-180° < 2x < 360°$. As cosine is symmetry about the y-axis, another solution is $-23.9°$. Hence, we have

$$2x = -\mathbf{23.9°, \ 23.9°, \ 336.1°}$$

Divide each solution by 2 as:

$$x = \frac{-\mathbf{23.9°}}{\mathbf{2}}, \ \frac{\mathbf{23.9°}}{\mathbf{2}}, \ \frac{\mathbf{336.1°}}{\mathbf{2}}$$

**Step 5:** The solutions of the equation in the range $-90° < x < 180°$ are

$$\therefore \theta = \mathbf{-12.0°, \ 12.0°, \ 168°}$$

## TABLE 5.9
### The $R$ Formulas

| Function | Sum | Difference |
|---|---|---|
| Sine | $a \sin x + b \cos x = R \sin(x + \alpha)$ | $a \sin x - b \cos x = R \sin(x - \alpha)$ |
| Cosine | $a \sin x + b \cos x = R \cos(x - \alpha)$ | $a \sin x - b \cos x = R \cos(x + \alpha)$ |

(5.5)

### 5.4.5 $R$ FORMULAS

In Chapter 4, we discussed $R$ formulas or identities and extended our coverage by examining the maximum/minimum value of the $R$ formulas. By way of a quick recap, $R$ formula is a trigonometric expression that is a sum or difference of sine and cosine functions. We notice that we can write (more technically transform) the sum of or difference of functions to a single function of sine or cosine, with a phase angle (or horizontal translation) and a vertical stretch. The formulas are reproduced in Table 5.9.

Again, our focus here is on using the above identities to solve trigonometric equations. It is a good time to try a couple of examples.

### Example 9

Solve the following trigonometric equations in the indicated intervals and sketch the graph to confirm your answer. Present your answer correct to 3 s.f. where necessary.

a) $2\sqrt{3} \sin x - 2 \cos x - 3 = 0$ ; $0 < x < 2\pi$
b) $3 \cos x - \sqrt{2} \sin x - 1 = 0$ ; $-180° < x < 180°$

What did you get? Find the solution below to double-check your answer.

### Solution to Example 9

a) $2\sqrt{3} \sin x - 2 \cos x - 3 = 0$ ; $0 < x < 2\pi$
**Solution**
**Step 1:** Let's simplify

$$2\sqrt{3} \sin x - 2 \cos x - 3 = 0$$

This implies that

$$2\sqrt{3} \sin x - 2 \cos x = 3$$
$$\sqrt{3} \sin x - \cos x = \frac{3}{2}$$

# Solving Trigonometric Equations

**Step 2:** We will use the $R$ formula to express the LHS in the sine function alone as:

$$\sqrt{3}\sin x - \cos x = R\sin(x+\beta)$$

For this

$$R = \sqrt{a^2 + b^2}$$
$$= \sqrt{\left(\sqrt{3}\right)^2 + (-1)^2} = 2$$
$$|\beta| = \left|\tan^{-1}\left(\frac{-1}{\sqrt{3}}\right)\right| = \frac{\pi}{6}$$

This is the second quadrant, the adjusted $\beta = -\frac{\pi}{6}$. Thus

$$\sqrt{3}\sin x - \cos x = R\sin(x-\beta) = 2\sin\left(x - \frac{\pi}{6}\right)$$

**Step 3:** We now need to substitute and simplify as:

$$2\sin\left(x - \frac{\pi}{6}\right) = \frac{3}{2}$$
$$\sin\left(x - \frac{\pi}{6}\right) = \frac{3}{4}$$

This implies that

$$x - \frac{\pi}{6} = \sin^{-1}\left(\frac{3}{4}\right) = \mathbf{0.848}$$

The new range is $-\frac{\pi}{6} < \left(x - \frac{\pi}{6}\right) < \frac{11}{6}\pi$.

**Step 4:** Sine is also positive in the second quadrant, which implies that

$$x - \frac{\pi}{6} = \pi - 0.848 = \mathbf{2.29}$$

**0.848** and **2.29** are the only values within the new range.

**Step 5:** The values of $x$ are:

$$x = \frac{\pi}{6} + 0.848 = \mathbf{1.37}$$

and

$$x = \frac{\pi}{6} + 2.29 = \mathbf{2.81}$$

**Step 6:** The solutions of the equation in the range $0 < \theta < 2\pi$ are

$$\therefore x = \mathbf{1.37, 2.81}$$

The graph is shown in Figure 5.24.

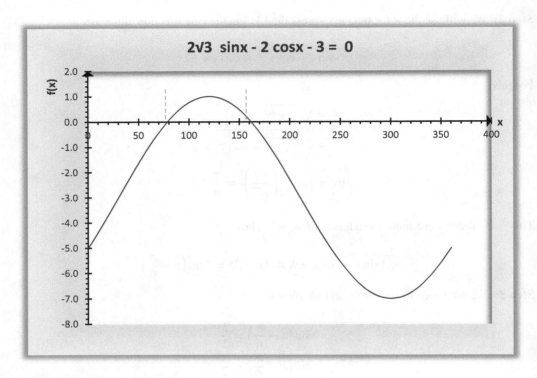

**FIGURE 5.24** Solution to Example 9(a).

b) $3\cos x - \sqrt{2}\sin x - 1 = 0$; $-180° < x < 180°$
**Solution**
**Step 1:** Let's simplify

$$3\cos x - \sqrt{2}\sin x - 1 = 0$$

This implies that

$$3\cos x - \sqrt{2}\sin x = 1$$

**Step 2:** We will use the $R$ formula to express the LHS in cosine function alone as:

$$3\cos x - \sqrt{2}\sin x = R\cos(x + \beta)$$

For this

$$R = \sqrt{a^2 + b^2}$$

$$= \sqrt{3^2 + \left(-\sqrt{2}\right)^2} = \sqrt{11}$$

$$|\beta| = \left|\tan^{-1}\left(\frac{-\sqrt{2}}{3}\right)\right| = 25.2°$$

This is the fourth quadrant, the adjusted $\beta = 25.2°$. Thus

$$\cos x - \sqrt{2}\sin x = \sqrt{11}\cos(x + 25.2°)$$

# Solving Trigonometric Equations

**Step 3:** We now need to substitute and simplify as:

$$\sqrt{11} \cos(x + 25.2°) = 1$$

$$\cos(x + 25.2°) = \frac{1}{\sqrt{11}}$$

This implies that

$$x + 25.2° = \cos^{-1}\left(\frac{1}{\sqrt{11}}\right) = \mathbf{72.5°}$$

The new range is $-154.8° < (x + 25.2°) < 205.2°$.

**Step 4:** Cosine is also positive in the fourth quadrant, but this is outside the range. Using periodicity $-360°$ will also take the solution outside the range. We can use its symmetry to determine one solution as:

$$x + 25.2° = -72.5°$$

$\mathbf{72.5°}$ and $\mathbf{-72.5°}$ are the only values within the new range. The values of $x$ are:

$$x = 72.5° - 25.2° = \mathbf{47.3°}$$

and

$$x = -72.5° - 25.2° = \mathbf{-97.7°}$$

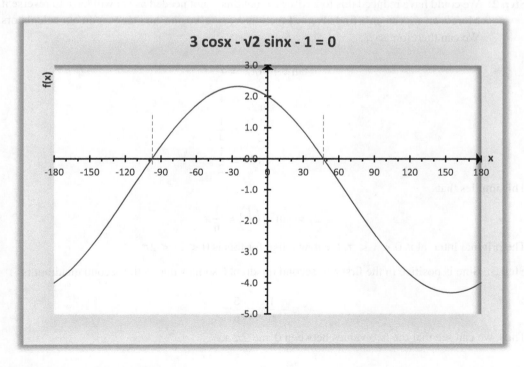

**FIGURE 5.25** Solution to Example 9(b).

**Step 5:** The solutions of the equation in the range $-180° < x < 180°$ are

$$\therefore x = -97.7°, \ 47.3°$$

Let's try one final example, though not particularly about a specific category of the identity.

### Example 10

Solve $\tan x + \cot x = 4$ in the intervals $0 < x < \pi$. Present your answer in exact form.

What did you get? Find the solution below to double-check your answer.

### Solution to Example 10

**Step 1:** We need to express this in one single trigonometric function. The LHS expression becomes:

$$\tan x + \cot x = \frac{\sin x}{\cos x} + \frac{\cos x}{\sin x} = \frac{\sin^2 x + \cos^2 x}{\cos x \sin x}$$
$$= \frac{1}{\frac{1}{2}(\sin 2x)} = \frac{2}{\sin 2x}$$

**Step 2:** We could have reduced this to $2 \operatorname{cosec} 2x$, but this is not needed as we will have to reverse it. Moreover, we can only find inverse functions of sine, cosine, and tangent on our calculators. We can therefore say,

$$\tan x + \cot x = 4$$
$$\frac{2}{\sin 2x} = 4$$
$$\frac{\sin 2x}{2} = \frac{1}{4}$$
$$\sin 2x = \frac{1}{2}$$

This implies that

$$2x = \sin^{-1}\left(\frac{1}{2}\right) = \frac{1}{6}\pi$$

The original interval is $0 < x < \pi$, the new range for this is $0 < 2x < 2\pi$.

**Step 3:** Sine is positive in the first and second quadrants, so its value in the second quadrant is

$$\pi - \frac{1}{6}\pi = \frac{5}{6}\pi$$

Thus, we can see that the two values between 0 and $2\pi$ are

$$2\theta = \frac{1}{6}\pi, \ \frac{5}{6}\pi$$

Solving Trigonometric Equations

Divide each answer by 2 to have the values of $x$ in the stated range as:

$$x = \frac{1}{12}\pi, \ \frac{5}{12}\pi$$

**Step 4:** The solutions of the equation in the range $0 < x < \pi$ are:

$$\therefore x = \frac{1}{12}\pi, \frac{5}{12}\pi$$

## 5.5 CHAPTER SUMMARY

1) A trigonometric equation (linear and non-linear) is an equation involving trigonometric ratios. As such, solving this type of equation requires that one is equipped with skills relating to trigonometric functions and identities, and how to solve equations in general.

2) Due to their periodicity, there are unlimited solutions for any given trigonometric equation. Therefore, we need to know the range of values (angles) required. A calculator will only give us one answer, usually in the first quadrant for positive values.

3) Unlike polynomials, we cannot foretell the number of solutions in trigonometric equations. The bigger the interval the more the solutions, but two solutions are generally expected for every 360-degree (or $2\pi$) interval.

4) There are three methods to solve trigonometric equations, namely:
    - Graph
    - CAST
    - Quadrant formula

5) **CAST** stands for **C**os, **A**ll, **S**in, and **T**an. CAST diagram is a diagram which divides $x$–$y$ plane into four equal parts, called quadrants.

6) Each of the trigonometric functions is positive in two quadrants and negative in the other two. They are all positive in the first quadrant.

7) Quadrant formulas are $180° - \theta$, $180° + \theta$, and $360° - \theta$ for the second, third, and fourth quadrant, respectively.

8) When the trigonometric functions are transformed, the interval of the solutions will change accordingly.

9) The three distinct transformations and how to solve problems involving each type.
    a) For $y = \sin ax$, $y = \cos ax$, $y = \tan ax$: This is simply a horizontal stretch, where $a$ is the factor of stretching. Consequently, we have that:
        - the intervals will be $a$ times **the original interval**, and
        - the number of solutions will be, in general, $a$ times the **original number of solutions**.

b) For $y = \sin(x \pm b)$, $y = \cos(x \pm b)$, $y = \tan(x \pm b)$: This is a horizontal translation, where $b$ is a positive real number and represents the amount of translation, left or right. Consequently, we have that:
- if the given interval is $\theta_1 \leq x \leq \theta_2$, the new interval will be $\theta_1 \pm b \leq (x \pm b) \leq \theta_2 \pm b$, and
- the number of solutions will mostly remain unchanged.

c) For $y = \sin(ax \pm b)$, $y = \cos(ax \pm b)$, $y = \tan(ax \pm b)$: This last type is a combination of types 1 and 2. We can say that if the given interval is $\theta_1 \leq x \leq \theta_2$, the new interval will be $a\theta_1 \pm b \leq (ax \pm b) \leq a\theta_2 \pm b$. Notice that we need to first multiply the limits (lower and upper) by the coefficient of the angle and then add/subtract the phase angle. This order is very essential.

\*\*\*\*

## 5.6 FURTHER PRACTICE

To access complementary contents, including additional exercises, please go to www.dszak.com.

# 6 The Binomial Expansion

*Learning Outcomes*

Once you have studied the content of this chapter, you should be able to:

- Explain Pascal's triangle and its derivation
- Compute factorials for $n$ items
- Determine the coefficients of a binomial expansion using $\binom{n}{r}$ or $^nC_r$
- Discuss the relationship between factorials and Pascal's triangle of coefficients
- Discuss the limitations to the application of Pascal's triangle
- Use Pascal's triangle to solve binomial expressions involving powers that are positive integers
- Use binomial expansion theorem for $(1+x)^n$ and $(a+bx)^n$ for any value of $n$
- State the validity of the expansion for rational power $n$
- Use binomial theorem to determine the approximate value of a rational number

## 6.1 INTRODUCTION

If we have understood the opening brackets method of solving algebraic expressions, then expressions such as $(2x+5)(4x-y)$, $(4x-3y)^2$, and even $(4x-3y)^3$ will be easy. But imagine that we are required to expand $(4x-3y)^{10}$. Did you take a deep breath? Perhaps you wondered about its purpose and how long this will take. Well, we may need this for good reasons, but it will take a pretty long time if we follow the 'conventional' way of opening the brackets. Of course, we also have software to do this. However, we equally have another technique in the absence of a calculating or processing device. This technique is known as **Binomial Expansion**, which is the focus of this chapter.

## 6.2 WHAT IS A BINOMIAL EXPRESSION?

'**Bi**' means two just like '**poly**' implies many. A binomial expression is an expression with just two terms, with a plus (+) or minus (−) between them. In general, it takes the following forms:

$$\boxed{a \pm bx^m} \text{ OR } \boxed{ax^m \pm by^m}$$

where $a$, $b$, and $m$ are rational numbers.

Here, $m$ is usually taken as 1, whilst $a$ and $b$ are (usually) integers but valid for any real number. That is:

$$\boxed{a \pm bx} \text{ OR } \boxed{ax \pm by} \tag{6.1}$$

However, what we will learn in this chapter is applicable to when $m \neq 1$. Through manipulation, we can apply the binomial theorem to trinomials (expressions with three terms) and others.

## 6.3 PASCAL'S TRIANGLE

Pascal's triangle is a pattern formed using the coefficients of the terms obtained when binomial expressions of different orders are simplified. To illustrate, let's take the binomial expression $ax + by$ and raise it to power $n$ as $(ax + by)^n$. Power $n$ ($n$ is called the order) is by principle any real number, but for Pascal's triangle, we are limited to positive integers. We will, for simplicity, consider that $a = b = 1$, hence we will expand $(x + y)^n$ for $n \geq 0$ as:

| $n$ | Binominal Expression | | | | | | |
|---|---|---|---|---|---|---|---|
| 0 | $(x+y)^0$ | $=$ | $1x^0y^0$ | | | | |
| 1 | $(x+y)^1$ | $=$ | $1x^1y^0$ | $+1x^0y^1$ | | | |
| 2 | $(x+y)^2$ | $=$ | $1x^2y^0$ | $+2x^1y^1$ | $+1x^0y^2$ | | |
| 3 | $(x+y)^3$ | $=$ | $1x^3y^0$ | $+3x^2y^1$ | $+3x^1y^2$ | $+1x^0y^3$ | |
| 4 | $(x+y)^4$ | $=$ | $1x^4y^0$ | $+4x^3y^1$ | $+6x^2y^2$ | $+4x^1y^3$ | $+1x^0y^4$ |
| 5 | $(x+y)^5$ | $=$ | $1x^5y^0$ | $+5x^4y^1$ | $+10x^3y^2$ | $+10x^2y^3$ | $+5x^1y^4$ | $+1x^0y^5$ |

Before we make any comments on the above, let's try the same with the expression $(1 + x)^n$ as:

| $n$ | Binominal Expression | | | | | | |
|---|---|---|---|---|---|---|---|
| 0 | $(1+x)^0$ | $=$ | $1x^0$ | | | | |
| 1 | $(1+x)^1$ | $=$ | $1x^0$ | $+1x^1$ | | | |
| 2 | $(1+x)^2$ | $=$ | $1x^0$ | $+2x^1$ | $+1x^2$ | | |
| 3 | $(1+x)^3$ | $=$ | $1x^0$ | $+3x^1$ | $+3x^2$ | $+1x^3$ | |
| 4 | $(1+x)^4$ | $=$ | $1x^0$ | $+4x^1$ | $+6x^2$ | $+4x^3$ | $+1x^4$ |
| 5 | $(1+x)^5$ | $=$ | $1x^0$ | $+5x^1$ | $+10x^2$ | $+10x^3$ | $+5x^4$ | $+1x^5$ |

We will now summarise the key points from the above as:

**Note 1** If the power is $n$, the number of terms is $n + 1$. For example, if the power is 3, the number of terms will be 4.

**Note 2** Sum of the powers of the variables $x$ and $y$ equals $n$ in each term of the expansion of $(x + y)^n$. For example, when $n = 4$, the sum of the powers of $x$ and $y$ will also be equal to 4 in each of the five terms. Consequently, if a term has $x^1$, then it must have $y^3$ to make the sum of the powers 4.

**Note 3** We generally like to arrange the results in ascending or descending order of the power of the variable $x$ or $y$. If this is followed, the power of the variable will increase/decrease by 1 when moving from one term to the immediate next.

# The Binomial Expansion

**Note 4** If Note 3 is followed, then the array of coefficients (of the terms) will form a pattern called **Pascal's triangle**. The array can be easily formed.

Now let's turn back to our solution of $(x+y)^n$ and $(1+x)^n$, for $n = 0, 1, 2, 3, 4, 5$, and extract the coefficients to form a pattern, as shown in Figure 6.1.

We've used the coefficients to form an array of numbers. An extra line, without coefficients, has been added (when $n = 6$), which will be completed shortly. A critical look at Figure 6.1 shows that:

**a)** The array of coefficients forms a pattern that can be enveloped in a triangle, hence the name Pascal's triangle. It is named after the French inventor and mathematician, Blaise Pascal.

**b)** The array is symmetrical. In other words, once you've obtained the first half of the coefficients, the second half is a lateral reflection of the first about the middle (or the line of symmetry).

**c)** It is arranged such that there is an empty unit space between two vertically and horizontally adjacent coefficients. The space is seen empty and greyed in Figure 6.1.

**d)** The array starts with 1 in the first row. This is followed with 1 twice in the second row, each is placed to the left and right of the 1 in the first row. Each of the subsequent rows starts and ends with 1.

**e)** To find a coefficient, follow a three-step approach:

**Step 1:** Look above the cell whose coefficient is to be found; this should be empty.
**Step 2:** Pick the number to the immediate left and immediate right of step 1. We've shown this for 2 in Figure 6.2.
**Step 3:** Add the two numbers in step 2; this becomes the unknown coefficient.

Note **e)** has now been applied to complete the triangle as shown in Figure 6.2. Did you get the gist?

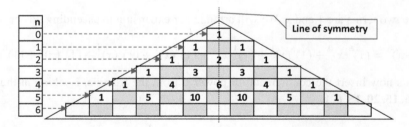

**FIGURE 6.1** Pascal's triangle illustrated.

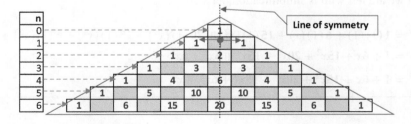

**FIGURE 6.2** Pascal's triangle illustrated (extended).

That's all for Pascal's triangle. Let's take a break and try a few examples.

### Example 1

Using Pascal's triangle, expand and completely simplify the following:

a) $(1+x)^6$  b) $(x-y)^5$  c) $(2x+3y)^3$

What did you get? Find the solution below to double-check your answer.

### Solution to Example 1

**HINT**

- Since we have Pascal's triangle drawn above, we will simply refer to it. Ordinarily, you may have to do this.
- Note that anything raised to power zero is **1**.
- We will also write power **1** without the number, i.e., $x^1$ will be written simply as $x$.

**a)** $(1+x)^6$
**Solution**
Let's go slowly here.

**Step 1:** $n = 6$, so we expect seven terms as:

$$(1+x)^6 = (\ )(\ )+(\ )(\ )+(\ )(\ )+(\ )(\ )+(\ )(\ )+(\ )(\ )+(\ )(\ )$$

**Step 2:** The two terms are 1 and $x$. We will arrange our expansion in ascending power of $x$ as:

$$(1+x)^6 = (1)^6(x)^0 + (1)^5(x)^1 + (1)^4(x)^2 + (1)^3(x)^3 + (1)^2(x)^4 + (1)^1(x)^5 + (1)^0(x)^6$$

**Step 3:** Let's now insert the coefficients for power 6 from the triangle. Looking at the array, it is $1, 6, 15, 20, 15, 6, 1$.

$$(1+x)^6 = \mathbf{1}(1)^6(x)^0 + \mathbf{6}(1)^5(x)^1 + \mathbf{15}(1)^4(x)^2 + \mathbf{20}(1)^3(x)^3 + \mathbf{15}(1)^2(x)^4 + \mathbf{6}(1)^1(x)^5 + \mathbf{1}(1)^0(x)^6$$

**Step 4:** All we are left with is simplification; let's go:

$$(1+x)^6 = \mathbf{1}(1)(1) + \mathbf{6}(1)(x) + \mathbf{15}(1)(x^2) + \mathbf{20}(1)(x^3) + \mathbf{15}(1)(x^4) + \mathbf{6}(1)(x^5) + \mathbf{1}(1)(x^6)$$
$$= 1 + 6x + 15x^2 + 20x^3 + 15x^4 + 6x^5 + 1x^6$$
$$\therefore (1+x)^6 = 1 + 6x + 15x^2 + 20x^3 + 15x^4 + 6x^5 + x^6$$

**b)** $(x-y)^5$
**Solution**
This is a bit different because of the sign, but the steps are the same. Consider it as $[x+(-y)]^5$. You get the gist? It is a very useful tip in this topic. Let's go.

# The Binomial Expansion

**Step 1:** $n = 5$, so we expect six terms: a pair of brackets for $x$ term and the other for $-y$ as:

$$(x - y)^5 = (\ )(\ ) + (\ )(\ ) + (\ )(\ ) + (\ )(\ ) + (\ )(\ ) + (\ )(\ )$$

**Step 2:** The two terms are $x$ and $-y$. We will arrange our expansion in ascending power of $y$ as:

$$(x - y)^5 = (x)^5(-y)^0 + (x)^4(-y)^1 + (x)^3(-y)^2 + (x)^2(-y)^3 + (x)^1(-y)^4 + (x)^0(-y)^5$$

**Step 3:** Let's now insert the coefficients for power 5 from the triangle. Looking at the array it is **1, 5, 10, 10, 5, 1**.

$$(x - y)^5 = \mathbf{1}(x)^5(-y)^0 + \mathbf{5}(x)^4(-y)^1 + \mathbf{10}(x)^3(-y)^2 + \mathbf{10}(x)^2(-y)^3 + \mathbf{5}(x)^1(-y)^4 + \mathbf{1}(x)^0(-y)^5$$

**Step 4:** All we are left with is simplification; let's go:

$$(x - y)^5 = \mathbf{1}\left(x^5\right)(1) + \mathbf{5}\left(x^4\right)(-y) + \mathbf{10}\left(x^3\right)\left(y^2\right) + \mathbf{10}\left(x^2\right)\left(-y^3\right) + \mathbf{5}(x)\left(y^4\right) + \mathbf{1}(1)\left(-y^5\right)$$
$$= 1x^5 - 5x^4y + 10x^3y^2 - 10x^2y^3 + 5xy^4 - 1y^5$$
$$\therefore (x - y)^5 = x^5 - 5x^4y + 10x^3y^2 - 10x^2y^3 + 5xy^4 - y^5$$

---

**c)** $(2x + 3y)^3$
**Solution**
This is of the form $ax + by$ where $a \neq b \neq 1$; it is all the same as we will see now. Let's go.

**Step 1:** $n = 3$, so we expect four terms as:

$$(2x + 3y)^3 = (\ )(\ ) + (\ )(\ ) + (\ )(\ ) + (\ )(\ )$$

**Step 2:** The two terms are $2x$ and $3y$. We will arrange our expansion in ascending power of $y$ as:

$$(2x + 3y)^3 = (2x)^3(3y)^0 + (2x)^2(3y)^1 + (2x)^1(3y)^2 + (2x)^0(3y)^3$$

**Step 3:** Let's now insert the coefficients for power 3 from the triangle. Looking at the array it is **1, 3, 3, 1**.

$$(2x + 3y)^3 = \mathbf{1}(2x)^3(3y)^0 + \mathbf{3}(2x)^2(3y)^1 + \mathbf{3}(2x)^1(3y)^2 + \mathbf{1}(2x)^0(3y)^3$$

**Step 4:** All we are left with is simplification; let's go:

$$(2x + 3y)^3 = \mathbf{1}\left(8x^3\right)(1) + \mathbf{3}\left(4x^2\right)(3y) + \mathbf{3}(2x)\left(9y^2\right) + \mathbf{1}(1)\left(27y^3\right)$$
$$= 8x^3 + 36x^2y + 54xy^2 + 27y^3$$
$$\therefore (2x + 3y)^3 = 8x^3 + 36x^2y + 54xy^2 + 27y^3$$

Another example to try.

## Example 2

The coefficient of $y^2$ in the expansion of $(ab^2 + by)^3$ is 192. Given that $b$ is half of $a$, determine the exact value of $a$ and $b$.

---

What did you get? Find the solution below to double-check your answer.

## Solution to Example 2

Looking at Pascal's array for power 3, we have $\mathbf{1, 3, 3, 1}$ as the coefficients of the terms. The third term will contain $y^2$. Thus, we have

$$3(ab^2)(by)^2 = 3(ab^2)(b^2y^2) = 3ab^4y^2$$

Therefore

$$3ab^4 = 192$$

But

$$a = 2b$$

Thus

$$3 \times 2b \times b^4 = 192$$
$$6b^5 = 192$$
$$b^5 = 32$$
$$b = \sqrt[5]{32} = 2$$

Hence

$$a = 2 \times 2 = 4$$
$$\therefore a = \mathbf{4}, b = \mathbf{2}$$

---

Here is another couple of examples to try.

## Example 3

Using Pascal's triangle, write down the expansion of the following in descending powers of $x$:

a) $\left(x^3 - \dfrac{1}{3x}\right)^3$  b) $(x^2 - x - 2)^3$

# The Binomial Expansion

What did you get? Find the solution below to double-check your answer.

## Solution to Example 3

**a)** $\left(x^3 - \dfrac{1}{3x}\right)^3$

**Solution**

Let's go slowly here

**Step 1:** $n = 3$, so we expect four terms as:

$$\left(x^3 - \dfrac{1}{3x}\right)^3 = (\ )(\ ) + (\ )(\ ) + (\ )(\ ) + (\ )(\ )$$

**Step 2:** The two terms are $x^3$ and $-\dfrac{1}{3x}$. We will arrange our expansion in descending powers of $x$ as:

$$\left(x^3 - \dfrac{1}{3x}\right)^3 = (x^3)^3\left(-\dfrac{1}{3x}\right)^0 + (x^3)^2\left(-\dfrac{1}{3x}\right)^1 + (x^3)^1\left(-\dfrac{1}{3x}\right)^2 + (x^3)^0\left(-\dfrac{1}{3x}\right)^3$$

**Step 3:** Let's now insert the coefficients for power 3 from the triangle. Looking at the array, it is $1, 3, 3, 1$.

$$\left(x^3 - \dfrac{1}{3x}\right)^3 = 1(x^3)^3\left(-\dfrac{1}{3x}\right)^0 + 3(x^3)^2\left(-\dfrac{1}{3x}\right)^1 + 3(x^3)^1\left(-\dfrac{1}{3x}\right)^2 + 1(x^3)^0\left(-\dfrac{1}{3x}\right)^3$$

**Step 4:** All we are left with is simplification; let's go:

$$\left(x^3 - \dfrac{1}{3x}\right)^3 = 1(x^3)^3\left(-\dfrac{1}{3x}\right)^0 + 3(x^3)^2\left(-\dfrac{1}{3x}\right)^1 + 3(x^3)^1\left(-\dfrac{1}{3x}\right)^2 + 1(x^3)^0\left(-\dfrac{1}{3x}\right)^3$$

$$= 1(x^9)(1) + 3(x^6)\left(-\dfrac{1}{3x}\right) + 3(x^3)\left(\dfrac{1}{9x^2}\right) + 1(1)\left(-\dfrac{1}{27x^3}\right)$$

$$= 1(x^9) + 3(-3x^5) + 3\left(\dfrac{x}{9}\right) + 1\left(-\dfrac{1}{27x^3}\right)$$

$$= 1x^9 - 9x^5 + \dfrac{1}{3}x - \dfrac{1}{27x^3}$$

$$\therefore \left(x^3 - \dfrac{1}{3x}\right)^3 = x^9 - 9x^5 + \dfrac{1}{3}x - \dfrac{1}{27x^3}$$

---

**b)** $(x^2 - x - 2)^3$

**Solution**

We will do this in two stages. As we are dealing with a trinomial, we will model it as a binomial by letting $y = x^2 - x$ then we have

$$(x^2 - x - 2)^3 = (y - 2)^3$$

Now, let's go

**Step 1:** $n = 3$, so we expect four terms as:

$$(y-2)^3 = (\ )(\ ) + (\ )(\ ) + (\ )(\ ) + (\ )(\ )$$

**Step 2:** The two terms are $y$ and $-2$. We will arrange our expansion in descending powers of $y$ as:

$$(y-2)^3 = (y)^3(-2)^0 + (y)^2(-2)^1 + (y)^1(-2)^2 + (y)^0(-2)^3$$

**Step 3:** Let's now insert the coefficients for power 3 from the triangle. Looking at the array, it is $1, 3, 3, 1$.

$$(y-2)^3 = \mathbf{1}(y)^3(-2)^0 + \mathbf{3}(y)^2(-2)^1 + \mathbf{3}(y)^1(-2)^2 + \mathbf{1}(y)^0(-2)^3$$

**Step 4:** All we are left with is simplification; let's go:

$$\begin{aligned}(y-2)^3 &= \mathbf{1}\left(y^3\right)(1) + \mathbf{3}\left(y^2\right)(-2) + \mathbf{3}(y)(4) + \mathbf{1}(1)(-8) \\ &= \mathbf{1}\left(y^3\right) + \mathbf{3}\left(-2y^2\right) + \mathbf{3}(4y) + \mathbf{1}(-8) \\ &= y^3 - 6y^2 + 12y - 8\end{aligned}$$

**Step 5:** Let's replace $y$ with $x^2 - x$, we have

$$= \left(x^2 - x\right)^3 - 6\left(x^2 - x\right)^2 + 12\left(x^2 - x\right) - 8$$

**Step 6:** Let expand the powers of $x^2 - x$ as

$$\begin{aligned}\left(x^2 - x\right)^2 &= x^4 - 2x^3 + x^2 \\ \left(x^2 - x\right)^3 &= 1(x^2)^3(-x)^0 + 3(x^2)^2(-x)^1 + 3(x^2)^1(-x)^2 + 1(x^2)^0(-x)^3 \\ &= x^6 - 3x^5 + 3x^4 - x^3\end{aligned}$$

**Step 7:** It is time to put our results in step 6 into step 5 and then simplify

$$\begin{aligned}\left(x^2 - x - 2\right)^3 &= \left(x^2 - x\right)^3 - 6\left(x^2 - x\right)^2 + 12\left(x^2 - x\right) - 8 \\ &= x^6 - 3x^5 + 3x^4 - x^3 - 6\left(x^4 - 2x^3 + x^2\right) + 12\left(x^2 - x\right) - 8 \\ &= x^6 - 3x^5 + 3x^4 - x^3 - 6x^4 + 12x^3 - 6x^2 + 12x^2 - 12x - 8 \\ &= x^6 - 3x^5 + 3x^4 - 6x^4 - x^3 + 12x^3 - 6x^2 + 12x^2 - 12x - 8 \\ \therefore \left(x^2 - x - 2\right)^3 &= x^6 - 3x^5 - 3x^4 + 11x^3 + 6x^2 - 12x - 8\end{aligned}$$

The Binomial Expansion

Another example to try.

### Example 4

Using Pascal's triangle, determine the value of $1.9^4$. Give your answer correct to 2 s.f.

What did you get? Find the solution below to double-check your answer.

### Solution to Example 4

**HINT**

Think about how the single number can be best split into two separate digits. Each represents a term in a binomial.

There are many ways of splitting 1.9 but think about the one that you can easily deal with their powers. For the current case, we will consider $1.9^4$ as $(2-0.1)^4$. Let's work this out as before:

**Step 1:** $n=4$, so we expect five terms as:

$$(2-0.1)^4 = ()() + ()() + ()() + ()() + ()()$$

**Step 2:** The two terms are 2 and $-0.1$. We will arrange our expansion in ascending powers of 0.1 as:

$$(2-0.1)^4 = (2)^4(-0.1)^0 + (2)^3(-0.1)^1 + (2)^2(-0.1)^2 + (2)^1(-0.1)^3 + (2)^0(-0.1)^4$$

**Step 3:** Let's now insert the coefficients for power 4. Looking at the array it is **1, 4, 6, 4, 1**.

$$(2-0.1)^4 = \mathbf{1}(2)^4(-0.1)^0 + \mathbf{4}(2)^3(-0.1)^1 + \mathbf{6}(2)^2(-0.1)^2 + \mathbf{4}(2)^1(-0.1)^3 + \mathbf{1}(2)^0(-0.1)^4$$

**Step 4:** All we are left with is simplification; let's go:

$$(2-0.1)^4 = \mathbf{1}(16)(1) + \mathbf{4}(8)(-0.1) + \mathbf{6}(4)(0.01) + \mathbf{4}(2)(-0.001) + \mathbf{1}(1)(0.0001)$$
$$= 16 - 3.2 + 0.24 - 0.008 + 0.0001$$
$$= 13.0321$$

$$\therefore \mathbf{1.9^4 = 13\,(2\text{ s.f.})}$$

To wrap up our discussion on Pascal's triangle, it is important to highlight the limitations and challenges to its usage, which include the following:

a) Unless provided, if we require power 7 for example, we need to find the coefficients for power 6 and all numbers < 6 down to 1.

b) It can only be used for positive integers of $n$. In other words, we cannot use it when $n$ is negative or a fraction, such as $(x-3)^{-4}$ or $(2x+y)^{\frac{1}{3}}$ respectively.

We will now explore how we can overcome these limitations and create a more universal and robust approach.

## 6.4 FACTORIALS

Factorial is denoted by an exclamation mark (!). Technically, it implies the product of all positive integers between a particular given integer and 1. Two things to note here are that factorials are generally computed for: (i) positive numbers and (ii) integers.

Given that $r$ is an integer, $r!$ is read as '$r$ factorial' or 'factorial $r$', and it is given by:

$$\boxed{r! = r \times (r-1) \times (r-2) \times (r-3) \times \cdots \times 3 \times 2 \times 1} \tag{6.2}$$

**OR**

$$\boxed{r! = 1 \times 2 \times 3 \times \cdots \times (r-3) \times (r-2) \times (r-1) \times r} \tag{6.3}$$

In other words, if we like we can start from the given number until 1 or from 1 until the given number. To make this clear, we've provided a few examples in Table 6.1.

But you need not go through the above process as there is a function key on most calculators. It is denoted as $x!$ where $x$ is the number.

Perhaps you noticed that 10! is not fully computed (or expanded) and looks a bit unusual. It is a good trick that is purposefully introduced here. If we stop on a number, then we can finish it off with the factorial symbol. Since the value of 6! is known, it will be used to simplify 10!, that is, $10! = 10 \times 9 \times 8 \times 7 \times 720 = 3,628,800$. 6! is however not the only place to stop. In fact, 10! can be expressed in any of the following ways:

$$10! = 10 \times 9!$$
$$= 10 \times 9 \times 8!$$
$$= 10 \times 9 \times 8 \times 7!$$
$$= \ldots$$
$$= 10 \times 9 \times 8 \times 7 \times 6 \times 5 \times 4 \times 3 \times 2 \times 1 = 3,628,800$$

**TABLE 6.1**
**Factorial Illustrated**

| Example | Read as | Solution (A) | Solution (B) |
|---|---|---|---|
| 3! | 3-factorial | $3! = 3 \times 2 \times 1 = \mathbf{6}$ | $3! = 1 \times 2 \times 3 = \mathbf{6}$ |
| 6! | 6-factorial | $6! = 6 \times 5 \times 4 \times 3 \times 2 \times 1 = \mathbf{720}$ | $6! = 1 \times 2 \times 3 \times 4 \times 5 \times 6 = \mathbf{720}$ |
| 10! | 10-factorial | $10! = \mathbf{10 \times 9 \times 8 \times 7 \times 6!}$ | $10! = \mathbf{6! \times 7 \times 8 \times 9 \times 10}$ |

# The Binomial Expansion

All done and dusted on factorials. It is now time to look at some examples.

### Example 5

Without using a calculator, simplify the following. Confirm your answer using a suitable calculator.

**a)** $4!$  **b)** $\dfrac{10!}{8!}$  **c)** $\dfrac{15!}{5!13!}$  **d)** $\dfrac{12!}{9!3!}$  **e)** $1!$  **f)** $0!$

What did you get? Find the solution below to double-check your answer.

### Solution to Example 5

**a)** $4!$
**Solution**
Let's go

$$4! = 4 \times 3 \times 2 \times 1 = 24$$
$$\therefore 4! = 24$$

**b)** $\dfrac{10!}{8!}$
**Solution**
Let's go

$$\dfrac{10!}{8!} = \dfrac{10 \times 9 \times 8!}{8!}$$
$$= 10 \times 9 = 90$$
$$\therefore \dfrac{10!}{8!} = 90$$

**c)** $\dfrac{15!}{5!13!}$
**Solution**
Let's go

$$\dfrac{15!}{5!13!} = \dfrac{15 \times 14 \times 13!}{(5 \times 4 \times 3 \times 2 \times 1) \times 13!} = \dfrac{15 \times 14 \times 13!}{(5 \times 4 \times 3 \times 2 \times 1) \times 13!}$$
$$= \dfrac{15 \times 14}{5 \times 4 \times 3 \times 2 \times 1} = \dfrac{7}{4}$$
$$\therefore \dfrac{15!}{5!13!} = \dfrac{7}{4}$$

**d)** $\frac{12!}{9!3!}$

**Solution**

Let's go

$$\frac{12!}{9!3!} = \frac{12 \times 11 \times 10 \times 9!}{9! \times (3 \times 2 \times 1)} = \frac{12 \times 11 \times 10 \times 9!}{9! \times (3 \times 2 \times 1)}$$

$$= \frac{12 \times 11 \times 10}{6} = 220$$

$$\therefore \frac{12!}{9!3!} = 220$$

---

**e)** $1!$

**Solution**

Obviously, we are starting with 1 and ending with 1. What do you think? Let's go

$$1! = 1$$

$$\therefore 1! = 1$$

---

**f)** $0!$

**Solution**

This is an incredibly tricky one! Wait for a second, what we've just seen is not a factorial, but rather an exclamation, because factorials are counted from a number down to 1 and 0 is not included.

There are few ways to show this, which include modification of the definition of factorial to $n! = \frac{(n+1)!}{n+1}$ to include zero. Therefore, we have:

$$0! = \frac{1!}{1} = 1$$

---

## 6.5 NOTATION FOR COMBINATION

What we have covered on factorials above is a stepping stone to unravelling the limitations of Pascal's triangle. We now need to introduce a useful notation, which is

$$\boxed{nC_r} \text{ OR } \boxed{\binom{n}{r}} \tag{6.4}$$

The above notations can be used interchangeably and represent a combination of **n** items taken **r** at a time, where the order or sequence of arrangement is irrelevant. It is premised on the fact that $n \geq r$.

# The Binomial Expansion

Let's say we have a team of five equally capable participants, and we need just two to represent the team in a competition. The number of choices we have is denoted by $\binom{5}{2}$. Let's use $A, B, C, D,$ and $E$ to represent the participants. The choices, or more technically the combinations, are given below:

$$AB \quad AC \quad AD \quad AE \quad BC \quad BD \quad BE \quad CD \quad CE \quad DE$$

This shows that there are ten different choices. The query here is, could we have obtained this without listing it 'manually'. The answer to this is yes indeed! This is by using the formula:

$$^nC_r = \binom{n}{r} = \frac{n!}{r!\,(n-r)!} \tag{6.5}$$

We know how to handle factorials, so let's try the above equation with our example as:

$$^5C_2 = \binom{5}{2} = \frac{5!}{2!\,(5-2)!} = \frac{5!}{2!3!} = \frac{5 \times 4 \times 3!}{(2 \times 1) \times 3!} = \frac{5 \times 4}{(2 \times 1)} = 10$$

Great! We obtained the same answer '10'. If we need four students out of five, this will be denoted by $^5C_4$ and the answer is 5.

$$^5C_4 = \binom{5}{4} = \frac{5!}{4!\,(5-4)!} = \frac{5!}{4!1!} = \frac{5 \times 4!}{4! \times 1} = 5$$

Like factorial, simply enter this on your calculator via the notation $^nC_r$. To perform this, follow these steps in order: $n \to second\ function \to nC_r \to r$. There is a sister operator called permutation and is denoted as $^nP_r$. It is not directly connected to our binomial expression, so it is fine to ignore it here.

## 6.6 RELATIONSHIP BETWEEN PASCAL'S TRIANGLE AND COMBINATION

By now we should have mastered factorials $n!$ and combination $^nC_r$. Let's start this section by finding the answer to the following: $5C_0, 5C_1, 5C_2, 5C_3, 5C_4, 5C_5$. What did you get? My answers are $1, 5, 10, 10, 5, 1$. Does the sequence look familiar? Yes, it does. It is the coefficients of the binomial expansion for $n = 5$. Really! That sounds cool! Well, try a new set, say $n = 6$, and check our Pascal's triangle on page 205 and see the values on Figure 6.3 or you can create yours if your $n$ cannot be found on page 205.

At this stage, it is obvious that:

a) we can obtain the coefficients of any binomial expansion, where $n$ is a positive integer without the need to create Pascal's triangle, and

b) we can obtain the coefficient of a single term in any expansion.

This is a great milestone and will be used in our subsequent work in this chapter.

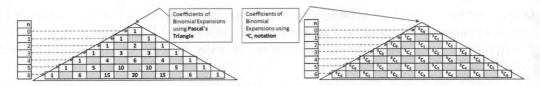

**FIGURE 6.3** Coefficient of binomial expansions using Pascal's triangle and $^nC_r$ compared.

## 6.7 BINOMIAL EXPANSIONS FOR $(1 \pm ax)^n$

Let's consider the expansion of the form $(1 \pm x)^n$ and $(1 \pm ax)^n$ for when $n$ is any rational number.

### 6.7.1 When $n$ Is a Positive Integer

- When $a = 1$

Recall that the coefficients of a binomial expansion can be obtained using $^nC_r$ notation. Hence, $(1 + x)^n$ can be written as:

$$(1+x)^n = \binom{n}{0}x^0 + \binom{n}{1}x^1 + \binom{n}{2}x^2 + \binom{n}{3}x^3 + \binom{n}{4}x^4 + \cdots + \binom{n}{n}x^n$$

You can choose to use $^nC_r$ instead of the $\binom{n}{r}$ notation and we have:

$$(1+x)^n = {^nC_0}x^0 + {^nC_1}x^1 + {^nC_2}x^2 + {^nC_3}x^3 + {^nC_4}x^4 + \cdots + {^nC_n}x^n$$

Have you noticed the pattern? The power of $x$ in any term is $r$. It makes it easy to recall, but we can tidy this up a bit. Note that $^nC_0 = {^nC_n} = 1$, $^nC_1 = n$, and $x^0 = 1$ and $x^1 = x$. Thus:

$$\boxed{(1+x)^n = 1 + nx + {^nC_2}x^2 + {^nC_3}x^3 + {^nC_4}x^4 + \cdots + x^n} \tag{6.6}$$

The above is the standard formula for $(1+x)^n$. If we replace $x$ with $-x$, we will have:

$$\boxed{(1-x)^n = 1 - nx + {^nC_2}x^2 - {^nC_3}x^3 + {^nC_4}x^4 + \cdots \pm x^n} \tag{6.7}$$

Notice that the formula is the same for both cases above except that the sign alternates in the latter, such that it becomes negative for every odd power and positive for even powers. The last term will be positive if $n$ is even otherwise negative.

- When $a \neq 1$

To expand $(1 + ax)^n$, we use the expansion of $(1 + x)^n$ and substitute each $x$ with $ax$ as:

$$(1+ax)^n = 1 + n(ax) + {^nC_2}(ax)^2 + {^nC_3}(ax)^3 + {^nC_4}(ax)^4 + \cdots + (ax)^n$$

Simplify the above as:

$$\boxed{(1+ax)^n = 1 + nax + {^nC_2}a^2x^2 + {^nC_3}a^3x^3 + {^nC_4}a^4x^4 + \cdots + a^nx^n} \tag{6.8}$$

# The Binomial Expansion

and by extension, we also have

$$(1 - ax)^n = 1 - nax + {}^nC_2 a^2 x^2 - {}^nC_3 a^3 x^3 + {}^nC_4 a^4 x^4 + \cdots \pm a^n x^n \qquad (6.9)$$

Notice that it is not different – we've only introduced $a$ with a similar power as $x$ for every given term.

Given $(1 + ax)^n$, we may be interested in a particular term. To derive a specific term, the $(r + 1)$th, the formula is:

$$ {}^nC_r (ax)^r $$

Now it is time to look at a couple of examples.

## Example 6

Using the binomial theorem, expand the following giving the first five terms in ascending power of $x$.

**a)** $(1 + x)^{12}$  **b)** $(1 - 2x)^8$

What did you get? Find the solution below to double-check your answer.

## Solution to Example 6

**HINT**

We just need to apply

$$(1 + ax)^n = 1 + nax + {}^nC_2 a^2 x^2 + {}^nC_3 a^3 x^3 + {}^nC_4 a^4 x^4 + \cdots + a^n x^n$$

Remember that we only need to show the first five terms.

**a)** $(1 + x)^{12}$
**Solution**
Let's get started

**Step 1:** Write down the formula

$$(1 + x)^{12} = 1 + nax + \binom{n}{2} a^2 x^2 + \binom{n}{3} a^3 x^3 + \binom{n}{4} a^4 x^4 + \ldots$$

**Step 2:** Substitute for $n = 12$ and $a = 1$.

$$(1 + x)^{12} = 1 + 12(1)x + \binom{12}{2}(1)^2 x^2 + \binom{12}{3}(1)^3 x^3 + \binom{12}{4}(1)^4 x^4 + \ldots$$

**Step 3:** It is time to simplify

$$(1 + x)^{12} = 1 + 12x + 66x^2 + 220x^3 + 495x^4 + \ldots$$
$$\therefore (1 + x)^{12} = 1 + 12x + 66x^2 + 220x^3 + 495x^4 + \ldots$$

**b)** $(1-2x)^8$

**Solution**

Let's get started

**Step 1:** Write down the formula

$$(1-2x)^8 = 1 + nax + \binom{n}{2}a^2x^2 + \binom{n}{3}a^3x^3 + \binom{n}{4}a^4x^4 + \ldots$$

**Step 2:** Substitute for $n = 8$ and $a = -2$.

$$(1-2x)^8 = 1 + 8(-2)x + \binom{8}{2}(-2)^2x^2 + \binom{8}{3}(-2)^3x^3 + \binom{8}{4}(-2)^4x^4 + \ldots$$

**Step 3:** It is time to simplify

$$(1-2x)^8 = 1 - 16x + (28)(4)x^2 + (56)(-8)x^3 + (70)(16)x^4 + \ldots$$
$$= 1 - 16x + 112x^2 - 448x^3 + 1120x^4 + \ldots$$
$$\therefore (1-2x)^8 = 1 - 16x + 112x^2 - 448x^3 + 1120x^4 + \ldots$$

---

Another example to try.

### Example 7

Using binomial theorem or otherwise, determine the coefficient of $x^3$ in the expansion $(x+1)^5(1-3x)^7$.

---

What did you get? Find the solution below to double-check your answer.

### Solution to Example 7

**HINT**

- Apply either of the formulas below to each term. Show up to the fourth term only since we want only the coefficient of $x^3$.

$$\boxed{(1+x)^n = 1 + nx + {}^nC_2 x^2 + {}^nC_3 x^3 + \ldots}$$

$$\boxed{(1+ax)^n = 1 + nax + {}^nC_2 a^2 x^2 + {}^nC_3 a^3 x^3 + \ldots}$$

- Appropriately multiply the results from above.

Let's get started with $(x+1)^5 = (1+x)^5$

Notice that we have switched the position of 1 and $x$. It is based on the commutative law of addition, and it also helps to write the expression in the same format as the formula.

# The Binomial Expansion

**Step 1:** Write down the formula

$$(1+x)^n = 1 + nx + \binom{n}{2}x^2 + \binom{n}{3}x^3 + \ldots$$

**Step 2:** Substitute for $n = 5$ and $a = 1$.

$$(1+x)^5 = 1 + 5x + \binom{5}{2}x^2 + \binom{5}{3}x^3 + \ldots$$

**Step 3:** It is time to simplify

$$(1+x)^5 = 1 + 5x + 10x^2 + 10x^3 + \ldots$$
$$\therefore (1+x)^5 = 1 + 5x + 10x^2 + 10x^3 + \ldots$$

Now let's deal with $(1-3x)^7$.

**Step 4:** Write down the formula

$$(1+ax)^n = 1 + nax + \binom{n}{2}a^2x^2 + \binom{n}{3}a^3x^3 + \ldots$$

**Step 5:** Substitute for $n = 7$ and $a = -3$.

$$(1-3x)^7 = 1 + 7(-3)x + \binom{7}{2}(-3)^2x^2 + \binom{7}{3}(-3)^3x^3 + \ldots$$

**Step 6:** It is time to simplify

$$(1-3x)^7 = 1 - 21x + (21)(9)x^2 + (35)(-27)x^3 + \ldots$$
$$\therefore (1-3x)^7 = 1 - 21x + 189x^2 - 945x^3 + \ldots$$

Now let's deal with $(x+1)^5(1-3x)^7$

**Step 7:** Write down the formula

$(x+1)^5(1-3x)^7$
$= (1 + 5x + 10x^2 + 10x^3 + \ldots)(1 - 21x + 189x^2 - 945x^3 + \ldots)$
$= 1(1 - 21x + 189x^2 - 945x^3 + \ldots) + 5x(1 - 21x + 189x^2 + \ldots)$
$\quad + 10x^2(1 - 21x + \ldots) + 10x^3(1 + \ldots)$
$= 1 - 21x + 189x^2 - 945x^3 + \cdots + 5x - 105x^2 + 945x^3 + \cdots + 10x^2 - 210x^3 + \cdots + 10x^3 \ldots$
$= 1 - 16x + 94x^2 - 200x^3 + \ldots$

$$\therefore \text{ The coefficient of } x^3 \text{ is } -200$$

Another example to try.

### Example 8

Using the binomial theorem, determine the term containing the indicated term of $x$. Give your answer in the simplest form.

a) $(1-x)^{10}$, with $[x^8]$

b) $(1+3x)^{13}$, with $[x^5]$

What did you get? Find the solution below to double-check your answer.

### Solution to Example 8

**HINT**

Apply

$$\binom{n}{r}(ax)^r$$

We can try to verify this using the formula below

$$(1+ax)^n = 1 + nax + {}^nC_2 a^2 x^2 + {}^nC_3 a^3 x^3 + {}^nC_4 a^4 x^4 + \cdots + a^n x^n$$

a) $(1-x)^{10}$, with $[x^8]$
**Solution**
Let's get started.

**Step 1:** Write down the formula for finding a term

$$\binom{n}{r}(ax)^r$$

**Step 2:** Substitute for $n = 10$, $r = 8$ and $a = -1$.

$$\binom{10}{8}(-x)^8$$

**Step 3:** It is time to simplify

$$45x^8$$

Thus, the term containing $x^8$ in $(1-x)^{10}$

$$45x^8$$

# The Binomial Expansion

**b)** $(1 + 3x)^{13}$, with $[x^5]$

**Solution**

Let's get started.

**Step 1:** Write down the formula for finding a term

$$\binom{n}{r}(ax)^r$$

**Step 2:** Substitute for $n = 13$, $r = 5$, and $a = 3$.

$$\binom{13}{5}(3x)^5$$

**Step 3:** It is time to simplify

$$(1287)(3^5)x^5 = 312741x^5$$

Thus, the term containing $x^5$ in $(1 + 3x)^{13}$ is

$$312741x^5$$

---

## 6.7.2 When $n$ Is Negative

We may wonder what is special about a negative number. Yes, there is nothing special about it, but our approach so far cannot take care of a negative value of $n$. This is because Pascal's triangle only has positive values of coefficients, and $\binom{n}{r}$ is not valid for $n < 0$. You can try this on your calculator, say $\binom{-5}{3}$.

We will thus need to modify our previous formula to eliminate the need for $\binom{n}{r}$. Before this, it is important to mention that when $n$ is positive, the resulting series is finite, i.e., we have a fixed number of terms. What we will have now is an infinite series, such that we can only show up to a particular term. Now let's go.

$$(1 + x)^n = 1 + nx + {}^nC_2 x^2 + {}^nC_3 x^3 + {}^nC_4 x^4 + \cdots + x^n$$
$$= 1 + nx + ({}^nC_2)x^2 + ({}^nC_3)x^3 + ({}^nC_4)x^4 + \cdots + x^n$$
$$= 1 + nx + \left(\frac{n!}{2!(n-2)!}\right)x^2 + \left(\frac{n!}{3!(n-3)!}\right)x^3 + \left(\frac{n!}{4!(n-4)!}\right)x^4 + \cdots$$

'Open up' (or evaluate) $n!$ in each numerator as:

$$= 1 + nx + \left(\frac{n(n-1)(n-2)!}{2!(n-2)!}\right)x^2 + \left(\frac{n(n-1)(n-2)(n-3)!}{3!(n-3)!}\right)x^3$$
$$+ \left(\frac{n(n-1)(n-2)(n-3)(n-4)!}{4!(n-4)!}\right)x^4 + \cdots + x^n$$

After cancelling out, we have:

$$= 1 + nx + \left(\frac{n(n-1)}{2!}\right)x^2 + \left(\frac{n(n-1)(n-2)}{3!}\right)x^3 + \left(\frac{n(n-1)(n-2)(n-3)}{4!}\right)x^4 + \cdots$$

We can therefore write a general formula which has no need for the combination notation as:

$$(1+x)^n = 1 + nx + \frac{n(n-1)}{2!}x^2 + \frac{n(n-1)(n-2)}{3!}x^3$$
$$+ \frac{n(n-1)(n-2)(n-3)}{4!}x^4 + ...$$
(6.10)

The above modified form has eliminated the need for $\binom{n}{r}$.

Let's summarise the main key points about this formula to ensure we get the pattern and can effectively use it.

**Note 1** The first and the second terms are identical to what we have in the first case when $n$ is a positive integer.

**Note 2** The last term cannot be obtained as the expansion is **infinite** as opposed to a **finite** expansion when $n$ is a positive integer.

**Note 3** If the denominator is $r!$, the power of $x$ or $ax$ will be $r$.

**Note 4** If the denominator is $r!$, there will be $r$ multiplicands in the numerator such that the first will be $n$ (positive or negative), the second will be one less than the first, and so on. To illustrate, if the binomial expression is $(1+x)^{-5}$ and the denominator of a term is $3!$, then the numerator will be $(-5 \times -6 \times -7)$. Notice that we have produced three numbers starting with $n = -5$ and subtracted 1 each time to have $-6$ and $-7$.

**Note 5** The above formula can be used for both positive and negative values of $n$.

**Note 6** We can now expand a fractional algebra, such as $\frac{1-x}{(3-2x)^5}$, provided that the denominator is a binomial expression. Otherwise, we will need to apply the partial fractions technique (see Chapter 7) first before the application of the binomial expansion technique.

For $(1 + ax)^n$, it follows the same pattern as $(1 + x)^n$ with a little modification as:

$$(1+ax)^n = 1 + nax + \frac{n(n-1)}{2!}a^2x^2 + \frac{n(n-1)(n-2)}{3!}a^3x^3$$
$$+ \frac{n(n-1)(n-2)(n-3)}{4!}a^4x^4 + ...$$
(6.11)

### 6.7.3 When $n$ Is a Fraction

Recall that $\sqrt[n]{(1+ax)} = (1+ax)^{\frac{1}{n}}$. When $n$ is a fraction, the approach will be the same as when $n$ is negative, hence the formulas shown before (Equations 6.10 and 6.11) remain the same. You will see this shortly in the worked examples.

# The Binomial Expansion

## 6.7.4 Validity of Binomial Expression

Now that we've covered all possible rational values of $n$, we should mention the validity of our expansions as follows:

**Rule 1** When $n > 1$, the binomial expression $(a + bx)^n$ is valid for all values of $x$.
**Rule 2** When $n < 1$ ($n$ is negative or fraction), the binomial expression $(a + bx)^n$ is valid only when

$$|bx| < a \text{ OR } |x| < \frac{a}{b}$$

for all values of $x$.

It is time to try examples to illustrate this.

### Example 9

Using the binomial theorem, expand the following up to (and including) the term in $x^3$. Also state the range of $x$ for which each expansion is valid.

a) $\dfrac{1}{(1+x)^4}$      b) $\sqrt{\left(1 - \dfrac{1}{3}x\right)}$      c) $\sqrt[5]{(1-4x)^2}$      d) $\dfrac{(1-x)^3}{(1+6x)^7}$

What did you get? Find the solution below to double-check your answer.

### Solution to Example 9

#### Hint

Firstly, we need to apply the relevant law(s) of indices to obtain $(1 + ax)^n$ and then apply the expansion below

$$(1 + ax)^n = 1 + nax + \frac{n(n-1)}{2!}a^2x^2 + \frac{n(n-1)(n-2)}{3!}a^3x^3 + \frac{n(n-1)(n-2)(n-3)}{4!}a^4x^4 + \ldots$$

We will use the above for all four variants $(1 \pm x)^n$ and $(1 \pm ax)^n$ by appropriate substitutions.

a) $\dfrac{1}{(1+x)^4}$
**Solution**

$$\frac{1}{(1+x)^4} = (1+x)^{-4}$$

Let's get started.

**Step 1:** Write down the formula

$$(1+ax)^n = 1 + nax + \frac{n(n-1)}{2!}a^2x^2 + \frac{n(n-1)(n-2)}{3!}a^3x^3 + \ldots$$

**Step 2:** Substitute for $n = -4$ and $a = 1$

$$(1+x)^n = 1 + nx + \frac{n(n-1)}{2!}x^2 + \frac{n(n-1)(n-2)}{3!}x^3 + \ldots$$

$$(1+x)^{-4} = 1 + (-4)x + \frac{-4(-4-1)}{2!}x^2 + \frac{-4(-4-1)(-4-2)}{3!}x^3 + \ldots$$

**Step 3:** Now simplify

$$(1+x)^{-4} = 1 - 4x + \frac{20}{2!}x^2 + \frac{-120}{3!}x^3 + \ldots$$
$$= 1 - 4x + 10x^2 - 20x^3 + \ldots$$
$$\therefore \frac{1}{(1+x)^4} \approx 1 - 4x + 10x^2 - 20x^3$$

$\frac{1}{(1+x)^4}$ is valid when $|x| < 1$

$$\therefore \textbf{Validity is } |x| < 1$$

---

**b)** $\sqrt{\left(1 - \frac{1}{3}x\right)}$

**Solution**

$$\sqrt{\left(1 - \frac{1}{3}x\right)} = \left(1 - \frac{1}{3}x\right)^{\frac{1}{2}}$$

Let's get started.

**Step 1:** Write down the formula

$$(1+ax)^n = 1 + nax + \frac{n(n-1)}{2!}a^2x^2 + \frac{n(n-1)(n-2)}{3!}a^3x^3 + \ldots$$

**Step 2:** Substitute for $n = \frac{1}{2}$ and $a = -\frac{1}{3}$

$$\left(1 - \frac{1}{3}x\right)^{\frac{1}{2}} = 1 + \left(\frac{1}{2}\right)\left(-\frac{1}{3}\right)x + \frac{\frac{1}{2}\left(\frac{1}{2}-1\right)}{2!}\left(-\frac{1}{3}\right)^2 x^2 + \frac{\frac{1}{2}\left(\frac{1}{2}-1\right)\left(\frac{1}{2}-2\right)}{3!}\left(-\frac{1}{3}\right)^3 x^3 + \ldots$$

# The Binomial Expansion

**Step 3:** Now simplify

$$\left(1-\frac{1}{3}x\right)^{\frac{1}{2}} = 1 - \frac{1}{6}x + \frac{\left(-\frac{1}{4}\right)}{2!}\left(\frac{1}{9}\right)x^2 + \frac{\frac{3}{8}}{3!}\left(-\frac{1}{27}\right)x^3 + \ldots$$

$$= 1 - \frac{1}{6}x + \left(-\frac{1}{8}\right)\left(\frac{1}{9}\right)x^2 + \left(\frac{1}{16}\right)\left(-\frac{1}{27}\right)x^3 + \ldots$$

$$= 1 - \frac{1}{6}x - \frac{1}{72}x^2 - \frac{1}{432}x^3 + \ldots$$

$$\therefore \sqrt{\left(1-\frac{1}{3}x\right)} \approx 1 - \frac{1}{6}x - \frac{1}{72}x^2 - \frac{1}{432}x^3$$

$\sqrt{\left(1-\frac{1}{3}x\right)}$ is valid when $\left|\frac{1}{3}x\right| < 1$

$$\frac{1}{3}|x| < 1 \text{ OR } |x| < 3$$

$$\therefore \textbf{Validity is } |x| < 3$$

---

**c)** $\sqrt[5]{(1-4x)^2}$

**Solution**

$$\sqrt[5]{(1-4x)^2} = (1-4x)^{\frac{2}{5}}$$

Let's get started.

**Step 1:** Write down the formula

$$(1+ax)^n = 1 + nax + \frac{n(n-1)}{2!}a^2x^2 + \frac{n(n-1)(n-2)}{3!}a^3x^3 + \ldots$$

**Step 2:** Substitute for $n = \frac{2}{5}$ and $a = -4$

$$(1-4x)^{\frac{2}{5}} = 1 + \left(\frac{2}{5}\right)(-4)x + \frac{\frac{2}{5}\left(\frac{2}{5}-1\right)}{2!}(-4)^2x^2 + \frac{\frac{2}{5}\left(\frac{2}{5}-1\right)\left(\frac{2}{5}-2\right)}{3!}(-4)^3x^3 + \ldots$$

**Step 3:** Now simplify

$$(1-4x)^{\frac{2}{5}} = 1 - \frac{8}{5}x + \frac{\left(-\frac{6}{25}\right)}{2!}(16)x^2 + \frac{\frac{48}{125}}{3!}(-64)x^3 + \ldots$$

$$= 1 - \frac{8}{5}x + \left(-\frac{3}{25}\right)(16)x^2 + \left(\frac{8}{125}\right)(-64)x^3 + \ldots$$

$$= 1 - \frac{8}{5}x - \frac{48}{25}x^2 - \frac{512}{125}x^3 + \ldots$$

$$\therefore \sqrt[5]{(1-4x)^2} \approx 1 - \frac{8}{5}x - \frac{48}{25}x^2 - \frac{512}{125}x^3$$

$\sqrt[5]{(1-4x)^2}$ is valid when $|4x| < 1$

$$4|x| < 1 \text{ OR } |x| < \frac{1}{4}$$

$$\therefore \textbf{Validity is } |x| < \frac{1}{4}$$

---

**d)** $\frac{(1-x)^3}{(1+6x)^7}$
**Solution**

$$\frac{(1-x)^3}{(1+6x)^7} = (1-x)^3(1+6x)^{-7}$$

Let's start with $(1-x)^3$

**Step 1:** Write down the formula

$$(1+x)^n = 1 + nx + \frac{n(n-1)}{2!}x^2 + \frac{n(n-1)(n-2)}{3!}x^3 + \ldots$$

**Step 2:** Substitute for $n = 3$ and $a = -1$. Alternatively, replace $x$ with $-x$

$$(1-x)^3 = 1 + 3(-x) + \frac{3(3-1)}{2!}(-x)^2 + \frac{3(3-1)(3-2)}{3!}(-x)^3$$

**Step 3:** Now simplify

$$(1-x)^3 = 1 - 3x + \frac{6}{2!}(x^2) + \frac{6}{3!}(-x^3)$$
$$= 1 - 3x + 3x^2 - x^3$$

Now we need $(1+6x)^{-7}$

**Step 4:** Write down the formula

$$(1+ax)^n = 1 + nax + \frac{n(n-1)}{2!}a^2x^2 + \frac{n(n-1)(n-2)}{3!}a^3x^3 + \ldots$$

**Step 5:** Substitute for $n = -7$ and $a = 6$.

$$(1+6x)^{-7} = 1 + (-7)(6)x + \frac{-7(-7-1)}{2!}(6)^2x^2 + \frac{-7(-7-1)(-7-2)}{3!}(6)^3x^3 + \ldots$$

**Step 6:** Now simplify

$$(1+6x)^{-7} = 1 - 42x + \frac{56}{2!}(36)x^2 + \frac{(-504)}{3!}(216)x^3 + \ldots$$
$$= 1 - 42x + (28)(36)x^2 + (-84)(216)x^3 + \ldots$$
$$= 1 - 42x + 1008x^2 - 18144x^3 + \ldots$$

Now we need $(1-x)^3(1+6x)^{-7}$.

**Step 7:** Multiply the two expressions

$$(1-x)^3(1+6x)^{-7} = (1 - 3x + 3x^2 - x^3)(1 - 42x + 1008x^2 - 18144x^3 + \ldots)$$

**Step 8:** Simplify

$$(1-x)^3(1+6x)^{-7}$$
$$= 1(1 - 42x + 1008x^2 - 18144x^3 + \ldots) - 3x(1 - 42x + 1008x^2 - 18144x^3 + \ldots)$$
$$+ 3x^2(1 - 42x + 1008x^2 - 18144x^3 + \ldots) - x^3(1 - 42x + 1008x^2 - 18144x^3 + \ldots)$$
$$= 1 - 42x + 1008x^2 - 18144x^3 + \cdots - 3x + 126x^2 - 3024x^3 + \ldots 3x^2 - 126x^3 + \cdots - x^3$$
$$= 1 - 45x + 1137x^2 - 21295x^3 + \ldots$$

$$\therefore \frac{(1-x)^3}{(1+6x)^7} \approx 1 - 45x + 1137x^2 - 21295x^3$$

$(1-x)^3$ is valid for all values of $x$, because $n$ is a positive integer and the expansion is therefore finite.
$(1+6x)^{-7}$ is valid when $|6x| < 1$

$$6|x| < 1$$
$$|x| < \frac{1}{6}$$

The validity of $\frac{(1-x)^3}{(1+6x)^7}$ is the overlap region of the two validities

$$\therefore \textbf{Validity is } |x| < \frac{1}{6}$$

---

## 6.8 BINOMIAL EXPANSION: $(p+q)^n$

In this section, we will look at the last possible case that we may encounter in binomial expansion. This takes the general form of $(p+q)^n$ and can fall under one of the following categories:

**Type 1**    $p$ and $q$ are numerical values, with a form $(a+b)^n$ where $a$ and $b$ are rational numbers. For example, $(2+3)^5$ is a simple indicial expression and $(\sqrt{3}-1)^2$ is a surdic expression, and do not fall under binomial expression.

**Type 2**    $p$ is an algebraic term $ax$ and $q$ is a numerical value $b$, which takes a general form $(ax+b)^n$; for example, $(2x+3)^5$. Alternatively, $p$ is a numerical value $a$ and $q$ is an algebraic term $bx$, which takes a general form $(a+bx)^n$. For example, $(5+2x)^{10}$. Both are the same and can be taken as one.

**Type 3**    $p$ is an algebraic term $ax$ and $q$ is another algebraic term $by$, such that $x \neq y$. This takes a general form $(ax+by)^n$; for example, $(2x+3y)^5$. If, however, $x = y$ this becomes similar to type 1 and is treated as a simple algebraic expression.

We will now consider two methods of dealing with Type 2 and Type 3 as follows.

## 6.8.1 METHOD 1

Given $(ax + by)^n$, the binomial expansion is

$$(ax + by)^n = \binom{n}{0}(ax)^n(by)^0 + \binom{n}{1}(ax)^{n-1}(by)^1 + \binom{n}{2}(ax)^{n-2}(by)^2 + \binom{n}{3}(ax)^{n-3}(by)^3 + \ldots$$
$$+ \binom{n}{n-1}(ax)^1(by)^{n-1} + \binom{n}{n}(ax)^0(by)^n$$

You should recognise this pattern from Pascal's triangle at the beginning of this chapter. Let's tidy the above up a bit as:

$$\boxed{\begin{array}{c}(ax + by)^n = (ax)^n + n(ax)^{n-1}(by)^1 + \binom{n}{2}(ax)^{n-2}(by)^2 + \binom{n}{3}(ax)^{n-3}(by)^3 + \\ \cdots + n(ax)^1(by)^{n-1} + (by)^n\end{array}}$$
(6.12)

To ensure that we can use the above formula for any rational value of $n$, including negative and fraction, let's get rid of the $^nC_r$ bit. Here we go:

$$\boxed{\begin{array}{c}(ax + by)^n = (ax)^n + n(ax)^{n-1}(by)^1 + \frac{n(n-1)}{2!}(ax)^{n-2}(by)^2 + \\ \frac{n(n-1)(n-2)}{3!}(ax)^{n-3}(by)^3 + \ldots\end{array}}$$
(6.13)

By appropriate substitution, we can say that:

$$\boxed{\begin{array}{c}(a + bx)^n = (a)^n + n(a)^{n-1}(bx)^1 + \frac{n(n-1)}{2!}(a)^{n-2}(bx)^2 \\ + \frac{n(n-1)(n-2)}{3!}(a)^{n-3}(bx)^3 + \ldots\end{array}}$$
(6.14)

## 6.8.2 METHOD 2

This approach is more appropriate for $(a + bx)^n$, but can also be applied to $(ax + by)^n$. Recall that

$$a + bx = a\left(1 + \frac{b}{a}x\right)$$

Thus,

$$(a + bx)^n = \left[a\left(1 + \frac{b}{a}x\right)\right]^n = a^n\left(1 + \frac{b}{a}x\right)^n$$

What this implies is that to find the expansion for $(a + bx)^n$, we will find the expansion for $\left(1 + \frac{b}{a}x\right)^n$ and multiply each term of the result by $a^n$ as follows

$$(a + bx)^n = a^n\left(1 + \frac{b}{a}x\right)^n$$
$$= a^n\left[1 + n\left(\frac{b}{a}\right)x + \frac{n(n-1)}{2!}\left(\frac{b}{a}\right)^2 x^2 + \frac{n(n-1)(n-2)}{3!}\left(\frac{b}{a}\right)^3 x^3\right.$$
$$\left.+ \frac{n(n-1)(n-2)(n-3)}{4!}\left(\frac{b}{a}\right)^4 x^4 + \ldots\right]$$

# The Binomial Expansion

You should be able to use the same trick shown above to $(ax+by)^n$. I guess we've now mastered the technique. It is time to try some examples.

## Example 10

Using the binomial theorem, expand the following up to (and including) the term indicated. State the range of values of $x$ for which the expansion is valid.

a) $(4x+2)^3$, up to $x^3$

b) $\sqrt{(5x+9)}$, up to $x^2$

c) $\dfrac{1}{(2-3x)^7}$, up to $x^2$

d) $(3x+4y)^5$, up to $x^5$

What did you get? Find the solution below to double-check your answer.

## Solution to Example 10

a) $(4x+2)^3$
**Solution**
Let's get started.

**Step 1:** Preliminary manipulation

$$(4x+2)^3 = (2+4x)^3$$
$$= \left[2\left(1+\frac{4}{2}x\right)\right]^3 = 2^3(1+2x)^3$$
$$= 8(1+2x)^3$$

**Step 2:** Write down the formula

$$(1+ax)^n = 1 + nax + \frac{n(n-1)}{2!}a^2x^2 + \frac{n(n-1)(n-2)}{3!}a^3x^3 + \ldots$$

**Step 3:** Substitute for $n=3$ and $a=2$.

$$(2+4x)^3 = 8(1+2x)^3$$
$$= 8\left[1 + 3(2)x + \frac{3(3-1)}{2!}(2)^2 x^2 + \frac{3(3-1)(3-2)}{3!}(2)^3 x^3\right]$$

**Step 4:** It is time to simplify

$$(2+4x)^3 = 8\left[1 + 6x + \frac{6}{2!}(4)x^2 + \frac{6}{3!}(8)x^3\right]$$
$$= 8\left[1 + 6x + \frac{6}{2!}(4)x^2 + \frac{6}{3!}(8)x^3\right]$$
$$= 8\left[1 + 6x + 12x^2 + 8x^3\right]$$
$$= 8 + 48x + 96x^2 + 64x^3$$
$$\therefore (4x+2)^3 = 8 + 48x + 96x^2 + 64x^3$$

$(4x + 2)^3$ is valid for all values of $x$, because $n$ is a positive integer and the expansion is therefore finite.

$$\therefore \textbf{Valid for all values of } x$$

**ALTERNATIVE METHOD**

**Step 1:** Write down the formula

$$(a + bx)^n = (a)^n + n(a)^{n-1}(bx)^1 + \frac{n(n-1)}{2!}(a)^{n-2}(bx)^2 + \frac{n(n-1)(n-2)}{3!}(a)^{n-3}(bx)^3$$

**Step 2:** Substitute for $n = 3$, $a = 2$, and $b = 4$.

$$(4x + 2)^3 = (2 + 4x)^3$$
$$= (2)^3 + 3(2)^{3-1}(4x)^1 + \frac{3(3-1)}{2!}(2)^{3-2}(4x)^2 + \frac{3(3-1)(3-2)}{3!}(2)^{3-3}(4x)^3$$

**Step 3:** It is time to simplify

$$(4x + 2)^3 = (2 + 4x)^3 = 8 + 3(2)^2(4x) + \frac{3(2)}{2!}(2)(4^2x^2) + \frac{3(2)(1)}{3!}(2)^0(4^3x^3)$$
$$= 8 + 48x + (3)(2)(16x^2) + (1)(1)(64x^3)$$
$$\therefore (4x + 2)^3 = 8 + 48x + 96x^2 + 64x^3$$

---

**b)** $\sqrt{(5x + 9)}$

**Solution**

Let's get started.

**Step 1:** Preliminary manipulation

$$\sqrt{(5x + 9)} = \sqrt{(9 + 5x)} = (9 + 5x)^{\frac{1}{2}}$$

**Step 2:** Write down the formula

$$(a + bx)^n = (a)^n + n(a)^{n-1}(bx)^1 + \frac{n(n-1)}{2!}(a)^{n-2}(bx)^2 + \ldots$$

**Step 3:** Substitute for $n = \frac{1}{2}$, $a = 9$, and $b = 5$.

$$(9 + 5x)^{\frac{1}{2}} = (9)^{\frac{1}{2}} + \left(\frac{1}{2}\right)(9)^{\left(\frac{1}{2}-1\right)}(5x)^1 + \frac{\frac{1}{2}\left(\frac{1}{2}-1\right)}{2!}(9)^{\left(\frac{1}{2}-2\right)}(5x)^2 + \ldots$$

# The Binomial Expansion

**Step 4:** It is time to simplify

$$(9+5x)^{\frac{1}{2}} = 3 + \left(\frac{1}{2}\right)(9)^{\left(-\frac{1}{2}\right)}(5x) + \frac{\left(-\frac{1}{4}\right)}{2!}(9)^{\left(-\frac{3}{2}\right)}(5^2 x^2) + \ldots$$

$$= 3 + \frac{1}{2}\left(\frac{1}{3}\right)(5x) - \frac{1}{8}\left(\frac{1}{27}\right)(25x^2) + \ldots$$

$$= 3 + \frac{5}{6}x - \frac{25}{216}x^2 + \ldots$$

$$\therefore \sqrt{(5x+9)} \approx 3 + \frac{5}{6}x - \frac{25}{216}x^2$$

$(9+5x)^{\frac{1}{2}}$ is valid when $|5x| < 9$

$$5|x| < 9$$

$$|x| < \frac{9}{5}$$

$$\therefore \textbf{Validity is } |x| < \frac{9}{5}$$

## ALTERNATIVE METHOD

**Step 1:** Preliminary manipulation

$$\sqrt{(5x+9)} = \sqrt{(9+5x)} = (9+5x)^{\frac{1}{2}}$$

$$= \left[9\left(1 + \frac{5}{9}x\right)\right]^{\frac{1}{2}} = 9^{\left(\frac{1}{2}\right)}\left(1 + \frac{5}{9}x\right)^{\frac{1}{2}}$$

$$= 3\left(1 + \frac{5}{9}x\right)^{\frac{1}{2}}$$

**Step 2:** Write down the formula

$$(1 + ax)^n = 1 + nax + \frac{n(n-1)}{2!}a^2x^2 + \ldots$$

**Step 3:** Substitute for $n = \frac{1}{2}$ and $a = \frac{5}{9}$.

$$\sqrt{(9+5x)} = 3\left(1 + \frac{5}{9}x\right)^{\frac{1}{2}}$$

$$= 3\left[1 + \frac{1}{2}\left(\frac{5}{9}\right)x + \frac{\frac{1}{2}\left(\frac{1}{2}-1\right)}{2!}\left(\frac{5}{9}\right)^2 x^2 + \ldots\right]$$

**Step 4:** It is time to simplify

$$\sqrt{(9+5x)} = 3\left[1 + \frac{5}{18}x + \frac{\frac{1}{2}\left(-\frac{1}{2}\right)}{2!}\left(\frac{25}{81}\right)x^2 + \ldots\right]$$

$$= 3\left[1 + \frac{5}{18}x - \frac{1}{8}\left(\frac{25}{81}\right)x^2 + \ldots\right]$$

$$= 3\left[1 + \frac{5}{18}x - \frac{25}{648}x^2 + \ldots\right]$$

$$= 3 + \frac{5}{6}x - \frac{25}{216}x^2 + \ldots$$

$$\therefore \sqrt{(9+5x)} \approx 3 + \frac{5}{6}x - \frac{25}{216}x^2$$

c) $\frac{1}{(2-3x)^7}$

**Solution**
Let's get started.

**Step 1:** Preliminary manipulation

$$\frac{1}{(2-3x)^7} = (2-3x)^{-7}$$

**Step 2:** Write down the formula

$$(a+bx)^n = (a)^n + n(a)^{n-1}(bx)^1 + \frac{n(n-1)}{2!}(a)^{n-2}(bx)^2 + \ldots$$

**Step 3:** Substitute for $n = -7$, $a = 2$ and $b = -3$.

$$(2-3x)^{-7} = (2)^{-7} + (-7)(2)^{-7-1}(-3x)^1 + \frac{-7(-7-1)}{2!}(2)^{-7-2}(-3x)^2 + \ldots$$

**Step 4:** It is time to simplify

$$(2-3x)^{-7} = \frac{1}{2^7} + (-7)(2)^{-8}(-3x)^1 + \frac{-7(-8)}{2!}(2)^{-9}(-3x)^2 + \ldots$$

$$= \frac{1}{128} - \frac{7}{2^8}(-3x) + (28)\left(\frac{1}{2^9}\right)(9x^2) + \ldots$$

$$= \frac{1}{128} + \frac{21}{256}x + \frac{63}{128}x^2 + \ldots$$

$$\therefore \frac{1}{(2-3x)^7} \approx \frac{1}{128} + \frac{21}{256}x + \frac{63}{128}x^2$$

# The Binomial Expansion

$\frac{1}{(2-3x)^7}$ is valid when $|3x| < 2$

$$3|x| < 2$$

$$|x| < \frac{2}{3}$$

$$\therefore \text{ Validity is } |x| < \frac{2}{3}$$

---

**d)** $(3x + 4y)^5$
**Solution**
Let's get started.
**Step 1:** Write down the formula

$$(ax + by)^n = \binom{n}{0}(ax)^n(by)^0 + \binom{n}{1}(ax)^{n-1}(by)^1 + \binom{n}{2}(ax)^{n-2}(by)^2 + \binom{n}{3}(ax)^{n-3}(by)^3 + \ldots$$

**Step 2:** Substitute for $n = 5$, $a = 3$, and $b = 4$.

$(3x + 4y)^5$
$= \binom{5}{0}(3x)^5(4y)^0 + \binom{5}{1}(3x)^{5-1}(4y)^1 + \binom{5}{2}(3x)^{5-2}(4y)^2 + \binom{5}{3}(3x)^{5-3}(4y)^3 + \binom{5}{4}(3x)^{5-4}(4y)^4$
$\quad + \binom{5}{5}(3x)^{5-5}(4y)^5$

**Step 3:** It is time to simplify

$(3x + 4y)^5$
$= (1)(3x)^5 + \binom{5}{1}(3x)^4(4y)^1 + \binom{5}{2}(3x)^3(4y)^2 + \binom{5}{3}(3x)^2(4y)^3 + \binom{5}{4}(3x)^1(4y)^4 + (1)(4y)^5$
$= (3^5 x^5) + (5)(3^4 x^4)(4y) + (10)(3^3 x^3)(4^2 y^2) + (10)(3^2 x^2)(4^3 y^3) + (5)(3x)(4^4 y^4) + (4^5 y^5)$
$= (3^5 x^5) + (5 \times 3^4 \times 4)(x^4 y) + (10 \times 3^3 \times 4^2)(x^3 y^2) + (10 \times 3^2 \times 4^3)(x^2 y^3) + (5 \times 3 \times 4^4)(xy^4)$
$\quad + (4^5 y^5)$
$= 243x^5 + (1620)(x^4 y) + (4320)(x^3 y^2) + (5760)(x^2 y^3) + (3840)(xy^4) + (1024 y^5)$
$= 243x^5 + 1620x^4 y + 4320x^3 y^2 + 5760x^2 y^3 + 3840xy^4 + 1024y^5$
$\therefore (3x + 4y)^5 = \mathbf{243x^5 + 1620x^4 y + 4320x^3 y^2 + 5760x^2 y^3 + 3840xy^4 + 1024y^5}$

$(3x + 4y)^5$ is valid for all values of $x$, and because $n$ is a positive integer, the expansion is therefore finite.

$$\therefore \text{ Validity is all values of } x$$

## 6.9 APPROXIMATION

We can use binomial expansion to find the approximate value of a number raised to a power $n$, and we have encountered this in Example 4. To do this, we use the expression $(a + bx)^n$ such that the value of $x$ is very small. This is because the term $x^r$ will be negligible at higher powers and can thus be ignored. In other words, we only need the first few terms of the expansion of $(a + bx)^n$ to approximate the value of the given number.

Let's try a few examples to finish this chapter.

### Example 11

Using the binomial theorem, determine the approximate value of the following. Present your answer correct to 3 d.p.

a) $0.993^{12}$ 	b) $\sqrt[3]{1.03}$ 	c) $\frac{500}{494}$

What did you get? Find the solution below to double-check your answer.

### Solution to Example 11

**a)** $0.993^{12}$
**Solution**
Let's get started.
**Step 1:** Preliminary manipulation

$$0.993^{12} = (1 - 0.007)^{12}$$
$$= (1 - 7x)^{12}$$

where $x = 0.001$

**Step 2:** Write down the formula

$$(1 + ax)^n = 1 + nax + \frac{n(n-1)}{2!}a^2x^2 + \frac{n(n-1)(n-2)}{3!}a^3x^3 + \ldots$$

**Step 3:** Substitute for $n = 12$ and $a = -7$.

$$(1 - 7x)^{12} = 1 + 12(-7)x + \frac{12(12-1)}{2!}(-7)^2x^2 + \frac{12(12-1)(12-2)}{3!}(-7)^3x^3 + \ldots$$

**Step 4:** It is time to simplify

$$(1 - 7x)^{12} = 1 - 84x + (66)(49)x^2 + 220(-343)x^3 + \ldots$$
$$= 1 - 84x + 3234x^2 - 75460x^3 + \ldots$$

**Step 5:** Now substitute for $x = 0.001$.

$$(1 - 7x)^{12} = 1 - 84(0.001) + 3234(0.001)^2 - 75460(0.001)^3 + \ldots$$
$$= 1 - 0.084 + 3.234 \times 10^{-3} - 7.546 \times 10^{-5} + \cdots \approx 0.91915854$$

$$\therefore 0.993^{12} \approx 0.919$$

**b)** $\sqrt[3]{1.03}$

**Solution**

Let's get started.

**Step 1:** Preliminary manipulation

$$\sqrt[3]{1.03} = (1.03)^{\frac{1}{3}} = (1 + 0.03)^{\frac{1}{3}}$$
$$= (1 + 3x)^{\frac{1}{3}}$$

where $x = 0.01$

**Step 2:** Write down the formula

$$(1 + ax)^n = 1 + nax + \frac{n(n-1)}{2!}a^2x^2 + \frac{n(n-1)(n-2)}{3!}a^3x^3 + \frac{n(n-1)(n-2)(n-3)}{4!}a^4x^4 + \ldots$$

**Step 3:** Substitute for $n = \frac{1}{3}$ and $a = 3$.

$$(1 + 3x)^{\frac{1}{3}} = 1 + \frac{1}{3}(3)x + \frac{\frac{1}{3}\left(\frac{1}{3} - 1\right)}{2!}(3)^2 x^2 + \frac{\frac{1}{3}\left(\frac{1}{3} - 1\right)\left(\frac{1}{3} - 2\right)}{3!}(3)^3 x^3$$
$$+ \frac{\frac{1}{3}\left(\frac{1}{3} - 1\right)\left(\frac{1}{3} - 2\right)\left(\frac{1}{3} - 3\right)}{4!}(3)^4 x^4 + \ldots$$

**Step 4:** It is time to simplify

$$(1 + 3x)^{\frac{1}{3}} = 1 + x + \left(-\frac{1}{9}\right)(9)x^2 + \left(\frac{5}{81}\right)(27)x^3 + \left(-\frac{10}{243}\right)(81)x^4 + \ldots$$
$$= 1 + x - x^2 + \frac{5}{3}x^3 - \frac{10}{3}x^4 + \ldots$$

**Step 5:** Now substitute for $x = 0.01$.

$$(1 + 3x)^{\frac{1}{3}} = 1 + (0.01) - (0.01)^2 + \frac{5}{3}(0.01)^3 - \frac{10}{3}(0.01)^4 + \ldots$$
$$= 1 + \frac{1}{10^2} - \frac{1}{10^4} + \frac{1}{6 \times 10^5} - \frac{1}{3 \times 10^7} + \cdots \approx 1.0099001633$$

$$\therefore \sqrt[3]{1.03} \approx 1.010$$

c) $\frac{500}{494}$

**Solution**
Let's get started.

**Step 1:** Preliminary manipulation

$$\frac{500}{494} = \frac{1000}{988} = \frac{1}{\left(\frac{988}{1000}\right)} = \frac{1}{0.988}$$

$$= (0.988)^{-1} = (1 - 0.012)^{-1}$$

$$= (1 - 6x)^{-1}$$

where $x = 0.002$

**Step 2:** Write down the formula

$$(1 + ax)^n = 1 + nax + \frac{n(n-1)}{2!}a^2x^2 + \frac{n(n-1)(n-2)}{3!}a^3x^3 + \ldots$$

**Step 3:** Substitute for $n = -1$ and $a = -6$.

$$(1 - 6x)^{-1} = 1 + (-1)(-6)x + \frac{-1(-1-1)}{2!}(-6)^2x^2 + \frac{-1(-1-1)(-1-2)}{3!}(-6)^3x^3 + \ldots$$

**Step 4:** It is time to simplify

$$(1 - 6x)^{-1} = 1 + 6x + 36x^2 + (-1)(-216)x^3 + \ldots$$

$$= 1 + 6x + 36x^2 + 216x^3 + \ldots$$

**Step 5:** Now substitute for $x = 0.002$.

$$(1 - 6x)^{-1} = 1 + 6(0.002) + 36(0.002)^2 + 216(0.002)^3 + \ldots$$

$$= 1 + \frac{12}{10^3} + \frac{144}{10^6} + \frac{1728}{10^9} + \cdots \approx 1.012145728$$

$$\therefore \frac{500}{494} \approx \mathbf{1.012}$$

## 6.10 CHAPTER SUMMARY

1) '**Bi**' means two just like '**poly**' implies many.
2) A binomial expression is an expression with just two terms and takes the following forms:

$$\boxed{a \pm bx^m} \text{ OR } \boxed{ax^m \pm by^m}$$

# The Binomial Expansion

where $a$, $b$, and $m$ are rational numbers.
$m$ is usually taken as 1 whilst $a$ and $b$ are (usually) integers but valid for any real number, that is:

$$\boxed{a \pm bx} \text{ OR } \boxed{ax \pm by}$$

3) Pascal's triangle is a pattern formed using the coefficients of the terms obtained when binomial expressions of different orders are simplified.

4) Factorial is denoted by an exclamation mark (!). Technically, it implies the product of all positive integers between a particular given integer and 1.

5) Factorials are generally computed for: (i) positive numbers, and (ii) integers.

6) Given that $r$ is an integer, $r!$ is read as '$r$ factorial' and it is given by:

$$\boxed{r! = r \times (r-1) \times (r-2) \times (r-3) \times \cdots \times 3 \times 2 \times 1}$$

**OR**

$$\boxed{r! = 1 \times 2 \times 3 \times \cdots \times (r-3) \times (r-2) \times (r-1) \times r}$$

7) Combination of $n$ items taken $r$ at a time, where the order of arrangement is irrelevant, is denoted by:

$$\boxed{{}^nC_r} \text{ OR } \boxed{\binom{n}{r}}$$

The above notations can be used interchangeably and are premised on the fact that $n \geq r$ such that:

$$^nC_r = \binom{n}{r} = \frac{n!}{r!(n-r)!}$$

8) The expansion of the form $(1 \pm x)^n$ and $(1 \pm ax)^n$ for when $n$ is a positive integer.

- When $a = 1$

$$\boxed{(1+x)^n = 1 + nx + {}^nC_2 x^2 + {}^nC_3 x^3 + {}^nC_4 x^4 + \cdots + x^n}$$

$$\boxed{(1-x)^n = 1 - nx + {}^nC_2 x^2 - {}^nC_3 x^3 + {}^nC_4 x^4 + \cdots \pm x^n}$$

- When $a \neq 1$

$$\boxed{(1+ax)^n = 1 + nax + {}^nC_2 a^2 x^2 + {}^nC_3 a^3 x^3 + {}^nC_4 a^4 x^4 + \cdots + a^n x^n}$$

$$\boxed{(1-ax)^n = 1 - nax + {}^nC_2 a^2 x^2 - {}^nC_3 a^3 x^3 + {}^nC_4 a^4 x^4 + \cdots \pm a^n x^n}$$

9) Given $(1 + ax)^n$, the $(r+1)$th term is determined using:

$$\boxed{{}^nC_r (ax)^r}$$

10) The expansion of the form $(1 \pm x)^n$ and $(1 \pm ax)^n$ when $n$ is a negative integer is given by:

$$(1+x)^n = 1 + nx + \frac{n(n-1)}{2!}x^2 + \frac{n(n-1)(n-2)}{3!}x^3 + \frac{n(n-1)(n-2)(n-3)}{4!}x^4 + ...$$

$$(1+ax)^n = 1 + nax + \frac{n(n-1)}{2!}a^2x^2 + \frac{n(n-1)(n-2)}{3!}a^3x^3 + \frac{n(n-1)(n-2)(n-3)}{4!}a^4x^4 + ...$$

11) When $n$ is a fraction, the approach will be the same as when $n$ is negative, hence the formula shown above remains the same.

****

## 6.11 FURTHER PRACTICE

To access complementary contents, including additional exercises, please go to www.dszak.com.

# 7 Partial Fractions

## Learning Outcomes

Once you have studied the content of this chapter, you should be able to:

- Discuss partial fractions
- Carry out basic operations involving partial fractions
- Explain the various types of partial fractions
- Convert improper algebraic fractions into partial fractions

## 7.1 INTRODUCTION

Adding or subtracting two or more fractions (whether they are numeric or algebraic fractions) to obtain a single fraction is apparently necessary. However, the inverse of this process (i.e., splitting a compound algebraic fraction into two or more simple fractions) might be obscure, although it has its applications in integration and binomial expansion among others. This chapter will discuss the concept and techniques of splitting a fraction into two or more simple ones.

## 7.2 WHAT IS A PARTIAL FRACTION

Let's consider two arithmetic fractions, $\frac{1}{3}$ and $\frac{2}{5}$. To add these fractions, we will find the LCM of their denominators (3 and 5) which is 15, and express both fractions such that they have the same denominators. Here is the full workings:

$$\frac{1}{3} + \frac{2}{5} = \frac{5}{15} + \frac{6}{15} = \frac{5+6}{15} = \frac{11}{15}$$

$$\therefore \frac{1}{3} + \frac{2}{5} = \frac{11}{15}$$

In the above example, we were given $\frac{1}{3}$ and $\frac{2}{5}$ and added them to obtain a single fraction $\frac{11}{15}$. Though not frequently required, is it possible that the single fraction $\frac{11}{15}$ is given and we are required to find two or more simple fractions that form this? There are indeed many options that can lead to this result (i.e., $\frac{11}{15}$). Obviously, one of the options is $\frac{1}{3}$ and $\frac{2}{5}$. Another one is $\frac{2}{15}$ and $\frac{3}{5}$, because $\frac{11}{15} = \frac{2}{15} + \frac{3}{5}$. Furthermore, we know that subtracting $\frac{1}{15}$ from $\frac{4}{5}$ will yield $\frac{11}{15}$, since $\frac{11}{15} = \frac{4}{5} - \frac{1}{15}$. The option is endless, and therefore, we are unable to establish what gave the result. Will the situation be the same with algebraic fractions? Let's start by subtracting one algebraic fraction from another simple fraction, similar to what we did with $\frac{1}{3}$ and $\frac{2}{5}$.

### Example 1

Express the following as a single fraction:

$$\frac{3}{x-1} - \frac{2}{x+5}$$

What did you get? Find the solution below to double-check your answer.

### Solution to Example 1

$$\frac{3}{x-1} - \frac{2}{x+5}$$

**Step 1:** We will try to make the denominators the same by multiplying the first term by $(x+5)$ and the second term by $(x-1)$. In each case, both the numerator and denominator should be multiplied with the same as:

$$\frac{3}{x-1} - \frac{2}{x+5} = \frac{3(x+5)}{(x-1)(x+5)} - \frac{2(x-1)}{(x-1)(x+5)}$$

**Step 2:** Now that the denominators are the same. This is our LCM. Take the LCM as the denominator for the single fraction and subtract the numerator of the second from the first one as:

$$= \frac{3(x+5) - 2(x-1)}{(x-1)(x+5)}$$

**Step 3:** Open the brackets, with particular attention given to the signs.

$$= \frac{3x + 15 - 2x + 2}{(x-1)(x+5)}$$

**Step 4:** Simplify the terms (or collect the like terms).

$$\therefore \frac{3}{x-1} - \frac{2}{x+5} = \frac{x+17}{(x-1)(x+5)}$$

In the above example, we have managed to subtract one algebraic fraction ($\frac{2}{x+5}$) from another ($\frac{3}{x-1}$) to obtain a single fraction, i.e., $\frac{x+17}{(x-1)(x+5)}$. We know that if $2 + 3 = 5$, then $5 = 2 + 3$. In a similar way, we can say that since

$$\frac{3}{x-1} - \frac{2}{x+5} = \frac{x+17}{(x-1)(x+5)}$$

then
$$\frac{x+17}{(x-1)(x+5)} = \frac{3}{x-1} - \frac{2}{x+5}$$

Whilst the above are pretty much the same. The subtle difference, however, is that in the first case, we seemingly started with two simple fractions and ended with a single fraction that represents the combination. This is fine. The second case denotes that we begin with a single algebraic fraction and then end up with two simple fractions, in which one is subtracted from the other. The resulting (simple) fractions are called **partial fractions** of the original single fraction. The process of achieving this is called **resolving into partial fractions** or **partial fraction decomposition**, which is the subject of this chapter.

## 7.3 RESOLVING INTO PARTIAL FRACTIONS

From the problem we looked at in the last worked example, we obtained $\frac{x+17}{(x-1)(x+5)}$ when $\frac{2}{x+5}$ is subtracted from $\frac{3}{x-1}$. Great! What we want to do now is to set out the rules and procedures of deriving the two latter simple fractions when we are given the former single fraction.

Note that $\frac{x+17}{(x-1)(x+5)}$ is the same as $\frac{x+17}{x^2+4x-5}$. To split the last fraction (or any other fraction) into two or more simpler (partial) fractions, we need to note the following:

**Note 1** Degree of the denominator

Check if the degree of the denominator is greater than that of the numerator, otherwise apply the long division method to simplify the algebraic fraction first. To illustrate:

a) To resolve $\frac{3x^2-5}{x^3-6x^2+8x}$ into partial fractions, we will proceed as normal using the relevant principle(s) explained below. This is because the expression in the numerator is of power 2 and that of the denominator is of power 3, i.e., the power of the denominator is greater than that of the numerator.

b) However, to resolve $\frac{3x^2-5}{x^2-6x+8}$ or $\frac{3x^3-2x+5}{x^2-6x+8}$ into partial fractions, we will first need to use the long division method to simplify the rational before proceeding to resolve into partial fractions. This is because the expression in the numerator is of power 2 and that of the denominator is of power 2 or 3 (i.e., the power of the denominator is equal to or less than that of the numerator).

In summary, our first check is to ensure that the degree of the algebraic expression in the numerator is less than that of the denominator. So, for a linear denominator, the numerator must be a numerical value, while for a quadratic denominator, the numerator can either be a linear algebraic expression or a number.

**Note 2** Factorising the denominator

Completely factorise the denominator where applicable. If the expression in the denominator is already in prime factors, proceed to resolving to partial fractions. For example, in $\frac{x+5}{(3x-1)(x-2)(x+1)}$, the denominator $(3x-1)(x-2)(x+1)$ is already expressed as a product of its prime factors. Similarly, the denominator of $\frac{3x^2-x-5}{(x^2+3)(x-1)}$ is already in its simplest form. This is because $(x^2+3)$, though a quadratic expression, cannot be simplified further. Also, remember that we generally require that the number of prime factors is equal to the degree of the polynomial expression in the denominator.

**Note 3** Nature of the factors in the denominator

The nature of the factors in the denominator determines the approach to be taken in resolving into partial fractions. We can generally classify the nature into five:

a) Non-repeated linear factors
b) Repeated linear factors
c) Irreducible non-repeated non-linear factors
d) Irreducible repeated non-linear factors
e) Improper fractions

The explanation of each class will come in the next section under their respective headings. Follow the method (or steps) given for each class.

**Note 4** The use of identity

Use the identity $\equiv$ sign instead of the equal sign $=$ to indicate that the expressions on both sides are equal for all possible values of the unknown variable.

## 7.4 TYPES OF PARTIAL FRACTIONS

As noted above, the method of splitting a complex algebraic fraction into simpler ones will be determined strictly by the denominator of the fraction. We will now consider each of the five categories and illustrate them with examples.

### 7.4.1 NON-REPEATED LINEAR FACTORS

This is when the denominator has linear factors or can be factorised into linear factors. In general, an expression with two or more linear terms can be split into partial fractions as:

$$\frac{f(x)}{(a_1x+b_1)(a_2x+b_2)(a_3x+b_3)\ldots(a_nx+b_n)} \equiv \frac{A}{a_1x+b_1} + \frac{B}{a_2x+b_2} + \frac{C}{a_3x+b_3} + \cdots + \frac{D}{a_nx+b_n}$$

(7.1)

where $a_1, a_2, a_3, \ldots, b_1, b_2, b_3, \ldots$ and $A, B, C, \ldots$ are numerical values.

$f(x)$ is the numerator of the original fraction, which is a polynomial of degree less than that of the product of the factors in the denominator. Note that only $A, B, C, \ldots$ are to be determined while others will be derived from the single fraction.

# Partial Fractions

For example:
$$\frac{x+17}{(x-1)(x+5)} \equiv \frac{A}{(x-1)} + \frac{B}{(x+5)}$$

and

$$\frac{2x-1}{(x+1)(2x-3)(x+5)} \equiv \frac{A}{(x+1)} + \frac{B}{(2x-3)} + \frac{C}{(x+5)}$$

From the above two examples, it becomes apparent that the key challenge is how to find $A$, $B$, and $C$. To do this, we can use one of the following three methods:

1) Substitution method
2) Equating coefficients method
3) Eliminating a denominator method

The above methods can be used in most cases, but experience will be a good guide in choosing the appropriate one. Do not worry if you're new to this, consider them in the order listed above. We shall illustrate these methods now.

## Example 2

Using the substitution method, split $\frac{x+17}{(x-1)(x+5)}$ into partial fractions.

What did you get? Find the solution below to double-check your answer.

## Solution to Example 2

$$\frac{x+17}{(x-1)(x+5)}$$

**Step 1:** The denominator of the compound fraction is already linear and in factored form. Hence, we have:

$$\frac{x+17}{(x-1)(x+5)} \equiv \frac{A}{x-1} + \frac{B}{x+5}$$

**Step 2:** The LCM of the right-hand side is $(x-1)(x+5)$. Using this, we have the simplified expression.

$$\frac{x+17}{(x-1)(x+5)} \equiv \frac{A(x+5)}{(x-1)(x+5)} + \frac{B(x-1)}{(x-1)(x+5)}$$
$$\frac{x+17}{(x-1)(x+5)} \equiv \frac{A(x+5) + B(x-1)}{(x-1)(x+5)}$$

**Step 3:** Since the denominators at the LHS and RHS are the same, it is then logical to conclude that the numerators are equivalent. Now we can equate them as:

$$x + 17 \equiv A(x+5) + B(x-1)$$

**Step 4:** The sign $\equiv$ implies that this expression is true for all values of $x$. It is however important to diligently choose the values to substitute for $x$. As a result, we will choose $x = 1$ and $x = -5$. We chose these because they will make a term of the expression equal to zero or the expression in the brackets zero. Thus:

- when $x = 1$, we have:

$$1 + 17 = A(1+5) + B(1-1)$$
$$18 = 6A$$
$$\therefore A = \frac{18}{6} = 3$$

**Step 5:** For this case, once we find $A$, we can choose any $x$ value to substitute in order to find $B$. However, using $x = -5$ will still be better. Thus:

- when $x = -5$, we have:

$$-5 + 17 = A(-5+5) + B(-5-1)$$
$$12 = -6B$$
$$\therefore B = \frac{12}{-6} = -2$$

**Step 6:** As these cannot be simplified further, the partial fractions are:

$$\frac{x+17}{(x-1)(x+5)} = \frac{3}{x-1} + \frac{-2}{x+5}$$

**Step 7:** Notice that $B = -2$ appears as though $B = 2$. This is because $\frac{-2}{x+5} = \frac{2}{-(x+5)} = -\frac{2}{x+5}$. We chose the last variant. Hence

$$\therefore \frac{x+17}{(x-1)(x+5)} = \frac{3}{x-1} - \frac{2}{x+5}$$

**CHECK**

Let's double-check by simplifying the RHS to see if we can obtain the LHS as:

$$\frac{3}{x-1} - \frac{2}{x+5} = \frac{3(x+5) - 2(x-1)}{(x-1)(x+5)}$$
$$= \frac{3x + 15 - 2x + 2}{(x-1)(x+5)}$$
$$= \frac{x+17}{(x-1)(x+5)}$$

**Correct!**

# Partial Fractions

Another example to try is the second method.

**Example 3**

Using the method of equating the coefficients, express $\dfrac{4-x}{3x^2-x-10}$ into partial fractions.

What did you get? Find the solution below to double-check your answer.

**Solution to Example 3**

$$\frac{4-x}{3x^2-x-10}$$

**Step 1:** Let's factorise the denominator. Thus, we have:

$$3x^2 - x - 10 = (3x+5)(x-2)$$

So, we can write the fraction as:

$$\frac{4-x}{3x^2-x-10} = \frac{4-x}{(3x+5)(x-2)}$$

**Step 2:** The above is now in prime factors form and is linear, so we have:

$$\frac{4-x}{(3x+5)(x-2)} \equiv \frac{A}{3x+5} + \frac{B}{x-2}$$

**Step 3:** The LCM of the right-hand side is $(3x+5)(x-2)$. We can proceed as we did in Example 2 above, but let's use cross-multiplication. Using this, we have the simplified expression

$$\frac{4-x}{(3x+5)(x-2)} \equiv \frac{A}{3x+5} + \frac{B}{x-2}$$
$$\equiv \frac{A(x-2) + B(3x+5)}{(3x+5)(x-2)}$$

**Step 4:** Open the brackets in the numerator and simplify as:

$$\frac{4-x}{(3x+5)(x-2)} = \frac{Ax - 2A + 3Bx + 5B}{(3x+5)(x-2)}$$
$$= \frac{Ax + 3Bx + 5B - 2A}{(3x+5)(x-2)}$$
$$= \frac{(A+3B)x + (5B-2A)}{(3x+5)(x-2)}$$

**Step 5:** Since the denominators at the LHS and RHS are the same, it is logical to conclude that the numerators are equivalent. Now we can equate them

$$4 - x \equiv (A+3B)x + (5B-2A)$$

**Step 6:** As the method suggests, we will now compare the coefficients and equate them. Thus:

- For $[x^2]$

This does not appear on either side (LHS and RHS). No further action is required.

- For $[x]$, we have

$$-x = (A + 3B)x$$

Divide both sides by $x$, we have:

$$-1 = A + 3B$$
$$\therefore A + 3B = -1 \qquad ---(i)$$

- For [constant], we have

$$4 = 5B - 2A$$
$$\therefore -2A + 5B = 4 \qquad ---(ii)$$

**Step 7:** We now have two simultaneous linear equations. To eliminate $A$, we will multiply equation (i) by 2 as

$$2A + 6B = -2 \qquad ---(iii)$$

Add equation (ii) and (iii) to have

$$11B = 2$$
$$\therefore B = \frac{2}{11}$$

From equation (i), we have:

$$A = -3B - 1$$
$$= -3\left(\frac{2}{11}\right) - 1$$
$$= -\frac{17}{11}$$

Therefore

$$A = -\frac{17}{11}, \quad B = \frac{2}{11}$$

**Step 8:** The partial fractions are:

$$\frac{4-x}{3x^2 - x - 10} = \frac{4-x}{(3x+5)(x-2)}$$
$$= \frac{\left(-\frac{17}{11}\right)}{3x+5} + \frac{\left(\frac{2}{11}\right)}{x-2}$$

Partial Fractions

Simplify the above to have

$$= \frac{-17}{11(3x+5)} + \frac{2}{11(x-2)}$$

$$= \frac{2}{11(x-2)} - \frac{17}{11(3x+5)}$$

$$\therefore \frac{4-x}{3x^2-x-10} = \frac{2}{11(x-2)} - \frac{17}{11(3x+5)}$$

Here is another example to try the third method.

### Example 4

Using the eliminating a denominator method, split $\frac{x^2+1}{(x-3)(x+2)(x+5)}$ into partial fractions.

What did you get? Find the solution below to double-check your answer.

### Solution to Example 4

$$\frac{x^2+1}{(x-3)(x+2)(x+5)}$$

**Step 1:** The denominator of the compound fraction is already in factored form and linear. Hence, we have:

$$\frac{x^2+1}{(x-3)(x+2)(x+5)} \equiv \frac{A}{x-3} + \frac{B}{x+2} + \frac{C}{x+5}$$

**Step 2:** To find $A$, multiply both sides by the denominator of $A$, i.e., $(x-3)$.

$$\frac{x^2+1}{(x-3)(x+2)(x+5)} \times (x-3) \equiv \frac{A(x-3)}{x-3} + \frac{B(x-3)}{x+2} + \frac{C(x-3)}{x+5}$$

Simplify to have:

$$\frac{x^2+1}{(x+2)(x+5)} \equiv A + \frac{B(x-3)}{x+2} + \frac{C(x-3)}{x+5}$$

To determine $A$, we need to ensure that the terms containing $B$ and $C$ on the RHS become zero. This is achieved by using $x = 3$ as:

$$\frac{(3)^2 + 1}{(3+2)(3+5)} = A + \frac{B(3-3)}{3+2} + \frac{C(3-3)}{3+5}$$

$$\frac{9+1}{(5)(8)} = A + \frac{B(0)}{5} + \frac{C(0)}{8}$$

$$\frac{10}{40} = A + 0 + 0$$

$$\frac{1}{4} = A$$

$$\therefore A = \frac{1}{4}$$

**Step 3:** We will repeat step 2 to find $B$, in this case by multiplying with $(x+2)$.

$$\frac{x^2 + 1}{(x-3)(x+2)(x+5)} \times (x+2) \equiv \frac{A(x+2)}{x-3} + \frac{B(x+2)}{x+2} + \frac{C(x+2)}{x+5}$$

Simplify to have:

$$\frac{x^2 + 1}{(x-3)(x+5)} \equiv \frac{A(x+2)}{x-3} + B + \frac{C(x+2)}{x+5}$$

To determine $B$, we need to ensure that the terms containing $A$ and $C$ on the RHS become zero. This is achieved by using $x = -2$, we have:

$$\frac{(-2)^2 + 1}{(-2-3)(-2+5)} = \frac{A(-2+2)}{-2-3} + B + \frac{C(-2+2)}{-2+5}$$

$$\frac{4+1}{(-5)(3)} = \frac{A(0)}{-5} + B + \frac{C(0)}{3}$$

$$\frac{5}{-15} = 0 + B + 0$$

$$-\frac{1}{3} = B$$

$$\therefore B = -\frac{1}{3}$$

**Step 4:** Again, we will repeat step 2 to find $C$, in this case by multiplying by $(x+5)$.

$$\frac{x^2 + 1}{(x-3)(x+2)(x+5)} \times (x+5) \equiv \frac{A(x+5)}{x-3} + \frac{B(x+5)}{x+2} + \frac{C(x+5)}{x+5}$$

Simplify to have:

$$\frac{x^2 + 1}{(x-3)(x+2)} \equiv \frac{A(x+5)}{x-3} + \frac{B(x+5)}{x+2} + C$$

# Partial Fractions

To determine $C$, we need to ensure that the terms containing $A$ and $B$ on the RHS become zero. This is achieved by using $x = -5$, we have:

$$\frac{(-5)^2 + 1}{(-5-3)(-5+2)} = \frac{A(-5+5)}{-5-3} + \frac{B(-5+5)}{-5+2} + C$$

$$\frac{25+1}{(-8)(-3)} = \frac{A(0)}{-8} + \frac{B(0)}{-3} + C$$

$$\frac{26}{24} = 0 + 0 + C$$

$$\frac{13}{12} = C$$

$$\therefore C = \frac{13}{12}$$

**Step 5:** The partial fractions are:

$$\frac{x^2 + 1}{(x-3)(x+2)(x+5)} \equiv \frac{\left(\frac{1}{4}\right)}{x-3} + \frac{\left(-\frac{1}{3}\right)}{x+2} + \frac{\left(\frac{13}{12}\right)}{x+5}$$

Simplify the above to have

$$= \frac{1}{4(x-3)} + \frac{-1}{3(x+2)} + \frac{13}{12(x+5)}$$

$$\therefore \frac{x^2 + 1}{(x-3)(x+2)(x+5)} = \frac{1}{4(x-3)} - \frac{1}{3(x+2)} + \frac{13}{12(x+5)}$$

---

## 7.4.2 Repeated Linear Factors

This is when the denominator has linear factors, but one or more of the factors are repeated. In other words, the power is not equal to 1. The general expression for three linear factors in which one of them is repeated is given below.

$$\boxed{\frac{f(x)}{(a_1x+b_1)(a_2x+b_2)(a_3x+b_3)^n} \equiv \frac{A}{a_1x+b_1} + \frac{B}{a_2x+b_2} + \frac{C_1}{a_3x+b_3} + \frac{C_2}{(a_3x+b_3)^2} + \frac{C_3}{(a_3x+b_3)^3} + \dots \frac{C_n}{(a_3x+b_3)^n}}$$

(7.2)

where $a_1, a_2, a_3, \dots, b_1, b_2, b_3, \dots, n$, and $A, B, C_1, C_2, C_3, \dots, C_n$ are numerical values.

$f(x)$ is the numerator of the original fraction, which is a polynomial of a degree less than that of the product of the factors in the denominator. Note that only $A, B, C_1, C_2, C_3, \dots, C_n$ are to be determined, while others will be known from the compound fraction. For example:

$$\frac{17}{(3x-5)^3} \equiv \frac{A}{(3x-5)} + \frac{B}{(3x-5)^2} + \frac{C}{(3x-5)^3}$$

and

$$\frac{x^2 - x + 1}{(5x+1)(x-3)^3} \equiv \frac{A}{(5x+1)} + \frac{B}{(x-3)} + \frac{C}{(x-3)^2} + \frac{D}{(x-3)^3}$$

Similarly

$$\frac{2x^3 + 1}{(x-1)^2(x+3)(x-7)^2} \equiv \frac{A}{(x-1)} + \frac{B}{(x-1)^2} + \frac{C}{(x+3)} + \frac{D}{(x-7)} + \frac{E}{(x-7)^2}$$

From the above three examples, it is apparent that $A$, $B$, $C$, $D$, and $E$ are the unknown constants to be obtained. To find these unknowns, we can use one of the following two methods:

1) Substitution method
2) Equating coefficients method

The eliminating a denominator method is not suitable for this current case; it is most appropriate for the first category only. Let's illustrate this class using the above two methods.

### Example 5

Using the substitution method, split $\frac{x+9}{(x+2)^2}$ into partial fractions.

What did you get? Find the solution below to double-check your answer.

### Solution to Example 5

$$\frac{x+9}{(x+2)^2}$$

**Step 1:** The denominator of the compound fraction is already in factored form and linear. Hence, we have:

$$\frac{x+9}{(x+2)^2} \equiv \frac{A}{x+2} + \frac{B}{(x+2)^2}$$

**Step 2:** The LCM of the right-hand side is $(x+2)^2$. Recall that the LCM of 2 and 4 is 4 and not 8. $(x+2)$ and $(x+2)^2$ are like 2 and $2^2$ (or 4), respectively, so that is why we chose $(x+2)^2$ as our LCM. Thus, we have:

$$\frac{x+9}{(x+2)^2} \equiv \frac{A(x+2) + B}{(x+2)^2}$$

**Step 3:** Since the denominators at the LHS and RHS are the same, we have:

$$x + 9 \equiv A(x+2) + B$$

# Partial Fractions

**Step 4:** Let's substitute now

- when $x = -2$, we have

$$-2 + 9 = A(-2+2) + B$$
$$7 = A(0) + B$$
$$\therefore B = 7$$

- when $x = -1$, we have

$$-1 + 9 = A(-1+2) + B$$
$$8 = A(1) + B$$
$$8 = A + 7$$
$$A = 8 - 7$$
$$\therefore A = 1$$

**Step 5:** The partial fractions are:

$$\frac{x+9}{(x+2)^2} \equiv \frac{1}{x+2} + \frac{7}{(x+2)^2}$$

**CHECK**

Let's double-check by simplifying the RHS to see if we can obtain the LHS as:

$$\frac{1}{x+2} + \frac{7}{(x+2)^2} = \frac{(x+2)+7}{(x+2)^2}$$
$$= \frac{x+9}{(x+2)^2}$$

**Correct!**

Here is another example to try the second method.

## Example 6

Using the method of equating the coefficients, express $\frac{x^2+3x-2}{(x-1)^3}$ in partial fractions.

What did you get? Find the solution below to double-check your answer.

## Solution to Example 6

$$\frac{x^2 + 3x - 2}{(x-1)^3}$$

**Step 1:** The denominator of the compound fraction is already in factored form and linear. Hence, we have:

$$\frac{x^2 + 3x - 2}{(x-1)^3} \equiv \frac{A}{x-1} + \frac{B}{(x-1)^2} + \frac{C}{(x-1)^3}$$

**Step 2:** The LCM of the right-hand side is $(x-1)^3$, so we have

$$\frac{x^2 + 3x - 2}{(x-1)^3} \equiv \frac{A(x-1)^2 + B(x-1) + C}{(x-1)^3}$$

**Step 3:** Open the brackets in the numerator and simplify as:

$$\frac{x^2 + 3x - 2}{(x-1)^3} \equiv \frac{A(x^2 - 2x + 1) + B(x-1) + C}{(x-1)^3}$$

$$= \frac{Ax^2 - 2Ax + A + Bx - B + C}{(x-1)^3}$$

$$= \frac{Ax^2 - (2A - B)x + A - B + C}{(x-1)^3}$$

**Step 4:** We now need to equate the numerators as:

$$x^2 + 3x - 2 \equiv Ax^2 - (2A - B)x + A - B + C$$

**Step 5:** We will now compare the coefficients and equate them. Thus:

- For $[x^2]$, we have

$$Ax^2 = x^2$$

Divide both sides by $x^2$, we have:

$$\therefore A = 1$$

- For $[x]$, we have

$$-(2A - B)x = 3x$$

Divide both sides by $x$, we have:

$$-(2A - B) = 3$$
$$-2A + B = 3$$
$$B = 3 + 2A = 3 + 2$$
$$\therefore B = 5$$

- For [constant], we have

$$A - B + C = -2$$
$$C = -2 - A + B$$
$$= -2 - 1 + 5$$
$$\therefore C = 2$$

**Step 6:** The partial fractions are:

$$\frac{x^2 + 3x - 2}{(x-1)^3} = \frac{1}{x-1} + \frac{5}{(x-1)^2} + \frac{2}{(x-1)^3}$$
$$\therefore \frac{x^2 + 3x - 2}{(x-1)^3} = \frac{1}{x-1} + \frac{5}{(x-1)^2} + \frac{2}{(x-1)^3}$$

The repeated factor(s) can be mixed with non-repeated factors. Let's try an example.

### Example 7

Using the method of equating the coefficients, express $\frac{x(x-3)}{(x+1)(2x+1)^2}$ in partial fractions.

What did you get? Find the solution below to double-check your answer.

### Solution to Example 7

$$\frac{x(x-3)}{(x+1)(2x+1)^2}$$

**Step 1:** The denominator of the compound fraction is already in factored form and linear. Hence, we have:

$$\frac{x(x-3)}{(x+1)(2x+1)^2} \equiv \frac{A}{x+1} + \frac{B}{2x+1} + \frac{C}{(2x+1)^2}$$

**Step 2:** The LCM of the right-hand side is $(x+1)(2x+1)^2$, so we have:

$$\frac{x(x-3)}{(x+1)(2x+1)^2} \equiv \frac{A(2x+1)^2 + B(x+1)(2x+1) + C(x+1)}{(x+1)(2x+1)^2}$$

**Step 3:** Open the brackets in the numerator and simplify as:

$$\frac{x(x-3)}{(x+1)(2x+1)^2} \equiv \frac{A(4x^2+4x+1)+B(2x^2+3x+1)+C(x+1)}{(x+1)(2x+1)^2}$$

$$= \frac{4Ax^2+4Ax+A+2Bx^2+3Bx+B+Cx+C}{(x+1)(2x+1)^2}$$

$$= \frac{(4A+2B)x^2+(4A+3B+C)x+(A+B+C)}{(x+1)(2x+1)^2}$$

**Step 4:** We now need to equate the numerators as:

$$x(x-3) \equiv (4A+2B)x^2+(4A+3B+C)x+(A+B+C)$$
$$x^2 - 3x \equiv (4A+2B)x^2+(4A+3B+C)x+(A+B+C)$$

**Step 5:** We will now compare the coefficients and equate them. Thus:

- For $[x^2]$, we have:

$$(4A+2B)x^2 = x^2$$

Divide both sides by $x^2$, we have:

$$4A + 2B = 1 \qquad -----\text{(i)}$$

- For $[x]$, we have:

$$(4A+3B+C)x = -3x$$

Divide both sides by $x$, we have:

$$4A + 3B + C = -3 \qquad -----\text{(ii)}$$

Equation (ii) – equation (i), we have

$$B + C = -4 \qquad -----\text{(iii)}$$

- For [constant], we have:

$$A + B + C = 0 \qquad -----\text{(iv)}$$

Multiply equation (iv) by 4, we have

$$4A + 4B + 4C = 0 \qquad -----\text{(v)}$$

Equation (v) – equation (i), we have:

$$2B + 4C = -1 \qquad -----\text{(vi)}$$

# Partial Fractions

**Step 6:** Solving for $A$, $B$, and $C$:

First, we will use equations containing only $B$ and $C$, i.e., equations (iii) and (vi)

$$B + C = -4 \quad ----- \text{(iii)}$$
$$2B + 4C = -1 \quad ----- \text{(vi)}$$

Let's use the substitution method to solve these simultaneous equations. From (iii), we have

$$B = -4 - C \quad ----- \text{(vii)}$$

Substitute (vii) in (vi), we have

$$2B + 4C = -1$$
$$2(-4 - C) + 4C = -1$$
$$-8 - 2C + 4C = -1$$
$$2C = -1 + 8$$
$$2C = 7$$
$$\therefore C = \frac{7}{2}$$

From equation (vii), we have

$$B = -4 - C$$
$$B = -4 - \frac{7}{2}$$
$$\therefore B = -\frac{15}{2}$$

From equation (i), we have

$$4A + 2B = 1$$
$$4A = 1 - 2B$$
$$= 1 - 2\left(-\frac{15}{2}\right)$$
$$= 1 + 15$$
$$= 16$$
$$\therefore A = 4$$

**Step 7:** The partial fractions are:

$$\frac{x(x-3)}{(x+1)(2x+1)^2} = \frac{4}{x+1} + \frac{\left(-\frac{15}{2}\right)}{2x+1} + \frac{\left(\frac{7}{2}\right)}{(2x+1)^2}$$

$$\therefore \frac{x(x-3)}{(x+1)(2x+1)^2} = \frac{4}{x+1} - \frac{15}{2(2x+1)} + \frac{7}{2(2x+1)^2}$$

**CHECK**

Let's double-check by simplifying the RHS to see if we can obtain the LHS as:

$$\frac{4}{x+1} - \frac{15}{2(2x+1)} + \frac{7}{2(2x+1)^2} = \frac{4 \times 2(2x+1)^2 - 15(x+1)(2x+1) + 7(x+1)}{2(x+1)(2x+1)^2}$$

$$= \frac{8(4x^2+4x+1) - 15(2x^2+3x+1) + 7(x+1)}{2(x+1)(2x+1)^2}$$

$$= \frac{(32x^2+32x+8) - (30x^2+45x+15) + (7x+7)}{2(x+1)(2x+1)^2}$$

$$= \frac{(32x^2-30x^2) + (32x-45x+7x) + (8-15+7)}{2(x+1)(2x+1)^2}$$

$$= \frac{(2x^2) + (-6x) + (0)}{2(x+1)(2x+1)^2} = \frac{2x^2 - 6x}{2(x+1)(2x+1)^2}$$

$$= \frac{2x(x-3)}{2(x+1)(2x+1)^2} = \frac{x(x-3)}{(x+1)(2x+1)^2}$$

**Correct!**

### 7.4.3 IRREDUCIBLE NON-REPEATED NON-LINEAR FACTORS

This is when the denominator has non-linear factors (e.g., quadratic, cubic) that are not repeated, provided the non-linear polynomials cannot be factorised or simplified further.

The general expression for two non-repeated non-linear factors which cannot be reduced further is given below.

$$\boxed{\frac{f(x)}{(a_1x^2 + b_1x + c_1)(a_2x^3 + b_2x^2 + c_2x + d_1)} \equiv \frac{Ax + B}{a_1x^2 + b_1x + c_1} + \frac{Cx^2 + Dx + E}{a_2x^3 + b_2x^2 + c_2x + d_1}}$$

(7.3)

where $a_1, a_2, b_1, b_2, c_1, c_2, d_1$ and $A, B, C, D,$ and $E$ are numerical values.

$f(x)$ is the numerator of the original fraction, which is a polynomial of a degree less than that of the product of the factors in the denominator. Note that only $A, B, C, D,$ and $E$ are to be determined while the others will be known from the complex fraction. For example

$$\frac{x^2 - x + 1}{(x^2 + x - 1)(x^3 + x^2 + x - 1)} \equiv \frac{Ax + B}{(x^2 + x - 1)} + \frac{Cx^2 + Dx + E}{(x^3 + x^2 + x - 1)}$$

and

$$\frac{2x - 1}{x(x + 1)(x^3 + 3)} \equiv \frac{A}{x} + \frac{B}{x + 1} + \frac{Cx^2 + Dx + E}{x^3 + 3}$$

Notice that even when some terms are missing in the denominator, the complete form of the polynomial is used in the numerator. From the above two examples, it is apparent that the key challenge is how to find $A, B, C, D$ and $E$. To do this, we can use one of the following methods.

# Partial Fractions

1) Substitution method
2) Equating coefficients method

To test if a quadratic expression can be factorised into its prime factors or not, use the discriminant $D$, where $D = b^2 - 4ac$. If $D$ is a perfect square, then we can factorise it otherwise the quadratic expression is not reducible.

Let's try an example.

## Example 8

Using the method of equating the coefficients, express $\dfrac{4x}{(3x-1)(2x^2+x-5)}$ in partial fractions.

What did you get? Find the solution below to double-check your answer.

## Solution to Example 8

$$\frac{4x}{(3x-1)(2x^2+x-5)}$$

**Step 1:** The denominator of the compound fraction has a linear and quadratic factor and the latter cannot be factorised further. Given $(2x^2 + x - 5)$, we can show that $b^2 - 4ac = 1^2 - 4(2)(-5) = 41$. Whilst $b^2 - 4ac > 0$, it is not a perfect square so cannot be factorised. Hence, we can say that:

$$\frac{4x}{(3x-1)(2x^2+x-5)} \equiv \frac{A}{3x-1} + \frac{Bx+C}{2x^2+x-5}$$

Notice that the numerator used for $2x^2 + x - 5$ is a linear expression. This is based on the stated rule (page 256).

**Step 2:** The LCM of the right-hand side is $(3x-1)(2x^2+x-5)$, but we will just cross-multiply, thus:

$$\frac{4x}{(3x-1)(2x^2+x-5)} \equiv \frac{A(2x^2+x-5)+(Bx+C)(3x-1)}{(3x-1)(2x^2+x-5)}$$

**Step 3:** We now need to equate the numerators as:

$$4x \equiv A(2x^2+x-5) + (Bx+C)(3x-1)$$

**Step 4:** Open the brackets in the numerator and simplify as:

$$4x \equiv (2Ax^2 + Ax - 5A) + (3Bx^2 - Bx + 3Cx - C)$$
$$\equiv (2A+3B)x^2 + (A-B+3C)x + (-5A-C)$$

**Step 5:** We will now compare the coefficients and equate them. Thus:

- For $[x^2]$, we have:

$$(2A + 3B)x^2 = 0x^2$$

Divide both sides by $x^2$, we have:

$$2A + 3B = 0 \qquad -----\text{(i)}$$

- For $[x]$, we have

$$(A - B + 3C)x = 4x$$

Divide both sides by $x$, we have:

$$A - B + 3C = 4 \qquad -----\text{(ii)}$$

- For [constant], we have

$$-5A - C = 0$$
$$5A + C = 0 \qquad -----\text{(iii)}$$

Multiply equation (iii) by 3, we have

$$15A + 3C = 0 \qquad -----\text{(iv)}$$

Equation (iv) – equation (ii), we have

$$14A + B = -4 \qquad -----\text{(v)}$$

**Step 6:** Solving for $A$, $B$, and $C$

Let's start with equations (i) and (v)

$$2A + 3B = 0 \qquad -----\text{(i)}$$
$$14A + B = -4 \qquad -----\text{(v)}$$

Multiply (i) by 7, we have

$$14A + 21B = 0 \qquad -----\text{(vi)}$$

Equation (vi) – equation (v), we have

$$20B = 4$$
$$\therefore B = \frac{1}{5}$$

From equation (i), we have

$$2A + 3B = 0$$
$$2A = -3B = -3\left(\frac{1}{5}\right) = -\frac{3}{5}$$
$$\therefore A = -\frac{3}{10}$$

Partial Fractions

From equation (iii), we have

$$5A + C = 0$$

$$C = -5A = -5\left(-\frac{3}{10}\right)$$

$$C = \frac{3}{2}$$

**Step 7:** The partial fractions are:

$$\frac{4x}{(3x-1)(2x^2+x-5)} \equiv \frac{\left(-\frac{3}{10}\right)}{3x-1} + \frac{\frac{1}{5}x + \frac{3}{2}}{2x^2+x-5}$$

$$= \frac{\frac{1}{10}(2x+15)}{2x^2+x-5} - \frac{\left(\frac{3}{10}\right)}{3x-1}$$

$$= \frac{2x+15}{10(2x^2+x-5)} - \frac{3}{10(3x-1)}$$

$$\therefore \frac{4x}{(3x-1)(2x^2+x-5)} \equiv \frac{2x+15}{10(2x^2+x-5)} - \frac{3}{10(3x-1)}$$

We've attempted a linear with a quadratic above; let's go for a linear and a cubic expression now. Here we go.

### Example 9

Using the substitution method, express $\frac{18}{(x-2)(x^3+1)}$ in partial fractions.

What did you get? Find the solution below to double-check your answer.

### Solution to Example 9

$$\frac{18}{(x-2)(x^3+1)}$$

**Step 1:** The denominator of the compound fraction has a linear and a cubic factor. Obviously, $(x^3+1)$ cannot be factorised further, hence we can say that

$$\frac{18}{(x-2)(x^3+1)} \equiv \frac{A}{x-2} + \frac{Bx^2 + Cx + D}{x^3+1}$$

Notice that the numerator for $x^3 + 1$ is a complete quadratic expression, though $(x^3 + 1)$ is missing $[x^2]$ and $[x]$ terms. This is based on the stated rule on page 256.

**Step 2:** The LCM of the right-hand side is $(x-2)(x^3+1)$, but we will just cross-multiply as in the previous example, thus:

$$\frac{18}{(x-2)(x^3+1)} \equiv \frac{A(x^3+1)+(x-2)(Bx^2+Cx+D)}{(x-2)(x^3+1)}$$

**Step 3:** We now need to equate the numerators as:

$$18 \equiv A(x^3+1)+(x-2)(Bx^2+Cx+D)$$

**Step 4:** We do not need to open the brackets; we will however carefully choose values of $x$ to substitute. Thus:

- For $[x = 2]$, we have

$$18 = A(x^3+1)+(x-2)(Bx^2+Cx+D)$$
$$18 = A(2^3+1)+(2-2)(B \times 2^2 + C \times 2 + D)$$
$$18 = A(9)+(0)(4B+2C+D)$$
$$18 = 9A$$
$$\therefore A = 2$$

- For $[x = 0]$, we have

$$18 = A(x^3+1)+(x-2)(Bx^2+Cx+D)$$
$$18 = A(0^3+1)+(0-2)(B \times 0^2 + C \times 0 + D)$$
$$18 = A(1)+(-2)(0+0+D)$$
$$18 = 2 - 2D$$
$$2D = -16$$
$$\therefore D = -8$$

- For $[x = -1]$, we have

$$18 = A(x^3+1)+(x-2)(Bx^2+Cx+D)$$
$$18 = A((-1)^3+1)+(-1-2)(B \times (-1)^2 + C \times -1 + D)$$
$$18 = A(-1+1)+(-3)(B-C+D)$$
$$18 = A(0)-3(B-C-8)$$
$$18 = -3(B-C-8)$$
$$-6 = B-C-8$$
$$-6+8 = B-C$$
$$B-C = 2 \quad -----(i)$$

Partial Fractions

- For $[x = 1]$, we have

$$18 = A(x^3 + 1) + (x - 2)(Bx^2 + Cx + D)$$
$$18 = A((1)^3 + 1) + (1 - 2)(B \times (1)^2 + C \times 1 + D)$$
$$18 = A(1 + 1) + (-1)(B + C + D)$$
$$18 = 2(2) - (B + C - 8)$$
$$18 = 4 - B - C + 8$$
$$18 - 4 - 8 = -B - C$$
$$6 = -B - C$$
$$B + C = -6 \quad ----- \text{(ii)}$$

Equation (i) + equation (ii), we have

$$2B = -4$$
$$B = \frac{-4}{2}$$
$$\therefore B = -2$$

From equation (ii), we have

$$B + C = -6$$
$$C = -B - 6 = 2 - 6$$
$$\therefore C = -4$$

**Step 5:** The partial fractions are:

$$\frac{18}{(x-2)(x^3+1)} = \frac{A}{x-2} + \frac{Bx^2 + Cx + D}{x^3+1}$$
$$= \frac{2}{x-2} + \frac{-2x^2 - 4x - 8}{x^3+1}$$
$$= \frac{2}{x-2} + \frac{-2(x^2 + 2x + 4)}{x^3+1}$$
$$= \frac{2}{x-2} - \frac{2(x^2 + 2x + 4)}{x^3+1}$$
$$\therefore \frac{18}{(x-2)(x^3+1)} = \frac{2}{x-2} - \frac{2(x^2 + 2x + 4)}{x^3+1}$$

**CHECK**

Let's double-check by simplifying the RHS to see if we can obtain the LHS as:

$$\frac{2}{x-2} - \frac{2(x^2+2x+4)}{x^3+1} = \frac{2(x^3+1) - 2(x-2)(x^2+2x+4)}{(x-2)(x^3+1)}$$

$$= \frac{(2x^3+2) - 2(x^3+2x^2+4x-2x^2-4x-8)}{(x-2)(x^3+1)}$$

$$= \frac{(2x^3+2) - 2(x^3-8)}{(x-2)(x^3+1)}$$

$$= \frac{2x^3+2-2x^3+16}{(x-2)(x^3+1)}$$

$$= \frac{18}{(x-2)(x^3+1)}$$

**Correct!**

**CHECK (Author's method)**

A quick way to check is to choose an appropriate value for $x$ and substitute on the LHS and RHS independently. The answer should be the same provided the working is correct (i.e., they are equal) AND that the chosen value is such that it is highly improbable to accidentally have equal results. For this case, let's say $x = 0.2$.

- RHS

$$\frac{2}{x-2} - \frac{2(x^2+2x+4)}{x^3+1} = \frac{2}{0.2-2} - \frac{2\left((0.2)^2+2(0.2)+4\right)}{(0.2)^3+1}$$

$$= -\frac{625}{63}$$

- LHS

$$\frac{18}{(x-2)(x^3+1)} = \frac{18}{(0.2-2)\left((0.2)^3+1\right)} = -\frac{625}{63}$$

Thus,

$$LHS = RHS$$

**Correct!**

# Partial Fractions

## ALTERNATIVE METHOD

In the above workings, we stated that $x^3 + 1$ cannot be factorised. This is not entirely true as $x^3 + 1$ can be factorised to obtain $(x+1)(x^2 - x + 1)$. This is because

$$a^3 + b^3 = (a+b)(a^2 - ab + b^2) \text{ AND } a^3 - b^3 = (a-b)(a^2 + ab + b^2)$$

Therefore

$$\frac{18}{(x-2)(x^3+1)} \equiv \frac{18}{(x-2)(x+1)(x^2-x+1)}$$

**Step 1:** Let's re-write the above as

$$\frac{18}{(x-2)(x+1)(x^2-x+1)} \equiv \frac{A}{x-2} + \frac{B}{x+1} + \frac{Cx+D}{x^2-x+1}$$

**Step 2:** The LCM of the right-hand side is $(x-2)(x+1)(x^2-x+1)$, thus:

$$\frac{18}{(x-2)(x+1)(x^2-x+1)}$$
$$\equiv \frac{A(x+1)(x^2-x+1) + B(x-2)(x^2-x+1) + (Cx+D)(x-2)(x+1)}{(x-2)(x+1)(x^2-x+1)}$$

**Step 3:** We now need to equate the numerators as:

$$18 \equiv A(x+1)(x^2-x+1) + B(x-2)(x^2-x+1) + (Cx+D)(x-2)(x+1)$$

**Step 4:** We do not need to open the brackets; we will however carefully choose values of $x$ to substitute. Thus:

- For $[x = 2]$, we have

$$18 = A(x+1)(x^2-x+1) + B(x-2)(x^2-x+1) + (Cx+D)(x-2)(x+1)$$
$$18 = A(2+1)(2^2-2+1) + B(2-2)(2^2-2+1) + (C(2)+D)(2-2)(2+1)$$
$$18 = A(3)(3) + B(0)(3) + (2C+D)(0)(3)$$
$$18 = 9A$$
$$\therefore A = 2$$

- For $[x = -1]$, we have

$$18 = A(x+1)(x^2 - x + 1) + B(x-2)(x^2 - x + 1) + (Cx+D)(x-2)(x+1)$$
$$18 = A(-1+1)\left((-1)^2 - (-1) + 1\right) + B(-1-2)\left((-1)^2 - (-1) + 1\right)$$
$$\quad + (C(-1) + D)(-1-2)(-1+1)$$
$$18 = A(0)(3) + B(-3)(3) + (-C+D)(-3)(0)$$
$$18 = -9B$$
$$\therefore B = -2$$

- For $[x = 0]$, we have

$$18 = A(x+1)(x^2 - x + 1) + B(x-2)(x^2 - x + 1) + (Cx+D)(x-2)(x+1)$$
$$18 = A(0+1)(0^2 - 0 + 1) + B(0-2)(0^2 - 0 + 1) + (C(0) + D)(0-2)(0+1)$$
$$18 = A(1)(1) + B(-2)(1) + (D)(-2)(1)$$
$$18 = A - 2B - 2D$$
$$18 = 2 - 2(-2) - 2D$$
$$18 = 6 - 2D$$
$$12 = -2D$$
$$\therefore D = -6$$

- For $[x = 1]$, we have

$$18 = A(x+1)(x^2 - x + 1) + B(x-2)(x^2 - x + 1) + (Cx+D)(x-2)(x+1)$$
$$18 = A(1+1)(1^2 - 1 + 1) + B(1-2)(1^2 - 1 + 1) + (C(1) + D)(1-2)(1+1)$$
$$18 = A(2)(1) + B(-1)(1) + (C+D)(-1)(2)$$
$$18 = 2A - B - 2(C+D)$$
$$18 = 2(2) - (-2) - 2(C-6)$$
$$18 = 4 + 2 - 2C + 12$$
$$\therefore C = 0$$

The partial fractions are:

$$\frac{18}{(x-2)(x+1)(x^2 - x + 1)} \equiv \frac{A}{x-2} + \frac{B}{x+1} + \frac{Cx+D}{x^2 - x + 1}$$
$$= \frac{2}{x-2} + \frac{-2}{x+1} + \frac{-6}{x^2 - x + 1}$$
$$= \frac{2}{x-2} - \frac{2}{x+1} - \frac{6}{x^2 - x + 1}$$
$$\therefore \frac{18}{(x-2)(x^3 + 1)} = \frac{2}{x-2} - \frac{2}{x+1} - \frac{6}{x^2 - x + 1}$$

# Partial Fractions

**NOTE**
The final answer in this alternative method is not an exact match of the previous one, but we can modify the latter as

$$\frac{2}{x-2} - \frac{2}{x+1} - \frac{6}{x^2-x+1} = \frac{2}{x-2} - \left(\frac{2}{x+1} + \frac{6}{x^2-x+1}\right)$$

$$= \frac{2}{x-2} - \left(\frac{2(x^2-x+1) + 6(x+1)}{(x+1)(x^2-x+1)}\right)$$

$$= \frac{2}{x-2} - \left(\frac{2x^2 - 2x + 2 + 6x + 6}{(x+1)(x^2-x+1)}\right)$$

$$= \frac{2}{x-2} - \left(\frac{2x^2 + 4x + 8}{(x+1)(x^2-x+1)}\right)$$

$$= \frac{2}{x-2} - \left(\frac{2(x^2 + 2x + 4)}{(x+1)(x^2-x+1)}\right)$$

$$= \frac{2}{x-2} - 2\left(\frac{(x^2 + 2x + 4)}{(x+1)(x^2-x+1)}\right)$$

$$= \frac{2}{x-2} - \frac{2(x^2 + 2x + 4)}{x^3 + 1}$$

Thus the answers are identical.

---

One more example for us to finish this section. Get ready as the example will be challenging.

## Example 10

Using any suitable method, express $\frac{x^3-6x-5}{(x^2+1)(x^2-x+2)}$ in partial fractions.

---

What did you get? Find the solution below to double-check your answer.

## Solution to Example 10

$$\frac{x^3 - 6x - 5}{(x^2 + 1)(x^2 - x + 2)}$$

**Step 1:** The denominator of the compound fraction has two quadratic factors. Obviously, $(x^2 + 1)$ cannot be factorised further and checking with $b^2 - 4ac$ will equally show that we cannot factorise $(x^2 - x + 2)$, hence we say that

$$\frac{x^3 - 6x - 5}{(x^2 + 1)(x^2 - x + 2)} \equiv \frac{Ax + B}{x^2 + 1} + \frac{Cx + D}{x^2 - x + 2}$$

**Step 2:** We will apply cross-multiplication, thus:

$$\frac{x^3 - 6x - 5}{(x^2+1)(x^2-x+2)} \equiv \frac{(Ax+B)(x^2-x+2) + (Cx+D)(x^2+1)}{(x^2+1)(x^2-x+2)}$$

**Step 3:** We now need to equate the numerators as:

$$x^3 - 6x - 5 \equiv (Ax+B)(x^2-x+2) + (Cx+D)(x^2+1)$$

**Step 4:** Open the brackets in the numerator and simplify as:

$$\begin{aligned} x^3 - 6x - 5 &\equiv (Ax+B)(x^2-x+2) + (Cx+D)(x^2+1) \\ &\equiv \left(Ax^3 - Ax^2 + 2Ax + Bx^2 - Bx + 2B\right) + \left(Cx^3 + Cx + Dx^2 + D\right) \\ &\equiv \left(Ax^3 + Cx^3\right) + \left(-Ax^2 + Bx^2 + Dx^2\right) + (2Ax - Bx + Cx) + (2B + D) \\ &\equiv (A+C)x^3 + (-A+B+D)x^2 + (2A-B+C)x + (2B+D) \end{aligned}$$

**Step 5:** We will now compare the coefficients and equate them. Thus:

- For $[x^3]$

We have

$$(A+C)x^3 = x^3$$

Divide both sides by $x^3$, we have:

$$A + C = 1 \qquad -----\text{(i)}$$

- For $[x^2]$, we have

$$(-A+B+D)x^2 = 0x^2$$

Divide both sides by $x^2$, we have:

$$-A + B + D = 0 \qquad -----\text{(ii)}$$

- For $[x]$, we have

$$(2A - B + C)x = -6x$$

Divide both sides by $x$, we have:

$$2A - B + C = -6 \qquad -----\text{(iii)}$$

- For [constant], we have

$$2B + D = -5 \qquad -----\text{(iv)}$$

## Partial Fractions

**Step 6:** We now need to solve for $A$, $B$, $C$, and $D$.

Let's quickly reproduce the four equations:

$$A + C = 1 \quad\quad\quad\quad ----- (i)$$

$$-A + B + D = 0 \quad\quad\quad\quad ----- (ii)$$

$$2A - B + C = -6 \quad\quad\quad\quad ----- (iii)$$

$$2B + D = -5 \quad\quad\quad\quad ----- (iv)$$

Matrix or a similar technique would be ideal in this case, but we will try solving these 'manually'. As $A$ and $B$ appear in three out of four equations, and $C$ and $D$ only in two out of four, we will form two equations containing only these two as follows:

- Equation (iii) – equation (i) to eliminate C

$$A - B = -7 \quad\quad\quad\quad ----- (v)$$

- Equation (iv) – equation (ii) to eliminate D

$$A + B = -5 \quad\quad\quad\quad ----- (vi)$$

- Equation (v) + equation (vi) to eliminate B

$$2A = -12$$
$$\therefore A = -6$$

From equation (vi), we have

$$A + B = -5$$
$$B = -5 - A = -5 - (-6) = -5 + 6$$
$$\therefore B = 1$$

From equation (i), we have

$$A + C = 1$$
$$C = 1 - A = 1 - (-6) = 1 + 6$$
$$\therefore C = 7$$

From equation (ii), we have

$$-A + B + D = 0$$
$$D = A - B = -6 - 1$$
$$\therefore D = -7$$

Thus, we have

$$A = -6,\ B = 1,\ C = 7,\ D = -7$$

**Step 7:** The partial fractions are:

$$\frac{x^3 - 6x - 5}{(x^2 + 1)(x^2 - x + 2)} \equiv \frac{Ax + B}{x^2 + 1} + \frac{Cx + D}{x^2 - x + 2}$$

$$= \frac{-6x + 1}{x^2 + 1} + \frac{7x - 7}{x^2 - x + 2}$$

$$= \frac{-(6x - 1)}{x^2 + 1} + \frac{7(x - 1)}{x^2 - x + 2}$$

$$= \frac{7(x - 1)}{x^2 - x + 2} - \frac{(6x - 1)}{x^2 + 1}$$

$$\therefore \frac{x^3 - 6x - 5}{(x^2 + 1)(x^2 - x + 2)} = \frac{7(x - 1)}{x^2 - x + 2} - \frac{(6x - 1)}{x^2 + 1}$$

**CHECK (Author's method)**

For this case, let's say $x = \frac{1}{3}$

- **RHS**

$$\frac{7(x - 1)}{x^2 - x + 2} - \frac{(6x - 1)}{x^2 + 1} = \frac{7\left(\frac{1}{3} - 1\right)}{\left(\frac{1}{3}\right)^2 - \frac{1}{3} + 2} - \frac{\left(6 \times \frac{1}{3} - 1\right)}{\left(\frac{1}{3}\right)^2 + 1}$$

$$= \frac{7\left(-\frac{2}{3}\right)}{\frac{1}{9} - \frac{1}{3} + 2} - \frac{(2 - 1)}{\frac{1}{9} + 1} = -\frac{141}{40}$$

- **LHS**

$$\frac{x^3 - 6x - 5}{(x^2 + 1)(x^2 - x + 2)} = \frac{\left(\frac{1}{3}\right)^3 - 6 \times \frac{1}{3} - 5}{\left(\left(\frac{1}{3}\right)^2 + 1\right)\left(\left(\frac{1}{3}\right)^2 - \frac{1}{3} + 2\right)}$$

$$= \frac{\frac{1}{27} - 2 - 5}{\left(\frac{1}{9} + 1\right)\left(\frac{1}{9} - \frac{1}{3} + 2\right)} = -\frac{141}{40}$$

Thus,

$$LHS = RHS$$

**Correct!**

---

We've had enough examples on this class of partial fractions. Off to the next!

# Partial Fractions

## 7.4.4 Irreducible Repeated Non-Linear Factors

This is when the denominator has non-linear factors and one or more of the factors are repeated. The general expression for two non-linear factors in which one of them is repeated is given below.

$$\frac{f(x)}{(a_1x^2+b_1x+c_1)(a_2x^3+b_2x^2+c_2x+d_1)^n} \equiv \frac{Ax+B}{a_1x^2+b_1x+c_1} + \frac{Cx^2+Dx+E}{a_2x^3+b_2x^2+c_2x+d_1} + \frac{Fx^2+Gx+H}{(a_2x^3+b_2x^2+c_2x+d_1)^2} + \cdots + \frac{Ix^2+Jx+K}{(a_2x^3+b_2x^2+c_2x+d_1)^n} \tag{7.4}$$

where $a_1$, $a_2$, $b_1$, $b_2$, $c_1$, $c_2$, and $d_1$ and $A$, $B$, $C$, $D$, $E$, $F$, $G$, and $H$ are numerical values.

$f(x)$ is the numerator of the original fraction, which is a polynomial of a degree less than that of the product of the factors in the denominator. Note that only $A$, $B, C$, $D, E, F, G, H, \ldots$ are to be determined, while others will be known from the compound fraction. For example:

$$\frac{2x}{x(x^3+x^2+x-1)^2} \equiv \frac{A}{x} + \frac{Bx^2+Cx+D}{(x^3+x^2+x-1)} + \frac{Ex^2+Fx+G}{(x^3+x^2+x-1)^2}$$

and

$$\frac{5}{(x-3)(5x^2+7)^3} \equiv \frac{A}{x-3} + \frac{Bx+C}{5x^2+7} + \frac{Dx+E}{(5x^2+7)^2} + \frac{Fx+G}{(5x^2+7)^3}$$

From the above two examples, it is apparent that the key challenge is how to find $A, B, C, D, E, F$, and $G$. To do this, we can use one of the following methods.

1) Substitution method
2) Equating coefficient method

Notice that this category is like a combination of category 2 and category 3, and it is as though it is not an independent category. Let's try an example for this.

### Example 11

Using any suitable method, express $\frac{x^3-5x-1}{(x^2+3)^2}$ in partial fractions.

What did you get? Find the solution below to double-check your answer.

### Solution to Example 11

$$\frac{x^3-5x-1}{(x^2+3)^2}$$

**Step 1:** The denominator of the compound fraction has two quadratic factors. Obviously, $(x^2 + 3)$ cannot be factorised further. We have:

$$\frac{x^3 - 5x - 1}{(x^2 + 3)^2} \equiv \frac{Ax + B}{x^2 + 3} + \frac{Cx + D}{(x^2 + 3)^2}$$

**Step 2:** LCM in this case is $(x^2 + 3)^2$, thus we have:

$$\frac{x^3 - 5x - 1}{(x^2 + 3)^2} \equiv \frac{(Ax + B)(x^2 + 3) + (Cx + D)}{(x^2 + 3)^2}$$

**Step 3:** We now need to equate the numerators as:

$$x^3 - 5x - 1 \equiv (Ax + B)(x^2 + 3) + (Cx + D)$$

**Step 4:** Open the brackets and simplify as:

$$\begin{aligned} x^3 - 5x - 1 &\equiv (Ax + B)(x^2 + 3) + (Cx + D) \\ &\equiv (Ax^3 + Bx^2 + 3Ax + 3B) + (Cx + D) \\ &\equiv (Ax^3) + (Bx^2) + (3Ax + Cx) + (3B + D) \\ &\equiv Ax^3 + Bx^2 + (3A + C)x + (3B + D) \end{aligned}$$

**Step 5:** We will now compare the coefficients and equate them. Thus:

- For $[x^3]$, we have

$$Ax^3 = x^3$$

Divide both sides by $x^3$, we have:

$$\therefore A = 1$$

- For $[x^2]$, we have

$$Bx^2 = 0x^2$$

Divide both sides by $x^2$, we have:

$$\therefore B = 0$$

- For $[x]$, we have

$$(3A + C)x = -5x$$

Divide both sides by $x$, we have:

$$3A + C = -5$$
$$C = -3A - 5 = -3(1) - 5 = -3 - 5$$
$$\therefore C = -8$$

# Partial Fractions

- For [constant], we have

$$3B + D = -1$$
$$D = -1 - 3B = -1 - 3(0) = -1$$
$$\therefore D = -1$$

So we have

$$A = 1, \ B = 0, \ C = -8, \ D = -1$$

**Step 6:** The partial fractions are:

$$\frac{x^3 - 5x - 1}{(x^2 + 3)^2} \equiv \frac{Ax + B}{x^2 + 3} + \frac{Cx + D}{(x^2 + 3)^2}$$

$$= \frac{x + 0}{x^2 + 3} + \frac{-8x - 1}{(x^2 + 3)^2}$$

$$= \frac{x}{x^2 + 3} + \frac{-(8x + 1)}{(x^2 + 3)^2}$$

$$= \frac{x}{x^2 + 3} - \frac{8x + 1}{(x^2 + 3)^2}$$

$$\therefore \frac{x^3 - 5x - 1}{(x^2 + 3)^2} = \frac{x}{x^2 + 3} - \frac{8x + 1}{(x^2 + 3)^2}$$

**CHECK (Author's method)**
For this case, let's say $x = -0.25$ (since we've not used a negative number thus far).

- RHS

$$\frac{x}{x^2 + 3} - \frac{8x + 1}{(x^2 + 3)^2} = \frac{(-0.25)}{(-0.25)^2 + 3} - \frac{8(-0.25) + 1}{\left[(-0.25)^2 + 3\right]^2}$$

$$= -\frac{4}{49} - \left(-\frac{256}{2401}\right) = \frac{60}{2401}$$

- LHS

$$\frac{x^3 - 5x - 1}{(x^2 + 3)^2} = \frac{(-0.25)^3 - 5(-0.25) - 1}{\left[(-0.25)^2 + 3\right]^2} = \frac{60}{2401}$$

Thus,

$$LHS = RHS$$

**Correct!**

For this example, the forward process can be easily done, so let's verify our answer using the conventional method.

$$\frac{x}{x^2+3} - \frac{8x+1}{(x^2+3)^2} = \frac{x(x^2+3) - 8x+1}{(x^2+3)^2}$$
$$= \frac{x^3+3x-8x+1}{(x^2+3)^2}$$
$$= \frac{x^3-5x+1}{(x^2+3)^2}$$

**Correct!**

This will do for this category; let's move to the last category.

### 7.4.5 IMPROPER FRACTIONS

The last category is when the degree of the numerator is equal to or greater than the degree of the denominator. This is called an improper fraction, and is similar to the numerical improper fraction such as $\frac{7}{3}$. Examples include:

$$\frac{x^2+1}{x^2-5x+6} \qquad \frac{x^3+2x-1}{(x-11)(2x+3)} \qquad \frac{x^4-2x+1}{(x^2-3)(x+1)^2}$$

In this case, we first need to divide the numerator by the denominator using the Remainder Theorem before we proceed to splitting it into partial fractions.

Given that $n(x)$ and $d(x)$ are the numerator and denominator, respectively, using the Remainder Theorem, we can simplify this into:

$$\boxed{\frac{n(x)}{d(x)} = Q(x) + \frac{R}{d(x)}} \tag{7.5}$$

where $Q(x)$ and $d(x)$ are the quotient and remainder, respectively. $Q(x)$ can be a number or a polynomial, but it must be of a lesser degree than both the dividend and the divisor. $\frac{R}{d(x)}$ is a proper fraction as the degree of the remainder $R$ should be less than that of the divisor $d(x)$. We can therefore choose a suitable method out of the three discussed to split it into partial fractions, depending on the nature of the resulting fraction.

A couple of examples will make this clearer (though a review of division of polynomials covered in *Foundation Mathematics for Engineers and Scientists with Worked Examples* by the same author may be helpful).

**Example 12**

Using a suitable method, split $\frac{3x^2-4x-5}{(x-3)(x+5)}$ into partial fractions.

# Partial Fractions

What did you get? Find the solution below to double-check your answer.

## Solution to Example 12

$$\frac{3x^2 - 4x - 5}{(x-3)(x+5)}$$

**Step 1:** As the power of the expression in the numerator is equal to that in the denominator, we need to reduce this using the long division. Let's get started with this.

We need to open the brackets in the denominator first, as the long division will require that, so we have

$$\frac{3x^2 - 4x - 5}{(x-3)(x+5)} = \frac{3x^2 - 4x - 5}{x^2 + 2x - 15}$$

Let's carry out this in the long division format as:

$$\begin{array}{r}
3\phantom{xxxxxxxxxx} \\
x^2+2x-15 \overline{\smash{\big)}\, 3x^2 - 4x - 5\phantom{)}} \\
3x^2 + 6x - 45\phantom{)} \\
\hline
-10x + 40\phantom{)}
\end{array}$$

Hence, we can write the rational fraction as:

$$\frac{3x^2 - 4x - 5}{(x-3)(x+5)} \equiv 3 + \frac{-10x + 40}{x^2 + 2x - 15}$$

$$\equiv 3 + \frac{40 - 10x}{(x-3)(x+5)}$$

The expression $\frac{40-10x}{(x-3)(x+5)}$ was not in the simplest form and we needed to apply partial fractions technique to it. Notice that this is now in the format of the first category, and we know how to handle this. Let's move to the next step.

**Step 2:** Splitting $\frac{40-10x}{(x-3)(x+5)}$ into partial fractions.

$$\frac{40 - 10x}{(x-3)(x+5)} \equiv \frac{A}{x-3} + \frac{B}{x+5}$$

Eliminating the denominator method will be quick here.

- To obtain $A$, multiply through by $(x-3)$

$$\frac{40 - 10x}{(x-3)(x+5)}(x-3) = \frac{A}{x-3}(x-3) + \frac{B}{x+5}(x-3)$$

Therefore

$$\frac{40 - 10x}{(x+5)} = A + \frac{B}{x+5}(x-3)$$

Using $x = 3$, we have

$$\frac{40 - 10(3)}{(3+5)} = A + \frac{B}{3+5}(3-3)$$

$$\frac{40 - 30}{8} = A + \frac{B}{8}(0)$$

$$\frac{10}{8} = A + 0$$

$$\therefore A = \frac{5}{4}$$

- To obtain $B$, multiply through by $(x+5)$

$$\frac{40 - 10x}{(x-3)(x+5)}(x+5) = \frac{A}{x-3}(x+5) + \frac{B}{x+5}(x+5)$$

$$\frac{40 - 10x}{(x-3)} = \frac{A}{x-3}(x+5) + B$$

Using $x = -5$, we have

$$\frac{40 - 10(-5)}{(-5-3)} = \frac{A}{-5-3}(-5+5) + B$$

$$\frac{40 + 50}{-8} = A(0) + B$$

$$\frac{90}{-8} = 0 + B$$

$$\therefore B = -\frac{45}{4}$$

Hence

$$\frac{40 - 10x}{(x-3)(x+5)} \equiv \frac{\left(\frac{5}{4}\right)}{x-3} + \frac{\left(-\frac{45}{4}\right)}{x+5}$$

$$= \frac{5}{4(x-3)} - \frac{45}{4(x+5)}$$

**Step 3:** Finally, we need to write the partial fraction

$$\frac{3x^2 - 4x - 5}{(x-3)(x+5)} \equiv 3 + \frac{-10x + 40}{x^2 + 2x - 15}$$

$$\equiv 3 + \frac{5}{4(x-3)} - \frac{45}{4(x+5)}$$

You can check using either the author's method or the conventional method.

---

One final example for us to try.

## Example 13

Using a suitable method, split $\frac{(x^2-1)^2}{x^3 - 2x^2 - 5x + 6}$ into partial fractions.

# Partial Fractions

What did you get? Find the solution below to double-check your answer.

**Solution to Example 13**

$$\frac{(x^2-1)^2}{x^3-2x^2-5x+6}$$

**Step 1:** As the power of the expression in the numerator is greater than that in the denominator, we need to reduce this using long division. Let's get started with this.

We need to open the brackets in the numerator first, as the long division will require that, so we have

$$\frac{(x^2-1)^2}{x^3-2x^2-5x+6} = \frac{x^4-2x^2+1}{x^3-2x^2-5x+6}$$

Note that $x^4 - 2x^2 + 1$ can be written in the complete polynomial format as $x^4 + 0x^3 - 2x^2 + 0x + 1$. Let's start as:

$$
\begin{array}{r}
x+2\phantom{xxxxxxxxxxxxxxxx} \\
x^3-2x^2-5x+6 \overline{\smash{\big)}\, x^4 + 0x^3 - 2x^2 + 0x + 1} \\
\underline{x^4 - 2x^3 - 5x^2 + 6x\phantom{+1}} \\
2x^3 + 3x^2 - 6x + 1 \\
\underline{2x^3 - 4x^2 - 10x + 12} \\
7x^2 + 4x - 11
\end{array}
$$

Hence, we can write the rational fraction as:

$$\frac{(x^2-1)^2}{x^3-2x^2-5x+6} \equiv \frac{x^4-2x^2+1}{x^3-2x^2-5x+6} \equiv x+2+\frac{7x^2+4x-11}{x^3-2x^2-5x+6}$$

Although experience can tell that this is not in the simplest form, because the denominator (i.e., $x^3 - 2x^2 - 5x + 6$) can still be factorised into prime factors. This is our next task; otherwise the above could be taken as the simplest partial fractions and our final answer.

**Step 2:** Factorise $x^3 - 2x^2 - 5x + 6$

Given that

$$f(x) = x^3 - 2x^2 - 5x + 6$$

If $x - a$ is a factor of $f(x)$, then

$$f(a) = 0$$

Because the constant in $f(x)$ is 6, we can try the following values of $a$: $\pm 1, \pm 2, \pm 3,$ and $\pm 6$.

- $a = 1$

$$f(1) = (1)^3 - 2(1)^2 - 5(1) + 6 = 1 - 2 - 5 + 6 = 0$$

Because $f(1) = 0$, we have that $x - 1$ is a factor. We need two more linear factors or one quadratic factor.

- $a = -1$

$$f(-1) = (-1)^3 - 2(-1)^2 - 5(-1) + 6 = (-1) - 2(1) + 5 + 6 = -1 - 2 + 5 + 6 \neq 0$$

Because $f(-1) \neq 0$, we can conclude that $x + 1$ is NOT a factor.

- $a = -2$

$$\begin{aligned} f(-2) &= (-2)^3 - 2(-2)^2 - 5(-2) + 6 \\ &= (-8) - 2(4) + 10 + 6 = -8 - 8 + 10 + 6 = 0 \end{aligned}$$

Because $f(-2) = 0$, we have that $x + 2$ is a factor. One more linear factor to go. Since we've identified $a = 1$ and $a = -2$, the third and the last one is $a = 3$. This is because the constant in $f(x)$ is 6, such that $6 = -1 \times 2 \times -3$ or $c = a_1 \times a_2 \times a_3$. In other words, $x - 3$ is a factor.

- $a = 3$

$$f(3) = (3)^3 - 2(3)^2 - 5(3) + 6 = 27 - 18 - 15 + 6 = 0$$

Because $f(3) = 0$, we have that $x - 3$ is a factor. Therefore, the complete factorisation is

$$x^3 - 2x^2 - 5x + 6 = (x - 3)(x - 1)(x + 2)$$

**ALTERNATIVE METHOD**

Alternatively, when we obtained the first factor we can then write that

$$x^3 - 2x^2 - 5x + 6 = (x - 1)(ax^2 + bx + c)$$

Comparing the coefficients – we do not really need to open the brackets on the RHS – we have

$$a = 1, \ c = 6$$

Note that $a = 1$ is obtained by multiplying $x$ with $ax^2$ and $c = 6$ is obtained when you multiply the last term in the first bracket with the last term in the second bracket, i.e., $(-1)(c)$. This can be done visually.

Equate the power of $[x^2]$ on the RHS, if open, with that on the LHS

$$\begin{aligned} bx^2 - ax^2 &= -2x^2 \\ b - a &= -2 \\ b &= -2 + a = -2 + 1 \\ \therefore b &= -1 \end{aligned}$$

Therefore, we can write the quadratic factor as:

$$x^3 - 2x^2 - 5x + 6 = (x - 1)(x^2 - x + 6)$$

It is easy to factorise now

$$x^3 - 2x^2 - 5x + 6 = (x - 1)(x - 3)(x + 2)$$

Partial Fractions

In summary, we now have

$$\frac{(x^2-1)^2}{x^3-2x^2-5x+6} \equiv \frac{x^4-2x^2+1}{x^3-2x^2-5x+6} \equiv x+2+\frac{7x^2+4x-11}{(x-3)(x-1)(x+2)}$$

**Step 3:** Splitting $\frac{7x^2+4x-11}{(x-3)(x-1)(x+2)}$ into partial fractions.

$$\frac{7x^2+4x-11}{(x-3)(x-1)(x+2)} \equiv \frac{A}{x-3} + \frac{B}{x-1} + \frac{C}{x+2}$$

Eliminating the denominator method will be quick here.

- To obtain $A$, multiply through with $(x-3)$

$$\frac{7x^2+4x-11}{(x-3)(x-1)(x+2)} \equiv \frac{A}{x-3} + \frac{B}{x-1} + \frac{C}{x+2}$$

$$\frac{7x^2+4x-11}{(x-3)(x-1)(x+2)}(x-3) \equiv \frac{A}{x-3}(x-3) + \frac{B}{x-1}(x-3) + \frac{C}{x+2}(x-3)$$

Therefore

$$\frac{7x^2+4x-11}{(x-1)(x+2)} \equiv A + \frac{B}{x-1}(x-3) + \frac{C}{x+2}(x-3)$$

Using $x = 3$, we have

$$\frac{7(3)^2+4(3)-11}{(3-1)(3+2)} \equiv A + \frac{B}{3-1}(3-3) + \frac{C}{3+2}(3-3)$$

$$\frac{63+12-11}{(2)(5)} \equiv A + \frac{B}{3-1}(0) + \frac{C}{3+2}(0)$$

$$\frac{64}{10} \equiv A$$

$$\therefore A = \frac{32}{5}$$

- To obtain $B$, multiply through by $(x-1)$

$$\frac{7x^2+4x-11}{(x-3)(x-1)(x+2)} \equiv \frac{A}{x-3} + \frac{B}{x-1} + \frac{C}{x+2}$$

$$\frac{7x^2+4x-11}{(x-3)(x-1)(x+2)}(x-1) \equiv \frac{A}{x-3}(x-1) + \frac{B}{x-1}(x-1) + \frac{C}{x+2}(x-1)$$

Therefore

$$\frac{7x^2+4x-11}{(x-3)(x+2)} \equiv \frac{A}{x-3}(x-1) + B + \frac{C}{x+2}(x-1)$$

Using $x = 1$, we have

$$\frac{7(1)^2+4(1)-11}{(1-3)(1+2)} \equiv \frac{A}{1-3}(1-1) + B + \frac{C}{1+2}(1-1)$$

$$\frac{7+4-11}{(-2)(3)} \equiv \frac{A}{1-3}(0) + B + \frac{C}{1+2}(0)$$

$$\frac{0}{-6} \equiv B$$

$$\therefore B = 0$$

- To obtain $C$, multiply through by $(x+2)$

$$\frac{7x^2 + 4x - 11}{(x-3)(x-1)(x+2)} \equiv \frac{A}{x-3} + \frac{B}{x-1} + \frac{C}{x+2}$$

$$\frac{7x^2 + 4x - 11}{(x-3)(x-1)(x+2)}(x+2) \equiv \frac{A}{x-3}(x+2) + \frac{B}{x-1}(x+2) + \frac{C}{x+2}(x+2)$$

Therefore

$$\frac{7x^2 + 4x - 11}{(x-3)(x-1)} \equiv \frac{A}{x-3}(x+2) + \frac{B}{x-1}(x+2) + C$$

Using $x = -2$, we have

$$\frac{7(-2)^2 + 4(-2) - 11}{(-2-3)(-2-1)} \equiv \frac{A}{-2-3}(-2+2) + \frac{B}{-2-1}(-2+2) + C$$

$$\frac{28 - 8 - 11}{(-5)(-3)} \equiv \frac{A}{-2-3}(0) + \frac{B}{-2-1}(0) + C$$

$$\frac{9}{15} \equiv C$$

$$\therefore C = \frac{3}{5}$$

Hence

$$\frac{7x^2 + 4x + 13}{(x-3)(x-1)(x+2)} \equiv \frac{\left(\frac{32}{5}\right)}{x-3} + \frac{(0)}{x-1} + \frac{\left(\frac{3}{5}\right)}{x+2} = \frac{32}{5(x-3)} + \frac{3}{5(x+2)}$$

**Step 4:** Finally, we need to write the partial fraction

$$\frac{(x^2-1)^2}{x^3 - 2x^2 - 5x + 6} \equiv \frac{x^4 - 2x^2 + 1}{x^3 - 2x^2 - 5x + 6} \equiv x + 2 + \frac{7x^2 + 4x - 11}{(x-3)(x-1)(x+2)}$$

$$\therefore \frac{(x^2-1)^2}{x^3 - 2x^2 - 5x + 6} \equiv x + 2 + \frac{32}{5(x-3)} + \frac{3}{5(x+2)}$$

This is a long one! It will be helpful to check, but this will be done using author's approach.

**CHECK (Author's method)**
For this case, let's say $x = -0.4$

- RHS

$$x + 2 + \frac{32}{5(x-3)} + \frac{3}{5(x+2)} = (-0.4 + 2) + \frac{32}{5(-0.4-3)} + \frac{3}{5(-0.4+2)}$$

$$= \left(\frac{8}{5}\right) + \left(-\frac{32}{17}\right) + \left(\frac{3}{8}\right) = \frac{63}{680}$$

- LHS

$$\frac{(x^2-1)^2}{x^3-2x^2-5x+6} = \frac{\left[(-0.4)^2-1\right]^2}{(-0.4)^3-2(-0.4)^2-5(-0.4)+6} = \frac{\left(\frac{441}{625}\right)}{\left(\frac{952}{125}\right)} = \frac{63}{680}$$

Thus,

$$LHS = RHS$$

**Correct!**

This brings us to the end of this chapter.

## 7.5 CHAPTER SUMMARY

1) The need to add or subtract two or more fractions to obtain a single fraction is apparently necessary. However, the inverse of this process (i.e., converting or splitting a compound algebraic fraction into two or more simple fractions) might be obscure.

2) The resulting (simple) fractions are called *partial fractions* of the original single fraction. The process of achieving this is called *resolving into partial fractions*.

3) The nature of the factors in the denominator determines the steps to be taken in resolving into partial fractions. We can generally classify the nature into five:
   - Non-repeated linear factors
   - Repeated linear factors
   - Irreducible non-repeated non-linear factors
   - Irreducible repeated non-linear factors
   - Improper fractions

4) The methods commonly used for resolving into partial fractions are:
   - Substitution method
   - Equating coefficients method
   - Eliminating a denominator method

\*\*\*\*

## 7.6 FURTHER PRACTICE

To access complementary contents, including additional exercises, please go to www.dszak.com.

# 8 Complex Numbers I

## Learning Outcomes

Once you have studied the content of this chapter, you should be able to:

- Explain the $j$ operator
- Explain the real and imaginary parts
- Apply fundamental principles of complex numbers
- Carry out addition, subtraction, and multiplication of complex numbers
- Discuss complex conjugates
- Carry out division of complex numbers using rationalising the denominator

## 8.1 INTRODUCTION

You probably have heard about numbers and their different types. The category we are interested in is their classification into real and imaginary numbers. Most of the numbers (e.g., integers, rational, and irrational numbers) you might have encountered are real. Imaginary numbers, on the other hand, are not very common except in science and engineering, and together with real numbers form the basis of complex numbers. They are called complex numbers by convention, not because they are difficult to understand or work with. The core differences between complex numbers and other numbers are few, and it only takes understanding a few tips about what constitutes complex numbers to spot these dissimilarities. In this chapter, we will look at the definition of $j$ or $i$ operator, evaluating complex numbers, expressing complex numbers in different forms, and converting from one form to another.

## 8.2 COMPLEX NUMBERS

### 8.2.1 $j$ Operator

$j$ is an operator that functions just like $\times$, $\sqrt{}$, $\cup$, %, or any other mathematical operator. Letter $j$ is used in engineering while $i$ is employed in other STEM (Science, Technology, Engineering, and Mathematics) subjects for vector analysis, though both denote the same numerical process. $i$ symbolises electric current and usually a complex number is applied in AC (alternating current) circuit analysis where electric current is also used. Consequently, it will be difficult to discern when $i$ is used as electric current or complex number operator. In this chapter, $j$ will be used.

DOI:10.1201/9781032665122-8

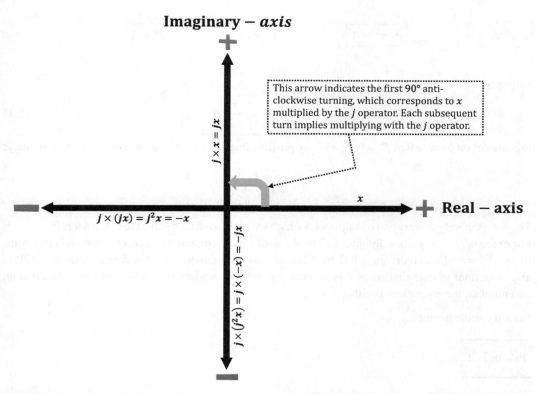

**FIGURE 8.1** Complex plane.

So what is $j$? The answer is simply

$$j = \sqrt{-1} \quad \text{OR} \quad j^2 = -1$$

In broader terms, $j$ is a 90° counter-clockwise rotation of a vector, usually taken from the positive $x$-axis. Notice that when such a rotation is performed twice, the vector will be on the same line but pointing towards the opposite direction. It is therefore not surprising that $j^2 = -1$. $j^2 = j \times j$, which is 90° turn followed by another 90° turn, making a total of 180°. In addition, when a 360° rotation is carried out on a vector, the vector will return to its original position and again $j^4 = 1$. We use $j^4$ because there are four 90° in 360°.

For example, when a vector of $x$ units (Figure 8.1), originally pointing East (that is, on the positive $x$-axis) is rotated counter-clockwise by 90°, the new vector can be regarded as $j * x$ units pointing northward or on the positive $y$-axis. If the same vector is rotated again in the same direction and with the same magnitude (i.e., 90° counter-clockwise), the new vector has a value of $j * j * x$ units, which is $-x$ units. It therefore follows that $j * j = j^2 = -1$.

The third and fourth rotations will produce vectors of values $j * j * j * x$ units and $j * j * j * j * x$ units, respectively. Obviously, after the third rotation, the vector will be pointing southward. Because it is on the negative $y$-axis, its value is the same in magnitude but opposite to the value of the vector pointing northward. Therefore, $j * j * j * x = -j * x$ or $j^3 = -j$. Similarly, the fourth rotation will be identical in magnitude and direction to the initial vector, thus $j * j * j * j * x = x$ or $j^4 = 1$. In summary, we have

$$\boxed{j = \sqrt{-1}} \qquad \boxed{j^2 = -1} \qquad \boxed{j^3 = -j} \qquad \boxed{j^4 = 1} \qquad \boxed{\frac{1}{j} = -j}$$

Note that

$$\boxed{\frac{1}{j} = \frac{1}{j} \times \frac{j}{j} = \frac{j}{j^2} = \frac{j}{-1} = -j} \qquad (8.1)$$

It is important to note that $j^n$, where $n$ is any positive integer, has only four possible values, namely:

$$1 \quad j \quad -1 \quad -j$$

They respectively correspond to values of $n$ which when divided by 4 leave the remainders 0, 1, 2, 3. For example, $j^{137} = j$ since dividing 137 by 4 leaves 1 as a remainder and this corresponds to $j$. Similarly, $j^{102} = -1$ since dividing 102 by 4 leaves 2 as a remainder and this corresponds to $-1$. It is also important to note that when $n$ is an even number, the answer is 1 or $-1$, and for when $n$ is an odd number, the equivalent is either $j$ or $-j$.

Let's try some examples.

### Example 1

Write the following in their simplest form.

a) $j^{40}$      b) $j^{15}$      c) $(3j)^3$      d) $(-3j)^5$      e) $j + 3j^3 - 2j^2 + 2j$

What did you get? Find the solution below to double-check your answer.

### Solution to Example 1

**HINT**

In this example, the key rule is to realise that $j^2 = -1$ and $j^4 = 1$. It is also important to apply the laws of indices when and where necessary.

**a) $j^{40}$**
**Solution**

$$j^{40} = (j^4)^{10} = (1)^{10} = 1$$

$$\therefore j^{40} = 1$$

**b) $j^{15}$**
**Solution**

$$j^{15} = j^{(4+4+4+2+1)} = j^4 \times j^4 \times j^4 \times j^2 \times j$$
$$= 1 \times 1 \times 1 \times -1 \times j$$
$$= -j$$

# Complex Numbers I

**c)** $(3j)^3$
**Solution**

$$(3j)^3 = 3^3 \times j^3 = 27 \times j^2 \times j$$
$$= 27 \times (-1) \times j = -27j$$
$$\therefore (3j)^3 = -27j$$

**d)** $(-3j)^5$
**Solution**

$$(-3j)^5 = (-3 \times j)^5 = (-3)^5(j)^5$$
$$= -243(j)^4 \cdot j = -243\,(1) \cdot j$$
$$\therefore (-3j)^5 = -243j$$

**e)** $j + 3j^3 - 2j^2 + 2j$
**Solution**

$$j + 3j^3 - 2j^2 + 2j = j + 3j^2 \times j - 2j^2 + 2j$$
$$= j + 3(-j) - 2(-1) + 2j$$
$$= j - 3j + 2 + 2j = 2$$
$$\therefore j + 3j^3 - 2j^2 + 2j = 2$$

So, the problem of the square roots of negative numbers is finally solved. We can now say, for example, that:

$$\sqrt{-9} = \sqrt{9 \times -1} = \sqrt{9} \times \sqrt{-1}$$
$$= \pm 3 \times j = \pm 3j$$

Let's try an example.

### Example 2

Without using a calculator, evaluate $\sqrt{-0.64}$.

What did you get? Find the solution below to double-check your answer.

### Solution to Example 2

$$\sqrt{-0.64} = \sqrt{-1 \times 0.64}$$
$$= \sqrt{-1} \times \sqrt{\frac{64}{100}} = j \times \frac{8}{10} = 0.8j$$
$$\therefore \sqrt{-0.64} = 0.8j$$

## NOTE

We can say that $\sqrt{\frac{64}{100}} = \pm\frac{8}{10} = \pm 0.8$. Therefore $\sqrt{-0.64} = \pm 0.8j$.

---

Before now we know that when discriminant $b^2 - 4ac < 0$, the quadratic equation cannot be solved or does not have real values. But with the above knowledge, it can be revealed that any quadratic equations can now be solved. Great! In general, we have

$$\boxed{\sqrt{-n} = \sqrt{-1 \times n} = \sqrt{-1} \times \sqrt{n} = j\sqrt{n}} \qquad (8.2)$$

Let's try some examples.

### Example 3

Use the quadratic formula to solve the following equations. Present the answers in the form $a + jb$, such that $a$ and $b$ are real numbers.

a) $x^2 + 9 = 0$     b) $3\alpha^2 - 2\alpha + 7 = 0$     c) $2t^2 + 4t + 6 = 0$

What did you get? Find the solution below to double-check your answer.

### Solution to Example 3

**a)** $x^2 + 9 = 0$
**Solution**
Remember that $x^2 + 9 = 0$ can be written as $x^2 + 0x + 9 = 0$. For this case, $a = 1$, $b = 0$, and $c = 9$. Let's substitute the values of $a$, $b$, and $c$ in the quadratic formula and then simplify.

$$x = \frac{-b \pm \sqrt{b^2 - 4ac}}{2a}$$

Here is the working:

$$x = \frac{-0 \pm \sqrt{0^2 - 4(1)(9)}}{2}$$

$$= \frac{\pm\sqrt{-36}}{2} = \pm\frac{6}{2}j = \pm 3j$$

Hence, the solutions to the equation are $x = 3j$ and $x = -3j$.

---

**b)** $3\alpha^2 - 2\alpha + 7 = 0$
**Solution**
For this case, $a = 3$, $b = -2$, and $c = 7$. Now substitute the values and simplify.

$$\alpha = \frac{2 \pm \sqrt{(-2)^2 - 4(3)(7)}}{2(3)} = \frac{2 \pm \sqrt{-80}}{6}$$

$$= \frac{2 \pm j4\sqrt{5}}{6} = \frac{1}{3}(1 \pm j2\sqrt{5})$$

# Complex Numbers I

Hence, the solutions to the equation are $\alpha = \frac{1}{3}(1+j2\sqrt{5})$ and $=\frac{1}{3}(1-j2\sqrt{5})$.

---

**c)** $2t^2 + 4t + 6 = 0$

**Solution**

For this case, $a = 2$, $b = 4$, and $c = 6$. Now substitute the values and then simplify. Here is the working:

$$t = \frac{-4 \pm \sqrt{4^2 - 4(2)(6)}}{2(2)}$$

$$= \frac{-4 \pm \sqrt{-32}}{4} = \frac{-4 \pm j4\sqrt{2}}{4}$$

$$\therefore t = -1 \pm j\sqrt{2}$$

Hence, the solutions to the equation are $t = -1 + j\sqrt{2}$ and $= -1 - j\sqrt{2}$.

---

Let's try another example.

### Example 4

Determine the roots of the equation $x^3 + 27 = 0$.

---

What did you get? Find the solution below to double-check your answer.

### Solution to Example 4

**HINT**

The above cubic equation (and similar ones) can be factorised using factor theorem or standard expressions:

$$a^3 + b^3 = (a+b)(a^2 - ab + b^2) \text{ and } a^3 - b^3 = (a-b)(a^2 + ab + b^2)$$

$x^3 + 27 = 0$ can be expressed as $x^3 + 3^3 = 0$, where $a = x$ and $b = 3$.
Thus

$$x^3 + 3^3 = (x+3)(x^2 - 3x + 9) = 0$$

Therefore, either

$$x + 3 = 0$$

which implies that

$$x = -3$$

or

$$x^2 - 3x + 9 = 0$$

this implies that

$$x = \frac{3 \pm \sqrt{(-3)^2 - 4(9)}}{2}$$

$$= \frac{3 \pm \sqrt{-27}}{2} = \frac{3 \pm j3\sqrt{3}}{2}$$

$$\therefore x = 1.5 \pm j1.5\sqrt{3}$$

Therefore, the roots of $x^3 + 27 = 0$ are

$$x = -3, \; x = 1.5 + j1.5\sqrt{3} \text{ and } x = 1.5 - j1.5\sqrt{3}.$$

Let's try one more example.

### Example 5

Determine the quadratic equation whose roots are $x = j3\sqrt{2}$ and $x = -j3\sqrt{2}$.

What did you get? Find the solution below to double-check your answer.

### Solution to Example 5

Let $\alpha$ and $\beta$ be the roots of a quadratic equation, it follows that

$$(x - \alpha)(x - \beta) = 0$$

or

$$x^2 - (\alpha + \beta)x + \alpha\beta = 0$$

Given that $x = j3\sqrt{2}$ and $x = -j3\sqrt{2}$, we have

$$x^2 - (\alpha + \beta)x + \alpha\beta = 0$$

this implies that

$$x^2 - \left(j3\sqrt{2} - j3\sqrt{2}\right)x + \left(j3\sqrt{2}\right)\left(-j3\sqrt{2}\right) = 0$$

$$x^2 - (0)x + \left(-j^2 \times 3^2 \times \left(\sqrt{2}\right)^2\right) = 0$$

$$x^2 - 0 + (1 \times 9 \times 2) = 0$$

$$\therefore x^2 + 18 = 0$$

**NOTE**

$x^2 + 18 = 0$ is one out of many (or infinite) equations that will have $x = j3\sqrt{2}$ and $x = -j3\sqrt{2}$ as their solutions. This is because any possible vertical stretch will produce the same roots. In general, the family of the equations that will produce these roots can be given as $k(x^2 + 18) = 0$, where $k$ is any real number.

---

To reiterate, whenever we take the square root of a number, we always have two answers. The two solutions are equal in magnitude but with opposite signs. Generally, $\pm x$ is used to denote this, where $x$ is the magnitude.

### 8.2.2 IMAGINARY NUMBER

A number such as $j3$ is called an imaginary number. A complex number is a number that is made up of two parts, the real and the imaginary. This is expressed as:

$$\boxed{z = x + jy}$$

where $x$ is the real part (or component) of the complex number ($z$), abbreviated as **Re** ($z$), and the imaginary part is $y$, shortened as **Im** ($z$). Note that the imaginary part does not include the $j$ operator itself.

Sometimes the $j$ operator is written first before the number, and the number is written before the $j$ operator at other times. Therefore, $2j$ can be written as $j2$, though the former seems more common.

Let's try some examples.

#### Example 6

State the real and imaginary components of the following complex numbers:

a) $z = 0.5 - 3j$  b) $z = j\sqrt{3}$  c) $z = 3\alpha + j(\beta - \gamma)$

What did you get? Find the solution below to double-check your answer.

#### Solution to Example 6

**HINT**

All you need to watch out for is the presence of the $j$ operator. If a term has the operator, it is the imaginary part, otherwise it is the real part.

**a)** $z = 0.5 - 3j$
**Solution**

$$\mathbf{Re}(z) = \mathbf{0.5}$$
$$\mathbf{Im}(z) = \mathbf{-3}$$

**b)** $z = j\sqrt{3}$
**Solution**

$$\mathbf{Re}(z) = 0$$
$$\mathbf{Im}(z) = \sqrt{3}$$

---

**c)** $z = 3\alpha + j(\beta - \gamma)$
**Solution**

$$\mathbf{Re}(z) = 3\alpha$$
$$\mathbf{Im}(z) = \beta - \gamma$$

---

Is there any number that is both real and imaginary? Yes, the answer is **0**. This is simply because $+0 = -0$ so we can regard $\sqrt{0}$ as real and $\sqrt{-0}$ as imaginary.

One other point that should be mentioned is that in engineering, the real part refers to the active (or in-phase) component and the imaginary is the reactive (or quadrature) part.

## 8.3 FUNDAMENTAL OPERATIONS OF COMPLEX NUMBERS

Addition, subtraction, multiplication, and division can be performed on complex numbers. The rules governing these operations are like other numbers, though with a minor manipulation.

### 8.3.1 ADDITION AND SUBTRACTION

To carry out addition or subtraction of two or more complex numbers, add/subtract the real parts and add/subtract the imaginary parts. Given that $z_1 = a_1 + jb_1$ and $z_2 = a_2 + jb_2$, then

$$\boxed{z_1 \pm z_2 = (a_1 \pm a_2) + j(b_1 \pm b_2)} \tag{8.3}$$

That is all we need to do. Let's try some examples.

### Example 7

Given that $Z_1 = 3 + 2j$, $Z_2 = 5 - j$, and $Z_3 = 1 + 7j$, simplify the following:

**a)** $Z_1 + Z_2$  **b)** $Z_2 - Z_3$  **c)** $Z_1 + Z_3$

# Complex Numbers I

What did you get? Find the solution below to double-check your answer.

## Solution to Example 7

**a)** $Z_1 + Z_2$
**Solution**

$$Z_1 + Z_2 = (3 + 2j) + (5 - j) = 3 + 2j + 5 - j$$
$$= 3 + 5 + 2j - j = 8 + j$$
$$\therefore Z_1 + Z_2 = 8 + j$$

---

**b)** $Z_2 - Z_3$
**Solution**

$$Z_2 - Z_3 = (5 - j) - (1 + 7j) = 5 - j - 1 - 7j$$
$$= 5 - 1 - j - 7j = 4 - 8j = 4(1 - 2j)$$
$$\therefore Z_2 - Z_3 = 4(1 - 2j)$$

---

**c)** $Z_1 + Z_3$
**Solution**

$$Z_1 + Z_3 = (3 + 2j) + (1 + 7j) = 3 + 2j + 1 + 7j$$
$$= 3 + 1 + 2j + 7j = 4 + 9j$$
$$\therefore Z_1 + Z_3 = 4 + 9j$$

---

### 8.3.2 Multiplication

Multiplication involving complex numbers is carried out in the same way as with numbers in arithmetic, but we need to note that $j \times j = -1$. In other words, if numbers with $j$ terms are multiplied, $j$ will disappear leaving a new term (now a real number) with a negative sign.

It is as simple as that. Let's try some examples.

## Example 8

Given that $Z_1 = 2 - 3j$, $Z_2 = 3 + 7j$, $Z_3 = 2 + j$, and $Z_4 = 2 + 3j$, simplify the following:

**a)** $3Z_1$    **b)** $jZ_2$    **c)** $Z_1 Z_4$    **d)** $(Z_2)^2$    **e)** $Z_1 Z_2 Z_3$

What did you get? Find the solution below to double-check your answer.

**Solution to Example 8**

**a)** $3Z_1$
**Solution**
In this example, a complex number is multiplied by a real number.

$$3Z_1 = 3 \times (2 - 3j)$$
$$= (3 \times 2) - (3 \times 3j) = 6 - 9j$$
$$\therefore 3Z_1 = 6 - 9j$$

**b)** $jZ_2$
**Solution**
In this example, a complex number is multiplied by an imaginary number.

$$jZ_2 = j \times (3 + 7j)$$
$$= (j \times 3) + (j \times 7j) = 3j + 7j^2 = 3j - 7$$
$$\therefore jZ_1 = -7 + 3j$$

**c)** $Z_1 Z_4$
**Solution**

$$Z_1 Z_4 = (2 - 3j)(2 + 3j)$$
$$= (4 + 6j - 6j + 9) = (4 + 9)$$
$$\therefore Z_1 Z_4 = 13$$

$Z_1$ and $Z_4$ are said to be conjugates because they differ only in the sign between the real part and the imaginary part; one is positive while the other is negative. Notice that the answer here is a real number. This is what happens when conjugate complex numbers are multiplied together. This is the same as conjugates in surds, which is covered in *Foundation Mathematics for Engineers and Scientists with Worked Examples* by the same author.

**d)** $(Z_2)^2$
**Solution**
In this example, a complex number is squared. This is the same as multiplying a complex number by itself.

$$(Z_2)^2 = Z_2 Z_2 = (3 + 7j)(3 + 7j)$$
$$= (9 + 21j + 21j - 49) = -40 + 42j = -2(20 - 21j)$$
$$\therefore Z_2 Z_2 = -2(20 - 21j)$$

**e)** $Z_1 Z_2 Z_3$

**Solution**

In this example, three complex numbers are multiplied together. We will use associative law here such that $Z_1 Z_2 Z_3 = Z_1 (Z_2 Z_3)$, i.e., we will multiply $Z_2$ and $Z_3$ together first and the result is multiplied with $Z_1$ as:

$$Z_1 Z_2 Z_3 = (2 - 3j)(3 + 7j)(2 + j)$$
$$= (6 + 14j - 9j + 21)(2 + j)$$
$$= (27 + 5j)(2 + j)$$
$$= 54 + 27j + 10j - 5 = 49 + 37j$$
$$\therefore Z_1 Z_2 Z_3 = 49 + 37j$$

### 8.3.3 Division

Division occurs in two forms, and each will be covered as follows.

#### 8.3.3.1 Division by Real Numbers

When a complex is divided by a real number (rational or irrational), the process is straightforward. The result will be another complex number with new real and imaginary parts. Let's try an example.

**Example 9**

Divide $3 + 2j$ by 5.

What did you get? Find the solution below to double-check your answer.

**Solution to Example 9**

$$3 + 2j \div 5 = \frac{3 + 2j}{5}$$

In this example, the denominator is a real number. Thus, using the fact that $\frac{a+b}{c} = \frac{a}{c} + \frac{b}{c}$, we will divide the real part and imaginary part by 5.

$$\frac{3 + 2j}{5} = \frac{3}{5} + \frac{2j}{5}$$
$$= 0.6 + 0.4j$$

Alternatively

$$\frac{3 + 2j}{5} = \frac{1}{5}(3 + 2j)$$

In the first solution, the problem is simplified giving the answer in the form $x + yj$. Alternatively, the problem could be simplified in a factored form.

### 8.3.3.2 Division by Complex Numbers

Division in this category is when:

**a)** a number is divided by a complex number, or

**b)** a complex number is divided by another complex number.

In either of the two above cases, division is carried out using the concept of rationalising the denominator. Recall that multiplying a surd with its conjugate results in a rational number. So, what about a complex conjugate?

Given a complex number $z = x + yj$, its complex conjugate is denoted as $\bar{z}$ such that $\bar{z} = x - yj$. Note that we have only changed the sign between the real and imaginary parts of the complex number to obtain a complex conjugate pair. $z^*$ is also used to indicate a complex conjugate.

If $z = x + yj$, then the following (Table 8.1) are true about a complex number and its conjugate.

**TABLE 8.1**
**Complex Conjugates Illustrated**

| Operation | Note |
| --- | --- |
| $z(\bar{z}) = x^2 + y^2$ | When conjugate complex numbers are multiplied together, the result is equal to the sum of the squares of the real and the imaginary parts. The result is therefore always a non-negative real number. It is also possible to obtain $-x^2 - y^2$ if the sign of the real part is changed to make the conjugate. In other words, $(x + yj)(-x + yj) = -x^2 - y^2$. |
| $\overline{(z_1 z_2)} = \overline{z_1} * \overline{z_2}$ | The product of two or more conjugates is equal to the product of their individual conjugates. |
| $z + \bar{z} = 2 * \mathbf{Re}(z)$ | The sum of a complex number and its conjugate is equal to its real part multiplied by 2. |
| $z - \bar{z} = 2j * \mathbf{Im}(z)$ | When a conjugate complex number is subtracted from its corresponding complex number, the result is the imaginary part multiplied by 2. |

**NOTE** The asterisk sign $*$ implies multiplication.

# Complex Numbers I

Let's try some examples.

## Example 10

Evaluate the following products of complex conjugates:

**a)** $(a - bj)(a + bj)$  **b)** $(1 + 5j)(1 - 5j)$

What did you get? Find the solution below to double-check your answer.

## Solution to Example 10

**a)** $(a - bj)(a + bj)$
**Solution**

$$(a - bj)(a + bj) = a^2 + abj - abj - b^2 j^2$$
$$= a^2 - b^2 j^2 = a^2 + b^2$$
$$\therefore (a - bj)(a + bj) = a^2 + b^2$$

**b)** $(1 + 5j)(1 - 5j)$
**Solution**

$$(1 + 5j)(1 - 5j) = 1 - 5j + 5j - 25j^2$$
$$= 1 - 25j^2 = 1 + 25 = 26$$
$$\therefore (1 + 5j)(1 - 5j) = 26$$

It can be observed from the above that multiplying a complex number with its complex conjugate gives a real number, it is therefore easy to render a complex denominator into a real number by multiplying both denominator and numerator by the conjugate of the denominator. Once this is done, dividing the numerator (complex number or another number) becomes easy. This process is called **rationalising the denominator**.

Let's try some examples.

## Example 11

Evaluate each of the following:

**a)** $\dfrac{1}{j^3 + 2j^8}$  **b)** $\dfrac{5 + 2j}{3 - j}$

What did you get? Find the solution below to double-check your answer.

## Solution to Example 11

**a)** $\frac{1}{j^3+2j^8}$

**Solution**

**Step 1:** First, we need to simplify the denominator. Note that $j^3 = -j$ and $j^4 = 1$, as noted on page 282. Therefore

$$\frac{1}{j^3+2j^8} = \frac{1}{-j+2} = \frac{1}{2-j}$$

**Step 2:** We need to rationalise the expression above using the conjugate of the denominator. That is, we need to multiply both the numerator and denominator by $2+j$.

$$\frac{1}{2-j} = \frac{1}{2-j} \times \frac{2+j}{2+j} = \frac{2+j}{(2-j)(2+j)}$$

**Step 3:** Open the brackets at the denominator which, because they are a conjugate pair, produces a real number.

$$\frac{2+j}{(2-j)(2+j)} = \frac{2+j}{2^2+1^2}$$
$$= \frac{2+j}{5} = \frac{2}{5} + \frac{j}{5}$$

Therefore

$$\frac{1}{j^3+2j^8} = 0.4 + 0.2j$$

---

**b)** $\frac{5+2j}{3-j}$

**Solution**

Given that

$$\frac{5+j2}{3-j}$$

**Step 1:** Let's rationalise the denominator by multiplying both the numerator and denominator by $3+j$.

$$\frac{5+j2}{3-j} = \frac{5+j2}{3-j} \times 1 = \frac{5+j2}{3-j} \times \frac{3+j}{3+j} = \frac{(5+j2)(3+j)}{(3-j)(3+j)}$$

**Step 2:** Open the brackets at the denominator which, because they are a conjugate pair, produces a real number.

$$= \frac{15+5j+6j-2}{3^2+1^2} = \frac{13+11j}{10} = \frac{13}{10} + \frac{11j}{10}$$

# Complex Numbers I

Therefore

$$\frac{5+j2}{3-j} = 1.3 + j1.1$$

Let's try another example.

## Example 12

Show that

$$\frac{\cos 2x + j\sin 2x}{\cos 3x + j\sin 3x} = \cos x - j\sin x$$

What did you get? Find the solution below to double-check your answer.

## Solution to Example 12

**Step 1:** Let's start with the LHS

$$\frac{\cos 2x + j\sin 2x}{\cos 3x + j\sin 3x}$$

**Step 2:** The complex conjugate of the denominator is $\cos 3x - j\sin 3x$, so we need to multiply both the numerator and denominator by this complex conjugate. In other words, we need to rationalise the expression. Thus,

$$\frac{\cos 2x + j\sin 2x}{\cos 3x + j\sin 3x} = \left\{\frac{\cos 2x + j\sin 2x}{\cos 3x + j\sin 3x}\right\} \times \left\{\frac{\cos 3x - j\sin 3x}{\cos 3x - j\sin 3x}\right\}$$

$$= \frac{(\cos 2x + j\sin 2x)(\cos 3x - j\sin 3x)}{(\cos 3x + j\sin 3x)(\cos 3x - j\sin 3x)}$$

**Step 3:** Let's simplify, starting with the numerator and then the denominator.

**Numerator**

$$(\cos 2x + j\sin 2x)(\cos 3x - j\sin 3x)$$
$$= \cos 2x \cos 3x - j\cos 2x \sin 3x + j\sin 2x \cos 3x + \sin 2x \sin 3x$$

Group this into real and imaginary parts, which gives:

$$= (\cos 2x \cos 3x + \sin 2x \sin 3x) + j(\sin 2x \cos 3x - \cos 2x \sin 3x)$$

Use trigonometric identities to simplify the above expression, which gives

$$= \cos(2x - 3x) + j\sin(2x - 3x)$$
$$= \cos(-x) + j\sin(-x)$$
$$= \cos x - j\sin x$$

**Denominator**

$$(\cos 3x + j\sin 3x)(\cos 3x - j\sin 3x) = \cos^2 3x + \sin^2 3x$$
$$= 1$$

**Step 4:** Now, it is time to evaluate the whole expression. Therefore

$$\frac{\cos 2x + j\sin 2x}{\cos 3x + j\sin 3x} = \frac{(\cos 2x + j\sin 2x)(\cos 3x - j\sin 3x)}{(\cos 3x + j\sin 3x)(\cos 3x - j\sin 3x)}$$
$$= \frac{\cos x - j\sin x}{1}$$

$$\therefore \frac{\cos 2x + j\sin 2x}{\cos 3x + j\sin 3x} = \cos x - j\sin x$$

Final example to try.

### Example 13

If $z = \frac{2+2j}{3-j}$, determine the real and the imaginary parts of the complex number $z - \frac{1}{z}$.

What did you get? Find the solution below to double-check your answer.

### Solution to Example 13

Given that

$$z = \frac{2+2j}{3-j}$$

**Step 1:** Let's determine the inverse $z$ (or $\frac{1}{z}$) as

$$\frac{1}{z} = \frac{3-j}{2+2j}$$

**Step 2:** Let's compute $z - \frac{1}{z}$ as:

$$z - \frac{1}{z} = \frac{2+2j}{3-j} - \frac{3-j}{2+2j}$$
$$= \frac{(2+2j)(2+2j) - (3-j)(3-j)}{(3-j)(2+2j)}$$
$$= \frac{(2+2j)^2 - (3-j)^2}{(3-j)(2+2j)}$$

Applying the difference of two squares, i.e., $a^2 - b^2 = (a+b)(a-b)$, for the numerator, we will have

$$= \frac{(2 + 2j + 3 - j)(2 + 2j - 3 + j)}{(3 - j)(2 + j2)}$$

$$= \frac{(5 + j)(-1 + 3j)}{(3 - j)(2 + 2j)}$$

Now open the brackets

$$= \frac{-8 + 14j}{8 + 4j}$$

Divide each term by 2

$$= \frac{-4 + 7j}{4 + 2j}$$

**Step 3:** Now it is time to rationalise using the complex conjugate of the denominator, thus

$$\frac{-4 + 7j}{4 + 2j} = \left\{\frac{-4 + 7j}{4 + 2j}\right\} \times \left\{\frac{4 - 2j}{4 - 2j}\right\} = \frac{(-4 + 7j)(4 - 2j)}{(4 + 2j)(4 - 2j)}$$

**Step 4:** Open the brackets at the denominator which, because they are a conjugate pair, produces a real number.

$$= \frac{-16 + 8j + 28j + 14}{4^2 + 2^2}$$

$$= \frac{-2 + 36j}{20} = \frac{-2}{20} + \frac{36j}{20}$$

$$= -0.1 + 1.8j$$

Therefore

$$z - \frac{1}{z} = -0.1 + 1.8j$$

Hence

**Re** $(z) = -0.1$

**Im** $(z) = 1.8$

## 8.4 FORMS OF COMPLEX NUMBERS

Complex numbers can be expressed in three main forms, namely Cartesian, polar, and exponential.

### 8.4.1 CARTESIAN FORM

The format of the complex number used so far is known as the Cartesian (or the rectangular) form, written as $x + yj$, where $x$ is the real part and $y$ is the imaginary part. This is the most commonly used form.

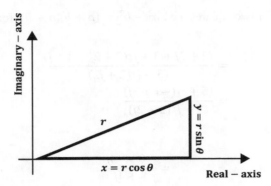

**FIGURE 8.2** Determining polar form illustrated.

### 8.4.2 POLAR FORM

The polar form is expressed as $z = r\angle\theta$, where $r$ is called the modulus (or magnitude) of the complex number, which is the length of the line joining the origin to the point representing the complex number. It is abbreviated as **mod** $z$ or $|z|$ and given by:

$$r = \sqrt{x^2 + y^2} \qquad (8.4)$$

On the other hand, $\theta$ is the angle between the positive real-axis and the line joining the complex number with the origin $(0, 0)$, as shown in Figure 8.2. The angle must be in the interval $-\pi < \theta \leq \pi$ (or $-180° < \theta \leq 180°$).

It follows from Figure 8.2 that

$$\theta = \tan^{-1}\left\{\frac{\text{imaginary part}}{\text{real part}}\right\} \quad \text{OR} \quad \theta = \tan^{-1}\left\{\frac{y}{x}\right\} \qquad (8.5)$$

Angle $\theta$ is generally called the argument of the complex number and is written as **arg**$(z)$.

Let's try some examples.

### Example 14

Determine the modulus and argument of the following complex numbers.

a) $z = 3 + 4j$      b) $z = 5 - 12j$

# Complex Numbers I

What did you get? Find the solution below to double-check your answer.

### Solution to Example 14

**a)** $z = 3 + 4j$
**Solution**

$$\text{mod } z = |z| = |3 + 4j|$$
$$= \sqrt{3^2 + 4^2}$$
$$= 5$$

and

$$\arg(z) = \arg(3 + 4j)$$
$$= \tan^{-1} \frac{4}{3}$$
$$= 53.13°$$

---

**b)** $z = 5 - 12j$
**Solution**

$$\text{mod } z = |z| = |5 - 12j|$$
$$= \sqrt{5^2 + (-12)^2}$$
$$= 13$$

and

$$\arg(z) = \arg(5 - 12j)$$
$$= \tan^{-1}\left(\frac{-12}{5}\right)$$
$$= -67.38°$$

---

Let's try another set of examples.

### Example 15

Given that $z_1 = 3 + 5j$, $z_2 = -4 + 7j$, $z_3 = -2 - 9j$, and $z_4 = 1 - j$, show that:

**a)** $z_1 \bar{z}_1 = |z_1|^2$    **b)** $z_2 \bar{z}_2 = |z_2|^2$    **c)** $z_3 \bar{z}_3 = |z_3|^2$    **d)** $z_4 \bar{z}_4 = |z_4|^2$

What did you get? Find the solution below to double-check your answer.

### Solution to Example 15

**a)** $z_1 \bar{z}_1 = |z_1|^2$
**Solution**
Given that $z_1 = 3 + 5j$, the conjugate is $\bar{z}_1 = 3 - 5j$

$$z_1 \bar{z}_1 = (3 + 5j)(3 - 5j) = 3^2 + 5^2$$
$$= 9 + 25$$
$$= 34$$

Also,

$$|z_1| = \sqrt{3^2 + 5^2}$$
$$= \sqrt{9 + 25}$$
$$= \sqrt{34}$$

Thus

$$|z_1|^2 = \left(\sqrt{34}\right)^2 = 34$$
$$\therefore z_1 \bar{z}_3 = |z_1|^2$$

---

**b)** $z_2 \bar{z}_2 = |z_2|^2$
**Solution**
Given that $z_2 = -4 + 7j$, the conjugate is $\bar{z}_2 = -4 - 7j$

$$z_2 \bar{z}_2 = (-4 + 7j)(-4 - 7j) = 4^2 + 7^2$$
$$= 16 + 49$$
$$= 65$$

Also,

$$|z_2| = \sqrt{4^2 + 7^2}$$
$$= \sqrt{16 + 49}$$
$$= \sqrt{65}$$

Thus

$$|z_2|^2 = \left(\sqrt{65}\right)^2 = 65$$
$$\therefore z_2 \bar{z}_2 = |z_2|^2$$

**c)** $z_3 \bar{z}_3 = |z_3|^2$

**Solution**

Given that $z_3 = -2 - 9j$, the conjugate is $\bar{z}_3 = -2 + 9j$

$$z_3 \bar{z}_3 = (-2 - 9j)(-2 + 9j) = 2^2 + 9^2$$
$$= 4 + 81$$
$$= 85$$

Also,

$$|z_3| = \sqrt{2^2 + 9^2}$$
$$= \sqrt{4 + 81}$$
$$= \sqrt{85}$$

Thus

$$|z_3|^2 = \left(\sqrt{85}\right)^2 = 85$$
$$\therefore z_3 \bar{z}_3 = |z_3|^2$$

---

**d)** $z_4 \bar{z}_4 = |z_4|^2$

**Solution**

Given that $z_4 = 1 - j$, the conjugate is $\bar{z}_4 = 1 - j$

$$z_4 \bar{z}_4 = (1 - j)(1 + j) = 1^2 + 1^2$$
$$= 1 + 1$$
$$= 2$$

Also,

$$|z_4| = \sqrt{1^2 + 1^2}$$
$$= \sqrt{1 + 1}$$
$$= \sqrt{2}$$

Thus

$$|z_4|^2 = \left(\sqrt{2}\right)^2 = 2$$
$$\therefore z_4 \bar{z}_4 = |z_4|^2$$

---

From the above examples, we can conclude that for any given complex $z = x + jy$, it follows that

$$\boxed{z\bar{z} = |z|^2} \qquad (8.6)$$

Another example to try.

## Example 16

Find the modulus and argument of $z = \frac{(7-j)(3+2j)}{(4-5j)^2}$.

What did you get? Find the solution below to double-check your answer.

## Solution to Example 16

### HINT

Here we need to simplify the expression until the complex number is in its simplest rectangular form.

$$z = \frac{(7-j)(3+2j)}{(4-5j)^2}$$

**Step 1:** We will start by simplifying the numerator and denominator separately.

**Numerator**

$$(7-j)(3+2j) = 21 + 14j - 3j + 2$$
$$= 23 + 11j$$

**Denominator**

$$(4-5j)^2 = (4-5j)(4-5j) = -9 - 40j$$

Thus,

$$z = \frac{(7-j)(3+2j)}{(4-5j)^2} = \frac{23+11j}{-9-40j}$$

**Step 2:** Rationalise the denominator by multiplying both the numerator and denominator by $-9 + 40j$.

$$z = \left\{\frac{23+11j}{-9-40j}\right\} \times \left\{\frac{-9+40j}{-9+40j}\right\} = \frac{(23+11j)(-9+40j)}{(-9-40j)(-9+40j)}$$

**Step 3:** Open the brackets at the denominator which, because they are a conjugate pair, produces a real number.

$$\frac{(23+11j)(-9+40j)}{(-9-40j)(-9+40j)} = \frac{-207 + 920j - 99j - 440}{(-9)^2 + 40^2}$$
$$= \frac{-647 + 821j}{1681}$$
$$= -\frac{647}{1681} + \frac{821}{1681}j$$

# Complex Numbers I

**Step 4:** Modulus

Let the modulus of the above complex number be $|z|$, therefore

$$|z| = \sqrt{\left(-\frac{647}{1681}\right)^2 + \left(\frac{821}{1681}\right)^2}$$

$$= \sqrt{\frac{650}{1681}}$$

$$= \mathbf{0.6218}$$

**Step 5:** Argument

If the argument is denoted by $\arg(z)$, therefore

$$\arg(z) = \tan^{-1}\left(-\frac{821}{647}\right)$$

$$= \mathbf{128.24°}$$

## 8.4.3 Exponential Form

The exponential form is represented as:

$$\boxed{z = re^{\pm j\theta}} \quad (8.7)$$

where $r$ is the modulus of $z$ and $\theta$ the argument of $z$. The angle $\theta$ here is measured in radians (unlike degrees in polar form).

The relationship between the different forms of complex numbers is summarised in Table 8.2.

**TABLE 8.2**
**Complex Number Forms**

| Expression | Name | Note |
|---|---|---|
| $z = x + jy$ | Rectangular or Cartesian | — |
| $z = r(\cos\theta \pm j\sin\theta)$ | Polar | This is also called a trigonometrical form in some textbooks, while some authors consider it as a rectangular form. |
| $z = r\angle \pm \theta$ | Shortened polar | It is simply a representation, which can be considered as a derivative of the exponential form except that its angle is measured in degrees while the former is in radians. |
| $z = re^{\pm j\theta}$ | Exponential | — |

## TABLE 8.3
### Operations of Complex Numbers Explained

| Expression | Note |
|---|---|
| $z_1 + z_2 = (x_1 + x_2) + j(y_1 + y_2)$ | **Note 1** Addition and subtraction carried out in rectangular form. |
| $z_1 - z_2 = (x_1 - x_2) + j(y_1 - y_2)$ | |
| $z_1 * z_2 = r_1 * r_2 \angle (\theta_1 + \theta_2)$ | **Note 2** Multiplication and division carried out in polar form. |
| $\frac{z_1}{z_2} = \frac{r_1}{r_2} \angle (\theta_1 - \theta_2)$ | **Note 3** The same can be done in exponential form. |

## 8.5 CONVERSION BETWEEN VARIOUS FORMS

Conversion from one form to another, particularly polar to rectangular and vice versa, is inevitable. A complex number $z = r \angle \theta$ in polar form has its rectangular form given as $z = x + jy$, such that

$$\boxed{x = r \cos \theta} \text{ and } \boxed{y = r \sin \theta} \tag{8.8}$$

Note that $r$ in polar form is the same as $r$ in the rectangular form. $\theta$ (in degree) in the rectangular form is a radian equivalent of the $\theta$ in exponential form.

It is important to sketch an Argand diagram to determine the quadrant of a complex number when converting between the various forms. This is because there are two possible values for angle $\theta$ between 0° and 360°.

It should also be noted that adding and subtracting complex numbers are easier in Cartesian form whilst multiplication and division can be easily evaluated in polar or exponential form.

Consider two complex numbers $z_1 = x_1 + jy_1 = r_1 \angle \theta_1$ and $z_2 = x_2 + jy_2 = r_2 \angle \theta_2$, the four operations are shown in Table 8.3.

The following relationship is also valid and can be proven.

$$\boxed{|z_1 * z_2| = |z_1| |z_2|} \quad \boxed{\arg(z_1 * z_2) = \arg(z_1) + \arg(z_2)}$$

$$\boxed{\left|\frac{z_1}{z_2}\right| = \frac{|z_1|}{|z_2|}} \quad \boxed{\arg\left(\frac{z_1}{z_2}\right) = \arg(z_1) - \arg(z_2)}$$

(8.9)

Let's try some examples.

### Example 17

Express the following complex numbers in polar form, presenting the answer in surd form where appropriate.

a) $z = 15$          b) $z = 6 - 7j$          c) $z = j$

Complex Numbers I

What did you get? Find the solution below to double-check your answer.

### Solution to Example 17

**HINT**

Conversion between various forms is very important and can be done quickly using a calculator. It is however essential that we are familiar with the principles of how this is 'manually' carried out.

**a)** $z = 15$
**Solution**
If $z = 15$, it follows that

$$r = \sqrt{15^2 + 0^2}$$
$$= 15$$

and

$$\theta = \tan^{-1}\frac{0}{15}$$
$$= 0°$$

Hence

$$z = 15 + j0$$
$$= 15\angle 0°$$

**b)** $z = 6 - 7j$
**Solution**
If $z = 6 - 7j$, it follows that

$$r = \sqrt{6^2 + (-7)^2}$$
$$= \sqrt{85}$$
$$= 9.22$$

and

$$\theta = \tan^{-1}\frac{-7}{6}$$
$$= -49.40°$$

Hence

$$z = 6 - 7j = 9.22\angle -49.40°$$

**c)** $z = j$
**Solution**
If $z = j$, it follows that

$$r = \sqrt{0^2 + 1^2}$$
$$= 1$$

and

$$\theta = \tan^{-1}\frac{1}{0}$$
$$= 90°$$

Hence

$$z = 0 + j$$
$$= 1\angle 90°$$

---

### Example 18

Express the following complex numbers in Cartesian form, presenting the answer in surd form where appropriate.

**a)** $z = 10(\cos 135° + j\sin 135°)$    **b)** $z = \sqrt{12}\,e^{-j\frac{\pi}{6}}$

---

What did you get? Find the solution below to double-check your answer.

### Solution to Example 18

**a)** $z = 10(\cos 135° + j\sin 135°)$
**Solution**

$$10(\cos 135° + j\sin 135°)$$

$$= 10\left(-\frac{\sqrt{2}}{2} + j\frac{\sqrt{2}}{2}\right)$$
$$= -5(\sqrt{2} - j\sqrt{2})$$

# Complex Numbers I

**b)** $z = \sqrt{12}e^{-j\frac{\pi}{6}}$
**Solution**

$$\sqrt{12}e^{-j\frac{\pi}{6}} = \sqrt{12}(\cos\frac{\pi}{6} - j\sin\frac{\pi}{6})$$
$$= \sqrt{12}(\cos 30° - j\sin 30°)$$
$$= \sqrt{12}\left(\frac{\sqrt{3}}{2} - j\frac{1}{2}\right)$$
$$= \frac{\sqrt{12}}{2}\left(\sqrt{3} - j\right)$$
$$= \sqrt{3}\left(\sqrt{3} - j\right)$$
$$= 3 - j\sqrt{3}$$

## Example 19

Express the following complex numbers in exponential form, presenting the answer in surd form where appropriate.

**a)** $z = 1 - j$       **b)** $z = 6 - 7j$

What did you get? Find the solution below to double-check your answer.

## Solution to Example 19

**a)** $z = 1 - j$
**Solution**

$$1 - j = r(\cos\theta + j\sin\theta)$$
$$= re^{j\theta}$$

where

$$r = \sqrt{1^2 + (-1)^2}$$
$$= \sqrt{2}$$

and

$$\theta = \tan^{-1}(-1)$$
$$= -45° = -\frac{\pi}{4}\text{ rad}$$

Hence

$$1 - j = \sqrt{2}\left(\cos\frac{\pi}{4} - j\sin\frac{\pi}{4}\right)$$
$$= \sqrt{2}e^{-j\frac{\pi}{4}}$$
$$= \sqrt{2}e^{-j0.7854}$$
$$\therefore 1 - j = \sqrt{2}e^{-j0.7854}$$

**b)** $z = 6 - 7j$
**Solution**

$$6 - 7j = r(\cos\theta + j\sin\theta)$$
$$= re^{j\theta}$$

where

$$r = \sqrt{6^2 + (-7)^2}$$
$$= \sqrt{85}$$

and

$$\theta = \tan^{-1}\left(\frac{-7}{6}\right) = -0.8622 \text{ rad}$$

Hence

$$6 - 7j = \sqrt{85}\left(\cos(0.8622) - j\sin(0.8622)\right)$$
$$= \sqrt{85}e^{-j0.8622}$$
$$= \sqrt{85}e^{-j0.8622}$$
$$\therefore 6 - 7j = \sqrt{85}e^{-j0.8622}$$

## 8.6 CHAPTER SUMMARY

1) $j$ is an operator that functions just like ×, $\sqrt{}$, ∪, %, or any other mathematical operator.
2) Letter $j$ is used in engineering while $i$ is employed in other STEM (Science, Technology, Engineering, and Mathematics) subjects for vector analysis, though both denote the same numerical process.
3) $j$ simply means

$$\boxed{j = \sqrt{-1}} \quad \text{OR} \quad \boxed{j^2 = -1}$$

# Complex Numbers I

4) $j$ is a 90° counter-clockwise rotation of a vector, usually taken from the positive $x$-axis. When such a rotation is performed twice, the vector will be on the same line but pointing towards the opposite direction. When a 360° rotation is carried out on a vector, the vector will return to its original position and again $j^4 = 1$. In summary, we have:

$$\boxed{j = \sqrt{-1}} \qquad \boxed{j^2 = -1} \qquad \boxed{j^3 = -j} \qquad \boxed{j^4 = 1} \qquad \boxed{\frac{1}{j} = -j}$$

5) A complex number is a number that is made up of two parts, the real and the imaginary. This is expressed as:

$$\boxed{z = x + jy}$$

where $x$ is the real part (or component) of the complex number ($z$) abbreviated as **Re**$(z)$ and the imaginary part is $y$, shortened as **Im**$(z)$.

6) Sometimes, the $j$ operator is written first before the number, and the number is written before the $j$ operator at other times. Therefore, $j2$ can be written as $2j$.

7) Addition, subtraction, multiplication, and division can be performed on complex numbers. The rules governing these operations are like other numbers, though with a minor manipulation.

8) Given a complex number $z = x + yj$, its complex conjugate is denoted as $\bar{z}$ such that $\bar{z} = x - yj$. Note that, generally, we only change the sign between the real and imaginary parts of the complex number to obtain a complex conjugate pair.

9) $z^*$ is also used to indicate a complex conjugate.

10) Multiplying a complex number with its complex conjugate gives a real number. Consequently, it is straightforward to convert a complex denominator into a real number by multiplying both denominator and numerator by the conjugate of the denominator. Once this is done, dividing the numerator (complex number or another number) becomes easy. This process is called **rationalising the denominator**.

11) For any given complex $z = x + yj$, it can be shown that

$$\boxed{z\bar{z} = |z|^2}$$

12) Writing the complex number as $x + yj$ is known as the Cartesian (or the rectangular) form, which is the most commonly used representation.

13) The polar form is expressed as $z = r\angle\theta$, where $r$ is called the modulus (or magnitude) of the complex number (or the length of the line joining the origin to the point representing the complex number). It is abbreviated as **mod** $z$ or $|z|$ and given by:

$$\boxed{r = \sqrt{x^2 + y^2}}$$

On the other hand, $\theta$ is the angle between the positive real-axis and the line joining the complex number with the origin. The angle must be in the interval $-\pi < \theta \leq \pi$ (or $-180° < \theta \leq 180°$). It follows that

$$\boxed{\theta = \tan^{-1}\left\{\frac{\text{imaginary part}}{\text{real part}}\right\}} \quad \text{OR} \quad \boxed{\theta = \tan^{-1}\left\{\frac{y}{x}\right\}}$$

Angle $\theta$ is generally called the argument of the complex number and is written as **arg**$(z)$.

14) The exponential form is represented as:

$$\boxed{z = re^{\pm j\theta}}$$

where $r$ is the modulus of $z$ and $\theta$ the argument of $z$. The angle $\theta$ here is measured in radians (unlike degrees in polar form).

15) Conversion from one form to another, particularly a polar to rectangular and vice versa, is inevitable. A complex number $z = r\angle\theta$ in polar form has its rectangular form given as $z = x + jy$, such that

$$\boxed{x = r\cos\theta} \quad \text{and} \quad \boxed{y = r\sin\theta}$$

16) $r$ in polar form is the same as $r$ in the rectangular form. $\theta$ (in degree) in the rectangular form is a radian equivalent of the $\theta$ in exponential form.

17) It is important to sketch an Argand diagram to determine the quadrant of a complex number when converting between the various forms.

18) The following relationship is valid:

$$\boxed{|z_1 * z_2| = |z_1||z_2|} \quad \boxed{\arg(z_1 * z_2) = \arg(z_1) + \arg(z_2)}$$

$$\boxed{\left|\frac{z_1}{z_2}\right| = \frac{|z_1|}{|z_2|}} \quad \boxed{\arg\left(\frac{z_1}{z_2}\right) = \arg(z_1) - \arg(z_2)}$$

\*\*\*\*

## 8.7 FURTHER PRACTICE

To access complementary contents, including additional exercises, please go to www.dszak.com.

# 9 Complex Numbers II

## Learning Outcomes

Once you have studied the content of this chapter, you should be able to:

- Determine the $n$th roots of complex numbers
- Determine the powers of trigonometric function and multiple angles
- Determine the logarithm of complex numbers

## 9.1 INTRODUCTION

In this chapter, we will continue our discussion on complex numbers, covering Argand diagrams, complex number equations, De Moivre's theorem, and logarithms of complex numbers, among others.

## 9.2 ARGAND DIAGRAMS

An Argand diagram, named after the French mathematician Jean-Robert Argand, is a geometrical plot of complex numbers on the $x$-$y$ Cartesian plane, also known as the complex plane. The $x$-axis (horizontal axis) represents the real parts and the $y$-axis (vertical axis) represents the imaginary parts of complex numbers. They are called the real axis and the imaginary axis, respectively.

To represent a complex number on an Argand diagram is very similar to the way a vector, specifically a position vector, will be represented. To illustrate, a complex $z = a + jb$ will be represented by a straight line which starts from the origin $(0,0)$ of the diagram to a point $(a,b)$ on the same diagram. Similarly, a complex number $w = c - jd$ will be represented by a straight line which starts from $(0,0)$ of the diagram to a point $(c,-d)$ on the same diagram. Both complex numbers are shown in Figure 9.1.

Let's try some examples on this.

### Example 1

Use the Argand diagram shown in Figure 9.2 to answer the following questions.

a) Write the complex numbers $z_1$, $z_2$, ..., $z_6$ in Cartesian form $z = x + jy$.

b) Use an Argand diagram to determine $z_7$ such that $z_1 + z_4$. Confirm this algebraically.

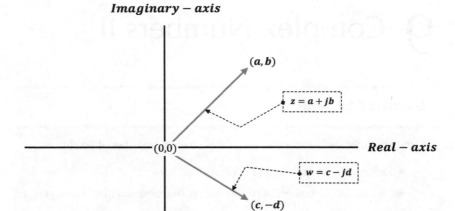

**FIGURE 9.1** Complex numbers illustrated on Argand diagram.

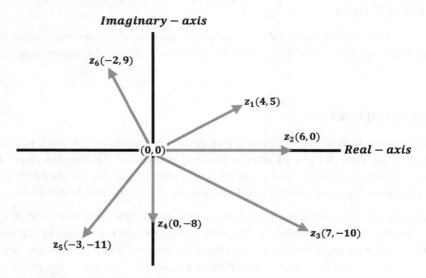

**FIGURE 9.2** Example 1 on Argand diagram.

What did you get? Find the solution below to double-check your answer.

### Solution to Example 1

**a)** $z_1, z_2, \ldots, z_6$
**Solution**

$$z_1 = 4 + j5$$
$$z_2 = 6 + j0$$
$$z_3 = 7 - j10$$

# Complex Numbers II

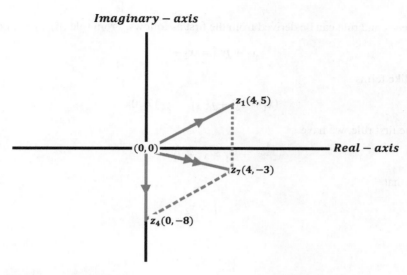

**FIGURE 9.3** Solution to Example 1(b).

$$z_4 = 0 - j8 = -j8$$
$$z_5 - 3 - j11$$
$$z_6 = -2 + j9$$

**NOTE**

You will notice that $z_2$ has no imaginary part or zero imaginary part; in fact, it's simply a real number. On the other hand, $z_4$ has only an imaginary part. However, both are still considered complex numbers.

**b)** $z_7 = z_1 + z_4$
**Solution** (Figure 9.3)

$$z_1 = 4 + j5$$
$$z_4 = (0 - j8)$$
$$z_7 = z_1 + z_4$$
$$= (4 + j5) + (0 - j8)$$
$$= 4 - j3$$
$$\therefore z_7 = 4 - j3$$

**NOTE**

Notice that the complex numbers are added graphically in the same way we add vectors (see Chapter 11).

## 9.3 EQUATIONS INVOLVING COMPLEX NUMBERS

When working with complex number equations, there are two rules to apply, namely:

**Rule 1** If a complex number is equal to zero, both the real part and the imaginary part must also be equal to zero. In other words, if $x + jy = 0$, then $x = 0$ and $y = 0$.

**Rule 2** If two complex numbers are equal, then their real parts and imaginary parts must be equal. For instance, given that $z_1 = x_1 + jy_1$ and $z_2 = x_2 + jy_2$ such that $z_1 = z_2$ then $x_1 = x_2$ and $y_1 = y_2$.

Note that the second rule can be derived from the first as follows. From rule (ii), $z_1 = z_2$ implies that

$$x_1 + jy_1 = x_2 + jy_2$$

Collect the like terms

$$(x_1 - x_2) + j(y_1 - y_2) = 0$$

Applying the first rule, we have

$$x_1 - x_2 = 0$$

this implies that

$$x_1 = x_2$$

and

$$y_1 - y_2 = 0$$

this gives

$$y_1 = y_2$$

as before.

Let's try some examples.

### Example 2

Determine the values of $x$ and $y$ given that $(1 - 3j)(1 + 3j)(0.5 + 0.6j) = x + jy$.

What did you get? Find the solution below to double-check your answer.

### Solution to Example 2

**Step 1:** Let us start by simplifying the left-hand side (LHS) as:

$$\begin{aligned}(1 - 3j)(1 + 3j)(0.5 + 0.6j) &= (1^2 + 3^2)(0.5 + 0.6j) \\ &= (1 + 9)(0.5 + 0.6j) \\ &= 10(0.5 + 0.6j) \\ &= 5 + 6j\end{aligned}$$

**Step 2:** Now we need to equate both sides as

$$5 + 6j = x + yj$$

Therefore

$$x = 5, \quad y = 6$$

# Complex Numbers II

### Example 3

Given that $(2x - y) + (x + 2y)j = 7 + 6j$, determine the values of $x$ and $y$ which satisfy this equation.

What did you get? Find the solution below to double-check your answer.

### Solution to Example 3

$$(2x - y) + (x + 2y)j = 7 + 6j$$

**Step 1:** Comparing both sides of the above equation, we have

$$2x - y = 7 \quad \text{-------(i)}$$

and

$$x + 2y = 6 \quad \text{------(ii)}$$

**Step 2:** Solve the simultaneous equations.

Using the elimination method, multiply equation (i) by 2 to obtain

$$4x - 2y = 14 \quad \text{------(iii)}$$

Add equations (ii) and (iii) to eliminate $y$,

$$5x = 20$$
$$x = \frac{20}{5}$$
$$\therefore x = 4$$

Substitute for $x$ in either equation (i) or (ii) to obtain the value of $y$. From equation (i), we have

$$y = 2x - 7$$
$$y = 2(4) - 7 = 1$$
$$\therefore y = 1$$

### Example 4

Determine the values of $x$ and $y$ which satisfy the complex equation:

$$\frac{2x - jy - 1}{x + j3y} = \frac{2 - j}{3}$$

What did you get? Find the solution below to double-check your answer.

### Solution to Example 4

Given that

$$\frac{2x - jy - 1}{x + j3y} = \frac{2 - j}{3} \quad ----(i)$$

**Step 1:** Multiplying both sides of the equation (i) by $3(x + j3y)$, we have

$$3(2x - jy - 1) = (2 - j)(x + j3y)$$

Simplifying this, we have

$$6x - j3y - 3 = 2x + j6y - jx + 3y$$
$$4x - 3y - 3 - j9y + jx = 0$$
$$(4x - 3y - 3) + j(x - 9y) = 0$$

**Step 2:** Equate the real part to zero and do the same for the imaginary part.

$$4x - 3y - 3 = 0$$
$$4x - 3y = 3 \quad -----(iii)$$

and

$$x - 9y = 0 \quad ------(iv)$$

**Step 3:** Solve the simultaneous equations.

Multiply equation (iv) by 4

$$4x - 36y = 0 \quad ------(v)$$

Subtract equation (v) from equation (iii)

$$33y = 3$$
$$\therefore y = \frac{1}{11}$$

From (iv)

$$x = 9y$$
$$\therefore x = \frac{9}{11}$$

Complex Numbers II

## Example 5

Show that $(5+j3)\,e^{(2+j2)x} + (5-j3)e^{(2-j2)x}$ is a real number if $x$ is also a real number.

What did you get? Find the solution below to double-check your answer.

## Solution to Example 5

Given that
$$(5+3j)\,e^{(2+2j)x} + (5-3j)\,e^{(2-2j)x}$$

**Step 1:** Let's simplify $(5+j3)\,e^{(2+j2)x}$

$$\begin{aligned}(5+j3)\,e^{2x} \times e^{j2x} &= (5+j3)\,e^{2x}\,(\cos 2x + j\sin 2x)\\&= \left(5e^{2x} + j3e^{2x}\right)(\cos 2x + j\sin 2x)\\&= \left(5e^{2x}\right)(\cos 2x + j\sin 2x) + \left(j3e^{2x}\right)(\cos 2x + j\sin 2x)\end{aligned}$$

**Step 2:** Let's simplify $(5-j3)e^{(2-j2)x}$

$$\begin{aligned}(5-j3)\,e^{2x} \times e^{-j2x} &= (5-j3)\,e^{2x}\,(\cos 2x - j\sin 2x)\\&= \left(5e^{2x} - j3e^{2x}\right)(\cos 2x - j\sin 2x)\\&= \left(5e^{2x}\right)[\cos 2x - j\sin 2x] - \left(j3e^{2x}\right)(\cos 2x - j\sin 2x)\end{aligned}$$

**Step 3:** Add the results from steps 1 and 2.

Therefore

$$\begin{aligned}(5+j3)\,e^{(2+j2)x} + (5-j3)\,e^{(2-j2)x} &= \{\left(5e^{2x}\right)[\cos 2x + j\sin 2x] + \left(j3e^{2x}\right)[\cos 2x + j\sin 2x]\}\\&\quad + \{\left(5e^{2x}\right)[\cos 2x - j\sin 2x] - \left(3je^{2x}\right)[\cos 2x - j\sin 2x]\}\\&= \{\left(5e^{2x}\right)[\cos 2x + j\sin 2x] + \left(5e^{2x}\right)[\cos 2x - j\sin 2x]\}\\&\quad + \{\left(j3e^{2x}\right)[\cos 2x + j\sin 2x] - \left(j3e^{2x}\right)[\cos 2x - j\sin 2x]\}\\&= \{\left(5e^{2x}\right)[\cos 2x + j\sin 2x + \cos 2x - j\sin 2x]\}\\&\quad + \{\left(j3e^{2x}\right)[\cos 2x + j\sin 2x - \cos 2x + j\sin 2x]\}\\&= \{\left(5e^{2x}\right)[2\cos 2x]\} + \{\left(j3e^{2x}\right)[j2\sin 2x]\}\\&= \{\left(10e^{2x} \times \cos 2x\right)\} + \{\left(j^2 6e^{2x}\sin 2x\right)\}\\&= \left(10e^{2x} \times \cos 2x\right) - \left(6e^{2x}\sin 2x\right)\\&= 2e^{2x}\,(5\cos 2x - 3\sin 2x)\end{aligned}$$

Since there is no $j$ component in the above answer, it follows that it's a real number because $x$ is also a real number.

## 9.4 PHASOR AND $j$ OPERATOR

In AC (alternating current) circuits, the current and voltage are either (i) in phase, which implies that the voltage and current reach maximum point, minimum point or any other point at the same time (ii) current leading the voltage, or (iii) current lagging the voltage, depending on the elements in the circuit.

In a purely resistive circuit, in which only resistors are connected, the voltage and the current are said to be in phase. In an $RL$ circuit, having resistance $R$ and inductance $L$, the voltage leads the current while current leads in an $RC$ (having resistance $R$ and capacitance $C$) circuit. The phase difference (lead or lag) is 90° in a purely capacitive and purely inductive circuit.

It is known that the angle between the $x$-axis and $y$-axis is 90°, so it is possible that if the positive $x$-axis is taken to represent the voltage in an $RC$ series circuit, the circuit current can be represented by the positive $y$-axis. Similarly, in an $RL$ circuit, the current can be represented by the negative $y$-axis while the voltage is kept on the positive $x$-axis.

Leading by 90° is akin to multiplying a quantity by $j$ while a 90° lagging is similar to multiplying a number by $-j$. In $RLC$ circuits, the lagging or leading factors vary; it is however possible to identify which element is leading or lagging by expressing both quantities in polar form.

The graphical representation of voltages and currents in an AC circuit is known as a **phasor diagram**, which is similar to an **Argand diagram**.

Before we move to examples, it is important to provide us with some useful formulas for AC circuits analysis as these will be referred.

$$\boxed{Z_{RC} = R - jX_c} \quad ---\text{ for } RC \text{ circuits} \tag{9.1}$$

for which

$$\boxed{X_c = \frac{1}{2\pi f C}} \tag{9.2}$$

and

$$\boxed{Z_{RL} = R + jX_L} \quad ---\text{ for } RL \text{ circuits} \tag{9.3}$$

for which

$$\boxed{X_L = 2\pi f L} \tag{9.4}$$

where

- $Z$ = impedance of the circuit
- $X_L$ = inductive reactance
- $C$ = capacitance
- $f$ = frequency
- $X_C$ = capacitive reactance
- $R$ = resistance
- $L$ = inductance

In general, the impedance in an $RLC$ circuit is given by

$$\boxed{Z = R + jX} \quad \text{OR} \quad \boxed{Z = R + j(X_L - X_c)} \tag{9.5}$$

# Complex Numbers II

Let's try some examples.

## Example 6

If $Z$ is the impedance of a series circuit, determine the resistance $R$ and series reactance $X$. Also determine the value of inductance or capacitance, given that the operating frequency is 60 Hz.

**a)** $Z = (5 - 7j)\ \Omega$        **b)** $Z = 20j\ \Omega$        **c)** $Z = 10\angle -30°\ \Omega$

What did you get? Find the solution below to double-check your answer.

## Solution to Example 6

### HINT

The presence of $j$ indicates a reactance. Positive is for inductive reactance while negative is for capacitive reactance. Inductive reactance ($X_L$) and capacitive reactance ($X_c$) are two types of reactance that depend on the presence of inductors and capacitors, respectively.

**a)** $Z = (5 - 7j)\ \Omega$
**Solution**

$$R = \text{Re}\ (5 - 7j)$$
$$= 5\ \Omega$$
$$X = \text{Im}\ (5 - 7j)$$
$$= -7j\ \Omega$$

Because the reactive part is negative, it is capacitive, i.e., $X = X_c$. Thus

$$X_c = \frac{1}{2\pi f C}$$

Re-arranging the formula, we have

$$C = \frac{1}{2\pi f X_c} = \frac{1}{2\pi\ (60)\ (7)}$$
$$\therefore C = 379\ \mu F$$

**b)** $Z = 20j\ \Omega$
**Solution**
Let's write this complex number in its standard (or full) format first as:

$$Z = 20j\ \Omega = (0 + 20j\ )\ \Omega$$

Thus

$$R = \text{Re}\ (0 + 20j)$$
$$= 0\ \Omega$$
$$X = \text{Im}\ (0 + 20j)$$
$$= j20\ \Omega$$

Because the reactive part is positive, it is inductive, i.e., $X = X_L$. Thus

$$X_L = 2\pi f L$$

Re-arranging the formula, we have

$$L = \frac{X_L}{2\pi f} = \frac{20}{2\pi (60)}$$

$$\therefore L = 53 \text{ mH}$$

---

**c)** $Z = 10\angle -30° \; \Omega$
**Solution**

$$10\angle -30° \; \Omega = 10\cos(-30°) + j10\sin(-30°)$$

$$= 8.66 - 5j$$
$$R = \text{Re}(8.66 - 5j)$$
$$= 8.66 \; \Omega$$
$$X = \text{Im}(8.66 - 5j)$$
$$= -5j \; \Omega$$

Because the reactive part is negative, it is capacitive, i.e., $X = X_c$. Thus

$$X_c = \frac{1}{2\pi f C}$$

Re-arranging the formula, we have

$$C = \frac{1}{2\pi f X_c} = \frac{1}{2\pi (60)(5)}$$

$$\therefore C = 531 \; \mu\text{F}$$

---

### Example 7

A current of $I = (5 - 3j)$ A flows in a series circuit when it is connected to a supply voltage $V = (15 + 18j)$ V. Determine the:

a) magnitude of its impedance  
b) phase angle between current and voltage  
c) state whether current leads or lags  
d) resistance  
e) reactance  
f) power consumed

# Complex Numbers II

What did you get? Find the solution below to double-check your answer.

## Solution to Example 7

**a) Magnitude of its impedance**
**Solution**

$$Z = \frac{V}{I}$$
$$= \frac{15 + 18j}{5 - 3j} = \frac{23.43\angle 50.19°}{5.83\angle -30.96°}$$
$$Z = 4.02\angle 81.15° \, \Omega$$

Therefore, the magnitude of the impedance is **4.02 Ω**

### ALTERNATIVE METHOD

We will use rationalising the denominator here

$$Z = \frac{15 + 18j}{5 - 3j} = \frac{15 + 18j}{5 - 3j} \times \frac{5 + 3j}{5 + 3j}$$
$$= \frac{(15 + 18j)(5 + 3j)}{(5 - 3j)(5 + 3j)} = \frac{75 + 45j + 90j + 54j^2}{5^2 - (3j)^2}$$
$$= \frac{75 + 135j - 54}{25 + 9} = \frac{21 + 135j}{34}$$
$$\therefore Z = \frac{21}{34} + \frac{135}{34}j$$

**NOTE**

$Z = \frac{21}{34} + \frac{135}{34}j$ is the same as $Z = 4.02\angle 81.15° \, \Omega$ if both are expressed in the same format.

---

**b) Phase angle between current and voltage**
**Solution**

$$V = 23.43\angle 50.19°$$
$$I = 5.83\angle -30.96°$$

Given that the phase angle is $\varphi$, then

$$\varphi = |50.19° - (-30.96°)|$$
$$= |50.19° + 30.96°| = 81.15°$$
$$\therefore \varphi = \mathbf{81.15°}$$

**c) State whether current leads or lags**
**Solution**
From (b), we can state that current lags the voltage and the lagging angle is $81.15°$.

$$\therefore \textbf{Current lags the voltage}$$

---

**d) Resistance**
**Solution**
Resistance is the active component of the impedance.

$$Z = 4.02 \angle 81.15°$$
$$= 0.618 + 3.97j \ \Omega$$
$$\text{Resistance} = \text{Re}(Z)$$
$$= \mathbf{0.618 \ \Omega}$$

Alternatively,

$$\text{Resistance} = r \cos \theta$$
$$= 4.02 \cos 81.15°$$
$$= \mathbf{0.618 \ \Omega}$$

---

**e) Reactance**
**Solution**

$$\text{Reactance} = \text{Im}(Z)$$
$$= \mathbf{3.97 \ \Omega}$$

Alternatively,

$$\text{Resistance} = r \sin \theta$$
$$= 4.02 \sin 81.15°$$
$$= \mathbf{3.97 \ \Omega}$$

---

**f) Power consumed**
**Solution**
Power consumed is the power loss in the conductor or the active power, which is

$$P = I^2 R$$
$$= (5.83)^2 (0.618)$$
$$= \mathbf{21.01 \ W}$$

---

### Example 8

If one of the three impedances in a delta connection, denoted as $Z_c$, is given by:

$$Z_c = \frac{Z_1 Z_2 + Z_2 Z_3 + Z_3 Z_1}{Z_3}$$

# Complex Numbers II

Determine $Z_c$ given that $Z_1 = (5 + 10j)\,\Omega$, $Z_2 = (3 - 8j)\,\Omega$, and $Z_3 = (7 + 5j)\,\Omega$. Express the answer in rectangular form correct to three significant figures.

What did you get? Find the solution below to double-check your answer.

### Solution to Example 8

**Step 1:** Let us express the impedances in both polar and rectangular forms.

$$Z_1 = (5 + 10j)\,\Omega = 11.18\angle 63.43°$$
$$Z_2 = (3 - 8j)\,\Omega = 8.54\angle -69.44°$$
$$Z_3 = (7 + 5j)\,\Omega = 8.60\angle 35.54°$$

**Step 2:** Now let us evaluate the terms in the original expression one by one

$$Z_1 Z_2 = (11.18\angle 63.43°)(8.54\angle -69.44°)$$
$$= 95.48\angle -6.01° = 94.96 - j10.00$$
$$Z_2 Z_3 = (8.54\angle -69.44°)(8.60\angle 35.54°)$$
$$= 73.44\angle -33.90° = 60.96 - j40.96$$
$$Z_3 Z_1 = (8.60\angle 35.54°)(11.18\angle 63.43°)$$
$$= 96.15\angle 98.97° = -14.99 + j94.97$$

Therefore

$$Z_1 Z_2 + Z_2 Z_3 + Z_3 Z_1 = (94.96 - 10.00j) + (60.96 - 40.96j) + (-14.99 + 94.97j)$$
$$= 140.93 + 44.01j = 147.64\angle 17.34°$$

**Step 3:** It is now time to put everything together, thus

$$Z_c = \frac{Z_1 Z_2 + Z_2 Z_3 + Z_3 Z_1}{Z_3}$$
$$= \frac{147.64\angle 17.34°}{8.60\angle 35.54°} = 17.2\angle -18.2°$$
$$Z_c = 16.3 - 5.37j$$

### Example 9

An AC circuit supplies three different loads of impedances $Z_1$, $Z_2$, and $Z_3$. Loads $Z_2$ and $Z_3$ are connected in parallel and the combination is in series with $Z_1$, with the latter closest to the AC source. Given that $Z_1 = (2 - 3j)\,\Omega$, $Z_2 = (3 - 4j)\,\Omega$, $Z_3 = (5 + 2j)\,\Omega$, and that the supply current is $I = (2 + j)$ A, determine the:

a) magnitude of the supply voltage,
b) magnitude of the supply voltage,
c) phase angle between the current and the voltage and state which one is leading. Comment on this result.

---

What did you get? Find the solution below to double-check your answer.

### Solution to Example 9

**a) Magnitude of the supply voltage**
**Solution**

**Step 1:** Draw the circuit diagram for clarity.

**Step 2:** Determine the total impedance $Z$ of the circuit.

$$Z = Z_1 + \frac{Z_2 Z_3}{Z_2 + Z_3}$$

So

$$Z_2 Z_3 = (3 - 4j)(5 + 2j) = 15 + 6j - 20j - 8j^2$$
$$= 15 + 6j - 20j - 8(-1) = 15 - 14j + 8$$
$$= 23 - 14j$$
$$Z_2 + Z_3 = (3 - 4j) + (5 + 2j) = 8 - 2j$$

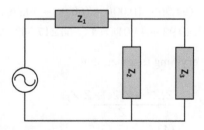

**FIGURE 9.4** Solution to Example 9.

# Complex Numbers II

Therefore

$$Z_1 + \frac{Z_2 Z_3}{Z_2 + Z_3} = (2 - 3j) + \frac{23 - 14j}{8 - 2j}$$

$$= (2 - 3j) + \frac{26.926\angle - 31.33°}{8.246\angle - 14.04°}$$

$$= (2 - 3j) + 3.265\angle - 17.29°$$

$$= (2 - 3j) + (3.117 - 0.970j)$$

$$= 5.117 - 3.970j$$

$$= 6.48\angle - 37.81°$$

**Step 3:** Determine the source voltage $V$.

$$V = IZ$$

Therefore,

$$V = (2 + j) \times (5.12 - 3.97j)$$

$$= (2.236\angle 26.57°)(6.479\angle - 37.79°)$$

$$= 14.49 \angle - 11.22°$$

The magnitude of the supply voltage is the modulus of the complex number above which is **14.5 V**.

**b) Magnitude of the supply current**
**Solution**

$$I = \frac{V}{Z}$$

$$= \frac{14.49 \angle - 11.22°}{6.48\angle - 37.81°}$$

$$= 2.236\angle 26.59°$$

**c) Phase angle between the current and the voltage and state which one is leading**
**Solution**

$$V = 14.49 \angle - 11.22°$$

$$I = 2.236\angle 26.59°$$

Given that the phase angle is $\varphi$, then

$$\varphi = |-11.22° - 26.59°|$$

$$= |-37.81°|$$

$$= 37.81°$$

$$\therefore \varphi = \mathbf{37.81°}$$

In this case, we can state that current leads the voltage with an angle of 37.81°.

## COMMENT

As the current is leading the voltage, the circuit is effectively capacitive. This is also evident in the value of the net impedance, i.e. $Z = 5.117 - 3.970j$. The negative sign in front of $X$ indicates that the reactance is capacitive in nature.

---

Let's try another example.

### Example 10

Two impedances $Z_1 = (8 + 10j)$ Ω and $Z_2 = (7 - 5j)$ Ω are connected in series to an AC power supply voltage of 240 V.

**a)** Determine the source current phasor (magnitude and phase angle).

**b)** If the two impedances were connected in parallel, determine the new source current phasor.

In each case, comment on the phase angle.

---

What did you get? Find the solution below to double-check your answer.

### Solution to Example 10

**a)** The source current phasor
**Solution**
Let $Z_t$ be the total impedance of the circuit. When connected in series (as shown in Figure 9.5)

$$Z_t = Z_1 + Z_2 = (8 + 10j) + (7 - 5j)$$
$$= 15 + 5j = 15.81 \angle 18.43°$$

From Ohm's law, the source current is

$$I = \frac{V}{Z_t}$$
$$= \frac{240 \angle 0°}{15.81 \angle 18.43°}$$
$$= 15.2 \angle -18.43° \, A$$

Hence, the current is **15.2 A** at **18.43°** lagging.

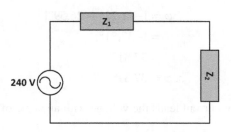

**FIGURE 9.5** Solution to Example 10(a).

# Complex Numbers II

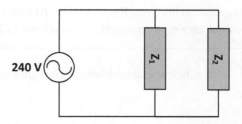

**FIGURE 9.6** Solution to Example 10(b).

**b)** The new source current phasor
**Solution**
When the impedances are connected in parallel (as shown in Figure 9.6), the total impedance is

$$Z_t = \frac{Z_1 Z_2}{Z_1 + Z_2}$$
$$= \frac{(8 + 10j)(7 - 5j)}{(8 + 10j) + (7 - 5j)}$$
$$= \frac{[12.81\angle 51.34°][8.60\angle -35.54°]}{15.81\angle 18.43°}$$
$$= \frac{110.17\angle 15.80°}{15.81\angle 18.43°}$$
$$= 6.97\angle -2.63°$$

Again, the circuit current is

$$I = \frac{V}{Z_t}$$
$$= \frac{240\angle 0°}{6.97\angle -2.63°}$$
$$= 34.43\angle 2.63° \text{ A}$$

In this parallel connection, the current is **34.43 A** at **2.63°** leading.

One final example to try.

### Example 11

The characteristic impedance (also known as natural impedance) $Z_0$ and the propagation coefficient $\gamma$ of a transmission line (lossy) are given by:

$$Z_0 = \sqrt{\frac{R + j\omega L}{G + j\omega C}}$$

and

$$\gamma = \sqrt{[(R + j\omega L)(G + j\omega C)]}$$

Determine $Z_0$ and $\gamma$ for a transmission line with $R = 15\,\Omega$, $L = 10$ mH, $G = 60\,\mu$S, $C = 0.5\,\mu$F, and $\omega = 1500$ rad/s. Present the final answer in rectangular form correct to 2 d.p.

---

What did you get? Find the solution below to double-check your answer.

### Solution to Example 11

Given that the characteristic impedance

$$Z_0 = \sqrt{\frac{R + j\omega L}{G + j\omega C}}$$

Substitute the values of $R, L, G, C$ and $\omega$ and simplify.

$$Z_0 = \sqrt{\frac{15 + j(1500)\left(10 \times 10^{-3}\right)}{60 \times 10^{-6} + j(1500)\left(0.5 \times 10^{-6}\right)}}$$

$$= \sqrt{\frac{15 + 15j}{10^{-6}(60 + 750j)}}$$

$$= \frac{1}{10^{-3}}\sqrt{\frac{15 + 15j}{60 + 750j}}$$

$$= 10^3 \sqrt{\frac{21.21\angle 45°}{752.40\angle 85.43}}$$

$$= 10^3 \sqrt{0.02819\angle - 40.43°}$$

$$= 10^3 [0.02819\angle - 40.43°]^{\frac{1}{2}}$$

$$= 10^3\, (0.02819)^{\frac{1}{2}} \angle\left(\frac{-40.43°}{2}\right)$$

$$= 10^3\, [0.1679\angle - 20.23°]$$

$$= 168\angle - 20.23°$$

$$\therefore Z_0 = 168\angle - 20.23° = 157.6 - 58.09j$$

Similarly, given that the propagation coefficient $\gamma$ of a transmission is

$$\gamma = \sqrt{[(R + j\omega L)(G + j\omega C)]}$$

substitute the values of $R, L, G, C,$ and $\omega$ and simplify.

$$\gamma = \sqrt{(21.21\angle 45°)\left[(752.40\angle 85.43) \times 10^{-6}\right]}$$

$$= 10^{-3}\sqrt{(21.21\angle 45°)(752.40\angle 85.43)}$$

$$= 10^{-3}\sqrt{15958.40\angle 130.43°}$$

# Complex Numbers II

$$= 10^{-3}[15958.40\angle 130.43°]^{\frac{1}{2}}$$

$$= 10^{-3}(15958.40)^{\frac{1}{2}}\angle\left(\frac{130.43°}{2}\right)$$

$$= 10^{-3}[126.33\angle 65.22°]$$

$$= 0.126\angle 65.22°$$

$$= \mathbf{0.05 + 0.11}\boldsymbol{j}$$

---

## 9.5 DE MOIVRE'S THEOREM

De Moivre's theorem is used to find the roots and powers of complex numbers. It states that if:

$$\boxed{z = r\angle\theta} \tag{9.6}$$

then

$$\boxed{z^n = [r\angle\theta]^n}$$
$$\boxed{= r^n \angle (n\theta)}$$
$$\boxed{= r^n(\cos n\theta + j\sin n\theta)} \tag{9.7}$$

This is because if

$$\boxed{z = re^{j\theta}} \tag{9.8}$$

then

$$\boxed{z^n = [re^{j\theta}]^n}$$
$$\boxed{= r^n e^{j(n\theta)}}$$
$$\boxed{= r^n \angle (n\theta)} \tag{9.9}$$

This theorem is valid for any real value of $n$ (i.e., positive, negative, whole, and fractional numbers). Let's try a couple of examples for $n$.

### Example 12

Using De Moivre's theorem, determine the value of the following in both polar and Cartesian forms:

**a)** $(3\angle 28°)^4$  **b)** $(2+j)^6$

The answer should be given in Cartesian form correct to 3 s.f.

What did you get? Find the solution below to double-check your answer.

### Solution to Example 12

**a)** $(3\angle 28°)^4$
**Solution**

$$(3\angle 28°)^4 = [3^4 \angle (4 \times 28°)]$$
$$= [81\angle(112°)]$$
$$\therefore (3\angle 28°)^4 = 81\angle 112°$$

Now in Cartesian form, we have

$$81\angle 112° = 81(\cos 112° + j\sin 112°)$$
$$= 81\cos 112° + j81\sin 112°$$
$$\therefore 81\angle 112° = -30.3 + j75.1$$

---

**b)** $(2+j)^6$
**Solution**
We need to convert this to polar form before we can use the theorem. Given $z = 2+j$, it follows that

$$r = \sqrt{2^2 + 1^2}$$
$$= \sqrt{5}$$

and

$$\theta = \tan^{-1}\frac{1}{2}$$
$$= 26.5651°$$

Hence

$$z = 2+j = \sqrt{5}\angle 26.5651°$$

Therefore

$$(2+j)^6 = \left(\sqrt{5}\angle 26.5651°\right)^6$$
$$= \left[\left(\sqrt{5}\right)^6 \angle (6 \times 26.5651°)\right]$$
$$= [125\angle(159.39°)]$$
$$\therefore (2+j)^6 = 125\angle 159°$$

Now in Cartesian form, we have

$$125\angle 159.39° = 125(\cos 159.39° + j\sin 159.39°)$$
$$= 125\cos 159.39° + j125\sin 159.39°$$
$$\therefore 125\angle 159.39° = -117 + j44.0$$

Complex Numbers II

### 9.5.1 ROOTS OF COMPLEX NUMBERS

When the square root of a complex number is taken, two complex numbers are produced. These roots will have the same magnitude $|z| = r$ such that their arguments only differ in sign. If one of the angles is $\theta$, the other will be $[\theta + (360°)/2]$ or $(\theta + 180°)$. Hence, the square roots of a complex number $z$ are $|z| \angle \theta$ and $|z| \angle (\theta + 180°)$.

You may wonder why angles $\theta$ and $(\theta + 180°)$ are said to be equal but only opposite in sign. Yes, they are. Note that $180°$ rotation is equal to $j^2$, which is in turn equal to $-1$. In fact, if the two roots are denoted by $z_1$ and $z_2$, it follows that:

$$\boxed{z_1 = |z| \angle \theta} \text{ AND } \boxed{z_2 = |z| \angle -\theta} \qquad (9.10)$$

or

$$\boxed{z_1 = |z|[\cos\theta + j\sin\theta]} \text{ AND } \boxed{z_2 = |z|[\cos\theta - j\sin\theta]} \qquad (9.11)$$

Similarly, the cube root of a complex number will have three roots, with the same modulus but different arguments, such that they are spaced by $[(360°)/3] = 120°$.

In general, the $n$th root of a complex number has $n$ roots, each having a magnitude of $|z| = r$. The first root has an angle $\theta = \arg(z)$ and others are symmetrically spaced from the first root and from each other by $(360°)/n$.

Since the $n$th root has $n$ complex numbers, it is sometimes required to find what is generally referred to as the principal root. This is the root that is closest to the positive $x$-axis, which can either be the first or the last root. If the first and last roots are equidistant from the positive real axis, the first root is conventionally taken as the principal root.

---

Let's try an example.

### Example 13

Determine the three cube roots of $z = 27(\cos 210° + j\sin 210°)$ and state which of them is the principal root. Also show the roots on a complex plane.

---

What did you get? Find the solution below to double-check your answer.

### Solution to Example 13

$$z = 27(\cos 210° + j\sin 210°)$$
$$= 27\angle 210°$$

Therefore

$$z^{\frac{1}{3}} = 27^{\frac{1}{3}} \angle {210°}/{3} = 3\angle 70°$$

For this case, the roots would be spaced at an angle of $\left(360°/3 = 120°\right)$. So let $\alpha$, $\beta$, and $\gamma$ be the first, second, and third roots, respectively. Thus, the first root is

$$\alpha = 3\angle 70°$$

and the two other roots are:

$$\beta = 3\angle(70° + 120°)$$
$$= 3\angle 190°$$
$$= 3\angle -170°$$

$$\gamma = 3\angle(190° + 120°)$$
$$= 3\angle 310°$$
$$= 3\angle -50°$$

or

$$\gamma = 3\angle(-170° + 120°)$$
$$= 3\angle -50°$$

Hence, the principal root (also the closest to the positive $x$-axis) is

$$\gamma = 3\angle -50°$$

The complex plane is shown in Figure 9.7.

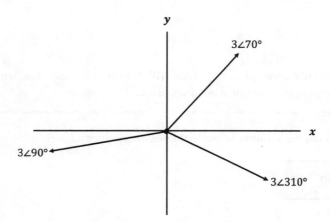

**FIGURE 9.7** Solution to Example 13.

# Complex Numbers II

**ALTERNATIVE METHOD**

Let the three cube roots of $z$ be denoted as $z_0$, $z_1$, and $z_2$ corresponding, respectively, to where $k = 0, 1,$ and 2. Thus, given that

$$z = 27(\cos 210° + j\sin 210°)$$

Therefore

$$z_0 = \sqrt[3]{27}\left[\cos\left(\frac{210°}{3} + \frac{2\pi \times 0}{3}\right) + j\sin\left(\frac{210°}{3} + \frac{2\pi \times 0}{3}\right)\right]$$
$$= 3(\cos 70° + j\sin 70°)$$
$$\therefore z_0 = 3\angle 70°$$

$$z_1 = \sqrt[3]{27}\left[\cos\left(\frac{210°}{3} + \frac{2\pi \times 1}{3}\right) + j\sin\left(\frac{210°}{3} + \frac{2\pi \times 1}{3}\right)\right]$$
$$= 3[\cos(70° + 120°) + j\sin(70° + 120°)]$$
$$= 3(\cos 190° + j\sin 190°)$$
$$= 3(\cos -170° + j\sin -170°)$$
$$= 3(\cos 170° - j\sin 170°)$$
$$\therefore z_1 = 3\angle -170°$$

$$z_2 = \sqrt[3]{27}\left[\cos\left(\frac{210°}{3} + \frac{2\pi \times 2}{3}\right) + j\sin\left(\frac{210°}{3} + \frac{2\pi \times 2}{3}\right)\right]$$
$$= 3[\cos(70° + 240°) + j\sin(70° + 240°)]$$
$$= 3(\cos 310° + j\sin 310°)$$
$$= 3(\cos -50° + j\sin -50°)$$
$$= 3(\cos 50° - j\sin 50°)$$
$$\therefore z_2 = 3\angle -50°$$

Let's try another example.

### Example 14

Determine the four fourth roots of $-81$ giving the answers in Cartesian form. Present the final answers in surd form.

What did you get? Find the solution below to double-check your answer.

**Solution to Example 14**

$$-81 = -81 + j0 = r\angle\theta$$

where

$$r = \sqrt{(-81)^2 + 0^2} \quad \text{and} \quad \theta = \tan^{-1}\left(\frac{0}{-81}\right)$$
$$= \sqrt{6561} = 81 \qquad\qquad\qquad = \tan^{-1}(-0)$$
$$\qquad\qquad\qquad\qquad\qquad\qquad = 180°$$

We chose $180°$ because the phasor is in the second-third quadrant. Therefore

$$-81 + j0 = 81\angle 180°$$

Therefore, the fourth root of $-81$ is:

$$(-81 + j0)^{\frac{1}{4}} = 81^{\frac{1}{4}}\angle\left(\frac{180°}{4}\right)$$
$$= 3\angle 45°$$

If $z_1, z_2, z_3$ and $z_4$ are the four roots of $-81$, then we have:

$z_1 = 3\angle 45°$  
$\quad = 3(\cos 45° + j\sin 45°)$  
$\quad = 3\left(\frac{\sqrt{2}}{2} + j\frac{\sqrt{2}}{2}\right)$  
$\quad = \frac{3\sqrt{2}}{2}(1 + j)$  

$z_3 = 3\angle(90° + 135°) = 3\angle 225°$  
$\quad = 3(\cos 225° + j\sin 225°)$  
$\quad = 3\left(-\frac{\sqrt{2}}{2} - j\frac{\sqrt{2}}{2}\right)$  
$\quad = -\frac{3\sqrt{2}}{2}(1 + j)$  

$z_2 = 3\angle(90° + 45°) = 3\angle 135°$  
$\quad = 3(\cos 135° + j\sin 135°)$  
$\quad = 3\left(-\frac{\sqrt{2}}{2} + j\frac{\sqrt{2}}{2}\right)$  
$\quad = \frac{3\sqrt{2}}{2}(-1 + j)$  

$z_4 = 3\angle(90° + 225°) = 3\angle 315°$  
$\quad = 3(\cos 315° + j\sin 315°)$  
$\quad = 3\left(\frac{\sqrt{2}}{2} - j\frac{\sqrt{2}}{2}\right)$  
$\quad = \frac{3\sqrt{2}}{2}(1 - j)$

# Complex Numbers II

A final example to try.

### Example 15

Determine the five fifth roots of $z = 2 + 2j$. Give the answers in both polar and exponential forms.

What did you get? Find the solution below to double-check your answer.

### Solution to Example 15

$$z = 2 + 2j = r\angle\theta$$

where

$$r = \sqrt{2^2 + 2^2} \quad \text{and} \quad \theta = \tan^{-1}\left(\frac{2}{2}\right) = \tan^{-1}(1)$$
$$= \sqrt{8} = 2\sqrt{2} \qquad\qquad\qquad\qquad = 45°$$

**Hence**

$$2 + 2j = 2\sqrt{2}\angle 45°$$

Therefore, the fifth root of $z$

$$(2 + j2)^{\frac{1}{5}} = \left(2\sqrt{2}\right)^{\frac{1}{5}} \angle \left(\frac{45°}{5}\right)$$
$$= 1.231\angle 9°$$

Since there are five roots, the spacing angle is

$$\frac{360°}{5} = 72°$$

If $z_1, z_2, z_3, z_4$, and $z_5$ are the five roots of $z$, then we have:

$z_1 = 1.231\angle 9°$ 
$= 1.231e^{j0.1571}$

$z_2 = 1.231\angle(9° + 72°)$
$= 1.231\angle 81°$
$= 1.231e^{j1.4137}$

$z_3 = 1.231\angle(81° + 72°)$
$= 1.231\angle 153°$
$= 1.231e^{j2.6704}$

$z_4 = 1.231\angle(153° + 72°)$
$= 1.231\angle 225°$
$= 1.231\angle -135°$
$= 1.231e^{-j2.3562}$

$z_5 = 1.231\angle(225° + 72°)$
$= 1.231\angle 297°$
$= 1.231\angle -63°$
$= 1.231e^{-j1.0996}$

## 9.5.2 POWERS OF TRIGONOMETRIC FUNCTIONS AND MULTIPLE ANGLES

Using De Moivre's theorem, it is possible to evaluate powers and multiple angles of sine and cosine functions. We know that

$$z^n = r^n(\cos n\theta + j \sin n\theta) \qquad (9.12)$$

If $r = 1$, then

$$\boxed{z^n = \cos n\theta + j \sin n\theta} \quad \text{-----(i)} \qquad (9.13)$$

Similarly,

$$\boxed{z^{-n} = \cos n\theta - j \sin n\theta} \quad \text{-----(ii)} \qquad (9.14)$$

From the last two equations, we can have two distinct results by:

adding $\qquad \boxed{z^n + z^{-n} = 2\cos n\theta} \quad \text{-----(iii)}$

subtracting $\qquad \boxed{z^n - z^{-n} = j2 \sin n\theta} \quad \text{-----(iv)} \qquad (9.15)$

When $n = 1$, we can also have

$$\boxed{z + z^{-1} = 2\cos\theta} \text{ AND } \boxed{z - z^{-1} = j2\sin\theta} \qquad (9.16)$$

It is also possible to find the values of **sin $n\theta$** and **cos $n\theta$** using

$$\boxed{\cos n\theta + j \sin n\theta = (\cos\theta + j \sin\theta)^n} \qquad (9.17)$$

When the right-hand side of the above equation is expanded (using binomial series) and like terms are collected together, $\sin n\theta$ would be equal to the imaginary part and $\cos n\theta$ equated to the real part. This is derived from De Moivre's theorem in conjunction with the rules used to solve complex number equations.

In summary, De Moivre's theorem is primarily used to determine the values of $\sin n\theta$, $\cos n\theta$, $\sin^n \theta$, and $\cos^n \theta$.

We can also state that

$$\boxed{Z = e^{j\theta} = \cos\theta + j\sin\theta} \quad \text{-------(v)} \qquad (9.18)$$

$$\boxed{Z^n = e^{jn\theta} = \cos n\theta + j\sin n\theta} \quad \text{-------(vi)} \qquad (9.19)$$

# Complex Numbers II

Also

$$\boxed{Z^{-1} = e^{-j\theta} = \cos\theta - j\sin\theta} \quad \text{-------(vii)}$$
(9.20)

$$\boxed{Z^{-n} = e^{-jn\theta} = \cos n\theta - j\sin n\theta} \quad \text{-------(viii)}$$
(9.21)

Add equations (v) and (vii), we can have

$$Z + Z^{-1} = e^{j\theta} + e^{-j\theta} = 2\cos\theta$$

which implies that

$$\boxed{\cos\theta = \frac{e^{j\theta} + e^{-j\theta}}{2}} \quad \text{-------(ix)}$$
(9.22)

$$\boxed{\cos n\theta = \frac{e^{jn\theta} + e^{-jn\theta}}{2}} \quad \text{-------(x)}$$
(9.23)

Subtract (vii) from (v), we have

$$Z - Z^{-1} = e^{j\theta} - e^{-j\theta} = 2j\sin\theta$$

which implies that

$$\boxed{\sin\theta = \frac{e^{j\theta} - e^{-j\theta}}{j2}} \quad \text{-------(xi)}$$
(9.24)

$$\boxed{\sin n\theta = \frac{e^{jn\theta} - e^{-jn\theta}}{j2}} \quad \text{-------(xii)}$$
(9.25)

The above can be used to express multiple and power of sine and cosine of angles using Euler–trigonometric relationship shown below.

**Euler–trigonometric relationship**

$$\cos^n\theta = (\cos\theta)^n = \left(\frac{e^{j\theta} + e^{-j\theta}}{2}\right)^n \quad \text{-----(xiii)}$$
(9.26)

$$\sin^n\theta = (\sin\theta)^n = \left(\frac{e^{j\theta} - e^{-j\theta}}{j2}\right)^n \quad \text{-----(xiv)}$$
(9.27)

Let's try a couple of examples.

### Example 16

Apply De Moivre's theorem to express $\cos^4\theta$ in terms of cosines of multiples of angle $\theta$.

What did you get? Find the solution below to double-check your answer.

### Solution to Example 16

We know that

$$z + \frac{1}{z} = 2\cos\theta \qquad -----\text{(i)}$$

$$z^n + \frac{1}{z^n} = 2\cos n\theta \qquad -----\text{(ii)}$$

From equation (i), we have that

$$(2\cos\theta)^4 = \left(z + \frac{1}{z}\right)^4 \qquad -----\text{(iii)}$$

Expand the RHS of equation (iii)

$$\left(z + \frac{1}{z}\right)^4 = z^4 + 4z^3 \cdot \frac{1}{z} + 6z^2 \cdot \frac{1}{z^2} + 4z \cdot \frac{1}{z^3} + \frac{1}{z^4}$$

$$= z^4 + 4z^2 + 6 + 4\frac{1}{z^2} + \frac{1}{z^4}$$

$$= z^4 + \frac{1}{z^4} + 4\left(z^2 + \frac{1}{z^2}\right) + 6 \qquad --\text{(iv)}$$

Use equation (ii) to express equation (iv) in terms of $\cos\theta$

$$\left(z + \frac{1}{z}\right)^4 = 2\cos 4\theta + 4(2\cos 2\theta) + 6$$

$$= 2\cos 4\theta + 8\cos 2\theta + 6$$

Based on equation (iii), we can then write that

$$(2\cos\theta)^4 = 2\cos 4\theta + 8\cos 2\theta + 6$$

this implies that

$$2^4 \cos^4\theta = 2\cos 4\theta + 8\cos 2\theta + 6$$

Divide through by 16

$$\cos^4\theta = \frac{1}{16}(2\cos 4\theta + 8\cos 2\theta + 6)$$

$$\therefore \cos^4\theta = \frac{1}{8}[(\cos 4\theta + 4\cos 2\theta + 3)]$$

# Complex Numbers II

**Example 17**

Expand $\sin 7\theta$ giving the answer in the simplest form involving only powers of $\sin \theta$.

What did you get? Find the solution below to double-check your answer.

**Solution to Example 17**

$$\cos 7\theta + j\sin 7\theta = (\cos \theta + j\sin \theta)^7 \quad ---(i)$$

Let

$$\cos \theta = a$$
$$\sin \theta = b$$

Substitute this in (i)

$$(\cos \theta + j\sin \theta)^7 = (a + jb)^7 \quad ------(ii)$$

Let us expand the RHS of equation (ii)

$$(a+jb)^7 = a^7 + 7a^6(jb) + 21a^5(jb)^2 + 35a^4(jb)^3 + 35a^3(jb)^4 + 21a^2(jb)^5 + 7a(jb)^6 + (jb)^7$$
$$= a^7 + j7a^6b - 21a^5b^2 - j35a^4b^3 + 35a^3b^4 + j21a^2b^5 - 7ab^6 - jb^7$$
$$= (a^7 - 21a^5b^2 + 35a^3b^4 - 7ab^6) + j(7a^6b - 35a^4b^3 + 21a^2b^5 - b^7)$$

Thus,

$$\cos 7\theta + j\sin 7\theta = (\cos \theta + j\sin \theta)^7$$
$$= (a^7 - 21a^5b^2 + 35a^3b^4 - 7ab^6) + j(7a^6b - 35a^4b^3 + 21a^2b^5 - b^7) \quad -----(iii)$$

For (iii) to be valid, according to our 'golden' rules, it follows that the real part on the LHS must be equal to the real part on the RHS, and the imaginary must be equated to its corresponding imaginary component. In our current case, we need the imaginary part only as we essentially want to evaluate $\sin 7\theta$. Thus, comparing both sides of (iii), we will have

$$\sin 7\theta = (7a^6b - 35a^4b^3 + 21a^2b^5 - b^7) \quad -(iv)$$

Now replace the value of $a$ and $b$ back into (iv), so we will have

$$\sin 7\theta = 7\cos^6\theta \sin\theta - 35\cos^4\theta\sin^3\theta + 21\cos^2\theta\sin^5\theta - \sin^7\theta$$
$$= 7\sin\theta(1-\sin^2\theta)^3 - 35\sin^3\theta(1-\sin^2\theta)^2 + 21\sin^5\theta(1-\sin^2\theta) - \sin^7\theta$$
$$= 7\sin\theta(1 - 3\sin^2\theta + 3\sin^4\theta - \sin^6\theta) - 35\sin^3\theta(1 - 2\sin^2\theta + \sin^4\theta)$$
$$\quad + 21\sin^5\theta - 21\sin^7\theta - \sin^7\theta$$
$$= 7\sin\theta - 21\sin^3\theta + 21\sin^5\theta - 7\sin^7\theta - 35\sin^3\theta + 70\sin^5\theta - 35\sin^7\theta$$
$$\quad + 21\sin^5\theta - 21\sin^7\theta - \sin^7\theta$$

$$\therefore \sin 7\theta = 7\sin\theta - 56\sin^3\theta + 112\sin^5\theta - 64\sin^7\theta$$

## 9.6 LOGARITHM OF A COMPLEX NUMBER

Sometimes, we need to find the logarithm of a complex number, particularly the natural logarithm, which is the logarithm of a complex number to the base of $e$. This is best determined if the complex number is in exponential form. Let us consider

$$z = re^{\pm j\theta} \qquad (9.28)$$

Take the natural logarithm of both sides, we have

$$\log_e z = \log_e re^{\pm j\theta} \qquad (9.29)$$

This is usually expressed as:

$$\ln z = \ln re^{\pm j\theta}$$
$$= \ln r + \ln e^{\pm j\theta}$$
$$= \ln r \pm j\theta \qquad (9.30)$$

Since $r = |z|$ and $\theta = \arg z$, the natural logarithm of a complex number $z$ can be expressed as:

$$\ln z = \ln |z| \pm j \arg z \qquad (9.31)$$

Let's try a couple of examples.

### Example 18

If $z_1 = 12(\cos 60° + j \sin 60°)$ and $z_2 = 5(\cos 300° + j \sin 300°)$, calculate:

**a)** $\ln z_1$          **b)** $\ln z_2$

Present the answer in Cartesian form correct to 3 d.p.

---

What did you get? Find the solution below to double-check your answer.

### Solution to Example 18

**a)** $\ln z_1$
**Solution**

$$z_1 = 12(\cos 60° + j \sin 60°)$$
$$= 12\angle 60°$$
$$= re^{j\theta}$$

where

$$r = 12$$

and

$$\theta = 60° = 1.047 \; rad$$
$$\therefore z_1 = 12e^{j1.0472}$$

# Complex Numbers II

Therefore,

$$\ln z_1 = \ln 12 e^{j1.0472}$$
$$= \ln 12 + j1.0472$$
$$= \mathbf{2.485 + j1.047}$$

**b)** $\ln z_2$
**Solution**

$$z_2 = 5(\cos 300° + j \sin 300°)$$
$$= 5(\cos 60° - j \sin 60°)$$
$$= 5\angle -60°$$
$$= re^{j\theta}$$

where

$$r = 5$$

and

$$\theta = -60° = -1.047 \; rad$$
$$\therefore z_2 = 5e^{-j1.047}$$

Therefore

$$\ln z_2 = \ln 5 e^{-j1.047}$$
$$= \ln 5 - j1.047$$
$$= \mathbf{1.609 - j1.047}$$

## 9.7 APPLICATIONS

To finish this chapter, we want to show two areas where complex numbers can be used, namely geometry and vector analysis. Let's try some examples to illustrate this.

### Example 19

Points $P$, $Q$, and $R$, on a complex plane, represent the complex numbers $2 - 7j$, $-2 + 5j$, and $-3 - 2j$, respectively. Represent this on a complex plane and show that triangle $PQR$ is isosceles.

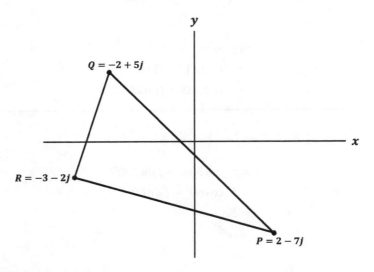

**FIGURE 9.8** Solution to Example 19.

What did you get? Find the solution below to double-check your answer.

### Solution to Example 19

#### HINT

An isosceles is a type of triangle having two sides of equal lengths.

As shown in the diagram in Figure 9.8, let

$$P = z_1 = 2 - 7j \qquad Q = z_2 = -2 + 5j \qquad R = z_3 = -3 - 2j$$

Also, let $\overline{PQ}$, $\overline{QR}$, and $\overline{RP}$ be the lengths of the three sides of the triangle $PQR$, it follows that

$$\begin{aligned}
\overline{PQ} &= |z_2 - z_1| \\
&= |(-2 + 5j) - (2 - 7j)| \\
&= |-4 + 12j| \\
&= \sqrt{(-4)^2 + 12^2} \\
&= \sqrt{160} \\
&= \mathbf{4\sqrt{10} \text{ units}}
\end{aligned} \qquad
\begin{aligned}
\overline{QR} &= |z_3 - z_2| \\
&= |(-3 - 2j) - (-2 + 5j)| \\
&= |-1 - 7j| \\
&= \sqrt{(-1)^2 + (-7)^2} \\
&= \sqrt{50} \\
&= \mathbf{5\sqrt{2} \text{ units}}
\end{aligned} \qquad
\begin{aligned}
\overline{RP} &= |z_1 - z_3| \\
&= |(2 - 7j) - (-3 - 2j)| \\
&= |5 - 5j| \\
&= \sqrt{(5)^2 + (-5)^2} \\
&= \sqrt{50} \\
&= \mathbf{5\sqrt{2} \text{ units}}
\end{aligned}$$

For triangle $PQR$, we have shown that $\overline{QR} = \overline{RP}$, therefore triangle PQR is isosceles.

# Complex Numbers II

## Example 20

Three coplanar forces $F_1$, $F_2$, and $F_3$ are acting on a point object. Force $F_1$ is 12 N acting eastward, $F_2$ is 20 N acting at an angle of 150° to force $F_1$ (measured in anti-clockwise direction from the line of action of $F_1$), and $F_3$ is 15 N acting southward. Using complex numbers, determine the magnitude and direction of the net force on the object.

What did you get? Find the solution below to double-check your answer.

## Solution to Example 20

### Hint

This question can be solved using other methods, including parallelogram and Lami's methods. However, we want to approach it from the complex number system route.

The three forces (shown in Figure 9.9) can be represented as complex number systems, with east taken as the reference axis.

Thus,

$F_1 = 12 + 0j$   $\quad F_2 = 20 \cos 150° + j20 \sin 150°$   $\quad F_3 = 15 \cos(-90°) + j15 \sin(-90°)$
$\phantom{F_1 = 12 + 0j \quad F_2} = -17.32 + 10j$   $\phantom{F_3} = 0 - 15j$

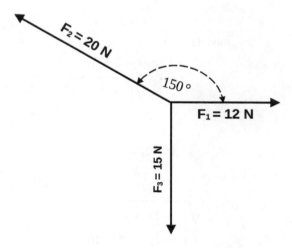

**FIGURE 9.9** Solution to Example 20.

Let $F_n$ represents the net force, therefore

$$\begin{aligned} F_n &= F_1 + F_2 + F_3 \\ &= (12 + 0j) + (-17.32 + 10j) + (0 - 15j) \\ &= -5.32 - 5j \\ &= 7.30\angle - 136.78° \text{ N} \end{aligned}$$

Hence, the net force is **7.30 N**, which is at an angle of **136.77°** measured clockwise from force $F_1$ or **223.23°** measured counter-clockwise from the same force.

Let's try another example.

### Example 21

Determine the equation of the locus of a point which moves in a complex plane such that $|z - 2| = 4$ where $z = x + jy$.

What did you get? Find the solution below to double-check your answer.

### Solution to Example 21

If $z = x + jy$, then

$$\begin{aligned} z - 2 &= x + jy - 2 \\ &= x - 2 + jy \\ \text{mod}\,(z - 2) &= |z - 2| \\ &= |x - 2 + jy| \\ &= \sqrt{(x - 2)^2 + y^2} \end{aligned}$$

Since

$$|z - 2| = 4$$

then

$$\begin{aligned} \sqrt{(x - 2)^2 + y^2} &= 4 \\ (x - 2)^2 + y^2 &= 16 \\ x^2 - 4x + 4 + y^2 &= 16 \end{aligned}$$

Hence, the equation of the locus, defined by $|z - 2| = 4$, is

$$x^2 + y^2 - 4x - 12 = 0$$

Complex Numbers II

Let's try another example.

## Example 22

Determine the equation of the locus of points which is given by $|z+j|^2 + 2|z-2|^2 = 0$.

What did you get? Find the solution below to double-check your answer.

## Solution to Example 22

If $z = x + jy$, then

$$z + j = x + jy + j \qquad\qquad z - 2 = x - 2 + jy$$
$$\phantom{z+j} = x + j(y+1) \qquad\qquad \phantom{z-2} = (x-2) + jy$$

Also

$$\begin{aligned}\text{mod } (z+j) &= |z+j| \\ &= |x+j(y+1)| \\ &= \sqrt{x^2 + (y+1)^2} \\ &= \sqrt{x^2 + y^2 + 2y + 1}\end{aligned} \qquad \begin{aligned}\text{mod } (z-2) &= |z-2| \\ &= |(x-2)+jy| \\ &= \sqrt{(x-2)^2 + y^2} \\ &= \sqrt{x^2 + y^2 - 4x + 4}\end{aligned}$$

Given that $|z + j2|^2 + 2|z - 2|^2 = 0$, it implies that

$$\left(\sqrt{x^2 + y^2 + 2y + 1}\right)^2 + 2\left(\sqrt{x^2 + y^2 - 4x + 4}\right)^2 = 0$$

Remove the roots, we have

$$(x^2 + y^2 + 2y + 1) + 2(x^2 + y^2 - 4x + 4) = 0$$
$$x^2 + y^2 + 2y + 1 + 2x^2 + 2y^2 - 8x + 8 = 0$$

Thus, the equation is a circle having a value of

$$3x^2 + 3y^2 - 8x + 2y + 9 = 0$$

Let's try another example.

## Example 23

Points $A$, $B$, $C$, and $D$, on an Argand diagram, represent the complex numbers $9+j$, $4+13j$, $-7+8j$, and $-3-4j$, respectively. Show that the quadrilateral formed by points $A$, $B$, $C$, and $D$ is neither a square nor a rectangle.

What did you get? Find the solution below to double-check your answer.

**Solution to Example 23**

Let

$$A = z_1 = 9 + j \qquad B = z_2 = 4 + j13 \qquad C = z_3 = -7 + j8 \qquad D = z_4 = -3 - 4j$$

Also, let $\overline{AB}, \overline{BC}, \overline{DC}$, and $\overline{AD}$ be the lengths of the four sides of the quadrilateral, it follows that

$$\overline{AB} = |z_2 - z_1|$$
$$= |(4 + 13j) - (9 + j)|$$
$$= |-5 + 12j|$$
$$= \sqrt{(-5)^2 + 12^2}$$
$$= \sqrt{169}$$
$$= 13 \text{ units}$$

$$\overline{BC} = |z_3 - z_2|$$
$$= |(-7 + 8j) - (4 + 13j)|$$
$$= |-11 - 5j|$$
$$= \sqrt{(-11)^2 + (-5)^2}$$
$$= \sqrt{146} \text{ units}$$

$$\overline{DC} = |z_3 - z_4|$$
$$= |(-7 + 8j) - (-3 - 4j)|$$
$$= |-4 + 12j|$$
$$= \sqrt{(-4)^2 + 12^2}$$
$$= \sqrt{160} \text{ units}$$

$$\overline{AD} = |z_4 - z_1|$$
$$= |(-3 - 4j) - (9 + j)|$$
$$= |-12 - 5j|$$
$$= \sqrt{(-12)^2 + (-5)^2}$$
$$= \sqrt{169}$$
$$= 13 \text{ units}$$

One of the conditions for $ABCD$ to be a square is that $\overline{AB} = \overline{BC} = \overline{DC} = \overline{AD}$. This is not the case here, so $ABCD$ is not a square. For $ABCD$ to be a rectangle, two pairs should be equal, i.e. $\overline{AB} = \overline{DC}$ and $\overline{BC} = \overline{AD}$, which is not the case here.

Let's try another example.

**Example 24**

Vectors $A$, $B$, and $C$ are given by $5\angle 40°$, $7\angle 135°$, and $2\angle 60°$, respectively. Presenting the answers in Cartesian form, determine the vectors represented by:

a) $A + 2B + 2C$ 

b) $3A - B - C$

What did you get? Find the solution below to double-check your answer.

### Solution to Example 24

First, let us convert the vectors to polar form since it is convenient to carry out addition and subtraction in this form.

$A = 5\angle 40°$  
$= 3.83 + 3.21j$

$B = 7\angle 135°$  
$= -4.95 + 4.95j$

$C = 2\angle 60°$  
$= 1 + 1.73j$

**a) $A + 2B + 2C$**
**Solution**

$$\begin{aligned} A + 2B + 2C &= (3.83 + 3.21j) + 2(-4.95 + 4.95j) + 2(1 + 1.73j) \\ &= (3.83 + 3.21j) + (-9.9 + 9.9j) + (2 + 3.46j) \\ &= (3.83 - 9.9 + 2) + (3.21 + 9.9 + 3.46)j \\ &= -4.07 + 16.57j \end{aligned}$$

$\therefore A + 2B + 2C = -4.07 + 16.57j$

**b) $3A - B - C$**
**Solution**

$$\begin{aligned} 3A - B - C &= 3(3.83 + 3.21j) - (-4.95 + 4.95j) - (1 + 1.73j) \\ &= (11.49 + 9.63j) - (-4.95 + 4.95j) - (1 + 1.73j) \\ &= (11.49 + 4.95 - 1) + (9.63 - 4.95 - 1.73)j \end{aligned}$$

$\therefore 3A - B - C = 15.44 + 2.95j$

## 9.8 CHAPTER SUMMARY

1) An Argand diagram is a geometrical plot of complex numbers on the x–y Cartesian plane, also known as the complex plane. The x-axis (horizontal axis) represents the real parts and the y-axis (vertical axis) represents the imaginary parts of complex numbers. They are called the real axis and the imaginary axis, respectively.

2) When working with complex number equations, there are two rules to apply:
   - If a complex number is equal to zero, both the real part and the imaginary part must also be equal to zero. In other words, if $x + jy = 0$, then $x = 0$ and $y = 0$.
   - If two complex numbers are equal, then their real parts and imaginary parts must be equal. For instance, given that $z_1 = x_1 + jy_1$ and $z_2 = x_2 + jy_2$ such that $z_1 = z_2$, then $x_1 = x_2$ and $y_1 = y_2$.

3) The graphical representation of voltages and currents in an AC circuit is known as a **phasor diagram**, which is like an **Argand diagram**.

4) A useful set of equations for AC circuits analysis include:

$$\boxed{Z_{RC} = R - jX_c} \quad ---- \text{ for } RC \text{ circuits}$$

for which

$$\boxed{X_c = \frac{1}{2\pi f C}}$$

and

$$\boxed{Z_{RL} = R + jX_L} \quad ---- \text{ for } RL \text{ circuits}$$

for which

$$\boxed{X_L = 2\pi f L}$$

where $Z$ = impedance of the circuit, $X_C$ = capacitive reactance, $X_L$ = inductive reactance, $R$ = resistance, $C$ = capacitance, $L$ = inductance, and $f$ = frequency.

5) In general, the impedance in an $RLC$ circuit is given by

$$\boxed{Z = R + jX} \quad \text{OR} \quad \boxed{Z = R + j\,(X_L - X_c)}$$

6) De Moivre's theorem is used to find the roots and powers of complex numbers. It states that if:

$$\boxed{z = r\angle\theta}$$

then

$$\boxed{z^n = [r\angle\theta]^n}$$
$$\boxed{= r^n \angle (n\theta)}$$
$$\boxed{= r^n(\cos n\theta + j\sin n\theta)}$$

This theorem is valid for any real value of $n$.

7) When the square root of a complex number is taken, two complex numbers are produced. These roots will have the same magnitude $|z| = r$, such that their arguments only differ in sign. If one of the angles is $\theta$, the other will be $[\theta + (360°)/2]$ or $(\theta + 180°)$. Hence, the square roots of a complex number $z$ are $|z|\angle\theta$ and $|z|\angle(\theta + 180°)$. In fact, if the two roots are denoted by $z_1$ and $z_2$, it follows that:

$$\boxed{z_1 = |z|\angle\theta} \text{ AND } \boxed{z_2 = |z|\angle -\theta}$$

or

# Complex Numbers II

$$z_1 = |z|[\cos\theta + j\sin\theta] \quad \text{AND} \quad z_2 = |z|[\cos\theta - j\sin\theta]$$

8) In general, the $n$th root of a complex number has $n$ roots, each having a magnitude of $|z| = r$. The first root has an angle $\theta = \arg(z)$ and the others are symmetrically spaced from the first root and from each other by $(360°)/n$.

9) Using De Moivre's theorem, it is possible to evaluate powers and multiple angles of sine and cosine functions. We know that:

$$z^n = r^n(\cos n\theta + j\sin n\theta)$$

If $r = 1$, then

$$z^n = \cos n\theta + j\sin n\theta$$

10) De Moivre's theorem is primarily used to determine the values of $\sin n\theta$, $\cos n\theta$, $\sin^n\theta$ and $\cos^n\theta$. We can write that:

$$e^{j\theta} = \cos\theta + j\sin\theta$$

$$e^{jn\theta} = \cos n\theta + j\sin n\theta$$

Also

$$e^{-j\theta} = \cos\theta - j\sin\theta$$

$$e^{-jn\theta} = \cos n\theta - j\sin n\theta$$

Hence

$$\cos\theta = \frac{e^{j\theta} + e^{-j\theta}}{2} \quad \text{AND} \quad \cos n\theta = \frac{e^{jn\theta} + e^{-jn\theta}}{2}$$

Similarly

$$\sin\theta = \frac{e^{j\theta} - e^{-j\theta}}{j2} \quad \text{AND} \quad \sin n\theta = \frac{e^{jn\theta} - e^{-jn\theta}}{j2}$$

11) Since $r = |z|$ and $\theta = \arg z$, the natural logarithm of a complex number $z$ can be expressed as:

$$\ln z = \ln|z| \pm j\arg z$$

\*\*\*\*

## 9.9 FURTHER PRACTICE

To access complementary contents, including additional exercises, please go to www.dszak.com.

# 10 Matrices

## Learning Outcomes

Once you have studied the content of this chapter, you should be able to:

- Explain the term matrix
- Discuss the types of matrices and their properties
- Perform addition and subtraction of matrices
- Perform scalar multiplication of a matrix
- Perform multiplication of two matrices
- Evaluate the determinant of a matrix
- Evaluate the adjoint of a square matrix
- Evaluate the inverse of a square matrix
- Use matrices to solve simultaneous equations

## 10.1 INTRODUCTION

Matrix is an essential area of mathematics that we, as human beings, use unknowingly. For example, the layout of groceries can be represented using a matrix and a particular product can be referenced using a matrix element. Matrices are effective in solving equations involving several unknown variables that would otherwise be difficult to solve using conventional methods, such as the substitution or elimination method. They are widely used in computing and programming. In this chapter, we will look at this important topic, covering its meaning, types, operations, and some of its applications.

## 10.2 MATRIX NOTATION

A matrix (plural: matrices) is defined as a rectangular array of numbers (or elements), consisting of rows and columns generally enclosed in square brackets [] or parentheses (); the use of either is not required if the matrix is 1 by 1. Examples of matrix include:

$$A = [3] \quad B = \begin{bmatrix} -1 & 0 & 2 \\ 8 & 12 & 5 \end{bmatrix} \quad C = \begin{bmatrix} 1 & 10 \\ -5 & 0 \\ -7 & 15 \end{bmatrix} \quad D = \begin{bmatrix} 1 & 2 & 3 \\ -5 & 10 & -15 \\ 3 & 6 & 9 \end{bmatrix}$$

Each number (or term) in the above matrices is called an **element** of the matrix. Therefore, changing the position of an element changes the entire matrix. We can say that 3 is an element of matrix **A**. Similarly, $-1$, $0$, $2$, $8$, $12$, and $5$ are all elements of matrix **B**. There is no mathematical connection between the elements of a matrix.

A matrix is generally represented by a bold capital letter and its elements with small letters. An element is completely identified by stating its row followed by its column. For example, in matrix **C**, $-5$ is an element in row 2 and column 1. In general, we can write matric **C** above as:

$$C = \begin{bmatrix} c_{11} & c_{12} \\ c_{21} & c_{22} \\ c_{31} & c_{32} \end{bmatrix}$$

where $c_{11} = 1$, $c_{12} = 10$, $c_{21} = -5$, $c_{22} = 0$, $c_{31} = -7$, and $c_{32} = 15$.

Similar to how points are identified in coordinate geometry and thus distinguish between its elements using $(x, y)$, we can equally use a general notation to state the location of an element in a matrix **A** as:

$$a_{ij}$$

where

- $i$ = row number
- $j$ = column number

We should also note the *size* (*order or dimension*) of a matrix, identified by stating the row and then the column, given as:

**Size (of a Matrix) = Row × Column = Row by Column**

By this, we can say that matrix **B** is **two by three**. Alternatively, we can write this as a **2 by 3** matrix or $2 \times 3$ matrix. Matrix **C** is $3 \times 2$ and matrix **D** is $3 \times 3$. Matrix **B** and **C**, though have the same number of elements, are not the same. As such, a matrix of $2 \times 3$ is not the same as a $3 \times 2$ matrix. Hence, the order is to state the number of rows and then the number of columns, the reverse is not valid.

## 10.3 TYPES OF MATRICES

Some matrices are important and are specifically identified with special names, as described below.

### 1) Column matrix

This is a matrix with a single (or one) column. The general dimension is $n \times 1$, where $n$ is the number of rows. Matrices **E** and **F** below are examples of column matrices, with dimensions of $2 \times 1$ and $3 \times 1$, respectively.

$$E = \begin{bmatrix} -2 \\ 4 \end{bmatrix} \qquad F = \begin{bmatrix} 3 \\ 2 \\ 5 \end{bmatrix}$$

To save space, matrix **F** can also be written as $\{3 \ \ 2 \ \ 5\}$ or $[3; 2; 5]$. Notice that in the first case, we used curly square brackets { } with a space between each element, and in the second case, we used [ ] with a semicolon (;) after each element. In general, the semicolon is used to terminate a row and a comma (or space) is used to separate elements in a row.

Let's consider a 3 by 2 matrix shown below.

$$Z = \begin{bmatrix} -1 & 7 \\ 2 & 4 \\ -6 & 3 \end{bmatrix}$$

Matrix **Z** can be written as $[-1, 7; 2, 4; -6, 3]$. The use of a comma shows the end of an element (provided that there is another element after it), and a semicolon at the end of a row (provided that there is a succeeding row). Let's imagine that we omit the comma and write **Z** as $[-17; 24; -63]$, this will be interpreted as a 3 by 1 matrix (i.e., a column matrix), and will be stored as:

$$Z = \begin{bmatrix} -17 \\ 24 \\ -63 \end{bmatrix}$$

To ensure that the above is not the case, one needs to create a space between the elements. Thus $[-1\,7; 2\,4; -6\,3]$ with a single space or $[-1\ \ 7; 2\ \ 4; -6\ \ 3]$ with a double space between elements will be correctly interpreted as:

$$\begin{bmatrix} -1 & 7 \\ 2 & 4 \\ -6 & 3 \end{bmatrix}$$

What if one omits the semicolon? Well, this is a tricky one. Let's first state that the system (or program) will only recognise the end of a row by seeing a semicolon, so let's say we omit the first semicolon and write $[-1, 72, 4; -6, 3]$. It will be considered as a matrix with two rows. The first row has three elements ($-1$, 72 and 4) and the second row has two elements ($-6$ and 3). This is inconsistent! Hence an error message will be displayed!

### 2) Row matrix

This is a matrix with a single (or one) row. It is also called a **line matrix**, and is the inverse of the column matrix. The general dimension is $1 \times n$, where $n$ is the number of columns. Matrices **S** and **R** shown below are examples of row matrices.

$$S = \begin{bmatrix} -2 & 4 \end{bmatrix} \qquad R = \begin{bmatrix} 3 & 2 & 5 \end{bmatrix}$$

### 3) Square matrix

This is a matrix with an equal number of rows and columns. Examples of square matrices are:

**2 × 2 square matrix**          **3 × 3 square matrix**          **4 × 4 square matrix**

$$A = \begin{bmatrix} -1 & -3 \\ 2 & 4 \end{bmatrix} \qquad B = \begin{bmatrix} 1 & 3 & 0 \\ 2 & 0 & 9 \\ 4 & 5 & -6 \end{bmatrix} \qquad C = \begin{bmatrix} 1 & 0 & 1 & 4 \\ 3 & 0 & 2 & 1 \\ 2 & 5 & 1 & 2 \\ 7 & 6 & 0 & 5 \end{bmatrix}$$

Square matrices are very important and central to many calculations in this topic. The determinant cannot be obtained except if a matrix is square.

### 4) Single (element) matrix

This is a matrix of a dimension $1 \times 1$, i.e., it has one element, one row, and one column. Examples include:

$$A = [2] \qquad B = [-3] \qquad C = [1.2]$$

### 5) Diagonal matrix

This is a type of square matrix where all the elements are zeros except the one on the diagonal line, generally called the leading diagonal (or principal diagonal). The diagonal matrix consists of all elements with $a_{ii}$, starting with $a_{11}$ on the top left (or 'north-west') to $a_{ii}$ on the bottom right (or 'south-east'), which can be covered with a backward slash (\). An example of a diagonal matrix is given below.

$$A = \begin{bmatrix} 1 & 0 & 0 \\ 0 & 3 & 0 \\ 0 & 0 & 7 \end{bmatrix}$$

The sum of the elements of the leading diagonal is called the trace of the same matrix and is denoted by $tr(A)$, if the matrix is given as A. For the above example, we have that

$$tr(A) = \sum_{i=1}^{n} a_{ii} = 1 + 3 + 7 = 11$$

Note that the trace of matrix only exists for square matrices (of which a diagonal matrix is a member), and the value can be any real number including zero.

Multiplying a diagonal matrix by another dimensionally compatible diagonal matrix results in a new diagonal matrix. An example for 3 by 3 matrices is such that if:

$$A = \begin{bmatrix} a_{11} & 0 & 0 \\ 0 & a_{22} & 0 \\ 0 & 0 & a_{33} \end{bmatrix} \qquad B = \begin{bmatrix} b_{11} & 0 & 0 \\ 0 & b_{22} & 0 \\ 0 & 0 & b_{33} \end{bmatrix}$$

Then

$$AB = \begin{bmatrix} a_{11} & 0 & 0 \\ 0 & a_{22} & 0 \\ 0 & 0 & a_{33} \end{bmatrix} \times \begin{bmatrix} b_{11} & 0 & 0 \\ 0 & b_{22} & 0 \\ 0 & 0 & b_{33} \end{bmatrix} = \begin{bmatrix} a_{11} \times b_{11} & 0 & 0 \\ 0 & a_{22} \times b_{22} & 0 \\ 0 & 0 & a_{33} \times b_{33} \end{bmatrix}$$

If only the elements below the leading diagonal are zero, this is called an **upper triangular matrix**, such as matrix C below. On the other hand, if the zero elements are the ones above the principal diagonal, the matrix is termed a **lower triangular matrix**, such as matrix D below.

$$C = \begin{bmatrix} a_{11} & a_{12} & a_{13} \\ 0 & a_{22} & a_{23} \\ 0 & 0 & a_{33} \end{bmatrix} \qquad D = \begin{bmatrix} b_{11} & 0 & 0 \\ b_{21} & b_{22} & 0 \\ b_{31} & b_{32} & b_{33} \end{bmatrix}$$

# Matrices

## 6) Unit (or identity) matrix

This is a diagonal matrix in which the elements along the diagonal line are ones (1). Examples of $2 \times 2$, $3 \times 3$, and $4 \times 4$ matrices are given below.

$$A = \begin{bmatrix} 1 & 0 \\ 0 & 1 \end{bmatrix} \quad B = \begin{bmatrix} 1 & 0 & 0 \\ 0 & 1 & 0 \\ 0 & 0 & 1 \end{bmatrix} \quad C = \begin{bmatrix} 1 & 0 & 0 & 0 \\ 0 & 1 & 0 & 0 \\ 0 & 0 & 1 & 0 \\ 0 & 0 & 0 & 1 \end{bmatrix}$$

An identity matrix is generally denoted with symbol **I**. You may also find that it's given as $I_n$, where $n$ stands for the order of the matrix. For example, $I_2$ and $I_5$ denote a 2 by 2 and a 5 by 5 identity matrices, respectively. Thus, matrix A, B, and C can be written as $I_2$, $I_3$, and $I_4$ respectively.

Unit matrices act like digit one (1) such that a matrix remains unchanged if it is multiplied (pre- or post-) by a similar identity matrix. That is, given matrices **A**, **B**, and **C** for example, we can write:

$$A \times I = I \times A = A, \quad B \times I = I \times B = B, \text{ and } C \times I = I \times C = C$$

Furthermore, when an identity matrix is multiplied by itself it gives the same result, $I \times I = I$, similar to multiplying one by one, i.e., $1 \times 1 = 1$. This (i.e., $I \times I = I$) will also be valid if the elements in the leading diagonal are $-1$ instead of 1, as in the case of matrix **D** shown below.

$$D = \begin{bmatrix} -1 & 0 & 0 \\ 0 & -1 & 0 \\ 0 & 0 & -1 \end{bmatrix}$$

## 7) Zero or null matrix

This is a matrix in which all the elements are zero; it is denoted by **0**. Examples are given below for a $2 \times 2$, $3 \times 4$, and $4 \times 4$ matrix.

$$A = \begin{bmatrix} 0 & 0 \\ 0 & 0 \end{bmatrix} \quad B = \begin{bmatrix} 0 & 0 & 0 & 0 \\ 0 & 0 & 0 & 0 \\ 0 & 0 & 0 & 0 \end{bmatrix} \quad C = \begin{bmatrix} 0 & 0 & 0 & 0 \\ 0 & 0 & 0 & 0 \\ 0 & 0 & 0 & 0 \\ 0 & 0 & 0 & 0 \end{bmatrix}$$

Although **A**, **B**, and **C** are null matrices, nevertheless, they are not equal, i.e., $A \neq B \neq C$, because they are dimensionally unequal. Pre- and post-multiplication of a matrix by a null matrix results in a null matrix, provided they are compatible for multiplication. This is like multiplying a number by digit zero. However, unlike in arithmetic, two non-zero matrices can be multiplied to get a zero matrix. Although unusual, we will see an example of this later in the worked examples.

## 8) Equal matrices

Two or more matrices are said to be equal if:

a) they have the same order, and
b) all their corresponding elements are equal.

For example, given matrices $A$ and $B$ such that $A = (4,6; 2,9)$ and $B = (4,6; 2,9)$, then we can say that $A = B$.

## 9) Singular matrix

This is a matrix whose determinant is zero. We will shortly discuss how to find the determinant of a matrix, but for now, it suffices to know that when the determinant of a matrix is zero it is called a singular matrix.

## 10) Transpose matrix

The transpose of a matrix is formed when the rows have been converted to respective columns and vice versa (i.e., rows and columns are interchanged). In this case, given a matrix **A** with a dimension of $m \times n$, the transpose of this is another matrix **B** with a dimension of $n \times m$. As matrix **B** is connected to **A**, it is written as:

$$\boxed{B = A^T} \quad \text{OR} \quad \boxed{B = A'} \quad \text{OR} \quad \boxed{B = \tilde{A}} \tag{10.1}$$

It can be shown that:

$$\boxed{(A \pm B)^T = A^T \pm B^T} \tag{10.2}$$

In other words, when two matrices are added/subtracted and then transposed or transposed and then added/subtracted, the resulting matrix will be the same. The principle is also the same for multiplication as shown below.

$$\boxed{(AB)^T = B^T A^T} \tag{10.3}$$

It is important to note that the order of multiplication is interchanged in the above. We will prove these identities in the worked examples.

If a square matrix is equal to its transpose, the matrix is *symmetric*.

$$\boxed{A = A^T} \quad \text{OR} \quad \boxed{a_{ij} = a_{ji}} \tag{10.4}$$

Examples include:

$$A = \begin{bmatrix} 7 & 4 \\ 4 & -5 \end{bmatrix} \quad \text{AND} \quad A^T = \begin{bmatrix} 7 & 4 \\ 4 & -5 \end{bmatrix}$$

$$B = \begin{bmatrix} 3 & 6 & -7 \\ 6 & -4 & 8 \\ -7 & 8 & 5 \end{bmatrix} \quad \text{AND} \quad B^T = \begin{bmatrix} 3 & 6 & -7 \\ 6 & -4 & 8 \\ -7 & 8 & 5 \end{bmatrix}$$

The principal diagonal acts as the line of symmetry or mirror line. On the other hand, if a square matrix is equal to its transpose multiplied by $k = -1$, the matrix is called a *skew-symmetric* matrix.

$$\boxed{A = -A^T} \quad \text{OR} \quad \boxed{a_{ij} = -a_{ji}} \tag{10.5}$$

Matrices **F** and **G** below are examples of *skew-symmetric* matrices.

$$F = \begin{bmatrix} 0 & -6 & -2 \\ 6 & 0 & -5 \\ 2 & 5 & 0 \end{bmatrix} \text{ and } G = \begin{bmatrix} 0 & 5 & 7 & 0 \\ -5 & 0 & 1 & 9 \\ -7 & -1 & 0 & 3 \\ 0 & -9 & -3 & 0 \end{bmatrix}$$

# Matrices

Let's pause here and try some examples.

## Example 1

Write the transpose of the following matrices.

a) $A = \begin{bmatrix} -2 \\ 5 \end{bmatrix}$
b) $B = \begin{bmatrix} 2.5 \\ -1.6 \\ 3.7 \end{bmatrix}$
c) $C = \begin{bmatrix} 0.3 & -1.2 \\ -0.6 & 0.5 \\ 2.1 & -0.7 \end{bmatrix}$
d) $D = \begin{bmatrix} 2 & 4 & -6 \\ 3 & 5 & 7 \\ -4 & 9 & -16 \end{bmatrix}$

What did you get? Find the solution below to double-check your answer.

## Solution to Example 1

a) $A^T = \begin{bmatrix} -2 & 5 \end{bmatrix}$
b) $B^T = \begin{bmatrix} 2.5 & -1.6 & 3.7 \end{bmatrix}$

c) $C^T = \begin{bmatrix} 0.3 & -0.6 & 2.1 \\ -1.2 & 0.5 & -0.7 \end{bmatrix}$
d) $D^T = \begin{bmatrix} 2 & 3 & -4 \\ 4 & 5 & 9 \\ -6 & 7 & -16 \end{bmatrix}$

## 10.4 MATRIX OPERATION

In this section, we will look at the basic operation of matrices, which includes addition, subtraction, and multiplication. The division of a matrix with a scalar is the same as multiplying with a fraction (or a reciprocal of the scalar). As for dividing a matrix with another matrix, this is not something that is explicitly carried out, but its concept is used in inverse matrices and when solving simultaneous equations, which will be covered later in this chapter.

### 10.4.1 ADDITION

Addition of two or more matrices is carried out by adding their corresponding elements, provided that their dimensions are the same, otherwise the operation is impossible. Notice that addition is both commutative and associative (i.e., the order of the operation is irrelevant) as shown below:

$$\boxed{A + B = B + A} \quad \text{AND} \quad \boxed{(A + B) + C = A + (B + C)} \tag{10.6}$$

Let's try some examples here.

## Example 2

Matrices **A**, **B**, and **C** are given as:

$$A = \begin{bmatrix} 1 & 3 \\ -2 & 6 \end{bmatrix} \quad B = \begin{bmatrix} 7 & -5 \\ 0 & 1 \end{bmatrix} \quad C = \begin{bmatrix} 7 & -6 \end{bmatrix}$$

Determine:

a) $A + B$             b) $A + C$             c) $B + C$

What did you get? Find the solution below to double-check your answer.

**Solution to Example 2**

a) $A + B$
**Solution**

$$A + B = \begin{bmatrix} 1 & 3 \\ -2 & 6 \end{bmatrix} + \begin{bmatrix} 7 & -5 \\ 0 & 1 \end{bmatrix}$$

$$= \begin{bmatrix} 1+7 & 3-5 \\ -2+0 & 6+1 \end{bmatrix}$$

$$\therefore A + B = \begin{bmatrix} 8 & -2 \\ -2 & 7 \end{bmatrix}$$

b) $A + C$
**Solution**

$$A + C = \begin{bmatrix} 1 & 3 \\ -2 & 6 \end{bmatrix} + \begin{bmatrix} 7 & -6 \end{bmatrix}$$

$$= \textit{Impossible because the dimensions are different}$$

c) $B + C$
**Solution**

$$B + C = \begin{bmatrix} 7 & -5 \\ 0 & 1 \end{bmatrix} + \begin{bmatrix} 7 & -6 \end{bmatrix}$$

$$= \textit{Impossible because the dimensions are not the same}$$

The last two examples are 'impossible' because the dimensions of the matrices are not the same. In other words, the matrices are not similar.

Another example to try.

**Example 3**

Matrices **A** and **B** are given as:

$$A = \begin{bmatrix} -3 & 0 \\ 2 & -1 \end{bmatrix} \quad B = \begin{bmatrix} 10 & -5 \\ 0 & 7 \end{bmatrix}$$

show that $(A + B)^T = A^T + B^T$.

# Matrices

What did you get? Find the solution below to double-check your answer.

## Solution to Example 3

**Solution**
Left-hand side

$$A + B = \begin{bmatrix} -3 & 0 \\ 2 & -1 \end{bmatrix} + \begin{bmatrix} 10 & -5 \\ 0 & 7 \end{bmatrix}$$

$$= \begin{bmatrix} -3+10 & 0-5 \\ 2+0 & -1+7 \end{bmatrix} = \begin{bmatrix} 7 & -5 \\ 2 & 6 \end{bmatrix}$$

$$\therefore (A + B)^T = \begin{bmatrix} 7 & 2 \\ -5 & 6 \end{bmatrix}$$

Right-hand side

$$A^T = \begin{bmatrix} -3 & 2 \\ 0 & -1 \end{bmatrix}$$

$$B^T = \begin{bmatrix} 10 & 0 \\ -5 & 7 \end{bmatrix}$$

$$A^T + B^T = \begin{bmatrix} -3 & 2 \\ 0 & -1 \end{bmatrix} + \begin{bmatrix} 10 & 0 \\ -5 & 7 \end{bmatrix}$$

$$= \begin{bmatrix} -3+10 & 2+0 \\ 0-5 & -1+7 \end{bmatrix}$$

$$\therefore A^T + B^T = \begin{bmatrix} 7 & 2 \\ -5 & 6 \end{bmatrix}$$

Thus

$$(A + B)^T = A^T + B^T$$

## 10.4.2 Subtraction

Same as addition, a matrix can only be subtracted from another if they have the same dimension.

Let's try some examples.

### Example 4

Matrices **A**, **B**, and **C** are given as:

$$A = \begin{bmatrix} 2 & -5 & 4 \\ -10 & 0 & 8 \end{bmatrix} \quad B = \begin{bmatrix} 2 & 9 \\ 0 & 12 \end{bmatrix} \quad C = \begin{bmatrix} -7 & 5 & 1 \\ 4 & 3 & 2 \end{bmatrix}$$

Determine:

a) $B - A$  
b) $A - C$  
c) $B - B$

What did you get? Find the solution below to double-check your answer.

**Solution to Example 4**

a) $B - A$
**Solution**

$$B - A = \begin{bmatrix} 2 & 9 \\ 0 & 12 \end{bmatrix} - \begin{bmatrix} 2 & -5 & 4 \\ -10 & 0 & 8 \end{bmatrix}$$
$$= Impossible$$

This is because they have different dimensions.

b) $A - C$
**Solution**

$$A - C = \begin{bmatrix} 2 & -5 & 4 \\ -10 & 0 & 8 \end{bmatrix} - \begin{bmatrix} -7 & 5 & 1 \\ 4 & 3 & 2 \end{bmatrix}$$
$$= \begin{bmatrix} 2-(-7) & -5-5 & 4-1 \\ -10-4 & 0-3 & 8-2 \end{bmatrix}$$
$$\therefore A - C = \begin{bmatrix} 9 & -10 & 3 \\ -14 & -3 & 6 \end{bmatrix}$$

c) $B - B$
**Solution**

$$B - B = \begin{bmatrix} 2 & 9 \\ 0 & 12 \end{bmatrix} - \begin{bmatrix} 2 & 9 \\ 0 & 12 \end{bmatrix}$$
$$= \begin{bmatrix} 2-2 & 9-9 \\ 0-0 & 12-12 \end{bmatrix}$$
$$\therefore B - B = \begin{bmatrix} 0 & 0 \\ 0 & 0 \end{bmatrix}$$

This is a zero or null matrix.

## 10.4.3 Scalar-Matrix Multiplication

When a matrix is multiplied by a scalar, each element is multiplied by this scalar factor. For example, given a 3 by 4 matrix **A**:

$$A = \begin{bmatrix} a_{11} & a_{12} & a_{13} & a_{14} \\ a_{21} & a_{22} & a_{23} & a_{24} \\ a_{31} & a_{32} & a_{33} & a_{34} \end{bmatrix}$$

If **A** is multiplied by a scalar factor $k$, where $k$ is any real number, the operation can be written as:

$$k \begin{bmatrix} a_{11} & a_{12} & a_{13} & a_{14} \\ a_{21} & a_{22} & a_{23} & a_{24} \\ a_{31} & a_{32} & a_{33} & a_{34} \end{bmatrix} = \begin{bmatrix} ka_{11} & ka_{12} & ka_{13} & ka_{14} \\ ka_{21} & ka_{22} & ka_{23} & ka_{24} \\ ka_{31} & ka_{32} & ka_{33} & ka_{34} \end{bmatrix}$$

# Matrices

The above operation is like 'expanding the brackets', and can be written in a short format as:

$$k[a_{ij}] = [ka_{ij}]$$

The reverse process can be performed. In other words, if there is a common factor among elements in a matrix, this can be taken out or 'factorised'.

$$[ka_{ij}] = k[a_{ij}]$$

$$\begin{bmatrix} ka_{11} & ka_{12} & ka_{13} & ka_{14} \\ ka_{21} & ka_{22} & ka_{23} & ka_{24} \\ ka_{31} & ka_{32} & ka_{33} & ka_{34} \end{bmatrix} = k \begin{bmatrix} a_{11} & a_{12} & a_{13} & a_{14} \\ a_{21} & a_{22} & a_{23} & a_{24} \\ a_{31} & a_{32} & a_{33} & a_{34} \end{bmatrix}$$

Let's try some examples.

### Example 5

A 3 by 4 matrix **A** is given as:

$$\begin{bmatrix} 3 & -2 & 5 & -1 \\ 1 & 0 & 2 & 4 \\ 0 & 7 & 6 & 8 \end{bmatrix}$$

Determine the following:

a) $3A$ 　　　　　　　　　　b) $-0.5A$ 　　　　　　　　　　c) $2\pi A$

What did you get? Find the solution below to double-check your answer.

### Solution to Example 5

**HINT**

In this example, we've tried to use parentheses () instead of square brackets [], as either can be used to express matrices.

**a) $3A$**
**Solution**

$$3A = 3 \begin{pmatrix} 3 & -2 & 5 & -1 \\ 1 & 0 & 2 & 4 \\ 0 & 7 & 6 & 8 \end{pmatrix}$$

$$= \begin{pmatrix} 3 \times 3 & 3 \times -2 & 3 \times 5 & 3 \times -1 \\ 3 \times 1 & 3 \times 0 & 3 \times 2 & 3 \times 4 \\ 3 \times 0 & 3 \times 7 & 3 \times 6 & 3 \times 8 \end{pmatrix}$$

$$\therefore 3A = \begin{pmatrix} 9 & -6 & 15 & -3 \\ 3 & 0 & 6 & 12 \\ 0 & 21 & 18 & 24 \end{pmatrix}$$

**b)** $-0.5A$
**Solution**

$$-0.5A = -0.5 \begin{bmatrix} 3 & -2 & 5 & -1 \\ 1 & 0 & 2 & 4 \\ 0 & 7 & 6 & 8 \end{bmatrix}$$

$$= \begin{bmatrix} -0.5 \times 3 & -0.5 \times -2 & -0.5 \times 5 & -0.5 \times -1 \\ -0.5 \times 1 & -0.5 \times 0 & -0.5 \times 2 & -0.5 \times 4 \\ -0.5 \times 0 & -0.5 \times 7 & -0.5 \times 6 & -0.5 \times 8 \end{bmatrix}$$

$$\therefore -0.5A = \begin{pmatrix} -1.5 & 1 & -2.5 & 0.5 \\ -0.5 & 0 & -1 & -2 \\ 0 & -3.5 & -3 & -4 \end{pmatrix}$$

---

**c)** $2\pi A$
**Solution**

$$2\pi A = 2\pi \begin{pmatrix} 3 & -2 & 5 & -1 \\ 1 & 0 & 2 & 4 \\ 0 & 7 & 6 & 8 \end{pmatrix}$$

$$= \begin{pmatrix} 2\pi \times 3 & 2\pi \times -2 & 2\pi \times 5 & 2\pi \times -1 \\ 2\pi \times 1 & 2\pi \times 0 & 2\pi \times 2 & 2\pi \times 4 \\ 2\pi \times 0 & 2\pi \times 7 & 2\pi \times 6 & 2\pi \times 8 \end{pmatrix}$$

$$\therefore 2\pi A = \begin{pmatrix} 6\pi & -4\pi & 10\pi & -2\pi \\ 2\pi & 0 & 4\pi & 8\pi \\ 0 & 14\pi & 12\pi & 16\pi \end{pmatrix}$$

---

Another example to try.

### Example 6

A 3 by 2 matrix B is given as:

$$\begin{bmatrix} 4 & -3 \\ 1 & 0 \\ -6 & 10 \end{bmatrix}$$

Express matrix **B** in the form of $k[a_{ij}]$, where $k = 2$ and $[a_{ij}]$ is a 3 by 2 matrix.

# Matrices

What did you get? Find the solution below to double-check your answer.

**Solution to Example 6**

$$B = \begin{bmatrix} 4 & -3 \\ 1 & 0 \\ -6 & 10 \end{bmatrix} = 2 \begin{bmatrix} \frac{4}{2} & -\frac{3}{2} \\ \frac{1}{2} & \frac{0}{2} \\ -\frac{6}{2} & \frac{10}{2} \end{bmatrix}$$

$$\therefore B = 2 \begin{bmatrix} 2 & -1.5 \\ 0.5 & 0 \\ -3 & 5 \end{bmatrix}$$

One final example to try in this section.

**Example 7**

Matrices **A**, **B**, and **C** are given as:

$$A = \begin{bmatrix} 2 & 1 & 0 \\ -3 & -2 & -1 \\ 2 & 0 & 4 \end{bmatrix} \quad B = \begin{bmatrix} 1 & 0 & 0 \\ -1 & 2 & 0 \\ -3 & -2 & 3 \end{bmatrix} \quad C = \begin{bmatrix} 1 & 3 & 5 \\ 0 & 1 & 7 \\ -5 & 0 & 9 \end{bmatrix}$$

Determine $3A + 2B - 5C$.

What did you get? Find the solution below to double-check your answer.

**Solution to Example 7**

Let's evaluate each term separately first, thus we have

$$3A = 3 \begin{bmatrix} 2 & 1 & 0 \\ -3 & -2 & -1 \\ 2 & 0 & 4 \end{bmatrix}$$

$$= \begin{bmatrix} 2 \times 3 & 1 \times 3 & 0 \times 3 \\ -3 \times 3 & -2 \times 3 & -1 \times 3 \\ 2 \times 3 & 0 \times 3 & 4 \times 3 \end{bmatrix}$$

$$= \begin{bmatrix} 6 & 3 & 0 \\ -9 & -6 & -3 \\ 6 & 0 & 12 \end{bmatrix}$$

$$2B = 2\begin{bmatrix} 1 & 0 & 0 \\ -1 & 2 & 0 \\ -3 & -2 & 3 \end{bmatrix}$$

$$= \begin{bmatrix} 1\times 2 & 0\times 2 & 0\times 2 \\ -1\times 2 & 2\times 2 & 0\times 2 \\ -3\times 2 & -2\times 2 & 3\times 2 \end{bmatrix}$$

$$= \begin{bmatrix} 2 & 0 & 0 \\ -2 & 4 & 0 \\ -6 & -4 & 6 \end{bmatrix}$$

$$5C = 5\begin{bmatrix} 1 & 3 & 5 \\ 0 & 1 & 7 \\ -5 & 0 & 9 \end{bmatrix}$$

$$= \begin{bmatrix} 1\times 5 & 3\times 5 & 5\times 5 \\ 0\times 5 & 1\times 5 & 7\times 5 \\ -5\times 5 & 0\times 5 & 9\times 5 \end{bmatrix}$$

$$= \begin{bmatrix} 5 & 15 & 25 \\ 0 & 5 & 35 \\ -25 & 0 & 45 \end{bmatrix}$$

Therefore

$$3A + 2B - 5C = 3\begin{bmatrix} 2 & 1 & 0 \\ -3 & -2 & -1 \\ 2 & 0 & 4 \end{bmatrix} + 2\begin{bmatrix} 1 & 0 & 0 \\ -1 & 2 & 0 \\ -3 & -2 & 3 \end{bmatrix} - 5\begin{bmatrix} 1 & 3 & 5 \\ 0 & 1 & 7 \\ -5 & 0 & 9 \end{bmatrix}$$

$$= \begin{bmatrix} 6 & 3 & 0 \\ -9 & -6 & -3 \\ 6 & 0 & 12 \end{bmatrix} + \begin{bmatrix} 2 & 0 & 0 \\ -2 & 4 & 0 \\ -6 & -4 & 6 \end{bmatrix} - \begin{bmatrix} 5 & 15 & 25 \\ 0 & 5 & 35 \\ -25 & 0 & 45 \end{bmatrix}$$

$$= \begin{bmatrix} (6+2-5) & (3+0-15) & (0+0-25) \\ (-9-2-0) & (-6+4-5) & (-3+0-35) \\ (6-6+25) & (0-4-0) & (12+6-45) \end{bmatrix}$$

$$= \begin{bmatrix} 3 & -12 & -25 \\ -11 & -7 & -38 \\ 25 & -4 & -27 \end{bmatrix}$$

$$\therefore 3A + 2B - 5C = \begin{bmatrix} 3 & -12 & -25 \\ -11 & -7 & -38 \\ 25 & -4 & -27 \end{bmatrix}$$

---

### 10.4.4 MATRIX-MATRIX MULTIPLICATION

Multiplying a matrix with another matrix has certain characteristics or 'compatibility requirements' that need to be noted and fulfilled. Let's say we want to multiply a matrix **A** by matrix **B**. If **A** is an $m \times n$ matrix and **B** is a $p \times q$ matrix, then we say:

**Note 1** $A \times B$ is not the same as $B \times A$ (i.e., $A \times B \neq B \times A$). In other words, multiplication is not commutative.

# Matrices

**Note 2** In $A \times B$, we say **A** pre-multiply **B** (or it is the first in the sequence) while **B** is said to post-multiply **A**. It is therefore important to know which one comes first.

**Note 3** For the operation $A \times B$ to be possible, we must have $n = p$. This means that the number of columns in matrix **A** must be equal to the number of rows in matrix **B**. Alternatively, we can say that the inner dimensions must be the same for both.

**Note 4** Operation $A \times B$ will result in a new matrix with a dimension of $m \times q$.

**Note 5** If both $A \times B$ and $B \times A$ are possible, it does not imply that $A \times B = B \times A$. In fact, this is a rare occurrence. Matrices A and B are said to commute if their product is commutative, i.e., $AB = BA$.

**Note 6** For the operation $B \times A$ to be possible, then $q = m$. In other words, the number of columns in matrix B must be equal to the number of rows in matrix A. Operation $B \times A$ will result in a new matrix with a dimension $p \times n$.

In summary, order is essential in multiplication. Provided the operation is valid, the following identities can be used in multiplication:

**Identity 1** Associative

$$\boxed{A(BC) = (AB)C} \tag{10.7}$$

**Identity 2** Distributive

$$\boxed{A(B+C) = AB + AC} \quad \text{AND} \quad \boxed{(A+B)(C+D) = AC + AD + BC + BD} \tag{10.8}$$

Before we proceed to examples, let's provide a few generic examples:

1) A compatible row matrix with a column matrix

$$\begin{bmatrix} a & b \end{bmatrix} \times \begin{bmatrix} c \\ d \end{bmatrix} = [ac + bd]$$

$$\begin{bmatrix} a & b & c \end{bmatrix} \times \begin{bmatrix} d \\ e \\ f \end{bmatrix} = [ad + be + cf]$$

2) A compatible row matrix with a square matrix

$$\begin{bmatrix} a & b \end{bmatrix} \times \begin{bmatrix} c & d \\ e & f \end{bmatrix} = \begin{bmatrix} \begin{bmatrix} a & b \end{bmatrix}\begin{bmatrix} c \\ e \end{bmatrix} & \begin{bmatrix} a & b \end{bmatrix}\begin{bmatrix} d \\ f \end{bmatrix} \end{bmatrix}$$
$$= \begin{bmatrix} ac + be & ad + bf \end{bmatrix}$$

$$\begin{bmatrix} a & b & c \end{bmatrix} \times \begin{bmatrix} d_1 & d_2 \\ e_1 & e_2 \\ f_1 & f_2 \end{bmatrix} = \begin{bmatrix} \begin{bmatrix} a & b & c \end{bmatrix}\begin{bmatrix} d_1 \\ e_1 \\ f_1 \end{bmatrix} & \begin{bmatrix} a & b & c \end{bmatrix}\begin{bmatrix} d_2 \\ e_2 \\ f_2 \end{bmatrix} \end{bmatrix}$$
$$= \begin{bmatrix} ad_1 + be_1 + cf_1 & ad_2 + be_2 + cf_2 \end{bmatrix}$$

3) A 2 by 2 matrix with another 2 by 2 matrix

$$\begin{bmatrix} a & b \\ c & d \end{bmatrix} \times \begin{bmatrix} p & r \\ s & t \end{bmatrix} = \begin{bmatrix} \begin{bmatrix} a & b \end{bmatrix}\begin{bmatrix} p \\ s \end{bmatrix} & \begin{bmatrix} a & b \end{bmatrix}\begin{bmatrix} r \\ t \end{bmatrix} \\ \begin{bmatrix} c & d \end{bmatrix}\begin{bmatrix} p \\ s \end{bmatrix} & \begin{bmatrix} c & d \end{bmatrix}\begin{bmatrix} r \\ t \end{bmatrix} \end{bmatrix}$$

$$= \begin{bmatrix} ap + bs & ar + bt \\ cp + ds & cr + dt \end{bmatrix}$$

That's not a straightforward operation, right? It requires more practice than addition and subtraction for one to get used to it. We will now look at a few examples.

### Example 8

Two matrices **A** and **B** are given below:

$$A = \begin{bmatrix} 2 & 1 \\ 3 & 5 \end{bmatrix} \quad B = \begin{bmatrix} -4 \\ 6 \end{bmatrix}$$

a) Determine $AB$  
b) Determine $BA$  
c) Given that I is a 2 by 2 matrix, show that $AI = IA = A$.

---

What did you get? Find the solution below to double-check your answer.

### Solution to Example 8

**a)** $AB$
**Solution**
**A** is $2 \times 2$ and **B** is $2 \times 1$. Hence, the number of columns in **A** is equal to the number of rows in **B**, so the operation is possible. The resulting matrix will be $2 \times 1$.

$$AB = \begin{bmatrix} 2 & 1 \\ 3 & 5 \end{bmatrix} \times \begin{bmatrix} -4 \\ 6 \end{bmatrix} = \begin{bmatrix} (2 \times -4) + (1 \times 6) \\ (3 \times -4) + (5 \times 6) \end{bmatrix}$$

$$= \begin{bmatrix} (-8) + (6) \\ (-12) + (30) \end{bmatrix}$$

$$\therefore AB = \begin{bmatrix} -2 \\ 18 \end{bmatrix}$$

---

**b)** $BA$
**Solution**

$$BA = \begin{bmatrix} -4 \\ 6 \end{bmatrix} \times \begin{bmatrix} 2 & 1 \\ 3 & 5 \end{bmatrix}$$

The number of columns in matrix **B** is not equal to the number of rows in **A**. Therefore, the operation $BA$ is not possible.

# Matrices

**c)** $AI = IA = A$
**Solution**
2 by 2 identity matrix is
$$\begin{bmatrix} 1 & 0 \\ 0 & 1 \end{bmatrix}$$

Thus:
$$AI = \begin{bmatrix} 2 & 1 \\ 3 & 5 \end{bmatrix} \times \begin{bmatrix} 1 & 0 \\ 0 & 1 \end{bmatrix}$$
$$= \begin{bmatrix} (2 \times 1) + (1 \times 0) & (2 \times 0) + (1 \times 1) \\ (3 \times 1) + (5 \times 0) & (3 \times 0) + (5 \times 1) \end{bmatrix}$$
$$= \begin{bmatrix} 2+0 & 0+1 \\ 3+0 & 0+5 \end{bmatrix}$$
$$= \begin{bmatrix} 2 & 1 \\ 3 & 5 \end{bmatrix} = A$$

Similarly,
$$IA = \begin{bmatrix} 1 & 0 \\ 0 & 1 \end{bmatrix} \times \begin{bmatrix} 2 & 1 \\ 3 & 5 \end{bmatrix}$$
$$= \begin{bmatrix} (1 \times 2) + (0 \times 3) & (1 \times 1) + (0 \times 5) \\ (0 \times 2) + (1 \times 3) & (0 \times 1) + (1 \times 5) \end{bmatrix}$$
$$= \begin{bmatrix} 2+0 & 1+0 \\ 0+3 & 0+5 \end{bmatrix}$$
$$= \begin{bmatrix} 2 & 1 \\ 3 & 5 \end{bmatrix} = A$$

Therefore,
$$AI = IA = A$$

Another example to try.

## Example 9

Matrices **A** and **B** are given as:
$$A = \begin{bmatrix} -3 & 0 \\ 2 & -1 \end{bmatrix}, B = \begin{bmatrix} 10 & -5 \\ 0 & 7 \end{bmatrix}$$

show that $(AB)^T = B^T A^T$.

What did you get? Find the solution below to double-check your answer.

### Solution to Example 9

Left-hand side

$$AB = \begin{bmatrix} -3 & 0 \\ 2 & -1 \end{bmatrix} \begin{bmatrix} 10 & -5 \\ 0 & 7 \end{bmatrix}$$

$$= \begin{bmatrix} (-3 \times 10 + 0 \times 0) & (-3 \times -5 + 0 \times 7) \\ (2 \times 10 + -1 \times 0) & (2 \times -5 + -1 \times 7) \end{bmatrix}$$

$$= \begin{bmatrix} -30 + 0 & 15 + 0 \\ 20 + 0 & -10 - 7 \end{bmatrix} = \begin{bmatrix} -30 & 15 \\ 20 & -17 \end{bmatrix}$$

$$\therefore (AB)^T = \begin{bmatrix} -30 & 20 \\ 15 & -17 \end{bmatrix}$$

Right-hand side

$$A^T = \begin{bmatrix} -3 & 2 \\ 0 & -1 \end{bmatrix}$$

$$B^T = \begin{bmatrix} 10 & 0 \\ -5 & 7 \end{bmatrix}$$

$$B^T A^T = \begin{bmatrix} 10 & 0 \\ -5 & 7 \end{bmatrix} \begin{bmatrix} -3 & 2 \\ 0 & -1 \end{bmatrix}$$

$$= \begin{bmatrix} (10 \times -3 + 0 \times 0) & (10 \times 2 + 0 \times -1) \\ (-5 \times -3 + 7 \times 0) & (-5 \times 2 + 7 \times -1) \end{bmatrix}$$

$$= \begin{bmatrix} -30 + 0 & 20 + 0 \\ 15 + 0 & -10 - 7 \end{bmatrix} = \begin{bmatrix} -30 & 20 \\ 15 & -17 \end{bmatrix}$$

$$\therefore B^T A^T = \begin{bmatrix} -30 & 20 \\ 15 & -17 \end{bmatrix}$$

Thus

$$(AB)^T = B^T A^T$$

Another example to try.

### Example 10

Two matrices **C**, $3 \times 2$, and **D**, $2 \times 3$, are:

$$C = \begin{bmatrix} 1 & -2 \\ -3 & 0 \\ 0 & 4 \end{bmatrix}, D = \begin{bmatrix} -1 & 0 & 2 \\ 5 & 3 & 0 \end{bmatrix}$$

Determine:

a) $CD$         b) $DC$

# Matrices

What did you get? Find the solution below to double-check your answer.

**Solution to Example 10**

**a)** $CD$
**Solution**
$C$ is $3 \times 2$ and $D$ is $2 \times 3$. Hence the number of columns in $C$ is equal to the number of rows in $D$, so the operation is possible. The resulting matrix will be 3 by 3.

$$CD = \begin{bmatrix} 1 & -2 \\ -3 & 0 \\ 0 & 4 \end{bmatrix} \times \begin{bmatrix} -1 & 0 & 2 \\ 5 & 3 & 0 \end{bmatrix}$$

$$= \begin{bmatrix} \{(1 \times -1) + (-2 \times 5)\} & \{(1 \times 0) + (-2 \times 3)\} & \{(1 \times 2) + (-2 \times 0)\} \\ \{(-3 \times -1) + (0 \times 5)\} & \{(-3 \times 0) + (0 \times 3)\} & \{(-3 \times 2) + (0 \times 0)\} \\ \{(0 \times -1) + (4 \times 5)\} & \{(0 \times 0) + (4 \times 3)\} & \{(0 \times 2) + (4 \times 0)\} \end{bmatrix}$$

$$= \begin{bmatrix} \{(-1) + (-10)\} & \{(0) + (-6)\} & \{(2) + (0)\} \\ \{(3) + (0)\} & \{(0) + (0)\} & \{(-6) + (0)\} \\ \{(0) + (20)\} & \{(0) + (12)\} & \{(0) + (0)\} \end{bmatrix}$$

$$\therefore CD = \begin{bmatrix} -11 & -6 & 2 \\ 3 & 0 & -6 \\ 20 & 12 & 0 \end{bmatrix}$$

**b)** $DC$
**Solution**
The number of columns in matrix $D$ is equal to the number of rows in $C$. The operation $DC$ is therefore possible. The resulting matrix will be 2 by 2.

$$DC = \begin{bmatrix} -1 & 0 & 2 \\ 5 & 3 & 0 \end{bmatrix} \times \begin{bmatrix} 1 & -2 \\ -3 & 0 \\ 0 & 4 \end{bmatrix}$$

$$= \begin{bmatrix} \{(-1 \times 1) + (0 \times -3) + (2 \times 0)\} & \{(-1 \times -2) + (0 \times 0) + (2 \times 4)\} \\ \{(5 \times 1) + (3 \times -3) + (0 \times 0)\} & \{(5 \times -2) + (3 \times 0) + (0 \times 4)\} \end{bmatrix}$$

$$= \begin{bmatrix} \{(-1) + (0) + (0)\} & \{(2) + (0) + (8)\} \\ \{(5) + (-9) + (0)\} & \{(-10) + (0) + (0)\} \end{bmatrix}$$

$$= \begin{bmatrix} (-1 + 0 + 0) & (2 + 0 + 8) \\ (5 - 9 + 0) & (-10 + 0 + 0) \end{bmatrix}$$

$$\therefore DC = \begin{bmatrix} -1 & 10 \\ -4 & -10 \end{bmatrix}$$

In this example, both options are possible, but the resulting matrices have different dimensions: 3 by 3 and 2 by 2.

Let's try another example.

### Example 11

Two $2 \times 2$ matrices, **E** and **F**, are given below:

$$E = \begin{bmatrix} -4 & 2 \\ 3 & -1 \end{bmatrix} \qquad F = \begin{bmatrix} 1 & 0 \\ -3 & 2 \end{bmatrix}$$

Determine:

a) $EF$  
b) $FE$

What did you get? Find the solution below to double-check your answer.

### Solution to Example 11

**a)** $EF$
**Solution**
**E** is $2 \times 2$ and **F** is $2 \times 2$. Hence the number of columns in **E** is equal to the number of rows in **F**, so the operation is possible. The resulting matrix will be $2 \times 2$.

$$EF = \begin{bmatrix} -4 & 2 \\ 3 & -1 \end{bmatrix} \times \begin{bmatrix} 1 & 0 \\ -3 & 2 \end{bmatrix}$$

$$= \begin{bmatrix} \{(-4 \times 1) + (2 \times -3)\} & \{(-4 \times 0) + (2 \times 2)\} \\ \{(3 \times 1) + (-1 \times -3)\} & \{(3 \times 0) + (-1 \times 2)\} \end{bmatrix}$$

$$= \begin{bmatrix} \{(-4) + (-6)\} & \{(0) + (4)\} \\ \{(3) + (3)\} & \{(0) + (-2)\} \end{bmatrix} = \begin{bmatrix} -4-6 & 0+4 \\ 3+3 & 0-2 \end{bmatrix}$$

$$\therefore EF = \begin{bmatrix} -10 & 4 \\ 6 & -2 \end{bmatrix}$$

**b)** $FE$
**Solution**
The number of columns in matrix **F** is equal to the number of rows in **E**. The operation **FE** is therefore possible. The resulting matrix will be $2 \times 2$.

$$FE = \begin{bmatrix} 1 & 0 \\ -3 & 2 \end{bmatrix} \times \begin{bmatrix} -4 & 2 \\ 3 & -1 \end{bmatrix}$$

$$= \begin{bmatrix} \{(1 \times -4) + (0 \times 3)\} & \{(1 \times 2) + (0 \times -1)\} \\ \{(-3 \times -4) + (2 \times 3)\} & \{(-3 \times 2) + (2 \times -1)\} \end{bmatrix}$$

$$= \begin{bmatrix} \{(-4) + (0)\} & \{(2) + (0)\} \\ \{(12) + (6)\} & \{(-6) + (-2)\} \end{bmatrix} = \begin{bmatrix} -4+0 & 2+0 \\ 12+6 & -6-2 \end{bmatrix}$$

$$\therefore FE = \begin{bmatrix} -4 & 2 \\ 18 & -8 \end{bmatrix}$$

# Matrices

Again, in this example, both options are possible, and the resulting matrices have the same size (2 by 2) but not the same elements, i.e., $F \times E \neq E \times F$.

## Example 12

Determine $A \times B$, such that:

$$A = \begin{bmatrix} 3 & 6 & 2 \\ 0 & 5 & 1 \\ 7 & 0 & -4 \end{bmatrix}, \quad B = \begin{bmatrix} 2 \\ -1 \\ 3 \end{bmatrix}$$

What did you get? Find the solution below to double-check your answer.

## Solution to Example 12

$$A \times B = \begin{bmatrix} 3 & 6 & 2 \\ 0 & 5 & 1 \\ 7 & 0 & -4 \end{bmatrix} \times \begin{bmatrix} 2 \\ -1 \\ 3 \end{bmatrix}$$

$$= \begin{bmatrix} (3 \times 2) + (6 \times -1) + (2 \times 3) \\ (0 \times 2) + (5 \times -1) + (1 \times 3) \\ (7 \times 2) + (0 \times -1) + (-4 \times 3) \end{bmatrix}$$

$$= \begin{bmatrix} (6) + (-6) + (6) \\ (0) + (-5) + (3) \\ (14) + (0) + (-12) \end{bmatrix}$$

$$= \begin{bmatrix} 6 \\ -2 \\ 2 \end{bmatrix}$$

$$\therefore A \times B = \begin{bmatrix} 6 \\ -2 \\ 2 \end{bmatrix}$$

One final example.

## Example 13

Three matrices **A**, **B**, and **C** are such that when **A** pre-multiply **B**, the result is **C**. Determine the value of elements $x$ and $y$ given that **A**, **B**, and **C** are:

$$A = \begin{bmatrix} 1 & 2 \\ 3 & -1 \end{bmatrix} \quad B = \begin{bmatrix} x \\ y \end{bmatrix} \quad C = \begin{bmatrix} -3 \\ 5 \end{bmatrix}$$

What did you get? Find the solution below to double-check your answer.

**Solution to Example 13**

The matrix equation is

$$AB = C$$

But

$$AB = \begin{bmatrix} 1 & 2 \\ 3 & -1 \end{bmatrix} \times \begin{bmatrix} x \\ y \end{bmatrix}$$

$$= \begin{bmatrix} (1 \times x) + (2 \times y) \\ (3 \times x) + (-1 \times y) \end{bmatrix}$$

$$= \begin{bmatrix} x + 2y \\ 3x - y \end{bmatrix}$$

Therefore

$$\begin{bmatrix} x + 2y \\ 3x - y \end{bmatrix} = \begin{bmatrix} -3 \\ 5 \end{bmatrix}$$

In this case, the corresponding element must be equal

$$x + 2y = -3 \quad -----(i)$$

and

$$3x - y = 5 \quad -----(ii)$$

Solving these equations simultaneously, we have

$$x = 1, \quad y = -2$$

## 10.5 DETERMINANT AND ITS PROPERTIES

The determinant of a matrix $A$, denoted as $|A|$ (or $\Delta$), is a number (scalar) that is calculated using the elements of $A$. This can only be computed for a square matrix.

Let's look at three categories of evaluating determinants.

### 10.5.1 SINGLE MATRIX (1 BY 1)

The determinant of a single matrix, which can be regarded as a 1 by 1 square matrix, is the same as its single element. For example, determinants of the following matrices:

$$A = [-2], \ B = [3], \ C = [0.6]$$

are

$$|A| = -2, \ |B| = 3, \ |C| = 0.6$$

## 10.5.2 Two-by-Two Matrix

The determinant of a 2 by 2 (or second-order) matrix is the product of the two elements on the leading diagonal minus the product of the two elements on the other diagonal. This is better illustrated as follows, using matrix $A$ such that:

$$A = \begin{bmatrix} a_{11} & a_{12} \\ a_{21} & a_{22} \end{bmatrix}$$

The determinant of $A$ is:

$$A = \begin{bmatrix} a_{11} & a_{12} \\ a_{21} & a_{22} \end{bmatrix}$$

$$\therefore |A| = (a_{11} \times a_{22}) - (a_{12} \times a_{21})$$

Let's look at a few examples.

### Example 14

Using the following matrices:

$$A = \begin{bmatrix} 10 & 20 \\ 15 & 5 \end{bmatrix}, B = \begin{bmatrix} 3 & -5 \\ 7 & -6 \end{bmatrix}, C = \begin{bmatrix} 0.5 & 1.2 \\ 0.8 & 2.2 \end{bmatrix}, D = \begin{bmatrix} j & -j \\ j-x & x \end{bmatrix},$$

$$E = \begin{bmatrix} 3\angle\left(\frac{1}{2}\pi\right) & 5\angle\left(-\frac{1}{3}\pi\right) \\ 2\angle\left(\frac{5}{6}\pi\right) & \sqrt{2}\angle\left(-\frac{1}{2}\pi\right) \end{bmatrix}$$

compute the determinant of:

a) $A$     b) $B$     c) $C$     d) $D$     e) $E$

What did you get? Find the solution below to double-check your answer.

### Solution to Example 14

**a)** $|A|$
**Solution**

$$|A| = \begin{vmatrix} 10 & 20 \\ 15 & 5 \end{vmatrix}$$
$$= (10 \times 5) - (20 \times 15)$$
$$= 50 - 300$$
$$\therefore |A| = -250$$

**b) $|B|$**
**Solution**

$$|B| = \begin{vmatrix} 3 & -5 \\ 7 & -6 \end{vmatrix}$$

$$= (3 \times -6) - (-5 \times 7)$$

$$= -18 - (-35)$$

$$\therefore |B| = 17$$

---

**c) $|C|$**
**Solution**

$$|C| = \begin{vmatrix} 0.5 & 1.2 \\ 0.8 & 2.2 \end{vmatrix}$$

$$= (0.5 \times 2.2) - (1.2 \times 0.8)$$

$$= 1.1 - 0.96$$

$$\therefore |C| = 0.14$$

---

**d) $|D|$**
**Solution**

$$|D| = \begin{vmatrix} j & -j \\ j-x & x \end{vmatrix}$$

$$= (j \times x) - (-j)(j-x)$$

$$= jx + j(j-x)$$

$$= jx + j^2 - jx$$

$$= j^2 = -1$$

$$\therefore |D| = -1$$

---

**e) $|E|$**
**Solution**

$$|E| = \begin{vmatrix} 3\angle\left(\frac{1}{2}\pi\right) & 5\angle\left(-\frac{1}{3}\pi\right) \\ 2\angle\left(\frac{5}{6}\pi\right) & \sqrt{2}\angle\left(-\frac{1}{2}\pi\right) \end{vmatrix}$$

$$= \left(3\angle\left(\frac{1}{2}\pi\right)\right)\left(\sqrt{2}\angle\left(-\frac{\pi}{2}\right)\right) - \left(5\angle\left(-\frac{1}{3}\pi\right)\right)\left(2\angle\left(\frac{5}{6}\pi\right)\right)$$

$$= \left(3 \times \sqrt{2}\right)\angle\left(\frac{1}{2}\pi - \frac{1}{2}\pi\right) - (5 \times 2)\angle\left(-\frac{1}{3}\pi + \frac{5}{6}\pi\right)$$

$$= \left(3\sqrt{2}\right)\angle(0) - (10)\angle\left(\frac{1}{2}\pi\right)$$

$$= 3\sqrt{2} - j10$$

$$\therefore |E| = 3\sqrt{2} - j10$$

### 10.5.3 Three-by-Three Matrix

Obtaining the determinant of a 3 by 3 and other higher square matrices is not as straightforward as that of a 2 by 2. To find the determinant of a 3 by 3, we need to follow these steps. Let's illustrate this using matrix **A**, given as:

$$A = \begin{bmatrix} a_{11} & a_{12} & a_{13} \\ a_{21} & a_{22} & a_{23} \\ a_{31} & a_{32} & a_{33} \end{bmatrix}$$

**Step 1:** Starting from the first element in the first row, assign positive (or +1) and then negative (or −1) and alternate this as you move across the first row. Continue this pattern to row 2 and then row 3. This is called the place value, and is illustrated below.

$$A = \begin{bmatrix} +a_{11} & -a_{12} & +a_{13} \\ -a_{21} & +a_{22} & -a_{23} \\ +a_{31} & -a_{32} & +a_{33} \end{bmatrix}$$

**Step 2:** Choose any row or column as your reference. It is customary to choose the first row. We will therefore choose row 1.

$$A = \begin{bmatrix} +a_{11} & -a_{12} & +a_{13} \\ -a_{21} & +a_{22} & -a_{23} \\ +a_{31} & -a_{32} & +a_{33} \end{bmatrix}$$

**Step 3:** Start with the first (or the very far left) element of the chosen row (i.e., row 1). Cross out all the elements in the column and row containing the first chosen element, and do the same for the two other elements in the same column or row. This is illustrated below:

$$|A| = a_{11} \begin{bmatrix} a_{11} & a_{12} & a_{13} \\ a_{21} & a_{22} & a_{23} \\ a_{31} & a_{32} & a_{33} \end{bmatrix} - a_{12} \begin{bmatrix} a_{11} & a_{12} & a_{13} \\ a_{21} & a_{22} & a_{23} \\ a_{31} & a_{32} & a_{33} \end{bmatrix} + a_{13} \begin{bmatrix} a_{11} & a_{12} & a_{13} \\ a_{21} & a_{22} & a_{23} \\ a_{31} & a_{32} & a_{33} \end{bmatrix}$$

**Step 4:** Once crossed out, there should be four elements left. This will give a nice 2 by 2 matrix, which is called the minor of the chosen element.

$$|A| = a_{11} \begin{vmatrix} a_{22} & a_{23} \\ a_{32} & a_{33} \end{vmatrix} - a_{12} \begin{vmatrix} a_{21} & a_{23} \\ a_{31} & a_{33} \end{vmatrix} + a_{13} \begin{vmatrix} a_{21} & a_{22} \\ a_{31} & a_{32} \end{vmatrix}$$

**Step 5:** Find the determinant of this matrix. Multiply each element by the place sign/value and compute the determinant of the newly formed 2 by 2 matrices. We have explained how to find the determinant of a 2 by 2 matrix above.

$$|A| = a_{11}(a_{22}a_{33} - a_{23}a_{32}) - a_{12}(a_{21}a_{33} - a_{23}a_{31}) + a_{13}(a_{21}a_{32} - a_{22}a_{31})$$

**Step 6:** Find the summation of these, which gives the determinant of the 3 by 3 matrix.

We've chosen row 1 to illustrate the determinant. Let's do the same with column 1 and summarise the steps quickly as:

$$|A| = a_{11}\begin{bmatrix} a_{11} & a_{12} & a_{13} \\ a_{21} & a_{22} & a_{23} \\ a_{31} & a_{32} & a_{33} \end{bmatrix} - a_{21}\begin{bmatrix} a_{11} & a_{12} & a_{13} \\ a_{21} & a_{22} & a_{23} \\ a_{31} & a_{32} & a_{33} \end{bmatrix} + a_{31}\begin{bmatrix} a_{11} & a_{12} & a_{13} \\ a_{21} & a_{22} & a_{23} \\ a_{31} & a_{32} & a_{33} \end{bmatrix}$$

$$|A| = a_{11}\begin{vmatrix} a_{22} & a_{23} \\ a_{32} & a_{33} \end{vmatrix} - a_{21}\begin{vmatrix} a_{12} & a_{13} \\ a_{32} & a_{33} \end{vmatrix} + a_{31}\begin{vmatrix} a_{12} & a_{13} \\ a_{22} & a_{23} \end{vmatrix}$$

$$|A| = a_{11}(a_{22}a_{33} - a_{23}a_{32}) - a_{21}(a_{12}a_{33} - a_{13}a_{32}) + a_{31}(a_{12}a_{23} - a_{13}a_{22})$$

With 3 by 3's, it means that there are six possible manipulation options. We have only illustrated two using the first row and first column. Others are second and third rows, and second and third columns, and they will be shown in the worked examples.

That's all about 3 by 3 determinant. Let's illustrate with examples.

### Example 15

A matrix **A** is given as:

$$A = \begin{bmatrix} 2 & -5 & 3 \\ 1 & -3 & 0 \\ 0 & 4 & -7 \end{bmatrix}$$

Find the determinant of A using all the six possible options.

What did you get? Find the solution below to double-check your answer.

### Solution to Example 15

**a) Using the first row**
**Solution**

$$|A| = 2\begin{bmatrix} 2 & -5 & 3 \\ 1 & -3 & 0 \\ 0 & 4 & -7 \end{bmatrix} - 5\begin{bmatrix} 2 & -5 & 3 \\ 1 & -3 & 0 \\ 0 & 4 & -7 \end{bmatrix} + 3\begin{bmatrix} 2 & -5 & 3 \\ 1 & -3 & 0 \\ 0 & 4 & -7 \end{bmatrix}$$

$$|A| = 2\begin{vmatrix} -3 & 0 \\ 4 & -7 \end{vmatrix} + 5\begin{vmatrix} 1 & 0 \\ 0 & -7 \end{vmatrix} + 3\begin{vmatrix} 1 & -3 \\ 0 & 4 \end{vmatrix}$$

$$|A| = 2\{(-3 \times -7) - (0 \times 4)\} + 5\{(1 \times -7) - (0 \times 0)\} + 3\{(1 \times 4) - (-3 \times 0)\}$$

$$|A| = 2(21 - 0) + 5(-7 - 0) + 3(4 - 0)$$

$$|A| = 2(21) + 5(-7) + 3(4)$$

$$= 42 - 35 + 12$$

$$|A| = 19$$

# Matrices

**b)** Using the first column
**Solution**

$$|A| = 2\begin{bmatrix} 2 & -5 & 3 \\  & -3 & 0 \\ 0 & 4 & -7 \end{bmatrix} - 1\begin{bmatrix} 2 & -5 & 3 \\  & -3 & 0 \\ 0 & 4 & -7 \end{bmatrix} + 0\begin{bmatrix} 2 & -5 & 3 \\  & -3 & 0 \\ 0 & 4 & -7 \end{bmatrix}$$

$$|A| = 2\begin{vmatrix} -3 & 0 \\ 4 & -7 \end{vmatrix} - 1\begin{vmatrix} -5 & 3 \\ 4 & -7 \end{vmatrix} + 0\begin{vmatrix} -5 & 3 \\ -3 & 0 \end{vmatrix}$$

$|A| = 2\{(-3 \times -7) - (0 \times 4)\} - 1\{(-5 \times -7) - (3 \times 4)\} + 0\{(-5 \times 0) - (3 \times -3)\}$

$|A| = 2(21 - 0) - 1(35 - 12) + 0(0 + 9)$

$|A| = 2(21) - 1(23) + 0(9)$

$\phantom{|A|} = 42 - 23 + 0$

$\therefore |A| = 19$

---

**c)** Using the second row
**Solution**

$$|A| = -1\begin{bmatrix} 2 & -5 & 3 \\  & -3 & 0 \\ 0 & 4 & -7 \end{bmatrix} - 3\begin{bmatrix} 2 & -5 & 3 \\  & -3 & 0 \\ 0 & 4 & -7 \end{bmatrix} - 0\begin{bmatrix} 2 & -5 & 3 \\  & -3 & 0 \\ 0 & 4 & -7 \end{bmatrix}$$

$$|A| = -1\begin{vmatrix} -5 & 3 \\ 4 & -7 \end{vmatrix} - 3\begin{vmatrix} 2 & 3 \\ 0 & -7 \end{vmatrix} - 0\begin{vmatrix} 2 & -5 \\ 0 & 4 \end{vmatrix}$$

$|A| = -1\{(-5 \times -7) - (3 \times 4)\} - 3\{(2 \times -7) - (3 \times 0)\} - 0\{(2 \times 4) - (-5 \times 0)\}$

$|A| = -1(35 - 12) - 3(-14 - 0) - 0(8 - 0)$

$|A| = -1(23) - 3(-14) - 0(8)$

$\phantom{|A|} = -23 + 42 - 0$

$\therefore |A| = 19$

---

**d)** Using the second column
**Solution**

$$|A| = 5\begin{bmatrix} 2 & -5 & 3 \\ 1 & -3 & 0 \\ 0 & 4 & -7 \end{bmatrix} - 3\begin{bmatrix} 2 & -5 & 3 \\ 1 & -3 & 0 \\ 0 & 4 & -7 \end{bmatrix} - 4\begin{bmatrix} 2 & -5 & 3 \\ 1 & -3 & 0 \\ 0 & 4 & -7 \end{bmatrix}$$

$$|A| = 5\begin{vmatrix} 1 & 0 \\ 0 & -7 \end{vmatrix} - 3\begin{vmatrix} 2 & 3 \\ 0 & -7 \end{vmatrix} - 4\begin{vmatrix} 2 & 3 \\ 1 & 0 \end{vmatrix}$$

$|A| = 5\{(1 \times -7) - (0 \times 0)\} - 3\{(2 \times -7) - (3 \times 0)\} - 4\{(2 \times 0) - (3 \times 1)\}$

$|A| = 5(-7 - 0) - 3(-14 - 0) - 4(0 - 3)$

$|A| = 5(-7) - 3(-14) - 4(-3)$

$\phantom{|A|} = -35 + 42 + 12$

$\therefore |A| = 19$

**e) Using the third row**
**Solution**

$$|A| = 0 \begin{bmatrix} 2 & -5 & 3 \\ 1 & -3 & 0 \\ 0 & 4 & -7 \end{bmatrix} - 4 \begin{bmatrix} 2 & -5 & 3 \\ 1 & -3 & 0 \\ 0 & 4 & -7 \end{bmatrix} - 7 \begin{bmatrix} 2 & -5 & 3 \\ 1 & -3 & 0 \\ 0 & 4 & -7 \end{bmatrix}$$

$$|A| = 0 \begin{vmatrix} -5 & 3 \\ -3 & 0 \end{vmatrix} - 4 \begin{vmatrix} 2 & 3 \\ 1 & 0 \end{vmatrix} - 7 \begin{vmatrix} 2 & -5 \\ 1 & -3 \end{vmatrix}$$

$$|A| = 0\{(-5 \times 0) - (3 \times -3)\} - 4\{(2 \times 0) - (3 \times 1)\} - 7\{(2 \times -3) - (-5 \times 1)\}$$

$$|A| = 0(0+9) - 4(0-3) - 7(-6+5)$$

$$|A| = 0(9) - 4(-3) - 7(-1)$$

$$= 0 + 12 + 7$$

$$\therefore |A| = 19$$

**f) Using the third column**
**Solution**

$$|A| = 3 \begin{bmatrix} 2 & -5 & 3 \\ 1 & -3 & 0 \\ 0 & 4 & -7 \end{bmatrix} - 0 \begin{bmatrix} 2 & -5 & 3 \\ 1 & -3 & 0 \\ 0 & 4 & -7 \end{bmatrix} - 7 \begin{bmatrix} 2 & -5 & 3 \\ 1 & -3 & 0 \\ 0 & 4 & -7 \end{bmatrix}$$

$$|A| = 3 \begin{vmatrix} 1 & -3 \\ 0 & 4 \end{vmatrix} - 0 \begin{vmatrix} 2 & -5 \\ 0 & 4 \end{vmatrix} - 7 \begin{vmatrix} 2 & -5 \\ 1 & -3 \end{vmatrix}$$

$$|A| = 3\{(1 \times 4) - (-3 \times 0)\} - 0\{(2 \times 4) - (-5 \times 0)\} - 7\{(2 \times -3) - (-5 \times 1)\}$$

$$|A| = 3(4-0) - 0(8-0) - 7(-6+5)$$

$$|A| = 3(4) - 0(8) - 7(-1)$$

$$= 12 - 0 + 7$$

$$\therefore |A| = 19$$

### 10.5.4 Determinants of a Higher-Order Matrix

For higher-size matrices (i.e., 4 by 4 or higher), it is obvious that the process will be cumbersome, and this will be better computed using software. However, let's illustrate this for a 4 by 4 matrix **B**, and this is the furthest we can attempt here in this chapter. Given a matrix **B** as:

$$B = \begin{bmatrix} b_{11} & b_{12} & b_{13} & b_{14} \\ b_{21} & b_{22} & b_{23} & b_{24} \\ b_{31} & b_{32} & b_{33} & b_{34} \\ b_{41} & b_{42} & b_{43} & b_{44} \end{bmatrix}$$

We will follow the steps highlighted for a 3 by 3 above. Let's assign place value or sign.

$$B = \begin{bmatrix} +b_{11} & -b_{12} & +b_{13} & -b_{14} \\ -b_{21} & +b_{22} & -b_{23} & +b_{24} \\ +b_{31} & -b_{32} & +b_{33} & -b_{34} \\ -b_{41} & +b_{42} & -b_{43} & +b_{44} \end{bmatrix}$$

Matrices

Notice that the 'signing' process looks different. We started the first row with a positive sign, the second with a negative and the next row with a positive sign, this continues to alternate. Choosing the first row (step 2), we can show step 3 as:

$$|B| = b_{11} \begin{bmatrix} \cancel{b_{11}} & \cancel{b_{12}} & \cancel{b_{13}} & \cancel{b_{14}} \\ \cancel{b_{21}} & b_{22} & b_{23} & b_{24} \\ \cancel{b_{31}} & b_{32} & b_{33} & b_{34} \\ \cancel{b_{41}} & b_{42} & b_{43} & b_{44} \end{bmatrix} - b_{12} \begin{bmatrix} \cancel{b_{11}} & \cancel{b_{12}} & \cancel{b_{13}} & \cancel{b_{14}} \\ b_{21} & \cancel{b_{22}} & b_{23} & b_{24} \\ b_{31} & \cancel{b_{32}} & b_{33} & b_{34} \\ b_{41} & \cancel{b_{42}} & b_{43} & b_{44} \end{bmatrix}$$

$$= + b_{13} \begin{bmatrix} \cancel{b_{11}} & \cancel{b_{12}} & \cancel{b_{13}} & \cancel{b_{14}} \\ b_{21} & b_{22} & \cancel{b_{23}} & b_{24} \\ b_{31} & b_{32} & \cancel{b_{33}} & b_{34} \\ b_{41} & b_{42} & \cancel{b_{43}} & b_{44} \end{bmatrix} - b_{14} \begin{bmatrix} \cancel{b_{11}} & \cancel{b_{12}} & \cancel{b_{13}} & \cancel{b_{14}} \\ b_{21} & b_{22} & b_{23} & \cancel{b_{24}} \\ b_{31} & b_{32} & b_{33} & \cancel{b_{34}} \\ b_{41} & b_{42} & b_{43} & \cancel{b_{44}} \end{bmatrix}$$

$$|B| = b_{11} \begin{vmatrix} b_{22} & b_{23} & b_{24} \\ b_{32} & b_{33} & b_{34} \\ b_{42} & b_{43} & b_{44} \end{vmatrix} - b_{12} \begin{vmatrix} b_{21} & b_{23} & b_{24} \\ b_{31} & b_{33} & b_{34} \\ b_{41} & b_{43} & b_{44} \end{vmatrix} + b_{13} \begin{vmatrix} b_{21} & b_{22} & b_{24} \\ b_{31} & b_{32} & b_{34} \\ b_{41} & b_{42} & b_{44} \end{vmatrix} - b_{14} \begin{vmatrix} b_{21} & b_{22} & b_{23} \\ b_{31} & b_{32} & b_{33} \\ b_{41} & b_{42} & b_{43} \end{vmatrix}$$

It is obvious before applying step 5 that we need to repeat steps 1 to 3 to evaluate the determinant of the newly formed 3 by 3. This will be a long-winded process, but it's possible.

### 10.5.5 Properties of Determinants

To finish this section, it is important to look at the key properties of a determinant, which will allow us to simplify complex matrix-based problems easily and efficiently. These include:

**Property 1** If all the elements in a particular row or column are zeros, then the determinant of the matrix is zero. For example, the determinants of the following are zero.

$$A = \begin{bmatrix} 0 & 12 \\ 0 & 13 \end{bmatrix} \quad B = \begin{bmatrix} 0 & 0 & 0 \\ 11 & -15 & 17 \\ -10 & 40 & 20 \end{bmatrix} \quad C = \begin{bmatrix} 0.1 & -0.2 & 0 \\ 1.8 & 1.5 & 0 \\ -0.8 & 4 & 0 \end{bmatrix} \quad D = \begin{bmatrix} 5 & 15 & -10 & 25 \\ 0 & 0 & 0 & 0 \\ -6 & 12 & -8 & 4 \\ 21 & 11 & 9 & 7 \end{bmatrix}$$

**Property 2** Given $k$ to be a real number, if the elements of any row or column are $k$ multiple of corresponding elements in another row or column, the determinant will be zero. Matrices **D**, **E**, and **F** below exhibit that property and therefore have zero determinants.

$$D = \begin{bmatrix} -2 & 3 \\ 4 & -6 \end{bmatrix}, \quad E = \begin{bmatrix} 10 & -6 & 18 \\ -14 & 8 & 12 \\ 5 & -3 & 9 \end{bmatrix}, \quad F = \begin{bmatrix} -2.5 & 1.5 & 0.7 & 2.5 \\ 0.2 & 0.5 & 1.2 & -0.2 \\ -0.4 & 1.4 & -0.8 & 0.4 \\ -1.4 & 1.1 & 0.9 & -1.4 \end{bmatrix}$$

- In **D**, column 2 is $-1.5$ of column 1, i.e., $k = -1.5$.
- In **E**, row 1 is double of row 3.
- In **F**, column 4 is $-1$ of column 1.

An extension to this is that if elements in any row (or column) are related to corresponding elements in another row (or column), the determinant will be zero. An example:

$$G = \begin{bmatrix} 4 & 0 & 8 \\ 7 & 3 & 11 \\ -5 & 2 & -12 \end{bmatrix}$$

In the above example, matrix $G$ is such that the elements in column 3 are formed by first doubling the corresponding elements in column 1 and then subtracting the corresponding elements in column 2.

**Property 3** If two rows (or columns) are identical, the determinant is zero. Example:

$$\begin{vmatrix} 2 & -5 & 7 \\ -6 & 3 & 8 \\ 2 & -5 & 7 \end{vmatrix} = 0$$

This is a special case of the second property where $k = 1$.

**Property 4** The determinant of a matrix is equal to the determinant of its transpose.

$$|A| = |A^T|$$

In the example below, $|A| = |B|$ because $B$ is the transpose of $A$ and vice versa.

$$A = \begin{bmatrix} 1 & 2 & 3 \\ 4 & 5 & 6 \\ 7 & 8 & 9 \end{bmatrix}, B = \begin{bmatrix} 1 & 4 & 7 \\ 2 & 5 & 8 \\ 3 & 6 & 9 \end{bmatrix}$$

It is important to add that the determinant of the product of two matrices, $A$ and $B$, is equal to the product of their individual determinant, provided $A$ and $B$ are square matrices with the same dimension. In other words,

$$\boxed{|AB| = |A||B| = |B||A|} \tag{10.9}$$

**Property 5** If two rows (or columns) are swap, the sign of the determinant will change from positive to negative and vice versa. Example:

$$\begin{vmatrix} 1 & 4 & 7 \\ 2 & 5 & 8 \\ 3 & 6 & 9 \end{vmatrix} = -\begin{vmatrix} 2 & 5 & 8 \\ 1 & 4 & 7 \\ 3 & 6 & 9 \end{vmatrix} = -\begin{vmatrix} 4 & 1 & 7 \\ 5 & 2 & 8 \\ 6 & 3 & 9 \end{vmatrix}$$

**Property 6** If all the element of a particular row (or column) is multiplied by a factor, the determinant is multiplied by the same factor. Example:

$$\begin{vmatrix} -3 & 12 \\ 6 & 0 \end{vmatrix} = 3\begin{vmatrix} -1 & 4 \\ 6 & 0 \end{vmatrix} = 3\begin{vmatrix} -1 & 12 \\ 2 & 0 \end{vmatrix}$$

# Matrices

## 10.6 INVERSE OF A SQUARE MATRIX AND ITS PROPERTIES

The inverse of a matrix **A** is written as $A^{-1}$ and pronounced as 'inverse A' or 'inverse of **A**' and not '**A** power −1' or 'per **A**'. In general, when a matrix is pre- or post-multiplied by its inverse, the product is an identity matrix of the same size. In other words,

$$AA^{-1} = A^{-1}A = I \qquad (10.10)$$

The following identities are also valid:

**Identity 1** The transpose of an inverse matrix is equal to the inverse of its transpose.

$$\left(A^{-1}\right)^T = \left(A^T\right)^{-1} \qquad (10.11)$$

To perform inversion and transposition on a particular matrix, the sequence is not essential, i.e., inverse and then transpose or transpose and then inverse the matrix.

**Identity 2** The inverse of the product of matrices is equal to the product of the matrices, but in reverse order.

$$(ABC)^{-1} = C^{-1}B^{-1}A^{-1} \qquad (10.12)$$

**Identity 3** The determinant of an inverse matrix is equal to the inverse of its determinant.

$$|A^{-1}| = |A|^{-1} \qquad (10.13)$$

Note that inverse matrix can only be determined for square and non-singular matrices; that is, non-square matrices have no inverses. Here, we will discuss two types, namely $2 \times 2$ and $3 \times 3$.

### 10.6.1 INVERSE OF A 2 × 2 MATRIX

There is a special formula for finding the inverse of a $2 \times 2$ matrix. Let us consider a matrix **A**, which is $2 \times 2$ as:

$$\begin{bmatrix} a & b \\ c & d \end{bmatrix} \qquad (10.14)$$

The inverse of **A**, written as $A^{-1}$, is given as:

$$A^{-1} = \frac{1}{ad - bc} \begin{bmatrix} d & -b \\ -c & a \end{bmatrix} \qquad (10.15)$$

The above shows that to find the inverse of a 2 × 2 matrix, we need the following steps:

**Step 1:** Interchange the two elements in the leading diagonal.
**Step 2:** Change the sign of the other elements.
**Step 3:** Write the new matrix formed by the modification in steps 1 and 2. These two steps can be remembered as '**Swap Sign**'.
**Step 4:** Find the determinant of the original matrix.
**Step 5:** Divide the modified matrix by the determinant of the original matrix.
**Step 6:** The resulting matrix from step 5 is the inverse of the original 2 × 2 matrix.

Let's try some examples.

## Example 16

Using the following matrices:

$$A = \begin{bmatrix} 1 & -3 \\ -2 & 5 \end{bmatrix} \quad B = \begin{bmatrix} 4 & 2 \\ 6 & 3 \end{bmatrix}$$

a) Find the inverse of $A$ or $A^{-1}$
b) Find $B^{-1}$
c) Show that $A^{-1}A = AA^{-1} = I$
d) Show that $(A^{-1})^T = (A^T)^{-1}$

What did you get? Find the solution below to double-check your answer.

## Solution to Example 16

**a)** $A^{-1}$
**Solution**

$$|A| = \begin{vmatrix} 1 & -3 \\ -2 & 5 \end{vmatrix}$$
$$= (1 \times 5) - (-3 \times -2)$$
$$= 5 - 6$$
$$\therefore |A| = -1$$

Therefore,

$$A^{-1} = \frac{\begin{bmatrix} 5 & 3 \\ 2 & 1 \end{bmatrix}}{-1}$$

$$\therefore A^{-1} = -\begin{bmatrix} 5 & 3 \\ 2 & 1 \end{bmatrix} = \begin{bmatrix} -5 & -3 \\ -2 & -1 \end{bmatrix}$$

## b) $B^{-1}$
**Solution**

$$|B| = \begin{vmatrix} 4 & 2 \\ 6 & 3 \end{vmatrix}$$
$$= (4 \times 3) - (2 \times 6)$$
$$= 12 - 12$$
$$\therefore |B| = 0$$

Therefore,

$$B^{-1} = \frac{\begin{bmatrix} 3 & -2 \\ -6 & 4 \end{bmatrix}}{0}$$
$$= No\ inverse$$

A matrix with a zero determinant is called a singular matrix; they do not have an inverse. Obviously, this is because $\frac{1}{0}$ is infinity.

---

## c) $A^{-1}A = AA^{-1} = I$
**Solution**
2 by 2 identity matrix is

$$\begin{bmatrix} 1 & 0 \\ 0 & 1 \end{bmatrix}$$

$$AA^{-1} = \begin{bmatrix} 1 & -3 \\ -2 & 5 \end{bmatrix} \times \begin{bmatrix} -5 & -3 \\ -2 & -1 \end{bmatrix}$$
$$= \begin{bmatrix} (1 \times -5) + (-3 \times -2) & (1 \times -3) + (-3 \times -1) \\ (-2 \times -5) + (5 \times -2) & (-2 \times -3) + (5 \times -1) \end{bmatrix}$$
$$= \begin{bmatrix} -5+6 & -3+3 \\ 10-10 & 6-5 \end{bmatrix}$$
$$= \begin{bmatrix} 1 & 0 \\ 0 & 1 \end{bmatrix} = I$$

Similarly,

$$A^{-1}A = \begin{bmatrix} -5 & -3 \\ -2 & -1 \end{bmatrix} \times \begin{bmatrix} 1 & -3 \\ -2 & 5 \end{bmatrix}$$
$$= \begin{bmatrix} (-5 \times 1) + (-3 \times -2) & (-5 \times -3) + (-3 \times 5) \\ (-2 \times 1) + (-1 \times -2) & (-2 \times -3) + (-1 \times 5) \end{bmatrix}$$
$$= \begin{bmatrix} -5+6 & 15-15 \\ -2+2 & 6-5 \end{bmatrix}$$
$$= \begin{bmatrix} 1 & 0 \\ 0 & 1 \end{bmatrix} = I$$

Therefore,
$$A^{-1}A = AA^{-1} = I$$

**d)** $\left(A^{-1}\right)^{T} = \left(A^{T}\right)^{-1}$
**Solution**

$$A^{-1} = \begin{bmatrix} -5 & -3 \\ -2 & -1 \end{bmatrix}$$

$$\therefore \left(A^{-1}\right)^{T} = \begin{bmatrix} -5 & -2 \\ -3 & -1 \end{bmatrix}$$

Also

$$A = \begin{bmatrix} 1 & -3 \\ -2 & 5 \end{bmatrix}$$

$$A^{T} = \begin{bmatrix} 1 & -2 \\ -3 & 5 \end{bmatrix}$$

$$|A^{T}| = \begin{vmatrix} 1 & -2 \\ -3 & 5 \end{vmatrix}$$

$$= (1 \times 5) - (-2 \times -3) = 5 - 6$$

$$= -1$$

In fact

$$|A^{T}| = |A| = -1$$

Therefore

$$\left(A^{T}\right)^{-1} = \frac{\begin{bmatrix} 5 & 2 \\ 3 & 1 \end{bmatrix}}{-1}$$

$$= -\begin{bmatrix} 5 & 2 \\ 3 & 1 \end{bmatrix} = \begin{bmatrix} -5 & -3 \\ -2 & -1 \end{bmatrix}$$

Hence,

$$\left|A^{-1}\right|^{T} = \left|A^{-T}\right|^{-1}$$

## 10.6.2 Inverse of any Square Matrix

Finding the inverse of higher-order matrices is similar to what was done when finding their determinants. We shall illustrate this using a 3 by 3 matrix. To proceed, we need to introduce and define a few terms using a 3 by 3 matrix **A** below.

$$A = \begin{bmatrix} 3 & -2 & 1 \\ 0 & 5 & 4 \\ -1 & 7 & 2 \end{bmatrix}$$

# Matrices

## 1) Minor

In matrix **A**, each element has a minor associated with it. The minor of each element is the determinant of the matrix that can be formed when the row and the column of the chosen element are crossed out. Remember that this is the element in question. Let's do this for element $a_{22}$, which is **5**.

**Step 1:** Cross out the row and column belonging to $a_{22} = 5$.

$$\begin{bmatrix} 3 & \cancel{-2} & 1 \\ \cancel{0} & \cancel{5} & \cancel{4} \\ -1 & \cancel{7} & 2 \end{bmatrix}$$

**Step 2:** Form a matrix from the left-over and let this be called **M**.

$$\therefore M = \begin{bmatrix} 3 & 1 \\ -1 & 2 \end{bmatrix}$$

**Step 3:** Find the determinant of M.

$$|M| = (3 \times 2) - (1 \times -1) = 6 + 1$$
$$= 7$$

**Step 4:** Hence the minor of element $a_{22}$ is 7.

## 2) Signed minor

A signed minor, known as a cofactor, is a minor that has been given a place sign value (which is either positive or negative) or has been multiplied by its place sign. There are two ways to get this:

**Option 1** Work out the determinant of matrix **A** by adding the sign value, starting from the top left element as shown below. The sign is shown in brackets.

$$A = \begin{bmatrix} (+)3 & (-)-2 & (+)1 \\ (-)0 & (+)5 & (-)4 \\ (+)-1 & (-)7 & (+)2 \end{bmatrix}$$

It is obvious that the sign value of $a_{22} = 5$ is positive; therefore, its minor remains unchanged as 7 and the cofactor of $a_{22} = 5$.

**Option 2** In this case, multiply the minor by the sign value $(-1)^{i+j}$, where $i$ and $j$ denotes its row and column number, respectively. Since $i = 2$ and $j = 2$ for element 5, thus its signed minor is

$$(-1)^{2+2} \times 7 = (-1)^4 \times 7 = 7$$

We can conclude that if $i + j$ equals an even number, then the signed minor is the same as minor, otherwise change the sign of the minor to get the signed minor.

## 3) Adjoint

It is written as **Adj(A)** for matrix **A**, which is obtained by replacing each of the elements in A with its signed minor (or cofactor), and then transpose the new matrix. The transpose matrix is the adjoint matrix of **A**.

Having known the above terms and how to compute them, we can now state the steps for finding the inverse of any higher-order matrix as follows:

**Step 1:** Find the determinant of the original matrix.
**Step 2:** Find the minors of all the elements of the original matrix.
**Step 3:** Compute their signed minors and replace each by its signed minor.
**Step 4:** Transpose the signed minors matrix to form the Adjoint matrix.
**Step 5:** Divide the Adjoint matrix by the determinant (provided it is not singular). The result is the inverse matrix of the original.

$$\boxed{A^{-1} = \frac{1}{|A|} [Adj\ (A)]} \tag{10.16}$$

It is a long process, but let's try an example.

### Example 17

A 3 by 3 matrix is given as:

$$A = \begin{bmatrix} 3 & -2 & 1 \\ 0 & 5 & 4 \\ -1 & 7 & 2 \end{bmatrix}$$

a) Determine the inverse of matrix A.    b) Show that $A^{-1}A = I$.

What did you get? Find the solution below to double-check your answer.

### Solution to Example 17

a) $A^{-1}$
**Solution**
**Step 1:** Find the determinant of **A**

$$|A| = 3 \begin{vmatrix} 5 & 4 \\ 7 & 2 \end{vmatrix} + 2 \begin{vmatrix} 0 & 4 \\ -1 & 2 \end{vmatrix} + 1 \begin{vmatrix} 0 & 5 \\ -1 & 7 \end{vmatrix}$$

$|A| = 3\{(5 \times 2) - (4 \times 7)\} + 2\{(0 \times 2) - (4 \times -1)\} + 1\{(0 \times 7) - (5 \times -1)\}$
$|A| = 3(10 - 28) + 2(0 + 4) + 1(0 + 5)$
$|A| = 3(-18) + 2(4) + (5) = -54 + 8 + 5$
$|A| = -41$

**Step 2:** Find the signed minor of each element.
$a_{11} = 3$

$$\begin{bmatrix} 3 & -2 & 1 \\ 0 & 5 & 4 \\ -1 & 7 & 2 \end{bmatrix}$$

# Matrices

$$(-1)^{i+j}\begin{vmatrix}5 & 4\\ 7 & 2\end{vmatrix} = (-1)^{1+1}\{(5\times 2)-(4\times 7)\}$$
$$= (-1)^2(10-28)$$
$$= -18$$

$a_{12} = -2$

$$\begin{bmatrix}3 & 2 & 1\\ 0 & 5 & 4\\ -1 & 7 & 2\end{bmatrix}$$

$$(-1)^{i+j}\begin{vmatrix}0 & 4\\ -1 & 2\end{vmatrix} = (-1)^{1+2}\{(0\times 2)-(4\times -1)\}$$
$$= (-1)^3(0+4)$$
$$= -4$$

$a_{13} = 1$

$$\begin{bmatrix}3 & 2 & 1\\ 0 & 5 & 4\\ -1 & 7 & 2\end{bmatrix}$$

$$(-1)^{i+j}\begin{vmatrix}0 & 5\\ -1 & 7\end{vmatrix} = (-1)^{1+3}\{(0\times 7)-(5\times -1)\}$$
$$= (-1)^4(0+5)$$
$$= 5$$

$a_{21} = 0$

$$\begin{bmatrix}3 & -2 & 1\\ 0 & 5 & 4\\ -1 & 7 & 2\end{bmatrix}$$

$$(-1)^{i+j}\begin{vmatrix}-2 & 1\\ 7 & 2\end{vmatrix} = (-1)^{2+1}\{(-2\times 2)-(1\times 7)\}$$
$$= (-1)^3(-4-7)$$
$$= 11$$

$a_{22} = 5$

$$\begin{bmatrix}3 & 2 & 1\\ 0 & 5 & 4\\ -1 & 7 & 2\end{bmatrix}$$

$$(-1)^{i+j}\begin{vmatrix}3 & 1\\ -1 & 2\end{vmatrix} = (-1)^{2+2}\{(3\times 2)-(1\times -1)\}$$
$$= (-1)^4(6+1)$$
$$= 7$$

$a_{23} = 4$

$$\begin{bmatrix} 3 & -2 & 1 \\ 0 & 5 & 4 \\ -1 & 7 & 2 \end{bmatrix}$$

$(-1)^{i+j} \begin{vmatrix} 3 & -2 \\ -1 & 7 \end{vmatrix} = (-1)^{2+3} \{(3 \times 7) - (-2 \times -1)\}$

$\qquad\qquad\qquad = (-1)^5 (21 - 2)$

$\qquad\qquad\qquad = -19$

$a_{31} = -1$

$$\begin{bmatrix} 3 & -2 & 1 \\ 0 & 5 & 4 \\ -1 & 7 & 2 \end{bmatrix}$$

$(-1)^{i+j} \begin{vmatrix} -2 & 1 \\ 5 & 4 \end{vmatrix} = (-1)^{3+1} \{(-2 \times 4) - (1 \times 5)\}$

$\qquad\qquad\qquad = (-1)^4 (-8 - 5)$

$\qquad\qquad\qquad = -13$

$a_{32} = 7$

$$\begin{bmatrix} 3 & -2 & 1 \\ 0 & 5 & 4 \\ -1 & 7 & 2 \end{bmatrix}$$

$(-1)^{i+j} \begin{vmatrix} 3 & 1 \\ 0 & 4 \end{vmatrix} = (-1)^{3+2} \{(3 \times 4) - (1 \times 0)\}$

$\qquad\qquad\qquad = (-1)^5 (12 - 0)$

$\qquad\qquad\qquad = -12$

$a_{33} = 2$

$$\begin{bmatrix} 3 & -2 & 1 \\ 0 & 5 & 4 \\ -1 & 7 & 2 \end{bmatrix}$$

$(-1)^{i+j} \begin{vmatrix} 3 & -2 \\ 0 & 5 \end{vmatrix} = (-1)^{3+3} \{(3 \times 5) - (-2 \times 0)\}$

$\qquad\qquad\qquad = (-1)^6 (15 - 0)$

$\qquad\qquad\qquad = 15$

**Step 3:** Form the matrix of the signed minors

$$\begin{bmatrix} -18 & -4 & 5 \\ 11 & 7 & -19 \\ -13 & -12 & 15 \end{bmatrix}$$

# Matrices

**Step 4:** Transpose the matrix of the signed minors to obtain Adjoint matrix

$$\text{Adjoint} = \begin{bmatrix} -18 & 11 & -13 \\ -4 & 7 & -12 \\ 5 & -19 & 15 \end{bmatrix}$$

**Step 5:** Form the inverse matrix by dividing the Adjoint by the determinant

$$A^{-1} = \frac{1}{-41}\begin{bmatrix} -18 & 11 & -13 \\ -4 & 7 & -12 \\ 5 & -19 & 15 \end{bmatrix}$$

$$= \frac{1}{41}\begin{bmatrix} 18 & -11 & 13 \\ 4 & -7 & 12 \\ -5 & 19 & -15 \end{bmatrix}$$

$$= \begin{bmatrix} 0.4390 & -0.2683 & 0.3171 \\ 0.0976 & -0.1707 & 0.2927 \\ -0.1220 & 0.4634 & -0.3659 \end{bmatrix}$$

**b)** $A^{-1}A = I$
**Solution**

$$A^{-1}A = \frac{1}{41}\begin{bmatrix} 18 & -11 & 13 \\ 4 & -7 & 12 \\ -5 & 19 & -15 \end{bmatrix} \times \begin{bmatrix} 3 & -2 & 1 \\ 0 & 5 & 4 \\ -1 & 7 & 2 \end{bmatrix}$$

$$= \frac{1}{41}\begin{bmatrix} \{(54)+(0)+(-13)\} & \{(-36)+(-55)+(91)\} & \{(18)+(-44)+(26)\} \\ \{(12)+(0)+(-12)\} & \{(-8)+(-35)+(84)\} & \{(4)+(-28)+(24)\} \\ \{(-15)+(0)+(15)\} & \{(10)+(95)+(-105)\} & \{(-5)+(76)+(-30)\} \end{bmatrix}$$

$$= \frac{1}{41}\begin{bmatrix} 41 & 0 & 0 \\ 0 & 41 & 0 \\ 0 & 0 & 41 \end{bmatrix}$$

$$= \begin{bmatrix} \frac{41}{41} & 0 & 0 \\ 0 & \frac{41}{41} & 0 \\ 0 & 0 & \frac{41}{41} \end{bmatrix}$$

$$= \begin{bmatrix} 1 & 0 & 0 \\ 0 & 1 & 0 \\ 0 & 0 & 1 \end{bmatrix} = I$$

$$\therefore A^{-1}A = I$$

## 10.7 SOLVING A SYSTEM OF LINEAR EQUATIONS

Substitution and elimination are well known methods of solving simultaneous equations. However, matrices provide a more robust and versatile approach, especially when the equations involve three or more unknown variables. In this section, we will limit ourselves to two and three equations involving two and three unknown variables, respectively. This can be extended and applied to any number of equations, provided there are as many equations as there are unknown variables, and the equations are linear and consistent.

### 10.7.1 CRAMER'S RULE

This is a useful rule for solving simultaneous equations and we will use it here for two and three equations, but it is valid for any number of equations.

#### 10.7.1.1 Simultaneous Equations in Two Unknowns

Let's say we have two equations as shown below:

$$a_1 x + b_1 y = c_1 \quad \text{(i)}$$
$$a_2 x + b_2 y = c_2 \quad \text{(ii)}$$

where $a_1$ and $a_2$ are the coefficients of the first variable, $b_1$ and $b_2$ the coefficients of the second unknown, and $c_1$ and $c_2$ are the constant terms.

There are a few ways to show this rule; we shall show two options here.

**Option 1** Follow these steps to determine the unknown variables $x$ and $y$

**Step 1:** Ensure that the equations are written in the format above, i.e. equations (i) and (ii).
**Step 2:** Form a 2 by 2 matrix of coefficients of the unknown as seen in the equations. Let's call this matrix **A** for ease of reference.

$$A = \begin{bmatrix} a_1 & b_1 \\ a_2 & b_2 \end{bmatrix}$$

**Step 3:** Find the determinant of **A**, and let's denote this as $\Delta$. Thus:

$$\Delta = \begin{vmatrix} a_1 & b_1 \\ a_2 & b_2 \end{vmatrix} = a_1 b_2 - a_2 b_1$$

**Step 4:** Replace the coefficients of $x$ in **A** with the constant terms. Find the determinant of this and let's denote it as $\Delta_x$. Thus:

$$\Delta_x = \begin{vmatrix} c_1 & b_1 \\ c_2 & b_2 \end{vmatrix} = c_1 b_2 - c_2 b_1$$

**Step 5:** Repeat step 4 for $y$ and let's denote it as $\Delta_y$. Thus:

$$\Delta_y = \begin{vmatrix} a_1 & c_1 \\ a_2 & c_2 \end{vmatrix} = a_1 c_2 - a_2 c_1$$

**Step 6:** To find the unknown variables $x$ and $y$, use:

$$x = \frac{\Delta_x}{\Delta}, \quad y = \frac{\Delta_y}{\Delta}$$

This is generally known as Cramer's rule.

**Option 2** Follow these steps to determine the unknown variables $x$ and $y$

**Step 1:** Ensure that the equations are written in the format above.
**Step 2:** Form a 2 by 2 matrix using the coefficients of the unknown as seen in the equations. Let's call this matrix **A** for ease of reference.

$$A = \begin{bmatrix} a_1 & b_1 \\ a_2 & b_2 \end{bmatrix}$$

**Step 3:** Find the determinant of **A** and let's denote this as $\Delta$. So:

$$\Delta = \begin{vmatrix} a_1 & b_1 \\ a_2 & b_2 \end{vmatrix} = a_1 b_2 - a_2 b_1$$

**Step 4:** Cross out the column containing $x$ and form a matrix of the coefficients and constant terms in the order seen. Find the determinant of this and let's denote it as $\Delta_x$.

$$a_1 x + b_1 y = c_1 \qquad \text{(i)}$$
$$a_2 x + b_2 y = c_2 \qquad \text{(ii)}$$

$$\Delta_x = \begin{vmatrix} b_1 & c_1 \\ b_2 & c_2 \end{vmatrix} = b_1 c_2 - b_2 c_1$$

**Step 5:** Repeat step 4 for $y$ and let's denote it as $\Delta_y$.

$$\Delta_y = \begin{vmatrix} a_1 & c_1 \\ a_2 & c_2 \end{vmatrix} = a_1 c_2 - a_2 c_1$$

**Step 6:** To find the unknown variables $x$ and $y$, use

$$x = -\frac{\Delta_x}{\Delta}, \quad y = \frac{\Delta_y}{\Delta}$$

Note the negative sign when finding the value of $x$ value. We can combine them as:

$$-\frac{x}{\Delta_x} = \frac{y}{\Delta_y} = \frac{1}{\Delta}$$

In another version of this second option, the equations will be arranged as

$$a_1 x + b_1 y + c_1 = 0 \qquad \text{(i)}$$
$$a_2 x + b_2 y + c_2 = 0 \qquad \text{(ii)}$$

before proceeding as above, but the result will be

$$\frac{x}{\Delta_x} = -\frac{y}{\Delta_y} = \frac{1}{\Delta}$$

Notice the change in the position of the negative sign. So in the first instance of this option, we have $(-, +, +)$ but in the revised version we have $(+, -, +)$. Take note of the sequence.

Let's look at an example using Option 1 above.

### Example 18

Using Cramer's rule, solve the simultaneous equations below.

$$x - 3y = -11 \quad \text{(i)}$$
$$4x + 2y = 5 \quad \text{(ii)}$$

What did you get? Find the solution below to double-check your answer.

### Solution to Example 18

$$\Delta = \begin{vmatrix} 1 & -3 \\ 4 & 2 \end{vmatrix} = (1 \times 2) - (4 \times -3) = 2 + 12 = \mathbf{14}$$

$$\Delta_x = \begin{vmatrix} -11 & -3 \\ 5 & 2 \end{vmatrix} = (-11 \times 2) - (5 \times -3) = -22 + 15 = \mathbf{-7}$$

$$\Delta_y = \begin{vmatrix} 1 & -11 \\ 4 & 5 \end{vmatrix} = (1 \times 5) - (4 \times -11) = 5 + 44 = \mathbf{49}$$

Therefore

$$x = \frac{\Delta_x}{\Delta} = \frac{-7}{14}$$
$$\therefore x = -\mathbf{0.5}$$

and

$$y = \frac{\Delta_y}{\Delta} = \frac{49}{14}$$
$$\therefore y = \mathbf{3.5}$$

Another example to try option 2.

### Example 19

Solve the simultaneous equations below using Option 2.

$$2x - 5y - 9 = 0 \quad \text{(i)}$$
$$3x + 7y + 1 = 0 \quad \text{(ii)}$$

# Matrices

What did you get? Find the solution below to double-check your answer.

**Solution to Example 19**

Remember to present the equation in the right format first. Re-arrange the equations as this:

$$2x - 5y = 9 \quad \text{(i)}$$
$$3x + 7y = -1 \quad \text{(ii)}$$

Thus

$$\Delta = \begin{vmatrix} 2 & -5 \\ 3 & 7 \end{vmatrix} = (2 \times 7) - (3 \times -5) = 14 + 15 = \mathbf{29}$$

$$\Delta_x = \begin{vmatrix} -5 & 9 \\ 7 & -1 \end{vmatrix} = (-5 \times -1) - (7 \times 9) = 5 - 63 = \mathbf{-58}$$

$$\Delta_y = \begin{vmatrix} 2 & 9 \\ 3 & -1 \end{vmatrix} = (2 \times -1) - (3 \times 9) = -2 - 27 = \mathbf{-29}$$

Therefore

$$x = -\frac{\Delta_x}{\Delta} = -\frac{-58}{29}$$
$$\therefore x = \mathbf{2}$$

and

$$y = \frac{\Delta_y}{\Delta} = \frac{-29}{29}$$
$$\therefore y = \mathbf{-1}$$

## 10.7.1.2 Simultaneous Equations in Three Unknowns

This is similar to the two equations covered above, except that we have one additional equation. Let's consider:

$$a_1 x + b_1 y + c_1 z = d_1 \quad \text{(i)}$$
$$a_2 x + b_2 y + c_2 z = d_2 \quad \text{(ii)}$$
$$a_3 x + b_3 y + c_3 z = d_3 \quad \text{(iii)}$$

where $a_1$, $a_2$, and $a_3$ are the coefficients of the first variable, $b_1$, $b_2$, and $b_3$ the coefficients of the second unknown, $c_1$, $c_2$, and $c_3$ the coefficients of the third unknown and $d_1$, $d_2$, and $d_3$ are the constant terms.

**Option 1** Follow these steps to determine the unknown variables $x$, $y$, and $z$.

**Step 1:** Ensure that the equations are written in the format above.

**Step 2:** Form a 3 by 3 matrix using the coefficients of the unknown as seen in the equations. Let's call this matrix **C** for ease of reference.

$$C = \begin{bmatrix} a_1 & b_1 & c_1 \\ a_2 & b_2 & c_2 \\ a_3 & b_3 & c_3 \end{bmatrix}$$

**Step 3:** Find the determinant of **C** and let's denote this as $\Delta$. Thus:

$$\Delta = \begin{vmatrix} a_1 & b_1 & c_1 \\ a_2 & b_2 & c_2 \\ a_3 & b_3 & c_3 \end{vmatrix} = a_1 \begin{vmatrix} b_2 & c_2 \\ b_3 & c_3 \end{vmatrix} - b_1 \begin{vmatrix} a_2 & c_2 \\ a_3 & c_3 \end{vmatrix} + c_1 \begin{vmatrix} a_2 & b_2 \\ a_3 & b_3 \end{vmatrix}$$

$$= a_1 (b_2 c_3 - b_3 c_2) - b_1 (a_2 c_3 - a_3 c_2) + c_1 (a_2 b_3 - a_3 b_2)$$

**Step 4:** Replace the coefficients of $x$ in **C** with the constant terms. Find the determinant of this and let's denote it as $\Delta_x$. Thus:

$$\Delta_x = \begin{vmatrix} d_1 & b_1 & c_1 \\ d_2 & b_2 & c_2 \\ d_3 & b_3 & c_3 \end{vmatrix} = d_1 \begin{vmatrix} b_2 & c_2 \\ b_3 & c_3 \end{vmatrix} - b_1 \begin{vmatrix} d_2 & c_2 \\ d_3 & c_3 \end{vmatrix} + c_1 \begin{vmatrix} d_2 & b_2 \\ d_3 & b_3 \end{vmatrix}$$

$$= d_1 (b_2 c_3 - b_3 c_2) - b_1 (d_2 c_3 - d_3 c_2) + c_1 (d_2 b_3 - d_3 b_2)$$

**Step 5:** Repeat step 4 for $y$ and let's denote it as $\Delta_y$. Thus:

$$\Delta_y = \begin{vmatrix} a_1 & d_1 & c_1 \\ a_2 & d_2 & c_2 \\ a_3 & d_3 & c_3 \end{vmatrix} = a_1 \begin{vmatrix} d_2 & c_2 \\ d_3 & c_3 \end{vmatrix} - d_1 \begin{vmatrix} a_2 & c_2 \\ a_3 & c_3 \end{vmatrix} + c_1 \begin{vmatrix} a_2 & d_2 \\ a_3 & d_3 \end{vmatrix}$$

$$= a_1 (d_2 c_3 - d_3 c_2) - d_1 (a_2 c_3 - a_3 c_2) + c_1 (a_2 d_3 - a_3 d_2)$$

**Step 6:** Repeat step 4 for $z$ and let's denote it as $\Delta_z$. Thus:

$$\Delta_z = \begin{vmatrix} a_1 & b_1 & d_1 \\ a_2 & b_2 & d_2 \\ a_3 & b_3 & d_3 \end{vmatrix} = a_1 \begin{vmatrix} b_2 & d_2 \\ b_3 & d_3 \end{vmatrix} - b_1 \begin{vmatrix} a_2 & d_2 \\ a_3 & d_3 \end{vmatrix} + d_1 \begin{vmatrix} a_2 & b_2 \\ a_3 & b_3 \end{vmatrix}$$

$$= a_1 (b_2 d_3 - b_3 d_2) - b_1 (a_2 d_3 - a_3 d_2) + d_1 (a_2 b_3 - a_3 b_2)$$

**Step 7:** To find the unknown variables $x$, $y$, and $z$, use:

$$x = \frac{\Delta_x}{\Delta}, \quad y = \frac{\Delta_y}{\Delta}, \quad z = \frac{\Delta_z}{\Delta}$$

**Option 2** Follow these steps to determine the unknown variables $x$, $y$, and $z$.

**Step 1:** Ensure that the equations are written in the format above.

**Step 2:** Form a 3 by 3 matrix using the coefficients of the unknown as seen in the equations. Let's call this matrix **C** for ease of reference.

# Matrices

$$C = \begin{bmatrix} a_1 & b_1 & c_1 \\ a_2 & b_2 & c_2 \\ a_3 & b_3 & c_3 \end{bmatrix}$$

**Step 3:** Find the determinant of **C** and let's denote this as $\Delta$. Thus:

$$\Delta = \begin{vmatrix} a_1 & b_1 & c_1 \\ a_2 & b_2 & c_2 \\ a_3 & b_3 & c_3 \end{vmatrix} = a_1 \begin{vmatrix} b_2 & c_2 \\ b_3 & c_3 \end{vmatrix} - b_1 \begin{vmatrix} a_2 & c_2 \\ a_3 & c_3 \end{vmatrix} + c_1 \begin{vmatrix} a_2 & b_2 \\ a_3 & b_3 \end{vmatrix}$$

$$= a_1(b_2c_3 - b_3c_2) - b_1(a_2c_3 - a_3c_2) + c_1(a_2b_3 - a_3b_2)$$

**Step 4:** Cross out the column containing $x$ and form a matrix of the coefficients and constant terms in the order seen. Find the determinant of this and let's denote it as $\Delta_x$.

$$\begin{aligned} \cancel{a_1 x} + b_1 y + c_1 z &= d_1 \\ \cancel{a_2 x} + b_2 y + c_2 z &= d_2 \\ \cancel{a_3 x} + b_3 y + c_3 z &= d_3 \end{aligned}$$

$$\Delta_x = \begin{vmatrix} b_1 & c_1 & d_1 \\ b_2 & c_2 & d_2 \\ b_3 & c_3 & d_3 \end{vmatrix} = b_1 \begin{vmatrix} c_2 & d_2 \\ c_3 & d_3 \end{vmatrix} - c_1 \begin{vmatrix} b_2 & d_2 \\ b_3 & d_3 \end{vmatrix} + d_1 \begin{vmatrix} b_2 & c_2 \\ b_3 & c_3 \end{vmatrix}$$

$$= b_1(c_2d_3 - c_3d_2) - c_1(b_2d_3 - b_3d_2) + d_1(b_2c_3 - b_3c_2)$$

**Step 5:** Repeat step 4 for $y$ and let's denote it as $\Delta_y$.

$$\Delta_y = \begin{vmatrix} a_1 & c_1 & d_1 \\ a_2 & c_2 & d_2 \\ a_3 & c_3 & d_3 \end{vmatrix} = a_1 \begin{vmatrix} c_2 & d_2 \\ c_3 & d_3 \end{vmatrix} - c_1 \begin{vmatrix} a_2 & d_2 \\ a_3 & d_3 \end{vmatrix} + d_1 \begin{vmatrix} a_2 & c_2 \\ a_3 & c_3 \end{vmatrix}$$

$$= a_1(c_2d_3 - c_3d_2) - c_1(a_2d_3 - a_3d_2) + d_1(a_2c_3 - a_3c_2)$$

**Step 6:** Repeat step 4 for $z$ and let's denote it as $\Delta_z$.

$$\Delta_z = \begin{vmatrix} a_1 & b_1 & d_1 \\ a_2 & b_2 & d_2 \\ a_3 & b_3 & d_3 \end{vmatrix} = a_1 \begin{vmatrix} b_2 & d_2 \\ b_3 & d_3 \end{vmatrix} - b_1 \begin{vmatrix} a_2 & d_2 \\ a_3 & d_3 \end{vmatrix} + d_1 \begin{vmatrix} a_2 & b_2 \\ a_3 & b_3 \end{vmatrix}$$

$$= a_1(b_2d_3 - b_3d_2) - b_1(a_2d_3 - a_3d_2) + d_1(a_2b_3 - a_3b_2)$$

**Step 7:** To find the unknown variables $x$, $y$ and $z$, use

$$x = \frac{\Delta_x}{\Delta}, \quad y = -\frac{\Delta_y}{\Delta}, \quad z = \frac{\Delta_z}{\Delta}$$

Note the negative sign when finding $y$ values. We can combine them as:

$$\frac{x}{\Delta_x} = -\frac{y}{\Delta_y} = \frac{z}{\Delta_z} = \frac{1}{\Delta}$$

Again, we can arrange the equations as

$$\begin{aligned} a_1 x + b_1 y + c_1 z + d_1 &= 0 & \text{(i)} \\ a_2 x + b_2 y + c_2 z + d_2 &= 0 & \text{(ii)} \\ a_3 x + b_3 y + c_3 z = d_3 &= 0 & \text{(iii)} \end{aligned}$$

before proceeding as above, but the result will be

$$\frac{x}{\Delta_x} = -\frac{y}{\Delta_y} = \frac{z}{\Delta_z} = -\frac{1}{\Delta}$$

Notice the change in the positions of the negative sign.

Let's look at an example using Option 1 above.

### Example 20

Using Cramer's rule, solve the simultaneous equations below.

$$7x - 2z = 1 \quad \text{(i)}$$
$$3x + 5y + z = -4 \quad \text{(ii)}$$
$$4x + y = 2 \quad \text{(iii)}$$

What did you get? Find the solution below to double-check your answer.

### Solution to Example 20

Re-write the equations as this:

$$7x + 0y - 2z = 1 \quad \text{(i)}$$
$$3x + 5y + z = -4 \quad \text{(ii)}$$
$$4x + y + 0z = 2 \quad \text{(iii)}$$

$$\Delta = \begin{vmatrix} 7 & 0 & -2 \\ 3 & 5 & 1 \\ 4 & 1 & 0 \end{vmatrix} = 7\begin{vmatrix} 5 & 1 \\ 1 & 0 \end{vmatrix} - 0\begin{vmatrix} 3 & 1 \\ 4 & 0 \end{vmatrix} - 2\begin{vmatrix} 3 & 5 \\ 4 & 1 \end{vmatrix}$$

$$= 7(5 \times 0 - 1 \times 1) - 0(3 \times 0 - 4 \times 1) - 2(3 \times 1 - 4 \times 5)$$
$$= 7(-1) - 0(-4) - 2(-17) = -7 - 0 + 34 = \mathbf{27}$$

$$\Delta_x = \begin{vmatrix} 1 & 0 & -2 \\ -4 & 5 & 1 \\ 2 & 1 & 0 \end{vmatrix} = 1\begin{vmatrix} 5 & 1 \\ 1 & 0 \end{vmatrix} - 0\begin{vmatrix} -4 & 1 \\ 2 & 0 \end{vmatrix} - 2\begin{vmatrix} -4 & 5 \\ 2 & 1 \end{vmatrix}$$

$$= 1(5 \times 0 - 1 \times 1) - 0(-4 \times 0 - 2 \times 1) - 2(-4 \times 1 - 2 \times 5)$$
$$= 1(-1) - 0(-2) - 2(-14) = -1 - 0 + 28 = \mathbf{27}$$

$$\Delta_y = \begin{vmatrix} 7 & 1 & -2 \\ 3 & -4 & 1 \\ 4 & 2 & 0 \end{vmatrix} = 7\begin{vmatrix} -4 & 1 \\ 2 & 0 \end{vmatrix} - 1\begin{vmatrix} 3 & 1 \\ 4 & 0 \end{vmatrix} - 2\begin{vmatrix} 3 & -4 \\ 4 & 2 \end{vmatrix}$$

$$= 7(-4 \times 0 - 2 \times 1) - 1(3 \times 0 - 4 \times 1) - 2(3 \times 2 - 4 \times -4)$$
$$= 7(-2) - 1(-4) - 2(22) = -14 + 4 - 44 = \mathbf{-54}$$

$$\Delta_z = \begin{vmatrix} 7 & 0 & 1 \\ 3 & 5 & -4 \\ 4 & 1 & 2 \end{vmatrix} = 7\begin{vmatrix} 5 & -4 \\ 1 & 2 \end{vmatrix} - 0\begin{vmatrix} 3 & -4 \\ 4 & 2 \end{vmatrix} + 1\begin{vmatrix} 3 & 5 \\ 4 & 1 \end{vmatrix}$$

$$= 7(5 \times 2 - 1 \times -4) - 0(3 \times 2 - 4 \times -4) + 1(3 \times 1 - 4 \times 5)$$
$$= 7(14) - 0(22) + 1(-17) = 98 - 0 - 17 = \mathbf{81}$$

Therefore

$$x = \frac{\Delta_x}{\Delta} = \frac{27}{27}$$
$$\therefore x = 1$$

$$y = \frac{\Delta_y}{\Delta} = \frac{-54}{27}$$
$$\therefore y = -2$$

and

$$z = \frac{\Delta_z}{\Delta} = \frac{81}{27}$$
$$\therefore z = 3$$

Hence, we have

$$x = 1, \ y = -2, \ z = 3$$

To round up on this method, one more example using Option 2.

## Example 21

Solve the simultaneous equations below using Option 2.

$$y - x - 2 = 0 \qquad \text{(i)}$$
$$z + 2x + 5 = 0 \qquad \text{(ii)}$$
$$7x + 9y - 3z = 0 \qquad \text{(iii)}$$

What did you get? Find the solution below to double-check your answer.

## Solution to Example 21

Re-write the equations as this:

$$-x + y + 0z = 2 \qquad \text{(i)}$$
$$2x + 0y + z = -5 \qquad \text{(ii)}$$
$$7x + 9y - 3z = 0 \qquad \text{(iii)}$$

$$\Delta = \begin{vmatrix} -1 & 1 & 0 \\ 2 & 0 & 1 \\ 7 & 9 & -3 \end{vmatrix} = -1\begin{vmatrix} 0 & 1 \\ 9 & -3 \end{vmatrix} - 1\begin{vmatrix} 2 & 1 \\ 7 & -3 \end{vmatrix} + 0\begin{vmatrix} 2 & 0 \\ 7 & 9 \end{vmatrix}$$
$$= -1(0 \times -3 - 9 \times 1) - 1(2 \times -3 - 7 \times 1) + 0(2 \times 9 - 7 \times 0)$$
$$= -1(-9) - 1(-13) + 0(18) = 9 + 13 + 0 = \mathbf{22}$$

$$\Delta_x = \begin{vmatrix} 1 & 0 & 2 \\ 0 & 1 & -5 \\ 9 & -3 & 0 \end{vmatrix} = 1\begin{vmatrix} 1 & -5 \\ -3 & 0 \end{vmatrix} - 0\begin{vmatrix} 0 & -5 \\ 9 & 0 \end{vmatrix} + 2\begin{vmatrix} 0 & 1 \\ 9 & -3 \end{vmatrix}$$
$$= 1(1 \times 0 - (-3) \times -5) - 0(0 \times 0 - 9 \times -5) + 2(0 \times -3 - 9 \times 1)$$
$$= 1(-15) - 0(45) + 2(-9) = -15 - 0 - 18 = \mathbf{-33}$$

$$\Delta_y = \begin{vmatrix} -1 & 0 & 2 \\ 2 & 1 & -5 \\ 7 & -3 & 0 \end{vmatrix} = -1\begin{vmatrix} 1 & -5 \\ -3 & 0 \end{vmatrix} - 0\begin{vmatrix} 2 & -5 \\ 7 & 0 \end{vmatrix} + 2\begin{vmatrix} 2 & 1 \\ 7 & -3 \end{vmatrix}$$
$$= -1(1 \times 0 - (-3) \times -5) - 0(2 \times 0 - 7 \times -5) + 2(2 \times -3 - 7 \times 1)$$
$$= -1(-15) - 0(35) + 2(-13) = 15 - 26 = \mathbf{-11}$$

$$\Delta_z = \begin{vmatrix} -1 & 1 & 2 \\ 2 & 0 & -5 \\ 7 & 9 & 0 \end{vmatrix} = -1\begin{vmatrix} 0 & -5 \\ 9 & 0 \end{vmatrix} - 1\begin{vmatrix} 2 & -5 \\ 7 & 0 \end{vmatrix} + 2\begin{vmatrix} 2 & 0 \\ 7 & 9 \end{vmatrix}$$
$$= -1(0 \times 0 - 9 \times -5) - 1(2 \times 0 - 7 \times -5) + 2(2 \times 9 - 7 \times 0)$$
$$= -1(45) - 1(35) - 2(18) = -45 - 35 + 36 = \mathbf{-44}$$

Therefore

$$x = \frac{\Delta_x}{\Delta} = \frac{-33}{22}$$
$$\therefore x = \mathbf{-1.5}$$

$$y = -\frac{\Delta_y}{\Delta} = -\frac{-11}{22}$$
$$\therefore y = \mathbf{0.5}$$

and

$$z = \frac{\Delta_z}{\Delta} = \frac{-44}{22}$$
$$\therefore z = \mathbf{-2}$$

Hence, we have

$$x = \mathbf{-1.5},\ y = \mathbf{0.5},\ z = \mathbf{-2}$$

---

### 10.7.2 MATRIX FORM

This method is based on two main principles that we have discussed before, which are:

1) if two matrices are equal, they must be of the same order and their corresponding elements must be equal, and

# Matrices

2) when a matrix is pre- or post-multiplied by its inverse, the result is an identity matrix.

The general expression for solving simultaneous equations using matrix form is:

$$\boxed{X = A^{-1}.b} \qquad (10.17)$$

where $A$ is the matrix formed from the coefficients of the variables in the equations, $X$ is the matrix of the unknown and $b$ the matrix of the constant terms, all arranged in a uniform order.

The above expression comes from the fact that if:

$$A.X = b$$

we can pre-multiply both sides by $A^{-1}$ to have

$$A^{-1}A.X = A^{-1}.b$$

which implies that

$$I.X = A^{-1}.b$$
$$X = A^{-1}.b$$

Because

$$A^{-1}A = AA^{-1} = I$$

If you ever have to post-multiply both sides, remember that there is a compatibility requirement for multiplication which will not be satisfied by the RHS. That is to say, $b.A^{-1}$ may not be possible.

Let's look at two cases here, but the method can be applied to any set of linear equations.

### 10.7.2.1 Simultaneous Equations in Two Unknowns

Let's consider these equations

$$a_1 x + b_1 y = c_1 \qquad \text{(i)}$$
$$a_2 x + b_2 y = c_2 \qquad \text{(ii)}$$

where $a_1$ and $a_2$ are the coefficients of the first variable, $b_1$ and $b_2$ the coefficients of the second unknown, and $c_1$ and $c_2$ are the constant terms.

Follow these steps to determine the unknown variables $x$ and $y$.

**Step 1:** Ensure that the equations are written in the format above.
**Step 2:** Form a matrix equation as shown below.

$$\begin{bmatrix} a_1 & b_1 \\ a_2 & b_2 \end{bmatrix} \begin{bmatrix} x \\ y \end{bmatrix} = \begin{bmatrix} c_1 \\ c_2 \end{bmatrix}$$

**Step 3:** Pre-multiply both sides by the matrix of coefficients.

$$\begin{bmatrix} a_1 & b_1 \\ a_2 & b_2 \end{bmatrix}^{-1} \cdot \begin{bmatrix} a_1 & b_1 \\ a_2 & b_2 \end{bmatrix} \begin{bmatrix} x \\ y \end{bmatrix} = \begin{bmatrix} a_1 & b_1 \\ a_2 & b_2 \end{bmatrix}^{-1} \cdot \begin{bmatrix} c_1 \\ c_2 \end{bmatrix}$$

$$I. \begin{bmatrix} x \\ y \end{bmatrix} = \begin{bmatrix} a_1 & b_1 \\ a_2 & b_2 \end{bmatrix}^{-1} \cdot \begin{bmatrix} c_1 \\ c_2 \end{bmatrix}$$

$$\begin{bmatrix} x \\ y \end{bmatrix} = \begin{bmatrix} a_1 & b_1 \\ a_2 & b_2 \end{bmatrix}^{-1} \cdot \begin{bmatrix} c_1 \\ c_2 \end{bmatrix}$$

The first two lines are not needed since we know that it will be a unity matrix.

**Step 4:** Find the inverse of the matrix of coefficients (we've shown this on page 256). Pre-multiply it with the matrix of constant terms. This should give you a column matrix of 2 by 1.

**Step 5:** Equate the left- and right-hand sides to obtain $x$ and $y$.

An example will be useful at this stage; let's go for it.

### Example 22

Using matrices, solve the simultaneous equations below.

$$3x + 4y - 13 = 0 \qquad \text{(i)}$$
$$6x - 7y + 7 = 0 \qquad \text{(ii)}$$

What did you get? Find the solution below to double-check your answer.

### Solution to Example 22

Remember to present the equation in the right format first. Re-arrange the equations as this:

$$3x + 4y = 13 \qquad \text{(i)}$$
$$6x - 7y = -7 \qquad \text{(ii)}$$

Write this in matrix form

$$\begin{bmatrix} 3 & 4 \\ 6 & -7 \end{bmatrix} \begin{bmatrix} x \\ y \end{bmatrix} = \begin{bmatrix} 13 \\ -7 \end{bmatrix}$$

Thus

$$\begin{bmatrix} x \\ y \end{bmatrix} = \begin{bmatrix} 3 & 4 \\ 6 & -7 \end{bmatrix}^{-1} \begin{bmatrix} 13 \\ -7 \end{bmatrix}$$

Let

$$A = \begin{bmatrix} 3 & 4 \\ 6 & -7 \end{bmatrix}$$

Matrices

$$|A| = \begin{vmatrix} 3 & 4 \\ 6 & -7 \end{vmatrix} = (3 \times -7) - (6 \times 4) = -21 - 24$$
$$= -45$$
$$A^{-1} = -\frac{1}{45} \begin{bmatrix} -7 & -4 \\ -6 & 3 \end{bmatrix}$$

Now

$$\begin{bmatrix} 3 & 4 \\ 6 & -7 \end{bmatrix}^{-1} \begin{bmatrix} 13 \\ -7 \end{bmatrix} = -\frac{1}{45} \begin{bmatrix} -7 & -4 \\ -6 & 3 \end{bmatrix} \begin{bmatrix} 13 \\ -7 \end{bmatrix}$$
$$= -\frac{1}{45} \begin{bmatrix} (-7 \times 13) + (-4 \times -7) \\ (-6 \times 13) + (3 \times -7) \end{bmatrix} = -\frac{1}{45} \begin{bmatrix} -91 + 28 \\ -78 - 21 \end{bmatrix}$$
$$= -\frac{1}{45} \begin{bmatrix} -63 \\ -99 \end{bmatrix} = \begin{bmatrix} 63/45 \\ 99/45 \end{bmatrix} = \begin{bmatrix} 1.4 \\ 2.2 \end{bmatrix}$$

Thus

$$\begin{bmatrix} x \\ y \end{bmatrix} = \begin{bmatrix} 1.4 \\ 2.2 \end{bmatrix}$$
$$\therefore x = 1.4, \; y = 2.2$$

### 10.7.2.2 Simultaneous Equations in Three Unknowns

Let's consider these equations:

$$a_1 x + b_1 y + c_1 z = d_1 \quad \text{(i)}$$
$$a_2 x + b_2 y + c_2 z = d_2 \quad \text{(ii)}$$
$$a_3 x + b_3 y + c_3 z = d_3 \quad \text{(iii)}$$

where $a_1$, $a_2$, and $a_3$ are the coefficients of the first variable, $b_1$, $b_2$, and $b_3$ the coefficients of the second unknown, $c_1$, $c_2$, and $c_3$ the coefficients of the third unknown and $d_1$, $d_2$, and $d_3$ are the constant terms.

Follow these steps to determine the unknown variables $x$, $y$, and $z$.

**Step 1:** Ensure that the equations are written in the format above.
**Step 2:** Form a matrix equation as shown below.

$$\begin{bmatrix} a_1 & b_1 & c_1 \\ a_2 & b_2 & c_2 \\ a_3 & b_3 & c_3 \end{bmatrix} \begin{bmatrix} x \\ y \\ z \end{bmatrix} = \begin{bmatrix} d_1 \\ d_2 \\ d_3 \end{bmatrix}$$

**Step 3:** Pre-multiply both sides by the inverse of the matrix of coefficients.

$$\begin{bmatrix} a_1 & b_1 & c_1 \\ a_2 & b_2 & c_2 \\ a_3 & b_3 & c_3 \end{bmatrix}^{-1} \cdot \begin{bmatrix} a_1 & b_1 & c_1 \\ a_2 & b_2 & c_2 \\ a_3 & b_3 & c_3 \end{bmatrix} \begin{bmatrix} x \\ y \\ z \end{bmatrix} = \begin{bmatrix} a_1 & b_1 & c_1 \\ a_2 & b_2 & c_2 \\ a_3 & b_3 & c_3 \end{bmatrix}^{-1} \cdot \begin{bmatrix} d_1 \\ d_2 \\ d_3 \end{bmatrix}$$

$$I. \begin{bmatrix} x \\ y \\ z \end{bmatrix} = \begin{bmatrix} a_1 & b_1 & c_1 \\ a_2 & b_2 & c_2 \\ a_3 & b_3 & c_3 \end{bmatrix}^{-1} \cdot \begin{bmatrix} d_1 \\ d_2 \\ d_3 \end{bmatrix}$$

$$\begin{bmatrix} x \\ y \\ z \end{bmatrix} = \begin{bmatrix} a_1 & b_1 & c_1 \\ a_2 & b_2 & c_2 \\ a_3 & b_3 & c_3 \end{bmatrix}^{-1} \cdot \begin{bmatrix} d_1 \\ d_2 \\ d_3 \end{bmatrix}$$

Again, the first two lines are not needed since we know that it will be a unity matrix.

**Step 4:** Find the inverse of the matrix of coefficients (we've shown this on page 385). Pre-multiply it with the matrix of constant terms. This should give you a column matrix of 3 by 1.

**Step 5:** Equate the left- and right-hand sides to obtain $x$, $y$ and $z$.

Again, an example will be useful here; let's go for it.

### Example 23

Using matrices, solve the simultaneous equations below.

$$x + 2y - z = 0 \quad \text{(i)}$$
$$3z - 7y + 4x + 1 = 0 \quad \text{(ii)}$$
$$5x - 3z + 4 = 0 \quad \text{(iii)}$$

What did you get? Find the solution below to double-check your answer.

### Solution to Example 23

Remember to present the equation in the right format first. Re-arrange the equations as this:

$$x + 2y - z = 0 \quad \text{(i)}$$
$$4x - 7y + 3z = -1 \quad \text{(ii)}$$
$$5x + 0y - 3z = -4 \quad \text{(iii)}$$

Write this in matrix form

$$\begin{bmatrix} 1 & 2 & -1 \\ 4 & -7 & 3 \\ 5 & 0 & -3 \end{bmatrix} \begin{bmatrix} x \\ y \\ z \end{bmatrix} = \begin{bmatrix} 0 \\ -1 \\ -4 \end{bmatrix}$$

Thus

$$\begin{bmatrix} x \\ y \\ z \end{bmatrix} = \begin{bmatrix} 1 & 2 & -1 \\ 4 & -7 & 3 \\ 5 & 0 & -3 \end{bmatrix}^{-1} \begin{bmatrix} 0 \\ -1 \\ -4 \end{bmatrix}$$

# Matrices

To find the inverse

$$A = \begin{bmatrix} 1 & 2 & -1 \\ 4 & -7 & 3 \\ 5 & 0 & -3 \end{bmatrix}$$

**Step 1:** Find the determinant of **A**.

$$|A| = \begin{vmatrix} 1 & 2 & -1 \\ 4 & -7 & 3 \\ 5 & 0 & -3 \end{vmatrix} = 1\begin{vmatrix} -7 & 3 \\ 0 & -3 \end{vmatrix} - 2\begin{vmatrix} 4 & 3 \\ 5 & -3 \end{vmatrix} - 1\begin{vmatrix} 4 & -7 \\ 5 & 0 \end{vmatrix}$$

$|A| = 1\{(-7 \times -3) - (0 \times 3)\} - 2\{(4 \times -3) - (5 \times 3)\} - 1\{(4 \times 0) - (5 \times -7)\}$
$|A| = 1(21 - 0) - 2(-12 - 15) - 1(0 + 35)$
$|A| = 1(21) - 2(-27) - 1(35) = 21 + 54 - 35$
$|A| = \mathbf{40}$

**Step 2:** Find the signed minor of each element.

- $a_{11} = 1$

$$\begin{vmatrix} 1 & 2 & -1 \\ 4 & -7 & 3 \\ 5 & 0 & -3 \end{vmatrix}$$

$$(-1)^{i+j}\begin{vmatrix} -7 & 3 \\ 0 & -3 \end{vmatrix} = (-1)^{1+1}\{(-7 \times -3) - (0 \times 3)\}$$
$$= (-1)^2 (21 - 0)$$
$$= \mathbf{21}$$

- $a_{12} = 2$

$$(-1)^{i+j}\begin{vmatrix} 4 & 3 \\ 5 & -3 \end{vmatrix} = (-1)^{1+2}\{(4 \times -3) - (5 \times 3)\}$$
$$= (-1)^3 (-12 - 15)$$
$$= \mathbf{27}$$

- $a_{13} = -1$

$$(-1)^{i+j}\begin{vmatrix} 4 & -7 \\ 5 & 0 \end{vmatrix} = (-1)^{1+3}\{(4 \times 0) - (5 \times -7)\}$$
$$= (-1)^4 (0 + 35)$$
$$= \mathbf{35}$$

- $a_{21} = 4$

$$(-1)^{i+j}\begin{vmatrix} 2 & -1 \\ 0 & -3 \end{vmatrix} = (-1)^{2+1}\{(2 \times -3) - (0 \times -1)\}$$
$$= (-1)^3 (-6 - 0)$$
$$= \mathbf{6}$$

- $a_{22} = -7$

$$(-1)^{i+j} \begin{vmatrix} 1 & -1 \\ 5 & -3 \end{vmatrix} = (-1)^{2+2} \{(1 \times -3) - (5 \times -1)\}$$
$$= (-1)^4 (-3 + 5)$$
$$= 2$$

- $a_{23} = 3$

$$(-1)^{i+j} \begin{vmatrix} 1 & 2 \\ 5 & 0 \end{vmatrix} = (-1)^{2+3} \{(1 \times 0) - (5 \times 2)\}$$
$$= (-1)^5 (0 - 10)$$
$$= 10$$

- $a_{31} = 5$

$$(-1)^{i+j} \begin{vmatrix} 2 & -1 \\ -7 & 3 \end{vmatrix} = (-1)^{3+1} \{(2 \times 3) - (-7 \times -1)\}$$
$$= (-1)^4 (6 - 7)$$
$$= -1$$

- $a_{32} = 0$

$$(-1)^{i+j} \begin{vmatrix} 1 & -1 \\ 4 & 3 \end{vmatrix} = (-1)^{3+2} \{(1 \times 3) - (4 \times -1)\}$$
$$= (-1)^5 (3 + 4)$$
$$= -7$$

- $a_{33} = -3$

$$(-1)^{i+j} \begin{vmatrix} 1 & 2 \\ 4 & -7 \end{vmatrix} = (-1)^{3+3} \{(1 \times -7) - (4 \times 2)\}$$
$$= (-1)^6 (-7 - 8)$$
$$= -15$$

**Step 3:** Form the matrix of the signed minors

$$\begin{bmatrix} 21 & 27 & 35 \\ 6 & 2 & 10 \\ -1 & -7 & -15 \end{bmatrix}$$

**Step 4:** Transpose the matrix of the signed minor to obtain Adjoint matrix

$$Adjoint = \begin{bmatrix} 21 & 6 & -1 \\ 27 & 2 & -7 \\ 35 & 10 & -15 \end{bmatrix}$$

# Matrices

**Step 5:** Form the inverse matrix by dividing the Adjoint by the determinant

$$A^{-1} = \frac{1}{40} \begin{bmatrix} 21 & 6 & -1 \\ 27 & 2 & -7 \\ 35 & 10 & -15 \end{bmatrix}$$

Now

$$\begin{bmatrix} 1 & 2 & -1 \\ 4 & -7 & 3 \\ 5 & 0 & -3 \end{bmatrix}^{-1} \begin{bmatrix} 0 \\ -1 \\ -4 \end{bmatrix} = \frac{1}{40} \begin{bmatrix} 21 & 6 & -1 \\ 27 & 2 & -7 \\ 35 & 10 & -15 \end{bmatrix} \begin{bmatrix} 0 \\ -1 \\ -4 \end{bmatrix}$$

$$= \frac{1}{40} \begin{bmatrix} (21 \times 0) + (6 \times -1) + (-1 \times -4) \\ (27 \times 0) + (2 \times -1) + (-7 \times -4) \\ (35 \times 0) + (10 \times -1) + (-15 \times -4) \end{bmatrix} = \frac{1}{40} \begin{bmatrix} 0 - 6 + 4 \\ 0 - 2 + 28 \\ 0 - 10 + 60 \end{bmatrix}$$

$$= \frac{1}{40} \begin{bmatrix} -2 \\ 26 \\ 50 \end{bmatrix} = \begin{bmatrix} -2/40 \\ 26/40 \\ 50/40 \end{bmatrix} = \begin{bmatrix} -0.05 \\ 0.65 \\ 1.25 \end{bmatrix}$$

Thus

$$\begin{bmatrix} x \\ y \\ z \end{bmatrix} = \begin{bmatrix} -0.05 \\ 0.65 \\ 1.25 \end{bmatrix}$$

$$\therefore x = -0.05, \ y = 0.65, \ z = 1.25$$

## 10.8 CHAPTER SUMMARY

1) A matrix (plural: matrices) is defined as an array of numbers (or elements), which consists of rows and columns enclosed in square brackets [] or parentheses ().

2) Each number in a matrix is called an **element** of the matrix. Therefore, changing the position of an element changes the entire matrix.

3) A matrix is generally represented by a bold capital letter and its elements with small letters.

4) An element is completely identified by stating its row followed by its column.

5) A general notation for stating the location of an element in a matrix $A$ as:

$$a_{ij}$$

where $i$ is the row number and $j$ the column number.

6) Some common types of matrix are:
   - **Column matrix**: This is a matrix with a single (or one) column. The general dimension is $n \times 1$, where $n$ is the number of rows.
   - **Row matrix**: This is a matrix with a single (or one) row. It is also called a **line matrix** and is the inverse of the column matrix. The general dimension is $1 \times n$, where $n$ is the number of columns.

- **Square matrix**: This is a matrix where the number of rows is the same as that of the columns i.e. $n \times n$ matrix.
- **Single (element) matrix**: This is a matrix of a dimension $1 \times 1$, i.e., it has one element, one row, and one column.
- **Diagonal matrix**: This is a type of square matrix where all elements are zeros, except the one on the diagonal line, generally called the leading diagonal. The diagonal consists of all elements with $a_{ii}$, starting with $a_{11}$ on the top left to $a_{ii}$ on the bottom right.
- **Unit (or identity) matrix**: This is a diagonal matrix in which the elements along the diagonal line are ones (1).
- **Zero or null matrix**: This is a matrix in which all the elements are zeros; it is denoted by **0**.
- **Singular matrix**: This is a matrix whose determinant is zero.
- **Transpose matrix**: Transpose of a matrix is formed when the rows have been converted to respective columns and vice versa (i.e., rows and columns are interchanged).

7) Two or more matrices are said to be equal if:
   - they have the same order, and
   - all their corresponding elements are equal.

8) Given a matrix **A** with a dimension of $m \times n$, the transpose of this is another matrix **B** with a dimension of $n \times m$. As matrix **B** is connected to **A**, it is written as:

$$\boxed{B = A^T} \text{ OR } \boxed{B = A'} \text{ OR } \boxed{B = \tilde{A}}$$

It can be shown that:

$$\boxed{(A + B)^T = A^T + B^T}$$

$$\boxed{(AB)^T = B^T A^T}$$

9) If a square matrix is equal to its transpose, the matrix is ***symmetric***.

$$\boxed{A = A^T} \text{ OR } \boxed{a_{ij} = a_{ji}}$$

10) If a square matrix is equal to its transpose multiplied by $k = -1$, the matrix is said to be or called ***skew-symmetric***.

$$\boxed{A = -A^T} \text{ OR } \boxed{a_{ij} = -a_{ji}}$$

11) Addition of two or more matrices is carried out by adding their corresponding elements, provided that their dimensions are the same, otherwise the operation is impossible. Subtraction of one matrix from another is carried out in a similar way.

12) Addition is both commutative and associative (i.e., the order of the operation is irrelevant).

$$\boxed{A + B = B + A} \text{ AND } \boxed{(A + B) + C = A + (B + C)}$$

13) When a matrix is multiplied by a scalar, each element is multiplied by this scalar factor.

14) Multiplication of a matrix by another matrix has certain 'compatibility requirements' that need to be noted and fulfilled.

15) $A \times B$ is not the same as $B \times A$ (i.e., $A \times B \neq B \times A$). In other words, multiplication is not commutative.

16) The determinant of a matrix $A$, denoted as $|A|$, is a number (scalar) that is calculated using the elements of $A$. This can only be computed for a square matrix.

17) The determinant of a 2 by 2 (or second-order) matrix is the product of the two elements on the leading diagonal minus the product of the two elements on the other diagonal.

18) The inverse of a 2 by 2 matrix $A$, written as $A^{-1}$, is given by:

$$A^{-1} = \frac{1}{ad - bc} \begin{bmatrix} d & -b \\ -c & a \end{bmatrix}$$

where $A$ is

$$\begin{bmatrix} a & b \\ c & d \end{bmatrix}$$

19) The inverse of a matrix $A$, denoted as $A^{-1}$, is given by:

$$A^{-1} = \frac{1}{|A|} [Adj\ (A)]$$

20) The general expression for solving simultaneous equations using matrix form is:

$$X = A^{-1}.b$$

\*\*\*\*

## 10.9 FURTHER PRACTICE

To access complementary contents, including additional exercises, please go to www.dszak.com.

# 11 Vectors

## Learning Outcomes

Once you have studied the content of this chapter, you should be able to:

- Discuss scalar and vector quantities
- Represent vector quantities using different notations
- Carry out basic operations involving vectors
- Explain the unit vectors in $x$-, $y$-, and $z$-directions
- Use Cartesian coordinates in two and three dimensions
- Use scalar and dot products of two vectors
- Understand the equation of a line in vector form
- Use vector analysis and techniques to solve problems in geometry
- Use vector equations to determine the intersection and angle between two lines

## 11.1 INTRODUCTION

Physical quantities are part of our daily activities. They include height, weight, time, speed, amount of substance, pressure, volume, and power among others. These quantities can be grouped using different criteria, and one such classification will be considered here in this chapter, as we discuss the meaning of a vector, its types, operations, and some applications.

## 11.2 SCALAR AND VECTOR

Physical quantities (simply referred to as quantities) can be divided into scalar and vector quantities. **Scalar** is any quantity that is completely identified or described by its magnitude (also known as amount or size) and associated units only. Examples include time, speed, height, temperature, pressure, power, money, and population. For these quantities, the knowledge about their size gives a full 'picture' of what the quantity means and their (potential) effect. Every physical quantity has this feature.

On the other hand, a **vector quantity** (or simply a vector) is a quantity which is completely identified by both its magnitude and direction (and associated units). Examples include velocity, displacement, force, electric current, stress, and momentum. We can say that velocity is speed in a particular direction and displacement is distance in a specific direction. Describing a scalar quantity with direction makes no sense; for example, a temperature of 2 degrees Celsius 'acting east' is not palatable to hear.

# Vectors

This classification is essential because adding and subtracting a scalar quantity follows the arithmetic principles that we all know, but the same is not true for a vector quantity. For example, if we have time $t_1 = 2s$ and $t_2 = 2s$, adding these will give 4s and subtracting one from the other equals 0 and nothing else. On the other hand, if we have forces $F_1 = 2N$ and $F_2 = 2N$, adding these two forces will give a value between $-4N$ and $4N$ (or $0N$ and $4N$ if we strictly mean addition). The precise answer depends on the exact direction of each of the two forces. How to determine this is what we are set to delve into.

## 11.3 FREE VECTOR

For analytical purposes, vectors are considered as free and referred to as '**free vectors**'. This implies that they can be anywhere such that the point of application is irrelevant. A vector applied at a point on a table in Figure 11.1 can be diagrammatically represented as though it is applied at a different point on the same table, provided that the direction remains unchanged.

Furthermore, a vector applied at one edge of a rectangle, say AB, is the same or can be considered as though is applied at the other edge (CD) of the same rectangle, as illustrated in Figure 11.2.

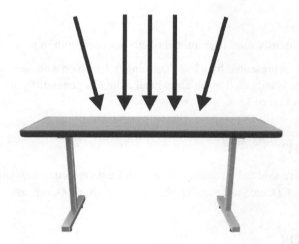

**FIGURE 11.1** Free vector illustrated (I).

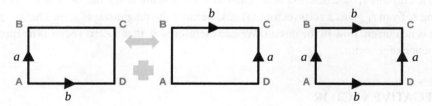

**FIGURE 11.2** Free vector illustrated (II).

**FIGURE 11.3** Vector representation.

## 11.4 VECTOR NOTATION

The common notations used for a vector include:

| | |
|---|---|
| $\mathbf{a}$ | This is a lowercase bold letter used to differentiate a vector from a scalar. |
| $\underline{a}$ | This is a lowercase letter with an underlined bar (or sometimes an underlined wavy bar is used). |
| $\bar{a}$ | This is a lowercase letter with an overbar. |
| $\overrightarrow{AB}$ | This is a 2-letter notation with an arrow pointing in the direction of the vector, starting from the point of application. |
| $\overline{AB}$ | This is a 2-letter notation starting from the point of application (or sometimes the bar is below $\underline{AB}$). |

The first notation is primarily used in print, but others are used both in print and in writing.

Graphically, a vector is represented by a line. The length, based on a chosen scale, denotes the magnitude and the direction is indicated by an arrowhead, which is generally placed at the end of the line or middle, as shown in Figure 11.3.

## 11.5 COLLINEARITY

Three or more points are said to be collinear if they all lie on the same straight line. It can be shown that if vectors $\overrightarrow{AB}$, $\overrightarrow{BC}$, $\overrightarrow{CD}$, etc., are parallel, then points $A$, $B$, $C$, $D$, etc., are collinear.

## 11.6 ZERO VECTOR

A **Zero** vector (also called a **Null** vector) is a vector whose magnitude is zero. For example, given two electric currents $I_1 = 2\,\text{mA}$ and $I_2 = 2\,\text{mA}$ that are flowing along the same branch of a circuit, subtracting $I_2$ from $I_1$ gives a zero vector of $0\,\text{mA}$, because the magnitude is zero. This simply means that there is no current flow in the circuit. We can therefore say that a Zero vector is a representation of the absence of a vector.

## 11.7 NEGATIVE VECTOR

The negation of a vector is a way of emphasising the direction feature of a vector. Given a vector $\mathbf{a}$, the negative vector of this is a vector that has the same magnitude as $\mathbf{a}$ but points in an opposite direction (i.e., 180° or say anti-phase). We indicate this with a sign change (ideally negative) or by flipping the direction of the arrow (Figure 11.4). In this case, the negative of $\mathbf{a}$ is given as $-\mathbf{a}$.

# Vectors

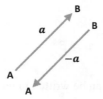

**FIGURE 11.4** Negative vector illustrated.

Unlike in scalar arithmetic, a negative vector does not imply deficiency. In other words, a vector $a$ can represent that which starts from the origin and points eastward while its negative $-a$ is the one pointing westward, also starting from the same origin. The reverse is also valid and possible.

Additionally, if the result of vector addition is negative (or a negative vector), it simply implies that it is a vector that acts in an opposite direction. For example, if a resultant force is obtained to be $F = -5\text{N}$ when we have taken North as a positive reference, then this means that the resultant is **5N** pointing southward. In this context, we can write that $\overrightarrow{AB} + \overrightarrow{BA} = 0$. This is because both have the same magnitude but different directions.

## 11.8 EQUALITY OF VECTORS

Two or more vectors are said to be equal if they all have the same magnitude and direction. If the lines in Figure 11.5 are equal and parallel, then the three vectors they represent are equal, i.e., $a = b = c$ or $\overrightarrow{XA} = \overrightarrow{YB} = \overrightarrow{ZC}$.

For non-zero, non-parallel vectors $a$ and $b$, if $\alpha a + \beta b = \lambda a + \mu b$ then $\alpha = \lambda$ and $\beta = \mu$. This is because:

$$\alpha a + \beta b = \lambda a + \mu b$$

which implies that

$$\alpha a - \lambda a = \mu b - \beta b$$
$$(\alpha - \lambda)a = (\mu - \beta)b$$

It follows that if $a$ and $b$ are not parallel and non-zero, then their coefficients must be zero for the equality to be valid. We thus have:

$$\alpha - \lambda = 0 \therefore \alpha = \lambda \quad \text{and} \quad \mu - \beta = 0 \therefore \mu = \beta$$

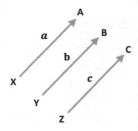

**FIGURE 11.5** Equality of vectors illustrated.

Therefore

$$\alpha = \lambda \qquad \mu = \beta$$

If two vectors $a$ and $b$ are parallel, one can be written in terms of the other as:

$$\boxed{a = \mu b} \quad \text{OR} \quad \boxed{b = \lambda a} \qquad (11.1)$$

where $\lambda$ and $\mu$ are non-zero scalars.

Let's illustrate the equality of vectors with an example.

### Example 1

$5u + (2x - 3y)v = (2x + y)u - 3v$ such that $u$ and $v$ are non-zero, non-parallel vectors. Calculate the values of the scalars $x$ and $y$.

What did you get? Find the solution below to double-check your answer.

### Solution to Example 1

Given that

$$5u + (2x - 3y)v = (2x + y)u - 3v$$

We will equate the coefficient of vector $u$ on the LHS with that on the RHS. The same will be applied to vector $v$. Therefore, we have:

$$2x + y = 5 \quad ----(i)$$
$$2x - 3y = -3 \quad ----(ii)$$

equation (i) – equation (ii), we have

$$4y = 8$$
$$y = \frac{8}{4} = 2$$

Substitute the value of $y$ in equation (i), we have

$$2x + 2 = 5$$
$$2x = 5 - 2$$
$$2x = 3$$
$$x = \frac{3}{2}$$
$$\therefore x = 1.5, \quad y = 2$$

# Vectors

## 11.9 UNIT VECTOR

A unit vector is a vector whose magnitude is unity, that is, equal to 1. For example, given two forces $F_1 = 3\,\text{N}$ and $F_2 = 2\,\text{N}$, if these forces act on the same line, subtracting $F_2$ from $F_1$ gives a unity vector of 1 N because the magnitude is unity. It will still be unity if the operation gives $-1\,\text{N}$ (i.e., when we reverse the subtrahend).

Follow these steps to find a unit vector in a particular direction.

**Step 1:** If not already given, determine the vector that acts in the stated direction. Let's say the vector is $a$.
**Step 2:** Determine the modulus of vector $a$, which is $|a|$.
**Step 3:** Divide vector $a$ by its modulus $|a|$. The resulting vector, denoted as $\hat{a}$, is a unit vector in the direction of $a$.

In general, a unit vector in the direction of $a$, written as 'hat' $a$, is given as:

$$\boxed{\hat{a} = \frac{a}{|a|}} \tag{11.2}$$

## 11.10 CARTESIAN REPRESENTATION OF VECTORS

We have standard unit vectors parallel to (or in the positive direction of) $x$, $y$, and $z$ axes. These vectors are at right angles to each other, commonly described as **mutually perpendicular**, and are denoted by $\hat{\imath}, \hat{\jmath}$, and $\hat{k}$, respectively, though the 'hat' is often omitted for these special unit vectors. For example, a vector of $3\hat{\imath}$ is a vector that is 3 units in length in the positive $x$-axis while $-5\hat{\jmath}$ is a vector that is 5 units in length in the negative $y$-axis. Similarly, $7\hat{k}$ is a vector that is 7 units in length in the positive $z$-axis.

Consider a vector $r$, which is entirely in 2D plane surface (Figure 11.6), the vector can be defined in terms of unit vectors as:

$$\boxed{r = x\hat{\imath} + y\hat{\jmath}}$$

If vector $r$ lies in space, then it is completely defined as:

$$\boxed{r = x\hat{\imath} + y\hat{\jmath} + z\hat{k}} \tag{11.3}$$

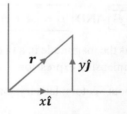

**FIGURE 11.6** Cartesian representation of vectors illustrated.

The column vector or matrix equivalents are:

$$\boxed{r = x\hat{i} + y\hat{j} = \begin{pmatrix} x \\ y \end{pmatrix}} \text{ AND } \boxed{r = x\hat{i} + y\hat{j} + z\hat{k} = \begin{pmatrix} x \\ y \\ z \end{pmatrix}} \quad (11.4)$$

where $x$, $y$, and $z$ are numerical values representing the magnitude of $r$ in each direction.

Note that a vector that lies entirely on the $x$-axis, $y$-axis, or $z$-axis can be represented as:

$$\begin{pmatrix} x \\ 0 \\ 0 \end{pmatrix}, \begin{pmatrix} 0 \\ y \\ 0 \end{pmatrix}, \begin{pmatrix} 0 \\ 0 \\ z \end{pmatrix} \quad (11.5)$$

As we will need this here too, remember that the coordinates of a point in two and three dimensions are, respectively, specified as:

$$\boxed{(x, y)} \text{ AND } \boxed{(x, y, z)} \quad (11.6)$$

## 11.11 MAGNITUDE OF A VECTOR

The magnitude of a vector (also known as modulus) is denoted using double vertical lines $||$, and it represents the size of a vector when the direction is not considered. Using the various notations for a vector (see page 410), the magnitude of a vector can thus be represented as: $|a|$, $|\underline{a}|$, $|\vec{a}|$, and $|\overrightarrow{AB}|$.

A vector and its negative have the same magnitude as:

$$\boxed{|a| = |-a|} \quad (11.7)$$

This is the same as taking the absolute value of numbers; for example, $|5| = |-5| = 5$. And $|5\,\text{mA}| = |-5\,\text{mA}| = 5\,\text{mA}$. In other words, magnitude is always a positive scalar.

Follow these steps to find the modulus of a vector

**Step 1:** Square the values of $x$, $y$, and $z$ (i.e. the coefficients of $i, j$, and $k$).
**Step 2:** Add the squares of $x$, $y$, and $z$, i.e., $x^2 + y^2 + z^2$.
**Step 3:** Take the square root of the answer in step 2.

In general, the magnitude is given by:

$$\boxed{|r| = |x\hat{i} + y\hat{j}| = \sqrt{x^2 + y^2}} \text{ AND } \boxed{|r| = |x\hat{i} + y\hat{j} + z\hat{k}| = \sqrt{x^2 + y^2 + z^2}} \quad (11.8)$$

Using Pythagoras' theorem will give us the magnitude in a two-dimension expression, and using the theorem twice will give us the three-dimension expression.

We've now covered important facts about vectors; let's try some examples.

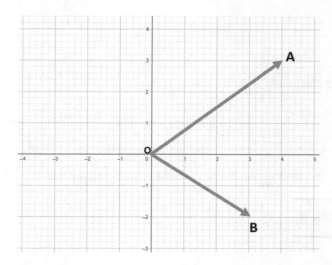

**FIGURE 11.7** Example 2.

### Example 2

$a$ and $b$ are vectors given by $\overrightarrow{OA}$ and $\overrightarrow{OB}$, as shown in the $x$–$y$ plane (Figure 11.7). Write $a$ and $b$ in Cartesian and matrix (or column) forms.

What did you get? Find the solution below to double-check your answer.

### Solution to Example 2

$$a = 4\hat{\imath} + 3\hat{\jmath}$$
$$= \begin{pmatrix} 4 \\ 3 \end{pmatrix}$$
$$b = 3\hat{\imath} - 2\hat{\jmath}$$
$$= \begin{pmatrix} 3 \\ -2 \end{pmatrix}$$

## Example 3

Determine the exact magnitude of the following vectors:

**a)** $a = \hat{i} - \sqrt{3}\hat{j}$        **b)** $b = 3(2\hat{i} - 6\hat{j} - 3\hat{k})$        **c)** $c = \begin{pmatrix} \sqrt{21} \\ -2 \end{pmatrix}$

What did you get? Find the solution below to double-check your answer.

## Solution to Example 3

**a)** $a = \hat{i} - \sqrt{3}\hat{j}$
**Solution**

$$|\hat{i} - \sqrt{3}\hat{j}| = \sqrt{(1)^2 + (-\sqrt{3})^2}$$
$$= \sqrt{1 + 3}$$
$$= \sqrt{4}$$
$$\therefore a = 2$$

**b)** $b = 3(2\hat{i} - 6\hat{j} - 3\hat{k})$
**Solution**

$$|3(2\hat{i} - 6\hat{j} - 3\hat{k})| = 3|2\hat{i} - 6\hat{j} - 3\hat{k}|$$
$$= 3\left(\sqrt{(2)^2 + (-6)^2 + (-3)^2}\right)$$
$$= 3\left(\sqrt{4 + 36 + 9}\right)$$
$$= 3\sqrt{49}$$
$$\therefore b = 21$$

Vectors

c) $c = \begin{pmatrix} \sqrt{21} \\ -2 \end{pmatrix}$

**Solution**

$$\left|\begin{pmatrix} \sqrt{21} \\ -2 \end{pmatrix}\right| = \sqrt{(\sqrt{21})^2 + (-2)^2}$$
$$= \sqrt{21 + 4}$$
$$= \sqrt{25}$$
$$\therefore c = 5$$

### Example 4

Determine a unit vector in the direction of the following vectors:

a) $p = 3\hat{i} - 4\hat{j}$

b) $q = 2\hat{i} - 3\hat{j} + 5\hat{k}$

What did you get? Find the solution below to double-check your answer.

### Solution to Example 4

a) $p = 3\hat{i} - 4\hat{j}$
**Solution**

$$|p| = \sqrt{(3)^2 + (-4)^2}$$
$$= \sqrt{9 + 16} = 5$$

$$\hat{p} = \frac{p}{|p|}$$
$$= \frac{3\hat{i} - 4\hat{j}}{5}$$
$$= \frac{1}{5}(3\hat{i} - 4\hat{j})$$
$$\therefore \hat{p} = \frac{3}{5}\hat{i} - \frac{4}{5}\hat{j}$$

**b)** $q = 2\hat{\imath} - 3\hat{\jmath} + 5\hat{k}$
**Solution**

$$|q| = \sqrt{(2)^2 + (-3)^2 + (5)^2}$$
$$= \sqrt{4 + 9 + 25}$$
$$= \sqrt{38}$$
$$\hat{q} = \frac{q}{|q|}$$
$$= \frac{2\hat{\imath} - 3\hat{\jmath} + 5\hat{k}}{\sqrt{38}}$$
$$= \frac{1}{\sqrt{38}}(2\hat{\imath} - 3\hat{\jmath} + 5\hat{k})$$
$$\therefore \hat{q} = \frac{2}{\sqrt{38}}\hat{\imath} - \frac{3}{\sqrt{38}}\hat{\jmath} + \frac{5}{\sqrt{38}}\hat{k}$$

---

## 11.12 POSITION VECTORS

Position vectors are used to specify the position of a point relative to a fixed reference, in two or three dimensions. If point $O$ in Figure 11.8 is chosen as the reference point or origin, then $\overrightarrow{OA}$ is the position vector of point $A$ relative to $O$ and $\overrightarrow{OB}$ is the position vector of $B$ relative to the same point. Since the reference is known and common to all, the above position vectors are written as $a$ and $b$, respectively. This is because it is customary to name a position vector with the finish point since the origin is common to all. In Cartesian notation, the origin is $(0, 0)$ for 2D and $(0, 0, 0)$ for 3D.

Vector $\overrightarrow{AB}$ in Figure 11.8 is the position vector of $B$ relative to $A$, and can be written in terms of $a$ and $b$ using the triangle law as:

$$\overrightarrow{AB} = \overrightarrow{AO} + \overrightarrow{OB} = \left(-\overrightarrow{OA}\right) + \overrightarrow{OB}$$

which implies that

$$\overrightarrow{AB} = -a + b = b - a$$

In other words, $\overrightarrow{AB}$ is the difference of the two position vectors.

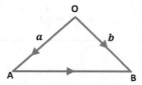

**FIGURE 11.8** Position vectors illustrated.

# Vectors

Let's try some examples.

### Example 5

The position vector of a point $P$ relative to the origin $O$ is $-\hat{i}-\hat{j}-3\hat{k}$. Determine the distance between point $P$ and the origin. Present your answer correct to 1 d.p.

What did you get? Find the solution below to double-check your answer.

### Solution to Example 5

**HINT**

Distance here is the length from the origin to point $P$, i.e. the magnitude of length OP.

$$\begin{aligned}|-\hat{i}-\hat{j}-3\hat{k}| &= \sqrt{(-1)^2+(-1)^2+(-3)^2} \\ &= \sqrt{1+1+9} = \sqrt{11} \\ &= 3.3\end{aligned}$$

### Example 6

$a$ and $b$ are position vectors of points $A$ and $B$ with respect to the origin $O$, as shown on the x–y plane below (Figure 11.9).

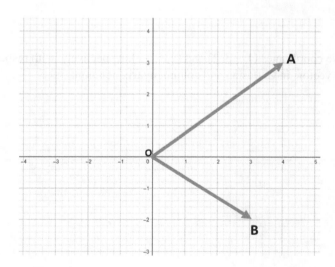

**FIGURE 11.9** Example 6.

a) Determine vector $\vec{AB}$ in Cartesian form.

b) Determine vector $\vec{BA}$ in matrix (or column) form.

---

What did you get? Find the solution below to double-check your answer.

## Solution to Example 6

**a) Vector $\vec{AB}$**
**Solution**

$$\vec{AB} = \vec{OB} - \vec{OA} = \mathbf{b} - \mathbf{a}$$
$$= (3\hat{\imath} - 2\hat{\jmath}) - (4\hat{\imath} + 3\hat{\jmath})$$
$$= -\hat{\imath} - 5\hat{\jmath}$$

---

**b) Vector $\vec{BA}$**
**Solution**

$$\vec{BA} = \vec{OA} - \vec{OB} = \mathbf{a} - \mathbf{b}$$
$$= \begin{pmatrix} 4 \\ 3 \end{pmatrix} - \begin{pmatrix} 3 \\ -2 \end{pmatrix}$$
$$= \begin{pmatrix} 1 \\ 5 \end{pmatrix}$$

**NOTE**

$$\vec{BA} = \vec{AB} = -(-i - 5j) = (i + 5j)$$

---

## Example 7

Points $A$ and $B$ have coordinates $(1, -3, 2)$ and $(2, 4, -3)$, respectively, with respect to the origin $O$. Find the following vectors in Cartesian and column matrix notation:

a) $\vec{OA}$

b) $\vec{OB}$

c) $\vec{AB}$

# Vectors

What did you get? Find the solution below to double-check your answer.

## Solution to Example 7

**HINT**

In this example, what we need is the coordinate $(x, y, z)$, and this should be translated to a vector equivalent in matrix form as $\begin{pmatrix} x \\ y \\ z \end{pmatrix}$ or Cartesian form as $xi + yj + zk$.

**a)** $\overrightarrow{OA}$
**Solution**

$$\overrightarrow{OA} = \hat{i} - 3\hat{j} + 2\hat{k}$$

$$\therefore \overrightarrow{OA} = \begin{pmatrix} 1 \\ -3 \\ 2 \end{pmatrix}$$

**b)** $\overrightarrow{OB}$
**Solution**

$$\overrightarrow{OB} = 2\hat{i} + 4\hat{j} - 3\hat{k}$$

$$\therefore \overrightarrow{OB} = \begin{pmatrix} 2 \\ 4 \\ -3 \end{pmatrix}$$

**c)** $\overrightarrow{AB}$
**Solution**

**Method 1** Cartesian form

$$\overrightarrow{AB} = \overrightarrow{OB} - \overrightarrow{OA}$$
$$= (2\hat{i} + 4\hat{j} - 3\hat{k}) - (\hat{i} - 3\hat{j} + 2\hat{k})$$
$$\therefore \overrightarrow{AB} = \hat{i} + 7\hat{j} - 5\hat{k}$$

**Method 2** Matrix form

$$\overrightarrow{AB} = \overrightarrow{OB} - \overrightarrow{OA}$$
$$= \begin{pmatrix} 2 \\ 4 \\ -3 \end{pmatrix} - \begin{pmatrix} 1 \\ -3 \\ 2 \end{pmatrix}$$
$$\therefore \overrightarrow{AB} = \begin{pmatrix} 1 \\ 7 \\ -5 \end{pmatrix}$$

## 11.13 ADDITION AND SUBTRACTION OF VECTORS

Vectors can be added and subtracted either analytically or by graphical method. Here, we will consider addition and subtraction of two vectors, but the principles can be used for any number of vectors.

### 11.13.1 ADDITION OF VECTORS

In Figure 11.10, there are two vectors: $a = \overrightarrow{AB}$ and $b = \overrightarrow{BC}$. These are called **components**. If we are to add these vectors together, we will have $a + b = \overrightarrow{AC}$. Vector $\overrightarrow{AC}$ is called the **resultant** of vectors $a$ and $b$. Notice that we use double arrows for resultant $\overrightarrow{AC}$. This is conventional, and it helps to differentiate it from component vectors.

Hence, we can write

$$\boxed{\overrightarrow{AC} = \overrightarrow{AB} + \overrightarrow{BC}} \tag{11.9}$$

If $\overrightarrow{AC} = c$, we can also write

$$\boxed{c = a + b} \tag{11.10}$$

The above is called the '*triangle law*' of vector addition. This law can be used repeatedly to find the resultant of any number of vectors in a polygon, for example.

Generally, a resultant is the shortest distance between the start and end point. The arrow of the components is such that the head of one will be followed by the tail of the next component in the same direction. The direction of the resultant is always to counterbalance the components. For example, if components follow a clockwise direction, the resultant will be anti-clockwise and vice versa, and this applies to any number of vectors.

In adding vectors, there are two laws that need to be considered and applied, namely:

1) Commutative Law

The law states that:

$$a + b = b + a$$

In other words, the order of adding two vectors is irrelevant.

This can be proved using a rectangle, square, parallelogram, or rhombus as shown in Figure 11.11.

The diagram for parallelogram (bottom right) is also called parallelogram law.

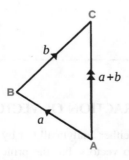

**FIGURE 11.10** Addition of vectors illustrated.

# Vectors

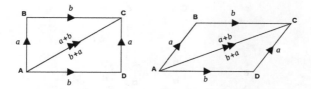

**FIGURE 11.11** Commutative law of vectors illustrated.

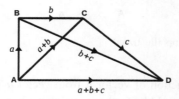

**FIGURE 11.12** Associative law of vectors illustrated.

2) Associative Law

Let's consider a trapezium $ABCD$ (Figure 11.12) with $\vec{AB}$, $\vec{BC}$, and $\vec{CD}$ being the components while $\vec{AD}$ the resultant.

To find the resultant, we use the triangle law twice. First, we take the triangle $ABC$ and obtain a resultant $\vec{AC} = \vec{AB} + \vec{BC} = a + b$. Now using triangle ACD, $\vec{AD} = \vec{AC} + \vec{CD} = (\vec{AB} + \vec{BC}) + \vec{CD}$. Thus, the resultant $\vec{AD}$ is

$$\vec{AD} = (a + b) + c$$

For the above case, we could have used triangle $BCD$ and then $ABD$. This will result in

$$\vec{AD} = a + (b + c)$$

Thus

$$(a + b) + c = a + (b + c)$$

The above is known as the *Associative Law* of vector addition, which shows that the order of adding three vectors is irrelevant.

## 11.13.2 Subtraction of Vectors

In Figure 11.13(a), we have two vectors added together to obtain $\vec{AC} = a + b$ as before. Recall that a negative vector is when the direction is rotated 180 degrees, but the magnitude remains unchanged. Let's create a negative vector of $b$ by rotation to produce $-b$ as shown in Figure 11.13(b). Our resultant is now $\vec{AC} = \vec{AB} + \vec{BC} = a + (-b) = a - b$.

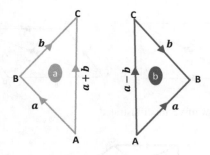

**FIGURE 11.13** Subtraction of vectors illustrated.

We can therefore say that subtraction of vector $b$ from vector $a$ is the same as (or equivalent to) addition of $a$ and negative $b$. It should be further noted that when going against the direction of the arrowhead on a vector, it means subtraction of that vector.

Let's try some examples.

### Example 8

Three vectors on a plane are given as $a = 2\hat{\imath} + 3\hat{\jmath}$, $b = \hat{\imath} - 2\hat{\jmath}$, and $c = -6\hat{\imath} + \hat{\jmath}$.

   a) Write vectors $a$, $b$, and $c$ in column (or matrix) form.
   b) Determine $a - 3b + c$.
   c) Determine the magnitude of $a - 3b + c$.
   d) Determine a unit vector in the direction of the vector in (b).

What did you get? Find the solution below to double-check your answer.

### Solution to Example 8

**a)** Column form
**Solution**

$a = 2\hat{\imath} + 3\hat{\jmath}$ $\qquad$ $b = \hat{\imath} - 2\hat{\jmath}$ $\qquad$ $c = -6\hat{\imath} + \hat{\jmath}$

$= \begin{pmatrix} 2 \\ 3 \end{pmatrix}$ $\qquad$ $= \begin{pmatrix} 1 \\ -2 \end{pmatrix}$ $\qquad$ $= \begin{pmatrix} -6 \\ 1 \end{pmatrix}$

**b)** $a - 3b + c$
**Solution**

$$a - 3b + c = \begin{pmatrix} 2 \\ 3 \end{pmatrix} - 3\begin{pmatrix} 1 \\ -2 \end{pmatrix} + \begin{pmatrix} -6 \\ 1 \end{pmatrix}$$

$$= \begin{pmatrix} 2 \\ 3 \end{pmatrix} - \begin{pmatrix} 3 \\ -6 \end{pmatrix} + \begin{pmatrix} -6 \\ 1 \end{pmatrix}$$

$$= \begin{pmatrix} 2 - 3 - 6 \\ 3 + 6 + 1 \end{pmatrix}$$

$$\therefore a - 3b + c = \begin{pmatrix} -7 \\ 10 \end{pmatrix}$$

**c) Magnitude**
**Solution**

$$|a - 3b + c| = \sqrt{(-7)^2 + (10)^2}$$
$$= \sqrt{49 + 100}$$
$$= \sqrt{149}$$

**d) Unit vector**
**Solution**
Let this be $\hat{d}$

$$\hat{d} = \frac{a - 3b + c}{|a - 3b + c|}$$
$$= \frac{-7\hat{\imath} + 10\hat{\jmath}}{\sqrt{149}}$$

This can also be written as

$$\hat{d} = -\frac{1}{\sqrt{149}}(7\hat{\imath} - 10\hat{\jmath})$$

Let's try another example.

### Example 9

if $a = \alpha\hat{\imath} + 2\hat{\jmath} + 4\hat{k}$ and $b = 4\hat{\imath} - 6\hat{\jmath} - 3\hat{k}$. Determine the possible values of $\alpha$ such that $|a + 2b| = 15$.

What did you get? Find the solution below to double-check your answer.

### Solution to Example 9

Given that

$$|a + 2b| = 15$$

We have

$$a + 2b = \begin{pmatrix} \alpha \\ 2 \\ 4 \end{pmatrix} + 2\begin{pmatrix} 4 \\ -6 \\ -3 \end{pmatrix}$$

$$= \begin{pmatrix} \alpha \\ 2 \\ 4 \end{pmatrix} + \begin{pmatrix} 8 \\ -12 \\ -6 \end{pmatrix}$$

$$= \begin{pmatrix} \alpha + 8 \\ -10 \\ -2 \end{pmatrix}$$

Now let's find the modulus of the above vector as:

$$|a + 2b| = \sqrt{(\alpha + 8)^2 + (-10)^2 + (-2)^2} = 15$$

$$\sqrt{(\alpha + 8)^2 + 104} = 15$$

$$(\alpha + 8)^2 + 104 = 225$$

$$(\alpha + 8)^2 = 121$$

$$\alpha + 8 = \pm 11$$

$$\alpha = -8 \pm 11$$

Thus

$$\alpha = -8 + 11 \quad \textbf{OR} \quad \alpha = -8 - 11$$
$$\alpha = 3 \quad\quad\quad\quad\quad\quad \alpha = -19$$

$$\therefore \alpha = 3 \text{ or} - 19$$

### ALTERNATIVE METHOD

$$(\alpha + 8)^2 = 121$$
$$\alpha^2 + 16\alpha + 64 = 121$$
$$\alpha^2 + 16\alpha - 57 = 0$$
$$(\alpha - 3)(\alpha + 19) = 0$$
$$\therefore \alpha = 3 \text{ or} - 19$$

# Vectors

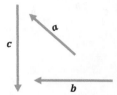

**FIGURE 11.14** Example 10.

Another example to try.

## Example 10

The diagram shown in Figure 11.14 shows three free vectors $a$, $b$, and $c$. Using a graphical method, find the resultant of these vectors.

What did you get? Find the solution below to double-check your answer.

## Solution to Example 10

If the resultant is denoted by $r$, then (Figure 11.15)

$$r = a + b + c$$

**NOTE**
This is only one option in many possible diagrams that can be formed using the vectors.

Another example to try.

## Example 11

Calculate the resultant of the polygon vector shown in Figure 11.16.

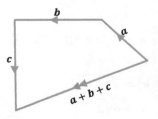

**FIGURE 11.15** Solution to Example 10.

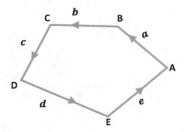

**FIGURE 11.16** Example 11.

What did you get? Find the solution below to double-check your answer.

### Solution to Example 11

$$\vec{AB} + \vec{BC} + \vec{CD} + \vec{DE} = \vec{AE}$$

Thus

$$a + b + c + d = -e$$

This implies that

$$a + b + c + d + e = \mathbf{0}$$

**NOTE**
Because the figure is closed and all the vectors follow the same direction, the resultant is zero.

---

A final example to try.

### Example 12

Calculate the magnitude of the resultant of vectors $u = -2\hat{\imath} + 5\hat{\jmath}$, $v = 7\hat{\imath} - \hat{\jmath}$, and $w = \hat{\imath} + 2\hat{\jmath}$.

---

What did you get? Find the solution below to double-check your answer.

### Solution to Example 12

$$u + v + w = (-2\hat{\imath} + 5\hat{\jmath}) + (7\hat{\imath} - \hat{\jmath}) + (\hat{\imath} + 2\hat{\jmath})$$
$$= 6\hat{\imath} + 6\hat{\jmath}$$
$$|u + v + w| = \sqrt{6^2 + 6^2}$$
$$= 6\sqrt{2}$$

# Vectors

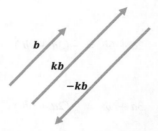

**FIGURE 11.17** Multiplication of a vector by a scalar illustrated.

## 11.14 MULTIPLICATION

### 11.14.1 Multiplication by a Scalar

A vector can be multiplied by a scalar (or number). When a non-zero scalar $k$ is used to multiply a vector $b$, the new vector will be $kb$ and will have a magnitude equal to $k|b|$ as shown in Figure 11.17. The two vectors will be parallel to each other but with different sizes. If $k$ is positive, $kb$ will be in the same direction as $b$ but if $k$ is negative, $kb$ will act in the opposite direction.

In other words, in the first case, we simply multiply a vector by a number. The second is when a vector multiplies another similar vector, and this is divided into two sub-divisions.

- $kb$ is obtained from $b$ by multiplying the latter by $k$ where $k > 1$.
- $-kb$ is obtained from $kb$ by inversion, i.e. multiplying with $k = -1$.

If $k > 1$ the size of the vector will be bigger, but if $0 < k < 1$ the magnitude of the resulting vector will be smaller than the original vector. In general, we can say that if vectors are scalar multiples of each other then they are parallel and vice versa.

Let's try an example.

**Example 13**

Show whether $2a + 6b$ and $3a + 9b$ are parallel or not.

What did you get? Find the solution below to double-check your answer.

**Solution to Example 13**

$2a + 6b$ and $3a + 9b$

$$2a + 6b = 2(a + 3b)$$

Also

$$3a + 9b = 3(a + 3b)$$

Therefore

$$2a + 6b = \frac{2}{3}(3a + 9b)$$

**OR**

$$3a + 9b = \frac{3}{2}(2a + 6b)$$

Therefore, the two vectors are

*Parallel*

---

### 11.14.2 SCALAR (OR DOT) PRODUCT

Given two vectors **a** and **b**, their scalar product (also called dot product) is denoted as **a.b** and is expressed as:

$$\boxed{a.b = |a|\,|b|\cos\theta} \qquad (11.11)$$

$\theta$ is the acute or obtuse angle between the two vectors when they are both pointing away from the origin or reference point, such that $0 \leq \theta \leq 180°$. In other words, when the tails of the two vectors meet at the origin.

**a.b** is read as 'a dot b' or 'a scalar b'. It is a 'dot' product because of the presence of the dot (and this must not be omitted) and it is 'scalar' because the result you obtain is always a scalar. This is true for vectors in two dimensions as well as three dimensions (Figure 11.18).

Since it is valid to write $|a|\,|b|\cos\theta$ as $|b|\,|a|\cos\theta$, it follows that if $|a|\,|b|\cos\theta = a.b$ then $|b|\,|a|\cos\theta = b.a$. Consequently, $a.b = b.a$, which implies that the dot product of two vectors is commutative. By re-arranging the formula above, we can determine the angle $\theta$ between two known vectors as:

$$\boxed{\cos\theta = \frac{a.b}{|a|\,|b|}} \qquad (11.12)$$

Following from Equation 11.11, there are two conditions of interest.

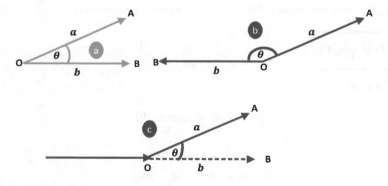

**FIGURE 11.18** Dot product of vectors illustrated.

# Vectors

**Condition 1** Vectors are perpendicular

This is when $\theta = 90°$. Since $\cos 90° = 0$ we have

$$\boxed{\begin{aligned} a.b &= |a|\,|b|\cos\theta \\ &= |a|\,|b| \times 0 \\ &= 0 \end{aligned}} \tag{11.13}$$

In other words, the scalar product of two perpendicular (or orthogonal, as sometimes called) vectors is zero. It is also true that either $|a|$, $|b|$, or both are zero. One of the consequences of this is:

$$\boxed{\hat{\imath}.\hat{\jmath} = \hat{\jmath}.\hat{\imath} = |\hat{\imath}|\,|\hat{\jmath}| \times 0 = 0} \quad \boxed{\hat{\imath}.\hat{k} = \hat{k}.\hat{\imath} = |\hat{\imath}|\,|\hat{k}| \times 0 = 0} \quad \boxed{\hat{\jmath}.\hat{k} = \hat{k}.\hat{\jmath} = |\hat{\jmath}|\,|k| \times 0 = 0} \tag{11.14}$$

**Condition 2** Vectors are parallel

This is when

- $\theta = 0$ (or when the vectors point in the same direction), and
- $\theta = 180°$ (or when the vectors point in opposite directions).

Since $\cos 0° = 1$ and $\cos 180° = -1$, we have

$$\boxed{a.b = |a|\,|b| \text{ for } \theta = 0} \quad \boxed{a.b = -|a|\,|b| \text{ for } \theta = 180°} \tag{11.15}$$

In other words, the scalar product of two parallel vectors is equal to the product of their magnitude. It is interesting that we are able to find the scalar product of parallel lines when they do not even intersect. This is because we are able to associate an angle to this. A particular case is when the vectors are parallel and equal, then we have $|a|\,|a|\cos\theta = |a|\,|a| \times 1 = |a|^2$. Consequently, we have

$$\boxed{\hat{\imath}.\hat{\imath} = |\hat{\imath}|\,|\hat{\imath}| \times 1 = |\hat{\imath}|^2 = 1} \quad \boxed{\hat{\jmath}.\hat{\jmath} = \hat{\jmath}.\hat{\jmath} = |\hat{\jmath}|\,|\hat{\jmath}| \times 1 = |\hat{\jmath}|^2 = 1}$$

$$\boxed{\hat{k}.\hat{k} = \hat{k}.\hat{k} = |\hat{k}|\,|k| \times 1 = |\hat{k}|^2 = 1} \tag{11.16}$$

We can use the above facts to derive a general formula for scalar products of two vectors, which can be used when the angle between them is unknown.

- **Two-Dimensional Vectors**

Let $a = a_1\hat{\imath} + a_2\hat{\jmath}$ and $b = b_1\hat{\imath} + b_2\hat{\jmath}$, then

$$\begin{aligned} a.b &= (a_1\hat{\imath} + a_2\hat{\jmath})(b_1\hat{\imath} + b_2\hat{\jmath}) \\ &= (a_1\hat{\imath}).(b_1\hat{\imath}) + (a_1\hat{\imath}).(b_2\hat{\jmath}) + (a_2\hat{\jmath}).(b_1\hat{\imath}) + (a_2\hat{\jmath}).(b_2\hat{\jmath}) \\ &= a_1b_1\hat{\imath}.\hat{\imath} + a_1b_2\hat{\imath}.\hat{\jmath} + a_2b_1\hat{\imath}.\hat{\jmath} + a_2b_2\hat{\jmath}.\hat{\jmath} \\ &= a_1b_1 \times 1 + a_1b_2 \times 0 + a_2b_1 \times 0 + a_2b_2 \times 1 \\ &= a_1b_1 + 0 + 0 + a_2b_2 \\ \therefore\; \boxed{a.b &= a_1b_1 + a_2b_2} \end{aligned} \tag{11.17}$$

In column or matrix form, if $a = \begin{pmatrix} a_1 \\ a_2 \end{pmatrix}$ and $b = \begin{pmatrix} b_1 \\ b_2 \end{pmatrix}$, then

$$\boxed{a.b = \begin{pmatrix} a_1 b_1 \\ a_2 b_2 \end{pmatrix}} \tag{11.18}$$

- **Three-Dimensional Vectors**

Let $a = a_1 \hat{i} + a_2 \hat{j} + a_3 \hat{k}$ and $b = b_1 \hat{i} + b_2 \hat{j} + b_3 \hat{k}$, then

$$\begin{aligned}
a.b &= (a_1 \hat{i} + a_2 \hat{j} + a_3 \hat{k})(b_1 \hat{i} + b_2 \hat{j} + b_3 \hat{k}) \\
&= (a_1 \hat{i}).(b_1 \hat{i}) + (a_1 \hat{i}).(b_2 \hat{j}) + (a_1 \hat{i}).(b_3 \hat{k}) + (a_2 \hat{j}).(b_1 \hat{i}) + (a_2 \hat{j}).(b_2 \hat{j}) + (a_2 \hat{j}).(b_3 \hat{k}) \\
&\quad + (a_3 \hat{k}).(b_1 \hat{i}) + (a_3 \hat{k}).(b_2 \hat{j}) + (a_3 \hat{k}).(b_3 \hat{k}) \\
&= a_1 b_1 \hat{i}.\hat{i} + a_1 b_2 \hat{i}.\hat{j} + a_1 b_3 \hat{i}.\hat{k} + a_2 b_1 \hat{i}.\hat{j} + a_2 b_2 \hat{j}.\hat{j} + a_2 b_3 \hat{j}.\hat{k} + a_3 b_1 \hat{i}.\hat{k} + a_3 b_2 \hat{j}.\hat{k} + a_3 b_3 \hat{k}.\hat{k} \\
&= a_1 b_1 \times 1 + a_1 b_2 \times 0 + a_1 b_3 \times 0 + a_2 b_1 \times 0 + a_2 b_2 \times 1 + a_2 b_3 \times 0 + a_3 b_1 \times 0 + a_3 b_2 \times 0 \\
&\quad + a_3 b_3 \times 1
\end{aligned}$$

$$\boxed{a.b = a_1 b_1 + a_2 b_2 + a_3 b_3} \tag{11.19}$$

In column or matrix form if, $a = \begin{pmatrix} a_1 \\ a_2 \\ a_3 \end{pmatrix}$ and $b = \begin{pmatrix} b_1 \\ b_2 \\ b_3 \end{pmatrix}$ then

$$\boxed{a.b = \begin{pmatrix} a_1 b_1 \\ a_2 b_2 \\ a_3 b_3 \end{pmatrix}} \tag{11.20}$$

The above short-hand formulas are very useful when computing scalar product where the angle is not given or not required.

Let's try some examples.

### Example 14

Simplify the following:

a) $(2\hat{i}).(\hat{k}) + (5\hat{j}).(3\hat{j}) - (4\hat{k}).(10\hat{i})$

b) $(2\hat{i}).(3\hat{i}) + (4\hat{i}).(-5\hat{j}) - (4\hat{j}).(7\hat{k}) + (-4\hat{k}).(3\hat{k})$

# Vectors

What did you get? Find the solution below to double-check your answer.

### Solution to Example 14

**a)** $(2\hat{\imath}).(\hat{k}) + (5\hat{\jmath}).(3\hat{\jmath}) - (4\hat{k}).(10\hat{\imath})$

**Solution**

$$(2\hat{\imath}).(\hat{k}) + (5\hat{\jmath}).(3\hat{\jmath}) - (4\hat{k}).(10\hat{\imath})$$
$$= (2 \times 1 \times \cos 90°) + (5 \times 3 \times \cos 0) - (4 \times 10 \times \cos 90°)$$
$$= (0) + (15) - (0)$$
$$= 15$$

**b)** $(2\hat{\imath}).(3\hat{\imath}) + (4\hat{\imath}).(-5\hat{\jmath}) - (4\hat{\jmath}).(7\hat{k}) + (-4\hat{k}).(3\hat{k})$

**Solution**

$$(2\hat{\imath}).(3\hat{\imath}) + (4\hat{\imath}).(-5\hat{\jmath}) - (4\hat{\jmath}).(7\hat{k}) + (-4\hat{k}).(3\hat{k})$$
$$= (2 \times 3 \times \cos 0) - (4 \times 5 \times \cos 90°) - (4 \times 7 \times \cos 90°) - (4 \times 3 \times \cos 0)$$
$$= (6) - (0) - (0) - (12)$$
$$= -6$$

### Example 15

if $a = 3\hat{\imath} + 2\hat{\jmath}$ and $b = \hat{\imath} - 3\hat{\jmath}$. Determine:

a) $a.b$

b) the angle between the two vectors. Present the answer correct to 1 d.p.

What did you get? Find the solution below to double-check your answer.

### Solution to Example 15

**a)** $a.b$

**Solution**

$$a.b = (3 \times 1) + (2 \times -3)$$
$$= (3) - (6)$$
$$\therefore a.b = -3$$

**b) The angle**
**Solution**

$$|a| = \sqrt{(3)^2 + (2)^2}$$
$$= \sqrt{13}$$
$$|b| = \sqrt{(1)^2 + (-3)^2}$$
$$= \sqrt{10}$$

$$a.b = |a||b|\cos\theta$$
$$\cos\theta = \frac{a.b}{|a||b|} = \frac{-3}{(\sqrt{13})(\sqrt{10})}$$

$$\theta = \cos^{-1}\left(\frac{-3}{(\sqrt{13})(\sqrt{10})}\right)$$

$$\therefore \theta = 105.3°$$

---

### Example 16

Given that $c = -3\hat{\imath} - 2\hat{\jmath} + 3\hat{k}$ and $d = -5\hat{\imath} + 6\hat{\jmath} - \hat{k}$, determine the following:

a) $c.d$  

b) the angle between the two vectors.

What did you get? Find the solution below to double-check your answer.

### Solution to Example 16

**a) $c.d$**
**Solution**

$$c.d = (-3 \times -5) + (-2 \times 6) + (3 \times -1)$$
$$= (15) + (-12) + (-3)$$
$$\therefore c.d = 0$$

**b) The angle**
**Solution**
Because the scalar product is zero, it implies that the vectors are perpendicular.

$$\therefore \theta = 90°$$

# Vectors

But let's check this analytically as:

$$|c| = \sqrt{(-3)^2 + (-2)^2 + (3)^2}$$
$$= \sqrt{22}$$
$$|d| = \sqrt{(-5)^2 + (6)^2 + (-1)^2}$$
$$= \sqrt{62}$$
$$c.d = |c||d|\cos\theta$$
$$\cos\theta = \frac{c.d}{|c||d|} = \frac{0}{(\sqrt{22})(\sqrt{62})}$$
$$\theta = \cos^{-1}\left(\frac{0}{(\sqrt{22})(\sqrt{62})}\right)$$
$$\therefore \theta = 90°$$

## Example 17

Given that $a = 3\hat{\imath} + k\hat{\jmath}$ and $b = 2\hat{\jmath} + \hat{k}$, determine two possible values of $k$ if the angle between the two vectors is 45°.

What did you get? Find the solution below to double-check your answer.

## Solution to Example 17

**Solution**
Given $a = 3\hat{\imath} + k\hat{\jmath}$ and $b = 2\hat{\jmath} + \hat{k}$, we have:

$$|a| = \sqrt{(3)^2 + (k)^2}$$
$$= \sqrt{9 + k^2}$$
$$|b| = \sqrt{(2)^2 + (1)^2}$$
$$= \sqrt{5}$$

Also

$$a.b = \begin{pmatrix} 3 \\ k \\ 0 \end{pmatrix} \cdot \begin{pmatrix} 0 \\ 2 \\ 1 \end{pmatrix} = 2k$$

$$\cos\theta = \cos 45° = \frac{1}{\sqrt{2}}$$

We have

$$a.b = |a||b|\cos\theta$$

$$2k = \left(\sqrt{9+k^2}\right)\left(\sqrt{5}\right)\frac{1}{\sqrt{2}}$$

$$2\sqrt{2}k = \left(\sqrt{9+k^2}\right)\left(\sqrt{5}\right)$$

Square both sides

$$2\sqrt{2}k = \left(\sqrt{9+k^2}\right)\left(\sqrt{5}\right)$$
$$8k^2 = 5\left(9+k^2\right)$$
$$8k^2 = 45 + 5k^2$$
$$3k^2 = 45$$
$$k^2 = 15$$
$$k = \pm\sqrt{15}$$
$$\therefore k = \pm\sqrt{15}$$

Another example to try.

### Example 18

Figure 11.19 shows two vectors $a$ and $b$, which represent forces of magnitude of 3 N and 5 N, respectively, in the direction shown by their arrow. Find the scalar product of these forces.

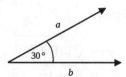

FIGURE 11.19  Example 18.

# Vectors

What did you get? Find the solution below to double-check your answer.

## Solution to Example 18

**Solution**

$$a.b = |a| |b| \cos 30°$$
$$= 3 \times 5 \times \cos 30°$$
$$= 15 \times \frac{\sqrt{3}}{2}$$
$$\therefore a.b = \frac{15\sqrt{3}}{2}$$

Another example to try.

## Example 19

The diagram in Figure 11.20 shows two vectors $p$ and $q$ which represent velocities of magnitude of $12 \text{ ms}^{-1}$ and $7 \text{ ms}^{-1}$, respectively, in the direction shown by their arrow. Determine the scalar product of these velocities.

What did you get? Find the solution below to double-check your answer.

## Solution to Example 19

The correct angle to use is the angle $\theta$ when the two vectors are pointing away from the common reference point $O$, which is shown in the re-drawn diagram in Figure 11.21. Therefore

$$p.q. = |p| |q| \cos \theta$$
$$= 12 \times 7 \times \cos(180° - 120°)$$
$$= 12 \times 7 \times \cos 60°$$
$$\therefore a.b = 42$$

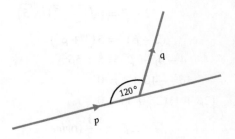

**FIGURE 11.20** Example 19.

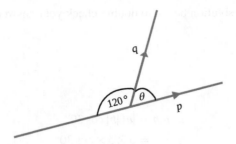

**FIGURE 11.21** Solution to Example 19.

Another example to try.

### Example 20

Determine the value of $\beta$ if vector $\boldsymbol{a} = \hat{\imath} + \beta\hat{\jmath}$ is parallel to vector $\boldsymbol{b} = 2\hat{\imath} - \hat{\jmath}$.

What did you get? Find the solution below to double-check your answer.

### Solution to Example 20

$$\boldsymbol{a}.\boldsymbol{b} = (1 \times 2) + (\beta \times -1)$$
$$= 2 - \beta$$

$$|\boldsymbol{a}| = \sqrt{(1)^2 + (\beta)^2}$$
$$= \sqrt{1 + \beta^2}$$

$$|\boldsymbol{b}| = \sqrt{(2)^2 + (-1)^2}$$
$$= \sqrt{5}$$

If the vectors are parallel, then

$$\boldsymbol{a}.\boldsymbol{b} = |\boldsymbol{a}|\,|\boldsymbol{b}|$$

which implies that

$$2 - \beta = \left(\sqrt{1+\beta^2}\right)\left(\sqrt{5}\right)$$
$$(2-\beta)^2 = 5(1+\beta^2)$$
$$4 - 4\beta + \beta^2 = 5 + 5\beta^2$$
$$4\beta^2 + 4\beta + 1 = 0$$
$$(2\beta + 1)(2\beta + 1) = 0$$
$$\therefore \beta = -\frac{1}{2} \; (twice)$$

# Vectors

Let's try another set of examples

### Example 21

Given that $a = \hat{\imath} - 3\hat{\jmath} + 5\hat{k}$, $b = 2\hat{\imath} + 3\hat{\jmath} + 4\hat{k}$, and $c = -2\hat{\imath} - \sqrt{6}\hat{\jmath} + \hat{k}$, show that:

a) $a.a = |a|^2$ 
b) $b.b = |b|^2$ 
c) $c.c = |c|^2$

What did you get? Find the solution below to double-check your answer.

### Solution to Example 21

a) $a.a = |a|^2$
**Solution**

$$a.a = (1 \times 1) + (-3 \times -3) + (5 \times 5)$$
$$= (1) + (9) + (25) = 35$$

Also

$$|a| = \sqrt{(1)^2 + (-3)^2 + (5)^2}$$
$$= \sqrt{1 + 9 + 25} = \sqrt{35}$$

Thus

$$|a|^2 = \left(\sqrt{35}\right)^2 = 35$$

$$\therefore a.a = |a|^2$$

b) $b.b = |b|^2$
**Solution**

$$b.b = (2 \times 2) + (3 \times 3) + (4 \times 4)$$
$$= (4) + (9) + (16) = 29$$

Also

$$|b| = \sqrt{(2)^2 + (3)^2 + (4)^2}$$
$$= \sqrt{4 + 9 + 16} = \sqrt{29}$$

Thus

$$|b|^2 = \left(\sqrt{29}\right)^2 = 29$$

$$\therefore b.b = |b|^2$$

c) $c.c = |c|^2$
**Solution**

$$c.c = (-2 \times -2) + \left(-\sqrt{6} \times -\sqrt{6}\right) + (1 \times 1)$$
$$= (4) + (6) + (1) = 11$$

Also

$$|c| = \sqrt{(-2)^2 + \left(-\sqrt{6}\right)^2 + (1)^2}$$
$$= \sqrt{4 + 6 + 1} = \sqrt{11}$$

Thus

$$|c|^2 = \left(\sqrt{11}\right)^2 = 11$$

$$\therefore c.c = |c|^2$$

---

From the above worked example, we can conclude that for any given vector $r$ in $xy$ plane or $xyz$ space, it follows that

$$\boxed{r.r = |r|^2} \tag{11.21}$$

Let's try another example.

### Example 22

Two vectors are given by $a = \hat{i} - x\hat{j} + 3\hat{k}$ and $b = -2\hat{i} + 5\hat{j} - x\hat{k}$. Determine the value of $x$ if $a.b = 14$.

---

What did you get? Find the solution below to double-check your answer.

### Solution to Example 22

$$a.b = \left(\hat{i} - x\hat{j} + 3\hat{k}\right).\left(-2\hat{i} + 5\hat{j} - x\hat{k}\right)$$
$$= (1)(-2) + (-x)(5) + (3)(-x)$$
$$= -2 - 5x - 3x$$
$$= -2 - 8x$$

But

$$a.b = 14$$

# Vectors

Thus

$$-2 - 8x = 14$$
$$-8x = 16$$
$$x = -2$$

Let's try one more example.

## Example 23

Given that $a = 3\hat{\imath} + 4\hat{\jmath} - 5\hat{k}$ and $b = \hat{\imath} - 3\hat{\jmath} + 4\hat{k}$, find two possible vectors that are perpendicular to both $a$ and $b$.

What did you get? Find the solution below to double-check your answer.

## Solution to Example 23

Let a perpendicular vector to $a$ and $b$ be $c$, such that

$$c = x\hat{\imath} + y\hat{\jmath} + z\hat{k}$$

Thus, we have the three-column vectors as

$$a = \begin{pmatrix} 3 \\ 4 \\ -5 \end{pmatrix}, b = \begin{pmatrix} 1 \\ -3 \\ 4 \end{pmatrix}, c = \begin{pmatrix} x \\ y \\ z \end{pmatrix}$$

Since $c$ is perpendicular to both $a$ and $b$, we have

$$a.c = 0$$

$$\begin{pmatrix} 3 \\ 4 \\ -5 \end{pmatrix} \cdot \begin{pmatrix} x \\ y \\ z \end{pmatrix} = 0$$

That's

$$3x + 4y - 5z = 0 \;----\;\text{(i)}$$

Similarly,

$$b.c = 0$$

$$\begin{pmatrix} 1 \\ -3 \\ 4 \end{pmatrix} \cdot \begin{pmatrix} x \\ y \\ z \end{pmatrix} = 0$$

That's

$$x - 3y + 4z = 0 \;----\;\text{(ii)}$$

For three unknowns, we need three equations, but we can assume that $z = 1$ for this case. Therefore we have from (i) that

$$3x + 4y - 5 \times 1 = 0$$
$$3x + 4y = 5 \quad ----\text{(iii)}$$

And from (ii), we have

$$x - 3y + 4 \times 1 = 0$$
$$x - 3y = -4 \quad ----\text{(iv)}$$

From (iv), we have

$$x = 3y - 4 \quad ----\text{(v)}$$

Substitute (v) in (iii), we have

$$3(3y - 4) + 4y = 5$$
$$9y - 12 + 4y = 5$$
$$13y = 17$$
$$\therefore y = \frac{17}{13}$$

From (v), we have

$$x = 3\left(\frac{17}{13}\right) - 4$$
$$\therefore x = -\frac{1}{13}$$

Hence

$$c = -\frac{1}{13}\hat{i} + \frac{17}{13}\hat{j} + \hat{k}$$

Let the other perpendicular vector to $a$ and $b$ be $d$. Since $c$ and $d$ must be parallel, it follows that

$$d = kc$$

where $k$ is any real number.

Let $k = 13$, then

$$d = 13c$$
$$= 13\left(-\frac{1}{13}\hat{i} + \frac{17}{13}\hat{j} + \hat{k}\right)$$
$$= -\hat{i} + 17\hat{j} + 13\hat{k}$$

The two perpendicular vectors are

$$c = -\frac{1}{13}\hat{i} + \frac{17}{13}\hat{j} + \hat{k}$$
$$d = -\hat{i} + 17\hat{j} + 13\hat{k}$$

Thus, we have the three-column vectors as

$$c = \begin{pmatrix} -\frac{1}{13} \\ \frac{17}{13} \\ 1 \end{pmatrix}, d = \begin{pmatrix} -1 \\ 17 \\ 13 \end{pmatrix}$$

**NOTE**
There are an infinite number of perpendicular vectors to both $a$ and $b$ since $k$ is any real number.

## 11.15 VECTOR EQUATION OF A LINE

The vector equation of a straight line, say $AR$ (Figure 11.22), passing through point $A$ and parallel to vector $b$ is given by

$$\boxed{r = a + tb} \tag{11.22}$$

where

- $r$ is the position vector of a point on line $AR$
- $a$ is the position vector of a fixed-point $A$
- $b$ is the vector (called direction vector) that is parallel to the line $AR$, such that $\overrightarrow{AR} = tb$
- $t$ is a variable scalar, which varies depending on the point on the line. Other letters, including $s$, $\lambda$, and $\mu$, can also be used.

The above is very similar to a Cartesian equation of a line $y = mx + c$ and can be used to find the vector equation. In this case, the $y$-intercept $c$ can be used to determine $a$, and the gradient $m$ is used to find $b$, or vice versa. The above is true for both 2D and 3D, except that the former has two components and the latter has three components.

Let's try a couple of examples.

### Example 24

Determine the vector equation of a straight line which passes through point $P$, with a position vector $3\hat{\imath} - \hat{\jmath} + 2\hat{k}$ and parallel to vector $\hat{\imath} - 5\hat{\jmath}$. Give the answer in both Cartesian and column notations.

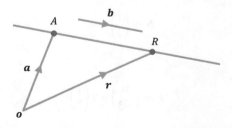

**FIGURE 11.22** Vector equation of a straight passing through points A and parallel to vector $b$ illustrated.

What did you get? Find the solution below to double-check your answer.

## Solution to Example 24

**Method 1** Cartesian form

$r = a + tb$
$= (3\hat{i} - \hat{j} + 2\hat{k}) + t(\hat{i} - 5\hat{j})$
$= 3\hat{i} - \hat{j} + 2\hat{k} + t\hat{i} - 5t\hat{j}$
$= 3\hat{i} + t\hat{i} - \hat{j} - 5t\hat{j} + 2\hat{k}$
$= (3+t)\hat{i} - (1+5t)\hat{j} + 2\hat{k}$

**Method 2** Matrix form

$r = a + tb$

$= \begin{pmatrix} 3 \\ -1 \\ 2 \end{pmatrix} + t \begin{pmatrix} 1 \\ -5 \\ 0 \end{pmatrix}$

$= \begin{pmatrix} 3 \\ -1 \\ 2 \end{pmatrix} + \begin{pmatrix} t \\ -5t \\ 0 \end{pmatrix}$

$= \begin{pmatrix} 3+t \\ -1-5t \\ 2 \end{pmatrix}$

## Example 25

$y = 2x - 3$ is the Cartesian equation of a straight line $l$. Write the vector equation of the line.

What did you get? Find the solution below to double-check your answer.

## Solution to Example 25

$$r = a + tb$$
$$y = mx + c$$
$$c = -3 = \begin{pmatrix} 0 \\ -3 \end{pmatrix}$$
$$\therefore a = \begin{pmatrix} 0 \\ -3 \end{pmatrix}$$
$$m = 2$$
$$\therefore b = \begin{pmatrix} 1 \\ 2 \end{pmatrix}$$

Therefore,

$$r = \begin{pmatrix} 0 \\ -3 \end{pmatrix} + t \begin{pmatrix} 1 \\ 2 \end{pmatrix}$$

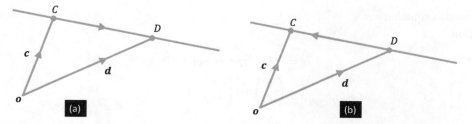

**FIGURE 11.23** Vector equation of a straight line passing through points C and D illustrated.

If we know two points along a line, as shown in Figure 11.23(a), such that $c$ is the position vector of point $C$ and $d$ is the position vector of point $D$, then the vector equation of the line is given as:

$$\boxed{r = c + t\overrightarrow{CD} = c + t(d - c)} \quad (11.23)$$

where $c$ is the position vector and $\overrightarrow{CD} = d - c$ is the direction vector.

If we flip the vector direction as shown in Figure 11.23(b), the direction vector becomes $\overrightarrow{DC} = c - d$ and we will have

$$\boxed{r = c + t(c - d)} \quad (11.24)$$

Note that we can replace the position vector with $d$ in the above two equations. The above equations will give different vector equations for the same line.

Let's try a few examples.

## Example 26

$5\hat{\imath} - 7\hat{\jmath}$ and $-\hat{\imath} + 2\hat{\jmath}$ are the position vectors of points $P$ and $Q$, respectively.

**a)** Determine the vector equation of a line connecting $P$ and $Q$.

**b)** Determine the Cartesian equation of the line through $P$ and $Q$ in the form $ax + by + c = 0$.

What did you get? Find the solution below to double-check your answer.

## Solution to Example 26

**a)** Vector equation
**Solution**

$$\begin{aligned} r &= p + t(q - p) \\ &= (5\hat{\imath} - 7\hat{\jmath}) + t[(-\hat{\imath} + 2\hat{\jmath}) - (5\hat{\imath} - 7\hat{\jmath})] \\ &= (5\hat{\imath} - 7\hat{\jmath}) + t[-\hat{\imath} + 2\hat{\jmath} - 5\hat{\imath} + 7\hat{\jmath}] \\ &= (5\hat{\imath} - 7\hat{\jmath}) + t[-6\hat{\imath} + 9\hat{\jmath}] \\ \therefore r &= (5\hat{\imath} - 7\hat{\jmath}) + t(-6\hat{\imath} + 9\hat{\jmath}) \end{aligned}$$

**b)** Cartesian equation
**Solution**

$$r = (5\hat{\imath} - 7\hat{\jmath}) + t(-6\hat{\imath} + 9\hat{\jmath})$$

$$= \begin{pmatrix} 5 \\ -7 \end{pmatrix} + t\begin{pmatrix} -6 \\ 9 \end{pmatrix}$$

$$m = -\frac{9}{6} = -\frac{3}{2}$$

$$y = mx + c$$

A point on the line is $\begin{pmatrix} 5 \\ -7 \end{pmatrix}$ which then has coordinates of $(5, -7)$. Thus

| Method 1 | Method 2 |
|---|---|
| $y = mx + c$ <br> $c = y - mx$ <br> $= -7 - \left(-\frac{3}{2}\right)(5) = \frac{1}{2}$ <br> Therefore <br> $y = -\frac{3}{2}x + \frac{1}{2}$ <br> $2y = -3x + 1$ <br> $3x + 2y - 1 = 0$ | $y - y_1 = m(x - x_1)$ <br> $y + 7 = -\frac{3}{2}(x - 5)$ <br> $2y + 14 = -3x + 15$ <br> $3x + 2y - 1 = 0$ |

**ALTERNATIVE METHOD**

$y$-intercept has coordinates of $(0, c)$. Hence,

$$\begin{pmatrix} 0 \\ c \end{pmatrix} = \begin{pmatrix} 5 \\ -7 \end{pmatrix} + t\begin{pmatrix} 6 \\ -9 \end{pmatrix} = \begin{pmatrix} 5 + 6t \\ -7 - 9t \end{pmatrix}$$

Equating the $\hat{\imath}$ component

$$0 = 5 + 6t$$

$$t = -\frac{5}{6}$$

Equating the $\hat{\jmath}$ component

$$c = -7 - 9t$$

$$c = -7 - 9\left(-\frac{5}{6}\right) = -7 + \frac{45}{6}$$

$$c = \frac{1}{2}$$

Hence

$$y = mx + c$$
$$y = -\frac{3}{2}x + \frac{1}{2}$$
$$2y = -3x + 1$$
$$3x + 2y - 1 = 0$$

**Example 27**

$r = \begin{pmatrix} 3 \\ 1 \\ 5 \end{pmatrix} + \lambda \begin{pmatrix} 1 \\ -1 \\ -2 \end{pmatrix}$ is a vector equation of a straight-line $AB$. Show whether or not point $C(4, 0, 3)$ lies on AB.

What did you get? Find the solution below to double-check your answer.

**Solution to Example 27**

$$r = \begin{pmatrix} 3 \\ 1 \\ 5 \end{pmatrix} + \lambda \begin{pmatrix} 1 \\ -1 \\ -2 \end{pmatrix}$$

$$= \begin{pmatrix} 3 + \lambda \\ 1 - \lambda \\ 5 - 2\lambda \end{pmatrix}$$

For $C(4, 0, 3)$ to be on the line, it implies that

$$\begin{pmatrix} 3 + \lambda \\ 1 - \lambda \\ 5 - 2\lambda \end{pmatrix} = \begin{pmatrix} 4 \\ 0 \\ 3 \end{pmatrix}$$

Equating $\hat{\imath}$ component

$$3 + \lambda = 4$$
$$\lambda = 1$$

Equating $\hat{\jmath}$ component

$$1 - \lambda = 0$$
$$\lambda = 1$$

Equating $\hat{k}$ component

$$5 - 2\lambda = 3$$
$$\lambda = 1$$

The value of $\lambda$ is the same in all cases so point $C$ lies on the line $AB$.

One final example to try.

## Example 28

$(a, 1, -1)$ are the coordinates of a point $R$ which lies on $r = (3\hat{\imath} - 2\hat{k}) + \mu(5\hat{\imath} - b\hat{\jmath} - \hat{k})$. Determine the values of $a$ and $b$.

What did you get? Find the solution below to double-check your answer.

## Solution to Example 28

$$r = (3\hat{\imath} - 2\hat{k}) + \mu(5\hat{\imath} - b\hat{\jmath} - \hat{k})$$

$$= \begin{pmatrix} 3 \\ 0 \\ -2 \end{pmatrix} + \mu \begin{pmatrix} 5 \\ -b \\ -1 \end{pmatrix}$$

$$= \begin{pmatrix} 3 + 5\mu \\ -\mu b \\ -2 - \mu \end{pmatrix}$$

For $(a, 1, -1)$ to be on the line, it implies that

$$\begin{pmatrix} 3 + 5\mu \\ -\mu b \\ -2 - \mu \end{pmatrix} = \begin{pmatrix} a \\ 1 \\ -1 \end{pmatrix}$$

Equating $\hat{k}$ component

$$-2 - \mu = -1$$
$$\mu = -1$$

Equating $\hat{\jmath}$ component

$$-\mu b = 1$$
$$b = \frac{1}{-\mu} = 1$$

Equating $\hat{\imath}$ component

$$3 + 5\mu = a$$
$$a = 3 - 5 = -2$$

Thus

$$a = -2, b = 1$$

Vectors

## 11.16 INTERSECTION OF TWO LINES

When two or more lines lie in the 2D $(x-y, x-z,$ or $y-z)$ plane or 3D space, one of these may happen:

**Case 1** The lines are parallel and never meet no matter how long they are (drawn or extended), such that the distance between them at all points remains the same. This can be demonstrated by showing that the vector equation of one is a scalar multiple of the other. For example, if two lines with vector equations $r = a + tb$ and $r = c + sd$ are parallel, then vector $b$ must be a multiple of vector $d$.

**Case 2** They intersect or cross each other at a single point. This condition can be demonstrated by equating the vector equations of the lines and showing that there is consistency. For example, if two lines with vector equations $r = a + \lambda b$ and $r = c + \mu d$ intersect, then $a + \lambda b = c + \mu d$ for a given value of $\mu$ and $\lambda$. Equate the $x$ components and $y$ components on both sides to find $\mu$ and $\lambda$. The coordinates of the point of intersection can be obtained by substituting $\mu$ or $\lambda$ in either equation, which should give the same answer.

**Case 3** They are neither parallel nor do they intersect. The lines in this case are called **skew**. This condition can be demonstrated by showing that the lines are neither parallel nor intersecting.

Cases 1 and 2 are common to both 2D and 3D, but case 3 only occurs in $xyz$ space. As a result, if two lines are in the same plane and not intersecting then they must be parallel, but this is not true if the same lines are in 3D space.

Let's try some examples.

### Example 29

Two lines $L_1$ and $L_2$ are, respectively, defined by vector equations $r = (\hat{\imath} - 3\hat{\jmath}) + t(2\hat{\imath} - 6\hat{\jmath} - 4\hat{k})$ and $r = (-2\hat{\imath} + \hat{\jmath} - 5\hat{k}) + s(3\hat{\imath} - 9\hat{\jmath} - 6\hat{k})$. Show whether these lines are parallel.

What did you get? Find the solution below to double-check your answer.

### Solution to Example 29

The general expression for vector equation is:

$$r = a + tb$$

The vector equations for the lines are

$L_1 : r = (\hat{\imath} - 3\hat{\jmath}) + t(2\hat{\imath} - 6\hat{\jmath} - 4\hat{k})$  $\qquad$ $L_2 : r = (-2\hat{\imath} + \hat{\jmath} - 5\hat{k}) + s(3\hat{\imath} - 9\hat{\jmath} - 6\hat{k})$

$= \begin{pmatrix} 1 \\ -3 \\ 0 \end{pmatrix} + t \begin{pmatrix} 2 \\ -6 \\ -4 \end{pmatrix}$ $\qquad\qquad\qquad\qquad$ $= \begin{pmatrix} -2 \\ 1 \\ -5 \end{pmatrix} + s \begin{pmatrix} 3 \\ -9 \\ -6 \end{pmatrix}$

For parallel lines, the direction vector of one line must be the scalar multiple of the other. If $b_1$ and $b_2$ are direction vectors for $L_1$ and $L_2$, respectively, then

$$b_1 = (2\hat{i} - 6\hat{j} - 4\hat{k}) \qquad b_2 = (3\hat{i} - 9\hat{j} - 6\hat{k})$$

$$= \begin{pmatrix} 2 \\ -6 \\ -4 \end{pmatrix} \qquad = \begin{pmatrix} 3 \\ -9 \\ -6 \end{pmatrix}$$

It can be shown that

$$\frac{3}{2} = \frac{-9}{-6} = \frac{-6}{-4} = 1.5$$

Thus

$$b_2 = 1.5 b_1$$

Hence, $L_1$ and $L_2$ are parallel.

### Example 30

The vector equations of two lines are given as $r = 2(4\hat{i} - \hat{j} + 3\hat{k}) + \lambda(-4\hat{i} - 2\hat{j} + 10\hat{k})$ and $r = 2(6\hat{i} + 2\hat{j} - 3\hat{k}) + \mu(2\hat{i} - \hat{j} - 4\hat{k})$.

**a)** Show that these lines intersect.

**b)** Find the coordinates of their point of intersection.

What did you get? Find the solution below to double-check your answer.

### Solution to Example 30

**a)** Show that these lines intersect.
**Solution**
Let's write the vectors in matrix form as:

$$r = (8\hat{i} - 2\hat{j} + 6\hat{k}) + \lambda(-4\hat{i} - 2\hat{j} + 10\hat{k})$$

$$= \begin{pmatrix} 8 \\ -2 \\ 6 \end{pmatrix} + \lambda \begin{pmatrix} -4 \\ -2 \\ 10 \end{pmatrix} = \begin{pmatrix} 8 - 4\lambda \\ -2 - 2\lambda \\ 6 + 10\lambda \end{pmatrix}$$

Also

$$r = (12\hat{i} + 4\hat{j} - 6\hat{k}) + \mu(2\hat{i} - \hat{j} - 4\hat{k})$$

$$= \begin{pmatrix} 12 \\ 4 \\ -6 \end{pmatrix} + \mu \begin{pmatrix} 2 \\ -1 \\ -4 \end{pmatrix} = \begin{pmatrix} 12 + 2\mu \\ 4 - \mu \\ -6 - 4\mu \end{pmatrix}$$

If the two lines intersect, then

$$\begin{pmatrix} 8 - 4\lambda \\ -2 - 2\lambda \\ 6 + 10\lambda \end{pmatrix} = \begin{pmatrix} 12 + 2\mu \\ 4 - \mu \\ -6 - 4\mu \end{pmatrix}$$

Equating $x$ components

$$8 - 4\lambda = 12 + 2\mu$$
$$8 - 12 = 4\lambda + 2\mu$$
$$4\lambda + 2\mu = -4$$
$$2\lambda + \mu = -2 \text{ ----- (i)}$$

Equating $y$ components

$$-2 - 2\lambda = 4 - \mu$$
$$-2 - 4 = 2\lambda - \mu$$
$$2\lambda - \mu = -6 \text{ ----- (ii)}$$

Let's solve the above two above equations. Add (i) and (ii)

$$4\lambda = -8$$
$$\boldsymbol{\lambda = -2}$$

From (i)

$$\mu = -2 - 2\lambda$$
$$\boldsymbol{\mu = 2}$$

Now we need to validate this on $z$ components. Equating $z$ components

$$6 + 10\lambda = -6 - 4\mu$$

This must be equal if the lines intersect

$$\text{LHS} = 6 + 10\lambda = 6 + 10(-2)$$
$$= -14$$
$$\text{RHS} = -6 - 4\mu = -6 - 4(2)$$
$$= -14$$

Since LHS equals RHS, then the two lines intersect.

---

**b)** Find the coordinates of their point of intersection.
**Solution**
To find the point of intersection, substitute in either equation.

- First equation

$$\begin{pmatrix} 8 - 4\lambda \\ -2 - 2\lambda \\ 6 + 10\lambda \end{pmatrix} = \begin{pmatrix} 8 - 4(-2) \\ -2 - 2(-2) \\ 6 + 10(-2) \end{pmatrix}$$

$$= \begin{pmatrix} 16 \\ 2 \\ -14 \end{pmatrix}$$

- Second equation

$$\begin{pmatrix} 12 + 2\mu \\ 4 - \mu \\ -6 - 4\mu \end{pmatrix} = \begin{pmatrix} 12 + 2(2) \\ 4 - (2) \\ -6 - 4(2) \end{pmatrix}$$

$$= \begin{pmatrix} 16 \\ 2 \\ -14 \end{pmatrix}$$

A final example.

## Example 31

The vector equations of two lines $L_1$ and $L_2$ are $r = (-\hat{i} - 2\hat{j} + 5\hat{k}) + t(2\hat{i} + \hat{j} + 7\hat{k})$ and $r = (2\hat{i} - 3\hat{k}) + s(-2\hat{i} + 3\hat{j} - \hat{k})$. Show that these lines can only be in 3D space.

What did you get? Find the solution below to double-check your answer.

## Solution to Example 31

If these lines can only be in 3D space, it implies that they are skew, i.e., neither parallel nor intersecting. The general expression for vector equation is

$$r = a + tb$$

- **Parallel?**

The vector equations for the lines in matrix form are

$$L_1 : r = (-\hat{i} - 2\hat{j} + 5\hat{k}) + t(2\hat{i} + \hat{j} + 7\hat{k})$$

$$= \begin{pmatrix} -1 \\ -2 \\ 5 \end{pmatrix} + t \begin{pmatrix} 2 \\ 1 \\ 7 \end{pmatrix}$$

$$L_2 : r = (2\hat{i} - 3\hat{k}) + s(-2\hat{i} + 3\hat{j} - \hat{k})$$

$$= \begin{pmatrix} 2 \\ 0 \\ -3 \end{pmatrix} + s \begin{pmatrix} -2 \\ 3 \\ -1 \end{pmatrix}$$

# Vectors

For parallel lines, the direction vector of one line must be the scalar multiple of the other. If $b_1$ and $b_2$ are direction vectors for $L_1$ and $L_2$, respectively, then

$$b_1 = (2\hat{i} + \hat{j} + 7\hat{k})$$
$$= \begin{pmatrix} 2 \\ 1 \\ 7 \end{pmatrix}$$
$$b_2 = (-2\hat{i} + 3\hat{j} - \hat{k})$$
$$= \begin{pmatrix} -2 \\ 3 \\ -1 \end{pmatrix}$$

It can be shown that

$$\frac{-2}{2} \neq -\frac{3}{1} \neq \frac{-1}{7}$$

Thus, $L_1$ and $L_2$ are not parallel, since $b_1$ is not a scalar multiple of $b_2$.

- **Intersecting?**

$$L_1 : r = (-\hat{i} - 2\hat{j} + 5\hat{k}) + t(2\hat{i} + \hat{j} + 7\hat{k})$$
$$= \begin{pmatrix} -1 \\ -2 \\ 5 \end{pmatrix} + t \begin{pmatrix} 2 \\ 1 \\ 7 \end{pmatrix} = \begin{pmatrix} -1 + 2t \\ -2 + t \\ 5 + 7t \end{pmatrix}$$

$$L_2 : r = (2\hat{i} - 3\hat{k}) + s(-2\hat{i} + 3\hat{j} - \hat{k})$$
$$= \begin{pmatrix} 2 \\ 0 \\ -3 \end{pmatrix} + s \begin{pmatrix} -2 \\ 3 \\ -1 \end{pmatrix} = \begin{pmatrix} 2 - 2s \\ 3s \\ -3 - s \end{pmatrix}$$

If the two lines intersect, then

$$\begin{pmatrix} -1 + 2t \\ -2 + t \\ 5 + 7t \end{pmatrix} = \begin{pmatrix} 2 - 2s \\ 3s \\ -3 - s \end{pmatrix}$$

Equating $x$ components

$$-1 + 2t = 2 - 2s$$
$$2t + 2s = 2 + 1$$
$$2t + 2s = 3 \text{ ----- (i)}$$

Equating $y$ components

$$-2 + t = 3s$$
$$t - 3s = 2 \text{ ----- (ii)}$$

Solving the above equations:

- (ii) ×2, we have

$$2t - 6s = 4 ----- (iii)$$

- (i)-(iii), we have

$$8s = -1$$
$$s = -\frac{1}{8}$$

From (ii)

$$t = 2 + 3s$$
$$t = \frac{13}{8}$$

Now we need to validate this on $z$ components. Equating $z$ components

$$5 + 7t = -3 - s$$

This must be equal if the lines intersect

$$\text{LHS} = 5 + 7t = 5 + 7\left(\frac{13}{8}\right)$$
$$= \frac{131}{8}$$
$$\text{RHS} = -3 - s = -3 - \left(-\frac{1}{8}\right)$$
$$= -\frac{23}{8}$$

LHS is not equal to RHS, then the two vector equations do not intersect. Therefore, the lines are skew since they are neither parallel nor intersect.

---

## 11.17 ANGLE BETWEEN TWO LINES

We can find the acute or obtuse angle $\theta$ between two lines from their vector equations. Let's say the vector equations of lines $L_1$ and $L_1$ are $r_1 = a_1 + \lambda b_1$ and $r_2 = a_2 + \mu b_2$, respectively. Since lines $L_1$ and $L_1$ are not parallel, the angle $\theta$ between their direction vectors $b_1$ and $b_2$ is given by:

$$\boxed{\cos \theta = \frac{b_1 . b_2}{|b_1| |b_2|}} \qquad (11.25)$$

Note that the angle here (Figure 11.24) is when the two direction vectors are pointing away from the common origin.

Like what was previously shown when dealing with the scalar product of two vectors, it can also be shown that two lines $L_1$ and $L_1$ are perpendicular (Figure 11.25) if the product of their direction vectors is zero, that is:

# Vectors

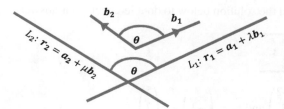

**FIGURE 11.24** Angle between two lines illustrated (1).

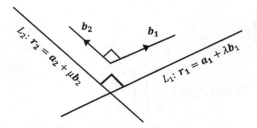

**FIGURE 11.25** Angle between two lines illustrated (2).

$$\boxed{b_1.b_2 = 0} \quad (11.26)$$

This is because for two perpendicular lines, $\theta = 90°$ and $\cos 90° = 0$, therefore:

$$\cos \theta = \frac{b_1.b_2}{|b_1||b_2|}$$

$$0 = \frac{b_1.b_2}{|b_1||b_2|}$$

$$0 = b_1.b_2$$

Also, when they are parallel then

$$b_1.b_2 = |b_1||b_2|$$

though the approach previously shown to check parallelism seems easier.

Let's try a few examples to finish this chapter.

### Example 32

A pair of vector equations represent two lines $L_1$ and $L_2$. Show whether these lines are parallel, perpendicular, or neither.

a) $r_1 = \begin{pmatrix} 3 \\ 2 \\ -11 \end{pmatrix} + s \begin{pmatrix} -2 \\ 3 \\ 4 \end{pmatrix}$ and $r_2 = \begin{pmatrix} 7 \\ -1 \\ 10 \end{pmatrix} + t \begin{pmatrix} 5 \\ 2 \\ 1 \end{pmatrix}$

b) $r_1 = (\hat{\jmath} - 9\hat{k}) + \lambda(\hat{\imath} - 5\hat{\jmath} - 4\hat{k})$ and $r_2 = (2\hat{\imath} - \hat{k}) + \mu(3\hat{\imath} - \hat{\jmath} + \hat{k})$

c) If they are neither, determine the angle between them. Present the answer correct to 1 d.p.

What did you get? Find the solution below to double-check your answer.

### Solution to Example 32

a) $r_1 = \begin{pmatrix} 3 \\ 2 \\ -11 \end{pmatrix} + s \begin{pmatrix} -2 \\ 3 \\ 4 \end{pmatrix}$ and $r_2 = \begin{pmatrix} 7 \\ -1 \\ 10 \end{pmatrix} + t \begin{pmatrix} 5 \\ 2 \\ 1 \end{pmatrix}$

**Solution**

$b_1 = \begin{pmatrix} -2 \\ 3 \\ 4 \end{pmatrix}$ and $b_2 = \begin{pmatrix} 5 \\ 2 \\ 1 \end{pmatrix}$

$$b_1 . b_2 = \begin{pmatrix} -2 \\ 3 \\ 4 \end{pmatrix} . \begin{pmatrix} 5 \\ 2 \\ 1 \end{pmatrix}$$
$$= (-2)(5) + (3)(2) + (4)(1)$$
$$= -10 + 6 + 4$$
$$= 0$$

Hence, the lines are perpendicular.

---

b) $r_1 = (\hat{\jmath} - 9\hat{k}) + \lambda (\hat{\imath} - 5\hat{\jmath} - 4\hat{k})$ and $r_2 = (2\hat{\imath} - \hat{k}) + \mu (3\hat{\imath} - \hat{\jmath} + \hat{k})$

**Solution**

$$r_1 = (\hat{\jmath} - 9\hat{k}) + \lambda (\hat{\imath} - 5\hat{\jmath} - 4\hat{k})$$

$$b_1 = \hat{\imath} - 5\hat{\jmath} - 4\hat{k} = \begin{pmatrix} 1 \\ -5 \\ -4 \end{pmatrix}$$

and

$$b_2 = 3\hat{\imath} - \hat{\jmath} + \hat{k} = \begin{pmatrix} 3 \\ -1 \\ 1 \end{pmatrix}$$

$$b_1 . b_2 = \begin{pmatrix} 1 \\ -5 \\ -4 \end{pmatrix} . \begin{pmatrix} 3 \\ -1 \\ 1 \end{pmatrix}$$
$$= (1)(3) + (-5)(-1) + (-4)(1)$$
$$= 3 + 5 - 4$$
$$= 4$$

Hence, the lines are not perpendicular. Now let's check for parallelism.

**Method 1**

It can be shown that

$$\frac{3}{1} \neq \frac{-1}{-5} \neq \frac{1}{-4}$$

Vectors

$b_2$ is not a scalar multiple of $b_1$, thus $L_1$ and $L_2$ are not parallel.

**Method 2**
For parallel lines

$$b_1 \cdot b_2 = |b_1| |b_2|$$
$$b_1 \cdot b_2 = 4$$

$$|b_1| |b_2| = \left(\sqrt{1^2 + (-5)^2 + (-4)^2}\right)\left(\sqrt{3^2 + (-1)^2 + 1^2}\right)$$
$$= \left(\sqrt{1 + 25 + 16}\right)\left(\sqrt{9 + 1 + 1}\right) = \sqrt{462}$$

But

$$\sqrt{462} \neq 4$$
$$b_1 \cdot b_2 \neq |b_1| |b_2|$$

Thus, the lines are not parallel either.

**c) Angle**
**Solution**

$$\cos \theta = \frac{b_1 \cdot b_2}{|b_1| |b_2|}$$
$$= \frac{4}{\sqrt{462}}$$
$$\theta = \cos^{-1}\left(\frac{4}{\sqrt{462}}\right)$$
$$\therefore \theta = \mathbf{79.3°}$$

## 11.18 CHAPTER SUMMARY

1) Physical quantities (simply referred to as quantities) can be divided into scalar and vector quantities.

2) A **scalar quantity** is any quantity that is completely identified or described by its magnitude (also known as amount or size) and associated units only. Examples include time, speed, height, temperature, pressure and power.

3) A **vector quantity** (or simply a vector) is a quantity which is completely identified by both its magnitude and direction (and associated units). Examples include velocity, displacement, force, electric current, stress, and momentum.

4) For analytical purposes, vectors are considered as free and referred to as '**free vectors**'. This implies that they can be anywhere in space and so the point of application is irrelevant.

5) Three or more points are said to be collinear if they all lie on the same straight line. It can be shown that if vectors $\overrightarrow{AB}, \overrightarrow{BC}, \overrightarrow{CD}$, etc., are parallel then points $A, B, C, D$, etc., are collinear.

6) A **Zero** vector (also called a **Null** vector) is a vector whose magnitude is zero.

7) The negation of a vector is a way of emphasising the direction feature of a vector. Given a vector $a$, the negative vector of this is a vector that has the same magnitude as $a$ but points in an opposite direction (i.e., 180° or say anti-phase). We indicate this with a sign change (ideally negative) or by flipping the direction of the arrow.

8) Two or more vectors are said to be equal if they all have the same magnitude and direction.

9) A unit vector is a vector whose magnitude is unity. In general, a unit vector in the direction of $a$ is given as:

$$\boxed{\hat{a} = \frac{a}{|a|}}$$

10) Consider a vector $r$, which is entirely in 2D plane surface, the vector can be defined in terms of unit vectors as:

$$\boxed{r = x\hat{i} + y\hat{j}}$$

If vector $r$ lies in space, then it is completely defined as:

$$\boxed{r = x\hat{i} + y\hat{j} + z\hat{k}}$$

The column vector or matrix equivalents are:

$$\boxed{r = x\hat{i} + y\hat{j} = \begin{pmatrix} x \\ y \end{pmatrix}} \text{ AND } \boxed{r = x\hat{i} + y\hat{j} + z\hat{k} = \begin{pmatrix} x \\ y \\ z \end{pmatrix}}$$

11) A vector that lies entirely on the $x$-axis, $y$-axis, or $z$-axis can be represented as:

$$\begin{pmatrix} x \\ 0 \\ 0 \end{pmatrix}, \begin{pmatrix} 0 \\ y \\ 0 \end{pmatrix}, \begin{pmatrix} 0 \\ 0 \\ z \end{pmatrix}$$

12) A vector and its negative have the same magnitude.

$$\boxed{|a| = |-a|}$$

13) In general, the magnitude is given by:

$$|r| = |x\hat{\imath} + y\hat{\jmath}| = \sqrt{x^2 + y^2} \quad \text{AND} \quad |r| = |x\hat{\imath} + y\hat{\jmath} + z\hat{k}| = \sqrt{x^2 + y^2 + z^2}$$

14) Given two vectors **a** and **b**, their scalar product (also called dot product) is denoted as **a.b** and is expressed as:

$$a.b = |a| |b| \cos \theta$$

By re-arranging the formula above, we can determine the angle $\theta$ between two known vectors as:

$$\cos \theta = \frac{a.b}{|a| |b|}$$

$$a.b = a_1 b_1 + a_2 b_2 \quad \text{AND} \quad a.b = a_1 b_1 + a_2 b_2 + a_3 b_3$$

Or

$$a.b = \begin{pmatrix} a_1 b_1 \\ a_2 b_2 \end{pmatrix} \quad \text{OR} \quad a.b = \begin{pmatrix} a_1 b_1 \\ a_2 b_2 \\ a_3 b_3 \end{pmatrix}$$

15) For any given vector **r** in $xy$ plane or $xyz$ space, it can be shown that

$$r.r = |r|^2$$

16) The vector equation of a straight line, say RS, passing through point $A$ and parallel to vector **b** is given by

$$r = a + tb$$

where

- **r** is the position vector of a point on line $RS$
- **a** is the position vector of a fixed-point $A$
- **b** is the vector that is parallel to the line $RS$
- **t** is a variable scalar, which varies depending on the point on the line. Other letters, including $s$, $\lambda$, and $\mu$, can also be used.

17) When we have two or more lines lie in 2D ($x-y$, $x-z$ or $y-z$) plane or 3D space, one of these may happen:
- The lines are parallel and never meet no matter how long they are (drawn) such that the distance between them at all points remains the same.
- They intersect or cross each other at a single point.
- They are neither parallel nor do they intersect. The lines in this case are called **skew**.

****

## 11.19 FURTHER PRACTICE

To access complementary contents, including additional exercises, please go to www.dszak.com.

# 12 Fundamentals of Differentiation

## Learning Outcomes

Once you have studied the content of this chapter, you should be able to:

- Discuss the concept of the gradient of a curve
- Determine the approximate gradient of the curve at a given point
- Use different notations of differentiation
- Determine differential coefficients of power and polynomial functions
- Find the derivative at a particular point on a curve
- Determine higher-order derivatives
- Find the derivative of trigonometric functions
- Differentiate exponential and logarithmic functions

## 12.1 INTRODUCTION

Calculus is a branch of mathematics with two distinctive parts, namely differentiation (or differential calculus) and integration (or integral calculus). Its application and relevance go beyond solving practical science and engineering problems, but also cover other sectors including business and medicine. As such, it is a very useful branch and shall be the last theme of this book. We will start this chapter with differentiation, covering the limits of functions, differentiation from the first principles, standard derivatives, and higher-order differentiation.

## 12.2 WHAT IS DIFFERENTIATION?

To put it in a very simple way, differentiation is a measure of the '**rate of change**'. In other words, it is a technique that is used to determine the size of change in one physical quantity (e.g., volume, distance, mass) when compared to an equivalent change in another physical quantity (e.g., time).

We have a few ways to denote differentiation, but $\frac{dy}{dx}$ (reads '***dee y dee x***') seems very popular. The variables $y$ and $x$ represent the two quantities (scalar or vector) being compared. For example, one may be interested in measuring how the volume of the water in a tank is changing over time. This is represented as $\frac{dV}{dt}$ (reads '***dee V dee t***'), where $V$ and $t$ are volume and time, respectively. The same can be said about $\frac{dv}{dt}$, which is the '**rate of change of velocity ($v$) with respect to time ($t$)**'. Notice the difference in $V$ (capital letter) for volume and $v$ (small letter) for velocity. You will soon find out that we typically shorten '*with respect to*' as '**wrt**'. This is the brief about the journey that we are set to begin, let's get on.

## 12.3 LIMITS

Limit (pl. limits) is linguistically understood to mean boundary or the greatest possible end, and it is a term that occurs frequently in science and engineering. This is the first time to encounter it in this book, and it is essential that we cover this concept as it is the basis upon which calculus is premised.

The limit is the boundary, and thus the limit of a function is the boundary of that function. For example, if a function of $x$ is denoted as $f(x)$, then the limit of this function is represented as:

$$\lim f(x)$$

This is read as '**limit of $f$ function of $x$**'. A more useful version is

$$\lim_{x \to \infty} f(x)$$

This is the '**limit of $f$ function of $x$ as $x \to \infty$**'. The sign $\to$ is read as '**tends to**' and technically implies that the left-hand variable $x$ approaches the right-hand variable or number. Table 12.1 shows further examples of the use of the limits.

Let's go back to our $\lim_{x \to \infty} f(x)$. What this means is that we are evaluating (or interested in knowing) the value of $f(x)$ as the independent variable $x$ becomes very large or infinite. We may also be interested in the limit of the function as the value of independent variable tends to other values, including zero.

To make this clear, let's say $f(x) = x^2 + x^{-5}$ and we are interested in knowing the value of this function when the value of $x$ approaches 3, i.e., $x \to 3$. The evaluation process is shown in Table 12.2. In (a), the values of $x$ are close to but less than 3. It can be said that $x$ approaches 3 from the left; this is denoted as $x \to 3^-$. Similarly, in (b), the values of $x$ are greater than but close to 3. It can be said that $x$ approaches 3 from the right and is denoted as $x \to 3^+$.

In both cases, (a) and (b), the value of the function $f(x)$ is approaching a particular value, which is 9 (correct to 1 s.f.). We can therefore write that:

$$\lim_{x \to 3} [f(x)] = \lim_{x \to 3^-} [f(x)] = \lim_{x \to 3^+} [f(x)] = 9$$

In our case, we will often be interested in either $x \to \infty$ or $x \to 0$, but the example is to show the concept in general. The limit is the approximation of the function, as the value of $x$ is very close to

**TABLE 12.1**
**The Use of the Limit Sign Illustrated**

| Example | Read as | Explanation |
|---|---|---|
| $x \to 2$ | $x$ tends to 2 | The value of the variable $x$ approaches a numerical value of 2. |
| $x \to n$ | $x$ tends to $n$ | The value of the variable $x$ approaches a variable $n$. |
| $n \to 0$ | $n$ tends to 0 | The value of the variable $n$ approaches 0. |

# Fundamentals of Differentiation

**TABLE 12.2**
**The Limit of $f(x)$ as $x$ Approaches 3 Illustrated**

| (a) | | (b) | |
|---|---|---|---|
| $x$ | $y = f(x)$ | $x$ | $y = f(x)$ |
| 2.990 | 8.94428 | 3.010 | 9.06415 |
| 2.991 | 8.95026 | 3.009 | 9.05814 |
| 2.992 | 8.95623 | 3.008 | 9.05212 |
| 2.993 | 8.96221 | 3.007 | 9.04612 |
| 2.994 | 8.96819 | 3.006 | 9.04011 |
| 2.995 | 8.97417 | 3.005 | 9.03411 |
| 2.996 | 8.98016 | 3.004 | 9.0281 |
| 2.997 | 8.98614 | 3.003 | 9.0221 |
| 2.998 | 8.99213 | 3.002 | 9.01611 |
| 2.999 | 8.99812 | 3.001 | 9.01011 |
| **3.000** | **9.00412** | **3.000** | **9.00412** |

3 but not equal to it. This is different from $f(3) = 3^2 + 3^{-5}$, which is when we evaluate the function at $x = 3$, as shown by the last row. The limit of a function is further illustrated in Table 12.3, as the variable tends to zero or infinity.

## 12.4 GRADIENT

Gradient is not a new concept to us, but we will advance our discussion on the term in this chapter.

### 12.4.1 STRAIGHT LINE

The gradient (or slope) of a straight line is given as '**delta y** (or $\Delta y$) **over delta x** (or $\Delta x$)', i.e., $\frac{\Delta y}{\Delta x}$. $\Delta$ is a Greek letter, generally used to denote a measurable change in a physical quantity as against $\delta$ – a lowercase delta – which denotes an infinitesimal (or an infinitely small) change in the quantity. $\Delta y$ is obtained by finding the difference in two $y$-coordinates; these values could be the initial and final or the first and second or second and fifth, and so on. The same can be said about $\Delta x$.

This implies that we can take any pair of points on a straight-line graph (Figure 12.1) to obtain a change in either $y$ or $x$. The value of the slope remains unchanged irrespective of the pair used in computing the change. In other words, the slope of a straight line is constant. Could the same be said about the slope of a non-linear function? Let's follow the discussion to get the answer to this.

## TABLE 12.3
### Examples of Limit of Function

| $f(x)$ | Limit | Answer | Note |
|---|---|---|---|
| $x^n + 1$ | $\lim\limits_{n \to 0} f(x)$ | 2 | This is because when $n$ approaches 0, $x^n \to x^0 \to 1$. |
| $x^n + 1$ | $\lim\limits_{n \to \infty} f(x)$ | • $x^n$, for $x > 1$<br>• 1, for $x < 1$<br>• 2, for $x = 1$ | • This is because when $n$ approaches $\infty$, $x^n$ becomes the dominant term.<br>• This is because when $n$ approaches $\infty$, $x^n$ becomes very small.<br>• This is because $1^\infty = 1$. |
| $\left(1 + \dfrac{1}{x}\right)^x$ | $\lim\limits_{x \to \infty} f(x)$ | $e \approx 2.718281\ldots$ | We will be using this soon. |
| $\sin \theta$ | $\lim\limits_{x \to 0} f(x)$ | $\theta$ | Provided that the angle is:<br>i) very small, and<br>ii) it is measured in radian.<br>We will be using this soon. |
| $\dfrac{\sin \theta}{\theta}$ | $\lim\limits_{x \to 0} f(x)$ | 1 | Provided that the angle is measured in radian.<br>We will be using this soon. |

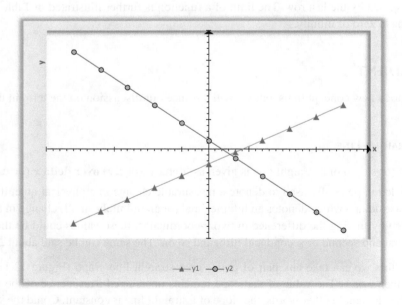

**FIGURE 12.1** Determining the gradient of a straight line illustrated.

Fundamentals of Differentiation

## 12.4.2 Non-linear Graph or Curve

Figure 12.2 is a plot of two non-linear functions, where $y_1 = f(x)$ represents a quadratic function, and $y_2 = g(x)$ represents a cubic function. Unlike a straight line, the gradient of these curves changes as we move from one point to another along the curve. For example, the gradient at point $A_1$ on $f(x)$ is different from that at points $A_2$ and $A_3$ on the same function. The same can be said about points $B_1$, $B_2$, and $B_3$ on $g(x)$. As a result, there are numerous, or even an infinite number of slopes for a given curve, depending on how much we zoom in.

The question of interest now is how do we find the gradient of a curve? For this, we will briefly discuss the common options available.

### 12.4.2.1 Drawing a Tangent

We will illustrate this graphical approach by determining the slope at point $x = 1.5$ on a graph of $y = x^2 - 1$. This method is illustrated in four steps below.

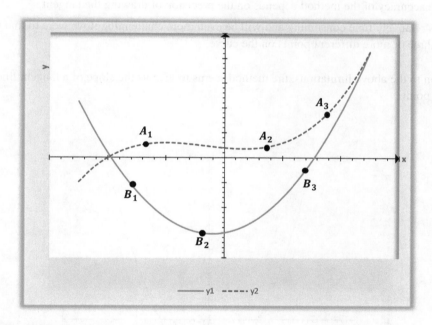

**FIGURE 12.2** Gradient of non-linear functions illustrated.

**Step 1:** Using a suitable method, draw the graph of the function $x^2 - 1$ if it is not already given (Figure 12.3).

**Step 2:** Locate the point by drawing a vertical line from point $x = 1.5$ until it reaches the curve drawn in Step 1 (Figure 12.4).

**Step 3:** Draw a tangent to the curve at the point of interest.

A tangent is a straight line that meets the curve at a right angle at the given point; it touches the curve but does not cross or cut or go through it. A very good approximation can be manually drawn (Figure 12.5).

**Step 4:** Determine the slope of the tangent using any two points on the tangent (Figure 12.6).

We have chosen points $A(1, 0)$ and $B(2, 3)$ and calculate the slope as:

$$Slope = \frac{\Delta y}{\Delta x} = \frac{y_2 - y_1}{x_2 - x_1}$$

$$= \frac{3 - 0}{2 - 1} = 3$$

The above four-step approach seems quick and easy, but there are some limitations:

**a)** The accuracy of the method depends on the precision of drawing the tangent.

**b)** It is relatively time consuming and will be even more challenging if we were to find the slope of three or more different points on the curve.

In addition to the above limitations, the method seems to give us the slope of a tangent line and not the given point.

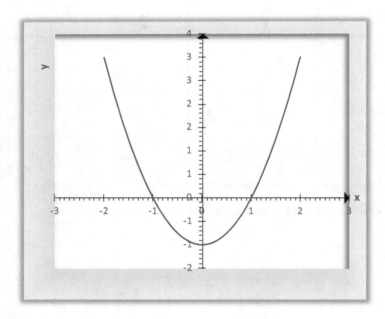

**FIGURE 12.3** Drawing tangent – Part I.

# Fundamentals of Differentiation

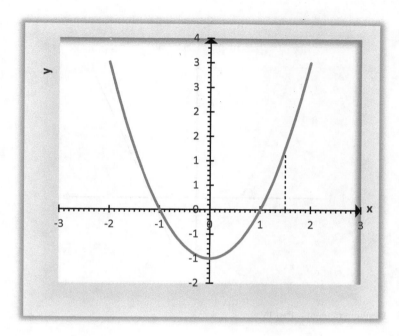

**FIGURE 12.4** Drawing tangent – Part II.

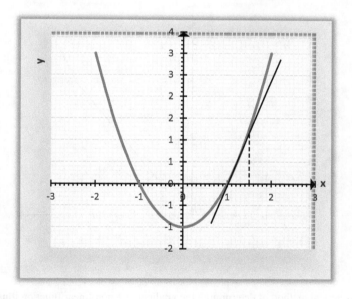

**FIGURE 12.5** Drawing tangent – Part III.

## 12.4.2.2 Numerical Method

For this method, we've chosen a slightly different function, which is $y = x^2 + 1$. Its plot is shown in Figure 12.7. Here we intend to find the gradient at point $A$ to the curve where $x = 0.5$.

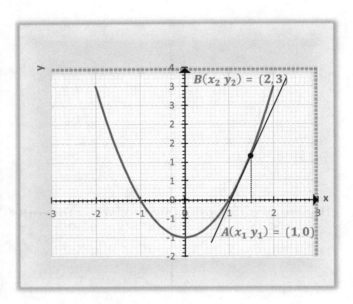

**FIGURE 12.6** Drawing tangent – Part IV.

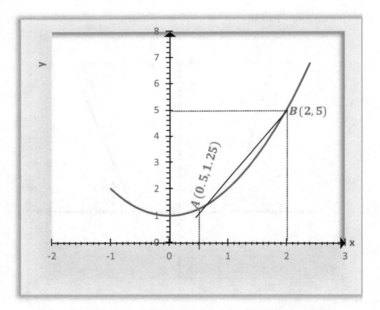

**FIGURE 12.7** Numerical method of determining the gradient of a non-linear function illustrated.

To do this, we will draw a straight line (or chord) $AB$ and find its gradient as:

$$\text{Slope}_{|\text{line } AB|} = m = \frac{\Delta y}{\Delta x} = \frac{y_2 - y_1}{x_2 - x_1}$$

$$= \frac{5 - 1.25}{2 - 0.5} = \frac{3.75}{1.5} = 2.5$$

The gradient is 2.5, which is, of course, the gradient of the line $AB$ and not the point $A$. AB is also not a tangent to the curve at point $A$. We will equally notice that the change in $x$, $\Delta x$ is 1.5. To improve the estimate, keep point $A$ fixed and move point $B$ closer and closer to $A$ along the curve, such that

# Fundamentals of Differentiation

$\Delta x$ is very small. The gradient obtained in this case will be very close to the gradient at point $A$. The result of the process is shown in Table 12.4. Point A, given as $(y_1, x_1)$, is fixed at $(0.5, 1.25)$. $y_1$ is found by substituting $x_1$ in $y = x^2 + 1$, i.e. $y_1 = 0.5^2 + 1 = 1.25$. $x_2$ is chosen arbitrarily, and the

**TABLE 12.4**
**Gradient of a Non-Linear Function Using Numerical Method Illustrated**

| $x_1$ | $y_1$ | $x_2$ | $y_2$ | $\Delta x$ | $\Delta y$ | $m$ |
|---|---|---|---|---|---|---|
| 0.5 | 1.25 | 0.550 | 1.30 | 0.05 | 0.05 | 1.05 |
| 0.5 | 1.25 | 0.548 | 1.30 | 0.05 | 0.05 | 1.05 |
| 0.5 | 1.25 | 0.546 | 1.30 | 0.05 | 0.05 | 1.05 |
| 0.5 | 1.25 | 0.544 | 1.30 | 0.04 | 0.05 | 1.04 |
| 0.5 | 1.25 | 0.542 | 1.29 | 0.04 | 0.04 | 1.04 |
| 0.5 | 1.25 | 0.540 | 1.29 | 0.04 | 0.04 | 1.04 |
| 0.5 | 1.25 | 0.538 | 1.29 | 0.04 | 0.04 | 1.04 |
| 0.5 | 1.25 | 0.536 | 1.29 | 0.04 | 0.04 | 1.04 |
| 0.5 | 1.25 | 0.534 | 1.29 | 0.03 | 0.04 | 1.03 |
| 0.5 | 1.25 | 0.532 | 1.28 | 0.03 | 0.03 | 1.03 |
| 0.5 | 1.25 | 0.530 | 1.28 | 0.03 | 0.03 | 1.03 |
| 0.5 | 1.25 | 0.528 | 1.28 | 0.03 | 0.03 | 1.03 |
| 0.5 | 1.25 | 0.526 | 1.28 | 0.03 | 0.03 | 1.03 |
| 0.5 | 1.25 | 0.524 | 1.27 | 0.02 | 0.02 | 1.02 |
| 0.5 | 1.25 | 0.522 | 1.27 | 0.02 | 0.02 | 1.02 |
| 0.5 | 1.25 | 0.520 | 1.27 | 0.02 | 0.02 | 1.02 |
| 0.5 | 1.25 | 0.518 | 1.27 | 0.02 | 0.02 | 1.02 |
| 0.5 | 1.25 | 0.516 | 1.27 | 0.02 | 0.02 | 1.02 |
| 0.5 | 1.25 | 0.514 | 1.26 | 0.01 | 0.01 | 1.01 |
| 0.5 | 1.25 | 0.512 | 1.26 | 0.01 | 0.01 | 1.01 |
| 0.5 | 1.25 | 0.510 | 1.26 | 0.01 | 0.01 | 1.01 |
| 0.5 | 1.25 | 0.508 | 1.26 | 0.01 | 0.01 | 1.01 |
| 0.5 | 1.25 | 0.506 | 1.26 | 0.01 | 0.01 | 1.01 |

corresponding $y_2$ using $y_2 = (x_2)^2 + 1$. In other words, we simply substitute $x_2$ in $y = x^2 + 1$ to find $y_2$. The computation follows this process for $x_3, x_4, x_5, x_6$, etc.

In the above, we've moved point $B$ from the $x$-coordinate 0.55 to 0.506. It can be observed that the limit of the gradient of the curve at $x = 0.5$, as $\Delta x$ tends to zero, is 1. In other words, the gradient of $f(x)$ is

$$\lim_{\Delta x \to 0} [x^2 + 1] = 1$$

In summary, we can find the approximate value of the gradient of a non-linear function at a given point numerically in this way:

**Step 1:** Evaluate $y_1$ by using $f(x_1)$. This gives the coordinates of the point at which the gradient is to be found, i.e., $(x_1, y_1)$.

**Step 2:** Choose another point $x_2$ such that $\Delta x$ (or $x_2 - x_1$) is very small or close to zero.

**Step 3:** Evaluate $y_2$ by using $f(x_2)$. This gives the coordinates of another point which is close to the given point, i.e., $(x_2, y_2)$.

**Step 4:** Find the slope $m$ using

$$\text{Slope} = \frac{\Delta y}{\Delta x} = \frac{y_2 - y_1}{x_2 - x_1}$$

### 12.4.2.3 Analytical Method

The above numerical method seems promising, but the 'smallness' of the change in $x$ is relative and may be subjective. In addition, there is a need to evaluate the function at $x_2$ to find the gradient. Unfortunately, the process will need to be repeated for every point on the curve that we need to find its gradient.

We now want to build on the above method and find a general procedure to obtain the gradient of a curve. We will start with the definition of a slope $m$ as:

$$m = \frac{\Delta y}{\Delta x} = \frac{y_2 - y_1}{x_2 - x_1}$$

Let the coordinates of the line $AB$ in Figure 12.8 (a re-production of Figure 12.7) be $A(x_1, y_1)$ and $B(x_2, y_2)$. Given that points $A$ and $B$ are on the curve, we can say that $\Delta x = x_2 - x_1$ or $x_2 = x_1 + \Delta x$. Given that $y = f(x)$, we have

$$y_1 = f(x_1)$$
$$y_2 = f(x_2) = f(x_1 + \Delta x)$$
$$\therefore \Delta y = f(x_1 + \Delta x) - f(x_1)$$

We can now write the slope as:

$$m = \frac{f(x_1 + \Delta x) - f(x_1)}{\Delta x}$$

Since $\Delta x$ needs to be very small as observed in the numerical method, we can replace $\Delta x$ with $\delta x$, thus

$$m = \frac{f(x_1 + \delta x) - f(x_1)}{\delta x}$$

# Fundamentals of Differentiation

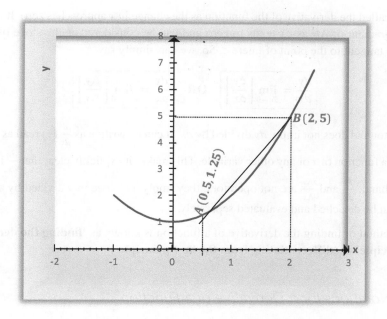

**FIGURE 12.8** Analytical method of determining the gradient of a non-linear function illustrated.

This is a more general representation of the gradient of a function.

### 12.4.2.4 Derivative of a Function

In the analytical method presented above, we noted that the slope is given as:

$$m = \frac{f(x_1 + \delta x) - f(x_1)}{\delta x}$$

$\delta x$, though infinitesimal, shows that the slope found using the above formula is for a line $AB$ (or tangent) and not a point of interest. The condition for a point is that $\delta x = 0$. However, if we were to substitute $\delta x = 0$ in the denominator, the $m$ becomes undefined and meaningless.

Let's make a final attempt to modify our approach. We will assume two arbitrary points on the curve $f(x)$. The first point $A$ has coordinates of $(x, y)$ and is the point whose gradient is to be found. The second point $B$ is that which is very close to point $A$, such that there is a small increment $\delta x$ in $x$ value and an equivalent increment $\delta y$ in $y$ value. Hence, the coordinates of $B$ are $(x + \delta x, y + \delta y)$.

Recall that we can write $y = f(x)$ and $y + \delta y = f(x + \delta x)$. Thus, $\delta y = f(x + \delta x) - f(x)$. The slope can now be given as:

$$m = \frac{\delta y}{\delta x} = \frac{f(x + \delta x) - f(x)}{\delta x}$$

If we assume that $\delta x \to 0$, using the concept of limits, we have

$$m_{\text{[at a given point]}} = \lim_{\delta x \to 0} \left[ \frac{\delta y}{\delta x} \right] = \lim_{\delta x \to 0} \left[ \frac{f(x + \delta x) - f(x)}{\delta x} \right]$$

The above is called the derivative of the function as the change in x approaches zero. It is the limiting value of the gradient of AB as $\delta x$ tends to zero and can be considered as the slope of the point of interest or the tangent to the point of interest. So, we can simply say

$$\boxed{\frac{dy}{dx} = \lim_{\delta x \to 0}\left[\frac{\delta y}{\delta x}\right]} \quad \text{OR} \quad \boxed{\frac{dy}{dx} = \lim_{\Delta x \to 0}\left[\frac{\Delta y}{\Delta x}\right]} \tag{12.1}$$

$\frac{dy}{dx}$ is an operator and does not imply $dy$ divided by $dx$. It can be written as $\frac{d}{dx}(y)$ read as '**dee dee x of y**', where y is a function of x or any other variable. This makes it explicitly clear that $\frac{dy}{dx}$ is an operator. On the other hand, $\frac{\delta y}{\delta x}$ and $\frac{\Delta y}{\Delta x}$ are not operators; they imply a change in y divided by a change in x, hence they can be detached and evaluated separately.

The above method of finding the derivative of a function is known as '**finding the derivative from the first principle**' or '**differentiating from the first principle**'.

### 12.4.2.5 Notations for a Derivative of a Function

The process of finding the derivative of a function is called **differentiation**. This is also called 'differential coefficient of a function *wrt* an independent variable' or 'the rate of change of function or gradient function' or simply $\frac{dy}{dx}$. It is a gradient function rather than a gradient, as it gives us a function that can be used to find the gradient at a given point on the curve.

In addition to $\frac{dy}{dx}$, there are other notations that are used for denoting derivatives and are given in Table 12.5.

The notations are the same and can be used interchangeably. They are called the 'first derivative'. We will come back soon to explain what it's meant by the second, third, etc., derivatives. Any of the notations (and similar ones) show that we are required to 'differentiate a function' or 'we've differentiated a function'.

**TABLE 12.5**
**Notations for a Derivative of a Function Illustrated**

| Notation | Read as |
|---|---|
| $\frac{dy}{dx}$ | dee y, dee x (or dee y by dee x or dee y over dee x) |
| $f'(x)$ | f-prime of x (or f-prime x) |
| $y'$ | y-prime |
| $\frac{d}{dx}(fx)$ | dee dee x of f function of x |
| $\dot{y}$ | y dot |

# Fundamentals of Differentiation

Here is a good time to look at some examples.

## Example 1

Determine the derivative of the following functions from the first principles.

**a)** $y = c$  **b)** $y = mx + c$  **c)** $y = x^2$  **d)** $y = x^3$

**e)** $y = ax^2 + bx + c$  **f)** $y = \frac{1}{x+3}$  **g)** $y = ax^n$

What did you get? Find the solution below to double-check your answer.

## Solution to Example 1

### HINT

- In this case, our main reference is

$$\frac{dy}{dx} = \lim_{\delta x \to 0}\left[\frac{\delta y}{\delta x}\right] = \lim_{\delta x \to 0}\left[\frac{f(x+\delta x)-f(x)}{\delta x}\right]$$

- Do not worry, you can choose $\delta$ or $\Delta$. We will always arrive at the same answer.
- We will also use the binomial theorem covered in Chapter 6.

$$(x+a)^n = (x)^n + n(x)^{n-1}(a)^1 + \frac{n(n-1)}{2!}(x)^{n-2}(a)^2 + \frac{n(n-1)(n-2)}{3!}(x)^{n-3}(a)^3 + \ldots$$

**a)** $y = c$
**Solution**

$$y = f(x) = c$$

Similarly

$$y + \delta y = f(x + \delta x)$$
$$= c$$

Therefore,

$$\delta y = f(x + \delta x) - f(x) = 0$$

This is because the graph is a straight line. Now we can evaluate the slope as:

$$\text{Slope} \approx \frac{\delta y}{\delta x} \approx \frac{0}{\delta x}$$
$$\approx 0$$

Thus, the derivative is

$$\frac{dy}{dx} = \lim_{\delta x \to 0}\left(\frac{\delta y}{\delta x}\right) = \lim_{\delta x \to 0}\left(\frac{0}{\delta x}\right)$$
$$= 0$$
$$\therefore \frac{dy}{dx} = \mathbf{0}$$

---

**b)** $y = mx + c$
**Solution**

$$y = f(x) = mx + c$$

Similarly

$$y + \delta y = f(x + \delta x)$$
$$= m(x + \delta x) + c$$
$$= mx + m\delta x + c$$

Therefore,

$$\delta y = f(x + \delta x) - f(x)$$
$$= [mx + m\delta x + c] - [mx + c]$$
$$= m\delta x$$

Now we can evaluate the slope as:

$$\text{Slope} \approx \frac{\delta y}{\delta x} \approx \frac{m\delta x}{\delta x}$$
$$\approx m$$

Thus, the derivative is

$$\frac{dy}{dx} = \lim_{\delta x \to 0}\left(\frac{\delta y}{\delta x}\right) = \lim_{\delta x \to 0}(m)$$
$$= m$$
$$\therefore \frac{dy}{dx} = \mathbf{m}$$

---

**c)** $y = x^2$
**Solution**

$$y = f(x) = x^2$$

Similarly

$$y + \delta y = f(x + \delta x)$$
$$= (x + \delta x)^2$$
$$= x^2 + 2x\delta x + (\delta x)^2$$

# Fundamentals of Differentiation

Therefore,

$$\delta y = f(x+\delta x) - f(x)$$
$$= \left[x^2 + 2x\delta x + (\delta x)^2\right] - \left[x^2\right]$$
$$= 2x\delta x + (\delta x)^2$$

Now we can evaluate the slope as:

$$\text{Slope} \approx \frac{\delta y}{\delta x} \approx \frac{2x\delta x + (\delta x)^2}{\delta x}$$
$$\approx 2x + \delta x$$

Thus, the derivative is

$$\frac{dy}{dx} = \lim_{\delta x \to 0} \left(\frac{\delta y}{\delta x}\right) = \lim_{\delta x \to 0} (2x + \delta x)$$
$$= 2x + 0$$
$$\therefore \frac{dy}{dx} = 2x$$

---

**d)** $y = x^3$

**Solution**

$$y = f(x) = x^3$$

Similarly

$$y + \delta y = f(x+\delta x)$$
$$= (x+\delta x)^3$$
$$= x^3 + 3x^2\delta x + 3x(\delta x)^2 + (\delta x)^3$$

Therefore,

$$\delta y = f(x+\delta x) - f(x)$$
$$= \left[x^3 + 3x^2\delta x + 3x(\delta x)^2 + (\delta x)^3\right] - \left[x^3\right]$$
$$= 3x^2\delta x + 3x(\delta x)^2 + (\delta x)^3$$

Now we can evaluate the slope as:

$$\text{Slope} \approx \frac{\delta y}{\delta x} \approx \frac{3x^2\delta x + 3x(\delta x)^2 + (\delta x)^3}{\delta x}$$
$$\approx 3x^2 + 3x\delta x + (\delta x)^2$$

Thus, the derivative is

$$\frac{dy}{dx} = \lim_{\delta x \to 0} \left(\frac{\delta y}{\delta x}\right)$$
$$= \lim_{\delta x \to 0} \left(3x^2 + 3x\delta x + (\delta x)^2\right)$$

As $\delta x \to 0$, $3x\delta x$ and $(\delta x)^2$ become relatively insignificant when compared to the term $3x^2$ and can both be ignored. Hence, we have

$$\lim_{\delta x \to 0} \left(3x^2 + 3x\delta x + (\delta x)^2\right) = 3x^2$$

$$\therefore \frac{dy}{dx} = 3x^2$$

---

e) $y = ax^2 + bx + c$
**Solution**

$$y = f(x) = ax^2 + bx + c$$

Similarly

$$y + \delta y = f(x + \delta x)$$
$$= a(x + \delta x)^2 + b(x + \delta x) + c$$
$$= ax^2 + 2ax\delta x + a(\delta x)^2 + bx + b\delta x + c$$

Therefore,

$$\delta y = f(x + \delta x) - f(x)$$
$$= \left[ax^2 + 2ax\delta x + a(\delta x)^2 + bx + b\delta x + c\right] - \left[ax^2 + bx + c\right]$$
$$= 2ax\delta x + a(\delta x)^2 + b\delta x$$
$$= \delta x \left(2ax + a\delta x + b\right)$$

Now we can evaluate the slope as:

$$\text{Slope} \approx \frac{\delta y}{\delta x} \approx \frac{\delta x (2ax + a\delta x + b)}{\delta x}$$
$$\approx 2ax + a\delta x + b$$

Thus, the derivative is

$$\frac{dy}{dx} = \lim_{\delta x \to 0} \left(\frac{\delta y}{\delta x}\right) = \lim_{\delta x \to 0} (2ax + a\delta x + b)$$
$$= 2ax + b$$

$$\therefore \frac{dy}{dx} = 2ax + b$$

---

f) $y = \frac{1}{x+3}$
**Solution**

$$y = f(x) = \frac{1}{x + 3}$$

Similarly

$$y + \delta y = f(x + \delta x)$$
$$= \frac{1}{(x + \delta x) + 3}$$

# Fundamentals of Differentiation

Therefore,

$$\delta y = f(x + \delta x) - f(x)$$
$$= \left[\frac{1}{(x + \delta x) + 3}\right] - \left[\frac{1}{x + 3}\right]$$
$$= \frac{(x + 3) - [(x + \delta x) + 3]}{[(x + \delta x) + 3][x + 3]}$$
$$= \frac{-\delta x}{(x + 3 + \delta x)(x + 3)}$$

Now we can evaluate the slope as:

$$\text{Slope} \approx \frac{\delta y}{\delta x} \approx \frac{\left[\frac{-\delta x}{(x+3+\delta x)(x+3)}\right]}{\delta x}$$
$$\approx \frac{-1}{(x + 3 + \delta x)(x + 3)}$$

Thus, the derivative is

$$\frac{dy}{dx} = \lim_{\delta x \to 0}\left(\frac{\delta y}{\delta x}\right)$$
$$= \lim_{\delta x \to 0}\left(\frac{-1}{(x + 3 + \delta x)(x + 3)}\right)$$
$$= \left(\frac{-1}{(x + 3 + 0)(x + 3)}\right) = \frac{-1}{(x + 3)(x + 3)}$$
$$\therefore \frac{dy}{dx} = -\frac{1}{(x + 3)^2}$$

This seems to be a long one!

---

**g) $y = ax^n$**
**Solution**

$$y = f(x) = ax^n$$

Similarly

$$y + \delta y = f(x + \delta x)$$
$$= a(x + \delta x)^n$$

We can use the binomial theorem, covered in Chapter 6, to expand $(x + \delta x)^n$. Thus, we have

$$y + \delta y = a\left[x^n + nx^{n-1}(\delta x)^1 + \frac{n(n-1)}{2!}x^{n-2}(\delta x)^2 + \frac{n(n-1)(n-2)}{3!}x^{n-3}(\delta x)^3 + \dots (\delta x)^n\right]$$
$$= ax^n + nax^{n-1}(\delta x)^1 + \frac{n(n-1)}{2!}ax^{n-2}(\delta x)^2 + \frac{n(n-1)(n-2)}{3!}ax^{n-3}(\delta x)^3 + \dots a(\delta x)^n$$

Therefore,

$$\delta y = f(x + \delta x) - f(x)$$
$$= \left[ax^n + nax^{n-1}(\delta x)^1 + \frac{n(n-1)}{2!}ax^{n-2}(\delta x)^2 + \frac{n(n-1)(n-2)}{3!}ax^{n-3}(\delta x)^3 + \dots a(\delta x)^n\right] - [ax^n]$$
$$= nax^{n-1}(\delta x)^1 + \frac{n(n-1)}{2!}ax^{n-2}(\delta x)^2 + \frac{n(n-1)(n-2)}{3!}ax^{n-3}(\delta x)^3 + \dots a(\delta x)^n$$

Now we can evaluate the slope as:

$$\text{Slope} \approx \frac{\delta y}{\delta x}$$

$$\approx \frac{nax^{n-1}(\delta x)^1 + \frac{n(n-1)}{2!}ax^{n-2}(\delta x)^2 + \frac{n(n-1)(n-2)}{3!}ax^{n-3}(\delta x)^3 + \ldots a(\delta x)^n}{\delta x}$$

$$\approx nax^{n-1} + \frac{n(n-1)}{2!}ax^{n-2}(\delta x)^1 + \frac{n(n-1)(n-2)}{3!}ax^{n-3}(\delta x)^2 + \ldots a(\delta x)^{n-1}$$

As $\delta x \to 0$, each term containing $\delta x$ becomes relatively insignificant when compared to the term $ax^{n-1}$ and can be ignored. Hence, we have

$$\lim_{\delta x \to 0} \left( nax^{n-1} + \frac{n(n-1)}{2!}ax^{n-2}(\delta x)^1 + \frac{n(n-1)(n-2)}{3!}ax^{n-3}(\delta x)^2 + \ldots a(\delta x)^{n-1} \right) = nax^{n-1}$$

$$\therefore \frac{dy}{dx} = nax^{n-1}$$

This last result is an important one, and we will come back to this shortly.

## 12.5 STANDARD DERIVATIVES

We have now been 'armed' with underpinning knowledge regarding differentiation and have managed to find derivatives of functions from first principles. The process can however be long and time consuming. What follows is the introduction of some standard derivatives of common functions, which will be followed by techniques to deal with complex and complicated functions.

### 12.5.1 POWER (OR POLYNOMIAL) FUNCTIONS

The most common standard derivative is the power function or polynomial function, which is given as:

$$\text{If } \boxed{f(x) = ax^n} \text{ then } \boxed{\frac{dy}{dx} = anx^{n-1}} \quad (12.2)$$

The above is valid for any real value of $n$ and $a$. If $a = 1$, we have a modified version

$$\text{If } \boxed{f(x) = x^n} \text{ then } \boxed{\frac{dy}{dx} = nx^{n-1}} \quad (12.3)$$

We've derived the formula above in Example 1(g) using the first principles. This standard or formula looks simple, but we need to understand it properly; otherwise, we will be limited by what it can be applied to. Here are the steps to follow in applying the formula:

**Step 1:** Arrange or simplify the function to be in a power form, i.e., $ax^n$.
**Step 2:** Multiply the function (i.e., $ax^n$) by the power (i.e., $n$).
**Step 3:** Reduce the power by 1 (or subtract 1 from the power).
**Step 4:** Simplify the resulting function.

# Fundamentals of Differentiation

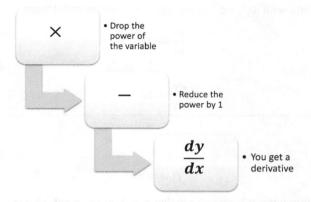

**FIGURE 12.9** Procedure for determining the derivative of a power function illustrated.

The above steps are illustrated in Figure 12.9.

The key to the derivative of a power function is '**drop down the power then reduce the power by 1**'. That's everything.

Let's try some examples.

## Example 2

Determine the derivative of the following functions with respect to $x$.

a) $y = x^2$   b) $y = x^{-4}$   c) $y = \frac{1}{5}x^{20}$   d) $y = 1.4x^{-5}$

e) $y = 6\sqrt{x}$   f) $y = \frac{1}{\sqrt[3]{x}}$   g) $y = -7x^{0.3}$   h) $y = \pi x$

i) $y = 5$   j) $y = x^{\sin \pi}$   k) $y = x^{-\frac{1}{5}}$   l) $y = \frac{x^7}{x^{-3}}$

What did you get? Find the solution below to double-check your answer.

## Solution to Example 2

**HINT**

Understanding the laws of indices will be helpful in differentiating some of the functions here and power functions in general.

**a)** $y = x^2$
**Solution**

$$f(x) = x^2$$

Comparing the function with $ax^n$, we can see that $a = 1$ and $n = 2$. Hence

$$\frac{dy}{dx} = 2x^{2-1}$$
$$= 2x^1 = 2x$$
$$\therefore \frac{dy}{dx} = 2x$$

**b)** $y = x^{-4}$
**Solution**

$$f(x) = x^{-4}$$

Comparing the function with $ax^n$, we can see that $a = 1$ and $n = -4$. Hence

$$\frac{dy}{dx} = -4x^{-4-1}$$
$$= -4x^{-5}$$
$$\therefore \frac{dy}{dx} = -4x^{-5}$$

**c)** $y = \frac{1}{5}x^{20}$
**Solution**

$$f(x) = \frac{1}{5}x^{20}$$

Comparing the function with $ax^n$, we can see that $a = \frac{1}{5}$ and $n = 20$. Hence

$$\frac{dy}{dx} = 20 \times \frac{1}{5}x^{20-1}$$
$$= 4x^{19}$$
$$\therefore \frac{dy}{dx} = 4x^{19}$$

You will agree that to do this from the first principles will take some time.

**d)** $y = 1.4x^{-5}$
**Solution**

$$f(x) = 1.4x^{-5}$$

Comparing the function with $ax^n$, we can see that $a = 1.4$ and $n = -5$. Hence

$$\frac{dy}{dx} = -5 \times 1.4x^{-5-1}$$
$$= -7x^{-6}$$
$$\therefore \frac{dy}{dx} = -7x^{-6}$$

For this, we could have written the final answer as $-\frac{7}{x^6}$ using one of the laws of indices.

# Fundamentals of Differentiation

**e)** $y = 6\sqrt{x}$
**Solution**

$$f(x) = 6\sqrt{x}$$

This is not in the power form, so we need to do something to present it in that form. To do this, we apply relevant law(s) of indices. We therefore have

$$f(x) = 6\sqrt{x} = 6x^{\frac{1}{2}}$$

Now comparing the function with $ax^n$, we can see that $a = 6$ and $n = \frac{1}{2}$. Hence

$$\frac{dy}{dx} = \frac{1}{2} \times 6x^{\left(\frac{1}{2}-1\right)}$$

$$= 3x^{-\frac{1}{2}}$$

$$\therefore \frac{dy}{dx} = 3x^{-\frac{1}{2}}$$

For this, we could have written the final answer as $\frac{3}{x^{\frac{1}{2}}}$ or $\frac{3}{\sqrt{x}}$ using one of the laws of indices.

---

**f)** $y = \frac{1}{\sqrt[3]{x}}$
**Solution**

$$f(x) = \frac{1}{\sqrt[3]{x}}$$

Again, this is not in the power form, so we need to modify it by using relevant law(s) of indices. Thus, we have

$$f(x) = \frac{1}{\sqrt[3]{x}} = \frac{1}{x^{\frac{1}{3}}} = x^{-\frac{1}{3}}$$

Comparing the function with $ax^n$, we can see that $a = 1$ and $n = -\frac{1}{3}$. Hence

$$\frac{dy}{dx} = -\frac{1}{3} \times x^{\left(-\frac{1}{3}-1\right)}$$

$$= -\frac{1}{3}x^{-\frac{4}{3}}$$

$$\therefore \frac{dy}{dx} = -\frac{1}{3}x^{-\frac{4}{3}}$$

For this, we could have written the final answer as $-\frac{1}{3x^{\frac{4}{3}}}$ or $-\frac{1}{3\sqrt[3]{x^4}}$ using relevant law(s) of indices.

**g)** $y = -7x^{0.3}$
**Solution**

$$f(x) = -7x^{0.3}$$

Comparing the function with $ax^n$, we can see that $a = -7$ and $n = 0.3$. Hence

$$\frac{dy}{dx} = 0.3 \times -7x^{0.3-1}$$
$$= -2.1x^{-0.7}$$
$$\therefore \frac{dy}{dx} = -2.1x^{-0.7}$$

---

**h)** $y = \pi x$
**Solution**

$$f(x) = \pi x = \pi x^1$$

Comparing the function with $ax^n$, we can see that $a = \pi$ and $n = 1$. Note that the absence of a power implies 1. Hence

$$\frac{dy}{dx} = 1 \times \pi x^{(1-1)}$$
$$= \pi x^0$$

$$\therefore \frac{dy}{dx} = \pi$$

---

**i)** $y = 5$
**Solution**

$$f(x) = 5$$

We need to change it into the power form, as it is not currently so. To do this, we will apply a special law of indices, i.e., $x^0 = 1$, as follows:

$$f(x) = 5 = 5 \times 1 = 5 \times x^0 = 5x^0$$

Great! Now comparing the function with $ax^n$, we can see that $a = 5$ and $n = 0$. Hence

$$\frac{dy}{dx} = 0 \times 5x^{(0-1)} = 0 \times 5x^{-1}$$
$$= 0$$

This is because anything multiplied by zero is zero.

$$\therefore \frac{dy}{dx} = 0$$

# Fundamentals of Differentiation

From this comes a general principle that

> *'the derivative of a constant is zero'*
>
> **OR**
>
> *'the rate of change of a constant is zero'*
>
> **OR**
>
> *'a constant does not change with respect to anything'*.

---

**j)** $y = x^{\sin \pi}$

**Solution**

$$f(x) = x^{\sin \pi}$$

This is in power form but not typical. Now comparing the function with $ax^n$, we can see that $a = 1$ and $n = \sin \pi$. Hence,

$$\frac{dy}{dx} = \sin \pi \times x^{(\sin \pi - 1)}$$

$$= \sin \pi . x^{(\sin \pi - 1)}$$

$$\therefore \frac{dy}{dx} = \sin \pi . x^{(\sin \pi - 1)}$$

Note that $\sin \pi$ is a number and should be treated as a constant in this.

---

**ALTERNATIVE METHOD**

We can simply say that $\sin \pi = 0$, therefore

$$f(x) = x^{\sin \pi}$$

$$= x^0 = 1$$

Thus

$$\frac{dy}{dx} = 0$$

---

**NOTE**

Note that if we simplify the first answer, we will also get zero as:

$$\sin \pi . x^{(\sin \pi - 1)} = 0 \times x^{0-1} = 0$$

---

**k)** $y = x^{-\frac{1}{5}}$

**Solution**

$$f(x) = x^{-\frac{1}{5}}$$

Comparing the function with $ax^n$, we can see that $a = 1$ and $n = -\frac{1}{5}$. Hence

$$\frac{dy}{dx} = -\frac{1}{5} \times x^{-\frac{1}{5}-1}$$

$$= -\frac{1}{5} x^{-\frac{6}{5}}$$

$$\therefore \frac{dy}{dx} = -\frac{1}{5} x^{-\frac{6}{5}}$$

That's the answer but we could have tidied that up a bit, using relevant laws of indices as:

$$-\frac{1}{5} x^{-\frac{6}{5}} = -\frac{1}{5} \times \frac{1}{x^{1.2}}$$

$$= -\frac{1}{5x^{1.2}}$$

$$\therefore \frac{dy}{dx} = -\frac{1}{5x^{1.2}}$$

---

l) $y = \frac{x^7}{x^{-3}}$
**Solution**

$$f(x) = \frac{x^7}{x^{-3}}$$

We first need to simplify this as:

$$f(x) = \frac{x^7}{x^{-3}} = x^7 \times x^3 = x^{10}$$

Great! Now comparing the function with $ax^n$, we can see that $a = 1$ and $n = 10$. Hence

$$\frac{dy}{dx} = 10 \times x^{10-1}$$

$$= 10x^9$$

$$\therefore \frac{dy}{dx} = 10x^9$$

---

Another set of examples to try.

## Example 3

A function is defined as $y = 5b^2 t^{-3} x^4$. Find the following derivatives.

a) $\frac{dy}{dx}$    b) $\frac{dy}{dt}$    c) $\frac{dy}{db}$    d) $\frac{dy}{dh}$    e) $\frac{dy}{dA}$

# Fundamentals of Differentiation

What did you get? Find the solution below to double-check your answer.

## Solution to Example 3

**HINT**

- The questions could be tricky so follow the **4**-step rule above and all should be fine.
- We will use *wrt* for 'with respect to'. It is a common phrase in this topic, and we are encouraged to use it.

**a)** $\frac{dy}{dx}$

**Solution**

$$y = 5b^2 t^{-3} x^4$$

As we are differentiating *wrt x*, the function can be re-written as:

$$y = \left(5b^2 t^{-3}\right) x^4$$

The above presentation is useful so that we can attain the general format. Now comparing the function with $ax^n$, we can see that $a = 5b^2 t^{-3}$ and $n = 4$. Hence

$$\frac{dy}{dx} = 4 \times \left(5b^2 t^{-3}\right) x^{4-1}$$
$$= 4\left(5b^2 t^{-3}\right) x^3$$
$$= 20 b^2 t^{-3} x^3$$
$$\therefore \frac{dy}{dx} = 20 b^2 t^{-3} x^3$$

---

**b)** $\frac{dy}{dt}$

**Solution**

$$y = 5b^2 t^{-3} x^4$$

As we are differentiating *wrt t*, the function can be re-written as:

$$y = \left(5b^2 x^4\right) t^{-3}$$

Comparing the function with $ax^n$, we can see that $a = 5b^2 x^4$ and $n = -3$. Hence

$$\frac{dy}{dt} = -3 \times \left(5b^2 x^4\right) t^{-3-1}$$
$$= -3 \left(5b^2 x^4\right) t^{-4}$$
$$= -15 b^2 x^4 t^{-4}$$
$$\therefore \frac{dy}{dt} = -15 b^2 x^4 t^{-4}$$

c) $\dfrac{dy}{db}$
**Solution**

$$y = 5b^2 t^{-3} x^4$$

As we are differentiating *wrt b*, the function can be re-written as:

$$y = \left(5t^{-3} x^4\right) b^2$$

Comparing the function with $ax^n$, we can see that $a = 5t^{-3}x^4$ and $n = 2$. Hence

$$\begin{aligned}\dfrac{dy}{db} &= 2 \times \left(5t^{-3}x^4\right) b^{2-1} \\ &= 2\left(5t^{-3}x^4\right) b^1 \\ &= 10b^1 t^{-3} x^4\end{aligned}$$

$$\therefore \dfrac{dy}{db} = 10bt^{-3}x^4$$

---

d) $\dfrac{dy}{dh}$
**Solution**

$$y = 5b^2 t^{-3} x^4$$

As we are differentiating *wrt h*, the function can be re-written as:

$$y = \left(5b^2 t^{-3} x^4\right) h^0$$

Comparing the function with $ax^n$, we can see that $a = 5b^2 t^{-3} x^4$ and $n = 0$. Hence

$$\begin{aligned}\dfrac{dy}{dh} &= 0 \times \left(5b^2 t^{-3} x^4\right) h^{0-1} \\ &= 0 \times \left(5b^2 t^{-3} x^4\right) h^{-1} \\ &= 0\end{aligned}$$

$$\therefore \dfrac{dy}{dh} = 0$$

---

e) $\dfrac{dy}{dA}$
**Solution**

$$y = 5b^2 t^{-3} x^4$$

As we are differentiating *wrt A*, the function can be re-written as:

$$y = \left(5b^2 t^{-3} x^4\right) A^0$$

# Fundamentals of Differentiation

Comparing the function with $ax^n$, we can see that $a = 5b^2t^{-3}x^4$ and $n = 0$. Hence

$$\frac{dy}{dA} = 0 \times (5b^2t^{-3}x^4)A^{0-1}$$
$$= 0 \times (5b^2t^{-3}x^4)A^{-1} = 0$$
$$\therefore \frac{dy}{dA} = 0$$

The last two examples show that when differentiating *wrt* a variable that is not in the expression of the function then it is as though we are differentiating a constant. Therefore, the derivative in this case is zero. Note this down.

Before we move to the next standard derivative, it is important to introduce a technique, called **'derivative of sum'**, also known as **'linearity rule of differentiation'**. This is used to find the derivative of a function which is made up of the summation of two or more separate functions or terms.

For example, $y = x^3 - 7x + 3$ is a function that consists of three other functions or terms, namely (i) $y_1 = x^3$, (ii) $y_2 = -7x$, and (iii) $y_3 = 3$. Well, you may start thinking that this is a polynomial of degree 3. Yes, it is, but it can be equally viewed as three functions. Again, you may say that $y_3 = 3$ is just a constant. You're right, but that does not 'deny' it being a function. Since we can plot it as a straight horizontal line passing through $y = 3$, it is a 'function'. The need for this technique becomes apparent when we have functions such as $3x + \sin \theta$ and $e^x + \sqrt{x^2}$ among others.

Now, back to our discussion: let's say we have $f(x)$, $U(x)$, $V(x)$, and $W(x)$, all functions of $x$. It follows that:

If $\boxed{f(x) = U(x) + V(x) + W(x)}$ then $\boxed{f'(x) = U'(x) + V'(x) + W'(x)}$ (12.4)

Alternatively,

If $\boxed{y = u + v + w}$ then $\boxed{\dfrac{dy}{dx} = \dfrac{du}{dx} + \dfrac{dv}{dx} + \dfrac{dw}{dx}}$ (12.5)

where $y$, $u$, $v$, and $w$ are functions of $x$.

We can write this rule in words as: **the derivative of a sum is the sum of the derivatives**.

This is indeed useful for finding the derivative of polynomial functions as well as other relevant functions that fall into this category or meet the criteria. One very quick note is that we will restrict the application to functions of powers here, as we haven't discussed other derivatives. However, the rule is more encompassing.

---

Let's try some examples.

### Example 4

Find the derivative of the following functions with respect to the indicated variable.

a) $y = 3x - 5$, $[x]$

b) $y = mx + c$, $[x]$

c) $y = 5t^2 - 11t + 3e$, $[t]$

d) $y = t - 13t^2 - t^3$, $[t]$

What did you get? Find the solution below to double-check your answer.

### Solution to Example 4

**HINT**

- Just follow the 4-step rule above and apply this to each term of the expression.
- We will also skip steps that are fundamental and have already been shown in previous examples.

**a)** $y = 3x - 5$
**Solution**

$$f(x) = 3x - 5 = 3x - 5x^0$$

Thus

$$\frac{dy}{dx} = 3x^0 - 0 \times 5x^{-1}$$
$$= 3 - 0$$
$$\therefore \frac{dy}{dx} = 3$$

---

**b)** $y = mx + c$
**Solution**

$$f(x) = mx + c = mx + cx^0$$

Thus

$$\frac{dy}{dx} = mx^0 + 0 \times cx^{-1}$$
$$= m + 0$$
$$\therefore \frac{dy}{dx} = m$$

This is the reason why the coefficient of $x$ in a linear expression $ax + b$ is usually taken as the slope of the line.

---

**c)** $y = 5t^2 - 11t + 3e$
**Solution**

$$f(x) = 5t^2 - 11t + 3e$$

Thus

$$\frac{dy}{dt} = 10t - 11$$
$$\therefore \frac{dy}{dt} = 10t - 11$$

# Fundamentals of Differentiation

**d)** $y = t - 13t^2 - t^3$
**Solution**

$$f(x) = t - 13t^2 - t^3$$

Thus

$$\frac{dy}{dt} = 1 - 26t - 3t^2$$

$$\therefore \frac{dy}{dt} = 1 - 26t - 3t^2$$

Since a derivative is usually a function of the gradient on the curve, it is essential to mention that when determining the gradient at a particular point, such as at $x = x_1$, we write:

$$\left.\frac{dy}{dx}\right|_{x=x_1} \tag{12.6}$$

Let's try an example.

## Example 5

Find the derivative of $y = x^5 - 2x^4 - 3x^3 + 7x^2 - x + 8$ at $x = 1$.

What did you get? Find the solution below to double-check your answer.

## Solution to Example 5

**Solution**

$$f(x) = x^5 - 2x^4 - 3x^3 + 7x^2 - x + 8$$

Thus

$$f'(x) = 5x^4 - 8x^3 - 9x^2 + 14x - 1$$

At $x = -1$, we have

$$f'(-1) = 5(-1)^4 - 8(-1)^3 - 9(-1)^2 + 14(-1) - 1$$
$$= 5 + 8 - 9 - 14 - 1 = -11$$

$$\therefore \left.\frac{dy}{dx}\right|_{x=-1} = -11$$

Let's try another example.

**Example 6**

Find the value of the gradient of the curve $y = \sqrt{x} - x^{-3} + 3x$ at point $P$, given that the $x$-coordinate at this point is 1. Present the answer in exact form.

---

What did you get? Find the solution below to double-check your answer.

**Solution to Example 6**

**Solution**

$$f(x) = \sqrt{x} - x^{-3} + 3x$$

We need to simplify this and present it in the required format as:

$$f(x) = \sqrt{x} - x^{-3} + 3x = x^{\frac{1}{2}} - x^{-3} + 3x$$

Thus, the gradient function is

$$f'(x) = \frac{1}{2}x^{-\frac{1}{2}} + 3x^{-4} + 3$$

At point $P$, we have $x = 1$, thus

$$f'(1) = \frac{1}{2}(1)^{-\frac{1}{2}} + 3(1)^{-4} + 3$$

$$= \frac{13}{2}$$

$$\therefore \left. \frac{dy}{dx} \right|_{x=1} = \frac{13}{2}$$

---

Let's try another example.

**Example 7**

Find the gradient of $y = x^2 \cdot \sqrt[3]{x} - \frac{x}{\sqrt[3]{x}}$ at point $A$ such that the coordinates of $A$ are given as (8, 124). Present the answer correct to 1 d.p.

# Fundamentals of Differentiation

What did you get? Find the solution below to double-check your answer.

**Solution to Example 7**

**Solution**

$$f(x) = x^2 \cdot \sqrt[3]{x} - \frac{x}{\sqrt[3]{x}}$$

We need to simplify this and present it in the required format as:

$$f(x) = x^2 \cdot \sqrt[3]{x} - \frac{x}{\sqrt[3]{x}}$$

$$= x^2 \cdot x^{\frac{1}{3}} - \frac{x}{x^{\frac{1}{3}}}$$

$$= x^{\left(2+\frac{1}{3}\right)} - x^{\left(1-\frac{1}{3}\right)}$$

$$= x^{\frac{7}{3}} - x^{\frac{2}{3}}$$

Thus, the gradient function is

$$f'(x) = \frac{7}{3}x^{\frac{4}{3}} - \frac{2}{3}x^{-\frac{1}{3}}$$

At point A (8, 124), we have that $x = 8$, thus we have

$$f'(8) = \frac{7}{3}(8)^{\frac{4}{3}} - \frac{2}{3}(8)^{-\frac{1}{3}}$$

$$= 37$$

$$\therefore \left.\frac{dy}{dx}\right|_{x=8} = 37.0 \ (1 \ d.p.)$$

---

Another example to try.

**Example 8**

The volume of a cylindrical shape is given by $V = \pi r^2 h$, where $V$ is the volume in cm$^3$, $r$ the radius in cm, and $h$ is the height in cm. The height is fixed at 10 cm in a batch production, such that the volume changes with respect to radius. Find the rate of change of volume *wrt* radius, when $r = 2$ cm. Present the answer in cm$^3$ correct to the nearest whole unit and comment on the answer.

What did you get? Find the solution below to double-check your answer.

**Solution to Example 8**

$$V = \pi r^2 h$$
$$\frac{dV}{dr} = 2\pi rh$$

At $r = 2$ **cm**, we have

$$\left.\frac{dV}{dr}\right|_{r=2 \text{ cm}} = 2\pi(2)(10)$$
$$= 40\pi \text{ cm}^3 \text{ per cm}$$
$$\therefore \left.\frac{dV}{dr}\right|_{r=2 \text{ cm}} = \mathbf{126 \text{ cm}^3 \text{ per cm}}$$

This means that for every 1 cm increase in radius length, the volume is increased by 126 cm³ approximately.

Another example to try.

**Example 9**

Simply $\frac{d}{dx}\left(x^2 + \sqrt{x}\right)$ at $x = 2$.

What did you get? Find the solution below to double-check your answer.

**Solution to Example 9**

Recall that one of the notations for the derivative is $\frac{d}{dx}(y)$. In other words, we are simply asked to find the derivative of the function $x^2 + \sqrt{x}$ at $x = 2$. Let's go

$$\frac{d}{dx}\left(x^2 + \sqrt{x}\right) = \frac{d}{dx}\left(x^2 + x^{\frac{1}{2}}\right)$$
$$= 2x + \frac{1}{2}x^{-\frac{1}{2}}$$

# Fundamentals of Differentiation

At $x = 2$, we have

$$\frac{d}{dx}\left(x^2 + \sqrt{x}\right) = 2(2) + \frac{1}{2}(2)^{-\frac{1}{2}}$$

$$= 4 + \frac{1}{2\sqrt{2}} = \frac{8\sqrt{2} + 1}{2\sqrt{2}}$$

$$= \frac{16 + \sqrt{2}}{4}$$

$$\therefore \frac{d}{dx}\left(x^2 + \sqrt{x}\right)\bigg|_{x=2} = \frac{16 + \sqrt{2}}{4}$$

## 12.5.2 Exponential Functions

For exponential functions, we have two cases.

**CASE 1** $e^x$

This can be regarded as a special situation where the base is a constant $e \approx 2.718...$. For this, we have that

$$\text{If } \boxed{y = e^x} \text{ then } \boxed{\frac{dy}{dx} = e^x} \qquad (12.7)$$

The above implies that a function is the same as its derivative. This is remarkable, and it's something special about the differential coefficient of $e^x$. The proof for this derivative is provided in Appendix A.

**CASE 2** $a^x$

This is a general case where the base is any value (constant or variable). We therefore have that

$$\text{If } \boxed{y = a^x} \text{ then } \boxed{\frac{dy}{dx} = a^x \ln a} \qquad (12.8)$$

Let's try some examples. Here we go.

### Example 10

Find the derivative of the following functions *wrt* the indicated variable.

a) $y = e^x$, $[x]$   b) $y = \sqrt{5}e^t$, $[t]$   c) $y = 0.5b^t$, $[t]$   d) $y = \frac{1}{6}a^t$, $[x]$

What did you get? Find the solution below to double-check your answer.

### Solution to Example 10

a) $y = e^x$, $[x]$

**Solution**

$$y = e^x$$

Thus

$$\frac{dy}{dx} = e^x$$

$$\therefore \frac{dy}{dx} = e^x$$

---

**b)** $y = \sqrt{5}e^t$, $[t]$
**Solution**

$$y = \sqrt{5}e^t$$

Thus

$$\frac{dy}{dt} = \sqrt{5}e^t$$

$$\therefore \frac{dy}{dt} = \sqrt{5}e^t$$

---

**c)** $y = 0.5b^t$, $[t]$
**Solution**

$$y = 0.5b^t$$

Thus

$$\frac{dy}{dt} = (0.5b^t) \ln b$$

$$\therefore \frac{dy}{dt} = \mathbf{0.5b^t \ln b}$$

---

**d)** $y = \frac{1}{6}a^t, [x]$
**Solution**

$$y = \frac{1}{6}a^t$$

We can modify this a bit to show how we arrive at the answer as:

$$y = \frac{1}{6}a^t = \frac{1}{6}a^t \times 1 = \left(\frac{1}{6}a^t\right) 1^x$$

Note that 1 to the power of anything is 1. In this case, we will consider $\frac{1}{6}a^t$ as a constant, thus

$$\frac{dy}{dx} = \left(\frac{1}{6}a^t\right) 1^x \times \ln 1$$

But $\ln 1 = 0$, hence

$$\frac{dy}{dx} = 0$$

$$\therefore \frac{dy}{dx} = \mathbf{0}$$

# Fundamentals of Differentiation

Here is another couple of examples to try.

### Example 11

Find the derivative of the following functions at the indicated value of the variable. Present the answer in exact form.

a) $y = 5x^{1.2} - 2e^x$, $[x = 32]$

b) $y = 2^x + x^2$, $[x = 3]$

What did you get? Find the solution below to double-check your answer.

### Solution to Example 11

a) $y = 5x^{1.2} - 2e^x$, $[x = 32]$
**Solution**

$$y = 5x^{1.2} - 2e^x$$

Thus

$$y' = 6x^{0.2} - 2e^x$$

At $x = 32$, we have

$$f'(32) = 6(32)^{0.2} - 2e^{32}$$
$$= 12 - 2e^{32}$$
$$\therefore \left.\frac{dy}{dx}\right|_{x=32} = 2(6 - e^{32})$$

b) $y = 2^x + x^2$, $[x = 3]$
**Solution**

$$y = 2^x + x^2$$

Thus

$$y' = 2^x \ln 2 + 2x$$

At $x = 3$, we have

$$f'(3) = 2^3 \ln 2 + 2 \times 3$$
$$= 8 \ln 2 + 6$$
$$\therefore \left.\frac{dy}{dx}\right|_{x=3} = 2(4\ln 2 + 3)$$

### 12.5.3 LOGARITHMIC FUNCTIONS

For logarithmic functions, we also have two cases.

**CASE 1**  $\log_e x$ or $\ln x$

This can be regarded as a special situation where the base is a value $e \approx 2.718\ldots$. We will have that

$$\text{If} \quad \boxed{y = \ln x} \quad \text{then} \quad \boxed{\frac{dy}{dx} = \frac{1}{x}} \quad (12.9)$$

**CASE 2**  $\log_a x$

This is when the base of the logarithmic function takes any real value or variable. We also have that

$$\text{If} \quad \boxed{y = \log_a x} \quad \text{then} \quad \boxed{\frac{dy}{dx} = \frac{1}{x \ln a}} \quad (12.10)$$

Let's try some examples.

### Example 12

Find the derivative of the following functions with respect to the indicated variable.

a) $y = 2 \ln x$, $[x]$     b) $y = \ln \sqrt{2}$, $[x]$     c) $y = \log_5 t$, $[\theta]$     d) $y = 7 \log_{\sqrt{3}} t$, $[t]$

What did you get? Find the solution below to double-check your answer.

### Solution to Example 12

**a)** $y = 2 \ln x$, $[x]$
**Solution**

$$y = 2 \ln x$$

Thus

$$\frac{dy}{dx} = 2 \times \frac{1}{x}$$

$$\therefore \frac{dy}{dx} = \frac{2}{x}$$

**b)** $y = \ln \sqrt{2}$, $[x]$
**Solution**

$$y = \ln \sqrt{2}$$

# Fundamentals of Differentiation

Thus

$$\frac{dy}{dx} = 0$$
$$\therefore \frac{dy}{dx} = 0$$

**c)** $y = \log_5 t$, $[\theta]$
**Solution**

$$y = \log_5 t$$

Thus

$$\frac{dy}{d\theta} = 0$$
$$\therefore \frac{dy}{d\theta} = 0$$

**d)** $y = 7\log_{\sqrt{3}} t$, $[t]$
**Solution**

$$y = 7\log_{\sqrt{3}} t$$

Thus

$$\frac{dy}{dt} = 7 \times \frac{1}{t \ln \sqrt{3}}$$
$$\therefore \frac{dy}{dt} = \frac{7}{t \ln \sqrt{3}}$$

Another set of examples to try.

## Example 13

Find the derivative of the following functions at the indicated value of the variable. Present the answer in exact form or correct to 2 decimal places.

**a)** $y = \ln x + \log x$, $[x = 5]$

**b)** $y = 4\log_3 x - 3^x - \ln v$, $[x = 1]$

What did you get? Find the solution below to double-check your answer.

## Solution to Example 13

**a)** $y = \ln x + \log x$, $[x = 5]$
**Solution**

$$y = \ln x + \log x$$

Thus
$$y' = \frac{1}{x} + \frac{1}{x \ln 10}$$

At $x = 5$, we have
$$y'(5) = \frac{1}{5} + \frac{1}{5 \ln 10}$$
$$= \frac{\ln 10 + 1}{5 \ln 10}$$
$$\therefore \left.\frac{dy}{dx}\right|_{x=5} = 0.29 \ (2 \ d.p.)$$

**b)** $y = 4 \log_3 x - 3^x - \ln v, \ [x = 1]$
**Solution**

$$y = 4 \log_3 x - 3^x - \ln v$$

Thus
$$y' = 4 \frac{1}{x \ln 3} - 3^x \ln 3 - 0$$
$$= \frac{4}{x \ln 3} - 3^x \ln 3$$
$$= \frac{4 - x(3^x)(\ln 3)^2}{x \ln 3}$$

At $x = 1$, we have
$$y'(1) = \frac{4 - 1 \times (3^1)(\ln 3)^2}{1 \times \ln 3}$$
$$= \frac{4 - 3(\ln 3)^2}{\ln 3}$$
$$\therefore \left.\frac{dy}{dx}\right|_{x=1} = 0.35 \ (2 \ d.p.)$$

## 12.5.4 Trigonometric Functions

In this section, we'll cover the derivative of the three fundamental trigonometric functions and their reciprocals. Relevant proofs are provided in Appendix A.

# Fundamentals of Differentiation

| Case 1 $\sin x$ | If | $y = \sin x$ | then | $\dfrac{dy}{dx} = \cos x$ | (12.11) |
|---|---|---|---|---|---|
| Case 2 $\cos x$ | If | $y = \cos x$ | then | $\dfrac{dy}{dx} = -\sin x$ | (12.12) |
| Case 3 $\tan x$ | If | $y = \tan x$ | then | $\dfrac{dy}{dx} = \sec^2 x$ | (12.13) |
| Case 4 $\sec x$ | If | $y = \sec x$ | then | $\dfrac{dy}{dx} = \sec x \tan x$ | (12.14) |
| Case 5 $\operatorname{cosec} x$ | If | $y = \operatorname{cosec} x$ | then | $\dfrac{dy}{dx} = -\operatorname{cosec} x \cot x$ | (12.15) |
| Case 6 $\cot x$ | If | $y = \cot x$ | then | $\dfrac{dy}{dx} = -\operatorname{cosec}^2 x$ | (12.16) |

Note that the derivative of the sine function is the cosine function, and the derivative of the cosine function is the negative of the sine function. This will be reversed in integration in Chapter 15. This is the case for all functions, but sine and cosine functions need particular attention.

Let's try some examples.

## Example 14

Find the derivative of the following functions *wrt* the indicated variable.

a) $y = \pi \sin x$, $[x]$

b) $y = \dfrac{\cos \theta}{\sqrt{6}}$, $[\theta]$

c) $y = 1.4 \tan \omega - \cos \omega$, $[\omega]$

d) $y = \varphi + \sec \varphi$, $[\varphi]$

e) $y = 3 \operatorname{cosec} \alpha + \sin x$, $[\alpha]$

f) $y = -\dfrac{\cot \beta}{\sqrt{3}}$, $[\beta]$

What did you get? Find the solution below to double-check your answer.

## Solution to Example 14

a) $y = \pi \sin x$, $[x]$
**Solution**

$$y = \pi \sin x$$

Thus
$$\frac{dy}{dx} = \pi \times \cos x$$
$$\therefore \frac{dy}{dx} = \pi \cos x$$

**b)** $y = \frac{\cos \theta}{\sqrt{6}}$, $[\theta]$
**Solution**

$$y = \frac{\cos \theta}{\sqrt{6}}$$
$$= \frac{1}{\sqrt{6}} \times \cos \theta$$

Thus
$$\frac{dy}{d\theta} = \frac{1}{\sqrt{6}} \times -\sin \theta$$
$$= -\frac{\sqrt{6}}{6} \times \sin \theta$$
$$\therefore \frac{dy}{d\theta} = -\frac{\sqrt{6} \sin \theta}{6}$$

**c)** $y = 1.4 \tan \omega - \cos \omega$, $[\omega]$
**Solution**

$$y = 1.4 \tan \omega - \cos \omega$$

Thus
$$\frac{dy}{d\omega} = 1.4 \tan \omega - \cos \omega$$
$$= 1.4 \sec^2 \omega - (-\sin \omega)$$
$$= 1.4 \sec^2 \omega + \sin \omega$$
$$\therefore \frac{dy}{d\omega} = \mathbf{1.4 \sec^2 \omega + \sin \omega}$$

**d)** $y = \varphi + \sec \varphi$, $[\varphi]$
**Solution**

$$y = \varphi + \sec \varphi$$

Thus
$$\frac{dy}{d\varphi} = 1 + \sec \varphi \tan \varphi$$
$$\therefore \frac{dy}{d\varphi} = \mathbf{1 + \sec \varphi \tan \varphi}$$

# Fundamentals of Differentiation

**e)** $y = 3 \csc \alpha + \sin x$, $[\alpha]$
**Solution**

$$y = 3 \csc \alpha + \sin x$$

Thus

$$\frac{dy}{d\alpha} = -3 \csc \alpha \cot \alpha + 0$$

$$\therefore \frac{dy}{d\alpha} = -3 \csc \alpha \cot \alpha$$

**NOTE**
The derivative of $\sin x$ *wrt* to $\alpha$ is zero.

---

**f)** $y = -\frac{\cot \beta}{\sqrt{3}}$, $[\beta]$
**Solution**

$$y = -\frac{\cot \beta}{\sqrt{3}} = -\frac{1}{\sqrt{3}} \times \cot \beta$$

Thus

$$\frac{dy}{d\beta} = -\frac{1}{\sqrt{3}} \times -\csc^2 \beta$$

$$= \frac{1}{\sqrt{3}} \times \csc^2 \beta$$

$$= \frac{\sqrt{3}}{3} \times \csc^2 \beta$$

$$\therefore \frac{dy}{d\beta} = \frac{\sqrt{3}}{3} \csc^2 \beta$$

---

Here is another couple of examples to try.

## Example 15

Find the derivative of the following functions at the indicated value of the variable. Present the answer in exact form.

**a)** $y = \sin x + \cos x$, $[x = \pi]$

**b)** $y = \sqrt{2} \tan \theta + \sqrt{3} \cot \theta$, $[\theta = 30°]$

What did you get? Find the solution below to double-check your answer.

### Solution to Example 15

**a)** $y = \sin x + \cos x$, $[x = \pi]$
**Solution**

$$y = \sin x + \cos x$$

Thus

$$y' = \cos x - \sin x$$

At $x = \pi$, we have

$$y'(\pi) = \cos \pi - \sin \pi$$
$$= -1 - 0 = -1$$
$$\therefore \left.\frac{dy}{dx}\right|_{x=\pi} = -1$$

---

**b)** $y = \sqrt{2}\tan\theta + \sqrt{3}\cot\theta$, $[\theta = 30°]$
**Solution**

$$y = \sqrt{2}\tan\theta + \sqrt{3}\cot\theta$$

Thus

$$y' = \sqrt{2}\sec^2\theta - \sqrt{3}\operatorname{cosec}^2\theta$$
$$= \frac{\sqrt{2}}{\cos^2\theta} - \frac{\sqrt{3}}{\sin^2\theta}$$

At $\theta = 30°$, we have

$$y'(30°) = \frac{\sqrt{2}}{\cos^2 30°} - \frac{\sqrt{3}}{\sin^2 30°}$$
$$= \frac{\sqrt{2}}{\left(\frac{\sqrt{3}}{2}\right)^2} - \frac{\sqrt{3}}{\left(\frac{1}{2}\right)^2} = \frac{\sqrt{2}}{\left(\frac{3}{4}\right)} - \frac{\sqrt{3}}{\left(\frac{1}{4}\right)}$$
$$= \frac{4\sqrt{2}}{3} - 4\sqrt{3} = \frac{4\sqrt{2} - 12\sqrt{3}}{3}$$
$$\therefore \left.\frac{dy}{d\theta}\right|_{\theta=30°} = \frac{4\left(\sqrt{2} - 3\sqrt{3}\right)}{3}$$

---

While we will be less concerned with inverse trigonometric functions in this book, we thought it is appropriate to provide the derivative of the inversion here for reference only (Table 12.6).

Fundamentals of Differentiation

**TABLE 12.6**
**Derivatives of Inverse Trigonometric Functions**

| Function | $\frac{dy}{dx}$ | Function | $\frac{dy}{dx}$ |
|---|---|---|---|
| $\sin^{-1}x$ | $\frac{1}{\sqrt{1-x^2}}$, $x \neq \pm 1$ | $\sec^{-1}x$ | $\frac{1}{|x|\sqrt{x^2-1}}$, $x \neq \pm 1, 0$ |
| $\cos^{-1}x$ | $\frac{-1}{\sqrt{1-x^2}}$, $x \neq \pm 1$ | $\csc^{-1}x$ | $\frac{-1}{|x|\sqrt{x^2-1}}$, $x \neq \pm 1, 0$ |
| $\tan^{-1}x$ | $\frac{1}{1+x^2}$ | $\cot^{-1}x$ | $\frac{-1}{1+x^2}$ |

## 12.6 HIGHER-ORDER DERIVATIVES

We've now covered the fundamentals of differentiation. It is our aim to finish off this chapter with a concept that is relevant to our work in the latter chapters when we start looking at the applications of differentiation. This is called the **higher derivatives**. It is a rather simple concept, and it implies finding the derivative of a derivative and another in a consecutive manner. The higher you go, the higher the number associated with the derivative.

The derivatives we've had so far are regarded as the **first derivative**, and if we take the derivative of the first derivative, we get the second derivative and when we take the derivative of the second one, we get the third, and so on. This will become easy with the examples to follow, but before that, here are some notations (Table 12.7) for the higher derivatives, given that $y = f(x)$.

As earlier noted for the first derivative, all other notations for higher derivatives are also operators and should be considered so.

**TABLE 12.7**
**Notations for Higher Derivatives Illustrated**

| Degree | Notation | Read as | Alternative Notations |
|---|---|---|---|
| First derivative | $\frac{dy}{dx}$ | Dee y dee x | $f'(x); \dot{y}; y'; \frac{d}{dx}(y)$ |
| Second derivative | $\frac{d^2y}{dx^2}$ | Dee two y dee x squared | $f''(x); \ddot{y}; y''; \frac{d}{dx}\left(\frac{dy}{dx}\right)$ |
| Third derivative | $\frac{d^3y}{dx^3}$ | Dee three y dee x cubed | $f'''(x); \dddot{y}; y'''; \frac{d}{dx}\left(\frac{d^2y}{dx^2}\right)$ |
| Fourth derivative | $\frac{d^4y}{dx^4}$ | Dee four y dee x quadrupled | $f'^v(x); \frac{d}{dx}\left(\frac{d^3y}{dx^3}\right)$ |
| nth derivative | $\frac{d^ny}{dx^n}$ | Dee n y dee x n | $\frac{d}{dx}\left(\frac{d^{n-1}y}{dx^{n-1}}\right)$ |

Let's try some examples to illustrate higher derivatives.

## Example 16

Find the first, second, and third derivatives of the following functions:

a) $y = x^3 + 5x - x^{-2}$
b) $s = 5t^2 - 11t + 3$
c) $y = \ln x$
d) $f(\theta) = \sin \theta$
e) $g(\varphi) = \cos \varphi$
f) $h(x) = e^x$

What did you get? Find the solution below to double-check your answer.

## Solution to Example 16

**HINT**

Nothing to worry about. We've used different notations to ensure that you are familiar with them. Feel free to choose from the options when you are solving problems.

a) $y = x^3 + 5x - x^{-2}$
**Solution**

$$y = x^3 + 5x - x^{-2}$$

Thus, the first derivative

$$\frac{dy}{dx} = 3x^2 + 5 + 2x^{-3}$$

differentiating $\frac{dy}{dx}$ to get the second derivative

$$\frac{d^2y}{dx^2} = 6x - 6x^{-4}$$

differentiating $\frac{d^2y}{dx^2}$ to get the third derivative

$$\frac{d^3y}{dx^3} = 6 + 24x^{-5}$$

Therefore

$$\frac{dy}{dx} = 3x^2 + 5 + 2x^{-3} \qquad \frac{d^2y}{dx^2} = 6x - 6x^{-4} \qquad \frac{d^3y}{dx^3} = 6 + 24x^{-5}$$

# Fundamentals of Differentiation

**b)** $s = 5t^2 - 11t + 3$
**Solution**

$$s = 5t^2 - 11t + 3$$

Thus, the first derivative

$$\dot{s} = 10t - 11$$

differentiating $\dot{s}$ to get the second derivative

$$\ddot{s} = 10$$

differentiating $\ddot{s}$ to get the third derivative

$$\dddot{s} = 0$$

Therefore

$$\dot{s} = 10t - 11 \qquad \ddot{s} = 10 \qquad \dddot{s} = 0$$

As you will see later in this book, if $s$ represents displacement (in m), then $\dot{s}$, $\ddot{s}$, and $\dddot{s}$ are the velocity (in m/s), acceleration (in m/s$^2$), and jerk or jolt (in m/s$^3$), respectively.

---

**c)** $y = \ln x$
**Solution**

$$y = \ln x$$

Thus, the first derivative

$$y' = \frac{1}{x} = x^{-1}$$

differentiating $y'$ to get the second derivative

$$y'' = -x^{-2} = -\frac{1}{x^2}$$

differentiating $y''$ to get the third derivative

$$y''' = 2x^{-3} = \frac{2}{x^3}$$

Therefore

$$y' = \frac{1}{x} \qquad y'' = -\frac{1}{x^2} \qquad y''' = \frac{2}{x^3}$$

**d)** $f(\theta) = \sin \theta$
**Solution**

$$f(\theta) = \sin \theta$$

Thus, the first derivative

$$f'(\theta) = \cos \theta$$

differentiating $f'(\theta)$ to get the second derivative

$$f''(\theta) = -\sin \theta$$

differentiating $f''(\theta)$ to get the third derivative

$$f'''(\theta) = -\cos \theta$$

Therefore

$$f'(\theta) = \cos \theta \quad f''(\theta) = -\sin \theta \quad f'''(\theta) = -\cos \theta$$

The cycle here is $S \to C \to -S \to -C$

---

**e)** $g(\varphi) = \cos \varphi$
**Solution**

$$g(\varphi) = \cos \varphi$$

Thus, the first derivative

$$g'(\varphi) = -\sin \varphi$$

differentiating $g'(\varphi)$ to get the second derivative

$$g''(\varphi) = -\cos \varphi$$

differentiating $g''(\varphi)$ to get the third derivative

$$g'''(\varphi) = \sin \varphi$$

Therefore

$$g'(\varphi) = -\sin \varphi \quad g''(\varphi) = -\cos \varphi \quad g'''(\varphi) = \sin \varphi$$

The cycle here is $C \to -S \to -C \to S$. It is therefore obvious that the cycle is continuous and can be represented as shown in Figure 12.10:

# Fundamentals of Differentiation

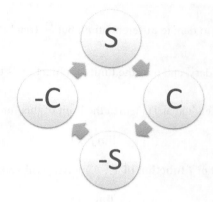

**FIGURE 12.10** Solution to Example 16(e).

**f)** $h(x) = e^x$
**Solution**

$$h(x) = e^x$$

Thus, the first derivative

$$h'(x) = e^x$$

differentiating $h'(x)$ to get the second derivative

$$h''(x) = e^x$$

differentiating $h''(x)$ to get the third derivative

$$h'''(x) = e^x$$

Therefore

$$h'(x) = e^x \qquad h''(x) = e^x \qquad h'''(x) = e^x$$

The function remains the same.

---

## 12.7 CHAPTER SUMMARY

1) Calculus is a branch of mathematics with two distinctive parts: differentiation and integration.
2) Differentiation is a measure of the '**rate of change**'. In other words, it is a technique that is used to determine the size of change in one physical quantity (e.g., volume, distance, mass) when compared to an equivalent change in another physical quantity (e.g., time).

3) We have a few ways to denote differentiation, but $\frac{dy}{dx}$ (reads '*dee y dee x*') seems very popular.

4) The limit is the boundary, and thus the limit of a function is the boundary of that function.

5) If a function of $x$ is denoted as $f(x)$, then the limit of this function is represented as:

$$\lim f(x)$$

This is read as '**limit of $f$ function of $x$**'. A more useful version is

$$\lim_{x \to \infty} f(x)$$

This is the '**limit of $f$ function of $x$ as $x \to \infty$**'. The sign $\to$ is read as '**tends to**' and technically implies that the left-hand variable $x$ approaches the right-hand variable or number.

6) Gradient (or slope) of a straight line is given by '**delta $y$ (or $\Delta y$) over delta $x$ (or $\Delta x$)**', i.e., $\frac{\Delta y}{\Delta x}$.

7) $\Delta$ is a Greek letter, generally used to denote a measurable change in a physical quantity as against $\delta$ – a lowercase delta – which denotes an infinitesimal (or an infinitely small) change in the quantity. $\Delta y$ is obtained by finding the difference in two $y$-coordinates; these values could be the initial and final or the first and second or second and fifth and so on. The same can be said about $\Delta x$.

8) Unlike a straight line, the gradient of a curve changes as we move from one point to another along the curve.

9) Derivative of the function as the change in $x$ approaches zero is written as:

$$\boxed{\frac{dy}{dx} = \lim_{\delta x \to 0} \left[ \frac{\delta y}{\delta x} \right]} \quad \text{OR} \quad \boxed{\frac{dy}{dx} = \lim_{\Delta x \to 0} \left[ \frac{\Delta y}{\Delta x} \right]}$$

10) $\frac{dy}{dx}$ is an operator and does not imply $dy$ divided by $dx$. It can be written as $\frac{d}{dx}(y)$ read as '**dee dee $x$ of $y$**', where $y$ is a function of $x$ or any other variable. This makes it explicitly clear that $\frac{dy}{dx}$ is an operator. On the other hand, $\frac{\delta y}{\delta x}$ and $\frac{\Delta y}{\Delta x}$ are not operators; they imply a change in $y$ divided by a change in $x$, hence they can be detached and evaluated separately.

11) A particular method of finding the derivative of a function is known as '**finding the derivative from the first principle**' or '**differentiating from the first principle**'.

12) The process of finding the derivative of a function is called **differentiation**. This is also called 'differential coefficient of a function *wrt* to an independent variable' or 'the rate of change of function or gradient function' or simply $\frac{dy}{dx}$. It is a gradient function rather than a gradient, as it gives us a function that can be used to find the gradient at a given point on the curve.

# Fundamentals of Differentiation

13) Since a derivative is usually a function of the gradient on the curve, it is essential to mention that when determining the gradient at a particular point such as at $x = x_1$, we write:

$$\frac{dy}{dx}\bigg|_{x=x_1}$$

14) Standard derivatives of common functions:

- Polynomial function or power function

    If $\boxed{f(x) = ax^n}$ then $\boxed{\dfrac{dy}{dx} = anx^{n-1}}$

    The above is valid for any real values of $n$ and $a$. If $a = 1$, we have a modified version

    If $\boxed{f(x) = x^n}$ then $\boxed{\dfrac{dy}{dx} = nx^{n-1}}$

- Exponential functions

    If $\boxed{y = e^x}$ then $\boxed{\dfrac{dy}{dx} = e^x}$

    If $\boxed{y = a^x}$ then $\boxed{\dfrac{dy}{dx} = a^x \ln a}$

- Logarithmic functions

    If $\boxed{y = \ln x}$ then $\boxed{\dfrac{dy}{dx} = \dfrac{1}{x}}$

    If $\boxed{y = \log_a x}$ then $\boxed{\dfrac{dy}{dx} = \dfrac{1}{x \ln a}}$

- Trigonometric functions and their reciprocals

    If $\boxed{y = \sin x}$ then $\boxed{\dfrac{dy}{dx} = \cos x}$

    If $\boxed{y = \cos x}$ then $\boxed{\dfrac{dy}{dx} = -\sin x}$

    If $\boxed{y = \tan x}$ then $\boxed{\dfrac{dy}{dx} = \sec^2 x}$

    If $\boxed{y = \sec x}$ then $\boxed{\dfrac{dy}{dx} = \sec x \tan x}$

    If $\boxed{y = \operatorname{cosec} x}$ then $\boxed{\dfrac{dy}{dx} = -\operatorname{cosec} x \cot x}$

    If $\boxed{y = \cot x}$ then $\boxed{\dfrac{dy}{dx} = -\operatorname{cosec}^2 x}$

15) The '**derivative of sum**' is used to find the derivative of a function which is made up of the summation of two or more separate functions or terms.

Let's say we have $f(x)$, $U(x)$, $V(x)$, and $W(x)$, all functions of $x$. It follows that:

If $\boxed{f(x) = U(x) + V(x) + W(x)}$ then $\boxed{f'(x) = U'(x) + V'(x) + W'(x)}$

Alternatively,

If $\boxed{y = u + v + w}$ then $\boxed{\dfrac{dy}{dx} = \dfrac{du}{dx} + \dfrac{dv}{dx} + \dfrac{dw}{dx}}$

where $y$, $u$, $v$, and $w$ are functions of $x$.

We can write this rule in words as: **the derivative of a sum is the sum of the derivatives**.

16) The **higher derivatives** imply finding the derivative of a derivative and another in a consecutive manner. The higher you go, the higher the number associated to the derivative..

\*\*\*\*

## 12.8 FURTHER PRACTICE

To access complementary contents, including additional exercises, please go to www.dszak.com.

# 13 Advanced Differentiation

*Learning Outcomes*

Once you have studied the content of this chapter, you should be able to:

- Determine the derivative of composite functions using the chain rule
- Determine the derivative of a product of two or more functions using the product rule
- Determine the derivative of rational functions using the quotient rule
- Determine the derivative of complex functions using a combination of rules
- Determine the derivative of implicit functions
- Determine the derivative of logarithmic functions
- Determine the derivative of parametric functions
- Use the inverse derivative

## 13.1 INTRODUCTION

In the preceding chapter, we went through the fundamentals of differential coefficients and focused on simple functions, which consists of one term and one process. We used the differentiation of the sum rule to deal with multiple-term functions. In practice, functions can be of multiple processes, products of two or more functions, division of functions, or combinations of these. These complex functions can be solved by applying further rules and techniques, as will be seen soon in this chapter. We will therefore cover the chain, product, and quotient rules, and discuss implicit, logarithmic, and parametric differentiation.

## 13.2 RULES OF DIFFERENTIATION

We will start off here by looking at the rule of:

a) multiple process,
b) multiplication, and
c) division.

Recall that the differential of the sum rule has catered for addition, and subtraction is also implied by extension.

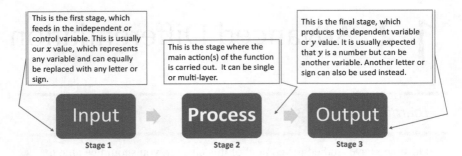

**FIGURE 13.1** Chain rule illustrated.

### 13.2.1 CHAIN RULE

The chain rule plays an important role in differentiation, and it is used to solve multiple, composite, or cascaded processes. This is commonly called '**function of a function**'.

Before we introduce the rule, let's quickly revise a function or a process. Figure 13.1 is a flow chart (or block diagram) of the three main stages of a function, which:

1) takes a value, generally denoted as $x$,
2) processes the value based on a pre-defined protocol or instruction(s), and finally
3) sends the processed value as an output $y$.

We write this process in a shorthand form as:

$$\boxed{y = f(x)} \tag{13.1}$$

The above simply implies that $y$ is a function of $x$ or the value of $y$ depends on the value of $x$, based on a pre-defined process. For example, $y = x^2 - 1$ is a function that depends on the value of $x$ to produce a related value of $y$. The process here consists of two instructions, namely:

a) Square the input (or multiply the input by itself).
b) Subtract 1 from the result obtained in (a).

The combination of these two instructions results in what is known as a quadratic function, and the result is sent to stage 3 as output $y$. This is a simple or single function, where the instructions are considered as a block of processes. On the other hand, we have situations where an input will undergo two or more independent processes to generate an output.

Under this, we will examine two different situations:

**Case 1** Function of a function

Imagine the result of the above process in stage 2, defined by $f(x) = x^2 - 1$, needs to be further processed by an instruction or a set of instructions, defined by $g(x) = \sin x$, before being sent as an output to stage 3. In this case, we have what we call (or can be regarded as) a **function of a function**. This is illustrated in Figure 13.2.

# Advanced Differentiation

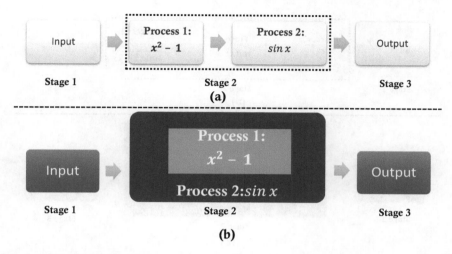

**FIGURE 13.2** Two levels of chain rule illustrated.

In (a), we have two processes: the input $x$ is fed into $x^2 - 1$, and the result is an input to a second process, $\sin(x^2 - 1)$. Once the sine function has been performed, the result is fed to stage 3, the output. The same can be visualised as illustrated in (b). In this case, the input from stage 1 is sent to the inner process of quadratic function, and the result of this process is passed to an outer process for further processing before being sent to stage 3, the output. This is simply a composite function, which can be denoted as $f[g(x)]$. Understanding the inner vs outer, as well as the first and second processes, is a pre-requisite for dealing with chain rule, which will be used to solve function of function problems.

The rule states that:

$$\text{If } \boxed{y = f(u) \text{ and } u = g(x)} \text{ then } \boxed{\frac{dy}{dx} = \frac{dy}{du} \times \frac{du}{dx}} \tag{13.2}$$

where $g(x)$ is the inner or first function to be evaluated and $f(u)$ is the outer or second function.

Other variants of the rule are:

$$\text{If } \boxed{y = f[g(x)]} \text{ then } \boxed{\frac{dy}{dx} = f'[g(x)] \times g'(x)} \tag{13.3}$$

and

$$\text{If } \boxed{y = [f(x)]^n} \text{ then } \boxed{\frac{dy}{dx} = n[f(x)]^{n-1} \times f'(x)} \tag{13.4}$$

Using our example above, we can say that $g(x) = x^2 - 1$ and $f(u) = \sin u$.

**Case 2** Function of a 'function of a function'

Sometimes, a function can be a cascade of more than two processes. Figure 13.3 is a three-process cascaded function.

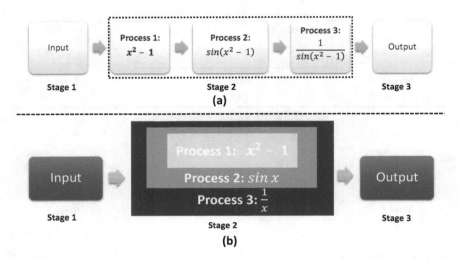

**FIGURE 13.3** Three levels of chain rule illustrated.

The chain rule for this can be stated as:

$$\text{If } \boxed{y = f(u),\ u = g(v),\ \text{and } v = h(x)} \text{ then } \boxed{\frac{dy}{dx} = \frac{dy}{du} \times \frac{du}{dv} \times \frac{dv}{dx}} \quad (13.5)$$

Following the above two cases, we should note the following:

**Note 1** A function of a function can be viewed as a loop within a loop.

**Note 2** The derivative of the overall function is the product of the derivative from each loop.

**Note 3** If there are $n$ loops or processes in stage 2, there will be $n$ derivatives in the differential coefficient. For example, in a function of a function, $n = 2$ and there are two derivatives to be multiplied, namely $\frac{dy}{du}$ and $\frac{du}{dx}$.

**Note 4** $n$ can be determined by the number of distinctly possible brackets. If one can distinctly apply $[\{(\ )\}]$, then there are three processes and it will be considered as a '**function of a function of a function**'.

**Note 5** Notice that the formula is such that the RHS can be cancelled out (as illustrated below) in order to obtain the LHS, though technically impossible as derivatives are operators.

$$\frac{dy}{dx} = \frac{dy}{\cancel{du}} \times \frac{\cancel{du}}{\cancel{dv}} \times \frac{\cancel{dv}}{dx} = \frac{dy}{dx}$$

# Advanced Differentiation

Let's try some examples.

## Example 1

Solve the following using the chain rule, and where possible, solve the same without using the rule.

**a)** $\frac{d}{dx}(x^2 + 2)^2$     **b)** $\frac{d}{dt}(3t - 1)^3$

What did you get? Find the solution below to double-check your answer.

## Solution to Example 1

**a)** $\frac{d}{dx}(x^2 + 2)^2$
**Solution**

$$y = (x^2 + 2)^2$$

**Chain Rule**
This is a function of a function. Let $u = x^2 + 2$ then $y = u^2$, therefore

$$\frac{du}{dx} = 2x \quad \text{and} \quad \frac{dy}{du} = 2u = 2(x^2 + 2)$$

Using the chain rule, we have that

$$\frac{dy}{du} = \frac{dy}{du} \times \frac{du}{dx}$$
$$= 2(x^2 + 2) \times 2x$$
$$= 4x(x^2 + 2)$$
$$\therefore \frac{dy}{dx} = 4x(x^2 + 2)$$

### ALTERNATIVE METHOD

Without the chain rule, we will need to expand the brackets

$$y = (x^2 + 2)^2$$
$$= x^4 + 4x^2 + 4$$

Differentiating $y$ wrt $x$, we have

$$\frac{dy}{dx} = 4x^3 + 8x$$

Factorise the differential coefficient, we have

$$= 4x(x^2 + 2)$$

$$\therefore \frac{dy}{dx} = 4x\left(x^2 + 2\right)$$

---

**b)** $\frac{d}{dt}(3t-1)^3$
**Solution**

$$y = (3t-1)^3$$

**Chain Rule**
This is a function of a function. Let $u = 3t - 1$ then $y = u^3$, therefore

$$\frac{du}{dt} = 3 \quad \text{and} \quad \frac{dy}{du} = 3u^2 = 3(3t-1)^2$$

Using the chain rule, we have that

$$\frac{dy}{dt} = \frac{dy}{du} \times \frac{du}{dt}$$

$$= 3(3t-1)^2 \times 3$$

$$= 9(3t-1)^2$$

$$\therefore \frac{dy}{dt} = 9(3t-1)^2$$

**ALTERNATIVE METHOD**

Without the chain rule, we will need to expand the brackets

$$y = (3t-1)^3$$

This can be expanded using the binomial theorem or another suitable approach. Doing this will give

$$y = (3t-1)^3$$
$$= (3t-1)(3t-1)(3t-1)$$
$$= (9t^2 - 6t + 1)(3t-1)$$
$$= 27t^3 - 9t^2 - 18t^2 + 6t + 3t - 1$$
$$= 27t^3 - 27t^2 + 9t - 1$$

Differentiating $y$ wrt $x$, we have

$$\frac{dy}{dx} = 81t^2 - 54t + 9$$

# Advanced Differentiation

Factorise the differential coefficient, we have

$$= 9(9t^2 - 6t + 1)$$
$$= 9(3t - 1)^2$$
$$\therefore \frac{dy}{dx} = 9(3t - 1)^2$$

From the above examples, we observe the following:

**Note 1** The answer is the same for either the chain rule or by expanding the brackets.
**Note 2** In (a), the chain rule appears relatively longer, but in (b), it was simple and relatively straightforward.

Whilst the expansion is possible here, and we could apply both methods, often the expansion is either impossible or too complicated to obtain, especially if $u$ is neither a power function nor a polynomial expression, e.g., $\sin(x^2 - 1)$. We shall try a new set of examples and determine how many we can try without the chain rule. Here we go.

## Example 2

Find the differential coefficient of the following functions *wrt x*.

a) $y = \cos^3 x$         b) $y = \cos x^3$         c) $y = e^{\tan x}$         d) $y = 2e^{1-3x}$

e) $y = \left(x^3 - x + \sqrt{3}\right)^{20}$    f) $y = \frac{1}{\sqrt{x^2 - 3x - 1}}$    g) $y = \operatorname{cosec}(3 - \ln x)$    h) $y = \log_2 e^x$

What did you get? Find the solution below to double-check your answer.

## Solution to Example 2

**HINT**

Once we are comfortable with this, we do not need to go through all the steps. All we need is to:

- Determine each 'block' and differentiate each.
- Substitute the inner block function in the derivative of the immediate next block.
- Multiply all the differential coefficients together.

**a)** $y = \cos^3 x$
**Solution**

$$y = \cos^3 x = (\cos x)^3$$

Let $u = \cos x$ then $y = u^3$, therefore

$$\frac{du}{dx} = -\sin x \quad \text{and} \quad \frac{dy}{du} = 3u^2 = 3(\cos x)^2 = 3\cos^2 x$$

Using the chain rule, we have that

$$\frac{dy}{dx} = \frac{dy}{du} \times \frac{du}{dx}$$
$$= 3\cos^2 x \times -\sin x$$
$$= -3\sin x \cos^2 x$$

$$\therefore \frac{d}{dx}\cos^3 x = -3\sin x \cos^2 x$$

---

**b)** $y = \cos x^3$
**Solution**

$$y = \cos x^3 = \cos(x^3)$$

Let $u = x^3$ then $y = \cos u$, therefore

$$\frac{du}{dx} = 3x^2 \quad \text{and} \quad \frac{dy}{du} = -\sin u = -\sin x^3$$

Using the chain rule, we have that

$$\frac{dy}{dx} = \frac{dy}{du} \times \frac{du}{dx}$$
$$= -\sin x^3 \times 3x^2$$
$$= -3x^2 . \sin x^3$$

$$\therefore \frac{d}{dx}\cos x^3 = -3x^2 . \sin x^3$$

---

**c)** $y = e^{\tan x}$
**Solution**

$$y = e^{\tan x}$$

Let $u = \tan x$ then $y = e^u$, therefore

$$\frac{du}{dx} = \sec^2 x \quad \text{and} \quad \frac{dy}{du} = e^u = e^{\tan x}$$

Using the chain rule, we have that

$$\frac{dy}{dx} = \frac{dy}{du} \times \frac{du}{dx}$$
$$= e^{\tan x} \times \sec^2 x$$
$$= \sec^2 x . e^{\tan x}$$

$$\therefore \frac{d}{dx} e^{\tan x} = \sec^2 x . e^{\tan x}$$

# Advanced Differentiation

**d)** $y = 2e^{1-3x}$
**Solution**

$$y = 2e^{1-3x}$$

Let $u = 1 - 3x$ then $y = 2e^u$, therefore

$$\frac{du}{dx} = -3 \quad \text{and} \quad \frac{dy}{du} = 2e^u = 2e^{1-3x}$$

Using the chain rule, we have that

$$\frac{dy}{dx} = \frac{dy}{du} \times \frac{du}{dx}$$
$$= 2e^{1-3x} \times -3$$
$$= -6e^{1-3x}$$

$$\therefore \frac{d}{dx} 2e^{1-3x} = -6e^{1-3x}$$

---

**e)** $y = \left(x^3 - x + \sqrt{3}\right)^{20}$
**Solution**

$$y = \left(x^3 - x + \sqrt{3}\right)^{20}$$

Although it is possible to expand this, avoid the expansion unless it is a stated requirement. Let $u = x^3 - x + \sqrt{3}$ then $y = u^{20}$, therefore

$$\frac{du}{dx} = 3x^2 - 1 \quad \text{and} \quad \frac{dy}{du} = 20u^{19} = 20\left(x^3 - x + \sqrt{3}\right)^{19}$$

Using the chain rule, we have that

$$\frac{dy}{dx} = \frac{dy}{du} \times \frac{du}{dx}$$
$$= 20\left(x^3 - x + \sqrt{3}\right)^{19} \times 3x^2 - 1$$
$$= 20\left(3x^2 - 1\right)\left(x^3 - x + \sqrt{3}\right)^{19}$$

$$\therefore \frac{d}{dx}\left(x^3 - x + \sqrt{3}\right)^{20} = 20\left(3x^2 - 1\right)\left(x^3 - x + \sqrt{3}\right)^{19}$$

---

**f)** $y = \dfrac{1}{\sqrt{x^2 - 3x - 1}}$
**Solution**

$$y = \frac{1}{\sqrt{x^2 - 3x - 1}}$$

We need to manipulate this a bit as follows

$$y = \frac{1}{(x^2 - 3x - 1)^{\frac{1}{2}}} = (x^2 - 3x - 1)^{-\frac{1}{2}}$$

Now it is in the right form. Let $u = x^2 - 3x - 1$ then $y = u^{-\frac{1}{2}}$, therefore

$$\frac{du}{dx} = 2x - 3 \quad \text{and} \quad \frac{dy}{du} = -\frac{1}{2}u^{-\frac{3}{2}} = -\frac{1}{2}(x^2 - 3x - 1)^{-\frac{3}{2}}$$

Using the chain rule, we have that

$$\frac{dy}{dx} = \frac{dy}{du} \times \frac{du}{dx}$$

$$= -\frac{1}{2}(x^2 - 3x - 1)^{-\frac{3}{2}} \times (2x - 3)$$

$$= -\frac{1}{2(x^2 - 3x - 1)^{\frac{3}{2}}} \times (2x - 3) = -\frac{2x - 3}{2\sqrt{(x^2 - 3x - 1)^3}}$$

$$\therefore \frac{d}{dx}\frac{1}{\sqrt{x^2 - 3x - 1}} = -\frac{2x - 3}{2\sqrt{(x^2 - 3x - 1)^3}} \quad \text{OR} \quad \frac{3 - 2x}{2\sqrt{(x^2 - 3x - 1)^3}}$$

**NOTE**
You will see later in this chapter that we can deal with this problem using the quotient rule.

**g)** $y = \operatorname{cosec}(3 - \ln x)$
**Solution**

$$y = \operatorname{cosec}(3 - \ln x)$$

Let $u = 3 - \ln x$ then $y = \operatorname{cosec} u$, therefore

$$\frac{du}{dx} = -\frac{1}{x} \quad \text{and} \quad \frac{dy}{du} = -\operatorname{cosec} u . \cot u = -\operatorname{cosec}(3 - \ln x) . \cot(3 - \ln x)$$

Using the chain rule, we have that

$$\frac{dy}{dx} = \frac{dy}{du} \times \frac{du}{dx}$$

$$= -\operatorname{cosec}(3 - \ln x) . \cot(3 - \ln x) \times -\frac{1}{x}$$

$$= \frac{\operatorname{cosec}(3 - \ln x) . \cot(3 - \ln x)}{x}$$

$$\therefore \frac{d}{dx}\operatorname{cosec}(3 - \ln x) = \frac{\operatorname{cosec}(3 - \ln x) . \cot(3 - \ln x)}{x}$$

**h)** $y = \log_2 e^x$
**Solution**

$$y = \log_2 e^x$$

Let $u = e^x$ then $y = \log_2 u$, therefore

$$\frac{du}{dx} = e^x \quad \text{and} \quad \frac{dy}{du} = \frac{1}{u . \ln 2} = \frac{1}{e^x . \ln 2}$$

# Advanced Differentiation

Using the chain rule, we have that

$$\frac{dy}{dx} = \frac{dy}{du} \times \frac{du}{dx}$$

$$= \frac{1}{e^x . \ln 2} \times e^x$$

$$= \frac{1}{\ln 2}$$

$$\therefore \frac{d}{dx} \log_2 e^x = \frac{1}{\ln 2}$$

**ALTERNATIVE METHOD**

$$y = \log_2 e^x$$

Using the power law of logarithm, we have

$$y = x \log_2 e = (\log_2 e) x$$

We need to recognise that $\log_2 e$ is a constant and should be differentiated in the same manner as $y = 3x$, hence

$$\frac{dy}{dx} = \log_2 e$$

Using a couple of laws, we can simplify the above as:

$$\frac{dy}{dx} = \log_2 e = \frac{\log_e e}{\log_e 2}$$

$$= \frac{1}{\log_e 2} = \frac{1}{\ln 2}$$

$$\therefore \frac{d}{dx} \log_2 e^x = \frac{1}{\ln 2}$$

Another set of examples to try. These are 'function of a function of a function' cases.

## Example 3

Find the differential coefficient of the following functions *wrt x*.

**a)** $y = \tan^4 (2x + 1)$  **b)** $y = \sin\left(e^{x^2}\right)$  **c)** $y = \sec^2 (1 - 5x)$  **d)** $y = \log_7 |\cos(\pi x)|$

What did you get? Find the solution below to double-check your answer.

## Solution to Example 3

### HINT

- The approach here is that we need to identify the three composite functions. The innermost is our $v$, then $u$ and the outermost (or the 'team leader') is our $y$. You may want to rearrange or re-present the function to give you the picture of a **3** cascaded function.
- Also, once you are comfortable, you can ignore certain steps. In this case, you just need to multiply the derivative of each of the blocks or layers as you see it.

**a)** $y = \tan^4(2x + 1)$
**Solution**

$$y = \tan^4(2x + 1)$$
$$= [\tan(2x + 1)]^4$$

Using the above format, $v = 2x + 1$, $u = \tan v$ and $y = u^4$, therefore

$$\frac{dv}{dx} = 2 \qquad \begin{aligned}\frac{du}{dv} &= \sec^2 v \\ &= \sec^2(2x+1)\end{aligned} \qquad \text{and} \qquad \begin{aligned}\frac{dy}{du} &= 4u^3 = 4(\tan v)^3 \\ &= 4\tan^3 v = 4\tan^3(2x+1)\end{aligned}$$

Using the chain rule, we have that

$$\frac{dy}{dx} = \frac{dy}{du} \times \frac{du}{dv} \times \frac{dv}{dx}$$
$$= 4\tan^3(2x+1) \times \sec^2(2x+1) \times 2$$
$$= 8\sec^2(2x+1)\tan^3(2x+1)$$
$$\therefore \frac{dy}{dx} = \mathbf{8\sec^2(2x+1)\tan^3(2x+1)}$$

---

**b)** $y = \sin\left(e^{x^2}\right)$
**Solution**

$$y = \sin\left(e^{x^2}\right)$$

Let $v = x^2$, $u = e^v$ and $y = \sin u$, therefore

$$\frac{dv}{dx} = 2x \qquad \begin{aligned}\frac{du}{dv} &= e^v \\ &= e^{x^2}\end{aligned} \qquad \text{and} \qquad \begin{aligned}\frac{dy}{du} &= \cos u \\ &= \cos\left(e^{x^2}\right)\end{aligned}$$

Using the chain rule, we have that

$$\frac{dy}{dx} = \frac{dy}{du} \times \frac{du}{dv} \times \frac{dv}{dx}$$
$$= \cos\left(e^{x^2}\right) \times e^{x^2} \times 2x$$
$$= 2x.e^{x^2}.\cos\left(e^{x^2}\right)$$
$$\therefore \frac{dy}{dx} = 2x.e^{x^2}.\cos\left(e^{x^2}\right)$$

---

**c)** $y = \sec^2(1 - 5x)$
**Solution**

$$y = \sec^2(1 - 5x)$$
$$= [\sec(1 - 5x)]^2$$

Using the above format, $v = 1 - 5x$, $u = \sec v$ and $y = u^2$, therefore

$$\frac{dv}{dx} = -5 \qquad \begin{aligned}\frac{du}{dv} &= \sec v.\tan v \\ &= \sec(1 - 5x).\tan(1 - 5x)\end{aligned} \qquad \text{and} \qquad \begin{aligned}\frac{dy}{du} &= 2u \\ &= 2\sec(1 - 5x)\end{aligned}$$

Using the chain rule, we have that

$$\frac{dy}{dx} = \frac{dy}{du} \times \frac{du}{dv} \times \frac{dv}{dx}$$
$$= 2\sec(1 - 5x) \times \sec(1 - 5x).\tan(1 - 5x) \times -5$$
$$= -10\sec^2(1 - 5x)\tan(1 - 5x)$$
$$\therefore \frac{dy}{dx} = -10\sec^2(1 - 5x)\tan(1 - 5x)$$

---

**d)** $y = \log_7|\cos(\pi x)|$
**Solution**

$$y = \log_7 \cos(\pi x)$$

Using the above format, $v = \pi x$, $u = \cos v$ and $y = \log_7 u$, therefore

$$\frac{dv}{dx} = \pi \qquad \begin{aligned}\frac{du}{dv} &= -\sin v \\ &= -\sin(\pi x)\end{aligned} \qquad \text{and} \qquad \begin{aligned}\frac{dy}{du} &= \frac{1}{u \ln 7} \\ &= \frac{1}{(\ln 7)\cos(\pi x)}\end{aligned}$$

Using the chain rule, we have that

$$\frac{dy}{dx} = \frac{dy}{du} \times \frac{du}{dv} \times \frac{dv}{dx}$$
$$= \frac{1}{(\ln 7)\cos(\pi x)} \times -\sin(\pi x) \times \pi$$
$$= -\frac{\pi \sin(\pi x)}{(\ln 7)\cos(\pi x)} = -\frac{\pi}{\ln 7}\tan(\pi x)$$
$$\therefore \frac{dy}{dx} = -\frac{\pi}{\ln 7}\tan(\pi x)$$

## 13.2.2 PRODUCT RULE

This rule is used when one function is multiplied by another function *wrt* the same variable, e.g., $y = \sin x \cdot (x^2 + 3)$. In this case, we will consider that $y$ is a function of $x$ which is a product of two functions, namely:

1) a sine function, $\sin x$, and
2) a quadratic function, $x^2 + 3$.

In the above example, if we choose $u$ and $v$ to represent $\sin x$ and $x^2 + 3$, respectively, then we can write $y = uv$. Also, just as we can write $y = f(x)$ to show that $y$ varies as $x$ or is a function of $x$, we can equally write $u = g(x)$ and $v = h(x)$ for the same purpose. Consequentially, it is a pre-requisite that $y$, $u$, and $v$ are functions of the same variable (i.e., $x$ in this case) to use the product rule. As a result, $y = \sin 2\pi (x^2 + 3)$ is not a product rule case since functions $\sin 2\pi$ and $x^2 + 3$ are not varying with respect to the same variable. In fact, this is a single function, i.e., quadratic function, since $\sin 2\pi$ is a constant once evaluated.

Product rule can be stated as:

$$\text{If } \boxed{y = uv, \text{ where } u = g(x) \text{ and } v = h(x)} \text{ then } \boxed{\frac{dy}{dx} = u\frac{dv}{dx} + v\frac{du}{dx}} \quad (13.6)$$

Since $y$, $u$, and $v$ are functions of $x$, we can write the rule in a short form as:

$$\boxed{y' = uv' + vu'} \quad (13.7)$$

where $y'$, $u'$ and $v'$ are derivatives of the respective functions.

Before we start looking at examples, let's outline the following steps to take in applying this rule:

**Step 1:** Identify that you have two functions that are multiplied together.
**Step 2:** Label the two functions as $u$ and $v$.
**Step 3:** Find the derivative of $u$ and call this $u'$.
**Step 4:** Do the same for $v$ and call this $v'$.
**Step 5:** Substitute $u$, $v$, $u'$, and $v'$ in $uv' + vu'$.
**Step 6:** Simplify the expression to obtain your derivative $y'$ or $\frac{dy}{dx}$.

The above steps are for two functions, but the same steps can be used when there are more than two functions in the product.

Let's try an example.

### Example 4

A function $y = f(x)$ is defined by $y = uvw$, where $u$, $v$, and $w$ are functions of $x$. Prove that the differential coefficient of $y$ wrt $x$ can be obtained using:

$$\frac{dy}{dx} = uv\frac{dw}{dx} + uw\frac{dv}{dx} + vw\frac{du}{dx}$$

# Advanced Differentiation

What did you get? Find the solution below to double-check your answer.

## Solution to Example 4

**Method 1** From the first principle

$$y = uvw$$

**Step 1:** Determine the change in $y$ or $y + \delta y$

$\delta x$ in $x$ will produce a corresponding $\delta y$, $\delta u$, $\delta v$, and $\delta w$ in $y$, $u$, $v$, and $w$, respectively. Therefore,

$$y + \delta y = f(x + \delta x)$$
$$= (u + \delta u)(v + \delta v)(w + \delta w)$$
$$= (uv + u\delta v + v\delta u + \delta u \delta v)(w + \delta w)$$
$$= uvw + uw\delta v + vw\delta u + w\delta u \delta v + uv\delta w + u\delta v \delta w + v\delta u \delta w + \delta u \delta v \delta w$$

It follows that

$$\delta y = [uvw + uw\delta v + vw\delta u + w\delta u \delta v + uv\delta w + u\delta v \delta w + v\delta u \delta w + \delta u \delta v \delta w] - [uvw]$$
$$= uw\delta v + vw\delta u + w\delta u \delta v + uv\delta w + u\delta v \delta w + v\delta u \delta w + \delta u \delta v \delta w$$

**Step 2:** Evaluate the slope as:

$$\text{Slope} \approx \frac{\delta y}{\delta x}$$
$$\approx \frac{uw\delta v + vw\delta u + w\delta u \delta v + uv\delta w + u\delta v \delta w + v\delta u \delta w + \delta u \delta v \delta w}{\delta x}$$
$$\approx uw\frac{\delta v}{\delta x} + vw\frac{\delta u}{\delta x} + w\delta u\frac{\delta v}{\delta x} + uv\frac{\delta w}{\delta x} + u\delta v\frac{\delta w}{\delta x} + v\delta u\frac{\delta w}{\delta x} + \delta u \delta v\frac{\delta w}{\delta x}$$

**Step 3:** Determine the derivative

$$\frac{dy}{dx} = \lim_{\delta x \to 0}\left(\frac{\delta y}{\delta x}\right)$$
$$= \lim_{\delta x \to 0}\left(uw\frac{\delta v}{\delta x} + vw\frac{\delta u}{\delta x} + w\delta u\frac{\delta v}{\delta x} + uv\frac{\delta w}{\delta x} + u\delta v\frac{\delta w}{\delta x} + v\delta u\frac{\delta w}{\delta x} + \delta u \delta v\frac{\delta w}{\delta x}\right)$$

As $\delta x \to 0$, $\frac{\delta w}{\delta x} \to \frac{dw}{dx}$, $\frac{\delta v}{\delta x} \to \frac{dv}{dx}$ and $\frac{\delta u}{\delta x} \to \frac{du}{dx}$, and $\delta u \to 0$, $\delta v \to 0$, and $\delta w \to 0$, hence

$$\frac{dy}{dx} = uw\frac{dv}{dx} + vw\frac{du}{dx} + w \times 0\frac{dv}{dx} + uv\frac{dw}{dx} + u \times 0\frac{dw}{dx} + v \times 0\frac{dw}{dx}$$

$$\therefore \frac{dy}{dx} = \frac{d}{dx}(uvw) = uw\frac{dv}{dx} + vw\frac{du}{dx} + uv\frac{dw}{dx}$$

**Method 2** Double application of the product rule

Alternatively, we can prove the same by applying the product rule formula twice as follows:

$$y = uvw$$
$$= u(vw)$$

**Step 1:** Product rule (first stage)

We will consider $u$ and $vw$, hence

$$\frac{d}{dx}y = u\frac{d}{dx}vw + vw\frac{du}{dx} \quad ---(i)$$

**Step 2:** Product rule (2nd stage)

Using the product rule again, we have

$$\frac{d}{dx}vw = v\frac{dw}{dx} + w\frac{dv}{dx} \quad ---(ii)$$

**Step 3:** Combining the two stages

Substituting (ii) in (i), we have

$$\frac{d}{dx}y = u\left[v\frac{dw}{dx} + w\frac{dv}{dx}\right] + vw\frac{du}{dx}$$
$$= uv\frac{dw}{dx} + uw\frac{dv}{dx} + vw\frac{du}{dx}$$
$$\therefore \frac{dy}{dx} = \frac{d}{dx}(uvw) = uw\frac{dv}{dx} + vw\frac{du}{dx} + uv\frac{dw}{dx}$$

The second method above can be easily used when $n$ functions are multiplied together by using the product rule $(n-1)$ times.

Another example to try.

## Example 5

A polynomial function is given as $y = (3x^2 - 5)(x^4 + 1)$. Using the product rule or otherwise, differentiate $y$ wrt $x$.

What did you get? Find the solution below to double-check your answer.

## Solution to Example 5

**Method 1** Using Product Rule

$$y = (3x^2 - 5)(x^4 + 1)$$

# Advanced Differentiation

Let $u = 3x^2 - 5$ and $v = x^4 + 1$, therefore

$$\frac{du}{dx} = 6x \quad \text{and} \quad \frac{dv}{dx} = 4x^3$$

Using the product rule, we have that

$$\begin{aligned}
\frac{dy}{dx} &= u\frac{dv}{dx} + v\frac{du}{dx} \\
&= (3x^2 - 5)(4x^3) + (x^4 + 1)(6x) \\
&= 12x^5 - 20x^3 + 6x^5 + 6x \\
&= 18x^5 - 20x^3 + 6x \\
&= 2x(9x^4 - 10x^2 + 3) \\
\therefore \frac{dy}{dx} &= 2x(9x^4 - 10x^2 + 3)
\end{aligned}$$

**Method 2** Using expansion

In this alternative method, we start by opening the brackets as:

$$\begin{aligned}
y &= (3x^2 - 5)(x^4 + 1) \\
&= 3x^6 - 5x^4 + 3x^2 - 5
\end{aligned}$$

This is now a polynomial function, thus

$$\begin{aligned}
y' &= 18x^5 - 20x^3 + 6x \\
&= 2x(9x^4 - 10x^2 + 3) \\
\therefore \frac{dy}{dx} &= 2x(9x^4 - 10x^2 + 3)
\end{aligned}$$

---

The above example highlights the possibility of using a different approach when a product rule question is presented. In fact, this question shows that the alternative method is quicker and easier. Yes, this seems true for this question, which is a product of two polynomial functions, i.e., $3x^2 - 5$ and $x^4 + 1$. In general, the function where the product rule is specially required to be used is when multiplying functions cannot be simplified (or fused) into one function. This will become apparently evident in the examples to follow shortly.

Another set of examples to try.

## Example 6

Find the differential coefficient of the following functions *wrt x*.

a) $y = (x^2 + x - 1)\tan x$

b) $y = \sin x . \cos x$

c) $y = 2x^5 \log x$

d) $y = e^x \cot x$

e) $y = \sqrt{x} . \ln x$

What did you get? Find the solution below to double-check your answer.

## Solution to Example 6

**a)** $y = (x^2 + x - 1)\tan x$
**Solution**

$$y = (x^2 + x - 1)\tan x$$

Let $u = x^2 + x - 1$ and $v = \tan x$, therefore

$$\frac{du}{dx} = 2x + 1 \quad \text{and} \quad \frac{dv}{dx} = \sec^2 x$$

Using the product rule, we have that

$$\frac{dy}{dx} = u\frac{dv}{dx} + v\frac{du}{dx}$$
$$= (x^2 + x - 1)(\sec^2 x) + (\tan x)(2x + 1)$$

There is less to simplify here

$$\therefore \frac{dy}{dx} = (x^2 + x - 1)\sec^2 x + (2x + 1)\tan x$$

---

**b)** $y = \sin x . \cos x$
**Solution**

$$y = \sin x . \cos x$$

Let $u = \sin x$ and $v = \cos x$, therefore

$$\frac{du}{dx} = \cos x \quad \text{and} \quad \frac{dv}{dx} = -\sin x$$

Using the product rule, we have that

$$\frac{dy}{dx} = u\frac{dv}{dx} + v\frac{du}{dx}$$
$$= \sin x . - \sin x + \cos x . \cos x$$
$$= \cos^2 x - \sin^2 x$$
$$= \cos 2x$$

$$\therefore \frac{dy}{dx} = \cos 2x$$

# Advanced Differentiation

**ALTERNATIVE METHOD**

$$y = \sin x \cdot \cos x$$

Using the double-angle formula, we have

$$y = \frac{1}{2}\sin 2x$$

Using the chain rule to differentiate the above, we have

$$y' = \left(\frac{1}{2}\cos 2x\right) \times 2$$
$$= \cos 2x$$
$$\frac{dy}{dx} = \cos 2x$$

---

**c)** $y = 2x^5 \log x$
**Solution**

$$y = 2x^5 \log x$$

Let $u = 2x^5$ and $v = \log x$, therefore

$$\frac{du}{dx} = 10x^4 \quad \text{and} \quad \frac{dv}{dx} = \frac{1}{x \ln 10}$$

Using the product rule, we have that

$$\frac{dy}{dx} = u\frac{dv}{dx} + v\frac{du}{dx}$$
$$= 2x^5 \cdot \frac{1}{x \ln 10} + \log x \cdot 10x^4$$
$$= 10x^4 \log x + \frac{2x^4}{\ln 10}$$
$$= 2x^4 \left(5\log x + \frac{1}{\ln 10}\right)$$
$$= 2x^4 \left(\log x^5 + \log e\right)$$
$$= 2x^4 \log ex^5$$
$$\therefore \frac{dy}{dx} = 2x^4 \log ex^5$$

---

**d)** $y = e^x \cot x$
**Solution**

$$y = e^x \cot x$$

Let $u = e^x$ and $v = \cot x$, therefore

$$\frac{du}{dx} = e^x \quad \text{and} \quad \frac{dv}{dx} = -\operatorname{cosec}^2 x$$

Using the product rule, we have that

$$\frac{dy}{dx} = u\frac{dv}{dx} + v\frac{du}{dx}$$
$$= e^x \cdot -\cosec^2 x + \cot x \cdot e^x$$
$$= e^x(\cot x - \cosec^2 x)$$
$$\therefore \frac{dy}{dx} = e^x(\cot x - \cosec^2 x)$$

---

**e)** $y = \sqrt{x} \cdot \ln x$
**Solution**

$$y = \sqrt{x} \cdot \ln x$$

Let $u = \sqrt{x}$ and $v = \ln x$, therefore

$$\frac{du}{dx} = \frac{1}{2}x^{-\frac{1}{2}} \quad \text{and} \quad \frac{dv}{dx} = \frac{1}{x}$$

Using the product rule, we have that

$$\frac{dy}{dx} = u\frac{dv}{dx} + v\frac{du}{dx}$$
$$= \sqrt{x} \cdot \frac{1}{x} + \ln x \cdot \frac{1}{2} x^{-\frac{1}{2}}$$
$$= x^{\frac{1}{2}} \cdot x^{-1} + \ln x \cdot \frac{1}{2x^{\frac{1}{2}}}$$
$$= x^{-\frac{1}{2}} + \frac{\ln x}{2\sqrt{x}} = \frac{1}{\sqrt{x}} + \frac{\ln x}{2\sqrt{x}}$$
$$= \frac{2 + \ln x}{2\sqrt{x}}$$
$$\therefore \frac{dy}{dx} = \frac{2 + \ln x}{2\sqrt{x}}$$

---

Before we leave this section, let's try cases where a function is a product of three and four functions. Here we go.

## Example 7

Find the differential coefficient of the following functions *wrt* x.

**a)** $y = e^x \cdot \sin x \cdot \ln x$  **b)** $y = (x^2 + 1)(x - 3)\log_6 x$  **c)** $y = x^2 \cdot e^x \cdot \tan x \cdot \ln x$

# Advanced Differentiation

What did you get? Find the solution below to double-check your answer.

## Solution to Example 7

**a)** $y = e^x \cdot \sin x \cdot \ln x$
**Solution**

$$y = e^x \cdot \sin x \cdot \ln x$$

Let $u = e^x$, $v = \sin x$, and $w = \ln x$, therefore

$$u' = e^x \quad v' = \cos x \quad \text{and} \quad w' = \frac{1}{x}$$

Using the product rule, we have that

$$y' = uvw' + uwv' + vwu'$$
$$= (e^x)(\sin x)\left(\frac{1}{x}\right) + (e^x)(\ln x)(\cos x) + (\sin x)(\ln x)(e^x)$$
$$= e^x \left(\frac{\sin x}{x} + \ln x \cdot \cos x + \ln x \cdot \sin x\right)$$

There is less to simplify here

$$\therefore \frac{dy}{dx} = e^x \left(\frac{\sin x}{x} + \ln x \cdot \cos x + \ln x \cdot \sin x\right)$$

**NOTE**
The functions $u$, $v$, and $w$ in this example are such that we cannot simplify to a product of two functions. Therefore, there is no way to cross-reference the solution except possibly by using the product rule twice. Let's look at another example where we have an apparent alternative method.

---

**b)** $y = (x^2 + 1)(x - 3) \log_6 x$
**Solution**

**Method 1 Product rule for three functions**

$$y = (x^2 + 1)(x - 3) \log_6 x$$

Let $u = x^2 + 1$, $v = x - 3$, and $w = \log_6 x$, therefore

$$u' = 2x \quad v' = 1 \quad \text{and} \quad w' = \frac{1}{x \ln 6}$$

Using the product rule, we have that

$$y' = uvw' + uwv' + vwu'$$
$$= (x^2 + 1)(x - 3)\left(\frac{1}{x \ln 6}\right) + (x^2 + 1)(\log_6 x)(1) + (x - 3)(\log_6 x)(2x)$$
$$= \frac{(x^2 + 1)(x - 3)}{x \ln 6} + (x^2 + 1) \log_6 x + 2x(x - 3) \log_6 x$$

$$= \frac{(x^2+1)(x-3)}{x\ln 6} + \log_6 x\left[(x^2+1) + 2x(x-3)\right]$$

$$= \frac{(x^2+1)(x-3)}{x\ln 6} + \log_6 x\left[x^2+1+2x^2-6x\right]$$

$$= \frac{(x^2+1)(x-3)}{x\ln 6} + \left(3x^2-6x+1\right)\log_6 x$$

$$\therefore \frac{dy}{dx} = \frac{(x^2+1)(x-3)}{x\ln 6} + \left(3x^2-6x+1\right)\log_6 x$$

**Method 2 Product rule for two functions**

For this case, we will simplify the function as:

$$y = (x^2+1)(x-3)\log_6 x$$
$$= (x^3 - 3x^2 + x - 3)\log_6 x$$

Let $u = x^3 - 3x^2 + x - 3$ and $v = \log_6 x$, therefore

$$v' = 3x^2 - 6x + 1 \quad \text{and} \quad w' = \frac{1}{x\ln 6}$$

Using the product rule, we have that

$$y' = uv' + vu'$$

$$= (x^3 - 3x^2 + x - 3)\left(\frac{1}{x\ln 6}\right) + (\log_6 x)(3x^2 - 6x + 1)$$

$$= \frac{x^3 - 3x^2 + x - 3}{x\ln 6} + \left(3x^2 - 6x + 1\right)\log_6 x$$

$$= \frac{(x^2+1)(x-3)}{x\ln 6} + \left(3x^2 - 6x + 1\right)\log_6 x$$

$$\therefore \frac{dy}{dx} = \frac{(x^2+1)(x-3)}{x\ln 6} + \left(3x^2 - 6x + 1\right)\log_6 x$$

as before.

---

**c)** $y = x^2 \cdot e^x \cdot \tan x \cdot \ln x$

**Solution**

We are considering a 4-function product here to illustrate its application, and you can extend it beyond this, though it is a rare requirement.

$$y = x^2 \cdot e^x \cdot \tan x \cdot \ln x$$

Let $u = x^2$, $v = e^x$, $w = \tan x$, and $z = \ln x$, therefore

$$u' = 2x \quad v' = e^x \quad w' = \sec^2 x \quad \text{and} \quad z' = \frac{1}{x}$$

Using the product rule, we have that

$$y' = uvwz' + uvzw' + uwzv' + vwzu'$$

$$= (x^2)(e^x)(\tan x)\left(\frac{1}{x}\right) + (x^2)(e^x)(\ln x)(\sec^2 x) + (x^2)(\tan x)(\ln x)(e^x) + (e^x)(\tan x)(\ln x)(2x)$$

$$= x \cdot e^x \cdot \tan x + x^2 \cdot e^x \cdot \ln x \cdot \sec^2 x + x^2 \cdot e^x \cdot \ln x \cdot \tan x + 2x \cdot e^x \cdot \ln x \cdot \tan x$$

$$= xe^x\left(\tan x + x \cdot \ln x \cdot \sec^2 x + x \cdot \ln x \cdot \tan x + 2\ln x \cdot \tan x\right)$$

# Advanced Differentiation

There is less to simplify here

$$\therefore \frac{dy}{dx} = xe^x \left( \tan x + x.\ln x.\sec^2 x + x.\ln x.\tan x + 2\ln x.\tan x \right)$$

## 13.2.3 Quotient Rule

We are now ready to take the last key rule of differentiation, called the **quotient rule**. It is not much different from the product rule, except that this is used for rational functions (i.e., one function is divided by another function) where it is either impossible to simplify it into a single function or practically difficult to do so.

The rule states that:

$$\text{If } \boxed{y = \frac{u}{v}, \text{ where } u = g(x) \text{ and } v = h(x)} \text{ then } \boxed{\frac{dy}{dx} = \frac{v\frac{du}{dx} - u\frac{dv}{dx}}{v^2}} \quad (13.8)$$

Since $y$, $u$, and $v$ are functions of $x$, we can write the rule in a compact form as:

$$\boxed{y' = \frac{vu' - uv'}{v^2}} \quad (13.9)$$

where $y'$, $u'$, and $v'$ are derivatives of the respective functions.

We should mention a couple of things about this rule, which makes it different from the product rule:

**Note 1** There is a negative sign between the two terms in the numerator.
**Note 2** The function in the denominator must be taken as $v$ when referring to the above (format of the) formula.

Before we start looking at examples, let's outline the steps to follow in applying this rule:

**Step 1:** Identify that you have two functions, namely dividend ($u$) and divisor ($v$). This should be expressed as $y = \frac{u}{v}$.
**Step 2:** Label the two functions as $u$ and $v$.
**Step 3:** Find the derivative of $u$ and call this $u'$.
**Step 4:** Do the same for $v$ and call this $v'$.
**Step 5:** Substitute $u$, $v$, $u'$, and $v'$ in $\frac{vu' - uv'}{v^2}$.
**Step 6:** Simplify the expression to obtain your derivative $y'$ or $\frac{dy}{dx}$.

Now it's time to try some examples.

## Example 8

Find the differential coefficient of the following functions wrt x.

a) $y = \dfrac{1}{x^5+3x+5}$
b) $y = \dfrac{5}{e^x}$

What did you get? Find the solution below to double-check your answer.

## Solution to Example 8

a) $y = \dfrac{1}{x^5+3x+5}$

**Solution**

**Method 1** Quotient rule

$$y = \dfrac{1}{x^5 + 3x + 5}$$

Let $u = 1$ and $v = x^5 + 3x + 5$, therefore

$$u' = 0 \quad \text{and} \quad v' = 5x^4 + 3$$

Using the quotient rule, we have that

$$y' = \dfrac{vu' - uv'}{v^2}$$

$$= \dfrac{(x^5 + 3x + 5) \times 0 - 1 \times (5x^4 + 3)}{(x^5 + 3x + 5)^2}$$

$$= \dfrac{-(5x^4 + 3)}{(x^5 + 3x + 5)^2}$$

There is less to simplify here

$$\therefore \dfrac{dy}{dx} = \dfrac{-(5x^4 + 3)}{(x^5 + 3x + 5)^2}$$

**Method 2** Chain rule

$$y = \dfrac{1}{x^5 + 3x + 5} = (x^5 + 3x + 5)^{-1}$$

If $u = x^5 + 3x + 5$ then $y = u^{-1}$, therefore

$$\dfrac{du}{dx} = 5x^4 + 3 \quad \text{and} \quad \dfrac{dy}{du} = -u^{-2} = -(x^5 + 3x + 5)^{-2}$$

# Advanced Differentiation

Using the chain rule, we have

$$\frac{dy}{dx} = \frac{dy}{du} \times \frac{du}{dx}$$
$$= -(x^5 + 3x + 5)^{-2} \times (5x^4 + 3)$$
$$= \frac{-(5x^4 + 3)}{(x^5 + 3x + 5)^2}$$
$$\therefore \frac{dy}{dx} = \frac{-(5x^4 + 3)}{(x^5 + 3x + 5)^2}$$

b) $y = \frac{5}{e^x}$

**Solution**

**Method 1** Quotient rule

$$y = \frac{5}{e^x}$$

Let $u = 5$ and $v = e^x$, therefore

$$\frac{du}{dx} = 0 \quad \text{and} \quad \frac{dv}{dx} = e^x$$

Using the quotient rule, we have that

$$y' = \frac{vu' - uv'}{v^2}$$
$$= \frac{(e^x) \times 0 - 5 \times (e^x)}{(e^x)^2}$$
$$= \frac{-5e^x}{e^{2x}} = \frac{-5}{e^x}$$
$$\therefore \frac{dy}{dx} = \frac{-5}{e^x} \quad \text{OR} \quad -5e^{-x}$$

**Method 2** Chain rule

$$y = \frac{5}{e^x} = 5e^{-x} = 5(e^x)^{-1}$$

If $u = e^x$ then $y = 5u^{-1}$, therefore

$$\frac{du}{dx} = e^x \quad \text{and} \quad \frac{dv}{dx} = -5u^{-2} = -5(e^x)^{-2} = -5e^{-2x}$$

Using the chain rule, we have that

$$\frac{dy}{dx} = \frac{dy}{du} \times \frac{du}{dx}$$
$$= -5e^{-2x} \times e^x$$
$$= -5e^{-x}$$
$$\therefore \frac{dy}{dx} = -5e^{-x}$$

We note from the above that there are cases where the quotient rule can be used, though alternative methods could prove more suitable. Also, in cases where there is an apparent division and the dividend (or numerator) is a constant, the quotient rule becomes

$$\boxed{y' = \frac{-uv'}{v^2}} \qquad (13.10)$$

Let's try another example.

### Example 9

A function of $t$ is given by $\frac{2t+1}{6t^2-7t-5}$. Show that the differential coefficient of $f(t)$ is $\frac{-3}{(3t-5)^2}$.

What did you get? Find the solution below to double-check your answer.

### Solution to Example 9

$$f(t) = \frac{2t+1}{6t^2 - 7t - 5}$$

**Method 1** Quotient rule

$$y = \frac{2t+1}{6t^2 - 7t - 5}$$

Let $u = 2t + 1$ and $v = 6t^2 - 7t - 5$, therefore

$$u' = 2 \quad \text{and} \quad v' = 12t - 7$$

Using the quotient rule, we have that

$$\begin{aligned} y' &= \frac{vu' - uv'}{v^2} \\ &= \frac{(6t^2 - 7t - 5) \times 2 - (2t+1) \times (12t - 7)}{(6t^2 - 7t - 5)^2} \\ &= \frac{2(6t^2 - 7t - 5) - (2t+1)(12t - 7)}{(6t^2 - 7t - 5)^2} \\ &= \frac{(12t^2 - 14t - 10) - (24t^2 - 2t - 7)}{(6t^2 - 7t - 5)^2} \\ &= \frac{-12t^2 - 12t - 3}{(6t^2 - 7t - 5)^2} \\ &= \frac{-3(4t^2 + 4t + 1)}{(6t^2 - 7t - 5)^2} \end{aligned}$$

# Advanced Differentiation

We can simplify the above expression as:

$$y' = \frac{-3(2t+1)(2t+1)}{[(2t+1)(3t-5)]^2}$$

$$= \frac{-3(2t+1)^2}{(2t+1)^2(3t-5)^2}$$

$$= \frac{-3}{(3t-5)^2}$$

$$\therefore \frac{dy}{dt} = \frac{-3}{(3t-5)^2}$$

**Method 2** Expansion and the chain rule

$$y = \frac{2t+1}{6t^2 - 7t - 5}$$

Let's factorise the denominator, we have

$$y = \frac{2t+1}{(2t+1)(3t-5)}$$

$$= \frac{1}{3t-5}$$

As this is now simplified, we will use the chain rule (for the same reasons we noted above), although the quotient rule can also be used.

$$y = \frac{1}{3t-5} = (3t-5)^{-1}$$

If $u = 3t - 5$ then $y = u^{-1}$, therefore

$$\frac{du}{dx} = 3 \quad \text{and} \quad \frac{dv}{dx} = -u^{-2} = -(3t-5)^{-2}$$

Using chain rule, we have that

$$\frac{dy}{dt} = \frac{dy}{du} \times \frac{du}{dt}$$

$$= -(3t-5)^{-2} \times 3$$

$$= \frac{-3}{(3t-5)^2}$$

There is less to simplify here.

$$\therefore \frac{dy}{dt} = \frac{-3}{(3t-5)^2}$$

For the above examples, we can avoid using the quotient rule. However, there are instances where the rational function cannot be simplified, and as a result, we can only differentiate them using the quotient rule. Let's try some examples to illustrate this.

## Example 10

Find the differential coefficient of the following functions *wrt x*.

a) $y = \dfrac{3x-7}{x^2+1}$  b) $y = \dfrac{\log_2 x}{\sec x}$  c) $y = \dfrac{e^x}{\sqrt{x^5}}$  d) $y = \dfrac{\cot x}{\ln x}$

What did you get? Find the solution below to double-check your answer.

## Solution to Example 10

**a)** $y = \dfrac{3x-7}{x^2+1}$
**Solution**

$$y = \frac{3x-7}{x^2+1}$$

Let $u = 3x - 7$ and $v = x^2 + 1$, therefore

$$u' = 3 \quad \text{and} \quad v' = 2x$$

Using the quotient rule, we have that

$$y' = \frac{vu' - uv'}{v^2}$$
$$= \frac{(x^2+1) \times 3 - (3x-7) \times 2x}{(x^2+1)^2}$$
$$= \frac{3x^2 + 3 - 6x^2 + 14x}{(x^2+1)^2} = \frac{-3x^2 + 14x + 3}{(x^2+1)^2}$$
$$\therefore \frac{dy}{dx} = \frac{-(3x^2 - 14x - 3)}{(x^2+1)^2}$$

**b)** $y = \dfrac{\log_2 x}{\sec x}$
**Solution**

$$y = \frac{\log_2 x}{\sec x}$$

Let $u = \log_2 x$ and $v = \sec x$, therefore

$$u' = \frac{1}{x \ln 2} \quad \text{and} \quad v' = \sec x . \tan x$$

# Advanced Differentiation

Using the quotient rule, we have that

$$y' = \frac{vu' - uv'}{v^2}$$

$$= \frac{(\sec x)\left(\frac{1}{x\ln 2}\right) - (\log_2 x)(\sec x . \tan x)}{(\sec x)^2}$$

$$= \frac{\sec x \left[\left(\frac{1}{x\ln 2}\right) - (\log_2 x)(\tan x)\right]}{(\sec x)(\sec x)}$$

$$= \frac{\left(\frac{1}{x\ln 2}\right) - (\log_2 x)(\tan x)}{\sec x}$$

$$= \frac{\left(\frac{1}{x\ln 2}\right)}{\sec x} - \frac{(\log_2 x)(\tan x)}{\sec x}$$

$$= \frac{1}{\left(\frac{1}{\cos x}\right)(x\ln 2)} - \frac{(\log_2 x)\left(\frac{\sin x}{\cos x}\right)}{\sec x}$$

$$= \frac{\cos x}{x\ln 2} - \frac{(\log_2 x)(\sin x . \sec x)}{\sec x}$$

$$= \frac{\cos x}{x\ln 2} - (\sin x)\log_2 x$$

$$= \frac{\cos x - x(\ln 2)(\sin x)\log_2 x}{x\ln 2}$$

$$\therefore \frac{dy}{dx} = \frac{\cos x - x(\ln 2)(\sin x)\log_2 x}{x\ln 2}$$

---

**c)** $y = \frac{e^x}{\sqrt{x^5}}$

**Solution**

$$y = \frac{e^x}{\sqrt{x^5}} = \frac{e^x}{x^{\frac{5}{2}}}$$

Let $u = e^x$ and $v = x^{\frac{5}{2}}$, therefore

$$u' = e^x \quad \text{and} \quad v' = \frac{5}{2}x^{\frac{3}{2}} = \frac{5}{2}\sqrt{x^3}$$

Using the quotient rule, we have that

$$y' = \frac{vu' - uv'}{v^2}$$

$$= \frac{\left(\sqrt{x^5}\right) \times e^x - (e^x) \times \frac{5}{2}\sqrt{x^3}}{\left(\sqrt{x^5}\right)^2}$$

$$= \frac{e^x\left(\sqrt{x^5} - 2.5\sqrt{x^3}\right)}{x^5}$$

Let's simplify this a bit further

$$= \frac{e^x\left[\left(\sqrt{x^2}\right)\left(\sqrt{x^3}\right) - 2.5\left(\sqrt{x^2}\right)\left(\sqrt{x}\right)\right]}{x^5}$$

$$= \frac{e^x\left[x\left(\sqrt{x^3}\right) - 2.5x\left(\sqrt{x}\right)\right]}{x^5}$$

$$= \frac{e^x \cdot x\left[\sqrt{x^3} - 2.5\sqrt{x}\right]}{x^5}$$

$$= \frac{e^x\left[\sqrt{x^3} - 2.5\sqrt{x}\right]}{x^4}$$

$$\therefore \frac{dy}{dx} = \frac{e^x\left[\sqrt{x^3} - 2.5\sqrt{x}\right]}{x^4}$$

---

**d)** $y = \dfrac{\cot x}{\ln x}$

**Solution**

$$y = \frac{\cot x}{\ln x}$$

Let $u = \cot x$ and $v = \ln x$, therefore

$$u' = -\csc^2 x \quad \text{and} \quad v' = \frac{1}{x}$$

Using the quotient rule, we have that

$$y' = \frac{vu' - uv'}{v^2}$$

$$= \frac{(\ln x) \times -\csc^2 x - (\cot x) \times \frac{1}{x}}{(\ln x)^2}$$

$$= \frac{-\ln x \cdot \csc^2 x - \frac{\cot x}{x}}{(\ln x)^2}$$

$$= \frac{-\frac{[x \cdot \ln x \cdot \csc^2 x + \cot x]}{x}}{(\ln x)^2}$$

$$= -\frac{x \cdot \ln x \cdot \csc^2 x + \cot x}{x \cdot (\ln x)^2}$$

$$\therefore \frac{dy}{dx} = -\frac{x \cdot \ln x \cdot \csc^2 x + \cot x}{x \cdot (\ln x)^2}$$

# Advanced Differentiation

## 13.2.4 COMBINED RULE

By the 'combined' rule, we intend to demonstrate cases where more than one rule would be needed for a question; in fact, this is common in general. At any rate, it is desirable to master the 'tricks' of the chain rule (and possibly the product rule too) such that one would not need to refer to any of the formulas when applying them. This is exactly what we will do in the worked examples to follow shortly. In other words, we will ignore presenting the chain rule, and sometimes the product rule will also be left out.

To summarise, given three functions $u$, $v$, and $w$, we would like to cover the following cases:

1) $y = \frac{uv}{w}$

2) $y = \frac{u}{vw}$

With the two cases mentioned above, $u$, $v$, and $w$ can be functions of functions, and there will be a need for the use of the chain rule. Let's try a couple of examples.

### Example 11

Find the differential coefficient of the following functions wrt $x$.

a) $y = \frac{x^3 \sin 4x}{e^{5x}}$

b) $y = \frac{(1-x)^2}{\ln 5x . \cos e^x}$

What did you get? Find the solution below to double-check your answer.

### Solution to Example 11

a) $y = \frac{x^3 \sin 4x}{e^{5x}}$

**Solution**

$$y = \frac{x^3 \sin 4x}{e^{5x}}$$

This is a complex expression where we will need the three rules, but the key rule is the quotient rule. Let $u = x^3 \sin 4x$ and $v = e^{5x}$. Apply both the chain and product rules to our $u$ to have

$$u' = (x^3)(4\cos 4x) + (3x^2)(\sin 4x)$$
$$= 4x^3 \cos 4x + 3x^2 \sin 4x$$

Also, apply the the chain rule to $v$ as:

$$v' = 5e^{5x}$$

Using the quotient rule, we have that:

$$y' = \frac{vu' - uv'}{v^2}$$

$$= \frac{\left(e^{5x}\right)\left(4x^3 \cos 4x + 3x^2 \sin 4x\right) - \left(x^3 \sin 4x\right)\left(5e^{5x}\right)}{\left(e^{5x}\right)^2}$$

$$= \frac{\left(4x^3 \cos 4x + 3x^2 \sin 4x\right) - \left(x^3 \sin 4x\right)(5)}{e^{5x}}$$

$$= \frac{4x^3 \cos 4x + 3x^2 \sin 4x - 5x^3 \sin 4x}{e^{5x}}$$

$$= \frac{x^2}{e^{5x}} \cdot [4x \cos 4x + 3 \sin 4x - 5x \sin 4x]$$

$$\therefore \frac{dy}{dx} = \frac{x^2}{e^{5x}} \cdot [4x \cos 4x + 3 \sin 4x - 5x \sin 4x]$$

**b)** $y = \frac{(1-x)^2}{\ln 5x \cdot \cos e^x}$

**Solution**

$$y = \frac{(1-x)^2}{\ln 5x \cdot \cos e^x}$$

This is a complex expression where we will need the three rules, but again the key rule is the quotient rule. Let $u = (1-x)^2$ and $v = \ln 5x \cdot \cos e^x$. Apply the chain rule to our $u$ to have

$$u' = 2(1-x) \times -1$$
$$= -2(1-x)$$

Also, apply both the chain and product rules to $v$ as:

$$v' = (\ln 5x)(e^x \cdot -\sin e^x) + \left(\frac{1}{x}\right)(\cos e^x)$$

$$= \frac{1}{x} \cdot \cos e^x - \ln 5x \cdot e^x \cdot \sin e^x$$

Using the quotient rule, we have that

$$y' = \frac{vu' - uv'}{v^2}$$

$$= \frac{(\ln 5x \cdot \cos e^x)(-2(1-x)) - (1-x)^2 \left(\frac{1}{x} \cdot \cos e^x - \ln 5x \cdot e^x \cdot \sin e^x\right)}{(\ln 5x \cdot \cos e^x)^2}$$

$$= \frac{-2(1-x)(\ln 5x \cdot \cos e^x) - \frac{1}{x}(1-x)^2 \cdot \cos e^x + (1-x)^2 \cdot \ln 5x \cdot e^x \cdot \sin e^x}{(\ln 5x \cdot \cos e^x)^2}$$

$$= \frac{(1-x)^2 \cdot \ln 5x \cdot e^x \cdot \sin e^x}{(\ln 5x \cdot \cos e^x)^2} - \frac{\frac{1}{x}(1-x)^2 \cdot \cos e^x}{(\ln 5x \cdot \cos e^x)^2} - \frac{2(1-x)(\ln 5x \cdot \cos e^x)}{(\ln 5x \cdot \cos e^x)^2}$$

$$= \frac{(1-x)^2 \cdot e^x \cdot \sin e^x}{\ln 5x \cdot \cos^2 e^x} - \frac{(1-x)^2}{x(\ln 5x)^2 \cdot \cos e^x} - \frac{2(1-x)}{\ln 5x \cdot \cos e^x}$$

# Advanced Differentiation

This expression is already in a simplified form, but we will tweak this a bit further to ensure that the answer looks like the one we will have later when we use a different method called logarithmic differentiation.

$$= \frac{(1-x)^2}{\ln 5x \cdot \cos e^x} \cdot \left[ \frac{e^x \cdot \sin e^x}{\cos e^x} - \frac{1}{x \ln 5x} - \frac{2}{(1-x)} \right]$$

$$\therefore \frac{dy}{dx} = \frac{(1-x)^2}{\ln 5x \cdot \cos e^x} \cdot \left[ \frac{e^x \cdot \sin e^x}{\cos e^x} - \frac{1}{x \ln 5x} - \frac{2}{(1-x)} \right]$$

## 13.3 IMPLICIT DIFFERENTIATION

In this section, there are two things to understand, namely implicit function and implicit differentiation, with the former necessitating the latter. So far, we've been dealing with functions where the LHS is $y$ (or a dependent variable) and the RHS is $x$ (or an independent variable), and as such, we write $y = f(x)$. This type of function, where $y$ can be expressed explicitly (i.e., exclusively or completely) in terms of $x$, is called an **explicit function**. Examples of this include $y = 3x - 5$, $y = \sin(2x + \alpha)$, $y = 3xe^{\tan x}$ and $y = \frac{1}{\sqrt{x^3 - 2x^2 + 5}}$.

However, there are cases where collecting the independent variable on one side and the dependent variable on the other is either difficult or impracticable, such functions are called **implicit functions**. $x^2 + y^2 - 4 = 0$ and $2xy - x^2 + y^2 = 6$ are examples of implicit functions. Instead of $y = f(x)$, we can express implicit functions as $f(x, y) = 0$ or $f(x, y) = g(x, y)$. The process of finding the derivatives of an implicit function is known as **implicit differentiation**.

To differentiate an implicit function is not different from that used for an explicit function, but it does require a few additional notes as:

**Note 1** First check if the expression can be simplified, such that $y$ is expressed explicitly in terms of $x$ (i.e., $y$ on one side and $x$ term(s) on another side). This is to establish that it is an implicit function and requires implicit differentiation.

**Note 2** Differentiate each term of the function on its own using the relevant differential standards.

**Note 3** If a term is purely in $x$ (or an independent variable), it should be differentiated as one would do in an explicit function.

**Note 4** If a term is purely in $y$ (or dependent variable), it should be differentiated using the chain rule since $y$ is a function of $x$. The general rule is $\frac{d}{dx}[y^n] = ny^{n-1} \cdot \frac{dy}{dx}$. For example, if a term is $3y^2$, its differential coefficient is $6y\frac{dy}{dx}$ and for $-3y$ it is $-3\frac{dy}{dx}$.

**Note 5** If a term is made up of both $x$ and $y$, it should be differentiated using the product rule, considering that $y$ is a function of $x$. In this case, we may view the independent (or $x$) as $u$, the dependent (or $y$) as $v$, and apply the product rule $y' = uv' + vu'$ to it. Still $y$ needs to be treated with the chain rule as per Note 3 above. The general rule is $\frac{d}{dx}[x^n y^n] = nx^{n-1}y^n + nx^n y^{n-1} \cdot \frac{dy}{dx}$. For example, if a term is $x^2 y^3$, its differential coefficient is $\left(2xy^3 + 3x^2 y^2 \frac{dy}{dx}\right)$ and for $-3xy^{-1}$ it is $\left(-3y^{-1} + 3xy^{-2}\frac{dy}{dx}\right)$.

These are the extra tips needed, and every other aspect is the same. Now let's try some examples.

### Example 12

Find the differential coefficient of the following functions *wrt x*.

a) $x^2 - y^2 - 4 = 0$     b) $3y - x^2 + 4 = 2xy$

What did you get? Find the solution below to double-check your answer.

### Solution to Example 12

a) $x^2 - y^2 - 4 = 0$
**Solution**
**Method 1 Simplification and explicit differentiation**

$$x^2 - y^2 - 4 = 0$$

Let's simplify this to make *y* the subject

$$x^2 - y^2 - 4 = 0$$
$$x^2 - 4 = y^2$$
$$y = \sqrt{x^2 - 4}$$
$$y = (x^2 - 4)^{\frac{1}{2}}$$

This is a function of function case, so let's use the chain rule as:

$u = x^2 - 4$ $\quad\Big|\quad$ $y = u^{\frac{1}{2}}$ $\quad\Big|\quad$ $\frac{du}{dx} = 2x$ $\quad\Big|\quad$ $\frac{dy}{du} = \frac{1}{2}u^{-\frac{1}{2}} = \frac{1}{2}(x^2 - 4)^{-\frac{1}{2}}$

Thus

$$\frac{dy}{dx} = \frac{dy}{du} \times \frac{du}{dx}$$
$$= \frac{1}{2}(x^2 - 4)^{-\frac{1}{2}} \times 2x$$
$$= \frac{2x}{2(x^2 - 4)^{\frac{1}{2}}}$$
$$= \frac{x}{(x^2 - 4)^{\frac{1}{2}}}$$
$$\therefore \frac{dy}{dx} = \frac{x}{\sqrt{x^2 - 4}}$$

# Advanced Differentiation

**Method 2  Implicit differentiation**

Differentiate each term one after the other and do check the tips above.

$$x^2 - y^2 - 4 = 0$$

Differentiating

$$2x - 2y\frac{dy}{dx} - 0 = 0$$

Now simplify as:

$$2y\frac{dy}{dx} = 2x$$

$$y\frac{dy}{dx} = x$$

$$\frac{dy}{dx} = \frac{x}{y}$$

Using the expression derived in method 1, we can further write the derivative as:

$$\frac{dy}{dx} = \frac{x}{\sqrt{x^2 - 4}}$$

---

**b)** $3y - x^2 + 4 = 2xy$

**Solution**

**Method 1  Simplification and explicit differentiation**

$$3y - x^2 + 4 = 2xy$$

Let's simplify this to make $y$ the subject

$$3y - 2xy = x^2 - 4$$

$$y(3 - 2x) = x^2 - 4$$

$$y = \frac{x^2 - 4}{3 - 2x}$$

Using the quotient rule, we have that

$$y' = \frac{vu' - uv'}{v^2}$$

$$= \frac{(3 - 2x)(2x) - (x^2 - 4)(-2)}{(3 - 2x)^2} = \frac{6x - 4x^2 + 2x^2 - 8}{(3 - 2x)^2}$$

$$= \frac{6x - 2x^2 - 8}{(3 - 2x)^2} = \frac{-2x^2 + 6x - 8}{(3 - 2x)^2} = \frac{-2(x^2 - 3x + 4)}{(3 - 2x)^2}$$

$$\therefore \frac{dy}{dx} = \frac{-2(x^2 - 3x + 4)}{(3 - 2x)^2}$$

## Method 2  Implicit differentiation

Differentiate each term one after the other and do check the tips above. Let's see

$$3y - x^2 + 4 = 2xy$$

Differentiating

$$3\frac{dy}{dx} - 2x + 0 = 2y + 2x\frac{dy}{dx}$$

Now simplify as:

$$3\frac{dy}{dx} - 2x\frac{dy}{dx} = 2y + 2x$$

$$(3 - 2x)\frac{dy}{dx} = 2(x + y)$$

$$\frac{dy}{dx} = \frac{2(x + y)}{(3 - 2x)}$$

Using the expression derived in method 1, we can further write the derivative as:

$$\frac{dy}{dx} = \frac{2x + 2\left(\frac{x^2-4}{3-2x}\right)}{(3-2x)} = \frac{\left[\frac{2x(3-2x)+2(x^2-4)}{3-2x}\right]}{(3-2x)}$$

$$= \frac{2x(3-2x) + 2(x^2-4)}{(3-2x)(3-2x)}$$

$$= \frac{6x - 4x^2 + 2x^2 - 8}{(3-2x)^2}$$

$$= \frac{6x - 2x^2 - 8}{(3-2x)^2} = \frac{-2x^2 + 6x - 8}{(3-2x)^2}$$

$$= \frac{-2(x^2 - 3x + 4)}{(3-2x)^2}$$

$$\therefore \frac{dy}{dx} = \frac{-2(x^2 - 3x + 4)}{(3-2x)^2}$$

From the above examples, it appears that in some cases it may be possible to re-arrange a seemingly involved or implicit function to an explicit one and differentiate. Yet differentiating implicitly looks easier and more compact. We will now consider where a function may prove impossible to write explicitly, and as such, we will need to apply implicit differentiation only. Let's try an example.

### Example 13

A function is defined implicitly as $4y^2 + x^2y - 3x = 0$, determine:

a) $\frac{dy}{dx}$ at $(1, -1)$
b) $\frac{d^2y}{dx^2}$ at $(-4, -1)$

# Advanced Differentiation

What did you get? Find the solution below to double-check your answer.

## Solution to Example 13

**a)** $\frac{dy}{dx}$ at $x = 1$ and $y = -1$

**Solution**

Differentiate each term one after the other.

$$4y^2 + x^2y - 3x = 0$$

Differentiating

$$8y\frac{dy}{dx} + 2xy + x^2\frac{dy}{dx} - 3 = 0$$

Now simplify as:

$$8y\frac{dy}{dx} + x^2\frac{dy}{dx} = 3 - 2xy$$

$$\frac{dy}{dx}(8y + x^2) = 3 - 2xy$$

$$\frac{dy}{dx} = \frac{3 - 2xy}{8y + x^2}$$

$$\therefore \frac{dy}{dx} = \frac{3 - 2xy}{x^2 + 8y}$$

At $(1, -1)$

$$\left.\frac{dy}{dx}\right|_{(1,-1)} = \frac{3 - 2(1)(-1)}{(1)^2 + 8(-1)} = \frac{3 + 2}{1 - 8} = -\frac{5}{7}$$

$$\therefore \left.\frac{dy}{dx}\right|_{(1,-1)} = -\frac{5}{7}$$

**b)** $\frac{d^2y}{dx^2}$ at $x = -4$ and $y = -1$

**Solution**

Differentiate the first derivative and apply the quotient rule

$$\frac{d^2y}{dx^2} = \frac{d}{dx}\left(\frac{3 - 2xy}{x^2 + 8y}\right)$$

$$= \frac{(x^2 + 8y)\left(-2y - 2x\frac{dy}{dx}\right) - (3 - 2xy)\left(2x + 8\frac{dy}{dx}\right)}{(x^2 + 8y)^2}$$

$$= \frac{-2(x^2 + 8y)\left(y + x\frac{dy}{dx}\right) - (3 - 2xy)\left(2x + 8\frac{dy}{dx}\right)}{(x^2 + 8y)^2}$$

We do not need to simplify the above further; we just need to substitute the coordinates. However, we need to start by evaluating the first derivative at $(-4, -1)$ as follows:

$$\left.\frac{dy}{dx}\right|_{(-4,-1)} = \frac{3 - 2(-4)(-1)}{(-4)^2 + 8(-1)}$$

$$= \frac{3 - 8}{16 - 8}$$

$$= -\frac{5}{8}$$

Thus, we can now determine the second derivation at $(-4, -1)$ by plugging in the values as

$$\left.\frac{d^2y}{dx^2}\right|_{(-4,-1)} = \frac{-2(16-8)\left(-1 - 4\left(-\frac{5}{8}\right)\right) - (3 - 2(-4)(-1))\left(2(-4) + 8\left(-\frac{5}{8}\right)\right)}{(16-8)^2}$$

$$= \frac{-2(8)\left(-1 + \frac{5}{2}\right) - (3-8)(-8-5)}{(8)^2}$$

$$= \frac{-16\left(\frac{3}{2}\right) + 5(-13)}{64}$$

$$= \frac{-24 - 65}{64}$$

$$= -\frac{89}{64}$$

$$\therefore \left.\frac{d^2y}{dx^2}\right|_{(-4,-1)} = -\frac{89}{64}$$

## 13.4 LOGARITHMIC DIFFERENTIATION

Earlier in this chapter, we looked at the product and quotient rules and how they can be used to determine the derivative of a function formed by multiplying or dividing two functions, and extending the product rule to three or more functions. It can be noted that real problems may be more involved or complicated, requiring the application of the two rules and also the third one (i.e., chain rule). Indeed, we can deal with such cases using these rules; however, sometimes this can be time consuming. There is another approach that is best suited for these cases, and this is known as **logarithmic differentiation**.

Fundamentally, logarithmic differentiation should be viewed as a process of dealing with complex derivatives, rather than a type of differentiation. We will illustrate this for two cases, which can be extended to similar functions.

**Case 1** $y = uvw$

This is a case where $u$, $v$, $w$, and $y$ are functions of $x$ (or another variable) such that $y$ is a product of $u$, $v$, and $w$. To differentiate $y = uvw$, follow these steps:

**Step 1:** Take the natural logarithmic of both sides. We chose natural logarithm instead of common logarithm (or any other) because it gives the simplest derivative. You will observe this later when we get to the worked examples.

$$\ln y = \ln(uvw)$$

# Advanced Differentiation

**Step 2:** Simplify the RHS using the multiplication law of logarithm.

$$\ln y = \ln(u) + \ln(v) + \ln(w)$$

**Step 3:** Take the derivative of both sides. Remember to apply implicit differentiation to the LHS ($\ln y$). Also remember to apply the chain rule to $u$, $v$, and $w$. We have:

$$\left(\frac{1}{y}\right)\frac{dy}{dx} = \left(\frac{1}{u}\right)\frac{du}{dx} + \left(\frac{1}{v}\right)\frac{dv}{dx} + \left(\frac{1}{w}\right)\frac{dw}{dx}$$

**Step 4:** Multiply both sides by $y$ to determine the differential coefficient of $y$ wrt $x$.

$$\boxed{\frac{dy}{dx} = \frac{d}{dx}(uvw) = y \cdot \left[\left(\frac{1}{u}\right)\frac{du}{dx} + \left(\frac{1}{v}\right)\frac{dv}{dx} + \left(\frac{1}{w}\right)\frac{dw}{dx}\right]} \quad (13.11)$$

or

$$\boxed{\frac{dy}{dx} = \frac{d}{dx}(uvw) = y \cdot \left[\left(\frac{1}{u}\right)u' + \left(\frac{1}{v}\right)v' + \left(\frac{1}{w}\right)w'\right]} \quad (13.12)$$

The final formula above is no different from what we obtained and used earlier. Let's show this in one go.

$$\frac{dy}{dx} = y\left[\left(\frac{1}{u}\right)\frac{du}{dx} + \left(\frac{1}{v}\right)\frac{dv}{dx} + \left(\frac{1}{w}\right)\frac{dw}{dx}\right]$$

$$= uvw\left[\left(\frac{1}{u}\right)\frac{du}{dx} + \left(\frac{1}{v}\right)\frac{dv}{dx} + \left(\frac{1}{w}\right)\frac{dw}{dx}\right]$$

$$= \left(\frac{uvw}{u}\right)\frac{du}{dx} + \left(\frac{uvw}{v}\right)\frac{dv}{dx} + \left(\frac{uvw}{w}\right)\frac{dw}{dx}$$

$$= (vw)\frac{du}{dx} + (uw)\frac{dv}{dx} + (uv)\frac{dw}{dx}$$

$$\boxed{\therefore \frac{d}{dx}(uvw) = (vw)u' + (uw)v' + (uv)w'} \quad (13.13)$$

Looking at the pattern of the logarithmic differentiation formula, we can extend it to a product of four functions of $y$ as:

$$\boxed{\frac{dy}{dx} = \frac{d}{dx}(uvwz) = y \cdot \left[\left(\frac{1}{u}\right)u' + \left(\frac{1}{v}\right)v' + \left(\frac{1}{w}\right)w' + \left(\frac{1}{z}\right)z'\right]} \quad (13.14)$$

**Case 2** $y = \frac{uv}{w}$

$u$, $v$, $w$, and $y$ are functions of $x$ (or another variable) such that $y$ is a product of $u$ and $v$ divided by $w$. To differentiate $y = \frac{uv}{w}$, follow these steps:

**Step 1:** Take the natural logarithmic of both sides. We choose natural log for the same reason given earlier.

$$\ln y = \ln\left(\frac{uv}{w}\right)$$

**Step 2:** Simplify the RHS using multiplication and division laws of logarithm

$$\ln y = \ln(u) + \ln(v) - \ln(w)$$

**Step 3:** Take the derivative of both sides. Remember to apply implicit differentiation to the LHS (ln y). Also remember to apply the chain rule to $u$, $v$, and $w$. We have

$$\left(\frac{1}{y}\right)\frac{dy}{dx} = \left(\frac{1}{u}\right)\frac{du}{dx} + \left(\frac{1}{v}\right)\frac{dv}{dx} - \left(\frac{1}{w}\right)\frac{dw}{dx}$$

**Step 4:** Multiply both sides by $y$ to determine the differential coefficient of $y$ wrt $x$.

$$\boxed{\frac{dy}{dx} = \frac{d}{dx}\left(\frac{uv}{w}\right) = y \cdot \left[\left(\frac{1}{u}\right)\frac{du}{dx} + \left(\frac{1}{v}\right)\frac{dv}{dx} - \left(\frac{1}{w}\right)\frac{dw}{dx}\right]} \quad (13.15)$$

or

$$\boxed{\frac{dy}{dx} = \frac{d}{dx}\left(\frac{uv}{w}\right) = y \cdot \left[\left(\frac{1}{u}\right)u' + \left(\frac{1}{v}\right)v' - \left(\frac{1}{w}\right)w'\right]} \quad (13.16)$$

We can easily show that $y = \frac{u}{vw}$.

$$\boxed{\frac{dy}{dx} = \frac{d}{dx}\left(\frac{uv}{w}\right) = y \cdot \left[\left(\frac{1}{u}\right)u' - \left(\frac{1}{v}\right)v' - \left(\frac{1}{w}\right)w'\right]} \quad (13.17)$$

**Case 3** $y = \frac{u}{vw}$

Following from case 2 above, we can easily show that $y = \frac{u}{vw}$.

$$\boxed{\frac{dy}{dx} = \frac{d}{dx}\left(\frac{uv}{w}\right) = y \cdot \left[\left(\frac{1}{u}\right)u' - \left(\frac{1}{v}\right)v' - \left(\frac{1}{w}\right)w'\right]} \quad (13.18)$$

It should be emphasised that this is a method of solving a complex problem rather than a rigid formula to be memorised or used.

Let's try some examples.

### Example 14

Find the differential coefficient of the following functions wrt $x$.

a) $y = e^x \cdot \sin x \cdot \ln x$     b) $y = (x^2 - 4) \log_3 x \cdot \tan^2 x$

What did you get? Find the solution below to double-check your answer.

### Solution to Example 14

a) $y = e^x \cdot \sin x \cdot \ln x$
**Solution**

$$y = e^x \cdot \sin x \cdot \ln x$$

We've solved this question previously, but let's try this same question using logarithmic differentiation.

# Advanced Differentiation

**Step 1:** Take the natural log of both sides and simplify.

$$\ln y = \ln(e^x \cdot \sin x \cdot \ln x)$$
$$= \ln(e^x) + \ln(\sin x) + \ln(\ln x)$$
$$= x \ln e + \ln(\sin x) + \ln(\ln x)$$
$$= x + \ln(\sin x) + \ln(\ln x)$$

**Step 2:** Apply implicit differentiation and simplify.

$$\ln y = x + \ln(\sin x) + \ln(\ln x)$$

$$\frac{1}{y} \cdot \frac{dy}{dx} = 1 + \frac{\cos x}{\sin x} + \frac{\frac{1}{x}}{\ln x} = 1 + \frac{\cos x}{\sin x} + \frac{1}{x \ln x}$$

$$\frac{dy}{dx} = y\left[1 + \frac{\cos x}{\sin x} + \frac{1}{x \ln x}\right] = e^x \cdot \sin x \cdot \ln x \left[1 + \frac{\cos x}{\sin x} + \frac{1}{x \ln x}\right]$$

$$= e^x \cdot \sin x \cdot \ln x + \frac{e^x \cdot \sin x \cdot \ln x \cdot \cos x}{\sin x} + \frac{e^x \cdot \sin x \cdot \ln x}{x \ln x}$$

$$= e^x \cdot \sin x \cdot \ln x + e^x \cdot \ln x \cdot \cos x + \frac{e^x \cdot \sin x}{x}$$

$$= e^x \left[\ln x \cdot \sin x + \ln x \cdot \cos x + \frac{\sin x}{x}\right]$$

$$\therefore \frac{dy}{dx} = e^x \left(\frac{\sin x}{x} + \ln x \cdot \cos x + \ln x \cdot \sin x\right)$$

**NOTE**
This is exactly what we obtained previously. Note that we carried out some simplification to ensure that the final answer here is a perfect match of the previous one, though this is not really a requirement.

---

**b)** $y = (x^2 - 4) \log_3 x \cdot \tan^2 x$

**Solution**

$$y = (x^2 - 4) \log_3 x \cdot \tan^2 x$$

We've not solved this question previously, but you can try to apply the previous approach and compare the answers.

**Step 1:** Take the natural log of both sides and simplify.

$$\ln y = \ln\left[(x^2 - 4) \log_3 x \cdot \tan^2 x\right]$$
$$= \ln(x^2 - 4) + \ln(\log_3 x) + \ln(\tan^2 x)$$
$$= \ln(x^2 - 4) + \ln(\log_3 x) + 2 \ln(\tan x)$$

**Step 2:** Apply implicit differentiation and simplify.

$$\ln y = \ln(x^2 - 4) + \ln(\log_3 x) + 2\ln(\tan x)$$

$$\frac{1}{y} \cdot \frac{dy}{dx} = \frac{2x}{x^2 - 4} + \frac{\frac{1}{x \ln 3}}{\log_3 x} + \frac{2\sec^2 x}{\tan x}$$

$$\frac{1}{y} \cdot \frac{dy}{dx} = \frac{2x}{x^2 - 4} + \frac{1}{x \cdot \ln 3 \cdot \log_3 x} + \frac{2\sec^2 x}{\tan x}$$

$$\frac{dy}{dx} = (x^2 - 4)\log_3 x \cdot \tan^2 x \left[ \frac{2x}{x^2 - 4} + \frac{1}{x \cdot \ln 3 \log_3 x} + \frac{2\sec^2 x}{\tan x} \right]$$

$$= \frac{2x \cdot (x^2 - 4)\log_3 x \cdot \tan^2 x}{x^2 - 4} + \frac{(x^2 - 4)\log_3 x \cdot \tan^2 x}{x \cdot \ln 3 \log_3 x} + \frac{2\sec^2 x \cdot (x^2 - 4)\log_3 x \cdot \tan^2 x}{\tan x}$$

$$= 2x \cdot \log_3 x \cdot \tan^2 x + \frac{(x^2 - 4) \cdot \tan^2 x}{x \ln 3} + 2\sec^2 x \cdot (x^2 - 4)\log_3 x \cdot \tan x$$

$$\therefore \frac{dy}{dx} = 2x \cdot \log_3 x \cdot \tan^2 x + \frac{(x^2 - 4) \cdot \tan^2 x}{x \ln 3} + 2(x^2 - 4) \cdot \log_3 x \cdot \tan x \cdot \sec^2 x$$

## Example 15

Find the differential coefficient of the following functions *wrt x*.

a) $y = \frac{x^3 \sin 4x}{e^{5x}}$

b) $y = \frac{(1-x)^2}{\ln 5x \cdot \cos e^x}$

What did you get? Find the solution below to double-check your answer.

## Solution to Example 15

a) $y = \frac{x^3 \sin 4x}{e^{5x}}$

**Solution**

$$y = \frac{x^3 \sin 4x}{e^{5x}}$$

We've solved this question previously, but let's try the same using logarithmic differentiation.

# Advanced Differentiation

**Step 1:** Take the natural log of both sides and simplify.

$$\ln y = \ln\left(\frac{x^3 \sin 4x}{e^{5x}}\right)$$
$$= \ln(x^3) + \ln(\sin 4x) - \ln(e^{5x})$$
$$= 3\ln(x) + \ln(\sin 4x) - 5x\ln(e)$$
$$= 3\ln(x) + \ln(\sin 4x) - 5x$$

**Step 2:** Apply implicit differentiation and simplify.

$$\ln y = 3\ln(x) + \ln(\sin 4x) - 5x$$
$$\frac{1}{y}\cdot\frac{dy}{dx} = \frac{3}{x} + \frac{4\cos 4x}{\sin 4x} - 5 = \frac{3}{x} + 4\cot 4x - 5$$
$$\frac{dy}{dx} = y\left[\frac{3}{x} + 4\cot 4x - 5\right]$$
$$= \frac{x^3 \sin 4x}{e^{5x}}\left[\frac{3}{x} + 4\cot 4x - 5\right]$$
$$\therefore \frac{dy}{dx} = \frac{x^3 \sin 4x}{e^{5x}}\left[\frac{3}{x} + 4\cot 4x - 5\right]$$

As obtained earlier.

---

**b)** $y = \frac{(1-x)^2}{\ln 5x \cdot \cos e^x}$

**Solution**

$$y = \frac{(1-x)^2}{\ln 5x \cdot \cos e^x}$$

We've solved this question previously, but let's attempt it again using logarithmic differentiation.

**Step 1:** Take the natural log of both sides and simplify.

$$\ln y = \ln\left(\frac{(1-x)^2}{\ln 5x \cdot \cos e^x}\right)$$
$$= \ln(1-x)^2 - \ln(\ln 5x) - \ln(\cos e^x)$$
$$= 2\ln(1-x) - \ln(\ln 5x) - \ln(\cos e^x)$$

**Step 2:** Apply implicit differentiation and simplify.

$$\ln y = 2\ln(1-x) - \ln(\ln 5x) - \ln(\cos e^x)$$
$$\frac{1}{y}\cdot\frac{dy}{dx} = \left(\frac{2}{1-x}\times -1\right) - \frac{\left(\frac{1}{x}\right)}{\ln 5x} - \frac{(-\sin e^x \times e^x)}{\cos e^x}$$
$$= -\frac{2}{1-x} - \frac{1}{x \cdot \ln 5x} + \frac{e^x \cdot \sin e^x}{\cos e^x}$$

$$= \frac{e^x . \sin e^x}{\cos e^x} - \frac{2}{1-x} - \frac{1}{x. \ln 5x}$$

$$\frac{dy}{dx} = y \left[ \frac{e^x . \sin e^x}{\cos e^x} - \frac{2}{1-x} - \frac{1}{x. \ln 5x} \right]$$

$$= \frac{(1-x)^2}{\ln 5x . \cos e^x} \left[ \frac{e^x . \sin e^x}{\cos e^x} - \frac{2}{1-x} - \frac{1}{x. \ln 5x} \right]$$

$$= \frac{(1-x)^2}{\ln 5x . \cos e^x} \left[ e^x . \tan e^x - \frac{2}{1-x} - \frac{1}{x. \ln 5x} \right]$$

$$\therefore \frac{dy}{dx} = \frac{(1-x)^2}{\ln 5x . \cos e^x} \left[ e^x . \tan e^x - \frac{2}{1-x} - \frac{1}{x. \ln 5x} \right]$$

as obtained earlier.

## 13.5 PARAMETRIC DIFFERENTIATION

We know that the graph of a function in $x$–$y$ plane is a locus of points connecting its coordinates $(x, y)$. Hitherto, we've observed that a single expression will have both $x$ and $y$ together, explicitly or implicitly. However, it is possible to define $x$ as a function of a new independent variable $t$, i.e., $x = f(t)$, and $y$ also as a function of the same variable $t$, $y = f(t)$. The new variable $t$ is called the **parameter** and the two equations $x = f(t)$ and $y = f(t)$ are called **parametric equations**, as they are based on the parameter $t$. That is to say that we've parametrised the variables $x$ and $y$. You will also find $\theta$ being used as a parameter.

Since $t$ is an independent variable, we can use a set of $t$ values to find $x$ values and corresponding $y$ values. The resulting $x$–$y$ coordinates can be used to draw the graph of $y = f(x)$. It is apparent from this that there is a relationship between $x$ and $y$, and we may therefore be interested in knowing the rate of change of $y$ wrt $x$. To do this, we need some tips as follows:

**Note 1** We need to understand that

$$\boxed{\frac{dy}{dx} = \frac{1}{\left(\frac{dx}{dy}\right)}} \quad \text{OR} \quad \boxed{\frac{dy}{dx} \times \frac{dx}{dy} = 1} \tag{13.19}$$

This is known as the derivative of the inverse function. The proof of this is provided in Appendix A.

**Note 2** We will need to apply a modified chain rule. Given a chain rule, $\frac{dy}{dx} = \frac{dy}{dt} \times \frac{dt}{dx}$, a modified version is

$$\boxed{\frac{dy}{dx} = \frac{dy}{dt} \times \frac{1}{\left(\frac{dx}{dt}\right)}} \tag{13.20}$$

**Note 3** We can also convert parametric equations to an equivalent Cartesian equation, by eliminating the parameter $t$. We can therefore differentiate the Cartesian equation, containing only $x$

# Advanced Differentiation

and $y$, using the standard method(s) of differentiation. The differential coefficient obtained here might seemingly be different from the one obtained using the above approach in Note 2, as the former is expressed in $t$ while the latter in $x$.

Let's try a couple of examples on inverse differentiation before we look at parametric differentiation.

## Example 16

For each of the following functions, determine $\frac{dx}{dy}$.

**a)** $y = x.\sqrt[5]{x}$  **b)** $y = x^2 - \sqrt{x}$

What did you get? Find the solution below to double-check your answer.

## Solution to Example 16

**a)** $y = x.\sqrt[5]{x}$
**Solution**

**Method 1  Using inverse derivative**

$$y = x.\sqrt[5]{x} = x.x^{\frac{1}{5}} = x^{\left(1+\frac{1}{5}\right)}$$
$$= x^{1.2}$$

Differentiating *wrt x*

$$\frac{dy}{dx} = 1.2x^{0.2}$$

Using inverse derivative, we have

$$\frac{dx}{dy} = \frac{1}{1.2x^{0.2}} = \frac{5}{6x^{\frac{1}{5}}}$$

$$\therefore \frac{dx}{dy} = \frac{5}{6\sqrt[5]{x}}$$

The above answer is in the form of $x$; we can express the same in the form of $y$ as:

$$\frac{dx}{dy} = \frac{5}{6x^{\frac{1}{5}}} = \frac{5}{6\left(y^{\frac{5}{6}}\right)^{\frac{1}{5}}}$$

$$= \frac{5}{6y^{\frac{1}{6}}}$$

$$\therefore \frac{dx}{dy} = \frac{5}{6\left(\sqrt[6]{y}\right)}$$

**Method 2 Re-arrangement**

$$y = x \cdot \sqrt[5]{x}$$
$$= x \cdot x^{\frac{1}{5}}$$
$$= x^{\frac{6}{5}}$$

Making $x$ the subject, we have:

$$x = y^{\frac{5}{6}}$$

Differentiating *wrt* $y$

$$\frac{dx}{dy} = \frac{5}{6} y^{-\frac{1}{6}}$$
$$= \frac{5}{6 y^{\frac{1}{6}}}$$
$$\therefore \frac{dx}{dy} = \frac{5}{6\left(\sqrt[6]{y}\right)}$$

The above answer is in the form of $y$; we can express the same in the form of $x$ as:

$$\frac{dx}{dy} = \frac{5}{6} y^{-\frac{1}{6}} = \frac{5}{6}\left(x^{\frac{6}{5}}\right)^{-\frac{1}{6}}$$
$$= \frac{5}{6} x^{-\frac{1}{5}} = \frac{5}{6 x^{\frac{1}{5}}}$$
$$\therefore \frac{dx}{dy} = \frac{5}{6 \sqrt[5]{x}}$$

**b)** $y = x^2 - \sqrt{x}$
**Solution**

$$y = x^2 - \sqrt{x}$$
$$= x^2 - x^{0.5}$$

Differentiating *wrt* $x$

$$\frac{dy}{dx} = 2x - 0.5 x^{-0.5}$$
$$= 2x - \frac{1}{2\sqrt{x}}$$
$$= \frac{4x\sqrt{x} - 1}{2\sqrt{x}}$$

# Advanced Differentiation

Using inverse derivative, we have:

$$\frac{dx}{dy} = \frac{1}{\left(\frac{4x\sqrt{x}-1}{2\sqrt{x}}\right)}$$

$$= \frac{2\sqrt{x}}{4x\sqrt{x}-1}$$

$$\therefore \frac{dx}{dy} = \frac{2\sqrt{x}}{4x\sqrt{x}-1}$$

The above answer is in the form of $x$; it is practically impossible or undesirable to express this in terms of $y$.

From the above, it is obvious that sometimes it is possible to re-arrange a function to switch the dependent and independent variables instead of using the inverse derivative formula. This is however not the case in all situations as apparent in the second example above (Example 16(b)).

Let's try some examples on inverse differentiation before we look at parametric differentiation.

## Example 17

Find the Cartesian equation of the curves given by the following parametric function equations.

a) $x = t - 5$, $y = t^2 + 1$

b) $x = \frac{3}{1-2t}$, $y = \frac{t}{2t-7} \left[ t \neq \frac{1}{2},\ t \neq \frac{7}{2} \right]$

c) $x = \sin\theta$, $y = \cos\theta$

d) $x = \sin\theta - 1$, $y = 2\cos\theta$

What did you get? Find the solution below to double-check your answer.

## Solution to Example 17

**a)** $x = t - 5$, $y = t^2 + 1$
**Solution**
Given

$$x = t - 5 \quad \text{-----(i)}$$

$$y = t^2 + 1 \quad \text{-----(ii)}$$

From (i), we can make the parameter $t$ the subject as

$$t = x + 5 \quad \text{-----(iii)}$$

Now substitute (iii) in (ii) and simplify, thus

$$y = (x+5)^2 + 1$$
$$= x^2 + 10x + 26$$

The Cartesian equation is therefore

$$y = x^2 + 10x + 26$$

---

**b)** $x = \dfrac{3}{1-2t}$, $y = \dfrac{t}{2t-7} \left[ t \neq \dfrac{1}{2},\ t \neq \dfrac{7}{2} \right]$

**Solution**
Given

$$x = \dfrac{3}{1-2t} \qquad\qquad \text{-----(i)}$$

$$y = \dfrac{t}{2t-7} \qquad\qquad \text{-----(ii)}$$

From (i), we can make the parameter $t$ the subject as

$$1 - 2t = \dfrac{3}{x}$$

$$-2t = \dfrac{3}{x} - 1$$

$$2t = 1 - \dfrac{3}{x} = \dfrac{x-3}{x}$$

$$t = \dfrac{x-3}{2x} \qquad\qquad \text{-----(iii)}$$

Now substitute (iii) in (ii) and simplify, thus

$$y = \dfrac{\left(\dfrac{x-3}{2x}\right)}{\left[2\left(\dfrac{x-3}{2x}\right) - 7\right]} = \dfrac{\left(\dfrac{x-3}{2x}\right)}{\left(\dfrac{x-3}{x} - 7\right)}$$

$$= \dfrac{\left(\dfrac{x-3}{2x}\right)}{\left(\dfrac{x-3-7x}{x}\right)} = \dfrac{\left(\dfrac{x-3}{2x}\right)}{\left(\dfrac{-6x-3}{x}\right)} = \dfrac{x-3}{2x} \times \dfrac{x}{-6x-3}$$

$$= \dfrac{x-3}{-2(6x+3)} = \dfrac{x-3}{-6(2x+1)}$$

$$= \dfrac{3-x}{6(2x+1)}$$

The Cartesian equation is therefore

$$y = \dfrac{3-x}{6(2x+1)}$$

---

**c)** $x = \sin\theta$, $y = \cos\theta$
**Solution**
Given

$$x = \sin\theta \qquad\qquad \text{-----(i)}$$

$$y = \cos\theta \qquad\qquad \text{-----(ii)}$$

# Advanced Differentiation

From (i), we have

$$(x)^2 = (\sin \theta)^2$$
$$x^2 = \sin^2 \theta \quad \text{----- (iii)}$$

From (ii), we have

$$(y)^2 = (\cos \theta)^2$$
$$y^2 = \cos^2 \theta \quad \text{----- (iv)}$$

Add equations (iii) and (iv), i.e., the LHS together and the RHS together, we have

$$x^2 + y^2 = \sin^2 \theta + \cos^2 \theta$$

Because $\sin^2 \theta + \cos^2 \theta = 1$, we have

$$x^2 + y^2 = 1$$

The Cartesian equation is therefore

$$x^2 + y^2 = 1$$

---

**d)** $x = \sin \theta - 1$, $y = 2 \cos \theta$

**Solution**
Given

$$x = \sin \theta - 1 \quad \text{----- (i)}$$

$$y = 2 \cos \theta \quad \text{----- (ii)}$$

From (i), we have

$$x + 1 = \sin \theta$$
$$(x + 1)^2 = (\sin \theta)^2$$
$$x^2 + 2x + 1 = \sin^2 \theta \quad \text{----- (iii)}$$

From (ii), we have

$$\frac{y}{2} = \cos \theta$$
$$\left(\frac{y}{2}\right)^2 = (\cos \theta)^2$$
$$\frac{y^2}{4} = \cos^2 \theta \quad \text{----- (iv)}$$

Add equations (iii) and (iv), i.e. LHS together and RHS together, we have

$$x^2 + 2x + 1 + \frac{y^2}{4} = \sin^2 \theta + \cos^2 \theta$$

Because $\sin^2\theta + \cos^2\theta = 1$, we have

$$x^2 + 2x + 1 + \frac{y^2}{4} = 1$$
$$4x^2 + 8x + 4 + y^2 = 4$$
$$4x^2 + 8x + y^2 = 0$$

The Cartesian equation is therefore

$$4x^2 + 8x + y^2 = 0$$

---

Now let's try some examples on parametric differentiation to finish this chapter.

## Example 18

Find the derivative of the following parametric equations.

**a)** $x = \sin 2\theta$, $y = \tan\theta$ 　　　　　　　　　　　　　　　　　　　　　**b)** $x = 3t - \ln t^2$, $y = e^{-5t}$

---

What did you get? Find the solution below to double-check your answer.

## Solution to Example 18

**a)** $x = \sin 2\theta$, $y = \tan\theta$
**Solution**
$x = \sin 2\theta$, $y = \tan\theta$
Differentiating

$$\frac{dx}{d\theta} = 2\cos 2\theta \quad \text{and} \quad \frac{dy}{d\theta} = \sec^2\theta$$

Using the chain rule, we have that

$$\frac{dy}{dx} = \frac{dy}{d\theta} \times \frac{d\theta}{dx}$$
$$= \sec^2\theta \times \frac{1}{2\cos 2\theta}$$
$$\therefore \frac{dy}{dx} = \frac{\sec^2\theta}{2\cos 2\theta}$$

---

**b)** $x = 3t - \ln t^2$, $y = e^{-5t}$
**Solution**
$x = 3t - \ln t^2$, $y = e^{-5t}$
Differentiating

$$\frac{dx}{dt} = 3 - \frac{2t}{t^2} = 3 - \frac{2}{t} = \frac{3t-2}{t} \quad \text{and} \quad \frac{dy}{dt} = -5e^{-5t}$$

# Advanced Differentiation

Using the chain rule, we have that

$$\frac{dy}{dx} = -5e^{-5t} \times \frac{t}{3t-2}$$

$$= -\frac{t \cdot 5e^{-5t}}{3t-2} = -\frac{5t}{(3t-2)e^{5t}}$$

$$\therefore \frac{dy}{dx} = -\frac{5t}{(3t-2)e^{5t}}$$

---

A final set of examples.

## Example 19

A function of $x$ is defined by the parametric equation $y = \tan\theta - \theta$, $x = 1 - \cos\theta$. Determine the following in the simplest form.

a) $\frac{dy}{dx}$ 

b) $\frac{d^2y}{dx^2}$

---

What did you get? Find the solution below to double-check your answer.

## Solution to Example 19

**a)** $\frac{dy}{dx}$

**Solution**

$y = \tan\theta - \theta$, $x = 1 - \cos\theta$

Differentiating

$$\frac{dy}{d\theta} = \sec^2\theta - 1 \qquad \text{and} \qquad \frac{dx}{d\theta} = \sin\theta$$

Using the chain rule, we have that

$$\frac{dy}{dx} = \frac{dy}{d\theta} \times \frac{d\theta}{dx}$$

$$= (\sec^2\theta - 1) \times \frac{1}{\sin\theta} = \frac{\sec^2\theta - 1}{\sin\theta}$$

$$= \frac{\tan^2\theta}{\sin\theta} = \frac{\left(\frac{\sin^2\theta}{\cos^2\theta}\right)}{\sin\theta} = \frac{\sin\theta}{\cos^2\theta}$$

$$\therefore \frac{dy}{dx} = \frac{\sin\theta}{\cos^2\theta}$$

**b)** $\frac{d^2y}{dx^2}$

**Solution**

The second derivative is a bit tricky for parametric equations. Let's go.

Differentiating

$$\frac{d^2y}{dx^2} = \frac{d}{dx}\left(\frac{dy}{dx}\right)$$

$$= \frac{d}{dx}\left(\frac{\sin\theta}{\cos^2\theta}\right)$$

But we will not be able to differentiate $\frac{\sin\theta}{\cos^2\theta}$ wrt $x$. We therefore need some manipulation as follows

$$\frac{d}{dx}\left(\frac{\sin\theta}{\cos^2\theta}\right) = \frac{d}{dx}\left(\frac{\sin\theta}{\cos^2\theta}\right) \times \frac{d\theta}{d\theta}$$

Although technically incorrect, we can say that $\frac{d\theta}{d\theta} = 1$. We will now switch the position to have

$$= \frac{d}{d\theta}\left(\frac{\sin\theta}{\cos^2\theta}\right) \times \frac{d\theta}{dx}$$

Great! We have the product of two differential coefficients, namely $\frac{d}{d\theta}\left(\frac{\sin\theta}{\cos^2\theta}\right)$ and $\frac{d\theta}{dx}$. For clarity, let's attempt each separately.

**Part 1**

$$\frac{d}{d\theta}\left(\frac{\sin\theta}{\cos^2\theta}\right) = \frac{(\cos^2\theta)(\cos\theta) - (\sin\theta)(-2\sin\theta\cos\theta)}{(\cos^2\theta)^2}$$

$$= \frac{\cos^3\theta + 2\cos\theta\sin^2\theta}{\cos^4\theta} = \frac{\cos\theta\left(\cos^2\theta + 2\sin^2\theta\right)}{\cos^4\theta}$$

$$= \frac{\cos^2\theta + 2\sin^2\theta}{\cos^3\theta}$$

**Part 2**

$$\frac{d\theta}{dx} = \frac{1}{\left(\frac{dx}{d\theta}\right)} = \frac{1}{\sin\theta}$$

It is time to combine the two parts

$$\frac{d^2y}{dx^2} = \frac{d}{d\theta}\left(\frac{\sin\theta}{\cos^2\theta}\right) \times \frac{d\theta}{dx}$$

$$= \left(\frac{\cos^2\theta + 2\sin^2\theta}{\cos^3\theta}\right)\left(\frac{1}{\sin\theta}\right)$$

$$= \frac{\cos^2\theta + 2\sin^2\theta}{\cos^3\theta \cdot \sin\theta}$$

$$\therefore \frac{d^2y}{dx^2} = \frac{\cos^2\theta + 2\sin^2\theta}{\cos^3\theta \cdot \sin\theta}$$

## 13.6 CHAPTER SUMMARY

1) The chain rule is used to solve multiple or cascaded process, which is also called **'function of a function'**. The rule states that:

$$\text{If } \boxed{y = f(u) \text{ and } u = g(x)} \quad \text{then} \quad \boxed{\frac{dy}{dx} = \frac{dy}{du} \times \frac{du}{dx}}$$

$$\text{If } \boxed{y = f(u), \ u = g(v), \text{ and } v = h(x)} \quad \text{then} \quad \boxed{\frac{dy}{dx} = \frac{dy}{du} \times \frac{du}{dv} \times \frac{dv}{dx}}$$

2) The product rule is used when one function is multiplied by another function *wrt* the same variable. This rule states that:

$$\text{If } \boxed{y = uv, \text{ where } u = g(x) \text{ and } v = h(x)} \quad \text{then} \quad \boxed{\frac{dy}{dx} = u\frac{dv}{dx} + v\frac{du}{dx}}$$

Since $y$, $u$, and $v$ are functions of $x$, we can write the rule in a short form as:

$$\boxed{y' = uv' + vu'}$$

where $y'$, $u'$, and $v'$ are derivatives of the respective functions.

3) The **quotient rule** is used when one function is divided by another function where it is either impossible to simplify it into a single function or practically difficult to do so. This rule states that:

$$\text{If } \boxed{y = \frac{u}{v}, \text{ where } u = g(x) \text{ and } v = h(x)} \quad \text{then} \quad \boxed{\frac{dy}{dx} = \frac{v\frac{du}{dx} - u\frac{dv}{dx}}{v^2}}$$

Since $y$, $u$, and $v$ are functions of $x$, we can write the rule in a short form as:

$$\boxed{y' = \frac{vu' - uv'}{v^2}}$$

4) In cases where there is an apparent division and the dividend (or numerator) is a constant, the quotient rule becomes

$$\boxed{y' = \frac{-uv'}{v^2}}$$

5) The process of finding the derivatives of an implicit function is known as **implicit differentiation**.

6) **Logarithmic differentiation** should be viewed as a process of dealing with complex derivatives rather than a type of differentiation.

7) It is possible to define $x$ as a function of a new independent variable $t$, i.e., $x = f(t)$ and $y$ also as a function of the same variable $t$, $y = f(t)$. The new variable $t$ is called the **parameter** and the two equations $x = f(t)$ and $y = f(t)$ are called **parametric equations**, as they are based on the parameter.

8) Inverse relationship is given by

$$\boxed{\frac{dy}{dx} = \frac{1}{\left(\frac{dx}{dy}\right)}} \text{ OR } \boxed{\frac{dy}{dx} \times \frac{dx}{dy} = 1}$$

\*\*\*\*

## 13.7 FURTHER PRACTICE

To access complementary contents, including additional exercises, please go to www.dszak.com.

# 14 Applications of Differentiation

*Learning Outcomes*

Once you have studied the content of this chapter, you should be able to:

- Determine velocity, acceleration, and jerk using differentiation
- Determine the equation of the tangent to a curve at a given point
- Determine the equation of the normal to a curve at a given point
- Evaluate increasing and decreasing functions
- Evaluate the stationary points and distinguish them
- Determine the maximum points, minimum points, and the point of inflexion
- Apply the stationary points to solve problems and sketch curves

## 14.1 INTRODUCTION

Having discussed the fundamentals of differentiation in the last two chapters, here we will introduce the common applications of differentiation, including mechanics, tangent and normal line equations, and stationary points.

## 14.2 RATE OF CHANGE

Differentiation is the rate of change of one physical quantity (an independent variable) with respect to another quantity (a dependent variable). The rate here is not restricted to time. In other words, the variables $x$ and $y$ in $\frac{dy}{dx}$ for $y = f(x)$ can represent any quantity. In this section, we will fix our $x$ variable as time and our $y$ variable will change according to the application. Generally, when this is the case, 'with respect to' is omitted.

### 14.2.1 Velocity

Velocity is the rate of change of displacement. It is a vector quantity, which means that it is completely identified by its magnitude (size) and direction. Its SI (**Système International or International System**) unit is metre per second, shortened as m/s or ms$^{-1}$. If $s = f(t)$, where $s$ and $t$ are position and time, respectively, velocity ($v$) is the differential coefficient of $s$ *wrt* $t$. This is expressed as:

$$\boxed{v = \frac{d}{dt}[s]} \tag{14.1}$$

In summary, velocity is the first derivative of position *wrt* time.

### 14.2.2 ACCELERATION

Acceleration is the rate of change of velocity, and it is also a vector quantity. Its SI unit is metre per second squared, shortened as m/s$^2$ or ms$^{-2}$. If $v = f(t)$, where $v$ and $t$ are velocity and time, respectively, acceleration ($a$) is the differential coefficient of $v$ wrt $t$. This is expressed as:

$$a = \frac{d}{dt}[v] \quad (14.2)$$

Alternatively, given that $s = f(t)$, where $s$ and $t$ are position and time, respectively, velocity is the second derivate of $s$ wrt $t$.

$$a = \frac{d}{dt}\left[\frac{ds}{dt}\right] = \frac{d^2s}{dt^2} \quad (14.3)$$

In summary, acceleration is the first derivative of velocity *wrt* time or the second derivative of position *wrt* time.

### 14.2.3 JERK

Jerk (also known as Jolt) is the rate of change of acceleration and a vector quantity. Its SI unit is metre per second cubed, shortened as m/s$^3$ or ms$^{-3}$. If $a = f(t)$, where $a$ and $t$ are acceleration and time, respectively, jerk ($j$) is the differential coefficient of $a$ wrt $t$. This is expressed as:

$$j = \frac{d}{dt}[a] \quad (14.4)$$

Alternatively, given that $s = f(t)$, where $s$ and $t$ are position and time, respectively, jerk is the third derivate of $s$ wrt $t$.

$$j = \frac{d}{dt}\left[\frac{d^2s}{dt^2}\right] = \frac{d^3s}{dt^3} \quad (14.5)$$

In summary, jerk is the first derivative of acceleration *wrt* time, the second derivative of velocity *wrt* time or the third derivative of position *wrt* time.

## 14.3 TANGENT AND NORMAL LINE EQUATION

When a straight-line, with a gradient $m$, passes through a point $(x_1, y_1)$, its equation is given by:

$$y - y_1 = m(x - x_1) \quad (14.6)$$

Now consider the graph of $y = f(x)$ shown in Figure 14.1. A straight line $L_1$ is drawn tangential to the curve at point A with coordinates $(x_1, y_1)$; $L_1$ is called the **tangent** at the given point. Similarly, another straight line $L_2$ is drawn perpendicularly to $L_1$ and passes through point A. This is called the **normal** at the same point $A$.

It is our desire here to apply differentiation to obtain the equation of the tangent $L_1$ and the normal $L_2$ at point $A$.

Applications of Differentiation

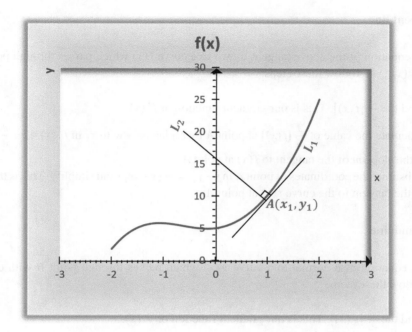

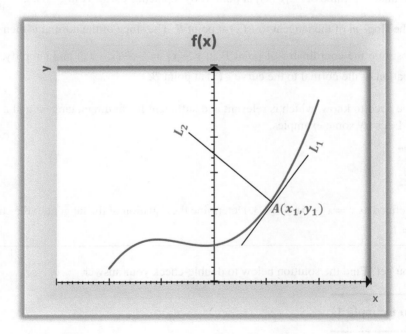

**FIGURE 14.1** Tangent and normal to a curve illustrated.

- **Tangent line**

To find the equation of the above tangent drawn to a curve $f(x)$ which passes through point $A$ with coordinates $(x_1, y_1)$, follow these steps:

**Step 1:** Find the $\frac{d}{dx}[f(x)]$. This is our gradient function or $f'(x)$.

**Step 2:** Calculate the value of $\frac{d}{dx}[f(x)]$ at point $A$ by replacing $x$ with $x_1$ in $f'(x)$, i.e., $\frac{dy}{dx}\bigg|_{x=x_1}$. This is the slope $m$ of the tangent to $f(x)$ at point $A$.

**Step 3:** Substitute the coordinates of point $A$ in $y - y_1 = m(x - x_1)$ and simplify. This is the equation of the tangent to the curve $f(x)$ at point $A$.

- **Normal line**

To find the equation of a normal to a curve $g(x)$ which passes through point $B$ with coordinates $(x_1, y_1)$, follow these steps:

**Step 1:** Find the $\frac{d}{dx}[g(x)]$. This is our gradient function or $g'(x)$.

**Step 2:** Calculate the value of $\frac{d}{dx}[g(x)]$ at point $B$ by replacing $x$ with $x_1$ in $g'(x)$, i.e., $\frac{dy}{dx}\bigg|_{x=x_1}$. This is the slope $m$ of the tangent to $g(x)$ at point $B$. The slope of the normal is then $-\frac{1}{m}$.

**Step 3:** Substitute the coordinates of point B in $y - y_1 = -\frac{1}{m}(x - x_1)$ and simplify. This is the equation of the normal to the curve $g(x)$ at point $B$.

This is all we need to know, which is relevant and sufficient for finding a tangent and a normal at a given point. Let's try some examples.

### Example 1

A curve is defined as $y = x^2 - 5x - \frac{1}{x} - 3$. Determine the equation of the tangent to the curve at point $A$ where $x = 1$.

What did you get? Find the solution below to double-check your answer.

### Solution to Example 1

$$y = x^2 - 5x - \frac{1}{x} - 3$$
$$= x^2 - 5x - x^{-1} - 3$$

Let's walk through the steps for this

Applications of Differentiation

**Step 1:** Determine the $y$-coordinate

Use the given equation to find the $y$ value at $x = 1$ as:

$$y = 1^2 - 5(1) - \frac{1}{1} - 3$$
$$= 1 - 5 - 1 - 3$$
$$= -8$$

The coordinates of point $A$ are $(x_1, y_1) = (1, -8)$,

**Step 2:** Find the differential coefficient of $y$ wrt $x$

$$\frac{dy}{dx} = 2x - 5 + x^{-2}$$
$$= 2x - 5 + \frac{1}{x^2}$$

**Step 3:** Calculate the $\frac{dy}{dx}$ at $x = 1$

$$\left.\frac{dy}{dx}\right|_{x=1} = 2(1) - 5 + \frac{1}{1^2}$$
$$= 2 - 5 + 1$$
$$= -2$$

Thus, the slope $m$ of the tangent at point $A$ is $-2$

**Step 4:** Substitute in the line equation

As we have a point $(1, -8)$ and a slope of $-2$, we will use

$$(y - y_1) = m(x - x_1)$$
$$(y - 1) = -2(x - 1)$$
$$y + 8 = -2x + 2$$
$$y + 8 + 2x - 2 = 0$$
$$y + 2x + 6 = 0$$
$$\therefore y + 2x + 6 = 0$$

**Example 2**

Determine the equation of the normal line to the curve $xy^2 - x^2y - 3y + \ln x = 5$ at point $P$ where $x = 1$ and for which $y$ is positive. Express the equation in the form $ax + by = c$ and state the value of $a$, $b$, and $c$.

What did you get? Find the solution below to double-check your answer.

**Solution to Example 2**

$$xy^2 - x^2y - 3y + \ln x = 5$$

Let's walk through the steps for this.

**Step 1:** Determine the $y$-coordinate

Use the function to find the $y$ value at $x = 1$

$$xy^2 - x^2y - 3y + \ln x = 5$$
$$1 \times y^2 - 1^2 \times y - 3y + \ln(1) = 5$$
$$y^2 - y - 3y + 0 - 5 = 0$$
$$y^2 - 4y - 5 = 0$$
$$(y - 5)(y + 1) = 0$$
$$\therefore y = 5,\ y = -1$$

From the above, we have two points, namely $(1, 5)$ and $(1, -1)$. Since point $P$ is a point where $y$ is positive, we can say that the coordinates of point $P$ are $(x_1, y_1) = (1, 5)$,

**Step 2:** Find the differential coefficient of $y$ wrt $x$

$$y^2 + 2xy \cdot \frac{dy}{dx} - 2xy - x^2 \cdot \frac{dy}{dx} - 3 \cdot \frac{dy}{dx} + \frac{1}{x} = 0$$
$$2xy \cdot \frac{dy}{dx} - x^2 \cdot \frac{dy}{dx} - 3 \cdot \frac{dy}{dx} = 2xy - y^2 - \frac{1}{x}$$
$$\frac{dy}{dx}(2xy - x^2 - 3) = 2xy - y^2 - \frac{1}{x}$$
$$= \frac{2x^2y - xy^2 - 1}{x}$$
$$\frac{dy}{dx} = \frac{2x^2y - xy^2 - 1}{x(2xy - x^2 - 3)}$$

**Step 3:** Calculate the $\frac{dy}{dx}$ at $(1, 5)$

$$\left.\frac{dy}{dx}\right|_{(1,5)} = \frac{2x^2y - xy^2 - 1}{x(2xy - x^2 - 3)}$$
$$= \frac{2(1)^2(5) - (1)(5)^2 - 1}{1 \times (2 \times 1 \times 5 - 1^2 - 3)}$$
$$= \frac{10 - 25 - 1}{10 - 1 - 3} = -\frac{16}{6} = -\frac{8}{3}$$

Thus, the slope $m$ of the tangent at point $A$ is $-\frac{8}{3}$.

# Applications of Differentiation

**Step 4:** Find the slope of the normal

If $m_1$ and $m_2$ represent the slopes of the tangent and normal, respectively, we can write that

$$m_1 \times m_2 = -1$$

Thus

$$m_2 = -\frac{1}{m_1} = -\frac{1}{\left(-\frac{8}{3}\right)}$$

$$\therefore m_2 = \frac{3}{8}$$

**Step 5:** Substitute in the straight-line equation

As we have a point $(1, 5)$ and a slope of $\frac{3}{8}$, we will use

$$(y - y_1) = m(x - x_1)$$

Now substituting, we have:

$$(y - 5) = \frac{3}{8}(x - 1)$$
$$8(y - 5) = 3(x - 1)$$
$$8y - 40 = 3x - 3$$
$$8y - 3x = 40 - 3$$
$$8y - 3x = 37$$
$$\therefore 3x - 8y = -37, \ a = 3, \ b = -8, \ c = -37$$

---

### Example 3

A curve is defined by parametric equations:

$$x = \frac{t+3}{e^t + 5} \qquad y = \frac{t^2 - 4}{e^t + 5}$$

Determine the equations of the tangent and normal at point $C$ on the curve such that $t = 0$. Express the answer in the form of $ax + by + c = 0$.

What did you get? Find the solution below to double-check your answer.

### Solution to Example 3

$$x = \frac{t+3}{e^t + 5} \qquad y = \frac{t^2 - 4}{e^t + 5}$$

Let's walk through the steps:

**Step 1:** Determine the $x$- and $y$-coordinates

$$t = 0$$
$$x = \frac{t+3}{e^t + 5} = \frac{0+3}{e^0 + 5}$$
$$x = \frac{3}{6} = \frac{1}{2}$$

Similarly,

$$y = \frac{t^2 - 4}{e^t + 5} = \frac{0 - 4}{e^0 + 5} = \frac{-4}{1 + 5}$$
$$y = \frac{-4}{6} = -\frac{2}{3}$$

The coordinates of point $C$ are $(x_1, y_1) = \left(\frac{1}{2}, -\frac{2}{3}\right)$,

**Step 2:** Find the differential coefficient of $y$ wrt $x$

This is a parametric differentiation, so let's go.

$$\frac{dx}{dt} = \frac{d}{dt}\left[\frac{t+3}{e^t + 5}\right] = \frac{(e^t + 5) - (t+3)e^t}{(e^t + 5)^2}$$

Similarly,

$$\frac{dy}{dt} = \frac{d}{dt}\left[\frac{t^2 - 4}{e^t + 5}\right] = \frac{(e^t + 5) \times 2t - (t^2 - 4)e^t}{(e^t + 5)^2}$$

Therefore

$$\frac{dy}{dx} = \frac{dy}{dt} \times \frac{dt}{dx}$$

$$= \left[\frac{(e^t + 5) \times 2t - (t^2 - 4)e^t}{(e^t + 5)^2}\right] \times \left[\frac{1}{\left(\frac{(e^t+5)-(t+3)e^t}{(e^t+5)^2}\right)}\right]$$

$$= \left[\frac{(e^t + 5) \times 2t - (t^2 - 4)e^t}{(e^t + 5)^2}\right] \times \left[\frac{(e^t + 5)^2}{(e^t + 5) - (t+3)e^t}\right]$$

$$= \frac{(e^t + 5) \times 2t - (t^2 - 4)e^t}{(e^t + 5) - (t+3)e^t}$$

Applications of Differentiation

**Step 3:** Calculate the $\frac{dy}{dx}$ at $t = 0$

Do not bother simplifying, as we will be substituting $t = 0$.

$$\begin{aligned}\frac{dy}{dx}\bigg|_{t=0} &= \frac{(e^0 + 5) \times 2 \times 0 - (0^2 - 4)e^0}{(e^0 + 5) - (0 + 3)e^0} \\ &= \frac{(1 + 5) \times 0 - (0 - 4) \times 1}{(1 + 5) - (0 + 3) \times 1} \\ &= \frac{(6) \times 0 - (-4) \times 1}{(6) - (3) \times 1} \\ &= \frac{0 + 4}{6 - 3} = \frac{4}{3}\end{aligned}$$

Thus, the slope $m$ of the tangent at point $C$ is $\frac{4}{3}$.

**Step 4:** Substitute into line equation to find the tangent

As we have a point $\left(\frac{1}{2}, -\frac{2}{3}\right)$ and a slope of $\frac{4}{3}$, we will use:

$$\begin{aligned}(y - y_1) &= m(x - x_1) \\ \left(y - \left(-\frac{2}{3}\right)\right) &= \frac{4}{3}\left(x - \frac{1}{2}\right) \\ y + \frac{2}{3} &= \frac{4}{3}x - \frac{2}{3} \\ 3y + 2 &= 4x - 2 \\ 3y + 2 - 4x + 2 &= 0 \\ -4x + 3y + 4 &= 0 \\ \therefore -4x + 3y + 4 &= 0\end{aligned}$$

**Step 5:** Substitute into line equation to find the normal

Recall that

$$\text{Slope of normal} = \frac{-1}{\text{slope of tangent}}$$

We can either find the slope of the normal using the above relationship or modify the line equation accordingly. We will use the latter here since we've demonstrated the first previously.

Now as we have a point $\left(\frac{1}{2}, -\frac{2}{3}\right)$ and $m = \frac{4}{3}$, we will use:

$$(y - y_1) = -\frac{1}{m}(x - x_1)$$

$$\left(y - \left(-\frac{2}{3}\right)\right) = -\frac{1}{\left(\frac{4}{3}\right)}\left(x - \frac{1}{2}\right)$$

$$\left(y - \left(-\frac{2}{3}\right)\right) = -\frac{3}{4}\left(x - \frac{1}{2}\right)$$

$$y + \frac{2}{3} = -\frac{3}{4}x + \frac{3}{8}$$

$$y + \frac{3}{4}x + \frac{2}{3} - \frac{3}{8} = 0$$

$$y + \frac{3}{4}x + \frac{2}{3} - \frac{3}{8} = 0$$

Multiply through by 24, which is the LCM of 3, 4, and 8.

$$24y + 18x + 16 - 9 = 0$$

$$24y + 18x + 7 = 0$$

$$\therefore 18x + 24y + 7 = 0$$

Note the subtle difference between the equations of the tangent and normal, even when the gradient is 1 and −1, respectively.

---

### Example 4

A curve is defined by parametric equations:

$$x = 3\tan 2\theta, \quad y = \sin 2\theta$$

Determine the equations of the tangent and normal at point $P$ on the curve such that $\theta = \frac{\pi}{12}$. Express the answer in the form of $y = mx + c$.

Applications of Differentiation

What did you get? Find the solution below to double-check your answer.

**Solution to Example 4**

$$x = 3\tan 2\theta, \quad y = \sin 2\theta$$

Let's walk through the steps:

**Step 1:** Determine the $(x, y)$ coordinates

$$\theta = \frac{\pi}{12}$$

$$x = 3\tan 2\theta = 3\tan 2\left(\frac{\pi}{12}\right) = 3\tan\left(\frac{\pi}{6}\right)$$

$$x = 3 \times \frac{\sqrt{3}}{3} = \sqrt{3}$$

Similarly,

$$y = \sin 2\theta = \sin 2\left(\frac{\pi}{12}\right) = \sin\left(\frac{\pi}{6}\right)$$

$$y = \frac{1}{2}$$

The coordinates of point $P$ are $(x_1, y_1) = \left(\sqrt{3}, \frac{1}{2}\right)$.

**Step 2:** Find the differential coefficient of $y$ wrt $x$

This is a parametric differentiation, so let's go

$$\frac{dx}{d\theta} = \frac{d}{d\theta}[3\tan 2\theta] = 6\sec^2 2\theta$$

Similarly,

$$\frac{dy}{d\theta} = \frac{d}{d\theta}[\sin 2\theta] = 2\cos 2\theta$$

Therefore

$$\frac{dy}{dx} = \frac{dy}{d\theta} \times \frac{d\theta}{dx}$$

$$= [2\cos 2\theta] \times \left[\frac{1}{6\sec^2 2\theta}\right]$$

$$= \left[\frac{\cos 2\theta}{3\sec^2 2\theta}\right]$$

$$= \frac{1}{3}\left[\frac{\cos 2\theta}{\left(\frac{1}{\cos 2\theta}\right)^2}\right]$$

$$= \frac{1}{3}(\cos 2\theta)^3$$

**Step 3:** Calculate the $\frac{dy}{dx}$ at $\theta = \frac{\pi}{12}$

$$\left.\frac{dy}{dx}\right|_{\theta=\frac{\pi}{6}} = \frac{1}{3}\left[\cos\left(2 \times \frac{\pi}{12}\right)\right]^3 = \frac{1}{3}\left[\cos\left(\frac{\pi}{6}\right)\right]^3$$

$$= \frac{1}{3}\left[\frac{\sqrt{3}}{2}\right]^3 = \frac{\sqrt{3}}{8}$$

Thus, the slope $m$ of the tangent at point P is $\frac{\sqrt{3}}{8}$.

**Step 4:** Substitute into line equation to find the equation of the tangent

As we have a point $\left(\sqrt{3}, \frac{1}{2}\right)$ and a slope of $\frac{\sqrt{3}}{8}$, we will use:

$$(y - y_1) = m(x - x_1)$$

which implies that

$$\left(y - \frac{1}{2}\right) = \frac{\sqrt{3}}{8}\left(x - \sqrt{3}\right)$$

$$8\left(y - \frac{1}{2}\right) = \sqrt{3}\left(x - \sqrt{3}\right)$$

$$8y - 4 = \sqrt{3}x - 3$$

$$8y = \sqrt{3}x + 1$$

$$\therefore y = \frac{\sqrt{3}}{8}x + \frac{1}{8}$$

**Step 5:** Substitute into line equation to find the equation of the normal

As we have a point $\left(\sqrt{3}, \frac{1}{2}\right)$ and a slope of $\frac{\sqrt{3}}{8}$, we will use:

$$(y - y_1) = -\frac{1}{m}(x - x_1)$$

which implies that

$$\left(y - \frac{1}{2}\right) = -\frac{8}{\sqrt{3}}\left(x - \sqrt{3}\right)$$

$$\left(y - \frac{1}{2}\right) = -\frac{8\sqrt{3}}{3}\left(x - \sqrt{3}\right)$$

$$y - \frac{1}{2} = -\frac{8\sqrt{3}}{3}x + \left(\frac{8\sqrt{3}}{3} \times \sqrt{3}\right)$$

$$y - \frac{1}{2} = -\frac{8\sqrt{3}}{3}x + 8$$

# Applications of Differentiation

Multiply through by 6, which is the LCM of 2 and 3.

$$6y - 3 = -16\sqrt{3}x + 48$$
$$16\sqrt{3}x + 6y - 3 - 48 = 0$$
$$16\sqrt{3}x + 6y - 51 = 0$$
$$\therefore \mathbf{16\sqrt{3}\,x + 6y - 51 = 0}$$

---

A final example to try.

### Example 5

Determine the angle between the straight line $y = 3x + 1$ and the parabola $y = x^2 + x - 2$ at the point of their intersection in the positive $x$-axis region. Present the angle correct to 1 d.p.

---

What did you get? Find the solution below to double-check your answer.

### Solution to Example 5

$$y = 3x + 1 \,,\, y = x^2 + x - 2$$

Let's walk through the steps:

**Step 1:** Determine the point of intersection(s) of the two functions

At the point of intersection, we have:

$$3x + 1 = x^2 + x - 2$$
$$x^2 - 2x - 3 = 0$$
$$(x - 3)(x + 1) = 0$$

Therefore

$$(x - 3) = 0 \qquad\qquad (x + 1) = 0$$
$$x - 3 = 0 \quad \text{OR} \quad x + 1 = 0$$
$$\therefore x = 3 \qquad\qquad \therefore x = -1$$

We can find the $y$-coordinate using either of the two equations, let's go:

When $x = 3$

$$y = 3x + 1$$
$$= 3(3) + 1 = 10$$
$$\therefore (\mathbf{3, 10})$$

When $x = -1$

$$y = 3x + 1$$
$$= 3(-1) + 1 = -2$$
$$\therefore (-1, -2)$$

There are two points of intersection, but we are interested in the positive $x$-region, thus we have $(3, 10)$ only.

**Step 2:** Find the differential coefficient of each function at $x = 3$.

- For $y = 3x + 1$

$$\frac{d}{dx}[3x + 1] = 3$$
$$\frac{d}{dx}[3x + 1]\bigg|_{x=3} = 3$$

- For $y = x^2 + x - 2$

$$\frac{d}{dx}[x^2 + x - 2] = 2x + 1$$
$$\frac{d}{dx}[x^2 + x - 2]\bigg|_{x=3} = 2(3) + 1 = 7$$

**Step 3:** Determine the angle between them

We will use the fact that

$$\tan\theta = m$$
$$\theta = \tan^{-1} m$$

- For $y = 3x + 1$

$$\theta_1 = \tan^{-1}(3) = 71.57°$$

- For $y = x^2 + x - 2$

$$\theta_1 = \tan^{-1}(7) = 81.87°$$

Hence, the angle $\theta$ between them is

$$\theta = |\theta_2 - \theta_1|$$
$$= |81.87° - 71.57°|$$
$$\therefore \theta = 10.3°$$

Applications of Differentiation

## 14.4 INCREASING AND DECREASING FUNCTIONS

A function $y = f(x)$ is said to be increasing, often within a particular interval, if $y$ increases as $x$ increases and vice versa. Figure 14.2(a) shows that $f(x)$ is increasing between $x = a$ and $x = b$, for $a < b$, which implies that $f(a) < f(b)$. Note that the slope is positive at all points between $a$ and $b$.

On the other hand, a function $y = g(x)$ is said to be decreasing, often in a particular interval, if $y$ increases as $x$ decreases and vice versa. Figure 14.2(b) shows that $g(x)$ is a decreasing function between $x = a$ and $x = b$. This is because as $x$ increases from $a$ to $b$, $y$ decreases accordingly, i.e., $a < b$ implies that $f(a) > f(b)$. Here the slope is negative at all points between $a$ and $b$.

The interval in both cases above can be written as $a \leq x \leq b$.

Often a function is both increasing and decreasing, but at different intervals. Let's consider a curve $y = h(x)$ which rises and falls as the value of $x$ changes from negative to positive (Figure 14.3).

For our discussion here, we've identified three portions, $A$, $B$, and $C$, to the right of $y$-axis, though we could have done the same to the left of the same axis. As the value of $x$ in the region $A$ increases, the value of $y$ decreases, the same holds for region $C$. Although the curve is in the positive $x$ region ($A$) and the negative $y$ region ($C$), it can however be noted that the gradient is negative in both cases, i.e. $m < 0$. This is a key and common criterion. The function $h(x)$ is therefore said to be decreasing in these two regions.

When we look at region $B$, it can be observed that as the value of $x$ increases the value of $y$ also increases. Although the curve is above the $x$-axis like region $A$, we note that the gradient is positive, i.e. $m > 0$. The function $h(x)$ is therefore said to be increasing in this region.

We can therefore summarise the above points as:

1) Increasing Function

$$\text{If } \boxed{h(x) \text{ is increasing}} \text{ then } \boxed{\frac{dy}{dx} > 0} \text{ OR } \boxed{\text{gradient is positive}} \quad (14.7)$$

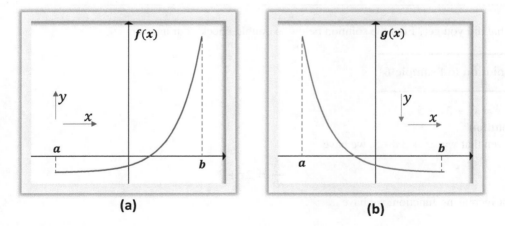

**FIGURE 14.2** (a) Increasing function and (b) decreasing function.

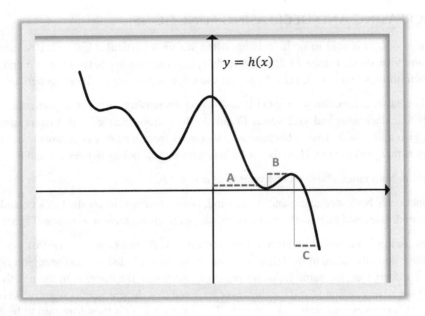

**FIGURE 14.3** A function that increases and decreases in a particular interval illustrated.

2) Decreasing Function

$$\text{If } \boxed{h(x) \text{ is decreasing}} \text{ then } \boxed{\frac{dy}{dx} < 0} \text{ OR } \boxed{\text{gradient is negative}} \quad (14.8)$$

It seems that we've said it all. Let's illustrate with examples.

### Example 6

Determine the range of values of $x$ for which each of the following functions is increasing.

a) $y = x^2 + 3x - 5$  
b) $y = 13 - x - 2x^2 - x^3$

What did you get? Find the solution below to double-check your answer.

### Solution to Example 6

a) $y = x^2 + 3x - 5$
**Solution**
Given that $y = x^2 + 3x - 5$, we have

$$\frac{dy}{dx} = 2x + 3$$

For increasing function, we have

$$\frac{dy}{dx} > 0$$

Applications of Differentiation

which implies that

$$2x + 3 > 0$$
$$2x > -3$$
$$x > -\frac{3}{2}$$

The above implies that the function $f(x)$ is increasing at any point when $x$ is greater than $-1.5$.

---

**b)** $y = 13 - x - 2x^2 - x^3$
**Solution**
Given that $y = 13 - x - 2x^2 - x^3$, we have

$$\frac{dy}{dx} = -1 - 4x - 3x^2$$

For increasing function, we have:

$$\frac{dy}{dx} > 0$$

which implies that

$$-1 - 4x - 3x^2 > 0$$
$$3x^2 + 4x + 1 < 0$$
$$(3x + 1)(x + 1) < 0$$
$$\therefore -1 < x < -\frac{1}{3}$$

The above implies that the function $f(x)$ is increasing at any point when $x$ is between $-1$ and $-\frac{1}{3}$.

---

Let's try another example.

**Example 7**

$v(t) = 2t^3 + \frac{1}{2}t^2 - 35t + 1$ represents the velocity of an object as a function of time.

   a) Determine the time interval when the object is accelerating.
   b) Determine the time interval when the object is decelerating.
   c) Show this on a graph.

What did you get? Find the solution below to double-check your answer.

## Solution to Example 7

### HINT

- Acceleration implies a positive or an increasing function.
- Deceleration implies a negative or a decreasing function.

**a) Accelerating**
**Solution**
Given that $v(t) = 2t^3 + \frac{1}{2}t^2 - 35t + 1$, we have

$$\frac{d}{dt}[v(t)] = 6t^2 + t - 35$$

During acceleration the function $v(t)$ will be increasing, thus

$$\frac{d}{dt}[v(t)] > 0$$
$$6t^2 + t - 35 > 0$$

The quadratic expression is $6t^2 + t - 35$, hence we solve as

$$6t^2 + t - 35 = 0$$
$$(2t + 5)(3t - 7) = 0$$

Therefore

$$2t + 5 = 0 \quad \text{OR} \quad 3t - 7 = 0$$
$$\therefore t = -2.5 \qquad \qquad \therefore t = \frac{7}{3}$$

We can conclude that the object is increasing (or accelerating) during the following time intervals

$$t < -2.5 \text{ and } t > \frac{7}{3}$$

**b) Decelerating**
**Solution**
During deceleration, the function $v(t)$ will be decreasing, thus

$$\frac{d}{dt}[v(t)] < 0$$
$$6t^2 + t - 35 < 0$$

As before, the roots of $6t^2 + t - 35 = 0$ are $t = -2.5$ or $t = \frac{7}{3}$.
We can conclude that the object is decreasing (or decelerating) during the following time intervals

$$-\frac{5}{2} < t < \frac{7}{3}$$

**c) Graph** (Figure 14.4 and Figure 14.5)

# Applications of Differentiation

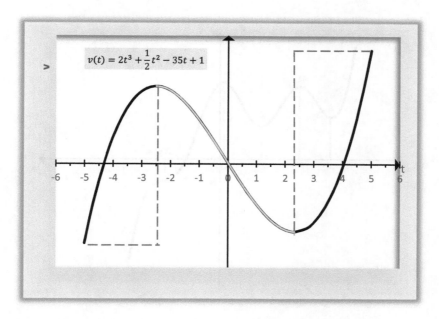

**FIGURE 14.4** Acceleration (as an increasing function of velocity) illustrated.

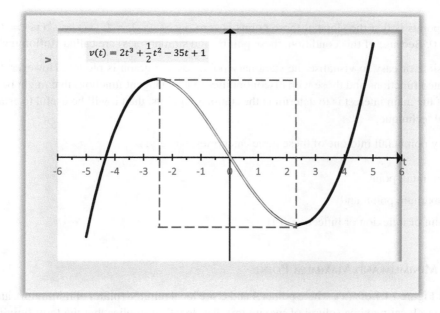

**FIGURE 14.5** Deceleration (as a decreasing function of velocity) illustrated.

## 14.5 STATIONARY POINTS

Figure 14.6 is a graph of $f(x)$, with a gradient function given as $f'(x)$. The gradient at a given point is obtained by evaluating $f'(x)$ using the value of $x$ at that point. For example, the gradient at points $A$, $B$, $C$, $D$, and $E$ are $f'(x_1), f'(x_2), f'(x_3), f'(x_4)$, and $f'(x_5)$, respectively. One thing that is common

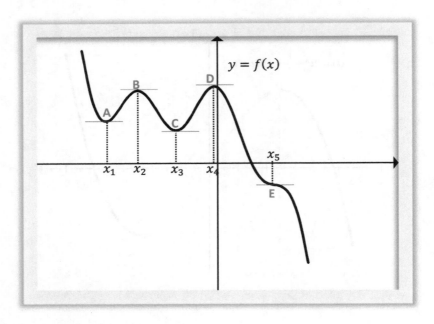

**FIGURE 14.6** Stationary points illustrated.

to these points is that the slope at these points is zero, i.e., $f'(x_1) = f'(x_2) = f'(x_3) = f'(x_4) = f'(x_5) = 0$. Because of this condition, these points (and similar ones) are called **stationary points**.

It is possible or easy to visualise the stationary points once a graph is plotted. However, there are complicated functions and those with discontinuities (e.g., a tangent function) that might be difficult to plot. If the main interest is to determine the stationary points, then it will be useful to establish an analytical technique.

Stationary points fall into one of these three categories, namely:

   a) Minimum point,
   b) Maximum point, and
   c) Point of inflexion or inflection (PI)

### 14.5.1 Minimum and Maximum Points

Consider Figure 14.7 of $f(x)$, where points $A$ and $C$ are local minima (plural of minimum), and points $B$ and $D$ are local maxima (plural of maximum). By 'local', it implies that the term 'minimum' or 'maximum' is related to a small region of the curve, otherwise we will only have one minimum point (e.g., point A) and one maximum point (e.g., point D) in the curve shown below. This also distinguishes them from the global minimum and maximum points of the function.

If we take a second look at the curve, we will notice that each stationary point ($A$, $B$, $C$, and $D$) has a pair of lines, one to the left and the other to the right of the point. Furthermore, the sign '$---$' indicates that the gradient of the curve in this region is negative and '$+++$' implies that the gradient of the curve in this region is positive. We can now summarise our observations as:

Applications of Differentiation

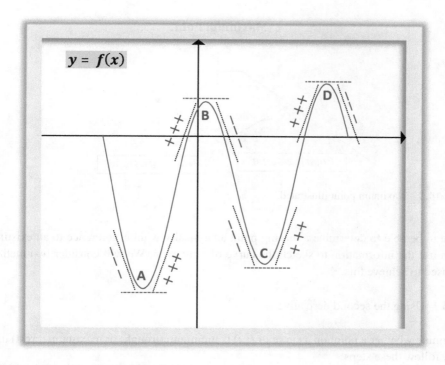

**FIGURE 14.7** Minimum and maximum points illustrated.

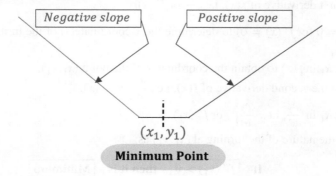

**FIGURE 14.8** Minimum point illustrated.

**Note 1** For each point, the gradient changes from either positive through zero to negative or from negative through zero to positive. Essentially, the curve switches its direction.

**Note 2** Points A and C switch signs from negative to positive and are called **minimum points**. It is also a point where the function changes from a decreasing function to an increasing function. This is schematically shown in Figure 14.8.

**Note 3** Points B and D switch signs from positive to negative and they are called **maximum points**. It is also a point where the function changes from being an increasing function to a decreasing function, as schematically shown in Figure 14.9.

**Note 4** Minima (A and C) and maxima (B and D) are also called the **turning points (TP)**. It should be added that turning points are sometimes used synonymously as stationary points. We can conclude that stationary is more general and unambiguous, whilst turning points can be used in a general sense or specific form.

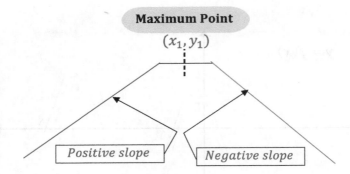

**FIGURE 14.9** Maximum point illustrated.

We need to be able to determine a turning point analytically without reference to an existing curve and then use the information to sketch the curve of a function. We will consider two methods that can be used to achieve this.

**Method 1** Using the second derivative

To determine whether a point on a curve of $f(x)$ is minimum (trough) or maximum (crest) using this method, follow these steps:

**Step 1:** Find the first derivative of $f(x)$, i.e., $\frac{dy}{dx}$ or $f'(x)$.

**Step 2:** Solve $\frac{dy}{dx} = 0$ (or $f'(x) = 0$) to determine the $x$-coordinate(s) of the turning point. Let this point be $x_1$.

**Step 3:** Substitute $x_1$ in $f(x)$ to obtain the coordinates of the point $(x_1, y_1)$.

**Step 4:** Determine the second derivative of $f(x)$, i.e., $\frac{d^2y}{dx^2}$ or $f''(x)$.

**Step 5:** Substitute $x_1$ in $\frac{d^2y}{dx^2}$, i.e., $\left.\frac{d^2y}{dx^2}\right|_{x_1}$ or $f''(x_1)$.

**Step 6:** Establish the nature of the turning at $(x_1, y_1)$ as:

$$\text{If } \boxed{f''(x_1) > 0} \text{ then it is } \boxed{\text{Minimum}} \tag{14.9}$$

and

$$\text{If } \boxed{f''(x_1) < 0} \text{ then it is } \boxed{\text{Maximum}} \tag{14.10}$$

**Method 2** Using change of gradient at stationary point

To determine whether a point on a curve of $f(x)$ is minimum (trough) or maximum (crest) using this method, follow these steps:

**Step 1:** Find the first derivative of $f(x)$, i.e., $\frac{dy}{dx}$ or $f'(x)$.

**Step 2:** Solve $\frac{dy}{dx} = 0$ or $f'(x) = 0$ to determine the $x$-coordinate of the turning point. Let this point be $x_1$.

# Applications of Differentiation

**Step 3:** Substitute $x_1$ in $f(x)$ to obtain the coordinates of the point $(x_1, y_1)$.

**Step 4:** Choose a point $x_1 - a$ to the left and $x_1 + a$ to the right of the stationary point $x_1$, where $a$ is a small positive real number. We say 'small' to ensure that the point chosen is as close as possible to $(x_1, y_1)$, otherwise the curve might have changed shape.

**Step 5:** Substitute $x_1 - a$ and $x_1 + a$ in the first derivative, i.e., $\left.\dfrac{dy}{dx}\right|_{x_1-a}$ and $\left.\dfrac{dy}{dx}\right|_{x_1+a}$.

**Step 6:** Establish the nature of the turning point at $(x_1, y_1)$ as:

$$\text{If } \boxed{f'(x_1 - a) < 0} \text{ AND } \boxed{f'(x_1 + a) > 0} \text{ then it is } \boxed{\text{Minimum}}$$
$$f'(a^-) < 0 \quad f'(a) = 0 \quad f'(a^+) > 0 \tag{14.11}$$

and

$$\text{If } \boxed{f'(x_1 - a) > 0} \text{ AND } \boxed{f'(x_1 + a) < 0} \text{ then it is } \boxed{\text{Maximum}}$$
$$f'(a^-) > 0 \quad f'(a) = 0 \quad f'(a^+) < 0 \tag{14.12}$$

Let's add a few points about the above two methods.

**Note 1** Instead of writing $f'(x_1 - a)$, we sometimes use $f'(x^-)$ or $f'(a^-)$ to mean a point to the left. Similarly, $f'(x^+)$ or $f'(a^+)$ is used instead of $f'(x_1 + a)$, to indicate a point to the right of the stationary point.

**Note 2** What is substituted in method 1 is the actual $x$-coordinate of the turning point and is substituted in $\dfrac{d^2y}{dx^2}$, but the $x$-values of the neighbouring points (left and right) of the stationary points are substituted in method 2 and is done in $\dfrac{dy}{dx}$.

The above process is also known as **determining the nature of the stationary point**.

It will be interesting to know what happens if:

a) $f'(x_1) \neq 0$
b) $f''(x_1) = 0$
c) $f'(x_1 - a) < 0$ AND $f'(x_1 + a) < 0$
d) $f'(x_1 - a) > 0$ AND $f'(x_1 + a) > 0$

The answer to the above will come in the next section when we discuss a point of inflexion (P-of-I), but for now, let's take a break and try some examples.

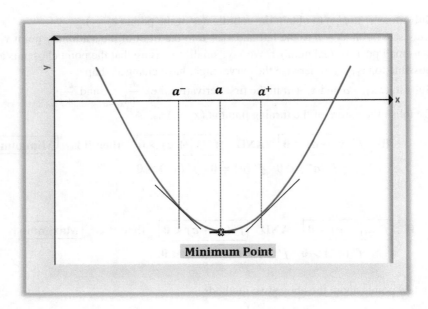

**FIGURE 14.10** Determining the minimum point using change of gradient at stationary point illustrated.

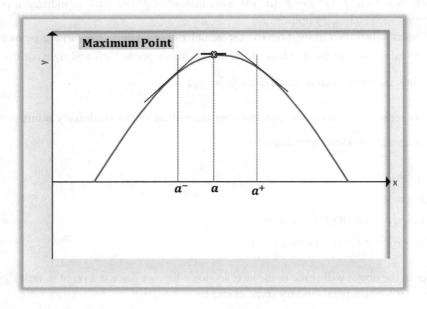

**FIGURE 14.11** Determining the maximum point using change of gradient at stationary point illustrated.

### Example 8

Determine the coordinates $(x_1, y_1)$ of the turning point on the following curves and state the nature.

**a)** $f(x) = x^2 - 7x + 5$  **b)** $g(x) = 13 - 8x - 5x^2$

Applications of Differentiation

What did you get? Find the solution below to double-check your answer.

### Solution to Example 8

**a)** $f(x) = x^2 - 7x + 5$

**Solution**

**Method 1** First derivative

Given that $f(x) = x^2 - 7x + 5$, we have

$$f'(x) = 2x - 7$$

At the turning point, we have

$$f'(x) = 0$$

which implies that

$$2x - 7 = 0$$
$$\therefore x = \frac{7}{2} = 3.5$$

This is the $x$-coordinate; the $y$-coordinate is found as:

$$f(x) = x^2 - 7x + 5$$
$$f(3.5) = \left(\frac{7}{2}\right)^2 - 7\left(\frac{7}{2}\right) + 5$$
$$= -\frac{29}{4} = -7.25$$

Hence, the coordinates are

$$(x_1, y_1) = \left(\frac{7}{2}, -\frac{29}{4}\right)$$

Remember that the above coordinates can also be found using the completing square method (covered in *Foundation Mathematics for Engineers and Scientists with Worked Examples* by the same author).

Let's determine the nature of the turning point. If $f'(x) = 2x - 7$, then

$$f''(x) = 2$$
$$\therefore f''\left(\frac{7}{2}\right) = 2 > 0$$

Since $f''(x) > 0$, there is a minimum at the turning point $\left(\frac{7}{2}, -\frac{29}{4}\right)$.

Notice that we evaluated $f''\left(\frac{7}{2}\right)$, but this is not needed in this situation, since $f''(x)$ is not a function but a numerical constant. This was shown to highlight the steps given above and we will subsequently omit it where necessary.

**Method 2** Change of gradient at the stationary point

You might have noticed that the first three steps are the same for both methods, and since we do have the answer for these steps, we will not be repeating them. Instead, we will start from step 4. The point to the left of $x_1 = \frac{7}{2}$ is $\frac{7}{2} - a$ and to the right is $\frac{7}{2} + a$, where $a$ is a small positive real number.

$$f'(x) = 2x - 7$$

Therefore

$$f'\left(\frac{7}{2} - a\right) = 2\left(\frac{7}{2} - a\right) - 7$$
$$= 7 - 2a - 7$$
$$= -2a < 0$$

Similarly,

$$f'\left(\frac{7}{2} + a\right) = 2\left(\frac{7}{2} + a\right) - 7$$
$$= 7 + 2a - 7$$
$$= 2a > 0$$

Since $f'\left(\frac{7}{2} - a\right) < 0$ AND $f'\left(\frac{7}{2} + a\right) > 0$, then we have

**Minimum**

**NOTE**
Instead of using $3.5 - a$ and $3.5 + a$, we could have used 3.4 and 3.6 respectively.

---

**b)** $g(x) = 13 - 8x - 5x^2$
**Solution**

**Method 1** First derivative

Given that $g(x) = 13 - 8x - 5x^2$, we have

$$g'(x) = -8 - 10x$$

At the turning point, we have

$$g'(x) = 0$$

Therefore

$$-10x - 8 = 0$$
$$\therefore x = -\frac{8}{10} = -0.8$$

This is the $x$-coordinate; the $y$-coordinate is found as:

$$g(x) = 13 - 8x - 5x^2$$
$$f(-0.8) = 13 - 8(-0.8) - 5(-0.8)^2$$
$$= \frac{81}{5} = 16.2$$

Hence, the coordinates are

$$(x_1, y_1) = (-0.8, 16.2)$$

If $g'(x) = -8 - 10x$, then

$$g''(x) = -10$$
$$\therefore g''(-0.8) = -10 < 0$$

Since $g''(x) < 0$, there is a maximum at the turning point $(-0.8, 16.2)$.

**Method 2** Change of gradient at the stationary point

The point to the left of $x_1 = -0.8$ is $-0.8 - a$ and to the right is $-0.8 + a$, where $a$ is a small positive real number.

$$g'(x) = -8 - 10x$$

Therefore

$$g'(-0.8 - a) = -8 - 10(-0.8 - a)$$
$$= -8 + 8 + 10a$$
$$= 10a > 0$$

Similarly,

$$g'(-0.8 + a) = -8 - 10(-0.8 + a)$$
$$= -8 + 8 - 10a$$
$$= -10a < 0$$

Since $g'(x^-) > 0$ AND $g'(x^+) < 0$, then we have:

**Maximum**

### ALTERNATIVE METHOD

In this example, we will try to use the following values.

If $x_1 = -0.8$, then we choose $x^- = -1.0$ and $x^+ = -0.6$. Notice that we chose a number to the left and another to the right of $x_1 = -0.8$. Now let's substitute as:

$$g'(x) = -8 - 10x$$

Therefore

$$g'(-1.0) = -8 - 10(-1.0)$$
$$= -8 + 10$$
$$= 2 > 0$$

Similarly,

$$g'(-0.6) = -8 - 10(-0.6)$$
$$= -8 + 6$$
$$= -2 < 0$$

Again, since $g'(x^-) > 0$ AND $g'(x^+) < 0$, then we have:

**Maximum**

**NOTE**
Here, we chose a point 0.2 to the left and right. For this case, this might be relatively small but perhaps too big in other situations. It is therefore important to keep this as small as practically possible (Figure 14.12).

Let's try another example.

## Example 9

A cubic polynomial function is given by $y = x^3 + 4x^2 - 3x - 9$.

a) Determine the coordinates of the stationary points.

b) State the nature of the stationary points.

c) Draw a graph showing the stationary points.

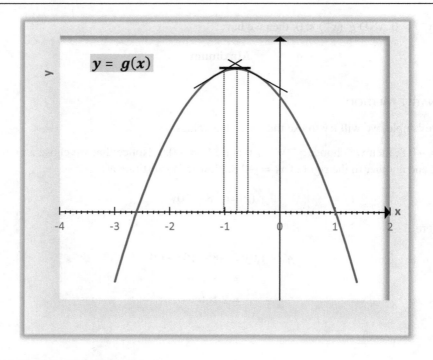

**FIGURE 14.12** Solution to Example 8(b).

Applications of Differentiation

What did you get? Find the solution below to double-check your answer.

**Solution to Example 9**

**a)** Stationary points
**Solution**
Given that $y = x^3 + 4x^2 - 3x - 9$, we have

$$y'(x) = 3x^2 + 8x - 3$$

At the turning point, we have

$$y'(x) = 0$$
$$\therefore 3x^2 + 8x - 3 = 0$$

Let's use quadratic formula here: $a = 3$, $b = 8$, and $c = -3$.

$$x = \frac{-b \pm \sqrt{b^2 - 4ac}}{2a}$$
$$= \frac{-8 \pm \sqrt{8^2 - 4(3)(-3)}}{2(3)}$$
$$= \frac{-8 \pm \sqrt{100}}{6} = \frac{-8 \pm 10}{6}$$
$$= \frac{-4 \pm 5}{3}$$

Therefore

$$x = \frac{-4+5}{3} \quad \text{OR} \quad x = \frac{-4-5}{3}$$
$$\therefore x = \frac{1}{3} \quad\quad\quad\quad \therefore x = -3$$

- When $x = \frac{1}{3}$

$$y\left(\frac{1}{3}\right) = \left(\frac{1}{3}\right)^3 + 4\left(\frac{1}{3}\right)^2 - 3\left(\frac{1}{3}\right) - 9$$
$$= -\frac{257}{27}$$

Thus, the coordinates are $(x_1, y_1) = \left(\frac{1}{3}, -\frac{257}{27}\right)$.

- When $x = -3$

$$y(-3) = (-3)^3 + 4(-3)^2 - 3(-3) - 9$$
$$= 9$$

Thus, the coordinates are $(x_2, y_2) = (-3, 9)$.

**b) Nature of the stationary points**
**Solution**
If $y'(x) = 3x^2 + 8x - 3$, then

$$y''(x) = 6x + 8$$

- When $x = \frac{1}{3}$

$$y''\left(\frac{1}{3}\right) = 6\left(\frac{1}{3}\right) + 8$$
$$= 10 > 0$$

Since $y''(x) > 0$, it is minimum at the turning point $\left(\frac{1}{3}, -\frac{257}{27}\right)$.

- When $x = -3$

$$y''\left(\frac{1}{3}\right) = 6(-3) + 8$$
$$= -10 < 0$$

Since $y''(x) < 0$, it is maximum at the turning point $(-3, 9)$.

**c) Plot**
**Solution** (Figure 14.13)

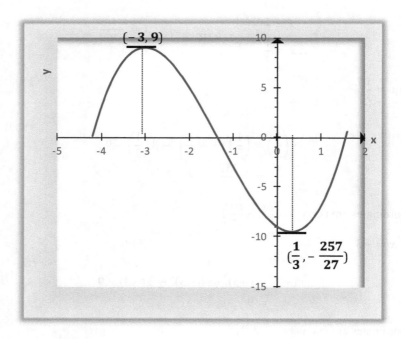

**FIGURE 14.13** Solution to Example 9(c).

# Applications of Differentiation

Let's try one final example.

## Example 10

A sinusoidal voltage signal is given by $v(t) = (\sin \omega t + 1)\ mV$. The signal operates at a constant angular frequency of $1000\pi$ rad/s.

**a)** Determine the time when the signal reaches its maximum/minimum voltage in the interval $0 \le t \le 2\pi$.

**b)** Determine the coordinates of the points in (a).

**c)** State the nature of the stationary points.

What did you get? Find the solution below to double-check your answer.

## Solution to Example 10

**a)** Stationary points
**Solution**
Given that $v(t) = \sin \omega t + 1$, we have

$$v'(t) = \omega \cos \omega t + 0 = \omega \cos \omega t$$

At the turning point, we have

$$v'(t) = 0$$

Therefore

$$\omega \cos \omega t = 0$$
$$\omega t = \cos^{-1}(0)$$
$$= \frac{\pi}{2}$$

This is the value in the first quadrant, the other value is

$$\omega t = 2\pi - \frac{\pi}{2} = \frac{3\pi}{2}$$

- When $\omega t = \frac{\pi}{2}$, we have

$$t = \frac{\pi}{2\omega} = \frac{\pi}{2 \times 1000\pi}$$
$$= \frac{1}{2000}s = 0.5\ ms$$
$$\therefore t_1 = 0.5\ ms$$

- When $\omega t = \frac{3\pi}{2}$, we have

$$t = \frac{3\pi}{2\omega} = \frac{3\pi}{2 \times 1000\pi}$$
$$= \frac{3}{2000}s = 1.5 \text{ ms}$$
$$\therefore t_2 = \mathbf{1.5 \text{ ms}}$$

Thus, we can say that the signal reaches its maximum/minimum at two points within the stated range.

---

**b) Coordinates**
**Solution**

- When $t = \mathbf{0.5 \text{ ms}}$

  $v(t) = \sin \omega t + 1$
  $v(0.5 \text{ ms}) = \sin\left(1000\pi \times 0.5 \times 10^{-3}\right) + 1$
  $= \sin\left(\frac{1}{2}\pi\right) + 1$
  $= 1 + 1 = 2 \text{ mV}$
  Thus, the coordinates are $(t_1, v_1) = $ **(0.5 ms, 2 mV)**

- When $t = \mathbf{1.5 \text{ ms}}$

  $v(t) = \sin \omega t + 1$
  $v(1.5 \text{ ms}) = \sin\left(1000\pi \times 1.5 \times 10^{-3}\right) + 1$
  $= \sin\left(\frac{3}{2}\pi\right) + 1$
  $= -1 + 1 = 0 \text{ mV}$
  Thus, the coordinates are $(t_2, v_2) = $ **(1.5 ms, 0 mV)**

---

**c) Nature of stationary point**
**Solution**
If $v'(t) = \omega \cos \omega t$, then

$$v''(t) = -\omega^2 \sin \omega t$$

- When $t = \mathbf{0.5 \text{ ms}}$

$$\therefore v''(0.5 \text{ ms}) = -(1000\pi)^2 \sin\left(1000\pi \times 0.5 \times 10^{-3}\right)$$
$$= -(1000\pi)^2 \sin\left(\frac{1}{2}\pi\right) = -(1000\pi)^2 < 0$$

Since $v''(t) < 0$, it is maximum at the turning point **(0.5 ms, 2 mV)**.

- When $t = \mathbf{1.5 \text{ ms}}$

$$\therefore v''(1.5 \text{ ms}) = -(1000\pi)^2 \sin\left(1000\pi \times 1.5 \times 10^{-3}\right)$$
$$= -(1000\pi)^2 \sin\left(\frac{3}{2}\pi\right) = -(1000\pi)^2 \times (-1) = (1000\pi)^2 > 0$$

Since $v''(t) > 0$, it is minimum at the turning point **(1.5 ms, 0 mV)**.

Applications of Differentiation

### 14.5.2 INFLEXION

This is the last type of stationary point, which is a bit different from the two we've covered. Let's start by stating what the point of inflexion is, as this will provide a clue to answering the four questions we asked at the end of the last section.

A **point of inflexion (PI)** is a point on a curve where there is a change in the magnitude of the gradient without a change in its sign, i.e., the curve changes from being concave to being convex to a particular direction or vice versa. In this situation, the slope of the curve does not change sign as one moves from left to right of the stationary point. In other words, if $m = +$ve it remains so and if it is negative, it will not change.

We can equally say that a point of inflexion exists at a point on a curve where the curve is generally increasing (positive) or decreasing (negative), but not changing from a positive to a negative slope. Figure 14.14 shows some possible examples of a point of inflexion.

To summarise, we can broadly divide a point of inflexion into two categories:

1) A generally increasing (or decreasing) function with a zero slope (i.e., $\frac{dy}{dx} = 0$) at the point of inflexion. This is the general case and the reason for including a point of inflexion in a stationary point.

2) A generally increasing (or decreasing) function with a non-zero slope (i.e., $\frac{dy}{dx} \neq 0$) at the point of inflexion. This is not generally intended when the point of inflexion is mentioned. Obviously, it is not a stationary point.

Based on this and subject to what we intend by 'turning', it can be said that the point of inflexion is not a turning point.

We now know that $\frac{dy}{dx} = 0$ is not a necessary condition to establish a point of inflexion. In fact, at the point of inflexion, $\frac{dy}{dx}$ can be negative, zero, or positive. Solving $\frac{dy}{dx} = 0$ will only detect the $x$-coordinate(s) of $y = f(x)$:

a) maximum point,
b) minimum point, and
c) point of inflexion if the slope is zero (Figure 14.14(a) and 14.14(b)). It will not reveal the point of inflexion for the cases in 14.14(b) and 14.14(d).

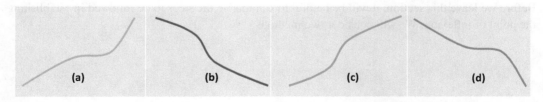

**FIGURE 14.14** Point of inflexion illustrated, showing rising PI (a) and (c) and falling PI (b) and (d).

There are two key measures that can be used to establish a point of inflexion, namely

**Condition 1:** Is there a point of inflexion?

We know that $f''(x) = 0$ (or undefined) confirms that there is neither a maximum nor a minimum at the given point, but does it suggest a point of inflexion? The answer is that it is a pre-condition for a point of inflexion, but it is not sufficient to confirm the presence of a point of inflexion. In summary $f''(x) = 0$ (or undefined) gives the $x$-coordinate(s) for a possible point (or points) of inflexion on the curve.

**Condition 2:** What is the nature of inflexion?

If condition 1 above is established, then there is a point of inflexion at point $x = x_1$ on the curve if the second condition is met. Condition 2 can be established using either of the two methods below.

**Method 1** Neighbourhood points (first derivative – test for no change in sign)

$$\text{If } \boxed{f'(x_1 - a) > 0} \text{ AND } \boxed{f'(x_1 + a) > 0} \text{ for } \boxed{\text{a rising function}} \quad (14.13)$$

OR

$$\text{If } \boxed{f'(x_1 - a) < 0} \text{ AND } \boxed{f'(x_1 + a) < 0} \text{ for } \boxed{\text{a falling function}} \quad (14.14)$$

**Method 2** Neighbourhood points (second derivative – test for change of sign)

$$\text{If } \boxed{f''(x_1 - a) < 0} \text{ AND } \boxed{f''(x_1 + a) > 0} \text{ for } \boxed{\text{a falling function}} \quad (14.15)$$

OR

$$\text{If } \boxed{f''(x_1 - a) > 0} \text{ AND } \boxed{f''(x_1 + a) < 0} \text{ for } \boxed{\text{a rising function}} \quad (14.16)$$

It should be mentioned that if we are only interested in whether a function is rising or falling at the point of inflexion $x_n$, we can find the third derivative of the function and subject it to the following test.

$$\boxed{f'''(x_n) > 0} \text{ for } \boxed{\text{a rising function}} \quad (14.17)$$

OR

$$\boxed{f'''(x_n) < 0} \text{ for } \boxed{\text{a falling function}} \quad (14.18)$$

Before we leave this section, it will be helpful to summarise the steps to be followed in establishing the point of inflexion. We will consider two methods.

# Applications of Differentiation

**Method 1** Test for no change in sign (using the first derivative)

To determine whether a point on a curve of $f(x)$ is a point of inflexion, follow these steps:

**Step 1:** Find the first derivative of $f(x)$, i.e., $\frac{dy}{dx}$ or $f'(x)$.

**Step 2:** Solve $\frac{dy}{dx} = 0$ or $f'(x) = 0$ to determine the x-coordinate of the turning point. Let this be $x_1$.

**Step 3:** Substitute $x_1$ in $f(x)$ to obtain the coordinates of the point $(x_1, y_1)$.

**Step 4:** Determine the second derivative of $f(x)$, i.e., $\frac{d^2y}{dx^2}$ or $f''(x)$.

**Step 5:** Substitute $x_1$ in $\frac{dy}{dx}$ or $f'(x)$, i.e., $\left.\frac{dy}{dx}\right|_{x_1}$ or $f'(x_1)$. If $\left.\frac{dy}{dx}\right|_{x_1} = 0$, then a pre-condition (or condition 1) for a point of inflexion at $x_1$ has been established. We need to proceed to the second condition.

**Step 6:** Use the following test to establish the existence of a point of inflexion at $x_1$:

   a) For a rising function, we must have:

$$\boxed{f'(x_1 - a) > 0} \quad \text{AND} \quad \boxed{f'(x_1 + a) > 0} \tag{14.19}$$

   b) For a falling function, we must have:

$$\boxed{f'(x_1 - a) < 0} \quad \text{AND} \quad \boxed{f'(x_1 + a) < 0} \tag{14.20}$$

**Method 2** Test for change of sign (using second derivative)

To determine whether a point on a curve of $f(x)$ is a point of inflexion, follow these steps:

**Step 1:** Find the first derivative of $f(x)$, i.e., $\frac{dy}{dx}$ or $f'(x)$.

**Step 2:** Solve $\frac{dy}{dx} = 0$ or $f'(x) = 0$ to determine the x-coordinate of the turning point. Let this be $x_1$.

**Step 3:** Substitute $x_1$ in $f(x)$ to obtain the coordinates of the point $(x_1, y_1)$.

**Step 4:** Determine the second derivative of $f(x)$, i.e., $\frac{d^2y}{dx^2}$ or $f''(x)$.

**Step 5:** Substitute $x_1$ in $\frac{d^2y}{dx^2}$ or $f''(x)$, i.e., $\left.\frac{d^2y}{dx^2}\right|_{x_1}$ or $f''(x_1)$. If $\left.\frac{d^2y}{dx^2}\right|_{x_1} = 0$, then a pre-condition (or condition 1) for a point of inflexion at $x_1$ has been established. We need to proceed to the second condition.

**Step 6:** Use the following test to establish the existence of a point of inflexion at $x_1$:

   a) For a rising function, we must have:

$$\boxed{f''(x_1 - a) > 0} \quad \text{AND} \quad \boxed{f''(x_1 + a) < 0} \tag{14.21}$$

   a) For a falling function, we must have:

$$\boxed{f''(x_1 - a) < 0} \quad \text{AND} \quad \boxed{f''(x_1 + a) > 0} \tag{14.22}$$

All we said above is to determine and establish the existence of a point of inflexion when the gradient is zero. What if $\frac{dy}{dx} \neq 0$ (i.e., when $\frac{dy}{dx} < 0$ and $\frac{dy}{dx} > 0$)? For this, the steps stated in the two methods above are still relevant but with a bit of tweak and this is as follows:

**Note 1** We will not solve $\frac{dy}{dx} = 0$ since this is not valid in this situation; we will instead use $\frac{d^2y}{dx^2} = 0$ and determine the possible points of inflexion. Let this be $x_1$.

**Note 2** Proceed in the first method, the substitution is carried out in $\frac{d^2y}{dx^2}$ while the same is done in $\frac{dy}{dx}$.

We can also conclude the existence of P-of-I if $\frac{dy}{dx} = 0$ and $\frac{d^2y}{dx^2} = 0$, but $\frac{d^3y}{dx^3} \neq 0$. This is for $y = x^3$ for example.

All done! Let's try an example to illustrate this.

### Example 11

Determine the coordinates of the points of inflexion, if any, on the graph of $f(x) = x^3 - 4x^2 - x + 5$.

---

What did you get? Find the solution below to double-check your answer.

### Solution to Example 11

Given that $f(x) = x^3 - 4x^2 - x + 5$, then we have

$$f'(x) = 3x^2 - 8x - 1$$
$$f''(x) = 6x - 8$$

A point of inflexion will occur, if it exists, when

$$f''(x) = 0$$

which implies that

$$6x - 8 = 0$$
$$\therefore x = \frac{4}{3}$$

Let's find the y-coordinate as:

$$f(x) = x^3 - 4x^2 - x + 5$$
$$f\left(\frac{4}{3}\right) = \left(\frac{4}{3}\right)^3 - 4\left(\frac{4}{3}\right)^2 - \left(\frac{4}{3}\right) + 5$$
$$= -\frac{29}{27}$$

Thus, the coordinates are $(x_1, y_1) = \left(\frac{4}{3}, -\frac{29}{27}\right)$.

Now, let $x^- = 1$ and $x^+ = \frac{5}{3}$.

# Applications of Differentiation

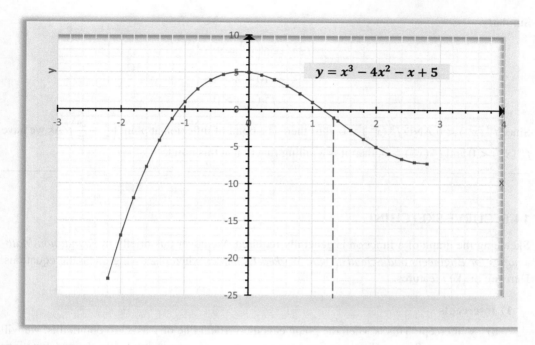

**FIGURE 14.15** Solution to Example 11.

**Method 1**

$$f'(x) = 3x^2 - 8x - 1$$
$$f'(x^-) = 3(1)^2 - 8(1) - 1$$
$$= 3 - 8 - 1 = -6$$
$$\therefore f'(x^-) < 0$$

Similarly

$$f'(x^+) = 3\left(\frac{5}{3}\right)^2 - 8\left(\frac{5}{3}\right) - 1$$
$$= \frac{25}{3} - \frac{40}{3} - 1 = -6$$
$$\therefore f'(x^+) < 0$$

Since $f'(x^-) < 0$ AND $f'(x^+) < 0$, thus there is a point of inflexion at point $\left(\frac{4}{3}, -\frac{29}{27}\right)$. As both conditions are negative, then it is a falling function at this point.

**Method 2**

$$f''(x) = 6x - 8$$
$$f''(x^-) = 6(1) - 8$$
$$= 6 - 8 = -2$$
$$\therefore f''(x^-) < 0$$

Similarly

$$f''(x^+) = 6\left(\frac{5}{3}\right) - 8$$
$$= 10 - 8 = 2$$
$$\therefore f''(x^+) > 0$$

Since $f''(x^-) < 0$ AND $f''(x^+) > 0$, thus there is a point of inflexion at point $\left(\frac{4}{3}, -\frac{29}{27}\right)$. As we have $f''(x^-) < 0$ and $f''(x^+) > 0$, then it is a falling function at this point.

---

## 14.6 CURVE SKETCHING

Sketching the graph of a function is generally required. We cover this briefly in *Foundation Mathematics for Engineers and Scientists with Worked Examples* when discussing quadratic equations. Here are the key features.

1) **Intercepts**

   a) *y*-intercept: This is when the graph cuts the vertical line or *y*-axis. To obtain this, we will use $x = 0$ in explicitly defined functions $y = f(x)$ or in implicitly defined functions $f(x, y) = f(x, y)$. In general, we expect a curve to cross the vertical axis once, at a point $(0, y_c)$, where $y_c$ is the point where the curve crosses the *y*-axis. However, a curve of an implicit function can cross the *y*-axis more than once.

   b) *x*-intercept: This is when the graph cuts the horizontal line or *x*-axis. To obtain this, we set $y = 0$ in explicitly defined functions $f(x)$ or in implicitly defined functions $f(x, y)$. The *x*-value or values obtained are called the roots or zeros of the function. In general, the number of *x*-values obtained is determined by the type of function. For a polynomial function, the number is the same as the degree of the polynomial. For a trigonometric function, this is primarily determined by the intervals and the periodicity (or angular frequency) of the function. If the roots of a function are $x_1, x_2, \ldots x_n$, the points of crossing the horizontal axis are $(x_1, 0), (x_2, 0), \ldots, (x_n, 0)$, respectively.

2) **Stationary points**

   We just covered this above. In summary, we need to determine the $(x, y)$ coordinates of the maximum and minimum points and the point of inflexion. This is particularly needed to show where the curve changes direction (concave or convex).

3) **General behaviour**

   In addition to the above points, there are other secondary factors that we need to consider when sketching a curve. It is not a condition that we check for each of the below when sketching the graph of a function. Nonetheless, it helps to have a better view and to obtain a more accurate shape.

   a) Symmetry: This is where a line can be drawn through the curve such that one part of the curve is a mirror of the other part. This line is called the line of symmetry. Generally, we expect only one line of symmetry where the graph is split into two. If this is a vertical line, where $x = n$ such that $n \in \mathbb{R}$, then we can say that the curve is symmetrical about the *y*-axis. This symmetry occurs when *x* is replaced with $-x$ in the $f(x)$ and can be represented as $f(x) = f(-x)$, e.g., $\cos\theta = \cos(-\theta)$.

# Applications of Differentiation

On the other hand, if this is a horizontal line, where $y = n$ such that $n \in \mathbb{R}$, then we say that the curve is symmetrical about the $x$-axis. This occurs when $y$ is replaced with $-y$ in $f(x)$ and can be represented as $f(x) = -f(x)$. For example, $x^2 + y^2 = 5$ is symmetrical about both the $x$-axis and the $y$-axis because we can replace either $x$ or $y$ or both with their respective negative value and the equation remains unchanged. The line of symmetry here is the diameter of the circle.

b) Interval (decreasing and increasing): Increasing and decreasing functions have been covered earlier in this chapter. We can examine this to confirm if the curve is progressing upward left and downward right (decreasing) or progressing upward right and downward left (increasing).

c) Asymptotes: Some functions are undefined at particular value(s) of $x$ or $y$, we can say that the curve is asymptotic at these point(s) of $x$ or $y$. A vertical asymptote is a vertical line, defined as $x = n$, for $n \in \mathbb{R}$, which the curve approaches but never touches. Similarly, a horizontal asymptote is a horizontal line, defined as $y = n$, for $n \in \mathbb{R}$, which the curve approaches but never touches. At these points, the curve is discontinuous. For example, a tangent function is discontinuous at angles $\theta = 90° \pm 180°n$ and these are the asymptotic lines. Similarly, a rational function $y = \frac{1}{x-1}$ is undefined when the denominator equals zero, i.e., $x - 1 = 0$, thus there is a vertical asymptote at $x = 1$. By re-arranging and making $x$ the subject, we can show that there is a horizontal asymptote at $y = 0$.

d) It is also helpful to check the behaviour of the curve at certain intervals.

e) Finally, knowing the default shape of certain functions (e.g., quadratic and cubic functions, trigonometric functions) will be immensely helpful in sketching a more accurate shape.

All covered! Let us try some examples.

## Example 12

Sketch the following functions.

a) $y = 3 - 2x$

b) $y = x^2 + 3x - 4$

c) $y = (1 - 2x)(x + 3)(x - 2)$

What did you get? Find the solution below to double-check your answer.

**Solution to Example 12**

**a)** $y = 3 - 2x$
**Solution**
We need the following information to be able to sketch the graph.

- **$x$-intercept**

When $y = 0$, we have

$$3 - 2x = 0$$
$$3 = 2x$$
$$\therefore x = 1.5$$

If this point is $A$, then the coordinates of $A$ are $(\mathbf{1.5, 0})$

- **$y$-intercept**

When $x = 0$, we have

$$y = 3 - 2(0)$$
$$\therefore y = 3$$

If this point is $B$, then the coordinates of $B$ are $(\mathbf{0, 3})$

- **Stationary point**

Knowing that this is a linear function, the two points $A$ and $B$ are sufficient to sketch this successfully. Nevertheless, let us apply this condition to it, so we differentiate

$$y = 3 - 2x$$
$$y' = -2$$

The first derivative is a numeric value, which implies that there is only one slope for this function at every point. Hence, this is a straight line (Figure 14.16).

---

**b)** $y = x^2 + 3x - 4$
**Solution**
We need the following information to be able to sketch the graph

- **$x$-intercept**

When $y = 0$, we have

$$x^2 + 3x - 4 = 0$$
$$(x - 1)(x + 4) = 0$$

# Applications of Differentiation

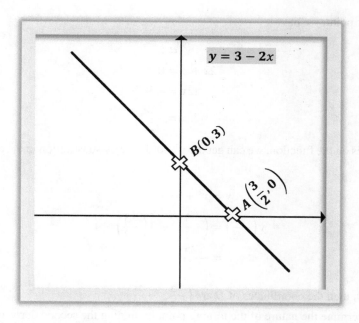

**FIGURE 14.16** Solution to Example 12(a).

Therefore

$$x - 1 \quad \text{OR} \quad x + 4 = 0$$
$$\therefore x = 1 \qquad \quad \therefore x = -4$$

If the points are A and B, then we have the following coordinates: $A(-4, 0)$ and $B(1, 0)$

- **y-intercept**

When $x = 0$, we have

$$y = (0)^2 + 3(0) - 4$$
$$\therefore y = -4$$

If this point is C, then the coordinates of C are $(0, -4)$

- **Stationary point**

Knowing that this is a quadratic function, we expect a single turning point. In fact, we know it is minimum since the coefficient of $x^2$ is positive, i.e. $a > 0$. The vertex (or the turning point) can also be obtained using completing the square method, but we will use differential calculus here. Now, differentiating

$$y = x^2 + 3x - 4$$
$$y' = 2x + 3$$

At the stationary point, we have

$$y' = 0$$
$$2x + 3 = 0$$
$$2x = -3$$
$$\therefore x = -\frac{3}{2}$$

Since this point is on the function, we can get the y-coordinate by substitution in the original function as:

$$y = x^2 + 3x - 4$$
$$y\left(-\frac{3}{2}\right) = \left(-\frac{3}{2}\right)^2 + 3\left(-\frac{3}{2}\right) - 4$$
$$= -\frac{25}{4}$$

If this point is $D$, then the coordinates of $D$ are $\left(-\frac{3}{2}, -\frac{25}{4}\right)$.

Now we can determine the nature of the turning point by finding the second derivative as:

$$y'' = 2$$

The value of $\frac{d^2y}{dx^2}$ at the turning point $x = -\frac{3}{2}$ is

$$y''\left(\frac{3}{2}\right) = 2 > 0$$

As this is positive, then there is a minimum at $\left(-\frac{3}{2}, -\frac{25}{4}\right)$.

- **Increasing/decreasing**

The function is increasing when

$$y' > 0$$
$$2x + 3 > 0$$
$$\therefore x > -\frac{3}{2}$$

It is decreasing when

$$y' < 0$$
$$2x + 3 < 0$$
$$\therefore x < -\frac{3}{2}$$

The above information is sufficient to accurately plot the graph (Figure 14.17).

Applications of Differentiation

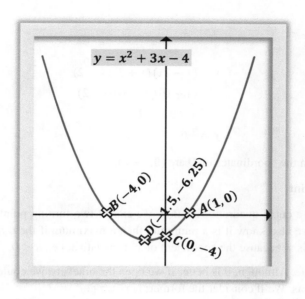

**FIGURE 14.17** Solution to Example 12(b).

c) $y = (1 - 2x)(x + 3)(x - 2)$
**Solution**
We need the following information to be able to sketch the graph

- **x-intercept**

When $y = 0$, we have

$$(1 - 2x)(x + 3)(x - 2) = 0$$

We're lucky that the expression is already factorised. Hence,

$$1 - 2x = 0$$
$$\therefore x = \frac{1}{2}$$

OR

$$x + 3 = 0$$
$$\therefore x = -3$$

OR

$$x - 2 = 0$$
$$\therefore x = 2$$

If the points are $A$, $B$, and $C$, then we have the following coordinates: $A(-3, 0)$, $B\left(\frac{1}{2}, 0\right)$, and $C(2, 0)$

- **y-intercept**

When $x = 0$, we have

$$y = (1 - 2x)(x + 3)(x - 2)$$
$$= (1 - 0)(0 + 3)(0 - 2)$$
$$= 1 \times 3 \times -2$$
$$\therefore y = -6$$

If this point is $D$, then the coordinates of $D$ are $(0, -6)$.

- **Stationary point**

Knowing that this is a cubic quadratic function, we expect two turning points (a maximum and a minimum). In fact, we also know it is a minimum then a maximum if the drawing is from the left side of the graph. This is because the coefficient of $x^3$ is negative, i.e., $a < 0$.

Now, let's differentiate. Although, it is better if we open the brackets, we could also use the product rule for triple functions. We'll consider the former; here we go.

$$y = (1 - 2x)(x + 3)(x - 2)$$
$$= (1 - 2x)(x^2 + x - 6)$$
$$= x^2 + x - 6 - 2x^3 - 2x^2 + 12x$$
$$= -2x^3 - x^2 + 13x - 6$$

Therefore

$$y' = -6x^2 - 2x + 13$$

At the stationary point, we have

$$y' = 0$$
$$-6x^2 - 2x + 13 = 0$$

Let's use quadratic formula to solve this. We know that $a = -6$, $b = -2$, and $c = 13$. Therefore,

$$x = \frac{-b \pm \sqrt{b^2 - 4ac}}{2a}$$
$$= \frac{2 \pm \sqrt{(-2)^2 - 4(-6)(13)}}{-12}$$
$$= \frac{2 \pm 2\sqrt{79}}{-12} = \frac{1 \pm \sqrt{79}}{-6}$$

Therefore

$$x_1 = -\frac{1 + \sqrt{79}}{6} = -1.65$$

**OR**

$$x_2 = \frac{-1 + 1\sqrt{79}}{6} = 1.31$$

# Applications of Differentiation

Since these points are on the function, we can get the y-coordinates by substitution in the original function as:

$$y = (1 - 2x)(x + 3)(x - 2)$$

At $x_1 = -\frac{1+1\sqrt{79}}{6}$

$$y\left(-\frac{1+\sqrt{79}}{6}\right) = \left[1 - 2\left(-\frac{1+\sqrt{79}}{6}\right)\right]\left[\left(-\frac{1+\sqrt{79}}{6}\right) + 3\right]\left[\left(-\frac{1+\sqrt{79}}{6}\right) - 2\right]$$
$$= -21.19$$

If this point is $E$, then the coordinates of $E$ are $(-1.65, -21.19)$

At $x_2 = \frac{-1+1\sqrt{79}}{6}$

$$y\left(\frac{-1+\sqrt{79}}{6}\right) = \left[1 - 2\left(\frac{-1+\sqrt{79}}{6}\right)\right]\left[\left(\frac{-1+\sqrt{79}}{6}\right) + 3\right]\left[\left(\frac{-1+\sqrt{79}}{6}\right) - 2\right]$$
$$= 4.82$$

If this point is $F$, then the coordinates of $F$ are $(1.31, 4.82)$

Now we can determine the nature of the turning points ($E$ and $F$) by finding the second derivative as:

$$y'' = -12x - 2$$

The value of $\frac{d^2y}{dx^2}$ at the turning point $E$ $x_1 = -1.65$ is

$$y''(-1.65) = -12(-1.65) - 2 = 17.8 > 0$$

As this is positive, then there is a minimum at $E(-1.65, -21.19)$.
Similarly, the value of $\frac{d^2y}{dx^2}$ at the turning point $F$ $x_1 = 1.31$ is

$$y''(1.31) = -12(1.31) - 2 = -17.72 < 0$$

As this is negative, then there is a maximum at $F(1.31, 4.82)$.

- **Increasing/decreasing**

The function is increasing when

$$y' > 0$$
$$-6x^2 - 2x + 13 > 0$$
$$6x^2 + 2x - 13 < 0$$
$$\therefore -1.65 < x < 1.31$$

It is decreasing when

$$y' < 0$$
$$-6x^2 - 2x + 13 < 0$$
$$6x^2 + 2x - 13 > 0$$
$$\therefore x < -1.65 \text{ OR } x > 1.31$$

The above information is sufficient to accurately plot the graph (Figure 14.18). Still, we can check the behaviour (Table 14.1) of the function beyond $x < -1.65$ and $x > 1.31$.

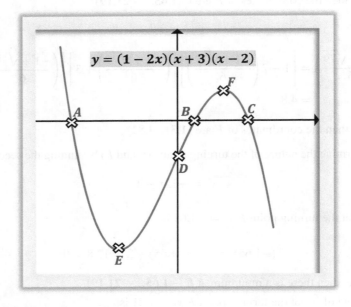

**FIGURE 14.18** Solution to Example 11(c).

**TABLE 14.1**
**Solution to Example 11(c)**

| Condition | Behaviour |
|---|---|
| $x \ll -1.65$ | At this instant, $-2x^3$ becomes the dominant term of $y = -2x^3 - x^2 + 13x - 6$. The cube of a negative value is negative; hence, we have $-2$ (negative) $=$ positive $= y$. In other words, the function remains positive above the horizontal axis. |
| $x \gg 1.31$ | At this instant, again $-2x^3$ dominates. The cube of a positive value is positive; hence, we have $-2$ (positive) $=$ negative $= y$. In other words, the function remains negative below the horizontal axis. |

# 14.7 CHAPTER SUMMARY

1) Velocity (a vector quantity) is the time rate of change of position. If $s = f(t)$, where $s$ and $t$ are position and time, respectively, velocity ($v$) is the differential coefficient of $s$ wrt $t$. This is expressed as:

$$v = \frac{d}{dt}[s]$$

2) Acceleration (also a vector quantity) is the time rate of change of velocity. If $v = f(t)$, where $v$ and $t$ are velocity and time, respectively, acceleration ($a$) is the differential coefficient of $v$ wrt $t$. This is expressed as:

$$a = \frac{d}{dt}[v]$$

Alternatively, given that $s = f(t)$, where $s$ and $t$ are position and time, respectively, velocity is the second derivate of $s$ wrt $t$.

$$a = \frac{d}{dt}\left[\frac{ds}{dt}\right] = \frac{d^2s}{dt^2}$$

3) Jerk (also known as Jolt) is the time rate of change of acceleration and is a vector quantity. If $a = f(t)$, where $a$ and $t$ are acceleration and time, respectively, jerk ($j$) is the differential coefficient of $a$ wrt $t$. This is expressed as:

$$j = \frac{d}{dt}[a]$$

Alternatively, given that $s = f(t)$, where $s$ and $t$ are position and time, respectively, jerk is the third derivate of $s$ wrt $t$.

$$a = \frac{d}{dt}\left[\frac{d^2s}{dt^2}\right] = \frac{d^3s}{dt^3}$$

4) A straight line $L_1$ drawn tangential to the curve at point $A$ is called the **tangent** at the given point. Similarly, a straight line $L_2$ drawn perpendicularly to $L_1$ and passes through point $A$ is called the **normal** at the same point $A$.

5) A function $y = f(x)$ is said to be increasing, often in a particular interval, if $y$ increases as $x$ increases and vice versa.

6) A function $y = g(x)$ is said to be decreasing, often in a particular interval, if $y$ increases as $x$ decreases and vice versa.

7) Stationary points fall into one of these categories, namely:
   - Minimum point,
   - Maximum point, and
   - Point of inflexion
8) A **point of inflexion** is a point on a curve where there is a change in the magnitude of the gradient without a change in its sign, i.e., the curve changes from being concave to convex to a particular direction or vice versa.

****

## 14.8 FURTHER PRACTICE

To access complementary contents, including additional exercises, please go to www.dszak.com.

# 15 Indefinite Integration

### Learning Outcomes

Once you have studied the content of this chapter, you should be able to:

- Explain integration as an inverse of differentiation
- Discuss the importance of the constant of integration
- Use the notations associated with integration
- Integrate the power of $x$
- Integrate standard functions
- Determine the equation of a curve when given its gradient and a point on the curve

## 15.1 INTRODUCTION

Integration is the second part of calculus and a direct inverse of differentiation (or anti-derivative). Integration is primarily used in applications that require finding the area and volume under a function. In fact, we can say that the formulas for finding areas of many standard shapes (triangle, rectangle, trapezium, circle, etc.) are or can be obtained using integration. We will, in this chapter, start with indefinite integration, covering the constant of integration, determining the integral of functions, and finding the equation of a curve.

## 15.2 INDEFINITE INTEGRAL

Integration can be either indefinite or definite. The former is the subject of discussion in this chapter, whilst the latter will be considered in the subsequent chapter. However, it will be helpful to set out the fundamental differences between the two.

**Note 1**  An indefinite integration is represented by $\int y \, dx$. The notation $\int$, which looks like an elongated $S$ (denoting sum), is called '**the integral of**'. It implies that the reverse of differentiation should be performed, just like the notation $\sqrt{\phantom{x}}$, is used to undo the square. $\int y \, dx$ is read as '**the integral of $y$ with respect to $x$**'. $y$ is the integrand and '$\boldsymbol{dx}$' shows the variable to which the integrand is being computed with respect to, which must be included otherwise the integration is meaningless. In this case, $x$ is the independent variable. If the independent variable is time ($t$), we will have $dt$ and if it is volume ($V$), we will have $dV$ and so on.

**Note 2**  On the other hand, a definite integration is represented by $\int_a^b y \, dx$. We say $a$ and $b$ are, respectively, the lower and upper limits of the integration. We shall come back to this soon.

**Note 3**  The operation of an indefinite integration is always a function of a variable, whilst the operation of a definite integration results in a numerical value or constant.

**Note 4** An indefinite integration is the anti-derivative or inverse of differentiation. On the other hand, a definite integration represents an area covered by the upper and lower limits.

**Note 5** A function that undergoes an indefinite integration can be recovered by taking the derivative of the integral function. However, if a definite integration is carried out on a function, it cannot be recovered by simply finding the differential coefficient of the result.

### 15.2.1 Constant of Integration

So far, we've been made aware that integration will undo what differentiation does. Let's investigate this concept further by finding the derivatives of the following quadratic functions $y_1, y_2, y_3$, and $y_4$, as detailed in Table 15.1.

From the functions in Table 15.1, of the form $ax^2 + bx + c$, we notice that all the four functions are the same but differ only in the value of the constant $c$. In other words, the value of $c$ in $y_1$ is 17 and in $y_2$, $y_3$, and $y_4$ are 8, 0, and $-20$, respectively. We equally note that the differential coefficient of all of them is the same, i.e., $6x - 5$. The derivative will remain the same as $6x - 5$ for any function of the above form such that only $c$ has changed. In fact, they represent a family of functions that have only been translated along the y-axis, i.e., moved up/down. The plots of the four functions, vertically translated, are shown in Figure 15.1.

The problem here is, if we were asked to integrate $6x - 5$, to recover the original function, how do we know which of the above four functions initially produced this integral? In other words, is the integral of $6x - 5$ with respect to $x$, technically written as $\int (6x - 5) \, dx$, equal to $3x^2 - 5x + 17$ or $3x^2 - 5x + 8$ or $3x^2 - 5x$ or $3x^2 - 5x - 20$ or even another family member? Obviously, we have lost the constant value in the process of finding the derivative, and we have no trace of what it was or a hint to inform us of what to do to recover it.

For the above reason, we always add a constant $C$ to the result of any integration. This is called the **constant of integration** or **arbitrary constant**, and it stands in place of a constant that has been lost when differentiating. It is common to use capital letter $C$, but it is not restricted. We shall also see other letters being used later in this chapter and subsequent ones.

Let's write our quadratic function family above as $3x^2 - 5x + C$, where $C$ represents any numerical constant. It follows that if:

$$\frac{d}{dx}(3x^2 - 5x + c) = 6x - 5$$

**TABLE 15.1**
**Constant of Integration Illustrated**

| Function $f(x)$ | Derivative $\frac{dy}{dx}$ | Function $f(x)$ | Derivative $\frac{dy}{dx}$ |
|---|---|---|---|
| $y_1 = 3x^2 - 5x + 17$ | $6x - 5$ | $y_2 = 3x^2 - 5x + 8$ | $6x - 5$ |
| $y_3 = 3x^2 - 5x$ | $6x - 5$ | $y_4 = 3x^2 - 5x - 20$ | $6x - 5$ |

# Indefinite Integration

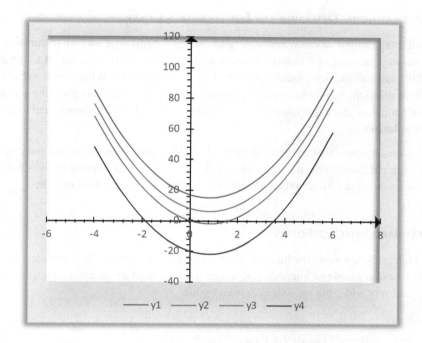

**FIGURE 15.1** A family of functions with the same derivatives illustrated.

then

$$\int (6x - 5)\,dx = 3x^2 - 5x + C$$

In other words, when we differentiate a function, the constant term disappears but when we integrate, a constant (i.e., constant of integration) appears. This is the gist! We have illustrated this in Figure 15.2.

As we've lost the original constant when a function is differentiated, adding the constant of integration is therefore mandatory in indefinite integration, otherwise the function cannot be 100 percent recovered. An expression with a constant of integration is called the **general solution**.

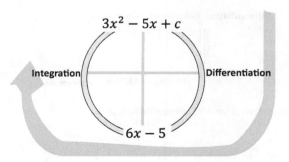

**FIGURE 15.2** Integration as an anti-derivative illustrated.

## 15.2.2 Integrating to Determine the Equation of a Curve

We discussed integrating a function and adding an arbitrary constant $C$, which means that the function represents any member of a family of functions. To find the value of constant $C$, we need to be given an additional information, usually a point $(x_1, y_1)$ on the curve. When this is substituted back into the general solution, we would be able to determine the value of $C$. This gives the actual function in a family of functions. An expression obtained, having replaced $C$ with a numerical value, is called a **particular solution**.

There is an essential aspect that needs to be made clear: How do we determine the integral of a function? In the example above, how do we know that $3x^2 - 5x + c$ is the integral of $6x - 5$ if we do not have the prior knowledge? Right, this is the journey that we are about to set out for.

## 15.2.3 Standard Integral Forms

By standard integrals, we mean the integrals of the common functions in their simplest forms. These are typically found in a table of integrals and can be used to integrate complex functions. Integration domain is seemingly wide, but we will take it bit by bit.

### 15.2.3.1 Integrating Power of $x$ or $x^n$ ($n \neq -1$)

This is used for power functions, and it states that

$$\text{If } \boxed{y = x^n} \quad \text{then} \quad \boxed{\int x^n dx = \frac{1}{n+1} x^{n+1} + C} \tag{15.1}$$

where $n$ is any real number excluding $-1$.

Here is the gist: for differentiation, we multiply then subtract, while in integration we add then divide. This is an exact inverse situation, as illustrated in Figure 15.3.

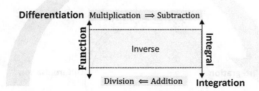

**FIGURE 15.3** Determining derivatives and integrals illustrated.

# Indefinite Integration

At this stage, let's try some examples.

## Example 1

Determine the integral of the following functions with respect to $x$.

**a)** $y = x^3$  **b)** $y = x^{-4}$  **c)** $y = x^{-1}$  **d)** $y = \sqrt{x}$  **e)** $y = \frac{1}{\sqrt[5]{x^3}}$  **f)** $y = x^{2.1}$

**g)** $y = x$  **h)** $y = 7$  **i)** $y = x^{\cos\frac{1}{6}\pi}$  **j)** $y = \frac{x^{-8}}{x^8}$  **k)** $y = x^{-\frac{2}{7}}$  **l)** $y = x^{\sqrt{2}}$

What did you get? Find the solution below to double-check your answer.

## Solution to Example 1

**HINT**

- Understanding the laws of indices will be helpful.
- Also, do not forget to add the constant of integration $C$.

**a)** $y = x^3$
**Solution**

$$y = x^3$$

$$\int y\, dx = \int x^3\, dx = \frac{1}{3+1} x^{3+1} + C$$

$$= \frac{1}{4} x^4 + C$$

$$\therefore \int x^3\, dx = \frac{1}{4} x^4 + C$$

**b)** $y = x^{-4}$
**Solution**

$$y = x^{-4}$$

$$\int y\, dx = \int x^{-4}\, dx = \frac{1}{-4+1} x^{-4+1} + C$$

$$= \frac{1}{-3} x^{-3} + C$$

$$\therefore \int x^{-4}\, dx = -\frac{1}{3} x^{-3} + C$$

**NOTE**

- For this, we could have written the final answer as $-\frac{1}{3x^3}$ using one of the laws of indices.
- One common error is to think that the power should be 'increased' from $-4$ to $-5$. This will be true for positive power $x^4$ where the power will go from 4 to 5. However, when the power is

−4, the new power will be −3 as this is an increase of 1 on the number line. It will be desirable to go with the formula.

**c)** $y = x^{-1}$
**Solution**

$$y = x^{-1}$$

$$\int y\, dx = \int x^{-1} dx = \frac{1}{-1+1} x^{-1+1} + C$$

$$= \frac{1}{0} x^0 + C = \frac{1}{0} + C = \infty + C$$

$$\therefore \int x^{-1} dx = ?$$

**NOTE**
Hold on a moment! It seems the formula states that it cannot be used when $n = -1$. But what do we do or how do we solve this type of problem? This will be covered by another integral standard that we will be looking at shortly.

**d)** $y = \sqrt{x}$
**Solution**

$$y = \sqrt{x}$$

This is not in the right format, so we need to transform this as:

$$y = \sqrt{x} = x^{\frac{1}{2}}$$

$$\int y\, dx = \int x^{\frac{1}{2}} dx = \frac{1}{\frac{1}{2}+1} x^{\left(\frac{1}{2}+1\right)} + C$$

$$= \frac{1}{\left(\frac{3}{2}\right)} x^{\frac{3}{2}} + C = \frac{2}{3} x^{\frac{3}{2}} + C$$

$$\therefore \int \sqrt{x}\, dx = \frac{2}{3} x^{\frac{3}{2}} + C$$

**NOTE**
For this, we could have written the final answer as $\frac{2}{3}\sqrt{x^3} + C$ using one of the laws of indices.

---

**ALTERNATIVE METHOD**

If you find it easier to work in decimal, we have:

$$\int y\, dx = \int x^{0.5} dx = \frac{1}{0.5+1} x^{(0.5+1)} + C$$

$$= \frac{1}{1.5} x^{1.5} + C$$

$$\therefore \int x^{0.5} dx = \frac{1}{1.5} x^{1.5} + C$$

# Indefinite Integration

**e)** $y = \dfrac{1}{\sqrt[5]{x^3}}$

**Solution**

$$y = \dfrac{1}{\sqrt[5]{x^3}}$$

This is not in the right format, so we need to apply relevant laws of indices as:

$$y = \dfrac{1}{\sqrt[5]{x^3}} = \dfrac{1}{x^{\frac{3}{5}}} = x^{-\frac{3}{5}}$$

$$\int y\,dx = \int x^{-\frac{3}{5}}\,dx = \dfrac{1}{-\frac{3}{5}+1}x^{\left(-\frac{3}{5}+1\right)} + C$$

$$= \dfrac{1}{\left(\frac{2}{5}\right)}x^{\frac{2}{5}} + C = \dfrac{5}{2}x^{\frac{2}{5}} + C$$

$$\therefore \int \dfrac{1}{\sqrt[5]{x^3}}\,dx = \dfrac{5}{2}x^{\frac{2}{5}} + C$$

**NOTE**

For this, we could have written the final answer as $\mathbf{2.5\sqrt[5]{x^2} + C}$ using one of the laws of indices.

---

**f)** $y = x^{2.1}$

**Solution**

$$y = x^{2.1}$$

$$\int y\,dx = \int x^{2.1}\,dx = \dfrac{1}{2.1+1}x^{2.1+1} + C$$

$$= \dfrac{1}{3.1}x^{3.1} + C = \dfrac{10}{31}x^{3.1} + C$$

$$\therefore \int x^{2.1}\,dx = \dfrac{10}{31}x^{3.1} + C$$

---

**g)** $y = x$

**Solution**

$$y = x = x^1$$

$$\int y\,dx = \int x\,dx = \dfrac{1}{1+1}x^{1+1} + C$$

$$= \dfrac{1}{2}x^2 + C$$

$$\therefore \int x\,dx = \dfrac{1}{2}x^2 + C$$

**h)** $y = 7$
**Solution**

$$y = 7 = 7x^0$$

$$\int y\,dx = \int 7x^0\,dx = \frac{1}{0+1}7x^{0+1} + C$$

$$= \frac{1}{1}7x + C$$

$$\therefore \int 7\,dx = 7x + C$$

**NOTE**
Recall that we used the style above when discussing differentiation. The same trick is used here too. A number or constant disappears when differentiated, while in integration a number comes with $x$ (or any independent variable used) attached to it.

---

**i)** $y = x^{\cos\frac{1}{6}\pi}$
**Solution**

$$y = x^{\cos\frac{1}{6}\pi}$$

This looks like the power format, except that we have $\cos\frac{1}{6}\pi$ as the power. Recall that the power $n$ can be any real number and $\cos\frac{1}{6}\pi$ is a real number. Hence, just consider it as a number.

$$\int y\,dx = \int x^{\cos\frac{1}{6}\pi}\,dx$$

$$= \frac{1}{\cos\frac{1}{6}\pi + 1}x^{\left(\cos\frac{1}{6}\pi + 1\right)} + C$$

$$\therefore \int x^{\cos\frac{1}{6}\pi}\,dx = \frac{1}{\cos\frac{1}{6}\pi + 1}x^{\left(\cos\frac{1}{6}\pi + 1\right)} + C$$

## Alternative Method

We can simplify the power as $\cos \frac{1}{6}\pi = \frac{\sqrt{3}}{2}$ and then proceed as:

$$\int y\,dx = \int x^{\cos \frac{1}{6}\pi}\,dx = \int x^{\frac{\sqrt{3}}{2}}\,dx$$

$$= \frac{1}{\frac{\sqrt{3}}{2}+1} x^{\left(\frac{\sqrt{3}}{2}+1\right)} + C$$

$$= \left(4 - 2\sqrt{3}\right) x^{\left(\frac{\sqrt{3}+2}{2}\right)} + C$$

$$\therefore \int x^{\cos \frac{1}{6}\pi}\,dx = \left(4 - 2\sqrt{3}\right) x^{\left(\frac{\sqrt{3}+2}{2}\right)} + C$$

### Note
This alternative method is even more complicated than the first method. This said, it will be required to know how to handle such a case without the need to simplify it first, as simplification might be impossible or difficult.

---

**j)** $y = \dfrac{x^{-8}}{x^8}$

**Solution**

$$y = \frac{x^{-8}}{x^8}$$

We need to apply indices here

$$y = \frac{x^{-8}}{x^8} = x^{-8} \cdot x^{-8} = x^{-16}$$

Good to go now as:

$$\int y\,dx = \int \frac{x^{-8}}{x^8}\,dx = \int x^{-16}\,dx$$

$$= \frac{1}{-16+1} x^{-16+1} + C$$

$$= \frac{1}{-15} x^{-15} + C$$

$$\therefore \int x^{-16}\,dx = -\frac{1}{15} x^{-15} + C \text{ OR } -\frac{1}{15x^{15}} + C$$

k) $y = x^{-\frac{2}{7}}$

**Solution**

$$y = x^{-\frac{2}{7}}$$

$$\int y\,dx = \int x^{-\frac{2}{7}}\,dx = \frac{1}{-\frac{2}{7}+1}x^{\left(-\frac{2}{7}+1\right)} + C$$

$$= \frac{1}{\left(\frac{5}{7}\right)}x^{\frac{5}{7}} + C = \frac{7}{5}x^{\left(\frac{5}{7}\right)} + C$$

$$\therefore \int x^{-\frac{2}{7}}\,dx = \frac{7}{5}x^{\frac{5}{7}} + C$$

---

l) $y = x^{\sqrt{2}}$

**Solution**

$$y = x^{\sqrt{2}}$$

$$\int y\,dx = \int x^{\sqrt{2}}\,dx = \frac{1}{\sqrt{2}+1}x^{\left(\sqrt{2}+1\right)} + C$$

$$= \left(\sqrt{2}-1\right)x^{\sqrt{2}+1} + C$$

When we rationalise $\frac{1}{\sqrt{2}+1}$ we will arrive at $\sqrt{2}-1$, but we can use our calculator here and leave the answer without rationalising.

$$\therefore \int x^{\sqrt{2}}\,dx = \left(\sqrt{2}-1\right)x^{\sqrt{2}+1} + C$$

---

We've managed to get started on integration. Before we try more examples, we need to talk about some fundamental rules that will help us with the integration of power functions and many other functions that we will be covering.

**Rule 1** A constant multiplier

The rule states that

$$\boxed{\int k.f(x)\,dx = k\int f(x)\,dx} \tag{15.2}$$

In other words, if a constant $k$ multiplies a function $f(x)$ and the product is integrated, the result is the same as the product of the constant and the integral of the function. Reverse is also possible. With this rule, it is possible to take out the constant from the integral notation. Let's illustrate this very quickly as:

$$\int ax^n\,dx = a\int x^n\,dx = a\left[\frac{1}{n+1}x^{n+1}\right] + C = \frac{a}{n+1}x^{n+1} + C$$

This rule allows us to modify our power function integral as:

If $\boxed{y = ax^n}$ for $n \neq -1$ then $\boxed{\int ax^n\,dx = \frac{a}{n+1}x^{n+1} + C}$ (15.3)

# Indefinite Integration

**Rule 2** Sum and difference of functions

This is also known as the '**linearity rule of integration**'; it states that:

$$\int [f(x) \pm g(x)]\, dx = \int f(x)\, dx \pm \int g(x)\, dx \quad (15.4)$$

In other words, the integral of the sum or difference of functions $f(x)$ and $g(x)$ is equal to the sum (or difference) of integrals of the functions. This will be useful for evaluating polynomials and functions that are made up of two or more terms, where the integral of each term will be taken in turn and summed up together.

Let's try some examples.

## Example 2

A function is defined as $y = b^{-4}r^3 x^4$. Determine the following:

a) $\int y\, dx$  b) $\int y\, dr$  c) $\int y\, db$  d) $\int y\, dt$  e) $\int y\, dA$

What did you get? Find the solution below to double-check your answer.

## Solution to Example 2

### HINT

These examples could be tricky, so do pay attention to some subtle details.

**a)** $\int y\, dx$
**Solution**

$$y = b^{-4} r^3 x^4$$

As we are integrating *wrt x*, the function can be re-written as:

$$y = (b^{-4} r^3) x^4$$

Thus, our constant $k$ here is $b^{-4} r^3$ and can be taken out of the integral notation as:

$$\int y\, dx = \int (b^{-4} r^3) x^4\, dx$$

$$= b^{-4} r^3 \int x^4\, dx$$

$$= b^{-4} r^3 \left[\frac{1}{4+1} x^{4+1}\right] + C$$

$$= b^{-4} r^3 \left[\frac{1}{5} x^5\right] + C$$

$$= \frac{b^{-4} r^3}{5} x^5 + C$$

## NOTE

Here, we did not multiply the constant of integration with $k$, the multiplier. We do not need to, and where it unavoidably happens for a reason, we may change the letter from $C$ to another letter. We can simplify the above and write the final answer as:

$$\therefore \int b^{-4} r^3 x^4 dx = \frac{r^3}{5b^4} x^5 + C$$

---

**b)** $\int y\, dr$
**Solution**

$$y = b^{-4} r^3 x^4$$

As we are integrating *wrt r*, the function can be re-written as:

$$y = \left(b^{-4} x^4\right) r^3$$

Thus, our constant $k$ here is $b^{-4} x^4$ and can be taken out of the integral notation. Thus we have

$$\int y\, dr = \int \left(b^{-4} x^4\right) r^3 dr$$

$$= b^{-4} x^4 \int r^3 dr$$

$$= b^{-4} x^4 \left[\frac{1}{3+1} r^{3+1}\right] + C$$

$$= b^{-4} x^4 \left[\frac{1}{4} r^4\right] + C$$

$$= \frac{b^{-4} x^4}{4} r^4 + C$$

$$\therefore \int b^{-4} r^3 x^4 dr = \frac{x^4}{4b^4} r^4 + C$$

---

**c)** $\int y\, db$
**Solution**

$$y = b^{-4} r^3 x^4$$

As we are integrating *wrt b*, the function can be re-written as:

$$y = \left(r^3 x^4\right) b^{-4}$$

Thus, our constant $k$ here is $r^3 x^4$ and can be taken out of the integral notation. Here we go.

$$\int y\, db = \int \left(r^3 x^4\right) b^{-4}\, db$$

$$= r^3 x^4 \int b^{-4}\, db$$

$$= r^3 x^4 \left[\frac{1}{-4+1} b^{-4+1}\right] + C$$

# Indefinite Integration

$$= r^3 x^4 \left[ \frac{1}{-3} b^{-3} \right] + C$$

$$= -\frac{r^3 x^4}{3} b^{-3} + C$$

$$\therefore \int b^{-4} r^3 x^4 \, db = -\frac{r^3 x^4}{3b^3} + C$$

**d)** $\int y \, dt$
**Solution**

$$y = b^{-4} r^3 x^4$$

As we are integrating *wrt* t, the function can be re-written as:

$$y = (b^{-4} r^3 x^4) \times 1$$
$$= (b^{-4} r^3 x^4) t^0$$

Notice that we have introduced $t^0$ in the expression, since $t^0 = 1$. We've chosen t as we are integrating *wrt* this variable and want to be able to map our solution with the standard integral given. It is also valid to introduce variable t when integrating the constant 1 here. As a result, our constant k here is $b^{-4} r^3 x^4$ and can be taken out of the integral notation. We therefore have

$$\int y \, dt = \int (b^{-4} r^3 x^4) t^0 \, dt$$

$$= b^{-4} r^3 x^4 \int t^0 \, dt$$

$$= b^{-4} r^3 x^4 \left[ \frac{1}{0+1} t^{0+1} \right] + C$$

$$= b^{-4} r^3 x^4 [t^1] + C$$

$$= b^{-4} r^3 x^4 t + C$$

$$\therefore \int b^{-4} r^3 x^4 \, dt = \frac{r^3 x^4 t}{b^4} + C$$

**e)** $\int y \, dA$
**Solution**

$$y = b^{-4} r^3 x^4$$

As we are integrating *wrt* A, the function can be re-written as:

$$y = (b^{-4} r^3 x^4) \times 1$$
$$= (b^{-4} r^3 x^4) A^0$$

Here, we introduced $A^0$ for the same reason mentioned above. Thus:

$$\int y \, db = \int (b^{-4} r^3 x^4) A^0 \, dA$$

$$= b^{-4} r^3 x^4 \int A^0 \, db$$

$$= b^{-4}r^3x^4 \left[\frac{1}{0+1}A^{0+1}\right] + C$$
$$= b^{-4}r^3x^4 \left[A^1\right] + C$$
$$= b^{-4}r^3x^4 A + C$$
$$\therefore \int b^{-4}r^3x^4 \, dA = \frac{r^3x^4 A}{b^4} + C$$

The last two examples show that when integrating with respect to a variable that is not in the expression of the function then it is as though we are integrating a constant. Therefore, the integral in this case will be the constant multiplied by the variable raised to power 1. Also, this underscores the significance and inevitability of placing $dx$, $dt$, etc. after the expression in integral notation. Note this down.

Let's now look at examples of the second rule on the 'integral of sum and difference'.

### Example 3

Solve the following integrals.

a) $\int (x^2 + 3x + 1) \, dx$ 
b) $\int (4t^3 - 9t^2 - 5) \, dt$
c) $\int (2\pi rh + 2\pi r^2) \, dr$ 
d) $\int (2\pi rh + 2\pi r^2) \, dh$

What did you get? Find the solution below to double-check your answer.

### Solution to Example 3

a) $\int (x^2 + 3x + 1) \, dx$
**Solution**
We will apply the second rule here to illustrate it.

$$\int (x^2 + 3x + 1) \, dx = \int x^2 \, dx + \int 3x \, dx + \int 1 \, dx$$
$$= \left(\frac{1}{2+1}x^{2+1} + c_1\right) + \left(\frac{3}{1+1}x^{1+1} + c_2\right) + \left(\frac{1}{0+1}x^{0+1} + c_3\right)$$
$$= \left(\frac{1}{3}x^3 + c_1\right) + \left(\frac{3}{2}x^2 + c_2\right) + \left(x^1 + c_3\right)$$
$$= \left(\frac{1}{3}x^3 + \frac{3}{2}x^2 + x\right) + (c_1 + c_2 + c_3)$$
$$= \frac{1}{3}x^3 + \frac{3}{2}x^2 + x + C$$
$$\therefore \int (x^2 + 3x - 1) \, dx = \frac{1}{3}x^3 + \frac{3}{2}x^2 + x + C$$

# Indefinite Integration

**NOTE**
We've collected the sub-functions in one place and the constants in one place. In general, we only need to show one arbitrary constant per integration. We obtained three constants here because we wanted to illustrate the second rule. This is not really needed, but we will ignore it moving forward and assume the rule allows us to do so.

---

**b)** $\int (4t^3 - 9t^2 - 5)\, dt$

**Solution**
In this case, the two rules will be needed; do spot their applications.

$$\int (4t^3 - 9t^2 - 5)\, dt = \left(\frac{4}{3+1}t^{3+1}\right) - \left(\frac{9}{2+1}t^{2+1}\right) - \left(\frac{5}{0+1}t^{0+1}\right) + C$$

$$= \left(\frac{4}{4}t^4\right) - \left(\frac{9}{3}t^3\right) - 5(t^1) + C$$

$$= t^4 - 3t^3 - 5t + C$$

$$\therefore \int (4t^3 - 9t^2 - 5)\, dt = t^4 - 3t^3 - 5t + C$$

---

**c)** $\int (2\pi r h + 2\pi r^2)\, dr$

**Solution**
In this case, the two rules will be needed. Also, we need to pay attention to the variable we are integrating with respect to, as there are a few letters in this expression. In the current case, our target is variable $r$, and we will maintain just one constant of integration from the start.

$$\int (2\pi r h + 2\pi r^2)\, dr = \int (2\pi r h)\, dr + \int (2\pi r^2)\, dr$$

We have applied the sum of integral rule. We can then apply the multiplier rule to have

$$= 2\pi h \int (r)\, dr + 2\pi \int (r^2)\, dr$$

Now integrate and just assign a single constant as:

$$= 2\pi h \left(\frac{r^2}{2}\right) + 2\pi \left(\frac{r^3}{3}\right) + C$$

We've skipped a few steps, but it is necessary as we need to get up to speed. Now let's tidy this up as:

$$= \pi h r^2 + \frac{2}{3}\pi r^3 + C$$

$$\therefore \int (2\pi r h + 2\pi r^2)\, dr = \pi h r^2 + \frac{2}{3}\pi r^3 + C$$

**d)** $\int \left(2\pi rh + 2\pi r^2\right) dh$

**Solution**

In the current case, we're integrating *wrt h*, so take note of this.

$$\int \left(2\pi rh + 2\pi r^2\right) dh = \int (2\pi rh) \, dh + \int (2\pi r^2) \, dh$$

$$= \int (2\pi r) \, h \, dh + \int (2\pi r^2) \, dh$$

$$= 2\pi r \int (h) \, dh + 2\pi r^2 \int dh$$

Note that $dh$ is the same as writing $1 \times dh$.

$$= 2\pi r \left(\frac{h^2}{2}\right) + 2\pi r^2 \left(\frac{h}{1}\right) + C$$

Now let's tidy this up

$$= \pi r h^2 + 2\pi r^2 h + C$$

$$\therefore \int \left(2\pi rh + 2\pi r^2\right) dh = \pi r h^2 + 2\pi r^2 h + C$$

**OR**

$$\therefore \int \left(2\pi rh + 2\pi r^2\right) dh = \pi r h (h + 2r) + C$$

---

More examples to try.

### Example 4

In each of the following cases, find the function $f(x)$ whose first derivative is stated below.

**a)** $ax - bx^{0.8} + x^{-5}$    **b)** $\frac{\omega}{x^2} - \alpha\sqrt{x} - \beta$

What did you get? Find the solution below to double-check your answer.

### Solution to Example 4

**HINT**

We know that integration will undo differentiation, so this is another way of asking us to integrate.

# Indefinite Integration

**a)** $ax - bx^{0.8} + x^{-5}$

**Solution**

$$\int (ax - bx^{0.8} + x^{-5})\, dx = a\left(\frac{x^2}{2}\right) - b\left(\frac{x^{1.8}}{1.8}\right) + \left(\frac{x^{-4}}{-4}\right) + C$$

$$= \frac{a}{2}x^2 - \frac{b}{1.8}x^{1.8} - \frac{1}{4}x^{-4} + C$$

$$= \frac{a}{2}x^2 - \frac{5b}{9}x^{1.8} - \frac{1}{4}x^{-4} + C$$

$$\therefore \int (ax - bx^{0.8} + x^{-5})\, dx = \frac{a}{2}x^2 - \frac{5}{9}bx^{1.8} - \frac{1}{4}x^{-4} + C$$

**b)** $\frac{\omega}{x^2} - \alpha\sqrt{x} - \beta$

**Solution**

$$\int \left(\frac{\omega}{x^2} - \alpha\sqrt{x} - \beta\right) dx = \int (\omega x^{-2} - \alpha x^{0.5} - \beta)\, dx$$

$$= \omega\left(\frac{x^{-1}}{-1}\right) - \alpha\left(\frac{x^{1.5}}{1.5}\right) - \beta(x) + C$$

$$= -\omega x^{-1} - \frac{\alpha}{1.5}x^{1.5} - \beta x + C$$

$$= -\omega x^{-1} - \frac{1}{1.5}\alpha x^{1.5} - \beta x + C$$

$$\therefore \int \left(\frac{\omega}{x^2} - \alpha\sqrt{x} - \beta\right) dx = -\omega x^{-1} - \frac{2}{3}\alpha x^{1.5} - \beta x + C$$

Another set of examples to try.

## Example 5

Solve the following integrals.

**a)** $\int x^3 \left(x + \frac{9}{x} - 4\right) dx$    **b)** $\int (3t + 5)(t - 2)\, dt$    **c)** $\int \left[(\sqrt{u} + 1)(\sqrt{u} - 1) - \frac{u - 5}{\sqrt{u}}\right] du$

What did you get? Find the solution below to double-check your answer.

### Solution to Example 5

**a)** $\int x^3 \left(x + \frac{9}{x} - 4\right) dx$

**Solution**

This does not look like what we have covered and definitely $\frac{9}{x}$ seems unsolvable. You're partially right, especially with regard to $\frac{9}{x}$. We will express this question differently by expanding and applying the sum-difference rule. Let's go.

$$\int x^3 \left(x + \frac{9}{x} - 4\right) dx = \int (x^4 + 9x^2 - 4x^3) \, dx$$

$$= \left(\frac{x^5}{5}\right) + 9\left(\frac{x^3}{3}\right) - 4\left(\frac{x^4}{4}\right) + C$$

$$= \frac{x^5}{5} + 3x^3 - x^4 + C$$

$$\therefore \int x^3 \left(x + \frac{9}{x} - 4\right) dx = \frac{1}{5}x^5 - x^4 + 3x^3 + C$$

**b)** $\int (3t + 5)(t - 2) dt$

**Solution**

Like the previous example, we need to expand this before proceeding.

$$\int (3t + 5)(t - 2) dt = \int (3t^2 - t - 10) \, dt$$

$$= 3\left(\frac{t^3}{3}\right) - \left(\frac{t^2}{2}\right) - 10\left(\frac{t^1}{1}\right) + C$$

$$= t^3 - \frac{t^2}{2} - 10t + C$$

$$\therefore \int (3t + 5)(t - 2) dt = t^3 - \frac{1}{2}t^2 - 10t + C$$

# Indefinite Integration

**c)** $\int \left[ (\sqrt{u}+1)(\sqrt{u}-1) - \frac{u-5}{\sqrt{u}} \right] du$

**Solution**

Let's expand this first.

$$\int \left[ (\sqrt{u}+1)(\sqrt{u}-1) - \frac{u-5}{\sqrt{u}} \right] du = \int \left[ (u-1) - \frac{u}{\sqrt{u}} - \frac{-5}{\sqrt{u}} \right] du$$

$$= \int \left[ u - 1 - u^{\frac{1}{2}} + 5u^{-\frac{1}{2}} \right] du$$

$$= \left( \frac{u^2}{2} \right) - (u) - \left( \frac{u^{\frac{3}{2}}}{\frac{3}{2}} \right) + 5 \left( \frac{u^{\frac{1}{2}}}{\frac{1}{2}} \right) + C$$

$$= \frac{u^2}{2} - u - \frac{2}{3} u^{\frac{3}{2}} + 10 u^{\frac{1}{2}} + C$$

$$\therefore \int \left[ (\sqrt{u}+1)(\sqrt{u}-1) - \frac{u-5}{\sqrt{u}} \right] du = \frac{u^2}{2} - u - \frac{2}{3} u^{\frac{3}{2}} + 10 u^{\frac{1}{2}} + C$$

---

Before moving to another standard integral, we need to illustrate how to find a particular solution of integration. Up to this point, we have only obtained general solutions, where the constant of integration remains as $C$. Let's go for an example.

### Example 6

The gradient of a curve $C$ is such that $f'(x) = \sqrt{x} \left( \frac{1-3x^{\frac{1}{2}}}{x} \right)$. Find the equation of the curve in the form $y = f(x)$, given that it passes through point $P(4, -1)$.

---

What did you get? Find the solution below to double-check your answer.

### Solution to Example 6

This will require the second rule.

$$y = \int f'(x)\, dx = \int \sqrt{x} \left( \frac{1 - 3x^{\frac{1}{2}}}{x} \right) dx$$

$$= \int x^{\frac{1}{2}} \left( \frac{1}{x} - \frac{3x^{\frac{1}{2}}}{x} \right) dx = \int \frac{x^{\frac{1}{2}}}{x} - \frac{3x^{\frac{1}{2}} \cdot x^{\frac{1}{2}}}{x} dx = \int x^{-\frac{1}{2}} - 3\, dx$$

$$= \left( \frac{x^{\frac{1}{2}}}{\frac{1}{2}} \right) - 3(x) + C$$

Therefore,
$$y = 2x^{\frac{1}{2}} - 3x + C$$

The above represents the general solution, so let's plug in the coordinates of point $P$ as follows:

$$-1 = 2(4)^{\frac{1}{2}} - 3(4) + C$$
$$-1 = 4 - 12 + C$$
$$\therefore C = 7$$

Now let's replace $C$ with its numerical value to find the equation of the curve as:

$$y = 2x^{\frac{1}{2}} - 3x + 7$$
$$\therefore f(x) = 2\sqrt{x} - 3x + 7$$

---

This has indeed been a good test of our understanding of integral standards; let's now take another one.

### 15.2.3.2 Integrating Rational $\frac{1}{x}$ and $\frac{1}{ax+b}$

Recall that we mentioned that the power integral standard is invalid when $n = -1$. We will now consider the rule that can be used for this, which states that

$$\boxed{y = \int \frac{1}{x} dx = \ln|x| + C} \tag{15.5}$$

Natural logarithmic is not valid for negative values. We've therefore used the modulus sign to ensure that only positive values of the logarithmic are taken.

A more general standard is when the denominator is a linear function of the form $(ax + b)$, i.e., $\frac{1}{ax+b}$. For this, we have two cases:

**Case 1** When $b = 0$ (i.e., $\frac{1}{ax}$)

$$\boxed{y = \int \frac{1}{ax} dx = \frac{1}{a} \ln|ax| + C} \tag{15.6}$$

Or simply

$$\boxed{y = \int \frac{1}{ax} dx = \frac{1}{a} \ln|x| + C} \tag{15.7}$$

# Indefinite Integration

Notice that we have removed the constant $a$ from the natural logarithm. This is because $\frac{d}{dx}(\ln x) = \frac{1}{x}$ and similarly $\frac{d}{dx}(\ln ax) = \frac{1}{ax} \times a = \frac{1}{x}$. The latter is based on the chain rule, but we can show that this is true using a different approach as:

$$\frac{d}{dx}(\ln ax) = \frac{d}{dx}(\ln a) + \frac{d}{dx}(\ln x)$$
$$= 0 + \frac{1}{x}$$
$$= \frac{1}{x}$$

Here we have simplified the expression using the product rule of logarithm and take the derivative. We can use a similar approach in integration as:

$$\int \frac{1}{ax} dx = \int \frac{1}{a}\left(\frac{1}{x}\right) dx$$
$$= \frac{1}{a}\int \frac{1}{x} dx = \frac{1}{a}\ln|x| + C$$

This is a case of considering $1/a$ as the coefficient, just like $a$ is the coefficient in $\int \frac{a}{x} dx$. In each case, the coefficient can be taken out of the integral sign to be a constant multiplier. The issue, however, is that if we simplify the first result of the integral, we will not recover the second result. Let's show this.

$$y = \int \frac{1}{ax} dx = \frac{1}{a}\ln|ax| + C$$

Using the logarithmic rule for the RHS as:

$$\frac{1}{a}(\ln|a| + \ln|x|) + C = \frac{1}{a}\ln|a| + \frac{1}{a}\ln|x| + C$$
$$= \frac{1}{a}\ln|x| + \frac{1}{a}\ln|a| + C$$

But $\frac{1}{a}\ln|a|$ is just a constant like $C$, the constant of integration. We can combine them together and represent them with $A$, another constant as:

$$\frac{1}{a}\ln|ax| + C = \frac{1}{a}\ln|x| + A$$

Changing from one constant to another is typical in integration. We will arrive at the same result using the two options, and you can try this with definite integration to confirm that the numerical value of both options is the same. There is another approach to get us the second option, but this will be covered in the next chapter.

**Case 2** When $b \neq 0$ (i.e., $\frac{1}{ax+b}$)

$$\boxed{y = \int \frac{1}{ax+b} dx = \frac{1}{a}\ln|ax+b| + C} \quad (15.8)$$

Let's make the following notes about the above two cases:

**Note 1** This is a function of a function style expression. The inner function must be linear.
**Note 2** We will integrate $\frac{1}{ax}$ or $\frac{1}{ax+b}$ as we do for $\frac{1}{x}$. We will then divide our answer with the derivative of the inner function $(ax+b)$ or simply $a$, whichever is easier to spot.

We will show later under the section on definite integration the two possible methods.

### 15.2.3.3 Integrating Exponential Functions $e^x$

The rule for this is

$$y = \int e^x dx = e^x + C \tag{15.9}$$

This is unique because it aligns with its derivative, but it's not surprising since we are looking for an antiderivative. If the base is not $e$, then we have

$$y = \int a^x dx = \frac{a^x}{\ln a} + C \tag{15.10}$$

What happens if the power is a linear function of the form $ax + b$, i.e., $e^{ax+b}$. For this, we have two cases:

**Case 1** When $b = 0$ (i.e., $e^{ax}$)

$$y = \int e^{ax} dx = \frac{1}{a} e^{ax} + C \tag{15.11}$$

**Case 2** When $b \neq 0$ (i.e., $e^{ax+b}$)

$$y = \int e^{ax+b} dx = \frac{1}{a} e^{ax+b} + C \tag{15.12}$$

Let's make the following notes:

**Note 1** This is a function of a function style expression. The inner function must be linear. In other words, this standard is not valid for use in the case of $e^{3x^2-1}$ for example.
**Note 2** We will integrate this $e^{ax}$ or $e^{ax+b}$ as we do for $e^x$. We will then divide our answer with the derivative of the inner function $(ax+b)$ or the coefficient of $x$.

# Indefinite Integration

Let's try some examples involving exponential and rational functions.

## Example 7

Evaluate the following:

a) $\int \left(3e^x + \frac{1}{3x}\right) dx$  b) $\int 6e^{1-2x} dx$  c) $\int \frac{4}{5x-1} dx$  d) $\int \left(\frac{1}{\sqrt{e^{x-2}}}\right) dx$

What did you get? Find the solution below to double-check your answer.

## Solution to Example 7

### HINT

In this example, we will go straight to the solution since we've illustrated how to apply the standard when there is a function of a function.

a) $\int \left(3e^x + \frac{1}{3x}\right) dx$
**Solution**

$$\int \left(3e^x + \frac{1}{3x}\right) dx = \int \left[3(e^x) + \frac{1}{3}\left(\frac{1}{x}\right)\right] dx$$

$$= 3(e^x) + \frac{1}{3} \ln |x| + C$$

$$\therefore \int \left(3e^x + \frac{1}{3x}\right) dx = 3e^x + \frac{1}{3} \ln |x| + C$$

b) $\int 6e^{1-2x} dx$
**Solution**

$$\int 6e^{1-2x} dx = 6 \int e^{1-2x} dx$$

$$= 6 \left[\frac{e^{1-2x}}{-2}\right] + C = \frac{6}{-2} [e^{1-2x}] + C$$

$$\therefore \int (6e^{1-2x} dx) dx = -3e^{1-2x} + C$$

c) $\int \frac{4}{5x-1} dx$
**Solution**

$$\int \frac{4}{5x-1} dx = 4 \int \frac{1}{5x-1} dx$$

$$= 4 \left[\frac{\ln |5x-1|}{5}\right] + C$$

$$\therefore \int \left(\frac{4}{5x-1}\right) dx = \frac{4}{5} (\ln |5x-1|) + C$$

**d)** $\int \left( \frac{1}{\sqrt{e^{x-2}}} \right) dx$

**Solution**

The expression will need to be simplified first before being mapped to the exponential integral standard above.

$$\int \left( \frac{1}{\sqrt{e^{x-2}}} \right) dx = \int \left[ \frac{1}{(e^{x-2})^{0.5}} \right] dx$$

$$= \int \left[ (e^{x-2})^{-0.5} \right] dx = \int e^{1-0.5x} dx$$

$$= \frac{1}{-0.5} e^{1-0.5x} + C = -2e^{1-0.5x} + C$$

$$\therefore \int \left( \frac{1}{\sqrt{e^{x-2}}} \right) dx = -2e^{1-0.5x} + C$$

### 15.2.3.4 Integrating Trigonometric Functions

Since integration is anti-derivative, it is easy to establish the integral of trigonometric functions.

**1)** $\sin x$, $\sin ax$, and $\sin(ax+b)$

The rule for this is

$$\boxed{y = \int \sin x \, dx = -\cos x + C} \tag{15.13}$$

when $x$ is replaced by a linear function, we have

$$\boxed{y = \int \sin ax \, dx = -\frac{1}{a} \cos ax + C} \quad \text{AND} \quad \boxed{y = \int \sin(ax+b) \, dx = -\frac{1}{a} \cos(ax+b) + C} \tag{15.14}$$

**2)** $\cos x$, $\cos ax$ and $\cos(ax+b)$

The rule for this is

$$\boxed{y = \int \cos x \, dx = \sin x + C} \tag{15.15}$$

when $x$ is replaced by a linear function, we have

$$\boxed{y = \int \cos ax \, dx = \frac{1}{a} \sin ax + C} \quad \text{AND} \quad \boxed{y = \int \cos(ax+b) \, dx = \frac{1}{a} \sin(ax+b) + C} \tag{15.16}$$

# Indefinite Integration

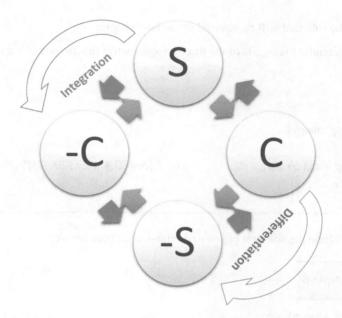

**FIGURE 15.4** Cyclic process of the sine and cosine functions illustrated.

The diagram (Figure 15.4) shows the cyclic process of the sine and cosine functions.

**3) $\tan x$, $\tan ax$, and $\tan(ax+b)$**

The rule for this is

$$\boxed{y = \int \tan x \, dx = \ln|\sec x| + C} \qquad (15.17)$$

When $x$ is replaced by a linear function, we have

$$\boxed{y = \int \tan ax \, dx = \frac{1}{a}\ln|\sec ax| + C} \quad \text{AND} \quad \boxed{y = \int \tan(ax+b)\, dx = \frac{1}{a}\ln|\sec(ax+b)| + C}$$
$$(15.18)$$

In the above integrals of the tangent function, we can remove the modulus sign and state that the integral is valid only for $-\frac{\pi}{2} < x < \frac{\pi}{2}$, as this is where the secant function is positive. This is because the natural logarithm of negative numbers yields complex numbers. The same for the integral of the secant function. As for cosecant and cotangent functions, the validity of the variable is $0 < x < \pi$.

Note that $\int \tan x \, dx$ can also be given as $-\ln|\cos x|$, by simplifying the above as

$$\ln|\sec x| = \ln\left|\frac{1}{\cos x}\right| = \ln\left|(\cos x)^{-1}\right| = -\ln|\cos x|$$

Alternatively, we can say that

$$\int \tan x \, dx = \int \frac{\sin x}{\cos x} \, dx = -\int (-\sin x) \times \frac{1}{\cos x} \, dx = -\ln|\cos x| + C$$

This is based on the rule that will be covered in the next chapter.

The above are the standard integrals of the three trigonometric functions. Let's try some examples.

### Example 8

Solve the following integrals.

a) $\int (\cos \varphi - \sin \varphi + \tan \varphi) \, d\varphi$

b) $\int [\cos 3\theta + 4 \sin (2\theta - 3)] \, d\theta$

c) $\int \left[\pi - 10 \tan \left(\frac{\pi}{3} - 2x\right)\right] dx$

What did you get? Find the solution below to double-check your answer.

### Solution to Example 8

a) $\int (\cos \varphi - \sin \varphi + \tan \varphi) \, d\varphi$
**Solution**

$$\int (\cos \varphi - \sin \varphi + \tan \varphi) \, d\varphi = \sin \varphi - (-\cos \varphi) + \ln |\sec \varphi| + C$$

$$= \sin \varphi + \cos \varphi + \ln |\sec \varphi| + C$$

$$\therefore \int (\cos \varphi - \sin \varphi + \tan \varphi) \, d\varphi = \sin \varphi + \cos \varphi + \ln |\sec \varphi| + C$$

b) $\int [\cos 3\theta + 4 \sin (2\theta - 3)] \, d\theta$
**Solution**

$$\int [\cos 3\theta + 4 \sin (2\theta - 3)] \, d\theta = \frac{1}{3} \sin 3\theta - 4 \times \frac{\cos (2\theta - 3)}{2} + C$$

$$= \frac{1}{3} \sin 3\theta - 2 \cos (2\theta - 3) + C$$

$$\therefore \int [\cos 3\theta + 4 \sin (2\theta - 3)] \, d\theta = \frac{1}{3} \sin 3\theta - 2 \cos (2\theta - 3) + C$$

c) $\int \left[\pi - 10 \tan \left(\frac{\pi}{3} - 2x\right)\right] dx$
**Solution**

$$\int \left[\pi - 10 \tan \left(\frac{\pi}{3} - 2x\right)\right] dx = \pi x - 10 \left(\frac{\ln \left|\sec \left(\frac{\pi}{3} - 2x\right)\right|}{-2}\right) + C$$

$$= \pi x + 5 \ln \left|\sec \left(\frac{\pi}{3} - 2x\right)\right| + C$$

$$\therefore \int \left[\pi - 10 \tan \left(\frac{\pi}{3} - 2x\right)\right] dx = \pi x + 5 \ln \left|\sec \left(\frac{\pi}{3} - 2x\right)\right| + C$$

# Indefinite Integration

## TABLE 15.2
### Standard Integrals

| $y$ | $\int y\,dx$ | $y$ | $\int y\,dx$ |
|---|---|---|---|
| $\sin x$ | $-\cos x + C$ | $\sin ax$ | $-\dfrac{1}{a}\cos ax + C$ |
| $\cos x$ | $\sin x + C$ | $\cos ax$ | $\dfrac{1}{a}\sin ax + C$ |
| $\tan x$ | $\ln|\sec x| + C$ | $\tan ax$ | $\dfrac{1}{a}\ln|\sec ax| + C$ |
| $\sec x$ | $\ln|\tan x + \sec x| + C$ | $\sec ax$ | $\dfrac{\ln|\tan ax + \sec ax|}{a} + C$ |
| $\operatorname{cosec} x$ | $-\ln|\cot x + \operatorname{cosec} x| + C$ | $\operatorname{cosec} ax$ | $-\dfrac{\ln|\cot ax + \operatorname{cosec} ax|}{a} + C$ |
| $\cot x$ | $\ln|\sin x| + C$ | $\cot ax$ | $\dfrac{\ln|\sin ax|}{a} + C$ |
| $\sec^2 x$ | $\tan x + C$ | $\sec^2 ax$ | $\dfrac{\tan ax}{a} + C$ |
| $\operatorname{cosec}^2 x$ | $-\cot x + C$ | $\operatorname{cosec}^2 ax$ | $-\dfrac{1}{a}\cot ax + C$ |
| $\sec x \tan x$ | $\sec x + C$ | $\sec ax \tan ax$ | $\dfrac{1}{a}\sec ax + C$ |
| $\operatorname{cosec} x \cot x$ | $-\operatorname{cosec} x + C$ | $\operatorname{cosec} ax \cot ax$ | $-\dfrac{1}{a}\operatorname{cosec} ax + C$ |
| $\dfrac{1}{\sqrt{1-x^2}}$ | $\sin^{-1} x + C$ | $\dfrac{-1}{x\sqrt{x^2-1}}$ | $\operatorname{cosec}^{-1} x + C$ |
| $\dfrac{-1}{\sqrt{1-x^2}}$ | $\cos^{-1} x + C$ | $\dfrac{1}{x\sqrt{x^2-1}}$ | $\sec^{-1} x + C$ |
| $\dfrac{1}{1+x^2}$ | $\tan^{-1} x + C$ | $\dfrac{-1}{1+x^2}$ | $\cot^{-1} x + C$ |

There is integration of trigonometric functions that would be difficult to do without relying on the derivatives of their inverse. Nevertheless, it is important that we know these integrals. It would therefore be helpful to provide a list of the common integral standards, usually called the **table of standard integrals** (Table 15.2). The main gist is that one needs to be able to map the relationship between a function and its integral and observe the transformation that ensues. We will also add others that, though we may not use them here, they are useful for integration.

Let's try some examples.

## Example 9

Solve the following:

a) $\int (\sec^2 x - \csc^2 x)\, dx$

b) $\int \left(\frac{1}{\cos^2 x} + \frac{1}{\sin^2 x}\right) dx$

c) $\int \left(\frac{\cos x}{\sin^2 x}\right) dx$

d) $\int \sec^2 x (1 + \cot^2 x)\, dx$

What did you get? Find the solution below to double-check your answer.

## Solution to Example 9

**a)** $\int \sec^2 x - \csc^2 x\, dx$
**Solution**

$$\int (\sec^2 x - \csc^2 x)\, dx = \tan x - (-\cot x) + C$$

$$= \tan x + \cot x + C$$

$$\therefore \int (\sec^2 x - \csc^2 x)\, dx = \tan x + \cot x + C$$

**NOTE**
We could have obtained $\int (\sec^2 x - \csc^2 x)\, dx = 2\csc 2x + C$. Starting with the expression in the integral, we have

$$\sec^2 x - \csc^2 x = \frac{1}{\cos^2 x} - \frac{1}{\sin^2 x} = \frac{\sin^2 x - \cos^2 x}{\cos^2 x \cdot \sin^2 x}$$

$$= \frac{-(\cos^2 x - \sin^2 x)}{(\sin x \cos x)^2} = \frac{-(\cos 2x)}{\left(\frac{1}{2}\sin 2x\right)^2}$$

$$= \frac{-\cos 2x}{\frac{1}{4}(\sin 2x)(\sin 2x)} = \frac{-4\cos 2x}{(\sin 2x)(\sin 2x)}$$

$$= -4\left(\frac{\cos 2x}{\sin 2x}\right)\left(\frac{1}{\sin 2x}\right) = -4\cot 2x \csc 2x$$

Thus

$$\int (\sec^2 x - \csc^2 x)\, dx = \int (-4 \cot 2x \csc 2x)\, dx$$

$$= -4 \int (\cot 2x \csc 2x)\, dx$$

$$= -4\left(-\frac{\csc 2x}{2}\right) + C$$

$$\therefore \int (\sec^2 x - \csc^2 x)\, dx = 2\csc 2x + C$$

# Indefinite Integration

**NOTE**

This example shows that the answer to integral can vary depending on the approach used, though they represent the same. We can equally take the first answer obtained and express it as $2\operatorname{cosec} 2x$. Also note that, if you use a calculating device to check your answer, do not be alarmed if it is very different from yours!

---

**b)** $\int \left( \dfrac{1}{\cos^2 x} + \dfrac{1}{\sin^2 x} \right) dx$

**Solution**

Notice that this is quite similar to the previous example, but let's go through it together as:

$$\int \left( \frac{1}{\cos^2 x} + \frac{1}{\sin^2 x} \right) dx = \int (\sec^2 x + \operatorname{cosec}^2 x)\, dx$$

$$= \tan x + (-\cot x) + C = \tan x - \cot x + C$$

$$\therefore \int \left( \frac{1}{\cos^2 x} + \frac{1}{\sin^2 x} \right) dx = \tan x - \cot x + C$$

By the same method, you should be able to show that $\int \left( \dfrac{1}{\cos^2 x} + \dfrac{1}{\sin^2 x} \right) dx = -2 \cot 2x + C$.

---

**c)** $\int \left( \dfrac{\cos x}{\sin^2 x} \right) dx$

**Solution**

$$\int \left( \frac{\cos x}{\sin^2 x} \right) dx = \int \left( \frac{\cos x}{\sin x} \times \frac{1}{\sin x} \right) dx$$

$$= \int (\cot x \cdot \operatorname{cosec} x)\, dx$$

$$= -\operatorname{cosec} x + C$$

$$\therefore \int \left( \frac{\cos x}{\sin^2 x} \right) dx = -\operatorname{cosec} x + C$$

---

**d)** $\int \sec^2 x (1 + \cot^2 x)\, dx$

**Solution**

$$\int \sec^2 x (1 + \cot^2 x)\, dx = \int \sec^2 x + \sec^2 x \cdot \cot^2 x\, dx$$

$$= \int \sec^2 x + \frac{1}{\cos^2 x} \cdot \frac{\cos^2 x}{\sin^2 x} dx$$

$$= \int \sec^2 x + \frac{1}{\sin^2 x} dx$$

$$= \int \sec^2 x + \operatorname{cosec}^2 x\, dx$$

$$= \tan x - \cot x + C$$

$$\therefore \int \sec^2 x (1 + \cot^2 x) = \tan x - \cot x + C$$

Another set of examples to try.

### Example 10

Determine $y$ in the following:

a) $\frac{dy}{dx} = 10 \sec 5x \tan 5x$

b) $\frac{dy}{d\theta} = \sec^2(7\theta - 1) + \csc^2(\pi - 5\theta)$

c) $\frac{dy}{d\omega} = \csc(2\omega t - \alpha) \cot(2\omega t - \alpha)$

d) $\frac{dy}{dx} = 5 \csc(2x - 1)$

e) $\frac{dy}{dx} = 10 \cot 5x - \sec(x + 6)$

What did you get? Find the solution below to double-check your answer.

### Solution to Example 10

**HINT**

This is another way to ask us to integrate. Let's illustrate this using question (a). Given that $\frac{dy}{dx} = \mathbf{10 \sec 5x \tan 5x}$ then (by multiplying both sides by $\boldsymbol{dx}$)

$$dy = (10 \sec 5x \tan 5x)\, dx$$

Taking the integral of both sides, we have

$$\int dy = \int (10 \sec 5x \tan 5x)\, dx$$

$$\therefore y = \int (10 \sec 5x \tan 5x)\, dx$$

Since

$$\int dy = \int 1 \times dy = y$$

We will not show this for each example; we are only illustrating this to show that the instruction is the integration of the RHS.

# Indefinite Integration

**a)** $\frac{dy}{dx} = 10 \sec 5x \tan 5x$

**Solution**

$$y = \int (10 \sec 5x \tan 5x)\, dx$$

$$= 10 \int (\sec 5x \tan 5x)\, dx$$

$$= 10 \left(\frac{\sec 5x}{5}\right) + C$$

$$\therefore \int (10 \sec 5x \tan 5x)\, dx = 2 \sec 5x + C$$

---

**b)** $\frac{dy}{d\theta} = \sec^2(7\theta - 1) + \operatorname{cosec}^2(\pi - 5\theta)$

**Solution**

$$y = \int \sec^2(7\theta - 1) + \operatorname{cosec}^2(\pi - 5\theta)\, d\theta$$

$$= \int \sec^2(7\theta - 1)\, d\theta + \int \operatorname{cosec}^2(\pi - 5\theta)\, d\theta$$

$$= \left(\frac{\tan(7\theta - 1)}{7}\right) + \left(\frac{-\cot(\pi - 5\theta)}{-5}\right) + C$$

$$\therefore \int \sec^2(7\theta - 1) + \operatorname{cosec}^2(\pi - 5\theta)\, d\theta = \frac{1}{7}\tan(7\theta - 1) + \frac{1}{5}\cot(\pi - 5\theta) + C$$

---

**c)** $\frac{dy}{d\omega} = \operatorname{cosec}(2\omega t - \alpha) \cot(2\omega t - \alpha)$

**Solution**

$$y = \int \operatorname{cosec}(2\omega t - \alpha) \cot(2\omega t - \alpha)\, d\omega$$

$$= \frac{-\operatorname{cosec}(2\omega t - \alpha)}{2t} + C$$

$$\therefore \int \operatorname{cosec}(2\omega t - \alpha) \cot(2\omega t - \alpha)\, d\omega = -\frac{1}{2t} \operatorname{cosec}(2\omega t - \alpha) + C$$

---

**d)** $\frac{dy}{dx} = 5 \operatorname{cosec}(2x - 1)$

**Solution**

$$y = \int 5 \operatorname{cosec}(2x - 1)\, dx$$

$$= 5 \int \operatorname{cosec}(2x - 1)\, dx$$

$$= -5 \left(\frac{\ln|\cot(2x - 1) + \operatorname{cosec}(2x - 1)|}{2}\right) + C$$

$$\therefore \int 5 \operatorname{cosec}(2x - 1)\, dx = -\frac{5}{2} \ln|\cot(2x - 1) + \operatorname{cosec}(2x - 1)| + C$$

e) $\frac{dy}{dx} = 10\cot 5x - \sec(x+6)$

**Solution**

$$y = \int 10\cot 5x - \sec(x+6)\, dx$$
$$= 10\int \cot 5x - \sec(x+6)\, dx$$
$$= \frac{10}{5}\ln|\sin 5x| - (\ln|\sec(x+6) + \tan(x+6)|) + C$$
$$= 2\ln|\sin 5x| - (\ln|\sec(x+6) + \tan(x+6)|) + C$$
$$= \ln\left|\sin^2 5x\right| - (\ln|\sec(x+6) + \tan(x+6)|) + C$$
$$= \ln\left|\frac{\sin^2 5x}{\sec(x+6) + \tan(x+6)}\right| + C$$
$$\therefore \int 10\cot 5x - \sec(x+6)\, dx = \ln\left|\frac{\sin^2 5x}{\sec(x+6) + \tan(x+6)}\right| + C$$

## 15.3 CHAPTER SUMMARY

1) Integration is a direct inverse of differentiation (or anti-derivative).

2) An indefinite integration is represented by $\int y\, dx$. The notation $\int$, which looks like an elongated $S$ (denoting sum), is called '**the integral of**'. It implies that the reverse of differentiation should be performed, just like the notation $\sqrt{\ }$ is used to undo the square. $\int y\, dx$ is read as '**the integral of $y$ with respect to $x$**'. The '$dx$' must be included otherwise the integration is meaningless. In this case, $x$ is the independent variable. If the independent variable is time ($t$), we will have $dt$ and if it is volume ($V$), we will have $dV$ and so on.

3) We always add a constant $C$ to the result of any integration. This is called the **constant of integration** or **arbitrary constant**, and it stands in place of a constant that has been lost when differentiating. It is common to use capital letter $C$, but other letters can be used.

4) An expression with a constant of integration is called the **general solution**.

5) An expression obtained, having replaced $C$ with a numerical value, is called a **particular solution**.

6) The standard integrals:

   a) Integrating powers of $x$ or $x^n$ ($n \neq -1$)

   If $\boxed{y = x^n}$ then $\boxed{\int x^n dx = \frac{1}{n+1}x^{n+1} + C}$

   If $\boxed{y = ax^n}$ for $n \neq -1$ then $\boxed{\int ax^n dx = \frac{a}{n+1}x^{n+1} + C}$

# Indefinite Integration

b) Sum and difference of functions

$$\int [f(x) \pm g(x)]\,dx = \int f(x)\,dx \pm \int g(x)\,dx$$

c) Integrating rational $\frac{1}{x}$ and $\frac{1}{ax+b}$

$$y = \int \frac{1}{x}\,dx = \ln|x| + C$$

$$y = \int \frac{1}{ax}\,dx = \frac{1}{a}\ln|ax| + C$$

Or simply

$$y = \int \frac{1}{ax}\,dx = \frac{1}{a}\ln|x| + C$$

$$y = \int \frac{1}{ax+b}\,dx = \frac{1}{a}\ln|ax+b| + C$$

d) Integrating exponential functions $e^x$

$$y = \int e^x\,dx = e^x + C$$

$$y = \int a^x\,dx = \frac{a^x}{\ln a} + C$$

$$y = \int e^{ax}\,dx = \frac{1}{a}e^{ax} + C$$

$$y = \int e^{ax+b}\,dx = \frac{1}{a}e^{ax+b} + C$$

e) $\sin x$, $\sin ax$, and $\sin(ax+b)$

$$y = \int \sin x\,dx = -\cos x + C$$

when $x$ is replaced by a linear function, we have

$$y = \int \sin ax\,dx = -\frac{1}{a}\cos ax + C \quad \text{AND}$$

$$y = \int \sin(ax+b)\,dx = -\frac{1}{a}\cos(ax+b) + C$$

f) $\cos x$, $\cos ax$, and $\cos(ax+b)$

$$y = \int \cos x\,dx = \sin x + C$$

when $x$ is replaced by a linear function, we have

$$\boxed{y = \int \cos ax \, dx = \frac{1}{a} \sin ax + C} \quad \text{AND}$$

$$\boxed{y = \int \cos(ax+b) \, dx = \frac{1}{a} \sin(ax+b) + C}$$

g) $\tan x$, $\tan ax$, and $\tan(ax+b)$

$$\boxed{y = \int \tan x \, dx = \ln|\sec x| + C}$$

when $x$ is replaced by a linear function, we have

$$\boxed{y = \int \tan ax \, dx = \frac{1}{a} \ln|\sec ax| + C} \quad \text{AND}$$

$$\boxed{y = \int \tan(ax+b) \, dx = \frac{1}{a} \ln|\sec(ax+b)| + C}$$

\*\*\*\*

## 15.4 FURTHER PRACTICE

To access complementary contents, including additional exercises, please go to www.dszak.com.

# 16 Definite Integration

*Learning Outcomes*

Once you have studied the content of this chapter, you should be able to:

- Discuss definite integrals
- Determine the area bounded by a curve and a line
- Determine the area bounded by two curves
- Discuss integrating to infinity

## 16.1 INTRODUCTION

This chapter will, in continuation of our discussion on integration, focus on definite integration. This is used to determine the area under a curve between two specified limits and the area bounded by two curves, linear or non-linear.

## 16.2 DEFINITE INTEGRAL

At the start of Chapter 15, we briefly introduced the definite integral as the second type of integration. Suppose we have a function $f(x)$ and we are required to integrate it from $x = a$ to $x = b$, we will represent this as:

$$\int_a^b f(x)\, dx$$

The above represents an area covered by the curve $f(x)$, $x = a$, $x = b$, and the horizontal axis as shown in Figure 16.1.

To determine the definite integral above, follow these steps:

**Step 1:** Find $\int f(x)\, dx$.
**Step 2:** Replace $x$ with the upper limit $b$ in the integral found in step 1. Do the same with the lower limit $a$.
**Step 3:** Subtract the lower limit value from the upper limit. This process will cancel out the constant of integration and as such we do not generally include the $C$ in definite integration. Hence, step 1 is done without $C$.
**Step 4:** The result of step 4 represents the area under the curve.

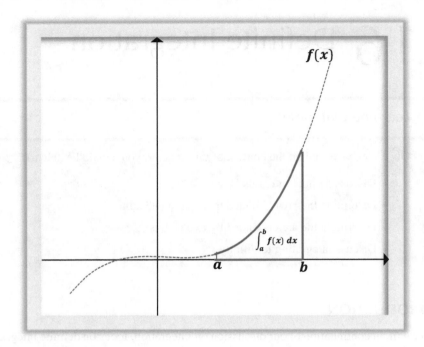

**FIGURE 16.1** Definite integral illustrated.

More compactly, we write this process as:

$$\int_a^b f(x)\,dx = \left[\int f(x)\,dx\right]_a^b = \left[\int f(x)\,dx\right]^b - \left[\int f(x)\,dx\right]^a \quad (16.1)$$

The choice of the square brackets [ ] is in line with the standard notation and should be used in this case. Subsequent calculations, including substitution of the limits, are generally carried out using round brackets ( ).

In Chapter 14, we learned that when we differentiate the function of position $s(t)$ with respect to time, we obtain a function of velocity $v(t)$. Similarly, if we differentiate $v(t)$, we get the function of acceleration. The reverse holds too. In other words, when we integrate the function of acceleration $a(t)$ with respect to time, we obtain a function of velocity. The same can be said about velocity: when it's integrated, we obtain displacement. What is more important is that as differentiation gives a slope, the integration gives us the area.

Given that $s$, $v$, $a$, and $j$ represent displacement, velocity, acceleration, and jerk, respectively, Table 16.1 provides a summary of the differentiation and integration of these physical quantities.

Consider an object which starts with an initial velocity of $u$ ms$^{-1}$ at point $A$ and accelerates uniformly until it reached a final velocity $v$ ms$^{-1}$. It maintained this velocity between point $B$ and point $C$ before decelerating to a rest position at point $D$. The journey is illustrated using a velocity–time graph, as shown in Figure 16.2. From our studies on mechanics, we know that the slope of $AB$ is the acceleration, and the slope of $CD$ is the deceleration. In calculus, this is the derivative of line $AB$ and $CD$. Similarly, the area under the curve $ABCD$ is used to find displacement in mechanics. If the curve is defined by a function $f(x)$, we can find displacement using the definite integral $\int_0^{t_D} f(x)\,dx$, where $t_D$ is the time at point $D$. We use 0 as our lower limit because the time at point A is zero, i.e. $t_A = 0$.

# Definite Integration

## TABLE 16.1
### Application of Calculus in Mechanics Illustrated

| Differentiation | | Integration | |
|---|---|---|---|
| $\frac{d}{dt}[s]$ | Velocity | $\int v \, dt$ | Displacement |
| $\frac{d}{dt}[v]$ | Acceleration | $\int a \, dt$ | Velocity |
| $\frac{d}{dt}[a]$ | Jerk | $\int j \, dt$ | Acceleration |

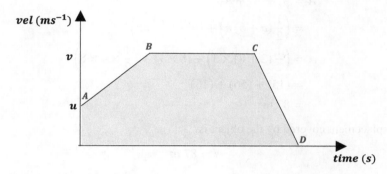

**FIGURE 16.2** Velocity–time graph analysed.

We will find this type of problem in mechanics, where the area under the curve represents a physical quantity. Often, this is limited to when we have straight lines that bound the area. Calculations typically involve splitting the shapes into multiple parts and using formulas for standard shapes, such as trapeziums, triangles, and rectangles. We will prove our use of definite integration to find such an area and compare this with the answer we get from the analytical or numerical calculation.

Let's try an example to illustrate this.

### Example 1

An object on a straight frictionless road starts with an initial velocity of 2 ms$^{-1}$ and accelerates uniformly for 3 s until it reaches a velocity of 8 ms$^{-1}$. The object maintains this velocity for 7 s before it starts to decelerate uniformly until it comes to rest after spending a total of 14 s in this journey. Determine the total distance covered by the object using:

a) graphical method
b) integration

What did you get? Find the solution below to double-check your answer.

### Solution to Example 1

a) Using graphical method
**Solution**
We first need to sketch the motion as shown in Figure 16.3.

The overall shape defined by the journey is not one of the standards and we've decided to divide this into three shapes with area $A_1$ (trapezium), $A_2$ (rectangle), and $A_3$ (triangle).

The area in $v - t$ graph represents the distance (ideally, we say the displacement). Therefore

$$A = A_1 + A_2 + A_3$$
$$= \left(\frac{1}{2}(a+b)h\right) + (bh) + \left(\frac{1}{2}bh\right)$$
$$= \left(\frac{1}{2}(2+8) \times 3\right) + (8 \times 7) + \left(\frac{1}{2} \times 4 \times 8\right)$$
$$= (15) + (56) + (16)$$
$$= 87 \text{ m}^2$$

Hence, the displacement covered by the object is

$$s = 87 \text{ m}$$

**NOTE**
We give A in $m^2$ because it represents area and s in $m$ as it is a displacement.

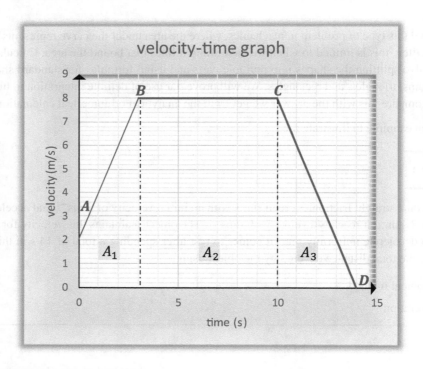

**FIGURE 16.3** Solution to Example 1(a).

Definite Integration

**b) Using integration**
**Solution**
We have re-reproduced the graph of the motion in Figure 16.4.

In this case, we note that the entire graph is formed from three different linear functions, similar to the three shapes in the first method described above. The functions are:

$$f_1(t) = 2t + 2$$
$$f_2(t) = 8$$
$$f_1(t) = -2t + 28$$

We now need to find the integral of each function within their limits, as follows.

**First function**

$$\int_0^3 f_1(t)\,dx = \int_0^3 (2t+2)\,dt = [t^2 + 2t]_0^3$$
$$= [t^2 + 2t]^3 - [t^2 + 2t]^0$$
$$= \{3^2 + 2(3)\} - \{0^2 + 2(0)\}$$
$$= \{15\} - \{0\}$$

$$\therefore \int_0^3 f_1(t)\,dt = 15$$

**Second function**

$$\int_3^{10} f_2(t)\,dt = \int_3^{10} (8)\,dt = [8t]_3^{10}$$
$$= [8t]^{10} - [8t]^3$$
$$= \{8(10)\} - \{8(3)\}$$
$$= \{80\} - \{24\}$$

$$\therefore \int_3^{10} f_2(t)\,dt = 56$$

**Third function**

$$\int_{10}^{14} f_3(t)\,dt = \int_{10}^{14} (-2t+28)\,dt = [-t^2 + 28t]_{10}^{14}$$
$$= [-t^2 + 28t]^{14} - [-t^2 + 28t]^{10}$$
$$= \{-14^2 + 28(14)\} - \{-10^2 + 28(10)\}$$
$$= \{196\} - \{180\}$$

$$\therefore \int_{10}^{14} f_3(t)\,dt = 16$$

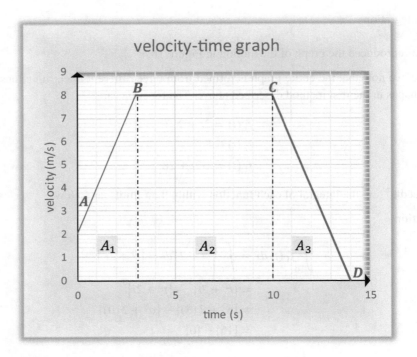

**FIGURE 16.4** Solution to Example 1(b).

Thus, the area covered by the function combined is

$$\int_0^{14} f(t)\,dt = \int_0^3 f_1(t)\,dt + \int_3^{10} f_2(t)\,dt + \int_{10}^{14} f_3(t)\,dt$$
$$= 15 + 56 + 16 = 87$$

$$\therefore \int_0^{14} f(t)\,dt = 87$$

### 16.2.1 AREA UNDER A CURVE

This is indeed the primary use of definite integration: to find the area covered by a shape for which we do not have a standard formula, even though we know the function that defines the curve. 'Under a curve' implies the area between the curve, the $x$-axis, and the two ordinates of the $y$-axis, defined by the upper and lower limits of integration. We just illustrated and proved its use for a compound function consisting of linear functions. Let's extend this to various functions.

Let's try some examples.

### Example 2

Evaluate the following definite integrals.

**a)** $\int_{-2}^{2} (x^2)\,dx$  
**b)** $\int_{3}^{5} \sqrt{x}\left(3 - \frac{1}{x} + 2\sqrt{x}\right)dx$

# Definite Integration

What did you get? Find the solution below to double-check your answer.

## Solution to Example 2

**a)** $\int_{-2}^{2} (x^2)\, dx$
**Solution**

$$\int_{-2}^{2} (x^2)\, dx = \left[\frac{x^3}{3}\right]_{-2}^{2}$$

$$= \left[\frac{x^3}{3}\right]^{2} - \left[\frac{x^3}{3}\right]^{-2}$$

$$= \left(\frac{2^3}{3}\right) - \left(\frac{(-2)^3}{3}\right)$$

$$= \left(\frac{8}{3}\right) - \left(-\frac{8}{3}\right) = \frac{8}{3} + \frac{8}{3} = \frac{16}{3}$$

$$\therefore \int_{-2}^{2} (x^2)\, dx = \frac{16}{3} \text{ unit}^2$$

This is shown in Figure 16.5.

**NOTE**

In this case, we can see that the function is symmetric about the y-axis. As such, the area in the region −2 to 0 is the same as that from 0 to 2. In other words, $\int_{-2}^{0} (x^2)\, dx = \int_{0}^{2} (x^2)\, dx$ or $\int_{-2}^{2} (x^2)\, dx = 2\left[\int_{-2}^{0} (x^2)\, dx\right] = 2\left[\int_{0}^{2} (x^2)\, dx\right]$. We can thus choose one and double the answer. This will be applicable in evaluating periodic functions, such as sine and cosine functions.

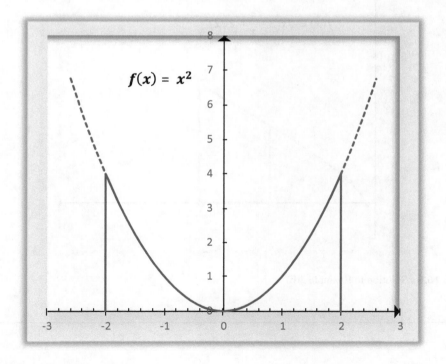

**FIGURE 16.5** Solution to Example 2(a).

**b)** $\int_3^5 \sqrt{x}\left(3 - \frac{1}{x} + 2\sqrt{x}\right) dx$

**Solution**

Let's quickly expand the function as:

$$\sqrt{x}\left(3 - \frac{1}{x} + 2\sqrt{x}\right) = x^{\frac{1}{2}}\left(3 - x^{-1} + 2x^{\frac{1}{2}}\right)$$

$$= 3x^{\frac{1}{2}} - x^{-\frac{1}{2}} + 2x$$

Thus

$$\int_3^5 \sqrt{x}\left(3 - \frac{1}{x} + 2\sqrt{x}\right) dx = \int_3^5 \left(3x^{\frac{1}{2}} - x^{-\frac{1}{2}} + 2x\right) dx$$

$$= \left[\frac{3x^{1.5}}{1.5} - \frac{x^{0.5}}{0.5} + x^2\right]_3^5 = \left[2x^{1.5} - 2x^{0.5} + x^2\right]_3^5$$

$$= \left[2x^{1.5} - 2x^{0.5} + x^2\right]^5 - \left[2x^{1.5} - 2x^{0.5} + x^2\right]^3$$

$$= \left\{2(5)^{1.5} - 2(5)^{0.5} + (5)^2\right\} - \left\{2(3)^{1.5} - 2(3)^{0.5} + (3)^2\right\}$$

$$= (42.8885\ldots) - (15.9282\ldots) = 26.96\ldots$$

$$\therefore \int_3^5 \sqrt{x}\left(3 - \frac{1}{x} + 2\sqrt{x}\right) dx = 27 \text{ unit}^2 \text{ (2 s.f.)}$$

This is shown in the graph in Figure 16.6.

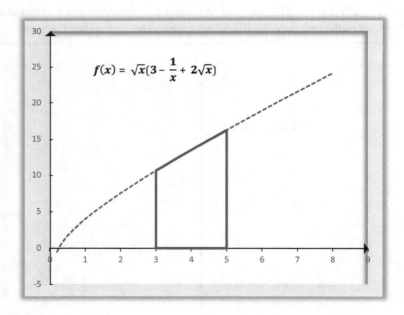

**FIGURE 16.6** Solution to Example 2(b).

# Definite Integration

We will now show that the result for the two possible integrals of $\int \frac{1}{ax} dx$ (i.e., $\int \frac{1}{ax} dx = \frac{1}{a} \ln ax + C = \frac{1}{a} \ln x + C$), as stated on page 632, is the same. Let's illustrate this.

### Example 3

Show that $\int_2^{10} \left(\frac{1}{5x}\right) dx = \frac{1}{5} \ln 5$ using the following two standards.

a) $\int \frac{1}{ax} dx = \frac{1}{a} \ln |x| + C$ 
b) $\int \frac{1}{ax} dx = \frac{1}{a} \ln |ax| + C$

What did you get? Find the solution below to double-check your answer.

### Solution to Example 3

a) $\int \frac{1}{ax} dx = \frac{1}{a} \ln |x| + C$

**Solution**

$$\int_2^{10} \left(\frac{1}{5x}\right) dx = \left[\frac{1}{5} \ln |x|\right]_2^{10}$$

$$= \left[\frac{1}{5} \ln |x|\right]^{10} - \left[\frac{1}{5} \ln |x|\right]^{2}$$

$$= \left(\frac{\ln 10}{5}\right) - \left(\frac{\ln 2}{5}\right)$$

$$= \frac{1}{5}(\ln 10 - \ln 2)$$

$$= \frac{1}{5}\left[\ln\left(\frac{10}{2}\right)\right] = \frac{1}{5} \ln 5$$

$$\therefore \int_2^{10} \left(\frac{1}{5x}\right) dx = \frac{1}{5} \ln 5$$

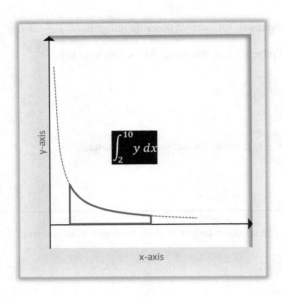

**FIGURE 16.7** Solution to Example 3(b).

**b)** $\int \frac{1}{ax} dx = \frac{1}{a} \ln |ax| + C$
**Solution**

$$\int_2^{10} \left(\frac{1}{5x}\right) dx = \left[\frac{1}{5} \ln |5x|\right]_2^{10}$$

$$= \left[\frac{1}{5} \ln |5x|\right]^{10} - \left[\frac{1}{5} \ln |5x|\right]^2$$

$$= \left(\frac{\ln 50}{5}\right) - \left(\frac{\ln 10}{5}\right)$$

$$= \frac{1}{5} (\ln 50 - \ln 10)$$

$$= \frac{1}{5} \left[\ln \left(\frac{50}{10}\right)\right] = \frac{1}{5} \ln 5$$

$$\therefore \int_2^{10} \left(\frac{1}{5x}\right) dx = \frac{1}{5} \ln 5$$

This is shown in Figure 16.7.

Another example to try.

## Example 4

Determine the area between the curve $f(x) = 15e^{2x}$, the x-axis, and the lines $x = -1$ and $x = 1$. Present the answer correct to 1 d.p.

# Definite Integration

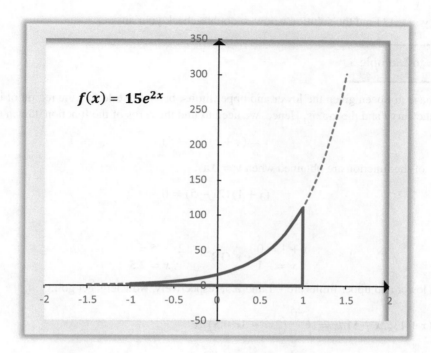

**FIGURE 16.8** Solution to Example 4.

What did you get? Find the solution below to double-check your answer.

### Solution to Example 4

**Solution**

$$\int_{x_1=-1}^{x_2=1} (15e^{2x}) \, dx = \left[ \frac{15e^{2x}}{2} \right]_{-1}^{1}$$

$$= \left[ \frac{15e^{2x}}{2} \right]^{1} - \left[ \frac{15e^{2x}}{2} \right]^{-1}$$

$$= \left( \frac{15e^{2}}{2} \right) - \left( \frac{15e^{-2}}{2} \right) = \frac{15}{2} \left( e^2 - \frac{1}{e^2} \right)$$

$$\therefore \int_{x_1=-1}^{x_2=1} (15e^{2x}) \, dx = 54.4 \text{ unit}^2$$

We've so far looked at situations where the curve is above the $x$-axis in the intervals of interest. What happens if the curve is below the graph? Let us investigate this.

### Example 5

Determine the area of the finite region $R$ bounded by the curve $y = (x+1)(2x-5)$ and the $x$-axis.

What did you get? Find the solution below to double-check your answer.

**Solution to Example 5**

Here we have not been given the lower and upper limits, but we're told that the region of interest is between the curve and the $x$-axis. Hence, we need to find the zeros of the function. Given that:

$$y = (x+1)(2x-5)$$

The zeros of the function are obtained when $y = 0$ as:

$$(x+1)(2x-5) = 0$$

Therefore

$$x + 1 = 0 \quad \text{OR} \quad 2x - 5 = 0$$
$$\therefore x = -1 \quad \quad \therefore x = 2.5$$

Thus, the lower and upper limits are $-1$ and $2.5$, respectively. Now let's get going.

$$\int_{-1}^{2.5} (x+1)(2x-5)\,dx = \int_{-1}^{2.5} \left(2x^2 - 3x - 5\right) dx$$

$$= \left[\frac{2x^3}{3} - \frac{3x^2}{2} - 5x\right]_{-1}^{2.5}$$

$$= \left[\frac{2x^3}{3} - \frac{3x^2}{2} - 5x\right]^{2.5} - \left[\frac{2x^3}{3} - \frac{3x^2}{2} - 5x\right]^{-1}$$

$$= \left\{\frac{2(2.5)^3}{3} - \frac{3(2.5)^2}{2} - 5(2.5)\right\} - \left\{\frac{2(-1)^3}{3} - \frac{3(-1)^2}{2} - 5(-1)\right\}$$

$$= \left(-\frac{275}{24}\right) - \left(\frac{17}{6}\right) = -\frac{343}{24}$$

**NOTE**

We obtained a negative value here; however, it is known that area is a scalar quantity, and therefore a negative value is invalid. When we look at the graph of this function (Figure 16.9), it is observed that the area of interest is below the $x$-axis and thus the negative value.

In cases like the above, we will ignore the negative sign and use the number. Technically, we say we've taken the absolute value of the answer and thus we have

$$\therefore \int_{-1}^{2.5} (2x^2 - 3x - 5)\,dx = \frac{343}{24}\ \text{unit}^2$$

It is important to note that the area evaluated is the one above the curve and below the $x$-axis, bounded to the left by the lower limit and to the right by the upper limit. This is a finite value. However, the area below the curve, which is bounded by the limits is infinite.

It would be helpful to state that $\int_{-1}^{3} (2x^2 - 3x - 5)\,dx$ represents the area that covers both below and above the $x$-axis, as shown in Figure 16.9.

We will deal with this in the next section.

Definite Integration

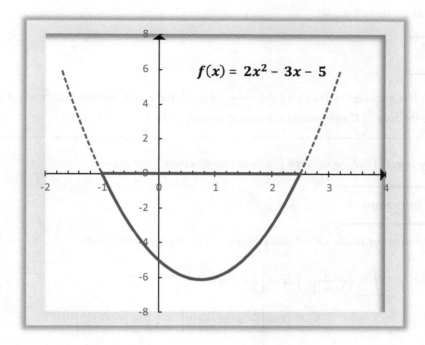

**FIGURE 16.9** Solution to Example 5 – Part I.

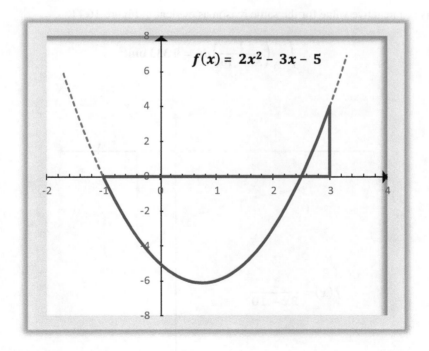

**FIGURE 16.10** Solution to Example 5 – Part II.

Let's look at another example.

### Example 6

The region $R$ is enclosed by the curve $y = \frac{1}{3x-10}$, $x \neq \frac{10}{3}$, the $x$-axis, and the lines $x = -3$ and $x = 1$. Determine the area of $R$ and present the answer correct to 3 s.f.

What did you get? Find the solution below to double-check your answer.

### Solution to Example 6

The lower and upper limits are $-3$ and $1$, respectively. We therefore have

$$\int_{-3}^{1} \left(\frac{1}{3x-10}\right) dx = \left[\frac{1}{3} \ln |3x-10|\right]_{-3}^{1}$$

$$= \left[\frac{1}{3} \ln |3x-10|\right]^{1} - \left[\frac{1}{3} \ln |3x-10|\right]^{-3}$$

$$= \left(\frac{1}{3} \ln |3(1)-10|\right) - \left(\frac{1}{3} \ln |3(-3)-10|\right)$$

$$= \frac{1}{3} \{(\ln |-7|) - (\ln |-19|)\} = -0.333$$

Again, this is a negative value for the same reason as evident in Figure 16.11.

$$\therefore \int_{-3}^{1} \left(\frac{1}{3x-10}\right) dx = 0.333 \text{ unit}^2$$

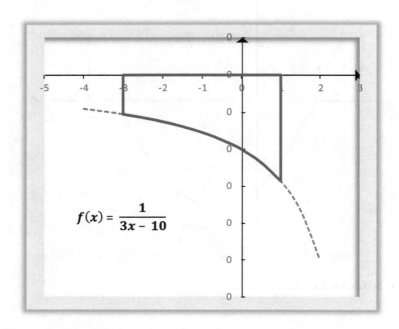

FIGURE 16.11  Solution to Example 6.

# Definite Integration

Before we move to another section on definite integration, it will be helpful to introduce an important rule. We previously covered two rules under indefinite integration; these rules also apply to definite integration. We will however need more rules as follows.

**Rule 1** Order of limits

The rule states that

$$\boxed{\int_a^c f(x)\,dx = -\int_c^a f(x)\,dx} \qquad (16.2)$$

In other words, swapping the limits (the upper limit becomes the lower limit and vice versa) changes the sign of the integral.

**Rule 2** Zero width interval

The rule states that

$$\boxed{\int_a^a f(x)\,dx = 0} \qquad (16.3)$$

In other words, the area covered by a line is zero.

**Rule 3** Decomposition of limits

The rule states that

$$\boxed{\int_a^c f(x)\,dx = \int_a^b f(x)\,dx + \int_b^c f(x)\,dx} \qquad (16.4)$$

This is because

$$\int_a^c f'(x)\,dx = f(c) - f(a)$$
$$= f(c) - f(a) + f(b) - f(b)$$
$$= f(c) - f(b) + f(b) - f(a)$$
$$= \underbrace{f(b) - f(a)}_{} + \underbrace{f(c) - f(b)}_{}$$
$$= \int_a^b f'(x)\,dx + \int_b^c f'(x)\,dx$$

In other words, the integration of a function between the lower limit $a$ and the upper limit $c$ is equal to the integral between point $a$ and an intermediate point $b$, plus the integral of the same function from the intermediate point $b$ and the original upper limit $c$. This intermediate point is such that $a < b < c$. This rule is illustrated with an example below.

$$\int_{-2}^{2} f(x)\,dx = \int_{-2}^{0} f(x)\,dx + \int_{0}^{2} f(x)\,dx$$

This makes sense since we're looking for an area between point $a$ and $c$. This rule is very important, especially when a part of a curve falls below and the other above the x-axis, which will be shown later. It can also be helpful to make the calculation simple, especially when we take one of the limits as zero in certain functions.

For now, let's try a few examples to prove the above rule.

### Example 7

In each of the following, show that the LHS integral is equal to the sum of the RHS integrals.

a) $\int_{-4}^{1} (x^2 + 3x - 4) \, dx = \int_{-4}^{0} (x^2 + 3x - 4) \, dx + \int_{0}^{1} (x^2 + 3x - 4) \, dx$

b) $\int_{-\frac{\pi}{2}}^{\frac{\pi}{2}} (\cos \theta) \, d\theta = \int_{-\frac{\pi}{2}}^{0} (\cos \theta) \, d\theta + \int_{0}^{\frac{\pi}{2}} (\cos \theta) \, d\theta$

c) $\int_{-\frac{\pi}{2}}^{\frac{\pi}{2}} (\sin \theta) \, d\theta = \int_{-\frac{\pi}{2}}^{0} (\sin \theta) \, d\theta + \int_{0}^{\frac{\pi}{2}} (\sin \theta) \, d\theta$

What did you get? Find the solution below to double-check your answer.

### Solution to Example 7

a) $\int_{-4}^{1} (x^2 + 3x - 4) \, dx = \int_{-4}^{0} (x^2 + 3x - 4) \, dx + \int_{0}^{1} (x^2 + 3x - 4) \, dx$
**Solution**
**LHS**

$$\int_{-4}^{1} (x^2 + 3x - 4) \, dx = \left[ \frac{x^3}{3} + \frac{3x^2}{2} - 4x \right]_{-4}^{1}$$

$$= \left[ \frac{x^3}{3} + \frac{3x^2}{2} - 4x \right]^{1} - \left[ \frac{x^3}{3} + \frac{3x^2}{2} - 4x \right]^{-4}$$

$$= \left\{ \frac{(1)^3}{3} + \frac{3(1)^2}{2} - 4(1) \right\} - \left\{ \frac{(-4)^3}{3} + \frac{3(-4)^2}{2} - 4(-4) \right\}$$

$$= \left( -\frac{13}{6} \right) - \left( \frac{56}{3} \right) = -\frac{125}{6}$$

$$\therefore \int_{-4}^{1} (x^2 + 3x - 4) \, dx = \frac{125}{6} \text{ unit}^2$$

# Definite Integration

**RHS (1)**

$$\int_{-4}^{0} (x^2 + 3x - 4)\, dx = \left[\frac{x^3}{3} + \frac{3x^2}{2} - 4x\right]_{-4}^{0}$$

$$= \left[\frac{x^3}{3} + \frac{3x^2}{2} - 4x\right]^{0} - \left[\frac{x^3}{3} + \frac{3x^2}{2} - 4x\right]^{-4}$$

$$= \left\{\frac{(0)^3}{3} + \frac{3(0)^2}{2} - 4(0)\right\}$$

$$- \left\{\frac{(-4)^3}{3} + \frac{3(-4)^2}{2} - 4(-4)\right\}$$

$$= (0) - \left(\frac{56}{3}\right) = -\frac{56}{3}$$

$$\therefore \int_{-4}^{0} (x^2 + 3x - 4)\, dx = \frac{56}{3} \text{ unit}^2$$

**RHS (2)**

$$\int_{0}^{1} (x^2 + 3x - 4)\, dx = \left[\frac{x^3}{3} + \frac{3x^2}{2} - 4x\right]_{0}^{1}$$

$$= \left[\frac{x^3}{3} + \frac{3x^2}{2} - 4x\right]^{1} - \left[\frac{x^3}{3} + \frac{3x^2}{2} - 4x\right]^{0}$$

$$= \left\{\frac{(1)^3}{3} + \frac{3(1)^2}{2} - 4(1)\right\}$$

$$- \left\{\frac{(0)^3}{3} + \frac{3(0)^2}{2} - 4(0)\right\}$$

$$= \left(-\frac{13}{6}\right) - (0) = -\frac{13}{6}$$

$$\therefore \int_{0}^{1} (x^2 + 3x - 4)\, dx = \frac{13}{6} \text{ unit}^2$$

Hence, for RHS we have

$$\therefore \int_{-4}^{0} (x^2 + 3x - 4)\, dx + \int_{0}^{1} (x^2 + 3x - 4)\, dx = \frac{56}{3} + \frac{13}{6} = \frac{125}{6} \text{ unit}^2$$

We have therefore shown that the LHS = RHS.

---

**b)** $\int_{-\frac{\pi}{2}}^{\frac{\pi}{2}} (\cos\theta)\, d\theta = \int_{-\frac{\pi}{2}}^{0} (\cos\theta)\, d\theta + \int_{0}^{\frac{\pi}{2}} (\cos\theta)\, d\theta$

**Solution**

**LHS**

$$\int_{-\frac{\pi}{2}}^{\frac{\pi}{2}} (\cos\theta)\, d\theta = [\sin\theta]_{-\frac{\pi}{2}}^{\frac{\pi}{2}}$$

$$= [\sin\theta]^{\frac{\pi}{2}} - [\sin\theta]^{-\frac{\pi}{2}}$$

$$= \left\{\sin\left(\frac{\pi}{2}\right)\right\} - \left\{\sin\left(-\frac{\pi}{2}\right)\right\}$$

$$= (1) - (-1) = 2$$

$$\therefore \int_{-\frac{\pi}{2}}^{\frac{\pi}{2}} (\cos\theta)\, d\theta = 2 \text{ unit}^2$$

**RHS(1)**

$$\int_{-\frac{\pi}{2}}^{0} (\cos\theta)\, d\theta = [\sin\theta]_{-\frac{\pi}{2}}^{0}$$

$$= [\sin\theta]^{0} - [\sin\theta]^{-\frac{\pi}{2}}$$

$$= \{\sin(0)\} - \left\{\sin\left(-\frac{\pi}{2}\right)\right\}$$

$$= (0) - (-1) = 1$$

$$\therefore \int_{-\frac{\pi}{2}}^{0} (\cos\theta)\, d\theta = 1 \text{ unit}^2$$

**RHS(2)**

$$\int_{0}^{\frac{\pi}{2}} (\cos\theta)\, d\theta = [\sin\theta]_{0}^{\frac{\pi}{2}}$$

$$= [\sin\theta]^{\frac{\pi}{2}} - [\sin\theta]^{0}$$

$$= \left\{\sin\left(\frac{\pi}{2}\right)\right\} - \{\sin(0)\}$$

$$= (1) - (0) = 1$$

$$\therefore \int_{0}^{\frac{\pi}{2}} (\cos\theta)\, d\theta = 1 \text{ unit}^2$$

Hence, for RHS we have

$$\therefore \int_{-\frac{\pi}{2}}^{0} (\cos\theta)\, dx + \int_{0}^{\frac{\pi}{2}} (\cos\theta)\, d\theta = 1 + 1 = 2 \text{ unit}^2$$

We have therefore shown that the LHS = RHS.

c) $\int_{-\frac{\pi}{2}}^{\frac{\pi}{2}} (\sin \theta) \, d\theta = \int_{-\frac{\pi}{2}}^{0} (\sin \theta) \, d\theta + \int_{0}^{\frac{\pi}{2}} (\sin \theta) \, d\theta$

**Solution**

**LHS**

$$\int_{-\frac{\pi}{2}}^{\frac{\pi}{2}} (\sin \theta) \, d\theta = [-\cos \theta]_{-\frac{\pi}{2}}^{\frac{\pi}{2}}$$
$$= [-\cos \theta]^{\frac{\pi}{2}} - [-\cos \theta]^{-\frac{\pi}{2}}$$
$$= \left\{-\cos \left(\frac{\pi}{2}\right)\right\} - \left\{-\cos \left(-\frac{\pi}{2}\right)\right\}$$
$$= (0) - (0) = 0$$
$$\therefore \int_{-\frac{\pi}{2}}^{\frac{\pi}{2}} (\sin \theta) \, d\theta = 0 \text{ unit}^2$$

It looks strange to say that the area covered is zero. Let's wait and see.

**RHS (1)**

$$\int_{-\frac{\pi}{2}}^{0} (\sin \theta) \, d\theta = [-\cos \theta]_{-\frac{\pi}{2}}^{0}$$
$$= [-\cos \theta]^{0} - [-\cos \theta]^{-\frac{\pi}{2}}$$
$$= \{-\cos (0)\} - \left\{-\cos \left(-\frac{\pi}{2}\right)\right\}$$
$$= (-1) - (0) = -1$$
$$\therefore \int_{-\frac{\pi}{2}}^{0} (\sin \theta) \, d\theta = 1 \text{ unit}^2$$

Hence, for RHS, we have

$$\therefore \int_{-\frac{\pi}{2}}^{0} (\sin \theta) \, dx + \int_{0}^{\frac{\pi}{2}} (\sin \theta) \, d\theta = 1 + 1 = 2 \text{ unit}^2$$

**RHS (2)**

$$\int_{0}^{\frac{\pi}{2}} (\sin \theta) \, d\theta = [-\cos \theta]_{0}^{\frac{\pi}{2}}$$
$$= [-\cos \theta]^{\frac{\pi}{2}} - [-\cos \theta]^{0}$$
$$= \left\{-\cos \left(\frac{\pi}{2}\right)\right\} - \{-\cos (0)\}$$
$$= (0) - (-1) = 1$$
$$\therefore \int_{0}^{\frac{\pi}{2}} (\sin \theta) \, d\theta = 1 \text{ unit}^2$$

Gush! $LHS \neq RHS$. Let's investigate this by looking at the graph of both sine and cosine in the $-\frac{\pi}{2} \leq \theta \leq \frac{\pi}{2}$ range as shown in Figure 16.12.

---

If you look at the graph in Figure 16.12, you will notice that for cosine, the entire function is above the $x$-axis. For the sine function, however, a part is above and another part is below the $x$-axis. In fact, it is equally split above and below the horizontal line. Hence, the areas cancel out, and this is where the application of the above rule is important. This can only be achieved if we have the information about the function or if we are requested to sketch the curve. Based on this, we can conclude that 2 units squared is the correct answer for both sine and cosine functions in this example.

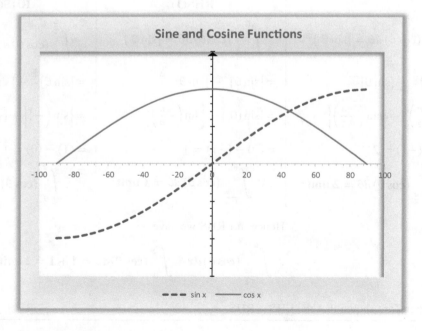

**FIGURE 16.12** Solution to Example 7(c).

Definite Integration

We have noted above that care must be taken when using definite integration between two given limits. It will be necessary or required in some cases to split the area using the rule above. While the rule splits the integral into two, considering only one intermediate point, we can extend this to create three or more integrals. This involves taking two or more intermediate points, as demonstrated below for three integrals.

$$\int_a^d f(x)\,dx = \int_a^b f(x)\,dx + \int_b^c f(x)\,dx + \int_c^d f(x)\,dx \qquad (16.5)$$

such that $a < b < c < d$.

The above will be useful, for example, where a curve switches from above to below and back to above the x-axis within the limits of interest.

### 16.2.2 Area of a Curve Above and Below x-Axis

From our example above, we can conclude that when a curve is above and below the x-axis in the region of upper and lower limits, we need to split the integral and evaluate each part separately. The portion above the horizontal axis should be evaluated and the same is done for the portion below the x-axis. The absolute values of the results are thereafter added together. Let's try some examples.

#### Example 8

In each of the following functions, determine the area bounded by the function.

a) $\int_{\frac{1}{6}\pi}^{\frac{5}{6}\pi} (2 \sin 3\theta)\, d\theta$

b) $\int_{\frac{1}{4}\pi}^{\frac{5}{6}\pi} (2 \sin 3\theta)\, d\theta$

What did you get? Find the solution below to double-check your answer.

#### Solution to Example 8

a) $\int_{\frac{1}{6}\pi}^{\frac{5}{6}\pi} (2 \sin 3\theta)\, d\theta$

**Solution**

It will be nice to sketch the graph if it is not given before proceeding. But let's assume everything is alright and proceed without plotting the graph as:

$$\int_{\frac{1}{6}\pi}^{\frac{5}{6}\pi} (2 \sin 3\theta)\, d\theta = \left[\frac{-2\cos 3\theta}{3}\right]_{\frac{1}{6}\pi}^{\frac{5}{6}\pi}$$

$$= \left[\frac{-2\cos 3\theta}{3}\right]^{\frac{5}{6}\pi} - \left[\frac{-2\cos 3\theta}{3}\right]^{\frac{1}{6}\pi}$$

$$= \left(\frac{-2\cos 3\left(\frac{5}{6}\pi\right)}{3}\right) - \left(\frac{-2\cos 3\left(\frac{1}{6}\pi\right)}{3}\right)$$

$$= \left(\frac{-2\cos\left(\frac{5}{2}\pi\right)}{3}\right) - \left(\frac{-2\cos\left(\frac{1}{2}\pi\right)}{3}\right)$$

$$= -\frac{2}{3}\left[\cos\left(\frac{5}{2}\pi\right) - \cos\left(\frac{1}{2}\pi\right)\right]$$

$$= -\frac{2}{3}[0 - 0] = 0$$

$$\therefore \int_{\frac{1}{6}\pi}^{\frac{5}{6}\pi} (2\sin 3\theta)\,d\theta = 0$$

This cannot be right. Now let us look at the graph, as shown in Figure 16.13.

Looking at the graph in Figure 16.13, we can see that there are three sections that form the area. Two regions are above the x-axis and are equal, and one is below the x-axis and is equal to the sum of the two above. Instead of looking for the three intervals (i.e., $\frac{1}{6}\pi \le \theta \le \frac{1}{3}\pi$, $\frac{1}{3}\pi \le \theta \le \frac{2}{3}\pi$, and $\frac{2}{3}\pi \le \theta \le \frac{5}{6}\pi$), we can take the area of the first section to the left and quadruple the answer. Let us do just that.

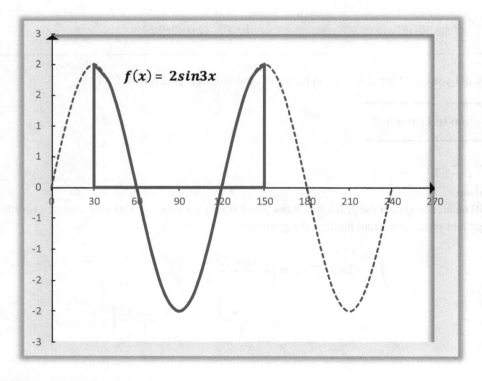

**FIGURE 16.13** Solution to Example 8(a).

# Definite Integration

$$\int_{\frac{1}{6}\pi}^{\frac{1}{3}\pi} (2\sin 3\theta)\, d\theta = \left[\frac{-2\cos 3\theta}{3}\right]_{\frac{1}{6}\pi}^{\frac{1}{3}\pi}$$

$$= \left[\frac{-2\cos 3\theta}{3}\right]^{\frac{1}{3}\pi} - \left[\frac{-2\cos 3\theta}{3}\right]^{\frac{1}{6}\pi}$$

$$= \left(\frac{-2\cos 3\left(\frac{1}{3}\pi\right)}{3}\right) - \left(\frac{-2\cos 3\left(\frac{1}{6}\pi\right)}{3}\right)$$

$$= \left(\frac{-2\cos(\pi)}{3}\right) - \left(\frac{-2\cos\left(\frac{1}{2}\pi\right)}{3}\right)$$

$$= \left(\frac{2}{3}\right) - (0) = \frac{2}{3}$$

We then need to multiply this by 4 to get $\frac{2}{3} \times 4 = \frac{8}{3}$

$$\therefore \int_{\frac{1}{6}\pi}^{\frac{5}{6}\pi} (2\sin 3\theta)\, d\theta = \frac{8}{3}\ \text{unit}^2$$

---

**b)** $\int_{\frac{1}{4}\pi}^{\frac{5}{6}\pi} (2\sin 3\theta)\, d\theta$

**Solution**

We have tried to change the interval here to highlight a crucial point in applying the decomposition rule. Again, we'll assume everything is alright and proceed without sketching the curve as:

$$\int_{\frac{1}{4}\pi}^{\frac{5}{6}\pi} (2\sin 3\theta)\, d\theta = \left[\frac{-2\cos 3\theta}{3}\right]_{\frac{1}{4}\pi}^{\frac{5}{6}\pi}$$

$$= \left[\frac{-2\cos 3\theta}{3}\right]^{\frac{5}{6}\pi} - \left[\frac{-2\cos 3\theta}{3}\right]^{\frac{1}{4}\pi}$$

$$= \left(\frac{-2\cos 3\left(\frac{5}{6}\pi\right)}{3}\right) - \left(\frac{-2\cos 3\left(\frac{1}{4}\pi\right)}{3}\right)$$

$$= \left(\frac{-2\cos\left(\frac{5}{2}\pi\right)}{3}\right) - \left(\frac{-2\cos\left(\frac{3}{4}\pi\right)}{3}\right)$$

$$= (0) - \left(\frac{\sqrt{2}}{3}\right) = -\frac{\sqrt{2}}{3}$$

Although this looks like we just need to remove the sign, and everything should be fine. But no, it's not just that. The area we found in the previous example is only slightly larger than the current one (Figure 16.14).

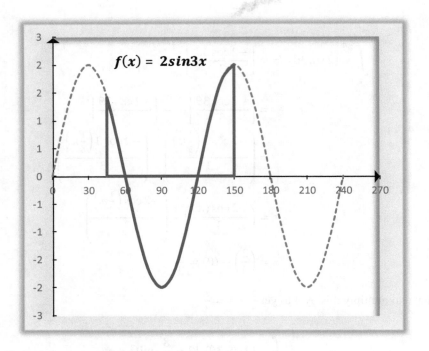

**FIGURE 16.14** Solution to Example 8(b).

For this case we will consider the three intervals.

$$\int_{\frac{1}{4}\pi}^{\frac{5}{6}\pi} (2\sin 3\theta)\, d\theta = \int_{\frac{1}{4}\pi}^{\frac{1}{3}\pi} (2\sin 3\theta)\, d\theta + \int_{\frac{1}{3}\pi}^{\frac{2}{3}\pi} (2\sin 3\theta)\, d\theta + \int_{\frac{2}{3}\pi}^{\frac{5}{6}\pi} (2\sin 3\theta)\, d\theta$$

**Section 1**

$$\int_{\frac{1}{4}\pi}^{\frac{1}{3}\pi} (2\sin 3\theta)\, d\theta = \left[\frac{-2\cos 3\theta}{3}\right]_{\frac{1}{4}\pi}^{\frac{1}{3}\pi}$$

$$= \left[\frac{-2\cos 3\theta}{3}\right]^{\frac{1}{3}\pi} - \left[\frac{-2\cos 3\theta}{3}\right]^{\frac{1}{4}\pi}$$

$$= \left(\frac{-2\cos 3\left(\frac{1}{3}\pi\right)}{3}\right) - \left(\frac{-2\cos 3\left(\frac{1}{4}\pi\right)}{3}\right)$$

$$= \left(\frac{-2\cos(\pi)}{3}\right) - \left(\frac{-2\cos\left(\frac{3}{4}\pi\right)}{3}\right)$$

$$= \left(\frac{2}{3}\right) - \left(\frac{\sqrt{2}}{3}\right) = \frac{2-\sqrt{2}}{3}$$

# Definite Integration

**Section 2**

$$\int_{\frac{1}{3}\pi}^{\frac{2}{3}\pi} (2\sin 3\theta)\, d\theta = \left[\frac{-2\cos 3\theta}{3}\right]_{\frac{1}{3}\pi}^{\frac{2}{3}\pi}$$

$$= \left[\frac{-2\cos 3\theta}{3}\right]^{\frac{2}{3}\pi} - \left[\frac{-2\cos 3\theta}{3}\right]^{\frac{1}{3}\pi}$$

$$= \left(\frac{-2\cos 3\left(\frac{2}{3}\pi\right)}{3}\right) - \left(\frac{-2\cos 3\left(\frac{1}{3}\pi\right)}{3}\right)$$

$$= \left(\frac{-2\cos(2\pi)}{3}\right) - \left(\frac{-2\cos(\pi)}{3}\right)$$

$$= \frac{-2}{3}[(\cos(2\pi)) - (\cos(\pi))]$$

$$= \frac{-2}{3}[(1) - (-1)]$$

$$= \frac{-2}{3}[1+1] = -\frac{4}{3}$$

It is a negative, hence

$$\int_{\frac{1}{3}\pi}^{\frac{2}{3}\pi} (2\sin 3\theta)\, d\theta = \frac{4}{3}$$

**Section 3**

$$\int_{\frac{2}{3}\pi}^{\frac{5}{6}\pi} (2\sin 3\theta)\, d\theta = \left[\frac{-2\cos 3\theta}{3}\right]_{\frac{2}{3}\pi}^{\frac{5}{6}\pi}$$

$$= \left[\frac{-2\cos 3\theta}{3}\right]^{\frac{5}{6}\pi} - \left[\frac{-2\cos 3\theta}{3}\right]^{\frac{2}{3}\pi}$$

$$= \left(\frac{-2\cos 3\left(\frac{5}{6}\pi\right)}{3}\right) - \left(\frac{-2\cos 3\left(\frac{2}{3}\pi\right)}{3}\right)$$

$$= \left(\frac{-2\cos\left(\frac{5}{2}\pi\right)}{3}\right) - \left(\frac{-2\cos(2\pi)}{3}\right)$$

$$= (0) - \left(-\frac{2}{3}\right) = \frac{2}{3}$$

We now need to add them up.

$$\int_{\frac{1}{4}\pi}^{\frac{5}{6}\pi} (2\sin 3\theta)\, d\theta = \int_{\frac{1}{4}\pi}^{\frac{1}{3}\pi} (2\sin 3\theta)\, d\theta + \int_{\frac{1}{3}\pi}^{\frac{2}{3}\pi} (2\sin 3\theta)\, d\theta + \int_{\frac{2}{3}\pi}^{\frac{5}{6}\pi} (2\sin 3\theta)\, d\theta$$

$$= \frac{2-\sqrt{2}}{3} + \frac{4}{3} + \frac{2}{3} = \frac{2-\sqrt{2}+4+2}{3}$$

$$= \frac{8-\sqrt{2}}{3}$$

$$\therefore \int_{\frac{1}{4}\pi}^{\frac{5}{6}\pi} (2\sin 3\theta)\, d\theta = \frac{8-\sqrt{2}}{3}\ \text{unit}^2$$

### 16.2.3 Area Bounded by a Curve and a Line

We may be interested in an area bounded by a curve and a straight line, such as the region **R** shown in Figure 16.15.

In Figure 16.15, a curve of $f(x)$ intersects a straight line of function $g(x)$ at two points $A(x_1, y_1)$ and $B(x_2, y_2)$. This is one of the ways a curve and a linear function can intersect, but the procedure for finding the bounded area is similar. We will consider two ways of finding area $R$.

**Method 1**

To find the area $R$, follow these steps:

**Step 1:** Solve for $x$ in $f(x) = g(x)$ to obtain $x$-coordinate of the points of intersection of the two functions. Let's denote these as $x_1$ and $x_2$, and they represent the lower and upper limits.

**Step 2:** Simplify the combined function $y = f(x) - g(x)$. Note that it is irrelevant which of the two is taken as subtrahend or minuend. The only consequence is that the area might be negative instead of positive.

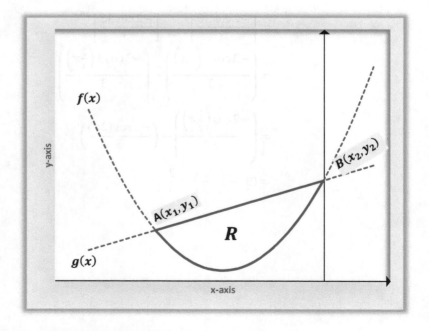

**FIGURE 16.15** Area bounded by a curve and a straight line illustrated.

# Definite Integration

**Step 3:** The area $R$ is found using:

$$\int_{x_1}^{x_2} \{y\}\, dx = \int_{x_1}^{x_2} \{f(x) - g(x)\}\, dx = \left[\int \{f(x) - g(x)\}\, dx\right]_{x_1}^{x_2} \quad (16.6)$$

This method can be used irrespective of whether the region is wholly or partially above or below the $x$-axis, provided that what is required is the entire area bounded by the two functions. Otherwise, method 2 below will be the only option.

## Method 2

To find the area $R$, follow these steps:

**Step 1:** Solve for $x$ in $f(x) = g(x)$ to obtain the $x$-coordinate of the points of intersection of the two functions. Let's denote these as $x_1$ and $x_2$, and they represent the lower and upper limits.

**Step 2:** Find the definite integrals of the two functions separately and then subtract as appropriate. Note that the subtrahend and minuend will be determined by the shape formed by the two functions.

**Step 3:** The area $R$ is found using:

$$\int_{x_1}^{x_2} \{g(x)\}\, dx - \int_{x_1}^{x_2} \{f(x)\}\, dx \quad (16.7)$$

This is illustrated in Figure 16.16.

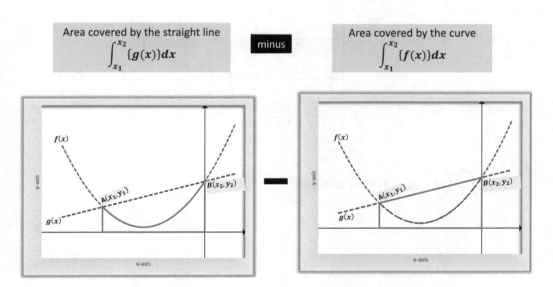

**FIGURE 16.16** Determining the area bounded by a curve and a straight line.

## Example 9

Determine the area of the finite region bounded by $y = x^2 + 1$ and $y = 3x + 5$.

What did you get? Find the solution below to double-check your answer.

## Solution to Example 9

We need to find the points of intersection of the curve and the straight line, which represent our limits. Given that $y = x^2 + 1$ and $y = 3x + 5$ implies that:

$$x^2 + 1 = 3x + 5$$
$$x^2 - 3x - 4 = 0$$
$$(x + 1)(x - 4) = 0$$

Therefore

$$x + 1 = 0 \quad \text{OR} \quad x - 4 = 0$$
$$x = -1 \quad \quad \quad x = 4$$

Hence, the lower and upper limits are $-1$ and $4$, respectively.

**Method 1**

$$\int_{-1}^{4} (x^2 - 3x - 4)\, dx = \left[\frac{x^3}{3} - \frac{3x^2}{2} - 4x\right]_{-1}^{4}$$

$$= \left[\frac{x^3}{3} - \frac{3x^2}{2} - 4x\right]^{4} - \left[\frac{x^3}{3} - \frac{3x^2}{2} - 4x\right]^{-1}$$

$$= \left(\frac{4^3}{3} - \frac{3(4)^2}{2} - 4(4)\right) - \left(\frac{(-1)^3}{3} - \frac{3(-1)^2}{2} - 4(-1)\right)$$

$$= \left(\frac{-56}{3}\right) - \left(\frac{13}{6}\right)$$

$$= -\frac{125}{6}$$

$$\therefore \int_{-1}^{4} (x^2 - 3x - 4)\, dx = 20.8 \text{ unit}^2$$

# Definite Integration

## Method 2

We need to find the area of the curve and the straight line separately. Let's start with the curve.

$$\int_{-1}^{4} (x^2 + 1)\, dx = \left[ \frac{x^3}{3} + x \right]_{-1}^{4}$$

$$= \left[ \frac{x^3}{3} + x \right]^{4} - \left[ \frac{x^3}{3} + x \right]^{-1}$$

$$= \left\{ \frac{(4)^3}{3} + (4) \right\} - \left\{ \frac{(-1)^3}{3} + (-1) \right\}$$

$$= \left( \frac{76}{3} \right) - \left( -\frac{4}{3} \right)$$

$$= \frac{80}{3}$$

$$\therefore \int_{-1}^{4} (x^2 + 1)\, dx = \frac{80}{3} \text{ unit}^2$$

Now the straight line.

$$\int_{-1}^{4} (3x + 5)\, dx = \left[ \frac{3x^2}{2} + 5x \right]_{-1}^{4}$$

$$= \left[ \frac{3x^2}{2} + 5x \right]^{4} - \left[ \frac{3x^2}{2} + 5x \right]^{-1}$$

$$= \left( \frac{3(4)^2}{2} + 5(4) \right) - \left( \frac{3(-1)^2}{2} + 5(-1) \right)$$

$$= (44) - \left( -\frac{7}{2} \right)$$

$$= \frac{95}{2}$$

$$\therefore \int_{-1}^{4} (3x + 5)\, dx = \frac{95}{2} \text{ unit}^2$$

Based on method 1 above, we can guess which one to subtract from which. Hence the area bounded is:

$$\int_{-1}^{4} (3x + 5)\, dx - \int_{-1}^{4} (x^2 + 1)\, dx = \frac{95}{2} - \frac{80}{3} = \frac{125}{6} \text{ unit}^2 = 20.8 \text{ unit}^2$$

This can be trickier in some cases and would require sketching the curves. For this example, the plot is shown in Figure 16.17 to confirm that the subtraction is right.

---

That's the gist for this.

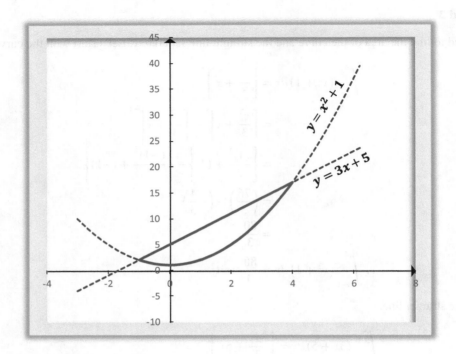

**FIGURE 16.17** Solution to Example 9.

### 16.2.4 AREA BOUNDED BY TWO CURVES

This is treated in the same way as when working with a curve and a straight line. Let's illustrate this too in the following examples.

**Example 10**

Determine the area of the area bounded by $y = x(x^2 - 2)$ and $y = x^2$ in the positive $x$-axis region.

What did you get? Find the solution below to double-check your answer.

**Solution to Example 10**

We need to find the points of intersection of the two curves, which represent our limits. Given that $y = x(x^2 - 2)$ and $y = x^2$ implies that:

$$x(x^2 - 2) = x^2$$
$$x^3 - 2x = x^2$$
$$x^3 - x^2 - 2x = 0$$
$$x(x^2 - x - 2) = 0$$
$$x(x + 1)(x - 2) = 0$$

Definite Integration

Therefore

$$x = 0 \quad \begin{array}{c} x+1=0 \\ x=-1 \end{array} \quad \textbf{OR} \quad \begin{array}{c} x-2=0 \\ x=2 \end{array}$$

$$x = -1, \; x = 0, \; x = 2$$

Hence, the two curves intersect at three points, namely $x = -1$, $x = 0$, and $x = 2$. Since we are only interested in the area of the positive $x$-axis, we will take only $x = 0$ and $x = 2$ as the lower and upper limits, respectively. Here we go.

$$\int_0^2 (x^3 - x^2 - 2x)\, dx = \left[\frac{x^4}{4} - \frac{x^3}{3} - x^2\right]_0^2$$

$$= \left[\frac{x^4}{4} - \frac{x^3}{3} - x^2\right]^2 - \left[\frac{x^4}{4} - \frac{x^3}{3} - x^2\right]^0$$

$$= \left(\frac{2^4}{4} - \frac{2^3}{3} - 2^2\right) - \left(\frac{0^4}{4} - \frac{0^3}{3} - 0^2\right)$$

$$= \left(\frac{-8}{3}\right) - (0) = -\frac{8}{3}$$

$$\therefore \int_0^2 (x^3 - x^2 - 2x)\, dx = \frac{8}{3}\; \textbf{unit}^2$$

This can be trickier in some cases and would require sketching the curves. For this question, the plot is shown in Figure 16.18 to confirm that the subtraction is right.

Although the region straddles the $x$-axis, the above method has accounted for this.

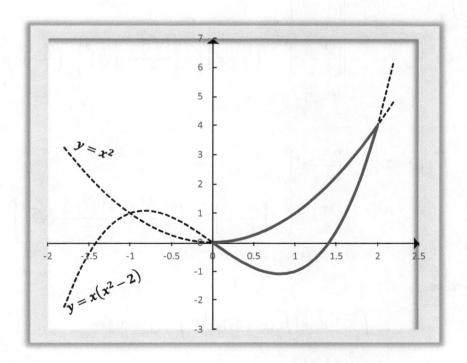

**FIGURE 16.18** Solution to Example 10.

## ALTERNATIVE METHOD

The region can be found as:

$$\int_0^2 (x^2)\,dx + \left|\int_0^{\sqrt{2}} (x^3 - 2x)\,dx\right| - \int_{\sqrt{2}}^2 (x^3 - 2x)\,dx$$

A couple of points here:

**a)** We use modulus sign to show that we need the absolute value, as the region of interest falls below the $x$-axis line.

**b)** The upper and lower limits are obtained by solving $x(x^2 - 2) = 0$.

So, let's see if it works.

First interval

$$\int_0^2 (x^2)\,dx = \left[\frac{x^3}{3}\right]_0^2$$

$$= \left[\frac{x^3}{3}\right]^2 - \left[\frac{x^3}{3}\right]^0 = \left(\frac{2^3}{3}\right) - \left(\frac{0^3}{3}\right)$$

$$= \frac{8}{3}$$

Second interval

$$\left|\int_0^{\sqrt{2}} (x^3 - 2x)\,dx\right| = \left|\left[\frac{x^4}{4} - x^2\right]_0^{\sqrt{2}}\right|$$

$$= \left|\left[\frac{x^4}{4} - x^2\right]^{\sqrt{2}} - \left[\frac{x^4}{4} - x^2\right]^0\right| = \left|\left(\frac{(\sqrt{2})^4}{4} - (\sqrt{2})^2\right) - \left(\frac{0^4}{4} - 0^2\right)\right|$$

$$= |(1 - 2) - (0)| = 1$$

Third interval

$$\int_{\sqrt{2}}^2 (x^3 - 2x)\,dx = \left[\frac{x^4}{4} - x^2\right]_{\sqrt{2}}^2$$

$$= \left[\frac{x^4}{4} - x^2\right]^2 - \left[\frac{x^4}{4} - x^2\right]^{\sqrt{2}} = \left(\frac{2^4}{4} - 2^2\right) - \left(\frac{(\sqrt{2})^4}{4} - (\sqrt{2})^2\right)$$

$$= (0) - (1 - 2) = 1$$

Therefore,

$$\int_0^2 (x^2)\,dx + \left|\int_0^{\sqrt{2}} (x^3 - 2x)\,dx\right| - \int_{\sqrt{2}}^2 (x^3 - 2x)\,dx$$

$$= \frac{8}{3} + 1 - 1 = \frac{8}{3}$$

# Definite Integration

## 16.2.5 Integrating to Infinity

Integrating to infinity is a situation where one of the two limits is infinity, i.e., ±∞. Essentially, when there is no boundary on one side, as shown below.

$$\int_{-\infty}^{a} \{y\}\,dx \quad \text{OR} \quad \int_{a}^{\infty} \{y\}\,dx \tag{16.8}$$

The above implies that we are interested in an area covered by the curve $y$ in the range $x \leq a$, (i.e., from $a$ and everything to the left) or in the range $x \geq a$, (i.e., from $a$ and everything to the right). In general, the area in either of these two cases is expected to be infinity itself. However, when the function being integrated tends to zero as $x$ becomes extremely small (Figure 16.19a) or extremely large (Figure 16.19b), then the area is finite and can thus be approximated.

Let's try an example.

### Example 11

Determine the area under the curve $y = 3e^{-2x}$ when $x \geq 1$. Present the answer in the exact form.

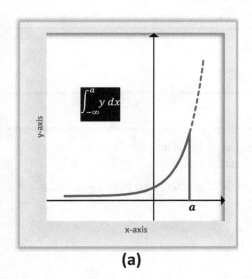
(a)

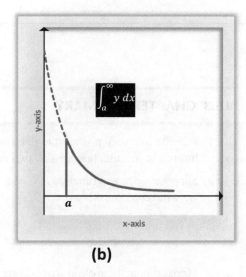
(b)

**FIGURE 16.19** Integrating to infinity illustrated: (a) from negative infinity to a given point, and (b) from a given point to positive infinity.

What did you get? Find the solution below to double-check your answer.

## Solution to Example 11

### HINT

Recall that when a number is divided by an infinity the result is zero. Also, multiplying an infinity with a constant $k$ results in another infinity, i.e., $k \times \infty = \infty$ for all positive real $k$ or $k \times \infty = -\infty$ for all negative real $k$.

The integral expression for this is:

$$\int_1^\infty 3e^{-2x}\, dx = \left[\frac{3e^{-2x}}{-2}\right]_1^\infty = \left[-\frac{3}{2e^{2x}}\right]_1^\infty$$

$$= \left[-\frac{3}{2e^{2x}}\right]^\infty - \left[-\frac{3}{2e^{2x}}\right]^1$$

$$= \left(-\frac{3}{2e^{2\times\infty}}\right) - \left(-\frac{3}{2e^{2\times 1}}\right)$$

$$= \left(-\frac{3}{\infty}\right) - \left(-\frac{3}{2e^2}\right) = (-0) + \frac{3}{2e^2}$$

$$= \frac{3}{2e^2}$$

$$\therefore \int_1^\infty 3e^{-2x}\, dx = \frac{3}{2e^2}\ \text{unit}^2$$

## 16.3 CHAPTER SUMMARY

1) Definite integration is used to determine the area under a curve between two specified limits and the area bounded by two curves, linear or non-linear.

2) Suppose we have a function $f(x)$ and we are required to integrate it from $x = a$ to $x = b$, we will represent this as:

$$\int_a^b f(x)\, dx$$

More compactly, we can write this process as:

$$\boxed{\int_a^b f(x)\, dx = \left[\int f(x)\, dx\right]_a^b = \left[\int f(x)\, dx\right]^b - \left[\int f(x)\, dx\right]^a}$$

Definite Integration

The choice of the square brackets [ ] is in line with the standard notation and should be used in this case. Subsequent calculations, including substitution of the limits, are generally carried out using round brackets ( ).

3) Given that $s$, $v$, $a$, and $j$ represent displacement, velocity, acceleration, and jerk, respectively, we have:

$$s = \int v\,dt \quad v = \int a\,dt \quad a = \int j\,dt$$

4) The rule of decomposition of limits states that:

$$\boxed{\int_a^c f(x)\,dx = \int_a^b f(x)\,dx + \int_b^c f(x)\,dx}$$

This intermediate point is such that $a < b < c$. Also by extension, we have

$$\boxed{\int_a^d f(x)\,dx = \int_a^b f(x)\,dx + \int_b^c f(x)\,dx + \int_c^d f(x)\,dx}$$

such that $a < b < c < d$.

5) When a curve is above and below the $x$-axis in the region of upper and lower limits, we need to split the integral and evaluate separately. The portion above the horizontal axis should be evaluated and the same is done for the portion below the $x$-axis. The absolute value of the results is thereafter added together.

6) If a curve of $f(x)$ intersects a straight line of function $g(x)$ at two points $A(x_1, y_1)$ and $B(x_2, y_2)$, the area $A$ is found using:

$$\boxed{\int_{x_1}^{x_2} \{y\}\,dx = \int_{x_1}^{x_2} \{f(x) - g(x)\}\,dx = \left[\int \{f(x) - g(x)\}\,dx\right]_{x_1}^{x_2}}$$

Alternatively

$$\boxed{\int_{x_1}^{x_2} \{y\}\,dx = \int_{x_1}^{x_2} \{g(x)\}\,dx - \int_{x_1}^{x_2} \{f(x)\}\,dx}$$

7) Integrating to infinity is a situation where one of the two limits is infinity, i.e., $\pm\infty$. Essentially, when there is no bound on one side. This is shown below.

$$\boxed{\int_{-\infty}^{a} \{y\}\, dx} \quad \textbf{OR} \quad \boxed{\int_{a}^{\infty} \{y\}\, dx}$$

The above implies that we are interested in an area covered by the curve $y$ in the range $x \leq a$ (i.e., from $a$ and everything to the left) or in the range $x \geq a$ (i.e., from $a$ and everything to the right). In general, the area in either of these two cases is expected to be infinity itself.

****

## 16.4 FURTHER PRACTICE

To access complementary contents, including additional exercises, please go to www.dszak.com.

# 17 Advanced Integration I

## Learning Outcomes

Once you have studied the content of this chapter, you should be able to:

- Integrate functions of the form $(ax + b)^n$
- Use the principle of reverse of the chain rule to solve complicated integrals
- Integrate functions of the form $\frac{f'(x)}{f(x)}$
- Integrate functions of the form $[f'(x)] \cdot [f(x)]$ and $[f'(x)] \cdot [f(x)]^n$
- Integrate rational expressions with linear denominators when it can be reduced to its partial fractions
- Use trigonometrical identities to solve complicated integrals
- Apply the technique of integration by substitution to resolve complicated expressions

## 17.1 INTRODUCTION

In the previous chapters, we delved into the world of integration and covered basic integral standards. We did this with both definite and indefinite integration. However, we encounter functions that need to be integrated but do not fit into any of the standard cases covered. We're lucky that a significant number of them share a similar form and become a family of functions that we can find a way of solving them – thanks to the fact that integration is the inverse of differentiation. In this chapter, we won't demonstrate how we've used differentiation to establish the integral of a function. Instead, we'll focus on solving these challenging families of functions.

## 17.2 INTEGRATION OF COMPLICATED FUNCTIONS

We are just about to kick start on taking common families of difficult functions that need integrating. They will be grouped under the heading that will explain the procedure to follow in resolving them. While we are making an effort to cover as many principles of solving complicated functions as possible, it will be impractical to cover all. Consequently, we may encounter functions that do not fit into any of the 'to-be-covered' families. In such cases, alternative methods, including numerical approaches, will be necessary.

### 17.2.1 Binomial of the Form $(ax+b)^n$

This rule is used for any function that takes the form of $(ax+b)^n$, and it states that

$$y = \int (ax+b)^n dx = \frac{1}{a(n+1)}(ax+b)^{n+1} + C \qquad (17.1)$$

where $n$ is any real number excluding $-1$.

There is a special rule for when $n = -1$, and this will be covered shortly. Let's make the following notes about the above rule:

**Note 1** This is a function of a function style expression. It is required that the inner function is linear. In other words, we cannot use this to integrate a function with the inner function being a quadratic or another higher order.

**Note 2** We will integrate this $(ax+b)^n$ as we do for $x^n$. We will then divide our answer with the derivative of the inner function $(ax+b)$ or the coefficient of $x$, whichever is easier to spot.

That's all. Let's try some examples.

#### Example 1

Determine $y$ in the following:

a) $\frac{dy}{dx} = (3x-5)^5$    b) $\frac{dy}{dx} = \frac{1}{(1-2x)^4}$    c) $\frac{dy}{dx} = (7-x)^{\frac{1}{2}}$    d) $\frac{dy}{dx} = \frac{10}{\sqrt[3]{5x+1}}$

What did you get? Find the solution below to double-check your answer.

#### Solution to Example 1

**a)** $\frac{dy}{dx} = (3x-5)^5$
**Solution**

$$\frac{dy}{dx} = (3x-5)^5$$

Let's take the integral of both sides with respect to $x$.

$$\int \frac{dy}{dx} dx = \int (3x-5)^5 dx$$

We can assume, though not technically correct, that $dx$ cancels out, and we have

$$\int dy = \int (3x-5)^5 dx$$

The integral of the left-hand side *wrt* $y$ is $y$. Hence, we have

$$y = \int (3x-5)^5 dx$$

# Advanced Integration I

**NOTE**
This is just to highlight how we've arrived at this conclusion and will be skipped in the subsequent examples.

The right-hand side matches the current binomial standard with $n = 5$ and $a = 3$. Thus

$$y = \int (3x-5)^5 dx$$

$$= \frac{(3x-5)^{5+1}}{3(5+1)} + C$$

$$= \frac{(3x-5)^6}{3(6)} + C$$

$$\therefore \int (3x-5)^5 dx = \frac{(3x-5)^6}{18} + C$$

**b)** $\frac{dy}{dx} = \frac{1}{(1-2x)^4}$
**Solution**

$$\frac{dy}{dx} = \frac{1}{(1-2x)^4}$$

which implies

$$y = \int \frac{1}{(1-2x)^4} dx$$

$$= \int (1-2x)^{-4} dx$$

In the above modified form, we can then say that $n = -4$ and $a = -2$, thus

$$y = \frac{(1-2x)^{-4+1}}{-2(-4+1)} + C$$

$$= \frac{(1-2x)^{-3}}{-2(-3)} + C$$

$$= \frac{(1-2x)^{-3}}{6} + C$$

$$\therefore \int \frac{1}{(1-2x)^4} dx = \frac{1}{6(1-2x)^3} + C$$

**c)** $\frac{dy}{dx} = (7-x)^{\frac{1}{2}}$
**Solution**

$$\frac{dy}{dx} = (7-x)^{\frac{1}{2}}$$

which implies

$$y = \int (7-x)^{\frac{1}{2}}$$

$n = \frac{1}{2}$ and $a = -1$, thus

$$y = \frac{(7-x)^{\frac{1}{2}+1}}{-1 \times \left(\frac{1}{2}+1\right)} + C$$

$$= -\frac{(7-x)^{\frac{3}{2}}}{\frac{3}{2}} + C$$

$$\therefore \int (7-x)^{\frac{1}{2}} dx = -\frac{2(7-x)^{\frac{3}{2}}}{3} + C$$

---

**d)** $\frac{dy}{dx} = \frac{10}{\sqrt[3]{5x+1}}$

**Solution**

$$\frac{dy}{dx} = \frac{10}{\sqrt[3]{5x+1}}$$

which implies

$$y = \int \frac{10}{\sqrt[3]{5x+1}} dx$$

$$= \int \frac{10}{(5x+1)^{\frac{1}{3}}} dx = \int 10(5x+1)^{-\frac{1}{3}} dx$$

$$= 10 \int (5x+1)^{-\frac{1}{3}} dx$$

In the above modified form, we can then say that $n = -\frac{1}{3}$ and $a = 5$, thus

$$y = 10 \left[ \frac{(5x+1)^{-\frac{1}{3}+1}}{5\left(-\frac{1}{3}+1\right)} \right] + C$$

$$= 10 \left[ \frac{(5x+1)^{\frac{2}{3}}}{5\left(\frac{2}{3}\right)} \right] + C$$

$$= 10 \left[ \frac{3(5x+1)^{\frac{2}{3}}}{5(2)} \right] + C$$

$$\therefore \int \frac{10}{\sqrt[3]{5x+1}} dx = 3(5x+1)^{\frac{2}{3}} + C$$

## 17.2.2 Linear Functions of the Form $f'(ax+b)$

The linear function rule is used when integrating a function of a linear function such that the inner function is a linear expression of the form $ax+b$, where $a$ and $b$ are real numbers and $a \neq 0$. In general, the function is of the form $f'(ax+b)$.

If $b \neq 0$, then we have $f'(ax+b)$. We can state that:

$$\boxed{\int f'(ax+b)\,dx = \frac{1}{a}f(ax+b) + C} \qquad (17.2)$$

and if $b = 0$, we have

$$\boxed{\int f'(ax)\,dx = \frac{1}{a}f(ax) + C} \qquad (17.3)$$

Here is the summary of this rule.

> *If a function shares the format with a standard integral but has an inner linear function, denote the inner function as u. Then determine the integral wrt to u when the inner function has been replaced with u and divide the answer with u' i.e. the derivative of u.*

Let's make the following notes about this standard:

**Note 1** Given that $y = f[u(x)]$, where $y$ is a function of $u$ and $u$ is in turn a function of $x$, we can say that the differentiation coefficient equals $f'(u) \times u'(x)$. In other words, the derivative of the function is the derivative of $y$ wrt $u$ (as a variable) **multiplied** by the derivative of $u$ wrt $x$.

**Note 2** For integration, given that $\emptyset = g[u(x)]$, where $\emptyset$ is a function of $u$ and $u$ is in turn a function of $x$, we say that the integration is equal to $\frac{\int g(u).dx}{u'(x)}$. In other words, the integral of the function is the integral of $\emptyset$ wrt $u$ (as a variable) **divided** by the derivative of $u$ wrt to $x$.

**Note 3** In both cases, we treat the inner function $u(x)$ as an independent variable and find the derivative (or integral) of the resulting function.

**Note 4** We also find the derivative of the inner function $u'(x)$ and we **multiply** it with the result in 3 if it is differentiation or **divide** if it is integration.

**Note 5** In integration, $u(x)$ must be a linear function, such that $u = ax + b$. However, for differentiation, the inner function can be any function. For example, we can differentiate or integrate a function $e^{2x-3}$ because the inner function $2x - 3$ is linear. However, it is not possible to integrate $e^{x^2-3x+1}$ or $e^{\sin x}$ using this rule, as the inner functions ($x^2 - 3x + 1$ and $\sin x$) are not linear. Nevertheless, both can be differentiated.

Some of the integral standards we outlined in Chapter 15 were based on this rule. In fact, $\int (ax+b)^n dx = \frac{1}{a(n+1)}(ax+b)^{n+1} + C$ that we just covered is a special form of this rule, as it is a function of a function expression.

Before we move to examples, the previously mentioned standards are re-produced here in Table 17.1. Although, we will stick to this rule and possibly compare the two, there will also be inevitable situations when we find ourselves making use of the standards.

## TABLE 17.1
### Standard Integrals for a Linear Expression of the Form $ax + b$

| $y$ | $\int y\,dx$ | $y$ | $\int y\,dx$ |
|---|---|---|---|
| $\dfrac{1}{ax+b}$ | $\dfrac{1}{a}\ln|ax+b| + C$ | $\dfrac{1}{ax}$ | $\dfrac{1}{a}\ln|ax| + C$ |
| $e^{ax+b}$ | $\dfrac{1}{a}e^{ax+b} + C$ | $e^{ax}$ | $\dfrac{1}{a}e^{ax} + C$ |
| $z^{ax+b}$ | $\dfrac{z^{ax+b}}{a\ln z} + C$ | $z^{ax}$ | $\dfrac{z^{ax}}{a\ln z} + C$ |
| $\ln(ax+b)$ | $\left(x + \dfrac{b}{a}\right)\ln(ax+b) - x + C$ | $\ln(ax)$ | $x\ln(ax) - x + C$ |
| $\sin(ax+b)$ | $-\dfrac{1}{a}\cos(ax+b) + C$ | $\sin(ax)$ | $-\dfrac{1}{a}\cos(ax) + C$ |
| $\cos(ax+b)$ | $\dfrac{1}{a}\sin(ax+b) + C$ | $\cos(ax)$ | $\dfrac{1}{a}\sin(ax) + C$ |
| $\tan(ax+b)$ | $\dfrac{1}{a}\ln|\sec(ax+b)| + C$ | $\tan(ax)$ | $\dfrac{1}{a}\ln|\sec(ax)| + C$ |
| $\sec(ax+b)$ | $\dfrac{1}{a}\ln|\tan(ax+b) + \sec(ax+b)| + C$ | $\sec(ax)$ | $\dfrac{1}{a}\ln|\tan(ax) + \sec(ax)| + C$ |
| $\text{cosec}(ax+b)$ | $-\dfrac{1}{a}\ln|\cot(ax+b) + \text{cosec}(ax+b)| + C$ | $\text{cosec}(ax)$ | $-\dfrac{1}{a}\ln|\cot(ax) + \text{cosec}(ax)| + C$ |
| $\cot(ax+b)$ | $\dfrac{1}{a}\ln|\sin(ax+b)| + C$ | $\cot(ax)$ | $\dfrac{1}{a}\ln|\sin(ax)| + C$ |
| $\sec^2(ax+b)$ | $\dfrac{1}{a}\tan(ax+b) + C$ | $\sec^2(ax)$ | $\dfrac{1}{a}\tan(ax) + C$ |
| $\text{cosec}^2(ax+b)$ | $-\dfrac{1}{a}\cot(ax+b) + C$ | $\text{cosec}^2(ax)$ | $-\dfrac{1}{a}\cot(ax) + C$ |
| $\sec(ax+b)\tan(ax+b)$ | $\dfrac{1}{a}\sec(ax+b) + C$ | $\sec(ax)\tan(ax)$ | $\dfrac{1}{a}\sec(ax) + C$ |
| $\text{cosec}(ax+b)\cot(ax+b)$ | $-\dfrac{1}{a}\text{cosec}(ax+b) + C$ | $\text{cosec}(ax)\cot(ax)$ | $-\dfrac{1}{a}\text{cosec}(ax) + C$ |

# Advanced Integration I

That's all. Let's try some examples.

## Example 2

Solve the following integrals.

a) $\int 3\sin(5x-1)\,dx$

b) $\int \operatorname{cosec}^2(7\varphi - \pi)\,d\varphi$

c) $\int 8e^{2x-3}\,dx$

d) $\int 5^{2-3x}\,dx$

e) $\int \sec\left(4\theta + \frac{1}{3}\pi\right)d\theta$

f) $\int 12\ln(4x-15)\,dx$

g) $\int \left(e^{3x} - e^{1-x}\right)^2 dx$

h) $\int \dfrac{3\sin\left(\frac{1}{2}x-5\right)-1}{\cos^2\left(\frac{1}{2}x-5\right)}\,dx$

What did you get? Find the solution below to double-check your answer.

## Solution to Example 2

### HINT

In this case, we will work through the problem with little explanation unless it is essentially necessary. Do pay attention to the subtle skills or techniques being used.

**a)** $\int 3\sin(5x-1)\,dx$
**Solution**

$$\int 3\sin(5x-1)\,dx = 3\int \sin(5x-1)\,dx$$

$$= 3\left[\frac{1}{5}\times -\cos(5x-1)\right] + C$$

$$\therefore \int 3\sin(5x-1)\,dx = -\frac{3}{5}\cos(5x-1) + C$$

**b)** $\int \operatorname{cosec}^2(7\varphi - \pi)\,d\varphi$
**Solution**

$$\int \operatorname{cosec}^2(7\varphi - \pi)\,d\varphi = \frac{1}{7}\times -\cot(7\varphi - \pi) + C$$

$$\therefore \int \operatorname{cosec}^2(7\varphi - \pi)\,d\varphi = -\frac{1}{7}\cot(7\varphi - \pi) + C$$

**c)** $\int 8e^{2x-3} dx$
**Solution**

$$\int 8e^{2x-3} dx = 8 \int e^{2x-3} dx$$
$$= 8\left[\frac{1}{2} \times e^{2x-3}\right] + C$$
$$\therefore \int 8e^{2x-3} dx = 4e^{2x-3} + C$$

---

**d)** $\int 5^{2-3x} dx$
**Solution**

$$\int 5^{2-3x} dx = -\frac{1}{3} \times \frac{5^{2-3x}}{\ln 5} + C$$
$$\therefore \int 5^{2-3x} dx = -\frac{5^{2-3x}}{3 \ln 5} + C$$

---

**e)** $\int \sec\left(4\theta + \frac{1}{3}\pi\right) d\theta$
**Solution**

$$\int \sec\left(4\theta + \frac{1}{3}\pi\right) d\theta = \frac{1}{4} \ln\left|\tan\left(4\theta + \frac{1}{3}\pi\right) + \sec\left(4\theta + \frac{1}{3}\pi\right)\right| + C$$
$$\therefore \int \sec\left(4\theta + \frac{1}{3}\pi\right) = \frac{1}{4} \ln\left|\tan\left(4\theta + \frac{1}{3}\pi\right) + \sec\left(4\theta + \frac{1}{3}\pi\right)\right| + C$$

---

**f)** $\int 12 \ln(4x - 15) dx$
**Solution**

$$\int 12 \ln(4x - 15) dx = 12 \int \ln(4x - 15) dx$$
$$= 12\left[\frac{(4x - 15) \ln(4x - 15) - 4x}{4}\right] + C$$
$$= 3\left[(4x - 15) \ln(4x - 15) - 4x\right] + C$$
$$\therefore \int 12 \ln(4x - 15) dx = 3\left[(4x - 15) \ln(4x - 15) - 4x\right] + C$$

---

**g)** $\int \left(e^{3x} - e^{1-x}\right)^2 dx$
**Solution**
The integral expression needs to be simplified first as:

$$\left(e^{3x} - e^{1-x}\right)^2 = \left(e^{3x}\right)^2 - 2\left(e^{3x}\right)\left(e^{1-x}\right) + \left(e^{1-x}\right)^2$$
$$= e^{6x} - 2e^{3x+1-x} + e^{2(1-x)}$$
$$= e^{6x} - 2e^{2x+1} + e^{2-2x}$$

# Advanced Integration I

Thus

$$\int (e^{3x} - e^{1-x})^2 \, dx = \int (e^{6x} - 2e^{2x+1} + e^{2-2x}) \, dx$$

$$= \frac{1}{6} \times e^{6x} - 2\left(\frac{1}{2} \times e^{2x+1}\right) + \left(-\frac{1}{2} \times e^{2-2x}\right) + C$$

$$= \frac{1}{6} e^{6x} - e^{2x+1} - \frac{1}{2} \times e^{2-2x} + C$$

$$\therefore \int (e^{3x} - e^{1-x})^2 \, dx = \frac{1}{6} e^{6x} - e^{2x+1} - \frac{1}{2} \times e^{2-2x} + C$$

---

h) $\int \dfrac{3\sin\left(\frac{1}{2}x - 5\right) - 1}{\cos^2\left(\frac{1}{2}x - 5\right)} dx$

**Solution**

The integral expression needs to be simplified first as:

$$\frac{3\sin\left(\frac{1}{2}x - 5\right) - 1}{\cos^2\left(\frac{1}{2}x - 5\right)} = \frac{3\sin\left(\frac{1}{2}x - 5\right)}{\cos^2\left(\frac{1}{2}x - 5\right)} - \frac{1}{\cos^2\left(\frac{1}{2}x - 5\right)}$$

$$= 3\left[\frac{\sin\left(\frac{1}{2}x - 5\right)}{\cos\left(\frac{1}{2}x - 5\right)}\right] \times \left[\frac{1}{\cos\left(\frac{1}{2}x - 5\right)}\right] - \frac{1}{\cos^2\left(\frac{1}{2}x - 5\right)}$$

$$= 3\tan\left(\frac{1}{2}x - 5\right)\sec\left(\frac{1}{2}x - 5\right) - \sec^2\left(\frac{1}{2}x - 5\right)$$

Thus

$$\int \frac{3\sin\left(\frac{1}{2}x - 5\right) - 1}{\cos^2\left(\frac{1}{2}x - 5\right)} dx = \int 3\tan\left(\frac{1}{2}x - 5\right)\sec\left(\frac{1}{2}x - 5\right) - \sec^2\left(\frac{1}{2}x - 5\right) dx$$

$$= 3\left[\frac{\sec\left(\frac{1}{2}x - 5\right)}{\frac{1}{2}}\right] - \left[\frac{\tan\left(\frac{1}{2}x - 5\right)}{\frac{1}{2}}\right] + C$$

$$= 6\sec\left(\frac{1}{2}x - 5\right) - 2\tan\left(\frac{1}{2}x - 5\right) + C$$

$$\therefore \int \frac{3\sin\left(\frac{1}{2}x - 5\right) - 1}{\cos^2\left(\frac{1}{2}x - 5\right)} dx = 6\sec\left(\frac{1}{2}x - 5\right) - 2\tan\left(\frac{1}{2}x - 5\right) + C$$

## 17.2.3 INTEGRATION OF THE FORM $\int \frac{f'(x)}{f(x)} dx$

Rational functions are generally difficult to handle, and their integration comes with a greater challenge. However, when the dividend (or numerator) is a derivative of the divisor (or denominator), then the integral is quite easy. The rule for this case states that:

$$y = \int \frac{f'(x)}{f(x)} dx = \ln |f(x)| + C \qquad (17.4)$$

This makes sense because when we differentiate $\ln |f(x)|$ wrt $x$, using the chain rule, we will arrive at $\frac{f'(x)}{f(x)}$. The main challenge in using this standard is to recognise that such a relationship exists. One will need to give special attention to a possible modification that would be necessary to get the form above. Let's try some examples.

### Example 3

Integrate the following functions wrt the indicated variables.

a) $\frac{2}{2x-7}$, $[x]$ 
b) $\frac{1}{5t-3}$, $[t]$ 
c) $\frac{4x-3}{2x^2-3x+1}$, $[x]$

What did you get? Find the solution below to double-check your answer.

### Solution to Example 3

**a)** $\frac{2}{2x-7}$, $[x]$
**Solution**
The numerator, 2, is the derivative of the denominator $2x - 7$, hence we have the $\frac{f'(x)}{f(x)}$ relation. Therefore,

$$\int \frac{2}{2x-7} dx = \ln |2x - 7| + C$$

$$\therefore \int \frac{2}{2x-7} dx = \ln |2x - 7| + C$$

This is straightforward.

**b)** $\frac{1}{5t-3}$, $[t]$
**Solution**
The numerator, 1, is NOT the derivative of the denominator $5t - 3$, and it seems as though we cannot use the $\frac{f'(x)}{f(x)}$ relation. We will show this shortly, but before that we can use the standard for integrating

# Advanced Integration I

rational functions $\frac{1}{ax+b}$ (Table 17.1). Therefore,

$$\int \frac{1}{5t-3} dt = \frac{\ln|5t-3|}{5} + C$$

To use the format $\frac{f'(x)}{f(x)}$, we need to tweak the expression as follows:

$$\int \frac{1}{5t-3} dt = \frac{5}{5} \int \frac{1}{5t-3} dt$$

$$= \frac{1}{5} \int \frac{5}{5t-3} dt$$

Now the format $\frac{f'(x)}{f(x)}$ is established. Proceed as per standard, and we have:

$$= \frac{1}{5} [\ln|5t-3|] + C$$

$$\therefore \int \frac{1}{5t-3} dt = \frac{\ln|5t-3|}{5} + C$$

Not as straightforward as the first, but equally good.

---

c) $\frac{4x-3}{2x^2-3x+1}$, $[x]$

**Solution**

$$\int \frac{4x-3}{2x^2-3x+1} dx = \ln|2x^2-3x+1| + C$$

$$\therefore \int \frac{4x-3}{2x^2-3x+1} dx = \ln|2x^2-3x+1| + C$$

---

Let's try another set of examples.

### Example 4

Evaluate the following integrals.

a) $\int \frac{3x-1}{(3x-5)(x+1)} dx$  
b) $\int \cot\theta \, d\theta$  
c) $\int \tan\theta \, d\theta$  
d) $\int \frac{2x(3x-2)}{(x+1)(x^2-2x+2)} dx$

e) $\int \frac{5\sec^2 2\theta}{\tan 2\theta - 3} d\theta$  
f) $\int \frac{e^{-2x}}{3e^{-2x}-5} dx$  
g) $\int \frac{\sin 3x + e^x}{3e^x - \cos 3x} dx$

What did you get? Find the solution below to double-check your answer.

## Solution to Example 4

**HINT**

In this case, we will work through the problem with little explanation unless it is essentially necessary. Do pay attention to the subtle skills or techniques being used.

**a)** $\int \frac{3x-1}{(3x-5)(x+1)} dx$

**Solution**

$$\int \frac{3x-1}{(3x-5)(x+1)} dx = \int \frac{3x-1}{3x^2 - 2x - 5} dx$$

$$= \frac{2}{2} \int \frac{3x-1}{3x^2 - 2x - 5} dx$$

$$= \frac{1}{2} \int \frac{2(3x-1)}{3x^2 - 2x - 5} dx$$

$$= \frac{1}{2} \int \frac{6x-2}{3x^2 - 2x - 5} dx$$

$$= \frac{1}{2} \ln |3x^2 - 2x - 5| + C$$

$$\therefore \int \frac{3x-1}{(3x-5)(x+1)} dx = \frac{1}{2} \ln |3x^2 - 2x - 5| + C$$

**b)** $\int \cot \theta \, d\theta$

**Solution**

$$\int \cot \theta \, d\theta = \int \frac{\cos \theta}{\sin \theta} d\theta$$

$$= \ln |\sin \theta| + C$$

$$\therefore \int \cot \theta \, d\theta = \ln |\sin \theta| + C$$

**c)** $\int \tan \theta \, d\theta$

**Solution**

$$\int \tan \theta \, d\theta = \int \frac{\sin \theta}{\cos \theta} d\theta$$

$$= -\int \frac{-\sin \theta}{\cos \theta} d\theta = -\ln |\cos \theta| + C$$

$$\therefore \int \tan \theta \, d\theta = -\ln |\cos \theta| + C$$

OR

$$\therefore \int \tan \theta \, d\theta = -\ln |\cos \theta| + C = \ln \left| \frac{1}{\cos \theta} \right| + C = \ln |\sec \theta| + C$$

These are two versions that you would find in the Table of Standard Integrals for the tangent function.

Let us quickly add a couple of notes here:

**Note 1** Sometimes we may need or be asked to express this standard (or any similar integral) in a single logarithmic term. Using the first answer for this, we will proceed as:

$$-\ln|\cos\theta| + C = -\ln|\cos\theta| + \ln k$$

$C$ is constant, hence $\ln k$, when $k$ is constant, is also a constant. Thus

$$-\ln|\cos\theta| + C = -\ln|\cos\theta| + \ln k$$
$$= \ln k - \ln|\cos\theta|$$
$$= \ln\left|\frac{k}{\cos\theta}\right|$$

Alternatively, we can avoid the fraction by stating that

$$-\ln|\cos\theta| + C = -\ln|\cos\theta| - \ln k$$
$$= -[\ln|\cos\theta| + \ln k]$$
$$= -\ln|k\cos\theta|$$

**Note 2** If we modify the last answer, we will arrive at $-\ln|k\cos\theta| = \ln\left|\frac{1}{k\cos\theta}\right|$. You may wonder that it does not look exactly like the first answer, i.e., $\ln\left|\frac{k}{\cos\theta}\right|$. You are quite right; however, in the first case, we replace $C$ with $\ln k$, whilst in the second case, it was replaced with $-\ln k$.

---

**d)** $\int \frac{2x(3x-2)}{(x+1)(x^2-2x+2)}\,dx$

**Solution**

$$\int \frac{2x(3x-2)}{(x+1)(x^2-2x+2)}\,dx = \int \frac{2x(3x-2)}{(x+1)(x^2-2x+2)}\,dx$$
$$= \int \frac{6x^2-4x}{x^3-x^2+2}\,dx = \int \frac{2(3x^2-2x)}{x^3-x^2+2}\,dx$$
$$= 2\int \frac{3x^2-2x}{x^3-x^2+2}\,dx$$
$$= 2\ln|(x^3-x^2+2)| + C$$

$$\therefore \int \frac{x(3x-2)}{(x+1)(x^2-2x+2)}\,dx = 2\ln|(x^3-x^2+2)| + C$$

e) $\int \frac{5\sec^2 2\theta}{\tan 2\theta - 3} d\theta$

**Solution**

$$\int \frac{5\sec^2 2\theta}{\tan 2\theta - 3} d\theta = \frac{2}{2} \int \frac{5\sec^2 2\theta}{\tan 2\theta - 3} d\theta$$

$$= \frac{5}{2} \int \frac{2\sec^2 2\theta}{\tan 2\theta - 3} d\theta$$

$$= \frac{5}{2} \ln|\tan 2\theta - 3| + C$$

$$\therefore \int \frac{5\sec^2 2\theta}{\tan 2\theta - 3} d\theta = \frac{5}{2} \ln|\tan 2\theta - 3| + C$$

f) $\int \frac{e^{-2x}}{3e^{-2x} - 5} dx$

**Solution**

$$\int \frac{e^{-2x}}{3e^{-2x} - 5} dx = -\frac{1}{6} \int \frac{-6e^{-2x}}{3e^{-2x} - 5} dx$$

$$= -\frac{1}{6} \ln|3e^{-2x} - 5| + C$$

$$\therefore \int \frac{e^{-2x}}{3e^{-2x} - 5} dx = -\frac{1}{6} \ln|3e^{-2x} - 5| + C$$

g) $\int \frac{\sin 3x + e^x}{3e^x - \cos 3x} dx$

**Solution**

$$\int \frac{\sin 3x + e^x}{3e^x - \cos 3x} dx = \frac{1}{3} \int \frac{3(\sin 3x + e^x)}{3e^x - \cos 3x} dx$$

$$= \frac{1}{3} \int \frac{3e^x + 3\sin 3x}{3e^x - \cos 3x} dx$$

$$= \frac{1}{3} \ln|3e^x - \cos 3x| + C$$

$$\therefore \int \frac{\sin 3x + e^x}{3e^x - \cos 3x} dx = \frac{1}{3} \ln|3e^x - \cos 3x| + C$$

### 17.2.4 Integration of the Form $\int [f'(x)].[f(x)] dx$ and $\int [f'(x)].[f(x)]^n dx$

This is another rule to integrate a product of two functions. It states that:

$$y = \int [f'(x)].[f(x)]^n dx = \frac{1}{n+1} [f(x)]^{n+1} + C \qquad (17.5)$$

where $n$ is any real number. If $n = 1$, we have a modified version as:

$$y = \int f'(x) f(x) dx = \frac{1}{2} [f(x)]^2 + C \qquad (17.6)$$

# Advanced Integration I

It seems the rule is self-explanatory, which is similar to the previous rule and based on the reverse chain rule. Nonetheless, let's provide these notes to guide us in applying it.

**Note 1** Establish that there is a product of two functions, say $f(x)$ and $g(x)$, where one is a derivative of the other. Say $g(x) = f'(x)$.

**Note 2** Based on Note 1, treat this as though you are integrating a power function, $\int x^n dx$.

Note that sometimes we need to modify the expression to ensure that the rule can be used. That's everything for now. Let's try some examples.

## Example 5

Solve the following:

a) $\int 3x^2(x^3 - 8)^4 dx$

b) $\int \frac{5x-3}{\sqrt{5x^2-6x+1}} dx$

c) $\int \frac{12}{e^{2x}(e^{-2x}-5)^3} dx$

d) $\int 8 \cos\theta \sin^3\theta \, d\theta$

e) $\int \cot x . \csc^2 x \, dx$

f) $\int \frac{\sec\theta \tan\theta}{1+\tan^2\theta} d\theta$

What did you get? Find the solution below to double-check your answer.

## Solution to Example 5

**a)** $\int 3x^2(x^3 - 8)^4 dx$
**Solution**
$f(x) = x^3 - 8, f'(x) = 3x^2$, $n = 4$, and $n + 1 = 5$. Hence

$$\int 3x^2(x^3 - 8)^4 dx = \frac{1}{5}(x^3 - 8)^5 + C$$

$$\therefore \int 3x^2(x^3 - 8)^4 dx = \frac{1}{5}(x^3 - 8)^5 + C$$

**b)** $\int \frac{5x-3}{\sqrt{5x^2-6x+1}} dx$
**Solution**
This is a rational function and looks like we cannot use the current rule. Let's modify this a bit as:

$$\int \frac{5x-3}{\sqrt{5x^2-6x+1}} dx = \int \frac{5x-3}{(5x^2-6x+1)^{\frac{1}{2}}} dx$$

$$= \int (5x-3)(5x^2-6x+1)^{-\frac{1}{2}} dx$$

The above shows that it's a product of two functions.

Hence, $f(x) = 5x^2 - 6x + 1$, $f'(x) = 10x - 6$, $n = -\frac{1}{2}$, and $n + 1 = \frac{1}{2}$. Thus

$$\int \frac{5x - 3}{\sqrt{5x^2 - 6x + 1}} dx = \frac{1}{2} \int 2(5x - 3)(5x^2 - 6x + 1)^{-\frac{1}{2}} dx$$

$$= \frac{1}{2} \int (10x - 6)(5x^2 - 6x + 1)^{-\frac{1}{2}} dx$$

$$= \frac{1}{2} \left[ \frac{(5x^2 - 6x + 1)^{\frac{1}{2}}}{\frac{1}{2}} \right] + C = (5x^2 - 6x + 1)^{\frac{1}{2}} + C$$

$$\therefore \int \frac{5x - 3}{\sqrt{5x^2 - 6x + 1}} dx = \sqrt{5x^2 - 6x + 1} + C$$

c) $\int \frac{12}{e^{2x}(e^{-2x} - 5)^3} dx$

**Solution**

This looks similar to the one above; let's modify this a bit as:

$$\int \frac{12}{e^{2x}(e^{-2x} - 5)^3} dx = \int 12 e^{-2x} (e^{-2x} - 5)^{-3} dx$$

Hence, $f(x) = e^{-2x} - 5$, $f'(x) = -2e^{-2x}$, $n = -3$ and $n + 1 = -2$. Thus

$$\int \frac{12}{e^{2x}(e^{-2x} - 5)^3} dx = \int 12 e^{-2x} (e^{-2x} - 5)^{-3} dx$$

$$= -6 \int -2e^{-2x} (e^{-2x} - 5)^{-3} dx$$

$$= -6 \left[ \frac{(e^{-2x} - 5)^{-2}}{-2} \right] + C = \frac{3}{(e^{-2x} - 5)^2} + C$$

$$\therefore \int \frac{12}{e^{2x}(e^{-2x} - 5)^3} dx = \frac{3}{(e^{-2x} - 5)^2} + C$$

d) $\int 8 \cos \theta \sin^3 \theta \, d\theta$

**Solution**

Let's modify this a bit as:

$$\int 8 \cos \theta \sin^3 \theta \, d\theta = 8 \int \cos \theta [\sin \theta]^3 d\theta$$

Hence, $f(\theta) = \sin \theta$, $f'(\theta) = \cos \theta$, $n = 3$, and $n + 1 = 4$. Thus

$$\int 8 \cos \theta \sin^3 \theta \, d\theta = 8 \int \cos \theta [\sin \theta]^3 d\theta$$

$$= 8 \left[ \frac{[\sin \theta]^4}{4} \right] + C = 2 \sin^4 \theta + C$$

$$\therefore \int 8 \cos \theta \sin^3 \theta \, d\theta = 2 \sin^4 \theta + C$$

# Advanced Integration I

**e)** $\int \cot x \cdot \operatorname{cosec}^2 x \, dx$

**Solution**

Hence, $f(x) = \cot x$, $f'(x) = -\operatorname{cosec}^2 x$, $n = 1$, and $n + 1 = 2$. Thus

$$\int \cot x \cdot \operatorname{cosec}^2 x \, dx = -\int [-\operatorname{cosec}^2 x][\cot x] \, dx$$

$$= -\left[\frac{[\cot x]^2}{2}\right] + C$$

$$= -\frac{1}{2}[\cot x]^2 + C$$

$$\therefore \int \cot x \cdot \operatorname{cosec}^2 x \, dx = -\frac{1}{2}\cot^2 x + C$$

**f)** $\int \frac{\sec\theta \tan\theta}{1+\tan^2\theta} \, d\theta$

**Solution**

Let's modify this a bit as:

$$\int \frac{\sec\theta \tan\theta}{1+\tan^2\theta} \, d\theta = \int \frac{\sec\theta \tan\theta}{\sec^2\theta} \, d\theta$$

$$= \int [\sec\theta \tan\theta] \cdot [\sec\theta]^{-2} \, d\theta$$

Hence, $f(x) = \sec\theta$, $f'(x) = \sec\theta \tan\theta$, $n = -2$, and $n + 1 = -1$. Thus

$$\int \frac{\sec\theta \tan\theta}{1+\tan^2\theta} \, d\theta = \int [\sec\theta \tan\theta] \cdot [\sec\theta]^{-2} \, d\theta$$

$$= \left[\frac{[\sec\theta]^{-1}}{-1}\right] + C$$

$$= -\frac{1}{\sec\theta} + C$$

$$\therefore \int \frac{\sec\theta \tan\theta}{1+\tan^2\theta} \, d\theta = -\frac{1}{\sec\theta} + C = -\cos\theta + C$$

**NOTE**

We could have simplified this further and avoid using this rule, as follows:

$$\int \frac{\sec\theta \tan\theta}{1+\tan^2\theta} \, d\theta = \int \frac{\sec\theta \tan\theta}{\sec^2\theta} \, d\theta$$

$$= \int \frac{\tan\theta}{\sec\theta} \, d\theta = \int \frac{\left(\frac{\sin\theta}{\cos\theta}\right)}{\left(\frac{1}{\cos\theta}\right)} \, d\theta$$

$$= \int \sin\theta \, d\theta = -\cos\theta + C$$

$$\therefore \int \frac{\sec\theta \tan\theta}{1+\tan^2\theta} \, d\theta = -\cos\theta + C$$

## 17.2.5 Integration of the Form $\int \left[\dfrac{du}{dx}\right] . [f'(u)] . dx$

This is used to find the integral of a product of two functions. It states that

$$y = \int \left[\dfrac{du}{dx}\right] . [f'(u)] \, dx = f(u) + C$$

This rule is based on the

a) function of a function rule, and that
b) integration is a direct inverse of differentiation, and vice versa.

Before we show this, let's quickly cover the following points, which are relevant to the use of this rule:

**Note 1** Establish that there is a product of two functions, say $f'(x)$ and $g(x)$.
**Note 2** One of the functions, $f'(x)$, is a function of function with inner function $u(x)$, such that $g(x) = u'(x)$.
**Note 3** Integrate $f'(u)$ to obtain $f(u)$, ignoring the fact it is a function of a function, then replace $u$.
**Note 4** Hence, $\int [g(x)] . [f'(x)] \, dx = f(u) + C$.

Recall that taking the integral will undo differentiation. Given a function of function $y = f(x)$, such that $y = f(u)$ and $u = g(x)$, then we can differentiate this function by applying the chain rule as:

$$\dfrac{dy}{dx} = \dfrac{dy}{du} \times \dfrac{du}{dx}$$

This is a forward process. Let us reverse this by integrating the result on the RHS *wrt* $x$. Remember that we need to apply the same to the LHS as:

$$\int \left[\dfrac{dy}{dx}\right] dx = \int \left(\left[\dfrac{dy}{du}\right] \times \left[\dfrac{du}{dx}\right]\right) dx$$

Swap the LHS and RHS of the equation, we have

$$\int \left(\left[\dfrac{dy}{du}\right] \times \left[\dfrac{du}{dx}\right]\right) dx = \int \left[\dfrac{dy}{dx}\right] dx$$

$$\int \left(\left[\dfrac{dy}{du}\right] \times \left[\dfrac{du}{dx}\right]\right) dx = \int dy$$

$$\int \left[\dfrac{du}{dx}\right] \times \left[\dfrac{dy}{du}\right] dx = y + C$$

$$\int \left[\dfrac{du}{dx}\right] \times [f'(u)] \, dx = f(u) + C$$

The subtle difference between this rule and the previous one is that, unlike in this where the function of function does not have power, in the previous rule, the power is what brought the function of a function (and thus the chain rule) into being.

That's all for this rule. Note that you can always differentiate the answer to confirm if the differential coefficient matches the original question.

# Advanced Integration I

Let's try some examples to illustrate.

### Example 6

Evaluate the following:

a) $\int 4e^{4x-5} dx$

b) $\int 2x \sin(x^2 - 1) dx$

c) $\int \sec^2 3\theta\, e^{\tan 3\theta} d\theta$

d) $\int 6 \tan(e^{2x}) e^{2x} dx$

e) $\int \operatorname{cosec} \theta \cot \theta\, e^{\pi - \operatorname{cosec} \theta} d\theta$

What did you get? Find the solution below to double-check your answer.

### Solution to Example 6

a) $\int 4e^{4x-5} dx$
**Solution**
We start with this question to illustrate this rule, though we could have solved this easily using a different method.

**Method 1** Using $\int e^{ax+b} dx = \frac{1}{a} e^{ax+b} + C$

$$\int 4e^{4x-5} dx = 4 \int e^{4x-5} dx$$
$$= 4\left[\frac{1}{4} e^{4x-5}\right] + C$$
$$\therefore \int 4e^{4x-5} dx = e^{4x-5} + C$$

**Method 2** Using $\int [g(x)][f'(x)] dx = f(u) + C$

Given that $\int 4e^{4x-5} dx$, our two functions of interest are 4 and $e^{4x-5}$. Thus, if $u = 4x - 5$, then $\frac{du}{dx} = 4$. If $f'(u) = e^u$, then $f(u) = e^u$. Hence

$$\int 4e^{4x-5} dx = e^u + C = e^{4x-5} + C$$
$$\therefore \int 4e^{4x-5} dx = e^{4x-5} + C$$

**NOTE**

1) When we integrate $f'(u)$ or $e^u$, we did not consider $u$ as a function of $x$, otherwise the result will be $\frac{1}{u'(x)} e^u$ and not simply $e^u$. This is a very important point to consider in this rule.

2) We consider 4 as a function here for illustration. For this rule, a function should have a variable included.

**b)** $\int 2x \sin(x^2 - 1) \, dx$

**Solution**

Two functions of interest are $2x$ and $\sin(x^2 - 1)$. Thus, if $u = x^2 - 1$, then $\frac{du}{dx} = 2x$. If $f'(u) = \sin u$, then $f(u) = -\cos u$. Hence

$$\int 2x \cdot \sin(x^2 - 1) \, dx = -\cos u + C$$

$$= -\cos(x^2 - 1) + C$$

$$\therefore \int 2x \sin(x^2 - 1) \, dx = -\cos(x^2 - 1) + C$$

---

**c)** $\int \sec^2 3\theta \, e^{\tan 3\theta} \, d\theta$

**Solution**

Two functions of interest are $\sec^2 3\theta$ and $e^{\tan 3\theta}$. Thus, if $u = \tan 3\theta$, then $\frac{du}{d\theta} = 3\sec^2 3\theta$. If $f'(u) = e^u$, then $f(u) = e^u$. For this case, we will modify the expression a bit before proceeding, hence:

$$\int \sec^2 3\theta \cdot e^{\tan 3\theta} \, d\theta = \frac{1}{3} \int 3\sec^2 3\theta \cdot e^{\tan 3\theta} \, d\theta$$

$$= \frac{1}{3}[e^u] + C = \frac{1}{3} e^{\tan 3\theta} + C$$

$$\therefore \int \sec^2 3\theta \, e^{\tan 3\theta} \, d\theta = \frac{1}{3} e^{\tan 3\theta} + C$$

---

**d)** $\int 6 \tan(e^{2x}) e^{2x} \, dx$

**Solution**

Two functions of interest are $e^{2x}$ and $\tan(e^{2x})$. Thus, if $u = e^{2x}$, then $\frac{du}{dx} = 2e^{2x}$. If $f'(u) = \tan u$, then $f(u) = \ln|\sec u|$. For this case, we will modify the expression a bit before proceeding, hence:

$$\int 6 \tan(e^{2x}) e^{2x} \, dx = 3 \int 2e^{2x} \cdot \tan(e^{2x}) \, dx$$

$$= 3[\ln|\sec u|] + C = 3\ln|\sec(e^{2x})| + C$$

$$\therefore \int 6 \tan(e^{2x}) e^{2x} \, dx = 3\ln|\sec(e^{2x})| + C$$

---

**e)** $\int \operatorname{cosec} \theta \cot \theta \, e^{\pi - \operatorname{cosec} \theta} \, d\theta$

**Solution**

Two functions of interest are $\operatorname{cosec} \theta \cot \theta$ and $e^{\pi - \operatorname{cosec} \theta}$. Thus, if $u = \pi - \operatorname{cosec} \theta$, then $\frac{du}{d\theta} = \operatorname{cosec} \theta \cot \theta$. If $f'(u) = e^u$, then $f(u) = e^u$. For this case, we do not need further modification, hence:

$$\int \operatorname{cosec} \theta \cot \theta \, e^{\pi - \operatorname{cosec} \theta} \, d\theta = e^u + C$$

$$= e^{\pi - \operatorname{cosec} \theta} + C$$

$$\therefore \int \operatorname{cosec} \theta \cot \theta \, e^{\pi - \operatorname{cosec} \theta} \, d\theta = e^{\pi - \operatorname{cosec} \theta} + C$$

# Advanced Integration I

## 17.3 INTEGRATION BY PARTIAL FRACTION

Partial fraction, covered in Chapter 7, is a great tool in integrating some rational expressions. An expression of the form $\frac{f(x)}{g(x)}$ can be split into partial fractions, such that the denominator of each of the resulting fractions is a linear function. Once split, we can apply the integral standards for rational expressions (i.e., $\frac{1}{x}$ and $\frac{1}{ax+b}$) to solve the problem. We can deduce that we only need to spot the expression that requires this approach and work through the steps. Let's try one or two examples.

### Example 7

Evaluate $\int \frac{6x-1}{3x^2-x-14} dx$, where $x > \frac{7}{3}$. Give your answer in the simplest form as a single logarithm.

What did you get? Find the solution below to double-check your answer.

### Solution to Example 7

**HINT**

- This example will be solved using partial fractions, but can easily be solved using $\int \frac{f'(x)}{f(x)} dx$.
- It is not immediately clear here why the question states $x > \frac{7}{3}$, but it will be in the second method.

$$\int \frac{6x-1}{3x^2 - x - 14} dx$$

**Method 1** Using $\int \frac{f'(x)}{f(x)} dx$

This expression is of the form $\int \frac{f'(x)}{f(x)} dx$ and we know that this should be

$$\int \frac{6x-1}{3x^2 - x - 14} dx = \ln|3x^2 - x - 14| + C$$

**Method 2** Using partial fractions

Let's use partial fractions and compare the answers. Here we go.

$$\frac{6x-1}{3x^2 - x - 14} = \frac{6x-1}{(3x-7)(x+2)}$$
$$\equiv \frac{A}{3x-7} + \frac{B}{x+2}$$
$$\equiv \frac{A(x+2) + B(3x-7)}{(3x-7)(x+2)}$$

Let's now equate the numerators of both sides as:

$$6x - 1 \equiv A(x+2) + B(3x-7)$$

Substituting $x = -2$, we have:

$$-12 - 1 = A(-2+2) + B(-6-7)$$
$$-13 = -13B$$
$$\therefore B = 1$$

Substituting $x = 0$, we have:

$$0 - 1 = A(0+2) + B(0-7)$$
$$-1 = 2A - 7B$$
$$-1 = 2A - 7(1)$$
$$2A = -1 + 7$$
$$\therefore A = \frac{6}{2} = 3$$

Hence

$$\frac{6x-1}{3x^2 - x - 14} = \frac{3}{3x-7} + \frac{1}{x+2}$$

Let's go back to the question, and we have

$$\int \frac{6x-1}{3x^2 - x - 14} dx = \int \left( \frac{3}{3x-7} + \frac{1}{x+2} \right) dx$$
$$= 3 \int \left( \frac{1}{3x-7} \right) dx + \int \left( \frac{1}{x+2} \right) dx$$
$$= 3 \left( \frac{1}{3} \ln|3x-7| \right) + \ln|x+2| + C$$
$$= \ln|3x-7| + \ln|x+2| + C$$

We can see here that when $x = \frac{7}{3}$, both $(3x-7)$ and $(x+2)$ are positive, so we can remove the modulus sign since we are sure that the value will always be positive. Thus:

$$\int \frac{6x-1}{3x^2 - x - 14} dx = \ln(3x-7) + \ln(x+2) + C$$

Combine this using the product rule of logarithms to have

$$\int \frac{6x-1}{3x^2 - x - 14} dx = \ln(3x-7)(x+2) + C$$

$$\therefore \int \frac{6x-1}{3x^2 - x - 14} dx = \ln[(3x-7)(x+2)] + C$$

# Advanced Integration I

If we open the brackets, we will obtain:

$$\int \frac{6x-1}{3x^2-x-14}dx = \ln(3x^2-x-14) + C$$

This is the same as what we obtained when we used $\int \frac{f'(x)}{f(x)}dx$ in the first method above.

---

Let's work through another example, but now a definite integral.

### Example 8

Evaluate $\int_3^4 \frac{5}{4x^2-25}dx$. Present the answer in the exact form.

---

What did you get? Find the solution below to double-check your answer.

### Solution to Example 8

$$\int_3^4 \frac{5}{4x^2-25}dx$$

This is not of the form $\int \frac{f'(x)}{f(x)}dx$, so let's use partial fractions. Thus:

$$\frac{5}{4x^2-25} = \frac{5}{(2x)^2-5^2} = \frac{5}{(2x+5)(2x-5)}$$

$$\equiv \frac{A}{2x+5} + \frac{B}{2x-5}$$

$$\equiv \frac{A(2x-5) + B(2x+5)}{(2x+5)(2x-5)}$$

Let's now equate the numerators of both sides as:

$$5 \equiv A(2x-5) + B(2x+5)$$

Substituting $x = -\frac{5}{2}$, we have

$$5 = A(-5-5) + B(5-5)$$

$$5 = -10A$$

$$\therefore A = -\frac{1}{2}$$

Substituting $x = \frac{5}{2}$, we have

$$5 = A(5-5) + B(5+5)$$
$$5 = 10B$$
$$\therefore B = \frac{1}{2}$$

Hence

$$\frac{5}{4x^2 - 25} = \frac{\left(-\frac{1}{2}\right)}{2x+5} + \frac{\left(\frac{1}{2}\right)}{2x-5}$$
$$= -\frac{1}{2}\left(\frac{1}{2x+5}\right) + \frac{1}{2}\left(\frac{1}{2x-5}\right)$$
$$= \frac{1}{2}\left(\frac{1}{2x-5}\right) - \frac{1}{2}\left(\frac{1}{2x+5}\right)$$

We therefore have

$$\int_3^4 \frac{5}{4x^2-25} dx = \int_3^4 \frac{1}{2}\left(\frac{1}{2x-5}\right) - \frac{1}{2}\left(\frac{1}{2x+5}\right) dx$$
$$= \left[\frac{1}{2}\left(\frac{1}{2}\ln|2x-5|\right) - \frac{1}{2}\left(\frac{1}{2}\ln|2x+5|\right)\right]_3^4$$
$$= \left[\frac{1}{4}\ln|2x-5| - \frac{1}{4}\ln|2x+5|\right]_3^4$$
$$= \left[\frac{1}{4}\ln|2x-5| - \frac{1}{4}\ln|2x+5|\right]^4 - \left[\frac{1}{4}\ln|2x-5| - \frac{1}{4}\ln|2x+5|\right]^3$$
$$= \left\{\frac{1}{4}\ln|2\times 4-5| - \frac{1}{4}\ln|2\times 4+5|\right\} - \left\{\frac{1}{4}\ln|2\times 3-5| - \frac{1}{4}\ln|2\times 3+5|\right\}$$
$$= \left(\frac{1}{4}\ln|3| - \frac{1}{4}\ln|13|\right) - \left(\frac{1}{4}\ln|1| - \frac{1}{4}\ln|11|\right)$$
$$= \frac{1}{4}(\ln 3 - \ln 13 - \ln 1 + \ln 11) = \frac{1}{4}\left[\ln\left(\frac{3\times 11}{13\times 1}\right)\right]$$
$$= \frac{1}{4}\ln\frac{33}{13}$$
$$\therefore \int_3^4 \frac{5}{4x^2-25} dx = \frac{1}{4}\ln\left(\frac{33}{13}\right)$$

---

## 17.4 Integration by Substitution

We have now covered rules to solve expressions for which we may not find a direct standard integral. Yet there are other functions, complicated or quasi-complicated, which are not members of the families we've just covered. In this regard, we rely on a technique called '**integration by substitution**'. Here we are talking about functional substitution rather than the numerical substitution that we use in definite integration for example.

Integration by substitution is based on and/or requires the chain rule. Here is a summary of the steps to follow when using this technique. Identify a (sub)function, which is a part of the whole expression, and denote this as $u$. What is taken as $u$ is either stated or chosen by experience.

# Advanced Integration I

**Step 1:** Find $\frac{du}{dx}$ and re-arrange to make $dx$ the subject of the formula.
**Step 2:** Substitute for $u$ and $dx$ in the original integral.
**Step 3:** Simplify the original integral so that it is only in $u$ and $du$.
**Step 4:** For a definite integral, the limits should be changed accordingly from $x_1$ to $u_1$ and $x_2$ to $u_2$.

Integration by substitution is more 'universal' and can be used for any of the cases covered so far. It is like a quadratic formula used to solve any quadratic equation. It's as simple as this. Let's try some examples.

## Example 9

Solve the following:

a) $\int 6xe^{(3x^2+1)}dx$

b) $\int \sin(3x)\cos^2(3x)\,dx$

c) $\int x^2(2-x^3)^2 dx$

d) $\int x^3(x^2+1)^2 dx$

e) $\int x\sqrt{x-3}\,dx$

f) $\int \frac{x-3}{x^2-6x-7}dx$

g) $\int \sin x \cdot \cos x\, dx$

What did you get? Find the solution below to double-check your answer.

## Solution to Example 9

**HINT**

Where it's relevant, we will attempt these examples using the substitution method as well as any of the following:

- $\int \left[\frac{du}{dx}\right][f'(u)]\,dx$
- $\int f'(x)[f(x)]^n\,dx$
- $\int \frac{f'(x)}{f(x)}dx$
- Partial fractions

**a)** $\int 6xe^{(3x^2+1)}dx$
**Solution**

**Method 1** Using substitution

We will take $u = 3x^2 + 1$, then $\frac{du}{dx} = 6x$. Thus, $dx = \frac{1}{6x}du$. Now, let's substitute in the original integral as:

$$\int 6xe^{(3x^2+1)}dx = \int 6xe^u \frac{1}{6x}du$$

$$= \int e^u du = e^u + C$$

$$= e^{(3x^2+1)} + C$$

$$\therefore \int 6xe^{(3x^2+1)}dx = e^{(3x^2+1)} + C$$

**Method 2** Using $\int \left[\frac{du}{dx}\right][f'(u)]dx$

Given that $\int 6x \cdot e^{(3x^2+1)} dx$, our two functions of interest are $6x$ and $e^{(3x^2+1)}$. Thus, if $u = 3x^2 + 1$, then $\frac{du}{dx} = 6x$. If $f'(u) = e^u$, then $f(u) = e^u$. Hence

$$\int 6xe^{(3x^2+1)} dx = e^u + C$$

$$= e^{3x^2+1} + C$$

$$\therefore \int 6xe^{(3x^2+1)} = e^{3x^2+1} + C$$

**NOTE**
This is a very simple one indeed. It is designed to illustrate the rule. We will take another 'simple' one.

---

**b)** $\int \sin 3x \cos^2 3x \, dx$
**Solution**

**Method 1** Using substitution

$$\int \sin 3x \cos^2 3x \, dx = \int \sin 3x [\cos 3x]^2 dx$$

We will take $u = \cos 3x$, then $\frac{du}{dx} = -3 \sin 3x$. Thus, $dx = -\frac{1}{3 \sin 3x}du$. Now, let's substitute in the original integral as:

$$\int \sin 3x \cos^2 3x \, dx = \int \sin 3x \, u^2 \cdot -\frac{1}{3 \sin 3x}du$$

$$= \int -\frac{1}{3}u^2 du = -\frac{1}{3}\left(\frac{u^3}{3}\right) + C$$

$$= -\frac{1}{9}\cos^3 3x + C$$

$$\therefore \int \sin 3x \cos^2 3x \, dx = -\frac{1}{9}\cos^3 3x + C$$

**Method 2** Using $\int f'(x) [f(x)]^n \, dx$

$f(x) = \cos 3x$, $f'(x) = -3 \sin 3x$, $n = 2$ and $n + 1 = 3$. Hence

$$\int \sin 3x \cdot \cos^2 3x \, dx = -\frac{1}{3} \int -3 \sin 3x [\cos 3x]^2 \, dx$$

$$= -\frac{1}{3} \left[ \frac{1}{3} \cos^3 3x \right] + C$$

$$= -\frac{1}{9} \cos^3 3x + C$$

$$\therefore \int \sin 3x \cos^2 3x \, dx = -\frac{1}{9} \cos^3 3x + C$$

c) $\int x^2 (2 - x^3)^2 \, dx$

**Solution**

Using substitution, we have $u = 2 - x^3$, then $\frac{du}{dx} = -3x^2$. Thus, $dx = -\frac{1}{3x^2} du$. Now, let's substitute in the original integral as:

$$\int x^2 (2 - x^3)^2 \, dx = \int x^2 \cdot u^2 \cdot -\frac{du}{3x^2}$$

$$= \int -\frac{1}{3} u^2 \, du = -\frac{1}{3} \int u^2 \, du = -\frac{1}{3} \left[ \frac{u^3}{3} \right] + C$$

$$= -\frac{1}{9} (2 - x^3)^3 + C$$

$$\therefore \int x^2 (2 - x^3)^2 \, dx = -\frac{1}{9} (2 - x^3)^3 + C$$

## ALTERNATIVE METHOD

Note that we could have used the $\int f'(x) [f(x)]^n \, dx$ technique for this. Also, we could have opened the brackets and then integrated, as shown below.

$$\int x^2 (2 - x^3)^2 \, dx = \int x^2 (4 - 4x^3 + x^6) \, dx$$

$$= \int 4x^2 - 4x^5 + x^8 \, dx$$

$$= \frac{4x^3}{3} - \frac{4x^6}{6} + \frac{x^9}{9} + C$$

$$= \frac{4x^3}{3} - \frac{2x^6}{3} + \frac{x^9}{9} + C$$

$$\therefore \int x^2 (2 - x^3)^2 \, dx = \frac{4x^3}{3} - \frac{2x^6}{3} + \frac{x^9}{9} + C$$

**NOTE**
Let's simplify the answer obtained using substitution as:

$$-\frac{1}{9}(2-x^3)^3 + C = -\frac{1}{9}(2-x^3)(2-x^3)^2 + C$$
$$= -\frac{1}{9}(2-x^3)(2-x^3)^2 + C = -\frac{1}{9}(2-x^3)(4-4x^3+x^6) + C$$
$$= -\frac{1}{9}(8 - 12x^3 + 6x^6 - x^9) + C$$
$$= -\frac{8}{9} + \frac{4x^3}{3} - \frac{2x^6}{3} + \frac{x^9}{9} + C = \frac{4x^3}{3} - \frac{2x^6}{3} + \frac{x^9}{9} + \left(C - \frac{8}{9}\right)$$

You may notice that the answers are slightly different, but recall that the arbitrary constant $C$ can be any numerical value. Thus, we can take it to be $C - \frac{8}{9}$ for the first case.

---

**d)** $\int x^3(x^2+1)^2 dx$

**Solution**
Unfortunately, we cannot use either of the two methods (i.e., $\int \left[\frac{du}{dx}\right][f'(u)]dx$ and $\int f'(x)[f(x)]^n dx$) covered earlier for the product of two functions. Hence, we will be using the substitution method only here. We have $u = x^2 + 1$, then $\frac{du}{dx} = 2x$. Thus, $dx = \frac{1}{2x}du$. Also, $x^2 = u - 1$. Now, let's substitute into the original integral as

$$\int x^3(x^2+1)^2 dx = \int x^3(u-1+1)^2 \frac{du}{2x} = \int x^3 u^2 \frac{du}{2x}$$
$$= \int \frac{1}{2}x^2 u^2 du = \frac{1}{2}\int (u-1)u^2 du$$
$$= \frac{1}{2}\int u^3 - u^2 du = \frac{1}{2}\left[\frac{1}{4}u^4 - \frac{1}{3}u^3\right] + C$$
$$= \frac{1}{8}(x^2+1)^4 - \frac{1}{6}(x^2+1)^3 + C$$
$$\therefore \int x^3(x^2+1)^2 dx = \frac{1}{8}(x^2+1)^4 - \frac{1}{6}(x^2+1)^3 + C$$

# Advanced Integration I

## ALTERNATIVE METHOD

We will open the brackets and integrate each term separately.

$$\int x^3(x^2+1)^2 dx = \int x^3(x^4+2x^2+1) dx$$

$$= \int (x^7 + 2x^5 + x^3) dx$$

$$= \frac{x^8}{8} + \frac{2x^6}{6} + \frac{x^4}{4} + C$$

$$= \frac{x^8}{8} + \frac{x^6}{3} + \frac{x^4}{4} + C$$

$$\therefore \int x^3(x^2+1)^2 dx = \frac{x^8}{8} + \frac{x^6}{3} + \frac{x^4}{4} + C$$

## NOTE
Although the answers look a bit different, but can be proved to be the same as we did on page 708.

---

e) $\int x\sqrt{x-3}\, dx$

**Solution**

Using the substitution method, we have $u = x - 3$, then $\frac{du}{dx} = 1$. Thus, $dx = du$. Also, $x = u + 3$.
Now, let's substitute in the original integral as:

$$\int x \cdot \sqrt{x-3}\, dx = \int (u+3)\sqrt{u}\, du$$

$$= \int (u+3)u^{0.5}\, du \int u^{1.5} + 3u^{0.5}\, du$$

$$= \frac{u^{2.5}}{2.5} + 3\frac{u^{1.5}}{1.5} + C$$

$$= \frac{2u^{2.5}}{5} + 2u^{1.5} + C$$

$$= \frac{2(x-3)^{2.5}}{5} + 2(x-3)^{1.5} + C$$

$$\therefore \int x\sqrt{x-3}\, dx = \frac{2(x-3)^{2.5}}{5} + 2(x-3)^{1.5} + C$$

### ALTERNATIVE METHOD

We can use a different substitution as follows:
$u = \sqrt{x-3} = (x-3)^{\frac{1}{2}}$, then $\frac{du}{dx} = \frac{1}{2}(x-3)^{-\frac{1}{2}} = \frac{1}{2\sqrt{x-3}}$. Thus, $dx = 2\sqrt{x-3}\,du$. Also, $u^2 = x - 3 \rightarrow x = u^2 + 3$. Now, let's substitute in the original integral as:

$$\int x\sqrt{x-3}\,dx = \int (u^2 + 3)\,u.2u\,du$$

$$= \int 2u^2(u^2+3)\,du = \int 2u^4 + 6u^2\,du$$

$$= \frac{2u^5}{5} + 6\frac{u^3}{3} + C = \frac{2u^5}{5} + 2u^3 + C$$

$$= \frac{2(\sqrt{x-3})^5}{5} + 2(\sqrt{x-3})^3 + C$$

$$\therefore \int x.\sqrt{x-3}\,dx = \frac{2(x-3)^{2.5}}{5} + 2(x-3)^{1.5} + C$$

### NOTE

For this example, substitution seems to be the only viable option. Also, the alternative method is a different version of (or approach to) substitution.

---

f) $\int \frac{x-3}{x^2-6x-7}\,dx$

**Solution**

For this example, we could have used two other methods (i.e., partial fraction and $\int \frac{f'(x)}{f(x)}\,dx$), but we will go for substitution to highlight its universality. We therefore have $u = x^2 - 6x - 7$, then $\frac{du}{dx} = 2x - 6 = 2(x-3)$. Thus, $dx = \frac{1}{2(x-3)}du$. Now, let's substitute in the original integral as:

$$\int \frac{x-3}{x^2-6x-7}\,dx = \int \frac{x-3}{u}\cdot\frac{1}{2(x-3)}\,du$$

$$= \int \frac{1}{2u}\,du = \frac{1}{2}\int \frac{1}{u}\,du$$

$$= \frac{1}{2}\ln u + C = \frac{1}{2}\ln(x^2 - 6x - 7) + C$$

$$\therefore \int \frac{x-3}{x^2-6x-7}\,dx = \frac{1}{2}\ln(x^2 - 6x - 7) + C$$

**g)** $\int \sin x \cos x \, dx$

**Solution**

Again, for this example we could have used $\int f(x) f'(x) \, dx$. However, we will go for substitution to highlight its universality. We will demonstrate two approaches based on what to choose as our $u$.

**Method 1** $u = \sin x$

Since $u = \sin x$, then $\frac{du}{dx} = \cos x$. Thus, $dx = \frac{1}{\cos x} du$. Now, let's substitute in the original integral as:

$$\int \sin x \cos x \, dx = \int u \cos x \frac{1}{\cos x} du$$

$$= \int u \, du = \frac{1}{2} u^2 + C$$

$$= \frac{1}{2} \sin^2 x + C$$

$$\therefore \int \sin x \cos x \, dx = \frac{1}{2} \sin^2 x + C$$

**Method 2** $u = \cos x$

Since $u = \cos x$, then $\frac{du}{dx} = -\sin x$. Thus, $dx = -\frac{1}{\sin x} du$. Now, let's substitute in the original integral as:

$$\int \sin x \cdot \cos x \, dx = \int u \sin x - \frac{1}{\sin x} du$$

$$= \int -u \cdot du = -\frac{1}{2} u^2 + C$$

$$= -\frac{1}{2} \cos^2 x + C$$

$$\therefore \int \sin x \cos x \, dx = -\frac{1}{2} \cos^2 x + C$$

**NOTE**

It is interesting that we have two different results, namely:

- $\frac{1}{2} \sin^2 x + C$
- $-\frac{1}{2} \cos^2 x + C$

But using the trigonometric identity, we will see that $\frac{1}{2} \sin^2 x = \frac{1}{2} - \frac{1}{2} \cos^2 x$. What this shows is that if the constant of integration in method 1 is $A$, that of method 2 is $A + \frac{1}{2}$. To confirm that the two answers are equal, try a definite integral. For example, we can confirm that $\int_{0.2}^{1.2} \sin x \cos x \, dx = \left[ \frac{1}{2} \sin^2 x \right]_{0.2}^{1.2} = \left[ -\frac{1}{2} \cos^2 x \right]_{0.2}^{1.2} = 0.4146$ (4 d.p.).

---

As earlier noted, when using the substitution method to solve a definite integral, the limits should be changed to $u$. This removes the need to replace $u$ with the function of $x$ in the final stage of the solution. To do this, just find $u(x_1)$ and $u(x_2)$, where $x_1$ and $x_2$ are the lower and upper limits, respectively, and $x_1 < x_2$.

During the change of limits, it is possible to see that $u_1 > u_2$, just swap them, i.e., make $u_1$ the new upper limit and $u_2$ the lower limits. For example, $\int_{x_1}^{x_2} f(x)\, dx$ becomes $\int_{u_2}^{u_1} g(u)\, du$. But if you decide to keep the limits, i.e., $\int_{u_1}^{u_2} g(u)\, du$, the answer will have a change of sign. We know that, for definite integrals, we are only after the absolute value and the sign can be ignored. Otherwise, there is nothing special in this case.

Let's try a couple of examples.

### Example 10

Use an appropriate substitution to determine the exact value of each of the following:

a) $\int_0^3 \dfrac{25x}{\sqrt{16-5x}}\, dx$ 
b) $\int_0^\pi \sin 2x\, (1 - \cos x)^4\, dx$

What did you get? Find the solution below to double-check your answer.

### Solution to Example 10

a) $\int_0^3 \dfrac{25x}{\sqrt{16-5x}}\, dx$

**Solution**

**Step 1:** Let's start with an indefinite integral.

$$\int \frac{25x}{\sqrt{16-5x}}\, dx$$

Using the substitution method, we have $u = 16 - 5x$, then $\frac{du}{dx} = -5$. Thus, $dx = -\frac{1}{5} du$. Also, $x = \frac{1}{5}(16 - u)$. Now, let's substitute as:

$$\int \frac{25x}{\sqrt{16-5x}}\, dx = 25 \int \frac{x}{\sqrt{16-5x}}\, dx$$

$$= 25 \int \frac{\frac{1}{5}(16-u)}{u^{\frac{1}{2}}} \times \left(-\frac{1}{5}\right) du = 25 \int -\frac{16-u}{25 u^{\frac{1}{2}}}\, du$$

$$= \int -\frac{16-u}{u^{\frac{1}{2}}}\, du = \int \frac{u-16}{u^{\frac{1}{2}}}\, du$$

$$= \int \frac{u}{u^{\frac{1}{2}}} - \frac{16}{u^{\frac{1}{2}}}\, du = \int u^{\frac{1}{2}} - 16 u^{-\frac{1}{2}}\, du$$

# Advanced Integration I

**Step 2:** Let's bring in definite integral.

$$\int_{x_1=0}^{x_2=3} \frac{25x}{\sqrt{16-5x}} dx = \int_{x_1=0}^{x_2=3} u^{\frac{1}{2}} - 16u^{-\frac{1}{2}} du$$

**Step 3:** Change the limits

Right, we need to change the limits from $x$ to $u$ as:

- When $x = 0$, we have

$$u_1 = 16 - 5(0) = 16$$

- When $x = 3$, we have

$$u_2 = 16 - 5(3) = 1$$

Notice that the limit is now from 16 to 1, i.e., from right to left. We can keep it as it is or swap it to 1 to 16 and stick a negative sign to the integral. It does not really matter, so we will keep it. Therefore

$$\int_{x_1=0}^{x_2=3} \frac{25x}{\sqrt{16-5x}} dx = \int_{u_1=16}^{u_2=1} u^{0.5} - 16u^{-0.5} . du = \left[ \frac{u^{1.5}}{1.5} - 16\frac{u^{0.5}}{0.5} \right]_{16}^{1}$$

$$= \left[ \frac{u^{1.5}}{1.5} - \frac{u^{0.5}}{0.5} \right]^{1} - \left[ \frac{u^{1.5}}{1.5} - \frac{u^{0.5}}{0.5} \right]^{16}$$

$$= \left( \frac{(1)^{1.5}}{1.5} - 16\frac{(1)^{0.5}}{0.5} \right) - \left( \frac{(16)^{1.5}}{1.5} - 16\frac{(16)^{0.5}}{0.5} \right)$$

$$= \left( -\frac{94}{3} \right) - \left( -\frac{256}{3} \right) = 54$$

$$\therefore \int_{x_1=0}^{x_2=3} \frac{25x}{\sqrt{16-5x}} dx = 54$$

## ALTERNATIVE METHOD

Alternatively, we can keep the limits in $x$ and change the $u$ back to function of $x$ as follows:

$$\int_{x_1=0}^{x_2=3} \frac{25x}{\sqrt{16-5x}} dx = \left[ \frac{u^{1.5}}{1.5} - 16\frac{u^{0.5}}{0.5} \right]_{u_1}^{u_2}$$

$$= \left[ \frac{(16-5x)^{1.5}}{1.5} - 16\frac{(16-5x)^{0.5}}{0.5} \right]_{0}^{3}$$

$$= \left( \frac{\{16 - 5(3)\}^{1.5}}{1.5} - 16 \frac{\{16 - 5(3)\}^{0.5}}{0.5} \right)$$
$$- \left( \frac{\{16 - 5(0)\}^{1.5}}{1.5} - 16 \frac{\{16 - 5(0)\}^{0.5}}{0.5} \right)$$
$$= \left( -\frac{94}{3} \right) - \left( -\frac{256}{3} \right) = 54$$
$$\therefore \int_{x_1=0}^{x_2=3} \frac{25x}{\sqrt{16 - 5x}} dx = 54$$

---

**b)** $\int_0^\pi \sin 2x (1 - \cos x)^4 dx$

**Solution**

**Step 1:** We first need to simplify the integral expression.

$$\sin 2x(1 - \cos x)^4 = 2 \sin x \cos x (1 - \cos x)^4$$

**Step 2:** Now let's start with indefinite integral

$$\int 2 \sin x \cos x (1 - \cos x)^4 dx$$

Using the substitution method, we have $u = 1 - \cos x$, then $\frac{du}{dx} = \sin x$. Thus, $dx = \frac{1}{\sin x} du$. Also, $\cos x = 1 - u$. Now, let's substitute as:

$$\int \sin 2x(1 - \cos x)^4 dx = \int 2 \sin x \cos x (1 - \cos x)^4 dx$$
$$= 2 \int \sin x \cos x (1 - \cos x)^4 dx$$
$$= 2 \int \sin x (1 - u) u^4 \frac{1}{\sin x} du$$
$$= 2 \int (1 - u) u^4 du = 2 \int u^4 - u^5 du$$

**Step 3:** Change of limits

We need to change the limits from $x$ to $u$ as:

- when $x = 0$, we have

$$u_1 = 1 - \cos x = 1 - \cos 0 = 1 - 1 = 0$$

- when $x = \pi$, we have

$$u_2 = 1 - \cos x = 1 - \cos \pi = 1 + 1 = 2$$

# Advanced Integration I

**Step 4:** Let's bring in the definite integral now, thus

$$\int_{x_1=0}^{x_2=\pi} \sin 2x\,(1-\cos x)^4\,dx = 2\int_{x_1=0}^{x_2=\pi} u^4 - u^5\,du$$

$$= 2\int_{u_1=0}^{u_2=2} u^4 - u^5\,du = 2\left[\frac{u^5}{5} - \frac{u^6}{6}\right]_0^2$$

$$= 2\left[\frac{u^5}{5} - \frac{u^6}{6}\right]^2 - 2\left[\frac{u^5}{5} - \frac{u^6}{6}\right]^0$$

$$= 2\left(\frac{2^5}{5} - \frac{2^6}{6}\right) - 2\left(\frac{0^5}{5} - \frac{0^6}{6}\right)$$

$$= 2\left(-\frac{64}{15}\right) - 2(0) = -\frac{128}{15}$$

$$\therefore \int_{x_1=0}^{x_2=\pi} \sin 2x\,(1-\cos x)^4\,dx = -\frac{128}{15}$$

---

## 17.5 CHAPTER SUMMARY

1) The rule of integrating a function that takes the form of $(ax+b)^n$ is:

$$\boxed{y = \int (ax+b)^n\,dx = \frac{1}{a(n+1)}(ax+b)^{n+1} + C}$$

where $n$ is any real number excluding $-1$.

2) The linear function rule is used when integrating a function of a linear function, such that the inner function is a linear expression of the form $ax+b$, where $a$ and $b$ are real numbers and $a \neq 0$. In general, the function is of the form $f'(ax+b)$. If $b \neq 0$, then we have $f'(ax+b)$. We can state that:

$$\boxed{\int f'(ax+b)\,dx = \frac{1}{a}f(ax+b) + C}$$

and if $b = 0$, we have

$$\boxed{\int f'(ax)\,dx = \frac{1}{a}f(ax) + C}$$

3) When the dividend (or numerator) of a rational function is a derivative of the divisor (or denominator) then the integral is given by:

$$\boxed{y = \int \frac{f'(x)}{f(x)}\,dx = \ln|f(x)| + C}$$

4) The rule to integrate a product of two functions $\int [f'(x)] \cdot [f(x)] dx$ and $\int [f'(x)] \cdot [f(x)]^n dx$ is given by:

$$y = \int [f'(x)][f(x)]^n dx = \frac{1}{n+1}[f(x)]^{n+1} + C$$

where $n$ is any real number. If $n = 1$, we have a modified version as:

$$y = \int f'(x) f(x) dx = \frac{1}{2}[f(x)]^2 + C$$

5) The rule to integrate a function of the form $\int \left[\frac{du}{dx}\right] \cdot [f'(u)] dx$ is given by:

$$y = \int \left[\frac{du}{dx}\right] [f'(u)] dx = f(u) + C$$

6) An expression of the form $\frac{f(x)}{g(x)}$ can be split into partial fractions, such that the denominator of each of the resulting fractions is a linear function. Once split, we can apply the integral standards for rational expressions (i.e., $\frac{1}{x}$ and $\frac{1}{ax+b}$) to solve the problem.

7) The technique called '**integration by substitution**' is used to integrate complicated or quasi-complicated, which are not members of the known families, and it seems to be more 'universal'.

\*\*\*\*

## 17.6 FURTHER PRACTICE

To access complementary contents, including additional exercises, please go to www.dszak.com.

# 18 Advanced Integration II

## Learning Outcomes

Once you have studied the content of this chapter, you should be able to:

- Apply parametric integration
- Apply the technique of integration by parts to solve the product of two functions
- Use trigonometrical identities to solve complex integrals

## 18.1 INTRODUCTION

In this chapter, we will continue our discussion on the integration of complicated functions, covering parametric integration, integration by parts, and integration of trigonometric functions.

## 18.2 PARAMETRIC INTEGRATION

Previously, we covered parametric differentiation, where a function $y = f(x)$ is defined parametrically using a third independent variable called a parameter, $t$ for example. As a result, $y = f(x)$ is given in two parametric equations $y = f(t)$ and $x = f(t)$. When we need to integrate a function that is defined parametrically, we will apply parametric integration, which is based on a chain rule similar to parametric differentiation. Like integration by substitution, it also requires changing the limits, and in addition, differentiating one of the parametric equations.

In short, we can say that if $y = f(x)$ is defined parametrically by $y = f(t)$ and $x = f(t)$, the indefinite integration is given by

$$\boxed{\int y\, dx = \int y \frac{dx}{dt} dt} \qquad (18.1)$$

and the definite integral is

$$\boxed{\int_{x_1}^{x_2} y\, dx = \int_{t_1}^{t_2} y \frac{dx}{dt} dt} \qquad (18.2)$$

Notice how we changed $dx$ to $dt$, as it is not possible to integrate $y = f(t)$ wrt $x$. Also, we changed the limits from $x_1$ and $x_2$ to $t_1$ and $t_2$, respectively, in definite integral.

Let's go for an example involving indefinite integration.

## Example 1

A curve is defined by the parametric equations $x = 5 + \cos^2 2\theta$ and $y = \sin^2 2\theta$. Determine $\int y\,dx$.

What did you get? Find the solution below to double-check your answer.

## Solution to Example 1

Given that $x = 5 + \cos^2 2\theta$ and $y = \sin^2 2\theta$, we have

$$\frac{dx}{d\theta} = (2\cos 2\theta) \times -2\sin 2\theta = -4\sin 2\theta \cdot \cos 2\theta$$

Thus

$$\int y\,dx = \int y \frac{dx}{d\theta} d\theta$$
$$= \int \left(\sin^2 2\theta\right)(-4\sin 2\theta \cdot \cos 2\theta)\,d\theta$$
$$= \int -4\cos 2\theta \sin^3 2\theta\,d\theta = -2 \int (2\cos 2\theta)(\sin 2\theta)^3 d\theta$$
$$= -2\left[\frac{1}{4}\sin^4 2\theta + C\right] = -\frac{1}{2}\sin^4 2\theta + C$$
$$\therefore \int y\,dx = -\frac{1}{2}\sin^4 2\theta + C$$

Another example on definite integral.

## Example 2

A curve is defined by parametric equations $x = 2t(t-2)$ and $y = 5t + 2$, where $t > 0$. Determine $\int_0^6 y\,dx$.

What did you get? Find the solution below to double-check your answer.

## Solution to Example 2

Given that $x = 2t(t-2)$ and $y = 5t + 2$, we have

$$\frac{d}{dt} 2t(t-2) = \frac{d}{dt}\left(2t^2 - 4t\right) = 4t - 4$$

Advanced Integration II

Let's change the limits too:

- For $x_1 = 0$, we have

$$2t(t-2) = 0$$

Therefore

$$2t = 0 \quad \text{OR} \quad t - 2 = 0$$
$$\therefore t = 0 \qquad\qquad \therefore t = 2$$

As $t > 0$, the valid answer is $t = 2$

- For $x_2 = 6$, we have

$$2t(t-2) = 6$$
$$2t^2 - 4t - 6 = 0$$
$$(2t+2)(t-3) = 0$$

Therefore

$$2t + 2 = 0 \quad \text{OR} \quad t - 3 = 0$$
$$\therefore t = -1 \qquad\qquad \therefore t = 3$$

As $t > 0$, the valid answer is $t = 3$. Thus

$$\int_{x_1=0}^{x_2=6} y\,dx = \int_{x_1=0}^{x_2=6} y\frac{dx}{dt}\,dt = \int_{t_1=2}^{t_2=3} (5t+2)(4t-4)\,dt$$

$$= \int_2^3 (20t^2 - 12t - 8)\,dt = \left[\frac{20}{3}t^3 - 6t^2 - 8t\right]_2^3$$

$$= \left[\frac{20}{3}t^3 - 6t^2 - 8t\right]^3 - \left[\frac{20}{3}t^3 - 6t^2 - 8t\right]^2$$

$$= \left(\frac{20}{3}(3)^3 - 6(3)^2 - 8(3)\right) - \left(\frac{20}{3}(2)^3 - 6(2)^2 - 8(2)\right)$$

$$= (180 - 54 - 24) - \left(\frac{160}{3} - 24 - 16\right)$$

$$= (102) - \left(\frac{40}{3}\right) = \frac{266}{3}$$

$$\therefore \int_{x_1=0}^{x_2=6} y\,dx = \frac{266}{3}\ \text{unit}^2$$

A final example involving definite integration.

## Example 3

A curve is defined by the parametric equations $x = 3t + 1$ and $y = t^2 + 5$. Determine $\int_{-2}^{7} y\,dx$.

What did you get? Find the solution below to double-check your answer.

## Solution to Example 3

**Method 1**

Given that $x = 3t + 1$ and $y = t^2 + 5$, we have

$$\frac{dx}{dt} = 3$$

Let's change the limits too:

- For $x_1 = -2$, we have

$$3t + 1 = -2$$
$$3t = -3$$
$$\therefore t = -1$$

- For $x_2 = 7$, we have

$$3t + 1 = 7$$
$$3t = 6$$
$$\therefore t = 2$$

Thus

$$\int_{x_1=-2}^{x_2=7} y\, dx = \int_{x_1=-2}^{x_2=7} y\frac{dx}{dt} dt$$
$$= \int_{t_1=-1}^{t_2=2} (t^2 + 5)(3)\, dt$$
$$= \int_{-1}^{2} (3t^2 + 15)\, dt$$
$$= \left[t^3 + 15t\right]_{-1}^{2}$$
$$= \left[t^3 + 15t\right]^{2} - \left[t^3 + 15t\right]^{-1}$$
$$= \left\{(2)^3 + 15(2)\right\} - \left\{(-1)^3 + 15(-1)\right\}$$
$$= (8 + 30) - (-1 - 15) = 38 + 16 = 54$$
$$\therefore \int_{x_1=-2}^{x_2=7} y\, dx = 54\ \text{unit}^2$$

**Method 2**

Given that $x = 3t + 1$, we have

$$3t = x - 1$$
$$t = \frac{1}{3}(x - 1)$$

Let's substitute $t$ in the $y(t)$ as:

$$y = t^2 + 5$$
$$= \left[\frac{1}{3}(x-1)\right]^2 + 5 = \frac{(x-1)^2}{9} + 5$$
$$= \frac{x^2 - 2x + 1}{9} + 5 = \frac{x^2}{9} - \frac{2}{9}x + \frac{1}{9} + 5$$
$$= \frac{x^2}{9} - \frac{2}{9}x + \frac{46}{9}$$

Thus

$$\int_{x_1=-2}^{x_2=7} y\,dx = \int_{x_1=-2}^{x_2=7} \left(\frac{x^2}{9} - \frac{2}{9}x + \frac{46}{9}\right) dx$$
$$= \left[\frac{1}{9}\left(\frac{x^3}{3}\right) - \frac{2}{9}\left(\frac{x^2}{2}\right) + \frac{46}{9}x\right]_{-2}^{7} = \left[\frac{x^3}{27} - \frac{x^2}{9} + \frac{46}{9}x\right]_{-2}^{7}$$
$$= \left[\frac{x^3}{27} - \frac{x^2}{9} + \frac{46}{9}x\right]^{7} - \left[\frac{x^3}{27} - \frac{x^2}{9} + \frac{46}{9}x\right]^{-2}$$
$$= \left\{\frac{(7)^3}{27} - \frac{(7)^2}{9} + \frac{46}{9}(7)\right\} - \left\{\frac{(-2)^3}{27} - \frac{(-2)^2}{9} + \frac{46}{9}(-2)\right\}$$
$$= \left(\frac{343}{27} - \frac{49}{9} + \frac{322}{9}\right) - \left(-\frac{8}{27} - \frac{4}{9} - \frac{92}{9}\right)$$
$$= \left(\frac{1162}{27}\right) - \left(-\frac{296}{27}\right) = 54$$
$$\therefore \int_{x_1=-2}^{x_2=7} y\,dx = 54 \text{ unit}^2$$

## 18.3 INTEGRATION BY PARTS (INTEGRATION OF PRODUCTS)

Integrating a product of two functions seems to be a tough one. Luckily, we have managed to identify common patterns and chose two lovely rules, namely:

1) $\int \left[\frac{du}{dx}\right] [f'(u)]\,dx$, and

2) $\int [f'(x)] \cdot [f(x)]^n\,dx$.

However, we realise that there are certain products which do not follow these two forms. As a result, we introduced the method of substitution as a more encompassing and robust approach, particularly to solve some rather complex products that do not fit the above two descriptions. These methods are based on the inverse chain rule, but unfortunately, they do not cover all product functions that we may need to integrate.

Following this, we are here with another method called **integration by parts**. It is distinctively known for being based on the inverse product rule and for using both differentiation and integration to solve a problem. It states that:

$$y = \int [u]\left[\frac{dv}{dx}\right]dx = uv - \int [v]\left[\frac{du}{dx}\right]dx \qquad (18.3)$$

It is often written in a compact format as:

$$y = \int [u][v']\,dx = uv - \int [v][u']\,dx \qquad (18.4)$$

where $v' = \frac{dv}{dx}$ and $u' = \frac{du}{dx}$.

Proof of this rule is provided in Appendix N.

Before we delve into worked examples, let's make the following notes to guide us further in our use of this technique:

**Note 1** Identify that there is a product of two functions, explicitly or implicitly expressed, and that it is impossible to use any of the techniques stated before.

**Note 2** One function is taken as $u$ and the other $v'$ or $\frac{dv}{dx}$. Use these to obtain $u'$ or $\frac{du}{dx}$ and $v$.

**Note 3** Substitute the four variables into the RHS expressions $\int uv'\,dx = uv - \int vu'\,dx$.

**Note 4** We do not need to add a constant of integration when writing $v$.

**Note 5** It is generally expected that the choice is made such that $u$ and $v'$ can, respectively, be differentiated and integrated to obtain a simpler result.

**Note 6** Based on Note 5, it is customarily known that the following priority order for $u$ should be used: Logarithmic function – Inverse trigonometric function – Algebraic function – Trigonometric function – Exponential function. This is collected in LIATE. It should be mentioned that LIAET is also valid, i.e., either a trigonometric function or an exponential function can be taken as $u$ first. The order is $\ln x$, inverse trigonometric functions ($sin^{-1}x$, $\cos^{-1} x$, and $\tan^{-1} x$), $x^n$, trigonometric functions ($\sin x$, $\cos x$, and $\tan x$), and then $e^x$.

**Note 7** Sometimes, a trigonometric function comes before $e^x$ in the order list. It's as though they are both on the same level of priority, just like multiplication and division in arithmetic operations.

**Note 8** Sometimes the integration by parts will need to be carried out two or more times.

# Advanced Integration II

It is time to look at some examples.

### Example 4

Solve the following:

a) $\int 3xe^x \, dx$  
b) $\int 6x \ln x \, dx$  
c) $\int \frac{\ln 3x}{5\sqrt{x}} \, dx$  
d) $\int -12x \sin(3x-1) \, dx$  
e) $\int \ln x \, dx$  
f) $\int 7x(8 - e^{-2x}) \, dx$  
g) $\int x^2 \ln x^3 \, dx$  
h) $\int \frac{5x-1}{\cos^2(3x+7)} \, dx$  
i) $\int 4x \operatorname{cosec}(1-2x) \cot(1-2x) \, dx$

What did you get? Find the solution below to double-check your answer.

### Solution to Example 4

**a)** $\int 3xe^x \, dx$  
**Solution**  
Using the priority order, we have $u = 3x$, then $\frac{du}{dx} = 3$. Similarly, $v' = e^x$, then $v = \int e^x \, dx = e^x$.  
Hence,

$$\int 3xe^x \, dx = uv - \int vu' \, dx$$

$$= 3xe^x - \int e^x \cdot 3 \, dx = 3x \, e^x - \int 3e^x \, dx$$

$$= 3x \, e^x - 3e^x + C$$

$$\therefore \int 3xe^x \, dx = 3e^x(x-1) + C$$

**b)** $\int 6x \ln x \, dx$  
**Solution**  
Using the priority order, we have $u = \ln x$, then $\frac{du}{dx} = \frac{1}{x}$. Similarly, $v' = 6x$, then $v = \int 6x \, dx = 3x^2$.  
Hence,

$$\int 6x \ln x \, dx = uv - \int vu' \, dx$$

$$= \ln x \cdot 3x^2 - \int 3x^2 \frac{1}{x} \, dx = 3x^2 \ln x - \int 3x \, dx$$

$$= 3x^2 \ln x - \frac{3}{2}x^2 + C$$

$$\therefore \int 6x \ln x \, dx = 3x^2 \left(\ln x - \frac{1}{2}\right) + C$$

c) $\int \frac{\ln 3x}{5\sqrt{x}} dx$

**Solution**
This does not explicitly look like a product of two functions, but can be modified to appear so:

$$\int \frac{\ln 3x}{5\sqrt{x}} dx = \int \ln 3x \cdot \frac{1}{5\sqrt{x}} dx = \int \ln 3x \cdot \frac{1}{5} x^{-0.5} dx$$

Using the priority order, we have $u = \ln 3x$, then $\frac{du}{dx} = \frac{1}{x}$. Similarly, $v' = \frac{1}{5} x^{-0.5}$, then $v = \int \frac{1}{5} x^{-0.5} dx = \frac{1}{5} \left( \frac{x^{0.5}}{0.5} \right) = \frac{2}{5} x^{0.5}$. Hence,

$$\int \frac{\ln 3x}{5\sqrt{x}} dx = uv - \int v u' dx$$

$$= \ln 3x \cdot \frac{2}{5} x^{0.5} - \int \frac{2}{5} x^{0.5} \cdot \frac{1}{x} dx = \frac{2}{5}\sqrt{x} \ln 3x - \int \frac{2}{5} x^{-0.5} dx$$

$$= \frac{2}{5}\sqrt{x} \ln 3x - \frac{2}{5} \left( \frac{x^{0.5}}{0.5} \right) + C = \frac{2}{5}\sqrt{x} \ln 3x - \frac{4}{5}\sqrt{x} + C$$

$$\therefore \int \frac{\ln 3x}{5\sqrt{x}} dx = \frac{2}{5}\sqrt{x} (\ln 3x - 2) + C$$

---

d) $\int -12x \sin(3x - 1) dx$

**Solution**
Using the priority order, we have $u = -12x$, then $\frac{du}{dx} = -12$. Similarly, $v' = \sin(3x - 1)$, then $v = \int \sin(3x - 1) dx = \frac{1}{3}(-\cos(3x - 1)) = -\frac{1}{3}\cos(3x - 1)$. Hence,

$$\int -12x \cdot \sin(3x - 1) dx = uv - \int v u' dx$$

$$= -12x \left\{ -\frac{1}{3} \cos(3x - 1) \right\} - \int -\frac{1}{3} \cos(3x - 1)(-12) dx$$

$$= 4x \cos(3x - 1) - \int 4 \cos(3x - 1) dx$$

$$= 4x \cos(3x - 1) - 4\left(\frac{1}{3} \sin(3x - 1)\right) + C = 4x \cos(3x - 1) - \frac{4}{3} \sin(3x - 1) + C$$

$$\therefore \int -12x \sin(3x - 1) = 4x \cos(3x - 1) - \frac{4}{3} \sin(3x - 1) + C$$

---

e) $\int \ln x \, dx$

**Solution**
This does not explicitly look like a product of two functions, but can be modified to appear so:

$$\int \ln x \, dx = \int 1 \cdot \ln x \, dx = \int x^0 \ln x \, dx$$

Using the priority order, we have $u = \ln x$, then $\frac{du}{dx} = \frac{1}{x}$. Similarly, $v' = x^0$, then $v = \int x^0 \, dx = x$. Hence,

$$\int \ln x \, dx = uv - \int vu' \, dx$$

$$= \ln x \cdot x - \int x \cdot \frac{1}{x} dx = x \ln x - \int 1 \, dx$$

$$= x \ln x - \int x^0 dx = x \ln x - x + C$$

$$\therefore \int \ln x \, dx = x \ln x - x + C$$

**f)** $\int 7x \left(8 - e^{-2x}\right) dx$

**Solution**

Using the priority order, we have $u = 7x$, then $\frac{du}{dx} = 7$. Similarly, $v' = 8 - e^{-2x}$, then $v = \int (8 - e^{-2x}) dx = 8x - \left(\frac{e^{-2x}}{-2}\right) = 8x + 0.5e^{-2x}$. Hence,

$$\int 7x \left(8 - e^{-2x}\right) dx = uv - \int v \frac{du}{dx} dx$$

$$= 7x \cdot \left(8x + 0.5e^{-2x}\right) - \int \left(8x + 0.5e^{-2x}\right) \cdot 7 \, dx$$

$$= 7x \left(8x + 0.5e^{-2x}\right) - 7 \int \left(8x + 0.5e^{-2x}\right) dx$$

$$= 7x \left(8x + 0.5e^{-2x}\right) - 7 \left[4x^2 + 0.5 \left(\frac{e^{-2x}}{-2}\right)\right] + C$$

$$= 7x \left(8x + 0.5e^{-2x}\right) - 7 \left[4x^2 - \frac{e^{-2x}}{4}\right] + C$$

$$= 56x^2 + 3.5x.e^{-2x} - 28x^2 + \frac{7}{4}e^{-2x} + C$$

$$= 28x^2 + 3.5x.e^{-2x} + \frac{7}{4}e^{-2x} + C$$

$$\therefore \int 7x \left(8 - e^{-2x}\right) dx = 28x^2 + \frac{7}{2}x.e^{-2x} + \frac{7}{4}e^{-2x} + C$$

**g)** $\int x^2 \ln x^3 \, dx$

**Solution**

Using the priority order, we have $u = \ln x^3$, then $\frac{du}{dx} = 3x^2 \times \frac{1}{x^3} = \frac{3}{x}$. Similarly, $v' = x^2$, then $v = \int x^2 dx = \frac{1}{3}x^3$. Hence,

$$\int x^2 \ln x^3 \, dx = uv - \int vu' \, dx$$

$$= \ln x^3 \cdot \frac{1}{3}x^3 - \int \frac{1}{3}x^3 \cdot \frac{3}{x} dx = \frac{1}{3}x^3 \cdot \ln x^3 - \int x^2 dx$$

$$= \frac{1}{3}x^3 \cdot \ln x^3 - \frac{1}{3}x^3 + C = \frac{1}{3}x^3 \left(\ln x^3 - 1\right) + C$$

$$\therefore \int x^2 \cdot \ln x^3 \, dx = \frac{1}{3}x^3 \left(\ln x^3 - 1\right) + C$$

**ALTERNATIVE METHOD**

Alternatively, we can attempt it this way

$$\int x^2 \cdot \ln x^3 \, dx = \int x^2 \cdot (3 \ln x) \, dx = \int 3x^2 \cdot \ln x \, dx$$

Using the priority order, we have $u = \ln x$, then $\frac{du}{dx} = \frac{1}{x}$. Similarly, $v' = 3x^2$, then $v = \int 3x^2 dx = x^3$. Hence,

$$\int 3x^2 \cdot \ln x \, dx = uv - \int vu' \, dx$$

$$= \ln x \cdot x^3 - \int x^3 \cdot \frac{1}{x} dx = x^3 \cdot \ln x - \int x^2 dx$$

$$= x^3 \cdot \ln x - \frac{1}{3}x^3 + C = \frac{1}{3}x^3 \left(3 \ln x - 1\right) + C$$

$$\therefore \int 3x^2 \cdot \ln x \, dx = \frac{1}{3}x^3 \left(3 \ln x - 1\right) + C$$

---

**h)** $\int \frac{5x-1}{\cos^2(3x+7)} dx$

**Solution**

This does not explicitly look like a product of two functions, but can be modified to appear so:

$$\int \frac{5x-1}{\cos^2(3x+7)} dx = \int (5x-1) \sec^2(3x+7) \, dx$$

Advanced Integration II

Using the priority order, we have $u = 5x - 1$, then $\frac{du}{dx} = 5$. Similarly, $v' = \sec^2(3x + 7)$, then $v = \int \sec^2(3x + 7)\, dx = \frac{1}{3}\tan(3x + 7)$. Hence,

$$\int \frac{5x - 1}{\cos^2(3x + 7)}\, dx = uv - \int vu'\, dx$$

$$= (5x - 1)\,\frac{1}{3}\tan(3x + 7) - \int \frac{1}{3}\tan(3x + 7).5\, dx$$

$$= \frac{1}{3}(5x - 1)\tan(3x + 7) - \frac{5}{3}\int \tan(3x + 7)\, dx$$

$$= \frac{1}{3}(5x - 1)\tan(3x + 7) - \frac{5}{3}\left[\frac{1}{3}\ln|\sec(3x + 7)|\right] + C$$

$$= \frac{1}{3}(5x - 1)\tan(3x + 7) - \frac{5}{9}\ln|\sec(3x + 7)| + C$$

$$\therefore \int \frac{5x - 1}{\cos^2(3x + 7)}\, dx = \frac{1}{3}(5x - 1)\tan(3x + 7) - \frac{5}{9}\ln|\sec(3x + 7)| + C$$

i) $\int 4x\,\text{cosec}(1 - 2x)\cot(1 - 2x)\, dx$

**Solution**

Using the priority order, we have $u = 4x$, then $\frac{du}{dx} = 4$. Similarly, $v' = \text{cosec}(1 - 2x)\cot(1 - 2x)$, then $v = \int \text{cosec}(1 - 2x)\cot(1 - 2x)\, dx = -\frac{1}{2}\text{cosec}(1 - 2x)$. Hence,

$$\int 4x\,\text{cosec}(1 - 2x)\cot(1 - 2x)\, dx$$

$$= uv - \int vu'\, dx$$

$$= 4x.\,-\frac{1}{2}\text{cosec}(1 - 2x) - \int -\frac{1}{2}\text{cosec}(1 - 2x).4\, dx$$

$$= -2x\,\text{cosec}(1 - 2x) - \int -2\,\text{cosec}(1 - 2x)\, dx$$

$$= -2x\,\text{cosec}(1 - 2x) + 2\int \text{cosec}(1 - 2x)\, dx$$

$$= -2x\,\text{cosec}(1 - 2x) + 2\left[-\frac{1}{2}\ln|\cot(1 - 2x) + \text{cosec}(1 - 2x)|\right] + C$$

$$= -2x\,\text{cosec}(1 - 2x) - \ln|\cot(1 - 2x) + \text{cosec}(1 - 2x)| + C$$

$$= -2x\,\text{cosec}(1 - 2x) - \ln|\cot(1 - 2x) + \text{cosec}(1 - 2x)| + C$$

$$\therefore \int 4x\,\text{cosec}(1 - 2x)\cot(1 - 2x)\, dx$$

$$= -2x\,\text{cosec}(1 - 2x) - \ln|\cot(1 - 2x) + \text{cosec}(1 - 2x)| + C$$

Another example to try.

### Example 5

The velocity of a particle changes at the rate defined by $4t.\ln(2t-3)$. Find the general expression for the velocity of the particle, such that $t \geq 2$.

---

What did you get? Find the solution below to double-check your answer.

### Solution to Example 5

From the above, we know that

$$\frac{dv}{dt} = 4t.\ln(2t-3)$$

$$\therefore v = \int 4t.\ln(2t-3)\,dt$$

Using the priority order, we have $u = \ln(2t-3)$, then $\frac{du}{dt} = \frac{2}{2t-3}$. Similarly, $v' = 4t$, then $v = \int 4t.dt = 2t^2$. Hence,

$$\int 4t.\ln(2t-3)\,dt = uv - \int v\frac{du}{dt}dt$$

$$= \ln(2t-3).2t^2 - \int 2t^2.\frac{2}{2t-3}dt$$

$$= 2t^2\ln(2t-3) - \int \frac{4t^2}{2t-3}dt$$

To simplify this $\int \frac{4t^2}{2t-3}dt$, we need the integration by partial fractions. Let's start by decomposing $\frac{4t^2}{2t-3}$.

$$
\begin{array}{r}
2t+3\phantom{000} \\
2t-3 \overline{\smash{)}\,4t^2\phantom{-6t00}} \\
\underline{4t^2 \phantom{0} -6t\phantom{00}} \\
6t\phantom{00} \\
\underline{6t \phantom{0} -9} \\
9\phantom{0}
\end{array}
$$

Therefore,

$$\int \frac{4t^2}{2t-3} dt = \int 2t + 3 + \frac{9}{2t-3} dt$$

$$= \int 2t\, dt + \int 3\, dt + 9 \int \frac{1}{2t-3} dt$$

Let's put all this back into where we stopped above as:

$$\int 4t.\ln(2t-3)\, dt = 2t^2 \ln(2t-3) - \int \frac{4t^2}{2t-3} dt$$

$$= 2t^2 \ln(2t-3) - \int 2t\, dt - \int 3\, dt - 9 \int \frac{1}{2t-3} dt$$

$$= 2t^2 \ln(2t-3) - t^2 - 3t - \frac{9}{2} \ln|2t-3| + C$$

Since $t \geq 2$, we may ignore the modulus sign since we are sure that the natural logarithm is positive at all times.

$$\therefore v = \int 4t.\ln(2t-3) = 2t^2 \ln(2t-3) - t^2 - 3t - \frac{9}{2} \ln(2t-3) + C$$

---

In the above examples, we've been able to use the formula $\int uv'\, dx = uv - \int vu'\, dx$ to solve each problem. In particular, the second part of the RHS expression (i.e., $\int vu'\, dx$) has been reduced to a single function. As a result, they were solved by simply referring to the table of integrals as/where applicable. However, there are situations where this is not possible; in other words, $\int vu'\, dx$ is still a product of two functions, though this might have been in a simpler form than the original problem $\int uv'\, dx$. For this case, we will apply the integration by parts repeatedly – two or more times – until we have a single function in $\int vu'\, dx$. Note that the constant of integration can only be one and should only be added to the last part of the integration. Let's try some examples.

### Example 6

Using integration by parts repeatedly, solve the following:

a) $\int \frac{3x^2}{e^{2x}} dx$

b) $\int x^2 \sin 4x\, dx$

c) $\int \frac{-6x^2}{(5-3x)^3} dx$

d) $\int (\ln 3x)^2 dx$

e) $\int e^{5x} . \cos x\, dx$

What did you get? Find the solution below to double-check your answer.

### Solution to Example 6

a) $\int \frac{3x^2}{e^{2x}} dx$

**Solution**

**Step 1:** Let's arrange this in a product-like form as:

$$\int \frac{3x^2}{e^{2x}} dx = \int 3x^2 e^{-2x} dx$$

**Step 2:** Using the priority order, we have $u = 3x^2$, then $\frac{du}{dx} = 6x$. Similarly, $v' = e^{-2x}$, then $v = \int e^{-2x} dx = -\frac{1}{2} e^{-2x}$. Hence,

$$\int 3x^2 e^{-2x} dx = uv - \int vu' dx$$

$$= 3x^2 \left(-\frac{1}{2} e^{-2x}\right) - \int -\frac{1}{2} e^{-2x} \cdot 6x \, dx$$

$$= -\frac{3x^2}{2e^{2x}} + \int 3x e^{-2x} dx$$

**Step 3:** In this case, we would apply integration by parts to $\int 3x \cdot e^{-2x} dx$. Although not a requirement, we will use slightly different notations for this second stage. Again, based on the priority order we have $u_1 = 3x$, then $\frac{du_1}{dx} = 3$. Similarly, $v'_1 = e^{-2x}$, then $v_1 = \int e^{-2x} dx = -\frac{1}{2} e^{-2x}$. Hence,

$$\int 3x \cdot e^{-2x} dx = u_1 v_1 - \int v_1 u'_1 dx$$

$$= 3x \cdot -\frac{1}{2} e^{-2x} - \int -\frac{1}{2} e^{-2x} \cdot 3 \, dx$$

$$= -\frac{3x}{2e^{2x}} + \frac{3}{2} \int e^{-2x} dx$$

$$= -\frac{3x}{2e^{2x}} + \left(\frac{3}{2}\right)\left(-\frac{1}{2} e^{-2x}\right) + C$$

$$= -\frac{3x}{2e^{2x}} - \frac{3}{4} e^{-2x} + C$$

**Step 4:** It is time to combine them, thus

$$\int 3x^2 e^{-2x} dx = -\frac{3x^2}{2e^{2x}} + \int 3x e^{-2x} dx$$

$$= -\frac{3x^2}{2e^{2x}} - \frac{3x}{2e^{2x}} - \frac{3}{4} e^{-2x} + C$$

$$\therefore \int \frac{3x^2}{e^{2x}} dx = -\frac{3x^2}{2e^{2x}} - \frac{3x}{2e^{2x}} - \frac{3}{4} e^{-2x} + C$$

# Advanced Integration II

**b)** $\int x^2 \sin 4x \, dx$

**Solution**

**Step 1:** Using the priority order, we have $u = x^2$, then $\frac{du}{dx} = 2x$. Similarly, $v' = \sin 4x$, then $v = \int \sin 4x \, dx = -\frac{1}{4}\cos 4x$. Hence,

$$\int x^2 \sin 4x \, dx = uv - \int v \frac{du}{dx} dx$$

$$= x^2\left(-\frac{1}{4}\cos 4x\right) - \int -\frac{1}{4}\cos 4x \cdot 2x \, dx$$

$$= -\frac{1}{4}x^2 \cos 4x + \int \frac{1}{2}x \cos 4x \, dx$$

**Step 2:** In this case, we would apply integration by parts to $\int \frac{1}{2} x \cos 4x \, dx$. Again, based on the priority order we have $u_1 = \frac{1}{2}x$, then $\frac{du_1}{dx} = \frac{1}{2}$. Similarly, $v_1' = \cos 4x$, then $v_1 = \int \cos 4x \, dx = \frac{1}{4}\sin 4x$. Hence,

$$\int \frac{1}{2}x \cdot \cos 4x \, dx = u_1 v_1 - \int v_1 u_1' \, dx$$

$$= \frac{1}{2}x \cdot \frac{1}{4}\sin 4x - \int \frac{1}{4}\sin 4x \cdot \frac{1}{2} dx$$

$$= \frac{1}{8}x \sin 4x - \frac{1}{8}\int \sin 4x \, dx$$

$$= \frac{1}{8}x \sin 4x - \frac{1}{8} \cdot -\frac{1}{4}\cos 4x + C$$

$$= \frac{1}{8}x \sin 4x + \frac{1}{32}\cos 4x + C$$

**Step 3:** It is time to combine them, thus

$$\int x^2 \sin 4x \, dx = -\frac{1}{4}x^2 \cos 4x + \int \frac{1}{2}x \cos 4x \, dx$$

$$= -\frac{1}{4}x^2 \cos 4x + \frac{1}{8}x \sin 4x + \frac{1}{32}\cos 4x + C$$

$$\therefore \int x^2 \sin 4x \, dx = -\frac{1}{4}x^2 \cos 4x + \frac{1}{8}x \sin 4x + \frac{1}{32}\cos 4x + C$$

**c)** $\int \frac{-6x^2}{(5-3x)^3} dx$

**Solution**

**Step 1:** Let's quickly re-arrange this as:

$$\int \frac{-6x^2}{(5-3x)^3} dx = \int -6x^2 (5-3x)^{-3} dx$$

**Step 2:** Using the priority order, we have $u = -6x^2$, then $\frac{du}{dx} = -12x$. Similarly, $v' = (5 - 3x)^{-3}$, then $v = \int (5 - 3x)^{-3} dx = \frac{1}{6}(5 - 3x)^{-2}$. Hence,

$$\int -6x^2 \cdot (5 - 3x)^{-3} dx = uv - \int v \frac{du}{dx} dx$$

$$= -6x^2 \cdot \frac{1}{6}(5 - 3x)^{-2} - \int \frac{1}{6}(5 - 3x)^{-2} \cdot -12x \, dx$$

$$= \frac{-x^2}{(5 - 3x)^2} + \int 2x \cdot (5 - 3x)^{-2} dx$$

**Step 3:** In this case, we would apply integration by parts to $\int 2x(5 - 3x)^{-2} dx$. Again, based on the priority order we have $u_1 = 2x$, then $\frac{du_1}{dx} = 2$. Similarly, $v_1' = (5 - 3x)^{-2}$, then $v_1 = \int (5 - 3x)^{-2} dx = \frac{1}{3}(5 - 3x)^{-1}$. Hence,

$$\int 2x(5 - 3x)^{-2} dx = u_1 v_1 - \int v_1 u_1' dx$$

$$= 2x \cdot \frac{1}{3}(5 - 3x)^{-1} - \int \frac{1}{3}(5 - 3x)^{-1} \cdot 2 \, dx$$

$$= \frac{2x}{3(5 - 3x)} - \frac{2}{3} \int \frac{1}{5 - 3x} dx$$

$$= \frac{2x}{3(5 - 3x)} - \frac{2}{3} \times -\frac{1}{3} \ln|5 - 3x| + C$$

$$= \frac{2x}{3(5 - 3x)} + \frac{2}{9} \ln|5 - 3x| + C$$

**Step 4:** It is time to combine them, thus

$$\int -6x^2(5 - 3x)^{-3} dx = \frac{-x^2}{(5 - 3x)^2} + \int 2x(5 - 3x)^{-2} dx$$

$$= \frac{-x^2}{(5 - 3x)^2} + \frac{2x}{3(5 - 3x)} + \frac{2}{9} \ln|5 - 3x| + C$$

$$\therefore \int -6x^2(5 - 3x)^{-3} dx = \frac{-x^2}{(5 - 3x)^2} + \frac{2x}{3(5 - 3x)} + \frac{2}{9} \ln|5 - 3x| + C$$

---

**d)** $\int (\ln 3x)^2 dx$

**Solution**

**Step 1:** Let's quickly re-arrange this as:

$$\int (\ln 3x)^2 dx = \int \ln^2 3x \, dx$$

**Step 2:** Using the priority order, we have $u = \ln^2 3x$, then $\frac{du}{dx} = 2(\ln 3x) \cdot \frac{1}{x} = \frac{2}{x} \ln 3x$. Similarly, $v' = 1$, then $v = \int 1 \, dx = x$. Hence,

$$\int \ln^2 3x \, dx = uv - \int vu' \, dx$$

$$= \ln^2 3x \cdot x - \int x \cdot \frac{2}{x} \ln 3x \, dx$$

$$= \ln^2 3x - \int 2 \ln 3x \, dx$$

**Step 3:** In this case, we would apply integration by parts to $\int 2 \ln 3x \, dx$. Again, based on the priority order we have $u_1 = \ln 3x$, then $\frac{du_1}{dx} = \frac{1}{3x} \cdot 3 = \frac{1}{x}$. Similarly, $v_1' = 2$, then $v_1 = \int 2 \, dx = 2x$. Hence

$$\int 2 \ln 3x \, dx = u_1 v_1 - \int v_1 u_1' \, dx$$

$$= \ln 3x \cdot 2x - \int 2x \cdot \frac{1}{x} dx$$

$$= 2x \ln 3x - \int 2 \, dx$$

$$= 2x \ln 3x - 2x + C$$

$$= 2x \ln 3x - 2x + C$$

**Step 4:** It is time to combine them, thus

$$\int \ln^2 3x \, dx = uv - \int vu' \, dx$$

$$= x\ln^2 3x - \int 2 \ln 3x \, dx$$

$$= x\ln^2 3x - (2x \ln 3x - 2x) + C$$

$$\therefore \int \ln^2 3x \, dx = x\ln^2 3x - 2x \ln 3x + 2x + C$$

---

e) $\int e^{5x} \cdot \cos x \, dx$
**Solution**

**Step 1:** Using the priority order, we have $u = e^{5x}$, then $\frac{du}{dx} = 5e^{5x}$. Similarly, $v' = \cos x$, then $v = \int \cos x \, dx = \sin x$. Hence,

$$\int e^{5x} \cdot \cos x \, dx = uv - \int vu' \, dx$$

$$= e^{5x} \cdot \sin x - \int \sin x \cdot 5e^{5x} \, dx$$

$$= e^{5x} \cdot \sin x - \int 5e^{5x} \cdot \sin x \, dx$$

**Step 2:** In this case, we would apply integration by parts to $\int 5e^{5x} \cdot \sin x \, dx$. Again, based on the priority order we have $u_1 = 5e^{5x}$, then $\frac{du_1}{dx} = 25e^{5x}$. Similarly, $v_1' = \sin x$, then $v_1 = \int \sin x \, dx = -\cos x$. Hence

$$\int 5e^{5x} \cdot \sin x \, dx = u_1 v_1 - \int v_1 u_1' \, dx$$

$$= 5e^{5x} \cdot -\cos x - \int -\cos x \cdot 25e^{5x} \, dx$$

$$= -5e^{5x} \cdot \cos x + \int 25e^{5x} \cdot \cos x \, dx$$

$$= -5e^{5x} \cdot \cos x + 25 \int e^{5x} \cdot \cos x \, dx$$

**Step 3:** Now the integral is still a product of two functions. Interestingly, it is exactly the question that we started with. If we were to repeat the method on this, we would end up with the same answer after two repetitions, and the cycle would continue. This is called a cyclic integration. Would we give up or what to do? Well, let's combine them first and take it a step at a time, thus

$$\int e^{5x} \cdot \cos x \, dx = uv - \int vu' \, dx$$

$$= e^{5x} \cdot \sin x - \int 5e^{5x} \cdot \sin x \, dx$$

$$= e^{5x} \cdot \sin x - \left[ -5e^{5x} \cdot \cos x + 25 \int e^{5x} \cdot \cos x \, dx \right]$$

$$= e^{5x} \cdot \sin x + 5e^{5x} \cdot \cos x - 25 \int e^{5x} \cdot \cos x \, dx$$

**Step 4:** Now let $I = \int e^{5x} \cdot \cos x \, dx$, therefore

$$I = e^{5x} \cdot \sin x + 5e^{5x} \cdot \cos x - 25I$$
$$I + 25I = e^{5x} \cdot \sin x + 5e^{5x} \cdot \cos x$$
$$26I = e^{5x} \cdot \sin x + 5e^{5x} \cdot \cos x$$

$$I = \frac{1}{26}(e^{5x} \cdot \sin x + 5e^{5x} \cdot \cos x) + C = \frac{e^{5x}}{26}(\sin x + 5\cos x) + C$$

$$\therefore \int e^{5x} \cdot \cos x \, dx = \frac{e^{5x}}{26}(\sin x + 5\cos x) + C$$

**NOTE**

If we swap the priority order between exponential and trigonometric functions by taking $u = \cos x$ and $v' = e^{5x}$, the answer will still be the same as above.

---

In the above examples, you will notice that we've kept to power of 2 (i.e., $ax^2$) where $a \in \mathbb{R}$. If the power were to be 3, the integration by parts will be carried out thrice, and if the power is 4 then we may need to use four loops of integration by parts, and so on. In a nutshell, if the power is $n$, where $n \in \mathbb{Z}^+$, the integration by parts will be carried out in $n$ times.

# Advanced Integration II

We've demonstrated integration by parts based on indefinite integrals, the same will be done for definite integrals. The specific formula for this is given in a compact format as:

$$y = \int_a^b [u.].[v']\,dx = [uv]_a^b - \int_a^b [v].[u']\,dx \qquad (18.5)$$

Let's try some examples.

## Example 7

Determine the exact value of the following:

a) $\int_1^2 10x^4 . \ln 7x \, dx$ 

b) $\int_{-1}^3 \frac{6x}{(x+2)^2}\,dx$ 

c) $\int_{\frac{\pi}{2}}^{\pi} x \left(\cos \frac{1}{3}x + 2\right) dx$

What did you get? Find the solution below to double-check your answer.

## Solution to Example 7

a) $\int_1^2 10x^4 . \ln 7x \, dx$
**Solution**

**Step 1:** Indefinite integral

Using the priority order, we have $u = \ln 7x$, then $\frac{du}{dx} = \frac{1}{x}$. Similarly, $v' = 10x^4$, then $v = \int 10x^4 dx = 2x^5$. Let's start with the indefinite integral

$$\int 10x^4 . \ln 7x \, dx = uv - \int v \frac{du}{dx} dx$$

$$= \ln 7x . 2x^5 - \int 2x^5 . \frac{1}{x} dx = 2x^5 . \ln 7x - \int 2x^4 \, dx$$

$$= 2x^5 . \ln 7x - \frac{2}{5}x^5 + C$$

**Step 2:** Definite integral

Thus,

$$\int_1^2 10x^4 . \ln 7x \, dx = \left[2x^5 . \ln 7x - \frac{2}{5}x^5\right]_1^2$$

$$= \left[2x^5 . \ln 7x - \frac{2}{5}x^5\right]^2 - \left[2x^5 . \ln 7x - \frac{2}{5}x^5\right]^1$$

$$= \left(2(2)^5 . \ln 7(2) - \frac{2}{5}(2)^5\right) - \left(2(1)^5 . \ln 7(1) - \frac{2}{5}(1)^5\right)$$

$$= \left(64\ln 14 - \frac{64}{5}\right) - \left(2\ln 7 - \frac{2}{5}\right) = 64\ln 14 - 2\ln 7 - \frac{64}{5} + \frac{2}{5}$$

$$= 64\ln 14 - 2\ln 7 - \frac{62}{5}$$

$$\therefore \int_1^2 10x^4 \cdot \ln 7x\, dx = 64\ln 14 - 2\ln 7 - \frac{62}{5}$$

---

**b)** $\int_{-1}^{3} \frac{6x}{(x+2)^2}\, dx$

**Solution**

**Step 1:** This looks a bit different, but we can apply the technique to it with a bit of modification as:

$$\int_{-1}^{3} \frac{6x}{(x+2)^2}\, dx = \int_{-1}^{3} 6x(x+2)^{-2}\, dx$$

**Step 2:** Indefinite integral

Using the priority order, we have $u = 6x$, then $\frac{du}{dx} = 6$. Similarly, $v' = (x+2)^{-2}$, then $v = \int (x+2)^{-2} dx = -(x+2)^{-1}$. Let's start with the indefinite integral

$$\int 6x(x+2)^{-2}\, dx = uv - \int v\frac{du}{dx}\, dx$$

$$= 6x \cdot -\frac{1}{x+2} - \int -\frac{1}{x+2} \cdot 6\, dx = \left(-\frac{6x}{x+2}\right) + \left(6\int \frac{1}{x+2}\, dx\right)$$

$$= \left(-\frac{6x}{x+2}\right) + (6\ln|x+2|) + C$$

**Step 3:** Definite integral

Thus,

$$\int_{-1}^{3} 6x(x+2)^{-2}\, dx = \left[6\ln|x+2| - \frac{6x}{x+2}\right]_{-1}^{3}$$

$$= \left[6\ln|x+2| - \frac{6x}{x+2}\right]^{3} - \left[6\ln|x+2| - \frac{6x}{x+2}\right]^{-1}$$

$$= \left(6\ln|3+2| - \frac{6\times 3}{3+2}\right) - \left(6\ln|-1+2| - \frac{6\times -1}{-1+2}\right)$$

$$= \left(6\ln|5| - \frac{18}{5}\right) - (6\ln|1| + 6)$$

$$= 6\ln|5| - \frac{18}{5} - 0 - 6$$

$$= 6\ln|5| - \frac{48}{5} = 6\left(\ln|5| - \frac{8}{5}\right)$$

$$\therefore \int_{-1}^{3} \frac{6x}{(x+2)^2}\, dx = 6\left(\ln|5| - \frac{8}{5}\right)$$

# Advanced Integration II

**ALTERNATIVE METHOD**

$$\int_{-1}^{3} 6x(x+2)^{-2} \, dx = [uv]_{-1}^{3} - \left[\int v \frac{du}{dx} dx\right]_{-1}^{3}$$

$$= \left[-\frac{6x}{x+2}\right]_{-1}^{3} - [-6\ln|x+2|]_{-1}^{3}$$

$$= \left\{\left[-\frac{6x}{x+2}\right]^{3} - \left[-\frac{6x}{x+2}\right]^{-1}\right\} - \left\{[-6\ln|x+2|]^{3} - [-6\ln|x+2|]^{-1}\right\}$$

$$= \left\{\left(-\frac{6\times 3}{3+2}\right) - \left(-\frac{6\times -1}{-1+2}\right)\right\} - \{(-6\ln|3+2|) - (-6\ln|-1+2|)\}$$

$$= \left(-\frac{18}{5} - 6\right) - (-6\ln 5 + 6\ln 1)$$

$$= -\frac{48}{5} + 6\ln 5$$

$$= 6\left(\ln 5 - \frac{8}{5}\right)$$

$$\therefore \int_{-1}^{3} \frac{6x}{(x+2)^{2}} dx = 6\left(\ln|5| - \frac{8}{5}\right)$$

The above alternative approach is based on the actual integral $\int_{a}^{b} [u.].[v'] \, dx = [uv]_{a}^{b} - \int_{a}^{b} [v].[u'] \, dx$, but can be long-winded, as seen in the case above.

---

**c)** $\int_{\frac{\pi}{2}}^{\pi} x\left(\cos \frac{1}{3}x + 2\right) dx$

**Solution**

**Step 1:** Indefinite integral

Using the priority order, we have $u = x$, then $\frac{du}{dx} = 1$. Similarly, $v' = \cos \frac{1}{3}x + 2$, then $v = \int \cos \frac{1}{3}x + 2 \, dx = \frac{\sin \frac{1}{3}x}{\frac{1}{3}} + 2x = 3\sin \frac{1}{3}x + 2x$. Let's start with the indefinite integral

$$\int x\left(\cos \frac{1}{3}x + 2\right) dx = uv - \int v \frac{du}{dx} dx$$

$$= x.\left(3\sin \frac{1}{3}x + 2x\right) - \int \left(3\sin \frac{1}{3}x + 2x\right).1 \, dx$$

$$= x\left(3\sin \frac{1}{3}x + 2x\right) - \int \left(3\sin \frac{1}{3}x + 2x\right) dx$$

$$= x\left(3\sin \frac{1}{3}x + 2x\right) - \left(-9\cos \frac{1}{3}x + x^{2}\right) + C$$

$$= 3x\sin \frac{1}{3}x + 2x^{2} + 9\cos \frac{1}{3}x - x^{2} + C$$

$$= 3x\sin \frac{1}{3}x + 9\cos \frac{1}{3}x + x^{2} + C$$

**Step 2:** Definite integral

Thus,

$$\int_{\frac{\pi}{2}}^{\pi} x\left(\cos\frac{1}{3}x + 2\right) dx = \left[3x\sin\frac{1}{3}x + 9\cos\frac{1}{3}x + x^2\right]_{\frac{\pi}{2}}^{\pi}$$

$$= \left[3x\sin\frac{1}{3}x + 9\cos\frac{1}{3}x + x^2\right]^{\pi} - \left[3x\sin\frac{1}{3}x + 9\cos\frac{1}{3}x + x^2\right]^{\frac{\pi}{2}}$$

$$= \left(3 \times \pi \sin\frac{1}{3}\pi + 9\cos\frac{1}{3}\pi + \pi^2\right)$$

$$\quad - \left(3 \times \frac{\pi}{2} \cdot \sin\frac{1}{3} \cdot \frac{\pi}{2} + 9\cos\frac{1}{3} \cdot \frac{\pi}{2} + \left(\frac{\pi}{2}\right)^2\right)$$

$$= \left(3\pi\sin\frac{1}{3}\pi + 9\cos\frac{1}{3}\pi + \pi^2\right) - \left(\frac{3\pi}{2} \cdot \sin\frac{\pi}{6} + 9\cos\frac{\pi}{6} + \frac{\pi^2}{4}\right)$$

$$= \left(3\pi \times \frac{\sqrt{3}}{2} + 9 \times \frac{1}{2} + \pi^2\right) - \left(\frac{3\pi}{2} \cdot \frac{1}{2} + 9 \times \frac{\sqrt{3}}{2} + \frac{\pi^2}{4}\right)$$

$$= \frac{3\pi\sqrt{3}}{2} + \frac{9}{2} + \pi^2 - \frac{3\pi}{4} - \frac{9\sqrt{3}}{2} - \frac{\pi^2}{4}$$

$$= \pi^2 - \frac{\pi^2}{4} + \frac{3\pi\sqrt{3}}{2} - \frac{3\pi}{4} + \frac{9}{2} - \frac{9\sqrt{3}}{2}$$

$$= \frac{3\pi^2}{4} + 3\pi\left(\frac{2\sqrt{3} - 1}{4}\right) + \frac{9}{2}\left(1 - \sqrt{3}\right)$$

$$\therefore \int_{\frac{\pi}{2}}^{\pi} x\left(\cos\frac{1}{3}x + 2\right) dx = \frac{3\pi^2}{4} + 3\pi\left(\frac{2\sqrt{3} - 1}{4}\right) + \frac{9}{2}\left(1 - \sqrt{3}\right)$$

## 18.4 INTEGRATION OF TRIGONOMETRIC FUNCTIONS

We have covered various techniques essential for dealing with complex integrals, including those involving trigonometric functions. However, there are trigonometric functions that cannot be integrated without being modified or simplified, using the trigonometric identities at our disposal. We will cover this under three headings, but this does not imply in any way that it's exhaustive.

### 18.4.1 POWER OF SINE AND COSINE FUNCTIONS

In this section, we will cover $\sin^n x$ and $\cos^n x$ for $2 \leq n \leq 5$, as $n = 1$ (i.e., $\sin x$ and $\cos x$) is already established and well known.

**CASE 1** $n = 2$: $\sin^2 x$ and $\cos^2 x$

For this case, we will need two out of the three variants of the cosine double-angle formula, namely

$$\cos 2x = 1 - 2\sin^2 x \quad \rightarrow \quad \sin^2 x = \frac{1}{2} - \frac{1}{2}\cos 2x$$

$$\cos 2x = 2\cos^2 x - 1 \quad \rightarrow \quad \cos^2 x = \frac{1}{2}\cos 2x + \frac{1}{2}$$

# Advanced Integration II

## TABLE 18.1
**Standard of Integrals for Powers of Sine and Cosine Functions $n = 2$**

| $\sin^2 x$ | $\cos^2 x$ |
|---|---|
| $\int \sin^2 x \, dx = \int \left( \frac{1}{2} - \frac{1}{2} \cos 2x \right) dx$ | $\int \cos^2 x \, dx = \int \left( \frac{1}{2} + \frac{1}{2} \cos 2x \right) dx$ |
| $= \frac{1}{2} x - \frac{1}{4} \sin 2x + C$ | $= \frac{1}{2} x + \frac{1}{4} \sin 2x + C$ |

The integrals for $n = 2$ are shown in Table 18.1.

**CASE 2** $n = 3$: $\sin^3 x$ and $\cos^3 x$

For this case, we will need the following trigonometric identity and technique:

- $\sin^2 x + \cos^2 x = 1$
- $\int f'(x) \cdot [f(x)]^n \, dx = \frac{1}{n+1} [f(x)]^{n+1} + C$

The integrals for $n = 3$ are shown in Table 18.2.

## TABLE 18.2
**Standard of Integrals for Powers of Sine and Cosine Functions $n = 3$**

| $\sin^3 x$ | $\cos^3 x$ |
|---|---|
| $\int \sin^3 x \, dx = \int \sin x . \sin^2 x \, dx$ | $\int \cos^3 x \, dx = \int \cos x . \cos^2 x \, dx$ |
| $= \int \sin x (1 - \cos^2 x) \, dx$ | $= \int \cos x (1 - \sin^2 x) \, dx$ |
| $= \int (\sin x - \sin x . \cos^2 x) \, dx$ | $= \int (\cos x - \cos x \sin^2 x) \, dx$ |
| $= \int \sin x \, dx - \int \sin x . \cos^2 x \, dx$ | $= \int \cos x \, dx - \int \cos x . \sin^2 x \, dx$ |
| $= \int \sin x \, dx + \int (-\sin x) . (\cos x)^2 \, dx$ | $= \int \cos x \, dx - \int (\cos x) . (\sin x)^2 \, dx$ |
| $= -\cos x + \frac{1}{2+1} (\cos x)^{2+1} + C$ | $= \sin x - \frac{1}{2+1} (\sin x)^{2+1} + C$ |
| $= -\cos x + \frac{1}{3} \cos^3 x + C$ | $= \sin x - \frac{1}{3} \sin^3 x + C$ |

## TABLE 18.3
### Standard of Integrals for Powers of Sine and Cosine Functions $n = 4$

| $\sin^4 x$ | $\cos^4 x$ |
|---|---|
| $\int \sin^4 x\, dx = \int \left(\sin^2 x\right)^2 dx$ | $\int \cos^4 x\, dx = \int \left(\cos^2 x\right)^2 dx$ |
| $= \int \left(\frac{1}{2} - \frac{1}{2}\cos 2x\right)^2 dx$ | $= \int \left(\frac{1}{2} + \frac{1}{2}\cos 2x\right)^2 dx$ |
| $= \int \left(\frac{1}{4} - \frac{1}{2}\cos 2x + \frac{1}{4}\cos^2 2x\right) dx$ | $= \int \left(\frac{1}{4} + \frac{1}{2}\cos 2x + \frac{1}{4}\cos^2 2x\right) dx$ |
| $= \int \left(\frac{1}{4} - \frac{1}{2}\cos 2x + \frac{1}{4}\left[\frac{1}{2}\cos 4x + \frac{1}{2}\right]\right) dx$ | $= \int \left(\frac{1}{4} + \frac{1}{2}\cos 2x + \frac{1}{4}\left[\frac{1}{2} + \frac{1}{2}\cos 4x\right]\right) dx$ |
| $= \int \left(\frac{1}{4} - \frac{1}{2}\cos 2x + \frac{1}{8}\cos 4x + \frac{1}{8}\right) dx$ | $= \int \left(\frac{1}{4} + \frac{1}{2}\cos 2x + \frac{1}{8}\cos 4x + \frac{1}{8}\right) dx$ |
| $= \int \left(\frac{3}{8} - \frac{1}{2}\cos 2x + \frac{1}{8}\cos 4x\right) dx$ | $= \int \left(\frac{3}{8} + \frac{1}{2}\cos 2x + \frac{1}{8}\cos 4x\right) dx$ |
| $= \frac{3}{8}x - \frac{1}{4}\sin 2x + \frac{1}{32}\sin 4x + C$ | $= \frac{3}{8}x + \frac{1}{4}\sin 2x + \frac{1}{32}\sin 4x + C$ |

**CASE 3** $n = 4$: $\sin^4 x$ and $\cos^4 x$

For this case, we will use the following cosine double-angle formula:

$$\cos 2x = 1 - 2\sin^2 x \quad \rightarrow \quad \sin^2 x = \frac{1}{2} - \frac{1}{2}\cos 2x$$

$$\cos 2x = 2\cos^2 x - 1 \quad \rightarrow \quad \cos^2 x = \frac{1}{2}\cos 2x + \frac{1}{2}$$

$$\cos 4x = 1 - 2\sin^2 2x \quad \rightarrow \quad \sin^2 2x = \frac{1}{2} - \frac{1}{2}\cos 4x$$

$$\cos 4x = 2\cos^2 2x - 1 \quad \rightarrow \quad \cos^2 2x = \frac{1}{2}\cos 4x + \frac{1}{2}$$

The integrals for $n = 4$ are shown in Table 18.3.

**CASE 4** $n = 5$: $\sin^5 x$ and $\cos^5 x$

For this case, we will use the following:

- $\sin^2 x + \cos^2 x = 1$
- $\int f'(x) \cdot [f(x)]^n dx = \frac{1}{n+1}[f(x)]^{n+1} + C$

The integrals for $n = 5$ are shown in Table 18.4.

What has been covered are not stand-alone formulas, but approaches to solve powers of common trigonometric functions up to power 5, which can still be applied to powers above 5.

Advanced Integration II

## TABLE 18.4
**Standard of Integrals for Powers of Sine and Cosine Functions $n = 5$**

| $\sin^5 x$ | $\cos^5 x$ |
|---|---|
| $\int \sin^5 x \, dx = \int \sin^4 x \cdot \sin x \, dx$ | $\int \cos^5 x \, dx = \int \cos^4 x \cdot \cos x \, dx$ |
| $= \int \left(\sin^2 x\right)^2 \cdot \sin x \, dx$ | $= \int \left(\cos^2 x\right)^2 \cdot \cos x \, dx$ |
| $= \int \left(1 - \cos^2 x\right)^2 \cdot \sin x \, dx$ | $= \int \left(1 - \sin^2 x\right)^2 \cdot \cos x \, dx$ |
| $= \int \left(1 - 2\cos^2 x + \cos^4 x\right) \cdot \sin x \, dx$ | $= \int \left(1 - 2\sin^2 x + \sin^4 x\right) \cdot \cos x \, dx$ |
| $= \int \left(\sin x - 2\cos^2 x \cdot \sin x + \cos^4 x \cdot \sin x\right) dx$ | $= \int \left(\cos x - 2\sin^2 x \cdot \cos x + \sin^4 x \cdot \cos x\right) dx$ |
| $= \int \sin x \, dx - 2 \int \cos^2 x \cdot \sin x \, dx$ | $= \int \cos x \, dx - 2 \int \sin^2 x \cdot \cos x \, dx$ |
| $\quad + \int \cos^4 x \cdot \sin x \, dx$ | $\quad + \int \sin^4 x \cdot \cos x \, dx$ |
| $= \int \sin x \, dx + 2 \int -\sin x (\cos x)^2 \, dx$ | $= \int \cos x \, dx - 2 \int \cos x \cdot (\sin x)^2 \, dx$ |
| $\quad - \int -\sin x (\cos x)^4 \, dx$ | $\quad + \int \cos x \cdot (\sin x)^4 \, dx$ |
| $= -\cos x + 2\left[\frac{1}{3}\cos^3 x\right] - \left[\frac{1}{5}\cos^5 x\right]$ | $= \sin x - 2\left[\frac{1}{3}\sin^3 x\right] + \left[\frac{1}{5}\sin^5 x\right]$ |
| $= -\cos x + \frac{2}{3}\cos^3 x - \frac{1}{5}\cos^5 x + C$ | $= \sin x - \frac{2}{3}\sin^3 x + \frac{1}{5}\sin^5 x + C$ |

### 18.4.2 Product of Sine and Cosine Functions

To integrate a function of a product of sine and cosine functions, we will use the following identities:

$2 \sin x \cos y = \sin(x+y) + \sin(x-y)$     **OR**     $\sin x \cos y = \frac{1}{2}\sin(x+y) + \frac{1}{2}\sin(x-y)$

$2 \cos x \sin y = \sin(x+y) - \sin(x-y)$     **OR**     $\cos x \sin y = \frac{1}{2}\sin(x+y) - \frac{1}{2}\sin(x-y)$

$2 \cos x \cos y = \cos(x+y) + \cos(x-y)$     **OR**     $\cos x \cos y = \frac{1}{2}\cos(x+y) + \frac{1}{2}\cos(x-y)$

$2 \sin x \sin y = \cos(x-y) - \cos(x+y)$     **OR**     $\sin x \sin y = \frac{1}{2}\cos(x-y) - \frac{1}{2}\cos(x+y)$

The product can be in one of the three following cases:

1) a product of sine and cosine
2) a product of sine and another sine
3) a product of cosine and another cosine

When the multiplier and multiplicand are the same function (sine or cosine), such as in 2 and 3, the argument must be different; otherwise, it will be a power case. To illustrate, $\sin 3x \cdot \sin 6x$ and $\cos x \cdot \cos 5x$ are examples of cases 2 and 3, respectively with arguments $3x$, $6x$, $x$, and $5x$. However, $\cos 2x \cdot \cos 2x$ is not to be treated as an example of case 3, as the argument is the same, i.e., $2x$, and the product can be written as $\cos^2 2x$.

### 18.4.3 OTHER TRIGONOMETRIC FUNCTIONS

This category includes all trigonometric functions that require special substitutions, including sine and cosine functions too. It can be powers of the functions, such as $\tan^2 x$, product of functions, and addition of a function with a numerical constant, such as $1 + \csc^2 x$. It could also come as a rational function. We have come across some of these cases when we discussed various techniques in this chapter or even the chapter before this. It is brought here to give it its due relevance as a family of integrable functions and emphasise the need to employ substitution of relevant trigonometric identities.

That's all for the family of trigonometric functions. Let's try some examples.

#### Example 8

Solve the following:

a) $\int \frac{\cos^2 x}{\sin^2 x} dx$
b) $\int (1 + \cos x)^2 dx$
c) $\int \sin 2x . \cos 2x \, dx$
d) $\int \sin^2 x . \cos^2 x \, dx$
e) $\int \sin 4x . \sin 6x \, dx$
f) $\int \cos x . \cos 5x \, dx$

What did you get? Find the solution below to double-check your answer.

#### Solution to Example 8

**HINT**

In this section and, obviously, the subsequent ones, we need to recall and apply the trigonometric identities.

a) $\int \frac{\cos^2 x}{\sin^2 x} dx$

**Solution**

$$\int \frac{\cos^2 x}{\sin^2 x} dx = \int \cot^2 x \, dx$$
$$= \int \csc^2 x - 1 \, dx$$

Because $\cot^2 x + 1 = \csc^2 x$, thus we have

$$\int \frac{\cos^2 x}{\sin^2 x} dx = \int \csc^2 x - 1 \, dx$$
$$= -\cot x - x + C$$
$$\therefore \int \frac{\cos^2 x}{\sin^2 x} dx = -x - \cot x + C$$

# Advanced Integration II

**b)** $\int (1 + \cos x)^2 dx$

**Solution**

$$\int (1 + \cos x)^2 dx = \int (1 + 2\cos x + \cos^2 x) dx$$

$$= \int \left(1 + 2\cos x + \frac{1}{2}\cos 2x + \frac{1}{2}\right) dx$$

$$= \int \left(\frac{3}{2} + 2\cos x + \frac{1}{2}\cos 2x\right) dx$$

Because $\cos 2x = 2\cos^2 x - 1$ which implies that $\cos^2 x = \frac{1}{2}\cos 2x + \frac{1}{2}$. Thus, we have

$$\int (1 + \cos x)^2 dx = \int \left(\frac{3}{2} + 2\cos x + \frac{1}{2}\cos 2x\right) dx$$

$$= \frac{3}{2}x + 2\sin x + \frac{1}{4}\sin 2x + C$$

$$\therefore \int (1 + \cos x)^2 dx = \frac{3}{2}x + 2\sin x + \frac{1}{4}\sin 2x + C$$

**c)** $\int \sin 2x . \cos 2x \, dx$

**Solution**

$$\int \sin 2x . \cos 2x \, dx = \frac{1}{2} \int \sin(2x + 2x) + \sin(2x - 2x) \, dx$$

$$= \frac{1}{2} \int \sin 4x \, dx$$

Because $2 \sin x \cos y = \sin(x + y) + \sin(x - y)$, thus we have

$$\int \sin 2x . \cos 2x \, dx = \frac{1}{2} \int \sin 4x \, dx$$

$$= \frac{1}{2}\left[-\frac{1}{4}\cos 4x\right] + C$$

$$\therefore \int \sin 2x . \cos 2x \, dx = -\frac{1}{8}\cos 4x + C$$

**NOTE**

In this question, we could have used $2 \sin x \cos y = \sin(x + y) - \sin(x - y)$ to suggest that cosine function is greater than the sine part. However, it will be irrelevant, as the second part of the identity will always give zero since the argument is the same for both sine and cosine in this question, i.e. $2x$. Take this a general rule for when the argument is the same for both.

**d)** $\int \sin^2 x . \cos^2 x \, dx$
**Solution**

$$\int \sin^2 x . \cos^2 x \, dx = \int \sin^2 x . (1 - \sin^2 x) \, dx$$

$$= \int \sin^2 x - \sin^4 x \, dx$$

$$= \int \left[\frac{1}{2} - \frac{1}{2}\cos 2x\right] - \left[\frac{3}{8} - \frac{1}{2}\cos 2x + \frac{1}{8}\cos 4x\right] dx$$

$$= \int \left(\frac{1}{8} - \frac{1}{8}\cos 4x\right) dx$$

Because $\cos 2x = 1 - 2\sin^2 x$, which implies that $\sin^2 x = \frac{1}{2} - \frac{1}{2}\cos 2x$, and we've demonstrated this for $\sin^2 x$ and $\sin^4 x$ on pages 739 and 740. Hence, we have

$$\int \sin^2 x . \cos^2 x \, dx = \int \left(\frac{1}{8} - \frac{1}{8}\cos 4x\right) dx$$

$$= \frac{1}{8}x - \frac{1}{32}\sin 4x + C$$

$$\therefore \int \sin^2 x . \cos^2 x \, dx = \frac{1}{8}x - \frac{1}{32}\sin 4x + C$$

**NOTE**
In this example, we could have substituted for $\sin^2 x$ and the result will still be the same. Give it a try.

---

**e)** $\int \sin 4x . \sin 6x \, dx$
**Solution**

$$\int \sin 4x . \sin 6x \, dx = \int \frac{1}{2}\cos(6x - 4x) - \frac{1}{2}\cos(4x + 6x) \, dx$$

$$= \int \frac{1}{2}\cos 2x - \frac{1}{2}\cos 10x \, dx$$

Because $\sin x \sin y = \frac{1}{2}\cos(x - y) - \frac{1}{2}\cos(x + y)$, thus we have

$$\int \sin 4x . \sin 6x \, dx = \frac{1}{2}\int \cos 2x - \cos 10x \, dx$$

$$= \frac{1}{2}\left[\frac{1}{2}\sin 2x - \frac{1}{10}\sin 10x\right] + C$$

$$\therefore \int \sin 4x . \sin 6x \, dx = \frac{1}{4}\sin 2x - \frac{1}{20}\sin 10x + C$$

---

**f)** $\int \cos x . \cos 5x \, dx$
**Solution**

$$\int \cos x . \cos 5x \, dx = \int \frac{1}{2}\cos(x + 5x) + \frac{1}{2}\cos(5x - x) \, dx$$

$$= \int \frac{1}{2}\cos 6x + \frac{1}{2}\cos 4x \, dx$$

# Advanced Integration II

Because $\cos x \cos y = \frac{1}{2}\cos(x+y) + \frac{1}{2}\cos(x-y)$, thus we have

$$\int \cos x \cdot \cos 5x\, dx = \frac{1}{2}\int \cos 6x + \cos 4x\, dx$$

$$= \frac{1}{2}\left[\frac{1}{6}\sin 6x + \frac{1}{4}\sin 4x\right] + C$$

$$\therefore \int \cos x \cdot \cos 5x\, dx = \frac{1}{12}\sin 6x + \frac{1}{8}\sin 4x + C$$

**NOTE**

In this example, we could have used $\frac{1}{2}\cos(x-5x) = \frac{1}{2}\cos(-4x)$, but this is the same as $\frac{1}{2}\cos(4x)$ due to the symmetry of the cosine function about the y-axis.

Let's try another set of examples.

### Example 9

Solve the following:

a) $\int \frac{4\sin 2x}{1-\cos^2 x}\, dx$ 

b) $\int (\tan x - \cot x)^2\, dx$ 

c) $\int (\sin x - \cos x)^2\, dx$

What did you get? Find the solution below to double-check your answer.

### Solution to Example 9

a) $\int \frac{4\sin 2x}{1-\cos^2 x}\, dx$

**Solution**

$$\int \frac{4\sin 2x}{1-\cos^2 x}\, dx = \int \frac{4\sin 2x}{\sin^2 x}\, dx$$

$$= \int \frac{4\sin 2x}{\frac{1}{2} - \frac{1}{2}\cos 2x}\, dx$$

Because $\cos 2x = 1 - 2\sin^2 x$ which implies that $\sin^2 x = \frac{1}{2} - \frac{1}{2}\cos 2x$. Notice here that $\frac{d}{dx}\left(\frac{1}{2} - \frac{1}{2}\cos 2x\right) = \sin 2x$. Here we will use $\int \frac{f'(x)}{f(x)}\, dx$, thus we have

$$\int \frac{4\sin 2x}{1-\cos^2 x}\, dx = 4\int \frac{\sin 2x}{\frac{1}{2} - \frac{1}{2}\cos 2x}\, dx$$

$$= 4\left[\ln\left(\frac{1}{2} - \frac{1}{2}\cos 2x\right)\right] + C = 4\left[\ln\left(\sin^2 x\right)\right]$$

$$\therefore \int \frac{4\sin 2x}{1-\cos^2 x}\, dx = 4\left[\ln(sin^2 x)\right] + C$$

**b)** $\int (\tan x - \cot x)^2 dx$
**Solution**

$$\int (\tan x - \cot x)^2 dx = \int (\tan^2 x - 2 \tan x \cot x + \cot^2 x) dx$$
$$= \int (\tan^2 x - 2 + \cot^2 x) dx$$
$$= \int [(\tan^2 x) - 2 + (\cot^2 x)] dx$$
$$= \int [(\sec^2 x - 1) - 2 + (\cosec^2 x - 1)] dx$$

Because $\tan^2 x + 1 = \sec^2 x$ and $\cot^2 x + 1 = \cosec^2 x$, thus we have

$$\int (\tan x - \cot x)^2 dx = \int (\sec^2 x + \cosec^2 x - 4) dx$$
$$= \tan x - \cot x - 4x + C$$

$$\therefore \int (\tan x - \cot x)^2 dx = \tan x - \cot x - 4x + C$$

---

**c)** $\int (\sin x - \cos x)^2 dx$
**Solution**

$$\int (\sin x - \cos x)^2 dx = \int (\sin^2 x - 2 \sin x \cos x + \cos^2 x) dx$$
$$= \int (\sin^2 x + \cos^2 x - 2 \sin x \cos x) dx$$
$$= \int (1 - 2 \sin x \cos x) dx$$
$$= \int 1 \, dx - \int 2 \sin x \cos x \, dx$$

At this point, we have two options using:

- $\int f'(x) . f(x) \, dx$, or
- $2 \sin x \cos x = \sin 2x$

**Option 1** $\int f'(x) . f(x) \, dx$

$$\int (\sin x - \cos x)^2 dx = \int 1 \, dx - \int 2 \sin x \cos x \, dx$$
$$= \int 1 \, dx - 2 \int \cos x \sin x \, dx$$
$$= x - 2 \left(\frac{1}{2}\sin^2 x\right) + C = x - \sin^2 x + C$$

Because $\frac{d}{dx}(\sin x) = \cos x$, we have

$$\therefore \int (\sin x - \cos x)^2 dx = x - \sin^2 x + C$$

# Advanced Integration II

**Option 2**  $2\sin x \cos x = \sin 2x$

$$\int (\sin x - \cos x)^2 dx = \int 1\, dx - \int 2\sin x \cos x\, dx$$
$$= \int 1\, dx - \int \sin 2x\, dx$$
$$= x - \left(-\frac{1}{2}\cos 2x\right) + C = x + \frac{1}{2}\cos 2x + C$$
$$\therefore \int (\sin x - \cos x)^2 dx = x + \frac{1}{2}\cos 2x + C$$

**NOTE**

We can simplify option 2 to obtain a similar answer as option 1 using the fact that $\cos 2x = 1 - 2\sin^2 x \rightarrow \frac{1}{2}\cos 2x = \frac{1}{2} - \sin^2 x$ as:

$$\int (\sin x - \cos x)^2 dx = x + \frac{1}{2}\cos 2x + C$$
$$= x + \frac{1}{2} - \sin^2 x + C$$
$$= x - \sin^2 x + \left(\frac{1}{2} + C\right)$$
$$\therefore \int (\sin x - \cos x)^2 dx = x - \sin^2 x + A$$

Note that we changed the arbitrary constant to $A$.

---

Let's try another set of examples.

### Example 10

Use an appropriate substitution to determine the exact value of each of the following:

a) $\int_{\pi/6}^{\pi/4} \tan^2\left(2x - \frac{\pi}{3}\right) dx$

b) $\int_{\frac{\pi}{2}}^{\pi} \sin 3x \cdot \cos 5x\, dx$

c) $\int_{\pi/6}^{\pi/3} \sec^2 x \cdot \operatorname{cosec}^2 x\, dx$

What did you get? Find the solution below to double-check your answer.

**Solution to Example 10**

**a)** $\int_{\pi/6}^{\pi/4} \tan^2(2x - \frac{\pi}{3}) \, dx$

**Solution**
Let's start off by using the identity $\tan^2 x + 1 = \sec^2 x$

$$\int_{\frac{\pi}{6}}^{\frac{\pi}{4}} \sec^2(2x - \frac{\pi}{3}) - 1 \, dx = \left[\frac{\tan(2x - \frac{\pi}{3})}{2} - x\right]_{\frac{\pi}{6}}^{\frac{\pi}{4}}$$

$$= \left[\frac{\tan(2x - \frac{\pi}{3})}{2} - x\right]^{\frac{\pi}{4}} - \left[\frac{\tan(2x - \frac{\pi}{3})}{2} - x\right]^{\frac{\pi}{6}}$$

$$= \left(\frac{\tan(2 \times \frac{\pi}{4} - \frac{\pi}{3})}{2} - \frac{\pi}{4}\right) - \left(\frac{\tan(2 \times \frac{\pi}{6} - \frac{\pi}{3})}{2} - \frac{\pi}{6}\right)$$

$$= \left(\frac{\sqrt{3}}{6} - \frac{\pi}{4}\right) - \left(0 - \frac{\pi}{6}\right) = \frac{\sqrt{3}}{6} - \frac{\pi}{4} + \frac{\pi}{6} = \frac{1}{12}\left(2\sqrt{3} - \pi\right)$$

$$\therefore \int_{\frac{\pi}{6}}^{\frac{\pi}{4}} \tan^2(2x - \frac{\pi}{3}) \, dx = \frac{1}{12}\left(2\sqrt{3} - \pi\right)$$

---

**b)** $\int_{\frac{\pi}{2}}^{\pi} \sin 3x \cdot \cos 5x \, dx$

**Solution**
Let's start with this

$$\cos x \sin y = \frac{1}{2}\sin(x+y) - \frac{1}{2}\sin(x-y)$$

$$\therefore \cos 5x \cdot \sin 3x = \frac{1}{2}\sin 8x - \frac{1}{2}\sin 2x$$

$$\int_{\frac{\pi}{2}}^{\pi} \sin 3x \cdot \cos 5x \, dx = \int_{\frac{\pi}{2}}^{\pi} \left(\frac{1}{2}\sin 8x - \frac{1}{2}\sin 2x\right) dx$$

$$= \left[-\frac{1}{16}\cos 8x + \frac{1}{4}\cos 2x\right]_{\frac{\pi}{2}}^{\pi}$$

$$= \left[\frac{1}{4}\cos 2x - \frac{1}{16}\cos 8x\right]_{\frac{\pi}{2}}^{\pi}$$

# Advanced Integration II

$$= \left[\frac{1}{4}\cos 2x - \frac{1}{16}\cos 8x\right]^{\pi} - \left[\frac{1}{4}\cos 2x - \frac{1}{16}\cos 8x\right]^{\frac{\pi}{2}}$$

$$= \left(\frac{1}{4}\cos 2\pi - \frac{1}{16}\cos 8\pi\right) - \left(\frac{1}{4}\cos \pi - \frac{1}{16}\cos 4\pi\right)$$

$$= \left(\frac{1}{4} - \frac{1}{16}\right) - \left(-\frac{1}{4} - \frac{1}{16}\right) = \frac{1}{2}$$

$$\therefore \int_{\frac{\pi}{2}}^{\pi} \sin 3x \cdot \cos 5x \, dx = 0.5$$

---

c) $\int_{\pi/6}^{\pi/3} \sec^2 x \cdot \csc^2 x \, dx$

**Solution**

In this example, we will use the following identities: $\tan^2 x + 1 = \sec^2 x$ and $\cot^2 x + 1 = \csc^2 x$. Let's start:

$$\sec^2 x \cdot \csc^2 x = (\tan^2 x + 1)(\cot^2 x + 1) = \tan^2 x \cdot \cot^2 x + \tan^2 x + \cot^2 x + 1$$

$$= 1 + \tan^2 x + \cot^2 x + 1$$

$$= \tan^2 x + \cot^2 x + 2$$

$$= (\sec^2 x - 1) + (\csc^2 x - 1) + 2$$

$$= \sec^2 x + \csc^2 x$$

Therefore

$$\int_{\pi/6}^{\pi/3} \sec^2 x \cdot \csc^2 x \, dx = \int_{\pi/6}^{\pi/3} \sec^2 x + \csc^2 x \, dx$$

$$= [\tan x - \cot x]_{\pi/6}^{\pi/3} = [\tan x - \cot x]_{\pi/6}^{\pi/3}$$

$$= \left[\tan x - \frac{1}{\tan x}\right]^{\pi/3} - \left[\tan x - \frac{1}{\tan x}\right]^{\pi/6}$$

$$= \left(\sqrt{3} - \frac{1}{\sqrt{3}}\right) - \left(\frac{\sqrt{3}}{3} - \frac{3}{\sqrt{3}}\right)$$

$$= \left(\frac{2\sqrt{3}}{3}\right) - \left(-\frac{2\sqrt{3}}{3}\right) = \frac{4\sqrt{3}}{3}$$

$$\therefore \int_{\pi/6}^{\pi/3} \sec^2 x \cdot \csc^2 x \, dx = \frac{4\sqrt{3}}{3}$$

## 18.5 CHAPTER SUMMARY

1) When we need to integrate a function that is defined parametrically, we will apply parametric integration, which is based on the chain rule like parametric differentiation. In short, we can say that if $y = f(x)$ is defined parametrically by $y = f(t)$ and $x = f(t)$, the indefinite integration is given by

$$\int y\, dx = \int y \cdot \frac{dx}{dt}\, dt$$

and the definite integral is

$$\int_{x_1}^{x_2} y\, dx = \int_{t_1}^{t_2} y \cdot \frac{dx}{dt}\, dt$$

Notice how we changed $dx$ to $dt$ as it is not possible to integrate $y = f(t)$ wrt $x$. Also, we changed the limits from $x_1$ and $x_2$ to $t_1$ and $t_2$, respectively, in definite integral.

2) **Integration by parts** is distinctively known for being based on the inverse product rule and for using both differentiation and integration to solve a problem. It states that:

$$y = \int [u.] \cdot \left[\frac{dv}{dx}\right] dx = uv - \int [v] \cdot \left[\frac{du}{dx}\right] dx$$

It is often written in a compact format as:

$$y = \int [u.] \cdot [v']\, dx = uv - \int [v] \cdot [u']\, dx$$

where $v' = \frac{dv}{dx}$ and $u' = \frac{du}{dx}$.

$$y = \int_a^b [u.] \cdot [v']\, dx = [uv]_a^b - \int_a^b [v] \cdot [u']\, dx$$

****

## 18.6 FURTHER PRACTICE

To access complementary contents, including additional exercises, please go to www.dszak.com.

# 19 Application of Integration

## Learning Outcomes

Once you have studied the content of this chapter, you should be able to:

- Estimate the areas covered by a function using numerical methods
- Use the trapezium rule to numerically estimate the value of definite integrals
- Use Simpson's rule to numerically estimate the value of definite integrals
- Estimate the error due to numerical integration
- Calculate the volume of revolution, given the Cartesian equation of the curve
- Calculate the volume of revolution, given the parametric equations of the curve
- Calculate the absolute and percentage errors in using numerical integration
- Increase the accuracy of the approximation methods

## 19.1 INTRODUCTION

Integration is an interesting part of calculus and, of course, comes with its challenges and complexities. In Chapters 17 and 18, we covered some advanced techniques involving integration, which represent a small fraction of the domain. We are here in this last chapter to go through more techniques, including numerical integration, mean values, and root mean square values of functions.

## 19.2 NUMERICAL INTEGRATION

We've spent time exploring methods and techniques to integrate complex functions that do not fit any of our already mentioned standards. We must, however, admit that there exist functions that cannot be solved using those techniques we covered in Chapters 17 and 18. Equally true is that covering the remaining techniques, including but not limited to reduction formulas, does not imply that all is done and dusted (i.e., we can integrate any function). Indeed, there are functions that we cannot integrate at all. As a result, we need to resort to a technique that we can use to solve any function, known as **numerical integration**.

This technique is used to find an area under a curve or between curves within the given limits. It is usually employed for functions that cannot be integrated. In other words, it is used for definite integration. The result from this is usually approximate and can be improved to reduce the error.

### 19.2.1 TRAPEZIUM RULE

This is the first of the techniques in numerical integration to be considered here and is one of the most popular. Before we go into the details of the trapezium (also called trapezoidal) rule, it is relevant to mention that this method uses trapezium, as is clearly apparent from its name.

Figure 19.1 shows three different orientations of a trapezium (a shape with a pair of parallel lines, which are connected by two non-parallel lines).

What is shown in Figure 19.1(a) is a mirror of Figure 19.1(b), and these are the two variants that are relevant in numerical integration.

Recall that the area of a trapezium is given by:

$$A = \frac{1}{2} h (l_1 + l_2) \tag{19.1}$$

where

- $l_1$ the length of the first parallel line (also denoted as $a$)
- $l_2$ the length of the second parallel line (also denoted as $b$)
- $h$ the length between the two parallel lines

The orientation shown in Figure 19.1(c) and similarly oriented trapeziums will not be encountered in this section. This is because the area of interest is the one below the curve, the $x$-axis, and the vertical lines representing the limits. On this basis, Figure 19.1(a) will be used for an increasing function or where a function is increasing, and Figure 19.1(b) for a decreasing function or where a function is decreasing.

Let's now consider how this rule works in Figure 19.2. The region $R$ in Figure 19.2(a) is a region under the curve $f(x)$, the $x$-axis, and the vertical lines at $x_1$ and $x_2$. To determine $R$, we need to evaluate $\int_{x_1}^{x_2} f(x)\ dx$. The width, representing the height $h$ in our formula, is given by:

$$h = \frac{x_2 - x_1}{7}$$

We have divided this by 7 because we require seven strips or trapezia.

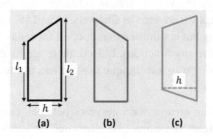

**FIGURE 19.1** Three possible orientations of a trapezium illustrated.

Application of Integration

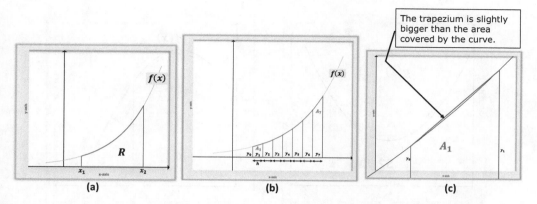

**FIGURE 19.2** Determining the area of equally spaced strips between the given limits illustrated.

In creating these strips, we need eight vertical lines represented as $y_0, y_1, y_2, \ldots, y_7$. If $A_1$ is the area of the first trapezium on the far left, then

$$A_1 = \frac{1}{2}h(y_0 + y_1)$$

In a similar manner, the area of the second trapezium $A_2$ is

$$A_2 = \frac{1}{2}h(y_1 + y_2)$$

The same is true for $A_3, A_4, A_5, A_6$, and $A_7$. Hence, we have

$$A = A_1 + A_2 + \cdots + A_7$$
$$= \frac{1}{2}h(y_0 + y_1) + \frac{1}{2}h(y_1 + y_2) + \cdots + \frac{1}{2}h(y_6 + y_7)$$
$$= \frac{1}{2}h[y_0 + 2(y_1 + y_2 + y_3 + y_4 + y_5 + y_6) + y_7]$$

We can then say that

$$\int_{x_1}^{x_2} [f(x)]\, dx \approx \frac{1}{2}h[y_0 + 2(y_1 + y_2 + y_3 + y_4 + y_5 + y_6) + y_7]$$

Notice that we have used the approximation sign $\approx$ to show that the answer we obtain numerically (the RHS) is not exact. In the current case, it's an overestimate, meaning that it's more than the actual value. When we look back at Figure 19.2(c), as we zoom in on the first trapezium, it's obvious that the area covered by the trapezium is a bit larger. This is generally the case for when a curve is convex to the x-axis as shown in Figure 19.1(a). Also, it's worth noting that a middle strip is actually a rectangular shape, not a trapezium. The greater the curvature, the more the overestimation. This is because the function $g(x)$ exhibits a vertical symmetry in this region. Therefore, we will use the trapezium rule for this case except for the middle strip, identified as S2 in Figure 19.3. We will use a special case of the trapezium formula, which is employed when the lengths of any two parallel lines are equal, i.e., $a = b$.

It's equally possible that the numerical integration results in an underestimate value. This occurs when the function is concave to the x-axis, as shown in Figure 19.3(a). We will soon show that the

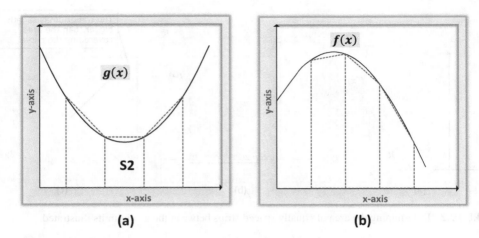

**FIGURE 19.3** Trapezium rule: (a) overestimation and (b) underestimation.

accuracy of the estimates is largely dependent on the number of strips taken. The more strips used, the greater the accuracy of the estimate.

It's necessary to have a general formula for any number of strips. Let's say, we need to find the area under the curve $f(x)$ between points $a$ and $b$ and choose $n$ strips. We express this as:

$$\int_a^b [f(x)] \, dx \approx \frac{1}{2}h \, [y_0 + 2(y_1 + y_2 + \ldots + y_{n-1}) + y_n] \qquad (19.2)$$

where $h = \frac{b-a}{n}$ and $y_i = f(a + ih)$, $0 \leq i \leq n$

Let's make the following notes about the above rule:

**Note 1**  The number of strips (or intervals) is one less than the number of ordinates (or $y$ values). In other words, if we choose $n$ strips, there will be $n + 1$ ordinates.

**Note 2**  Each strip represents a trapezium.

**Note 3**  The first and the last ordinates, $y_0$ and $y_n$, are multiplied by one while all other intermediate ordinates are multiplied by two.

**Note 4**  If $x_i$ is the value of $x$ at a corresponding ordinate $y_0$, it follows that $x_0 = a$, $x_1 = a + h$, $x_2 = a + 2h$, …, $x_{n-1} = a + (n-1)h$, $x_n = b$, or $x_n = a + nh$. In other words, to obtain the $x$-values, start with the lower limit $a$ and keep adding $h$ each time until you reach $x$-value equal to the upper limit $b$.

**Note 5**  To know whether the trapezium rule will overestimate or underestimate, plot the function and see if it is concave or convex to the $x$-axis in the region of interest. A concave shape results in an underestimate, while a convex shape implies an overestimate. If, however, it's a combination of the two, then we cannot determine this visually.

# Application of Integration

Do the following to determine the area under a curve using the trapezium rule:

**Step 1:** Decide on the number of strips, if not already given or stated.

**Step 2:** Determine $h$ using $h = \frac{1}{n}(b - a)$, where $a$ and $b$ are the limits of integration. $h$ is called the height or width of the strip.

**Step 3:** Set out the $x$-values in steps of $h$, starting with the lower limit as: $x_0 = a$, $x_1 = a + h$, $x_2 = a + 2h$, ..., $x_n = b$.

**Step 4:** Determine the ordinates (or $y$ values), by substituting the corresponding $x$-values in the original function $f(x)$. In other words, $y_0 = f(x_0)$, $y_1 = f(x_1)$, $y_2 = f(x_2)$, $y_3 = f(x_3)$, ....

**Step 5:** Substitute these values accordingly in $\frac{1}{2}h\,[y_0 + 2(y_1 + y_2 + ... + y_{n-1}) + y_n]$.

That's all. Let's try some examples.

## Example 1

$R$ is the region covered by $f(x) = e^{2x} - 2x$, the $x$-axis, and the vertical lines at $x = -3$ and $x = 1$.

a) Evaluate $I = \int_{-3}^{1} [e^{2x} - 2x]\,dx$.

b) Use the trapezium rule with four strips to estimate the value of $\int_{-3}^{1} [e^{2x} - 2x]\,dx$.

c) Use the trapezium rule with eight strips to estimate the value of $\int_{-3}^{1} [e^{2x} - 2x]\,dx$.

d) Comment on your answers to parts a, b, and c, and state whether this is an overestimate or underestimate.

Present your answers correct to 2 d.p.

What did you get? Find the solution below to double-check your answer.

## Solution to Example 1

**a)** $I = \int_{-3}^{1} [e^{2x} - 2x]\,dx$

**Solution**

$$I = \int_{-3}^{1} [e^{2x} - 2x]\,dx = \left[\frac{e^{2x}}{2} - x^2\right]_{-3}^{1}$$

$$= \left[\frac{e^{2x}}{2} - x^2\right]^{1} - \left[\frac{e^{2x}}{2} - x^2\right]^{-3}$$

$$= \left(\frac{e^2}{2} - 1^2\right) - \left(\frac{e^{-6}}{2} - (-3)^2\right)$$

$$= \left(\frac{e^2}{2} - 1\right) - \left(\frac{1}{2e^6} - 9\right) = \frac{e^2}{2} - \frac{1}{2e^6} + 8$$

$$= \frac{e^8 + 16e^6 - 1}{2e^6} = 11.69$$

$$\therefore I = \int_{-3}^{1} [e^{2x} - 2x]\,dx = 11.69 \text{ unit}^2 \text{ (2 d.p.)}$$

## TABLE 19.1
### Solution to Example 1(b)

|       | $y_0$ | $y_1$ | $y_2$ | $y_3$ | $y_4$ |
|-------|-------|-------|-------|-------|-------|
| $x$   | $-3$  | $-2$  | $-1$  | $0$   | $1$   |
| $f(x)$| 6.002 | 4.018 | 2.135 | 1.000 | 5.389 |

**b) Four strips**
**Solution**
$n = 4$, we have

$$h = \frac{1}{4}(1-(-3)) = \frac{1}{4}(1+3) = 1$$

$$\int_{-3}^{1} \left[e^{2x} - 2x\right] dx \approx \frac{1}{2}h\left[y_0 + 2(y_1 + y_2 + y_3) + y_4\right]$$

$$\approx \frac{1}{2}\left[y_0 + 2(y_1 + y_2 + y_3) + y_4\right]$$

It will be better to use a table for computing the ordinates. This is shown in Table 19.1.

Notice that we start with $a = -3$ and add $h = 1$ each time. Hence

$$\int_{-3}^{1} \left[e^{2x} - 2x\right] dx \approx \frac{1}{2}\left[6.002 + 2(4.018 + 2.135 + 1.000) + 5.389\right]$$

$$\approx \frac{1}{2}(25.697) \approx 12.85$$

$$\therefore \int_{-3}^{1} \left[e^{2x} - 2x\right] dx \approx \mathbf{12.85 \text{ unit}^2 \text{ (2 d.p.)}}$$

**c) Eight strips**
**Solution**
$n = 8$, we have

$$h = \frac{1}{8}(1-(-3)) = \frac{1}{8}(1+3) = 0.5$$

$$\int_{-3}^{1} \left[e^{2x} - 2x\right] dx \approx \frac{1}{2}h\left[y_0 + 2(y_1 + y_2 + y_3 + y_4 + y_5 + y_6 + y_7) + y_8\right]$$

$$\approx \frac{1}{4}\left[y_0 + 2(y_1 + y_2 + y_3 + y_4 + y_5 + y_6 + y_7) + y_8\right]$$

Using $h = 0.5$, the $x$-values and corresponding ordinates are shown in Table 19.2.

# Application of Integration

**TABLE 19.2**
**Solution to Example 1(c)**

|   | $y_0$ | $y_1$ | $y_2$ | $y_3$ | $y_4$ | $y_5$ | $y_6$ | $y_7$ | $y_8$ |
|---|---|---|---|---|---|---|---|---|---|
| $x$ | $-3$ | $-2.5$ | $-2$ | $-1.5$ | $-1$ | $-0.5$ | 0 | 0.5 | 1 |
| $f(x)$ | 6.002 | 5.007 | 4.018 | 3.050 | 2.135 | 1.368 | 1.000 | 1.718 | 5.389 |

Hence

$$\int_{-3}^{1} \left[e^{2x} - 2x\right] dx \approx \frac{1}{4} \left[6.002 + 2\left(5.007 + 4.018 + 3.050 + 2.135 + 1.368 \right.\right.$$
$$\left.\left. + 1.000 + 1.718\right) + 5.389\right]$$
$$\approx \frac{1}{4}(47.983) \approx 12.00$$
$$\therefore \int_{-3}^{1} \left[e^{2x} - 2x\right] dx \approx 12.00 \text{ unit}^2 \text{ (2 d.p.)}$$

**d) Comments**
**Solution**

- When we doubled the number of strips to 8, the estimate was much better, suggesting a direct proportionality between accuracy and the number of strips.
- In both cases, there is an overestimation, obvious from the numerical values. We can also see in Figure 19.4 that it's convex in the region $R$ between $x = -3$ and $x = 1$.

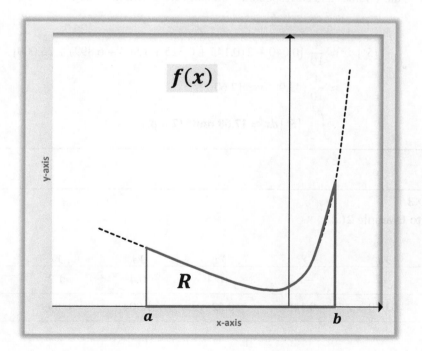

**FIGURE 19.4** Solution to Example 1(d).

Let's try another example.

### Example 2

Use the trapezium rule with six ordinates to estimate the following:

a) $\int_{-2}^{2} [5^x]\, dx$

b) $\int_{1}^{3} \left[\dfrac{x}{2x+1}\right] dx$

Present your answer correct to 2 d.p.

---

What did you get? Find the solution below to double-check your answer.

### Solution to Example 2

a) $\int_{-2}^{2} [5^x]\, dx$
**Solution**
Six ordinates imply 5 strips, thus $n = 5$. We also have $a = -2$ and $b = 2$. Therefore

$$h = \frac{1}{5}(2 - (-2)) = \frac{4}{5} = 0.8$$

$$\int_{-2}^{2} [5^x]\, dx \approx \frac{1}{2}h\,[y_0 + 2(y_1 + y_2 + y_3 + y_4) + y_5]$$

$$\approx \frac{4}{10}\,[y_0 + 2(y_1 + y_2 + y_3 + y_4) + y_5]$$

Using $h = \frac{4}{5}$, the $x$-values and corresponding ordinates are shown in Table 19.3.
Hence

$$\int_{-2}^{2} [5^x]\, dx \approx \frac{4}{10}\,[0.040 + 2(0.145 + 0.525 + 1.904 + 6.899) + 25.000]$$

$$\approx \frac{4}{10}(43.986) \approx 17.60$$

$$\therefore \int_{-2}^{2} [5^x]\, dx \approx \mathbf{17.60}\ \text{unit}^2\ \text{(2 d.p.)}$$

---

**TABLE 19.3**
**Solution to Example 2(a)**

|      | $y_0$ | $y_1$ | $y_2$ | $y_3$ | $y_4$ | $y_5$ |
|------|-------|-------|-------|-------|-------|-------|
| $x$  | $-2$  | $-1.2$ | $-0.4$ | $0.4$ | $1.2$ | $2$ |
| $f(x)$ | 0.040 | 0.145 | 0.525 | 1.904 | 6.899 | 25.000 |

# Application of Integration

**TABLE 19.4**
**Solution to Example 2(b)**

|   | $y_0$ | $y_1$ | $y_2$ | $y_3$ | $y_4$ | $y_5$ |
|---|---|---|---|---|---|---|
| $x$ | 1 | 1.4 | 1.8 | 2.2 | 2.6 | 3 |
| $f(x)$ | 0.333 | 0.368 | 0.391 | 0.407 | 0.419 | 0.429 |

b) $\int_1^3 \left[\dfrac{x}{2x+1}\right] dx$

**Solution**

Six ordinates imply 5 strips, thus $n = 5$. We also have $a = 1$ and $b = 3$. Therefore

$$h = \frac{1}{5}(3-1) = \frac{2}{5} = 0.4$$

$$\int_1^3 \left[\frac{x}{2x+1}\right] dx \approx \frac{1}{2}h \left[y_0 + 2(y_1 + y_2 + y_3 + y_4) + y_5\right]$$

$$\approx \frac{2}{10} \left[y_0 + 2(y_1 + y_2 + y_3 + y_4) + y_5\right]$$

Using $h = \dfrac{2}{5}$, the $x$-values and corresponding ordinates are shown in Table 19.4.
Hence

$$\int_1^3 \frac{x}{2x+1} dx \approx \frac{2}{10} \left[0.333 + 2(0.368 + 0.391 + 0.407 + 0.419) + 0.429\right]$$

$$\approx \frac{2}{10}(3.932) \approx 0.79$$

$$\therefore \int_1^3 \frac{x}{2x+1} dx \approx 0.79 \text{ unit}^2 \text{ (2 d.p.)}$$

## 19.2.2 Simpson's Rule

This is another approximation technique, which is named after an English mathematician, Thomas Simpson. Like the trapezium method, Simpson's rule is also based on dividing the region into equal strips. Unlike the trapezium rule, however, Simpson's rule can only be used for an even number of strips, such as 2, 4, 6, etc., strips.

Simpson's rule is also called Simpson's 1/3 rule or Simpson's first rule. Others are Simpson's 3/8 rule (or Simpson's second rule) and Simpson's third rule. Simpson's first rule (or simply Simpson's rule) is given as:

$$\int_a^b [f(x)] \, dx \approx \frac{1}{3}h \left[y_0 + 4y_1 + 2y_2 + 4y_3 + 2y_4 + \cdots + 2y_{n-2} + 4y_{n-1} + y_n\right] \quad (19.3)$$

**OR**

$$\int_a^b [f(x)] \, dx \approx \frac{1}{3}h \left[(y_0 + y_n) + 4(y_1 + y_3 + \ldots y_{n-1}) + 2(y_2 + y_4 + \ldots y_{n-2})\right] \quad (19.4)$$

where $h = \frac{b-a}{n}$ and $y_i = f(a+ih)$, $0 \le i \le n$

Let's quickly make the following comments about the formula:

**Note 1** We can only use an even number of intervals or strips. As a result, there will always be an odd number of ordinates.

**Note 2** The first and last ordinates are multiplied by one.

**Note 3** The odd-numbered ordinates are multiplied by 4. If you choose $y_0, y_1, y_2, \ldots, y_n$ style, then the odd ordinates are $y_1, y_3, y_5, \ldots, y_{n-1}$.

**Note 4** The even-numbered ordinates are multiplied by 2. If you choose $y_0, y_1, y_2, \ldots, y_n$ style, then they are $y_2, y_4, y_6, \ldots, y_{n-2}$.

Whilst it is possible to prove the above formula, this is however not our objective here.

As usual, let's try some examples.

### Example 3

$R$ is the region covered by $f(x) = e^{2x} - 2x$, the $x$-axis and the vertical lines at $x = -3$ and $x = 1$.

a) Evaluate $I = \int_{-3}^{1} [e^{2x} - 2x] \, dx$.

b) Use Simpson's rule with four strips to estimate the value of $\int_{-3}^{1} [e^{2x} - 2x] \, dx$.

c) Use Simpson's rule with eight strips to estimate the value of $\int_{-3}^{1} [e^{2x} - 2x] \, dx$.

d) Comment on your answers to parts a, b, and c, and whether this is an overestimate or underestimate.

Present your answers correct to 2 d.p.

What did you get? Find the solution below to double-check your answer.

### Solution to Example 3

**HINT**

Note that we have solved this using the trapezium method (see Example 1).

## Application of Integration

**a)** $I = \int_{-3}^{1} [e^{2x} - 2x] \, dx$

**Solution**

$$I = \int_{-3}^{1} [e^{2x} - 2x] \, dx = \left[ \frac{e^{2x}}{2} - x^2 \right]_{-3}^{1}$$

$$= \left[ \frac{e^{2x}}{2} - x^2 \right]^{1} - \left[ \frac{e^{2x}}{2} - x^2 \right]^{-3}$$

$$= \left( \frac{e^2}{2} - 1^2 \right) - \left( \frac{e^{-6}}{2} - (-3)^2 \right)$$

$$= \left( \frac{e^2}{2} - 1 \right) - \left( \frac{1}{2e^6} - 9 \right) = \frac{e^2}{2} - \frac{1}{2e^6} + 8$$

$$= \frac{e^8 + 16e^6 - 1}{2e^6} = 11.69$$

$$\therefore I = \int_{-3}^{1} [e^{2x} - 2x] \, dx = \mathbf{11.69 \ unit^2 \ (2 \ d.p.)}$$

**b) Four strips**

**Solution**

$n = 4$, we have

$$h = \frac{1}{4}(1 - (-3)) = \frac{1}{4}(1 + 3) = 1$$

$$\int_{-3}^{1} [e^{2x} - 2x] \, dx \approx \frac{1}{3} h \left[ (y_0 + y_4) + 4(y_1 + y_3) + 2(y_2) \right]$$

$$\approx \frac{1}{3} \left[ (y_0 + y_4) + 4(y_1 + y_3) + 2(y_2) \right]$$

Using $h = 1$, the $x$-values and corresponding ordinates are shown in Table 19.5.
Hence

$$\int_{-3}^{1} [e^{2x} - 2x] \, dx \approx \frac{1}{3} \left[ (6.002 + 5.389) + 4(4.018 + 1.000) + 2(2.135) \right]$$

$$\approx \frac{1}{3} \left[ (11.391) + (20.072) + (4.270) \right]$$

$$\approx \frac{1}{3} (35.733) \approx 11.911$$

$$\therefore \int_{-3}^{1} [e^{2x} - 2x] \, dx \approx \mathbf{11.91 \ unit^2 \ (2 \ d.p.)}$$

**TABLE 19.5**
**Solution to Example 3(b)**

|      | $y_0$ | $y_1$ | $y_2$ | $y_3$ | $y_4$ |
|------|-------|-------|-------|-------|-------|
| $x$  | $-3$  | $-2$  | $-1$  | $0$   | $1$   |
| $f(x)$ | 6.002 | 4.018 | 2.135 | 1.000 | 5.389 |

**TABLE 19.6**
Solution to Example 3(c)

|   | $y_0$ | $y_1$ | $y_2$ | $y_3$ | $y_4$ | $y_5$ | $y_6$ | $y_7$ | $y_8$ |
|---|-------|-------|-------|-------|-------|-------|-------|-------|-------|
| $x$    | $-3$  | $-2.5$ | $-2$  | $-1.5$ | $-1$  | $-0.5$ | 0     | 0.5   | 1     |
| $f(x)$ | 6.002 | 5.007 | 4.018 | 3.050 | 2.135 | 1.368 | 1.000 | 1.718 | 5.389 |

**c) Eight strips**
**Solution**
$n = 8$, we have

$$h = \frac{1}{8}(1-(-3)) = \frac{1}{8}(1+3) = 0.5$$

$$\int_{-3}^{1} \left[e^{2x} - 2x\right] dx \approx \frac{1}{3}h\left[(y_0 + y_8) + 4(y_1 + y_3 + y_5 + y_7) + 2(y_2 + y_4 + y_6)\right]$$

$$\approx \frac{1}{6}\left[(y_0 + y_8) + 4(y_1 + y_3 + y_5 + y_7) + 2(y_2 + y_4 + y_6)\right]$$

Using $h = 0.5$, the $x$-values and corresponding ordinates are shown in Table 19.6. Hence

$$\int_{-3}^{1} \left[e^{2x} - 2x\right] dx \approx \frac{1}{6}\left[(6.002 + 5.389) + 4(5.007 + 3.050 + 1.368 + 1.718)\right.$$

$$\left. + 2(4.018 + 2.135 + 1.000)\right]$$

$$\approx \frac{1}{6}\left[(11.391) + (44.572) + (14.306)\right]$$

$$\approx \frac{1}{6}(70.269) \approx 11.71$$

$$\therefore \int_{-3}^{1} \left[e^{2x} - 2x\right] dx \approx \mathbf{11.71 \text{ unit}^2 \text{ (2 d.p.)}}$$

**d) Comments**
**Solution**

- When we doubled the number of strips to 8, the estimate was much better, suggesting a direct proportionality between accuracy and the number of strips.
- In both cases, there is an overestimation. This is however better than what we observed with the trapezium method for the same function (Table 19.7).

# Application of Integration

**TABLE 19.7**
**Solution to Example 3(d)**

| Method | $n = 4$ | $n = 8$ |
|---|---|---|
| Exact | 11.69 | 11.69 |
| Trapezium | 12.85 | 12.00 |
| Simpson | 11.91 | 11.71 |

Let's try a couple of examples to compare the two methods side by side. Here we go.

## Example 4

Use (i) the trapezium rule and (ii) Simpson's rule, with seven ordinates, to estimate the following:

a) $\int_0^{\pi/2} [\sin 2\theta]\, d\theta$    b) $\int_{-\pi/6}^{\pi/6} [\sec^2 \theta]\, d\theta$

Present your answer in exact form or correct to 4 s.f.

What did you get? Find the solution below to double-check your answer.

## Solution to Example 4

**a)** $\int_0^{\pi/2} [\sin 2\theta]\, d\theta$
**Solution**

### i) Trapezium rule

Seven ordinates imply 6 strips, thus $n = 6$. We also have $a = 0$ and $b = \pi/2$. Therefore

$$h = \frac{1}{6}(\pi/2 - 0) = \frac{\pi}{12}$$

$$\int_0^{\pi/2} [\sin 2\theta]\, d\theta \approx \frac{1}{2}h\, [y_0 + 2(y_1 + y_2 + y_3 + y_4 + y_5) + y_6]$$

$$\approx \frac{\pi}{24} [y_0 + 2(y_1 + y_2 + y_3 + y_4 + y_5) + y_6]$$

## TABLE 19.8
### Solution to Example 4(a)

|   | $y_0$ | $y_1$ | $y_2$ | $y_3$ | $y_4$ | $y_5$ | $y_6$ |
|---|---|---|---|---|---|---|---|
| $\theta$ | 0 | $\pi/12$ | $\pi/6$ | $\pi/4$ | $\pi/3$ | $5\pi/12$ | $\pi/2$ |
| $f(\theta)$ | 0 | 0.5 | $\sqrt{3}/2$ | 1 | $\sqrt{3}/2$ | 0.5 | 0 |

Using $h = \frac{\pi}{12}$, the $x$-values and corresponding ordinates are shown in Table 19.8.
Hence

$$\int_0^{\pi/2} [\sin 2\theta]\, d\theta \approx \frac{\pi}{24}\left[0 + 2\left(0.5 + \sqrt{3}/2 + 1 + \sqrt{3}/2 + 0.5\right) + 0\right]$$

$$\approx \frac{\pi}{24}\left(4 + 2\sqrt{3}\right) \approx \frac{\pi}{12}\left(2 + \sqrt{3}\right) = 0.9770$$

$$\therefore \int_0^{\pi/2} [\sin 2\theta]\, d\theta \approx \frac{\pi}{12}\left(2 + \sqrt{3}\right) \approx \mathbf{0.9770 \text{ unit}^2 \text{ (4 s.f.)}}$$

### ii) Simpson's rule

We will use the same table and $h$ value.

$$\int_0^{\pi/2} [\sin 2\theta]\, d\theta \approx \frac{1}{3}h[(y_0 + y_6) + 4(y_1 + y_3 + y_5) + 2(y_2 + y_4)]$$

$$\approx \frac{\pi}{36}[(y_0 + y_6) + 4(y_1 + y_3 + y_5) + 2(y_2 + y_4)]$$

Hence

$$\int_0^{\pi/2} [\sin 2\theta]\, d\theta \approx \frac{\pi}{36}\left[(0 + 0) + 4(0.5 + 1 + 0.5) + 2\left(\sqrt{3}/2 + \sqrt{3}/2\right)\right]$$

$$\approx \frac{\pi}{36}\left[(0) + (8) + \left(2\sqrt{3}\right)\right]$$

$$\approx \frac{\pi}{36}\left(8 + 2\sqrt{3}\right) \approx 1.000$$

$$\therefore \int_0^{\pi/2} [\sin 2\theta]\, d\theta \approx \mathbf{1.000 \text{ unit}^2 \text{ (4 s.f.)}}$$

The exact value for this integral, correct to 4 s.f., is $I = \int_0^{\pi/2} [\sin 2\theta]\, d\theta = 1$. This implies that Simpson's rule is more accurate.

# Application of Integration

**TABLE 19.9**
Solution to Example 4(b)

|  | $y_0$ | $y_1$ | $y_2$ | $y_3$ | $y_4$ | $y_5$ | $y_6$ |
|---|---|---|---|---|---|---|---|
| $\theta$ | $-\pi/6$ | $-\pi/9$ | $-\pi/18$ | 0 | $\pi/18$ | $\pi/9$ | $\pi/6$ |
| $f(\theta)$ | 1.3333 | 1.1325 | 1.0311 | 1.0000 | 1.0311 | 1.1325 | 1.3333 |
|  | 1.3333 | 1.1325 | 1.0311 | 1.0000 | 1.0311 | 1.1325 | 1.3333 |

**b)** $\int_{-\pi/6}^{\pi/6} [\sec^2\theta]\, d\theta$
**Solution**

### i) Trapezium rule

Seven ordinates imply 6 strips, thus $n = 6$. We also have $a = -\pi/6$ and $b = \pi/6$, therefore

$$h = \frac{1}{6}(\pi/6 - (-\pi/6)) = \frac{1}{6}(\pi/6 + \pi/6) = \frac{\pi}{18}$$

$$\int_{-\pi/6}^{\pi/6} [\sec^2\theta]\, d\theta \approx \frac{1}{2}h\,[y_0 + 2(y_1 + y_2 + y_3 + y_4 + y_5) + y_6]$$

$$\approx \frac{\pi}{36}[y_0 + 2(y_1 + y_2 + y_3 + y_4 + y_5) + y_6]$$

Using $h = \frac{\pi}{18}$, the $x$-values and corresponding ordinates are shown in Table 19.9.
Hence

$$\int_{-\pi/6}^{\pi/6} [\sec^2\theta]\, d\theta \approx \frac{\pi}{36}[1.3333 + 2(1.1325 + 1.0311 + 1.0000 + 1.0311 + 1.1325) + 1.3333]$$

$$\approx \frac{\pi}{36}(13.3210) \approx 1.162$$

$$\therefore \int_{-\pi/6}^{\pi/6} [\sec^2\theta]\, d\theta \approx \mathbf{1.162\ unit^2}\ \text{(4 s.f.)}$$

### ii) Simpson's rule

We will use the same table and $h$ value.

$$\int_{-\pi/6}^{\pi/6} [\sec^2\theta]\, d\theta \approx \frac{1}{3}h\,[(y_0 + y_6) + 4(y_1 + y_3 + y_5) + 2(y_2 + y_4)]$$

$$\approx \frac{\pi}{54}[(y_0 + y_6) + 4(y_1 + y_3 + y_5) + 2(y_2 + y_4)]$$

Hence

$$\int_{-\pi/6}^{\pi/6} [\sec^2\theta]\, d\theta \approx \frac{\pi}{54}[(1.3333 + 1.3333) + 4(1.1325 + 1.0000 + 1.1325)$$

$$+ 2(1.0311 + 1.0311)]$$

$$\approx \frac{\pi}{54}[(2.6666) + (13.06) + (4.1244)]$$

$$\approx \frac{\pi}{54}(19.851) \approx 1.155$$

$$\therefore \int_{-\pi/6}^{\pi/6} [\sec^2\theta]\, d\theta \approx 1.155 \text{ unit}^2 \text{ (4 s.f.)}$$

The exact value for this integral, correct to 4 s.f., is $I = \int_{-\pi/6}^{\pi/6} [\sec^2\theta]\, d\theta = 1.155$. This implies once again that Simpson's rule is more accurate.

---

### 19.2.3 Calculating Error

We've seen above that the numerical integration provides an estimate of integrals, and is particularly useful when it's difficult or impossible to find the exact value of an integral. As a result, we may be interested in determining the error due to this method, provided we have the information of what the exact value is. There are four ways to determine this, namely:

**1) Absolute Error (AE)**

This is used to determine the actual value of deviation, and is given as:

$$\boxed{AE = \text{Exact value} - \text{Approximate value}} \qquad (19.5)$$

For this, a positive AE (or AE > 0) implies an underestimate, and a negative value shows an overestimate.

**2) Relative Error (RE)**

This is computed using:

$$\boxed{RE = |\text{Exact value} - \text{Approximate value}|} \qquad (19.6)$$

Note the use of the modulus | | sign. Hence, the answer will always be positive (or greater than 1), and it does not matter which of the two is the minuend or subtrahend. In other words, RE is used when we are simply interested in the difference between the exact and estimate. With this method, it's not possible to determine if the process has resulted in an underestimate or an overestimate.

**3) Absolute Percentage Error (APE)**

This is given by:

$$\boxed{APE = \frac{\text{Exact value} - \text{Approximate value}}{\text{Exact value}} \times 100\%} \qquad (19.7)$$

# Application of Integration

**4) Relative Percentage Error (RPE)**

This is given by:

$$\boxed{\text{RPE} = \left|\frac{\text{Exact value} - \text{Approximate value}}{\text{Exact value}}\right| \times 100\%} \tag{19.8}$$

Let's try some examples.

## Example 5

A function $f(x)$ shown in Figure 19.5 is defined by the equation $y = \frac{4x}{x^2+2}$. The region J is bounded by $f(x)$, the x-axis, and the vertical lines at $x = 1$ and $x = 3$.

a) Determine the exact area of J.

b) Use the trapezium rule to estimate the area J when:

  i) $n = 2$
  ii) $n = 3$
  iii) $n = 4$
  iv) $n = 5$
  v) $n = 10$

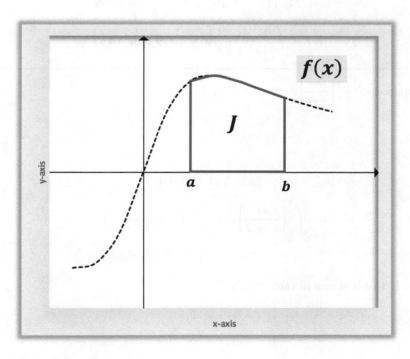

**FIGURE 19.5** Example 5.

c) Determine the absolute error in each case in (b) and state whether it's an underestimate or overestimate. Comment on your results.

d) Determine the relative percentage error in each case of (b) and comment on your results.

Present your answers correct to 4 s.f.

---

What did you get? Find the solution below to double-check your answer.

### Solution to Example 5

**a) Exact area**
**Solution**
For this case, we have $a = 1$ and $b = 3$. Thus

$$I = \int_1^3 \left[\frac{4x}{x^2+2}\right] dx = \int_1^3 2\left[\frac{2x}{x^2+2}\right] dx$$

This is a case of $\int \frac{f'(x)}{f(x)} dx$, thus

$$I = \left[2 \ln(x^2 + 2)\right]_1^3$$
$$= \left[2 \ln(x^2 + 2)\right]^3 - \left[2 \ln(x^2 + 2)\right]^1$$
$$= \{2 \ln(3^2 + 2)\} - \{2 \ln(1^2 + 2)\}$$
$$= (2 \ln 11) - (2 \ln 3) = 2 \ln\left(\frac{11}{3}\right) = 2.599$$

$$\therefore I = \int_1^3 \left[\frac{4x}{x^2+2}\right] dx = 2.599 \text{ unit}^2 \text{ (4 s.f.)}$$

---

**b) Approximation for $\int_1^3 \left[\frac{4x}{x^2+2}\right] dx$**
**Solution**

i) $n = 2$

$$h = \frac{1}{2}(3 - 1) = 1$$

$$\int_1^3 \left[\frac{4x}{x^2+2}\right] dx \approx \frac{1}{2} h \, [y_0 + 2y_1 + y_2]$$

$$\approx \frac{1}{2} [y_0 + 2y_1 + y_2]$$

because $h = 1$. This is shown in Table 19.10.

## TABLE 19.10
### Solution to Example 5(b) – Part I

|      | $y_0$  | $y_1$  | $y_2$  |
|------|--------|--------|--------|
| $x$    | 1      | 2      | 3      |
| $f(x)$ | 1.3333 | 1.3333 | 1.0909 |

Hence

$$\int_1^3 \left[\frac{4x}{x^2+2}\right] dx \approx \frac{1}{2}\left[1.3333 + 2(1.3333) + 1.0909\right]$$

$$\approx \frac{1}{2}(5.0908) \approx 2.5454$$

$$\therefore \int_1^3 \left[\frac{4x}{x^2+2}\right] dx \approx 2.545 \text{ unit}^2 \text{ (4 s.f.)}$$

ii) $n = 3$

$$h = \frac{1}{3}(3-1) = \frac{2}{3}$$

$$\int_1^3 \left[\frac{4x}{x^2+2}\right] dx \approx \frac{1}{2}h\left[y_0 + 2(y_1 + y_2) + y_3\right]$$

$$\approx \frac{1}{3}\left[y_0 + 2(y_1 + y_2) + y_3\right]$$

because $h = \frac{2}{3}$. This is shown in Table 19.11.

Hence

$$\int_1^3 \left[\frac{4x}{x^2+2}\right] dx \approx \frac{1}{3}\left[1.3333 + 2(1.3953 + 1.2537) + 1.0909\right]$$

$$\approx \frac{1}{3}(7.7222) \approx 2.57407$$

$$\therefore \int_1^3 \left[\frac{4x}{x^2+2}\right] dx \approx 2.574 \text{ unit}^2 \text{ (4 s.f.)}$$

## TABLE 19.11
### Solution to Example 5(b) – Part II

|      | $y_0$  | $y_1$  | $y_2$  | $y_3$  |
|------|--------|--------|--------|--------|
| $x$    | 1      | 5/3    | 7/3    | 3      |
| $f(x)$ | 1.3333 | 1.3953 | 1.2537 | 1.0909 |

**TABLE 19.12**
Solution to Example 5(b) – Part III

|      | $y_0$  | $y_1$  | $y_2$  | $y_3$  | $y_4$  |
|------|--------|--------|--------|--------|--------|
| $x$  | 1      | 1.5    | 2      | 2.5    | 3      |
| $f(x)$ | 1.3333 | 1.4118 | 1.3333 | 1.2121 | 1.0909 |

**iii)** $n = 4$

$$h = \frac{1}{4}(3-1) = \frac{1}{2}$$

$$\int_1^3 \left[\frac{4x}{x^2+2}\right] dx \approx \frac{1}{2}h \,[y_0 + 2(y_1 + y_2 + y_3) + y_4]$$

$$\approx \frac{1}{4}\,[y_0 + 2(y_1 + y_2 + y_3) + y_4]$$

because $h = \frac{1}{2}$. This is shown in Table 19.12.

Hence

$$\int_1^3 \left[\frac{4x}{x^2+2}\right] dx \approx \frac{1}{4}\,[1.3333 + 2(1.4118 + 1.3333 + 1.2121) + 1.0909]$$

$$\approx \frac{1}{4}(10.3386) \approx 2.58465$$

$$\therefore \int_1^3 \left[\frac{4x}{x^2+2}\right] dx \approx \mathbf{2.585 \text{ unit}^2} \text{ (4 s.f.)}$$

**iv)** $n = 5$

$$h = \frac{1}{5}(3-1) = \frac{2}{5}$$

$$\int_1^3 \left[\frac{4x}{x^2+2}\right] dx \approx \frac{1}{2}h \,[y_0 + 2(y_1 + y_2 + y_3 + y_4) + y_5]$$

$$\approx \frac{1}{5}\,[y_0 + 2(y_1 + y_2 + y_3 + y_4) + y_5]$$

because $h = \frac{2}{5}$. This is shown in Table 19.13.

Hence

$$\int_1^3 \left[\frac{4x}{x^2+2}\right] dx \approx \frac{1}{5}\,[1.3333 + 2(1.4141 + 1.3740 + 1.2865 + 1.1872) + 1.0909]$$

$$\approx \frac{1}{5}(12.9478) \approx 2.590$$

$$\therefore \int_1^3 \left[\frac{4x}{x^2+2}\right] dx \approx \mathbf{2.590 \text{ unit}^2} \text{ (4 s.f.)}$$

**TABLE 19.13**
Solution to Example 5(b) – Part IV

|   | $y_0$ | $y_1$ | $y_2$ | $y_3$ | $y_4$ | $y_5$ |
|---|---|---|---|---|---|---|
| $x$ | 1 | $7/5$ | $9/5$ | $11/5$ | $13/5$ | 3 |
| $f(x)$ | 1.3333 | 1.4141 | 1.3740 | 1.2865 | 1.1872 | 1.0909 |

**TABLE 19.14**
Solution to Example 5(b) – Part V

|   | $y_0$ | $y_1$ | $y_2$ | $y_3$ | $y_4$ | $y_5$ | $y_6$ | $y_7$ | $y_8$ | $y_9$ | $y_{10}$ |
|---|---|---|---|---|---|---|---|---|---|---|---|
| $x$ | 1 | 1.2 | 1.4 | 1.6 | 1.8 | 2 | 2.2 | 2.4 | 2.6 | 2.8 | 3 |
| $f(x)$ | 1.3333 | 1.3953 | 1.4141 | 1.4035 | 1.3740 | 1.3333 | 1.2865 | 1.2371 | 1.1872 | 1.1382 | 1.0909 |

v) $n = 10$

$$h = \frac{1}{10}(3-1) = \frac{1}{5}$$

$$\int_1^3 \left[\frac{4x}{x^2+2}\right] dx \approx \frac{1}{2}h\ [y_0 + 2(y_1 + y_2 + y_3 + y_4 + y_5 + y_6 + y_7 + y_8 + y_9) + y_{10}]$$

$$\approx \frac{1}{10}\ [y_0 + 2(y_1 + y_2 + y_3 + y_4) + y_5]$$

because $h = \frac{1}{5}$. This is shown in Table 19.14.
Hence

$$\int_1^3 \left[\frac{4x}{x^2+2}\right] dx \approx \frac{1}{10}\ [1.3333 + 2(1.3953 + 1.4141 + 1.4035 + 1.3740 + 1.3333$$

$$+ 1.2865 + 1.2371 + 1.1872 + 1.1382) + 1.0909]$$

$$\approx \frac{1}{10}(25.9626) \approx 2.596$$

$$\therefore \int_1^3 \left[\frac{4x}{x^2+2}\right] dx \approx \mathbf{2.596\ unit^2}\ \textbf{(4 s.f.)}$$

**TABLE 19.15**
**Solution to Example 5(c)**

| n  | Estimate Value | Exact Value | Absolute Error | Outcome       |
|----|----------------|-------------|----------------|---------------|
| 2  | 2.545          | 2.599       | 0.054          | Underestimate |
| 3  | 2.574          | 2.599       | 0.025          | Underestimate |
| 4  | 2.585          | 2.599       | 0.014          | Underestimate |
| 5  | 2.590          | 2.599       | 0.009          | Underestimate |
| 10 | 2.596          | 2.599       | 0.003          | Underestimate |

**c)** Absolute Error
**Solution**
Absolute Error is calculated using the formula below.

$$AE = \text{Exact value} - \text{Approximate value}$$

This has been calculated for each of the different strips and summarised in Table 19.15.

- A general decrease in absolute error as the number of strips increases.
- They are all positive, which implies an underestimate. This agrees with the shape of $f(x)$, which is concave to the $x$-axis between $x = 1$ and $x = 3$.

**d)** Relative Percentage Error
**Solution**
Relative Percentage Error is calculated using the formula below.

$$RPE = \left| \frac{\text{Exact value} - \text{Approximate value}}{\text{Exact value}} \right| \times 100 \%$$

This has been calculated for each of the different strips and summarised in Table 19.16.

- A general decrease in percentage error as the number of strips increases. The graph in Figure 19.6 shows that this trend will follow an exponential decay, and as $n \to \infty$, error tends to zero.
- The deviation is less than 3 %, which is quite remarkable.

# Application of Integration

**TABLE 19.16**
**Solution to Example 5(d)**

| n | Estimate Value | Exact Value | Relative Percentage Error (%) |
|---|---|---|---|
| 2 | 2.545 | 2.599 | 2.078 |
| 3 | 2.574 | 2.599 | 0.962 |
| 4 | 2.585 | 2.599 | 0.539 |
| 5 | 2.590 | 2.599 | 0.346 |
| 10 | 2.596 | 2.599 | 0.115 |

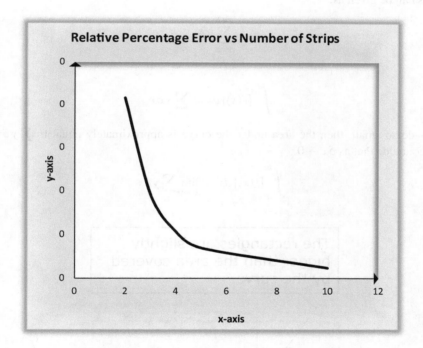

**FIGURE 19.6** Solution to Example 5(c) – Part II.

## 19.3 APPLICATIONS OF INTEGRATION

There are several applications of integration, not only in engineering and science but also in other sectors of human endeavours. We will only consider a few here.

### 19.3.1 Areas and Volumes

#### 19.3.1.1 Area

Recall that an area covered by a function $f(x)$, the $x$-axis, and the limits $x = a$ and $x = b$ can be written as a definite integral:

$$\int_a^b [f(x)]\, dx$$

We may also note, from our earlier discussion in this chapter, that we can approximate the area by taking several strips (or trapezia) from the region of interest. Furthermore, we established the fact that the error can be minimised by increasing the number of strips. Suppose we divide the region in this case into rectangular strips with equal width of $\delta x$ and length $y$, as shown in Figure 19.7. The area of each strip will be $y\, dx$, and the area of the region under the curve is the sum of the area of the strips. This can be given as:

$$\sum_{x=a}^{x=b} y\, \delta x$$

Thus

$$\int_a^b [f(x)]\, dx \approx \sum_{x=a}^{x=b} y\, \delta x$$

If $\delta x$ is made so small, then the area under the curve is approximately equal to $\sum y\, \delta x$. We can therefore conclude that as $\delta x \to 0$

$$\int_a^b [f(x)]\, dx = \lim_{\delta x \to 0} \sum y\, \delta x$$

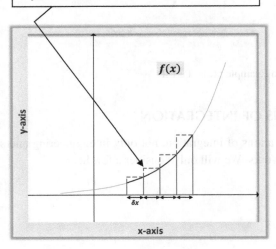

**FIGURE 19.7** Summation of the area of the strips under the curve.

# Application of Integration

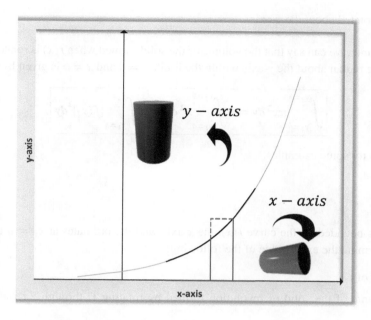

**FIGURE 19.8** Volume of solid of revolution illustrated.

## 19.3.1.2 Volumes of Solids of Revolution

Now let's revisit our discussion on the area above and take a step further. Consider Figure 19.8, where one single strip (or rectangle) is shown rotated through one revolution ($2\pi$ or 360°) about: (i) the $x$-axis and (ii) the $y$-axis. For each strip, a 3D shape is formed that is approximately cylindrical.

There are two instances of interest, namely:

**Case 1** When rotated about the $x$-axis

The volume of each cylinder is $Ah$, where $A$ is the cross-sectional area and $h$ is the height of the cylinder. For this case, the radius is $y$ and height is $\delta x$. Hence, the volume $V$ of each strip is

$$V_{\text{strip}} = Ah = \pi y^2 \delta x$$

The total volume is

$$\sum \pi y^2 \delta x \approx \int_a^b \pi y^2 \delta x$$

As $\delta x \to 0$, we have that

$$\lim_{\delta x \to 0} \sum \pi y^2 \delta x = \int_a^b \pi y^2 \, dx$$

The volume of the solid formed when $f(x)$ is rotated through one revolution or $2\pi$ radian about $x$-axis within the limits $x = a$ and $x = b$ is given by:

$$\boxed{\int_{x_1=a}^{x_2=b} \pi y^2 \, dx = \pi \int_{x_1=a}^{x_2=b} y^2 \, dx = \pi \int_{x_1=a}^{x_2=b} [f(x)]^2 \, dx} \tag{19.9}$$

**Case 2** When rotated about the y-axis

In the same manner, we can say that the volume of the solid formed when $f(x)$ is rotated through one revolution or $2\pi$ radian about the y-axis within the limits $x = a$ and $x = b$ is given by:

$$\boxed{\int_{y_1=a}^{y_2=b} \pi x^2 \, dy = \pi \int_{y_1=a}^{y_2=b} x^2 \, dy = \pi \int_{y_1=a}^{y_2=b} [f(y)]^2 \, dy} \qquad (19.10)$$

That's all. Let's try some examples.

## Example 6

The region $C$ is bounded by the curve $f(x)$, the x-axis, and the ordinates at $x_1 = a$ and $x_2 = b$. In each case, determine the exact value of the following:

i) The area of $C$.

ii) The volume of the solid of revolution formed by rotating $C$ through $2\pi$ radians about the x-axis.

a) $f(x) = x(x+2)$, $[x_1 = 1, x_2 = 3]$ 

b) $f(x) = \operatorname{cosec} x$, $\left[x_1 = \pi/4, x_2 = \pi/3\right]$

c) $f(x) = \frac{1}{\sqrt{x}} + \sqrt{x}$, $[x_1 = 1, x_2 = 4]$

d) $f(x) = 6 \sec 2x \cdot \tan 2x$, $\left[x_1 = 0, x_2 = \pi/6\right]$

What did you get? Find the solution below to double-check your answer.

## Solution to Example 6

**a)** $f(x) = x(x+2)$, $[x_1 = 1, x_2 = 3]$
**Solution**

i) Area

$$I = \int_1^3 x(x+2) \, dx = \int_1^3 x^2 + 2x \, dx = \left[\frac{x^3}{3} + x^2\right]_1^3$$

$$= \left[\frac{x^3}{3} + x^2\right]^3 - \left[\frac{x^3}{3} + x^2\right]^1 = \left(\frac{3^3}{3} + 3^2\right) - \left(\frac{1^3}{3} + 1^2\right)$$

$$= (18) - \left(\frac{4}{3}\right) = \frac{50}{3}$$

$$\therefore I = \int_1^3 x(x+2) \, dx = \frac{50}{3} \, \text{unit}^2$$

ii) Volume (x-axis)

$$y = x(x+2) = x^2 + 2x$$

Therefore

$$\begin{aligned} y^2 &= (x^2 + 2x)^2 \\ &= x^4 + 4x^3 + 4x^2 \end{aligned}$$

Now let's determine the volume.

$$V_{x-axis} = \pi \int_1^3 y^2 \, dx = \pi \int_1^3 (x^4 + 4x^3 + 4x^2) \, dx$$

$$= \pi \left[ \frac{x^5}{5} + x^4 + \frac{4x^3}{3} \right]_1^3 = \pi \left[ \frac{x^5}{5} + x^4 + \frac{4x^3}{3} \right]^3 - \pi \left[ \frac{x^5}{5} + x^4 + \frac{4x^3}{3} \right]^1$$

$$= \pi \left( \frac{3^5}{5} + 3^4 + \frac{4(3)^3}{3} \right) - \pi \left( \frac{1^5}{5} + 1^4 + \frac{4(1)^3}{3} \right)$$

$$= \pi \left( \frac{828}{5} \right) - \pi \left( \frac{38}{15} \right) = \frac{2446}{15} \pi$$

$$\therefore V = \pi \int_1^3 (x^4 + 4x^3 + 4x^2) \, dx = \frac{2446}{15} \pi \text{ unit}^3$$

---

**b)** $f(x) = \operatorname{cosec} x$, $\left[ x_1 = \pi/4, \, x_2 = \pi/3 \right]$

**Solution**

i) Area

$$I = \int_{\pi/4}^{\pi/3} \operatorname{cosec} x \, dx = -[\ln |\cot x + \operatorname{cosec} x| \,]_{\pi/4}^{\pi/3}$$

$$= -[\ln |\cot x + \operatorname{cosec} x|]^{\pi/3} + [\ln |\cot x + \operatorname{cosec} x|]^{\pi/4}$$

$$= -\left[ \ln \left| \frac{1}{\tan x} + \frac{1}{\sin x} \right| \right]^{\pi/3} + \left[ \ln \left| \frac{1}{\tan x} + \frac{1}{\sin x} \right| \right]^{\pi/4}$$

$$= -\left( \ln \left| \frac{1}{\tan(\pi/3)} + \frac{1}{\sin(\pi/3)} \right| \right) + \left( \ln \left| \frac{1}{\tan(\pi/4)} + \frac{1}{\sin(\pi/4)} \right| \right)$$

$$= -\left( \ln \left| \frac{1}{\sqrt{3}} + \frac{1}{(\sqrt{3}/2)} \right| \right) + \left( \ln \left| \frac{1}{1} + \frac{1}{(\sqrt{2}/2)} \right| \right)$$

$$= -\left( \ln |\sqrt{3}| \right) + \left( \ln |1 + \sqrt{2}| \right)$$

$$\therefore I = \int_{\pi/4}^{\pi/3} \operatorname{cosec} x \, dx = \left( \ln |1 + \sqrt{2}| \right) - \left( \ln |\sqrt{3}| \right) \text{ unit}^2$$

ii) Volume ($x$-axis)

$$y = \operatorname{cosec} x$$

Therefore

$$y^2 = (\operatorname{cosec} x)^2$$
$$= \operatorname{cosec}^2 x$$

Now let's determine the volume.

$$V_{x-axis} = \pi \int_{\pi/4}^{\pi/3} y^2\, dx = \pi \int_{\pi/4}^{\pi/3} \operatorname{cosec}^2 x\, dx$$

$$= \pi[-\cot x\,]_{\pi/4}^{\pi/3} = \pi[-\cot x]^{\pi/3} - \pi[-\cot x]^{\pi/4}$$

$$= \pi\left[-\frac{1}{\tan x}\right]^{\pi/3} - \pi\left[-\frac{1}{\tan x}\right]^{\pi/4}$$

$$= \pi\left(-\frac{1}{\tan(\pi/3)}\right) - \pi\left(-\frac{1}{\tan(\pi/4)}\right)$$

$$= \pi\left(-\frac{1}{\sqrt{3}}\right) - \pi\left(-\frac{1}{1}\right) = -\frac{1}{\sqrt{3}}\pi + \pi = \pi\left(\frac{3-\sqrt{3}}{3}\right)$$

$$\therefore V = \pi \int_{\pi/4}^{\pi/3} \operatorname{cosec}^2 x\, dx = \pi\left(\frac{3-\sqrt{3}}{3}\right) \text{ unit}^3$$

---

c) $f(x) = \frac{1}{\sqrt{x}} + \sqrt{x}$, $[x_1 = 1, x_2 = 4]$

**Solution**

i) Area

$$I = \int_1^4 \frac{1}{\sqrt{x}} + \sqrt{x}\, dx = \int_1^4 x^{-0.5} + x^{0.5}\, dx = \left[\frac{x^{0.5}}{0.5} + \frac{x^{1.5}}{1.5}\right]_1^4$$

$$= \left[2\sqrt{x} + \frac{2}{3}x^{1.5}\right]_1^4 = \left[2\sqrt{x} + \frac{2}{3}x^{1.5}\right]^4 - \left[2\sqrt{x} + \frac{2}{3}x^{1.5}\right]^1$$

$$= \left(2\sqrt{4} + \frac{2}{3}(4)^{1.5}\right) - \left(2\sqrt{1} + \frac{2}{3}(1)^{1.5}\right)$$

$$= \left(4 + \frac{16}{3}\right) - \left(2 + \frac{2}{3}\right) = \left(\frac{28}{3}\right) - \left(\frac{8}{3}\right) = \frac{20}{3}$$

$$\therefore I = \int_1^4 \frac{1}{\sqrt{x}} + \sqrt{x}\, dx = \frac{20}{3} \text{ unit}^2$$

ii) Volume ($x$-axis)

$$y = \frac{1}{\sqrt{x}} + \sqrt{x} = x^{-0.5} + x^{0.5}$$

Therefore

$$y^2 = (x^{-0.5} + x^{0.5})^2$$
$$= (x^{-0.5})^2 + 2(x^{-0.5})(x^{0.5}) + (x^{0.5})^2$$
$$= x^{-1} + 2 + x = \frac{1}{x} + 2 + x$$

Now let's determine the volume.

$$V_{x-axis} = \pi \int_1^4 y^2 \, dx = \pi \int_1^4 \left(\frac{1}{x} + 2 + x\right) dx$$
$$= \pi \left[\ln x + 2x + \frac{1}{2}x^2\right]_1^4$$
$$= \pi \left[\ln x + 2x + \frac{1}{2}x^2\right]^4 - \pi \left[\ln x + 2x + \frac{1}{2}x^2\right]^1$$
$$= \pi \left(\ln 4 + 2(4) + \frac{1}{2}(4)^2\right) - \pi \left(\ln 1 + 2(1) + \frac{1}{2}(1)^2\right)$$
$$= \pi (\ln 4 + 8 + 8) - \pi \left(0 + 2 + \frac{1}{2}\right) = \pi (\ln 4 + 16) - \pi \left(\frac{5}{2}\right)$$
$$= \pi \left(\ln 4 + 16 - \frac{5}{2}\right) = \pi (\ln 4 + 13.5)$$

$$\therefore V = \pi \int_1^4 \left(\frac{1}{x} + 2 + x\right) dx = \pi (\ln 4 + 13.5) \text{ unit}^3$$

---

**d)** $f(x) = 6 \sec 2x . \tan 2x$, $\left[x_1 = 0, \ x_2 = \pi/6\right]$
**Solution**

i) Area

$$I = \int_0^{\pi/6} 6 \sec 2x . \tan 2x \, dx = \left[6 \times \frac{1}{2} \sec(2x)\right]_0^{\pi/6}$$

$$= [3 \sec(2x)]^{\pi/6} + [3 \sec(2x)]^0 = \left[\frac{3}{\cos 2x}\right]^{\pi/6} - \left[\frac{3}{\cos 2x}\right]^0$$

$$= \left(\frac{3}{\cos\left(2 \times \pi/6\right)}\right) - \left(\frac{3}{\cos(2 \times 0)}\right) = \left(\frac{3}{\cos \pi/3}\right) - \left(\frac{3}{\cos 0}\right)$$

$$= \left(\frac{3}{1/2}\right) - \left(\frac{3}{1}\right) = 6 - 3 = 3$$

$$\therefore I = \int_0^{\pi/6} 6 \sec 2x . \tan 2x \, dx = 3 \text{ unit}^2$$

ii) Volume (x-axis)

$$y = 6\sec 2x . \tan 2x$$

Therefore

$$y^2 = (6\sec 2x . \tan 2x)^2$$
$$= 36\sec^2 2x . \tan^2 2x$$

Now let's determine the volume.

$$V_{x-axis} = \pi \int_0^{\pi/6} y^2 \, dx = \pi \int_0^{\pi/6} 36\sec^2 2x . \tan^2 2x \, dx$$

$$= 18\pi \int_0^{\pi/6} (2\sec^2 2x) . \tan^2 2x \, dx$$

$$= 18\pi \left[ \frac{1}{3} \tan^3 2x \right]_0^{\pi/6} = 6\pi [\tan^3 2x]_0^{\pi/6}$$

$$= 6\pi [\tan^3 2x]^{\pi/6} - 6\pi [\tan^3 2x]^0$$

$$= 6\pi \left( \left( \tan \left( 2 \times \pi/6 \right) \right)^3 \right) - 6\pi \left( (\tan(2 \times 0))^3 \right)$$

$$= 6\pi \left( \left( \tan (\pi/3) \right)^3 \right) - 6\pi \left( (\tan(0))^3 \right)$$

$$= 6\pi \left( (\sqrt{3})^3 \right) - 6\pi \left( (0)^3 \right) = 6\pi \left( 3\sqrt{3} \right) = 18\pi\sqrt{3}$$

$$\therefore V = \pi \int_0^{\pi/6} 36\sec^2 2x . \tan^2 2x \, dx = 18\pi\sqrt{3} \text{ unit}^3$$

---

Another example to try.

### Example 7

The region $R$ is bounded by a curve with equation $y = \tan 2\theta$, the x-axis, and the vertical lines at $\theta = 0$ and $\theta = \pi/6$.

a) Determine the exact area of $R$.

b) Determine the exact volume of the solid formed when the region $R$ is rotated through $360°$ about the x-axis.

# Application of Integration

What did you get? Find the solution below to double-check your answer.

## Solution to Example 7

**a) Area**
**Solution**

$$Area = I = \int_0^{\pi/6} [\tan 2\theta]\, d\theta = \left[\frac{1}{2} \ln |\sec(2\theta)|\right]_0^{\pi/6}$$

$$= \left[\frac{1}{2} \ln |\sec(2\theta)|\right]^{\pi/6} - \left[\frac{1}{2} \ln |\sec(2\theta)|\right]^0$$

$$= \left(\frac{1}{2} \ln \left|\sec\left(2 \times \pi/6\right)\right|\right) - \left(\frac{1}{2} \ln |\sec(2 \times 0)|\right)$$

$$= \frac{1}{2}\left(\ln \left|\frac{1}{\cos(\pi/3)}\right|\right) - \frac{1}{2}\left(\ln \left|\frac{1}{\cos(0)}\right|\right) = \frac{1}{2}(\ln|2|) - \frac{1}{2}(\ln|1|)$$

$$= 0.5 \ln 2 - 0 = 0.5 \ln 2$$

$$\therefore Area = I = \int_0^{\pi/6} [\tan 2\theta]\, d\theta = 0.5 \ln 2 \text{ unit}^2$$

**b) Volume (rotation about the $x$-axis)**
**Solution**

$$V_{x-axis} = \pi \int_0^{\pi/6} y^2\, d\theta$$

$y = \tan 2\theta \rightarrow y^2 = \tan^2 2\theta$, but we have $\tan^2\theta + 1 = \sec^2\theta$. Thus

$$V_{x-axis} = \pi \int_0^{\pi/6} y^2\, d\theta = V_{x-axis} = \pi \int_0^{\pi/6} [\tan^2 2\theta]\, d\theta$$

$$= \pi \int_0^{\pi/6} [\sec^2 2\theta - 1]\, d\theta = \pi \left[\frac{1}{2}\tan 2\theta - \theta\right]_0^{\pi/6}$$

$$= \pi \left[\frac{1}{2}\tan 2\theta - \theta\right]^{\pi/6} - \pi \left[\frac{1}{2}\tan 2\theta - \theta\right]^0$$

$$= \pi \left(\frac{\tan(2 \times \frac{\pi}{6})}{2} - \frac{\pi}{6}\right) - \pi \left(\frac{\tan(2 \times 0)}{2} - 0\right)$$

$$= \pi \left(\frac{\tan(\frac{\pi}{3})}{2} - \frac{\pi}{6}\right) - \pi \left(\frac{\tan(0)}{2} - 0\right)$$

$$= \pi\left(\frac{\sqrt{3}}{2} - \frac{\pi}{6}\right) - \pi\left(\frac{0}{2} - 0\right) = \pi\left(\frac{3\sqrt{3} - \pi}{6}\right)$$

$$\therefore V_{x-\text{axis}} = \pi \int_0^{\frac{\pi}{6}} \left[\tan^2 2\theta\right] d\theta = \pi\left(\frac{3\sqrt{3} - \pi}{6}\right) \text{ unit}^3$$

Another example to try.

### Example 8

Find the exact volume of revolution generated when the area enclosed by $y = \sqrt{3x^2 + 1}$ is rotated through 360° about the y-axis between the vertical lines $x_1 = 0$ and $x_2 = 4$ and the x-axis.

What did you get? Find the solution below to double-check your answer.

### Solution to Example 8

$$V_{y-\text{axis}} = \pi \int_{y_1=a}^{y_2=b} x^2 \, dy$$

Let's change the limits as:

- when $x_1 = 0$, $y = \sqrt{3(0)^2 + 1} = 1$
- when $x_1 = 4$, $y = \sqrt{3(4)^2 + 1} = 7$

Also, given that $y = \sqrt{3x^2 + 1}$, we have

$$y^2 = 3x^2 + 1$$

$$x^2 = \frac{1}{3}(y^2 - 1)$$

Now let's get to business as:

$$V_{y-\text{axis}} = \pi \int_{y_1=a}^{y_2=b} x^2 \, dy = \pi \int_1^7 \frac{1}{3}(y^2 - 1) \, dy$$

$$= \frac{1}{3}\pi \int_1^7 (y^2 - 1) \, dy = \frac{1}{3}\pi \left[\frac{1}{3}y^3 - y\right]_1^7$$

$$= \frac{1}{3}\pi \left[\frac{1}{3}y^3 - y\right]^7 - \frac{1}{3}\pi \left[\frac{1}{3}y^3 - y\right]^1$$

# Application of Integration

$$= \frac{1}{3}\pi\left(\frac{1}{3}(7)^3 - 7\right) - \frac{1}{3}\pi\left(\frac{1}{3}(1)^3 - 1\right)$$

$$= \frac{1}{3}\pi\left(\frac{343}{3} - 7\right) - \frac{1}{3}\pi\left(\frac{1}{3} - 1\right)$$

$$= \frac{1}{3}\pi\left(\frac{322}{3}\right) - \frac{1}{3}\pi\left(-\frac{2}{3}\right) = \frac{1}{3}\pi\left(\frac{322+2}{3}\right)$$

$$= \frac{1}{3}\pi(108) = 36\pi$$

$$\therefore V_{y-\text{axis}} = \pi\int_{1}^{7}\frac{1}{3}(y^2 - 1)\,dy = 36\pi \text{ unit}^3$$

---

Sometimes the function can be defined with parametric equations. This is not a problem, just apply parametric integration as we did in the previous chapter and that's all. Let's try an example.

### Example 9

A curve $K$ is defined parametrically as $x = 2t + 5$ and $y = t - 3$. A region $J$ is bounded by $K$, the $x$-axis, and the vertical lines at $x = 1$ and $x = 3$. Find the volume of revolution the solid generated by rotating $J$ $2\pi$ radians about the $x$-axis.

---

What did you get? Find the solution below to double-check your answer.

### Solution to Example 9

**Method 1**

Given $x = 2t + 5$ and $y = t - 3$, we have

$$\frac{dx}{dt} = 2$$

Let's change the limits too:
For $x_1 = 1$, we have

$$2t + 5 = 1$$
$$2t = -4$$
$$\therefore t = -2$$

For $x_2 = 3$, we have

$$2t + 5 = 3$$
$$2t = -2$$
$$\therefore t = -1$$

Thus

$$y = t - 3$$
$$\therefore y^2 = (t-3)^2 = t^2 - 6t + 9$$

$$V_{x\text{-axis}} = \pi \int_{x_1=1}^{x_2=3} y^2\, dx = \pi \int_{t_1=-2}^{t_2=-1} y^2 \cdot \frac{dx}{dt}\, dt$$

$$= \pi \int_{t_1=-2}^{t_2=-1} (t-3)^2 \times 2\, dt$$

$$= \pi \int_{t_1=-2}^{t_2=-1} 2\left(t^2 - 6t + 9\right) dt$$

$$= 2\pi \int_{t_1=-2}^{t_2=-1} \left(t^2 - 6t + 9\right) dt$$

$$= 2\pi \left[\frac{t^3}{3} - 3t^2 + 9t\right]_{-2}^{-1}$$

$$= 2\pi \left[\frac{t^3}{3} - 3t^2 + 9t\right]^{-1} - 2\pi \left[\frac{t^3}{3} - 3t^2 + 9t\right]^{-2}$$

$$= 2\pi \left(\frac{(-1)^3}{3} - 3(-1)^2 + 9(-1)\right) - 2\pi \left(\frac{(-2)^3}{3} - 3(-2)^2 + 9(-2)\right)$$

$$= 2\pi \left(-\frac{1}{3} - 3 - 9\right) - 2\pi \left(-\frac{8}{3} - 12 - 18\right) = 2\pi \left(-\frac{37}{3}\right) - 2\pi \left(-\frac{98}{3}\right)$$

$$= 2\pi \left(-\frac{37}{3}\right) + 2\pi \left(\frac{98}{3}\right) = 2\pi \left(-\frac{37}{3} + \frac{98}{3}\right) = \frac{122}{3}\pi$$

$$\therefore V = \pi \int_{x_1=1}^{x_2=3} y^2\, dx = \frac{122}{3}\pi \text{ unit}^3$$

**ALTERNATIVE METHOD**

Given $x = 2t + 5$, we have

$$2t = x - 5$$
$$t = \frac{1}{2}(x - 5)$$

Let's substitute $t$ in the $y(t)$ as:

$$y = t - 3$$
$$= \frac{1}{2}(x - 5) - 3 = \frac{x-5}{2} - \frac{6}{2}$$
$$= \frac{x - 11}{2} = \frac{x}{2} - \frac{11}{2}$$

Therefore

$$y^2 = \left(\frac{x}{2} - \frac{11}{2}\right)^2$$
$$= \frac{x^2}{4} - \frac{11x}{2} + \frac{121}{4}$$

# Application of Integration

Now let's determine the volume.

$$V_{x-\text{axis}} = \pi \int_{x_1=1}^{x_2=3} y^2\, dx = \pi \int_{x_1=1}^{x_2=3} \left(\frac{x^2}{4} - \frac{11x}{2} + \frac{121}{4}\right) dx$$

$$= \pi \left[\frac{x^3}{12} - \frac{11x^2}{4} + \frac{121}{4}x\right]_1^3$$

$$= \pi \left[\frac{x^3}{12} - \frac{11x^2}{4} + \frac{121}{4}x\right]^3 - \pi \left[\frac{x^3}{12} - \frac{11x^2}{4} + \frac{121}{4}x\right]^1$$

$$= \pi \left(\frac{(3)^3}{12} - \frac{11(3)^2}{4} + \frac{121}{4}(3)\right) - \pi \left(\frac{(1)^3}{12} - \frac{11(1)^2}{4} + \frac{121}{4}(1)\right)$$

$$= \pi \left(\frac{27}{12} - \frac{99}{4} + \frac{363}{4}\right) - \pi \left(\frac{1}{12} - \frac{11}{4} + \frac{121}{4}\right) = \pi \left(\frac{273}{4}\right) - \pi \left(\frac{331}{12}\right)$$

$$= \pi \left(\frac{273}{4} - \frac{331}{12}\right) = \frac{122}{3}\pi$$

$$\therefore\therefore V = \pi \int_{x_1=1}^{x_2=3} y^2\, dx = \frac{122}{3}\pi \text{ unit}^3$$

---

## 19.3.2 MEAN VALUES

We've studied how to calculate the area under a curve $f(x)$ between two points $x_1 = a$ and $x_2 = b$ using definite integration. We've equally covered methods to estimate the area of unusual functions and noted that the ordinates are generally different in values. Now, imagine that we can find one ordinate (call this $y_m$) such that

$$y_m = \frac{y_0 + y_1 + y_2 + \cdots + y_n}{n+1}$$

where $n$ is the number of strips.

The mean value (commonly called average) of a function $f(x)$ between the interval $a$ and $b$ is therefore a value $y_m$ such that when it's multiplied with the interval $b - a$, it gives the area under the curve (Figure 19.9).

For the mean value, we have that

$$(b-a)y_m = \int_a^b f(x)\, dx$$

Therefore,

$$\boxed{y_m = \frac{1}{b-a}\int_a^b [f(x)]\, dx} \qquad (19.11)$$

In other words, the mean value is equal to the area divided by the interval (or the difference in the limits). That's everything on this for now.

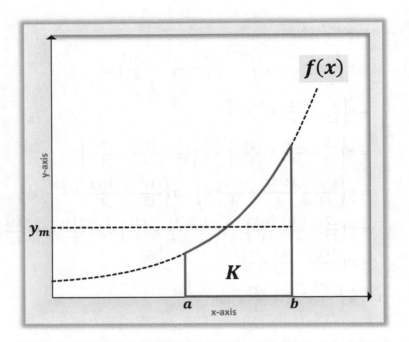

**FIGURE 19.9** Mean value of a function illustrated.

### 19.3.3 Root Mean Square (RMS) Value

It will soon be shown that the average (or mean) value of a sinusoidal signal (or sine wave) is always zero when taking over a period of one revolution (or $2\pi$ radian). To circumvent this, an alternative mean value is used in sectors (e.g., energy, communication) where sine waves are employed. This is called the **root mean square (RMS)** value, and is given by

$$\boxed{y_{\text{RMS}} = \sqrt{\frac{1}{b-a} \int_a^b [f(x)]^2 \, dx}} \tag{19.12}$$

Essentially, to find RMS perform 'RMS' in the reverse order. In other words:

- **S**: square the function $[f(x)]^2$
- **M**: take the mean of the result $\frac{1}{b-a} \int_a^b [f(x)]^2 \, dx$
- **R**: take the (square) root of the result $\sqrt{\frac{1}{b-a} \int_a^b [f(x)]^2 \, dx}$

# Application of Integration

This brings us to the end of the application that we are set to cover here. Let's try some examples.

### Example 10

In the stated limits or intervals for each of the following functions, find the:

i) mean value, and

ii) root mean square value.

a) $y = \sin \theta$, $a = 0$, $b = \pi$

b) $y = \cos \theta$, $0 \le \theta \le 2\pi$

c) $y = 3x^2 - 5$, $x_1 = 2$, $x_2 = 5$

d) $v = 40 \cos 20\pi t$, $0 \le t \le \frac{1}{40}$

What did you get? Find the solution below to double-check your answer.

### Solution to Example 10

**a)** $y = \sin \theta$, $a = 0$, $b = \pi$
**Solution**

**i) Mean value**

$$y_m = \frac{1}{b-a} \int_a^b y \, d\theta = \frac{1}{\pi - 0} \int_0^\pi \sin \theta \, d\theta$$

$$= \frac{1}{\pi} [-\cos \theta]_0^\pi = \frac{1}{\pi} [-\cos \theta]^\pi - \frac{1}{\pi} [-\cos \theta]^0$$

$$= \frac{1}{\pi} \{(-\cos \pi) - (-\cos 0)\} = \frac{1}{\pi} \{(1) - (-1)\}$$

$$= \frac{1}{\pi} (2) = \frac{2}{\pi} = 0.6366 \text{ (4 s.f.)}$$

$$\therefore y_m = \frac{1}{\pi - 0} \int_0^\pi \sin \theta \, d\theta = \frac{2}{\pi} = 0.6366 \text{ (4 s.f.)}$$

**ii) Root Mean Square value**

$$y_{\text{RMS}} = \sqrt{\frac{1}{b-a} \int_a^b y^2 \, d\theta}$$

We therefore have

$$(y_{\text{RMS}})^2 = \frac{1}{b-a} \int_a^b y^2 \, d\theta = \frac{1}{\pi - 0} \int_0^\pi \sin^2 \theta \, d\theta$$

$$= \frac{1}{\pi} \int_0^\pi \frac{1}{2} (1 - \cos 2\theta) \, d\theta = \frac{1}{2\pi} \int_0^\pi (1 - \cos 2\theta) \, d\theta$$

$$= \frac{1}{2\pi} \left[ \theta - \frac{1}{2} \sin 2\theta \right]_0^\pi = \frac{1}{2\pi} \left[ \theta - \frac{1}{2} \sin 2\theta \right]^\pi - \frac{1}{2\pi} \left[ \theta - \frac{1}{2} \sin 2\theta \right]^0$$

$$= \frac{1}{2\pi} \left\{ \left( \pi - \frac{1}{2} \sin 2\pi \right) - \left( 0 - \frac{1}{2} \sin 2 \times 0 \right) \right\} = \frac{1}{2\pi} \{(\pi - 0) - (0 - 0)\} = \frac{1}{2}$$

Thus

$$y_{RMS} = \sqrt{\frac{1}{2}} = \frac{\sqrt{2}}{2} = 0.7071 \text{ (4 s.f.)}$$

$$\therefore y_{RMS} = \sqrt{\frac{1}{b-a}\int_a^b \sin^2\theta\, d\theta} = \frac{\sqrt{2}}{2} = 0.7071 \text{ (4 s.f.)}$$

---

**b)** $y = \cos\theta,\ 0 \le \theta \le 2\pi$
**Solution**

   **i) Mean value**

$$y_m = \frac{1}{b-a}\int_a^b y\, d\theta$$

$$= \frac{1}{2\pi - 0}\int_0^{2\pi} \cos\theta\, d\theta$$

$$= \frac{1}{2\pi}[\sin\theta]_0^{2\pi} = \frac{1}{2\pi}[\sin\theta]^{2\pi} - \frac{1}{\pi}[\sin\theta]^{0}$$

$$= \frac{1}{2\pi}\{(\sin 2\pi) - (\sin 0)\} = \frac{1}{2\pi}\{(0) - (0)\}$$

$$= \frac{1}{2\pi}(0) = 0$$

$$\therefore y_m = \frac{1}{2\pi - 0}\int_0^{2\pi} \cos\theta\, d\theta = 0$$

   **ii) Root mean square value**

$$y_{RMS} = \sqrt{\frac{1}{b-a}\int_a^b y^2\, d\theta}$$

We therefore have

$$(y_{RMS})^2 = \frac{1}{b-a}\int_a^b y^2\, d\theta = \frac{1}{2\pi - 0}\int_0^{2\pi} \cos^2\theta\, d\theta$$

$$= \frac{1}{2\pi}\int_0^{2\pi} \frac{1}{2}(1 + \cos 2\theta)\, d\theta = \frac{1}{4\pi}\int_0^{2\pi} (1 + \cos 2\theta)\, d\theta$$

$$= \frac{1}{4\pi}\left[\theta + \frac{1}{2}\sin 2\theta\right]_0^{2\pi} = \frac{1}{4\pi}\left[\theta + \frac{1}{2}\sin 2\theta\right]^{2\pi} - \frac{1}{4\pi}\left[\theta + \frac{1}{2}\sin 2\theta\right]^0$$

$$= \frac{1}{4\pi}\left\{\left(2\pi + \frac{1}{2}\sin 4\pi\right) - \left(0 + \frac{1}{2}\sin 2\times 0\right)\right\}$$

$$= \frac{1}{4\pi}\{(2\pi + 0) - (0 + 0)\} = \frac{1}{2}$$

Thus

$$y_{RMS} = \sqrt{\frac{1}{2}} = \frac{1}{\sqrt{2}} = 0.7071 \text{ (4 s.f.)}$$

$$\therefore y_{RMS} = \sqrt{\frac{1}{b-a}\int_a^b \sin^2\theta\, d\theta} = 0.7071 \text{ (4 s.f.)}$$

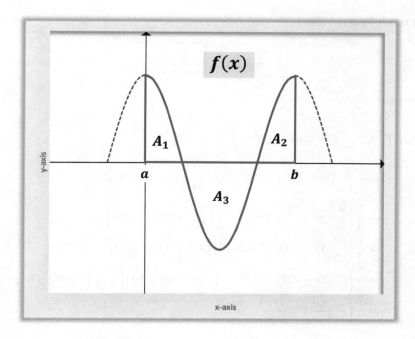

**FIGURE 19.10** Solution to Example 10(b).

### NOTE

1) The mean value is 0.6366 when taken over a half revolution, but zero when taken over a full cycle. However, the RMS value is the same at 0.7071 for both half a cycle and full cycle.

2) The reason for mean value being zero over a full revolution is illustrated in Figure 19.10. It is observed that the area in the interval $a \leq \theta \leq b$ has three parts: two above the $x$-axis in the positive region and one below the $x$-axis in the negative region. The algebraic sum of the area is zero, i.e., $A_1 + A_2 + A_3 = 0$.

3) On this basis, RMS value is preferred over mean value, since it has immunity against the interval used.

---

c) $y = 3x^2 - 5$, $x_1 = 2$, $x_2 = 5$
**Solution**

   i) **Mean value**

$$y_m = \frac{1}{b-a}\int_a^b y\,dx = \frac{1}{5-2}\int_2^5 (3x^2 - 5)\,dx$$

$$= \frac{1}{3}[x^3 - 5x]_2^5 = \frac{1}{3}[x^3 - 5x]^5 - \frac{1}{3}[x^3 - 5x]^2$$

$$= \frac{1}{3}\{(5^3 - 5\times 5) - (2^3 - 5\times 2)\} = \frac{1}{3}\{(100) - (-2)\}$$

$$= \frac{1}{3}(100 + 2) = 34$$

$$\therefore y_m = \frac{1}{5-2}\int_2^5 (3x^2 - 5)\,dx = 34$$

## ii) Root mean square value

$$y_{RMS} = \sqrt{\frac{1}{b-a}\int_a^b y^2\, dx}$$

We therefore have

$$(y_{RMS})^2 = \frac{1}{b-a}\int_a^b y^2\, dx = \frac{1}{5-2}\int_2^5 (3x^2-5)^2\, dx$$

$$= \frac{1}{3}\int_2^5 (9x^4 - 30x^2 + 25)\, dx = \frac{1}{3}\left[\frac{9}{5}x^5 - 10x^3 + 25x\right]_2^5$$

$$= \frac{1}{3}\left[\frac{9}{5}x^5 - 10x^3 + 25x\right]^5 - \frac{1}{3}\left[\frac{9}{5}x^5 - 10x^3 + 25x\right]^2$$

$$= \frac{1}{3}\left\{\left(\frac{9}{5}(5)^5 - 10(5)^3 + 25(5)\right) - \left(\frac{9}{5}(2)^5 - 10(2)^3 + 25(2)\right)\right\}$$

$$= \frac{1}{3}\left\{(5625 - 1250 + 125) - \left(\frac{288}{5} - 80 + 50\right)\right\} = \frac{1}{3}\left\{(4500) - \left(\frac{138}{5}\right)\right\}$$

$$= \frac{1}{3}\left(\frac{22362}{5}\right) = \frac{7454}{5}$$

Thus

$$y_{RMS} = \sqrt{\frac{7454}{5}} = 38.61 \ (4 \text{ s.f.})$$

$$\therefore y_{RMS} = \sqrt{\frac{1}{3}\int_2^5 (9x^4 - 30x^2 + 25)\, dx} = 38.61 \ (4 \text{ s.f.})$$

---

**d)** $v = 40\cos 20\pi t,\ 0 \le t \le \frac{1}{40}$

**Solution**

### i) Mean value

$$y_m = \frac{1}{b-a}\int_a^b y\, dt = \frac{1}{\frac{1}{40} - 0}\int_0^{1/40} 40\cos 20\pi t\, dt$$

$$= 40\int_0^{1/40} 40\cos 20\pi t\, dt$$

$$= 40 \times 40 \int_0^{1/40} \cos 20\pi t\, dt = 1600\left[\frac{1}{20\pi}\sin 20\pi t\right]_0^{1/40}$$

$$= 1600\left[\frac{1}{20\pi}\sin 20\pi t\right]^{1/40} - 1600\left[\frac{1}{20\pi}\sin 20\pi t\right]^0$$

$$= 1600\left\{\left(\frac{1}{20\pi}\sin 20\pi \times \frac{1}{40}\right) - \left(\frac{1}{20\pi}\sin 20\pi \times 0\right)\right\}$$

# Application of Integration

$$= 1600\left\{\left(\frac{1}{20\pi}\sin\frac{1}{2}\pi\right) - \left(\frac{1}{20\pi}\sin 0\right)\right\} = 1600\left\{\left(\frac{1}{20\pi}\right) - (0)\right\}$$

$$= \frac{1600}{20\pi} = \frac{80}{\pi}$$

$$\therefore y_m = 40\int_0^{1/40} 40\cos 20\pi t\, dt = \frac{80}{\pi}$$

### ii) Root mean square value

$$y_{RMS} = \sqrt{\frac{1}{b-a}\int_a^b y^2\, dt}$$

We therefore have

$$(y_{RMS})^2 = \frac{1}{\frac{1}{40} - 0}\int_0^{1/40} y^2\, dt = 40\int_0^{1/40}(40\cos 20\pi t)^2\, dt$$

$$= 40\int_0^{1/40} 40^2 \cos^2 20\pi t\, dt = 40^3 \int_0^{1/40} \cos^2 20\pi t\, dt$$

$$= 40^3 \int_0^{1/40} \frac{1}{2}(1 + \cos 40\pi t)\, dt = 20\times 40^2 \int_0^{1/40}(1 + \cos 40\pi t)\, dt$$

$$= 20\times 40^2 \left[t + \frac{1}{40\pi}\sin 40\pi t\right]_0^{1/40}$$

$$= 20\times 40^2 \left[t + \frac{1}{40\pi}\sin 40\pi t\right]^{1/40} - 20\times 40^2\left[t + \frac{1}{40\pi}\sin 40\pi t\right]^0$$

$$= 20\times 40^2 \left\{\left(\frac{1}{40} + \frac{1}{40\pi}\sin 40\pi \times \frac{1}{40}\right) - \left(0 + \frac{1}{40\pi}\sin 40\pi \times 0\right)\right\}$$

$$= 20\times 40^2 \left\{\left(\frac{1}{40} + 0\right) - (0+0)\right\} = 20\times 40^2\left(\frac{1}{40}\right) = 20\times 40 = 800$$

Thus

$$y_{RMS} = \sqrt{800} = 20\sqrt{2} = 28.28\ (4\text{ s.f.})$$

$$\therefore y_{RMS} = \sqrt{40\int_0^{1/40}(40\cos 20\pi t)^2\, dt} = 20\sqrt{2} = 28.28\ (4\text{ s.f.})$$

## 19.4 CHAPTER SUMMARY

1) There are functions that we cannot integrate at all. As a result, we need to resort to a technique that we can use to solve any function, and this is known as **numerical integration**. This technique is used to find an area under a curve or between curves within a given limit, usually employed for functions that cannot be integrated. In other words, it's used for complex integration. The result from this is usually approximate and can be improved to reduce the error.

2) The area of a trapezium is given by:

$$A = \frac{1}{2}h(l_1 + l_2)$$

3) The trapezium rule is given by:

$$\int_a^b [f(x)]\, dx \approx \frac{1}{2}h\, [y_0 + 2(y_1 + y_2 + \ldots + y_{n-1}) + y_n]$$

where $h = \frac{b-a}{n}$ and $y_i = f(a + ih)$, $0 \leq i \leq n$

4) Simpson's rule is another approximation technique, which is named after an English mathematician, Thomas Simpson. Like the trapezium method, this rule is also based on dividing the region into equal strips. Unlike the trapezium rule, however, Simpson's rule can only be used for an even number of strips, such as 2, 4, 6, etc., strips.

5) Simpson's rule is given by:

$$\int_a^b [f(x)]\, dx \approx \frac{1}{3}h\, [y_0 + 4y_1 + 2y_2 + 4y_3 + 2y_4 + \cdots + 2y_{n-2} + 4y_{n-1} + y_n]$$

**OR**

$$\int_a^b [f(x)]\, dx \approx \frac{1}{3}h\, [(y_0 + y_n) + 4(y_1 + y_3 + \ldots y_{n-1}) + 2(y_2 + y_4 + \ldots y_{n-2})]$$

where $h = \frac{b-a}{n}$ and $y_i = f(a + ih)$, $0 \leq i \leq n$.

6) There are four ways to determine the error due to trapezium and Simpson's rules:

- Absolute Error (AE): This is used to determine the actual value of deviation and is given as:

$$\text{AE} = \text{Exact value} - \text{Approximate value}$$

  For this, a positive AE (or AE > 0) implies an underestimate, and a negative value shows an overestimate.

- Relative Error (RE): This is computed using

$$\text{RE} = |\text{Exact value} - \text{Approximate value}|$$

Note the use of the modulus | | sign. Hence, the answer will always be positive (or greater than 1), and it does not matter which of the two is the minuend or subtrahend.

- Absolute Percentage Error (APE): This is given by

$$\boxed{APE = \frac{\text{Exact value} - \text{Approximate value}}{\text{Exact value}} \times 100\,\%}$$

- Relative Percentage Error (RPE): This is given by

$$\boxed{RPE = \left|\frac{\text{Exact value} - \text{Approximate value}}{\text{Exact value}}\right| \times 100\,\%}$$

7) The mean value of a function is given by:

$$\boxed{y_m = \frac{1}{b-a}\int_a^b [f(x)]\,dx}$$

8) The root mean square (RMS) value is given by

$$\boxed{y_{RMS} = \sqrt{\frac{1}{b-a}\int_a^b [f(x)]^2\,dx}}$$

\*\*\*\*

## 19.5 FURTHER PRACTICE

To access complementary contents, including additional exercises, please go to www.dszak.com.

# Appendix

## (A) PROOF OF THE DERIVATIVE OF $e^x$

We will use the infinite series expansion of $e^x$

$$e^x = 1 + \frac{x}{1!} + \frac{x^2}{2!} + \frac{x^3}{3!} + \frac{x^4}{4!} + \frac{x^5}{5!} + \ldots$$

**Method 1**

$$y = e^x$$

Using the above-stated expansion

$$y = 1 + \frac{x}{1!} + \frac{x^2}{2!} + \frac{x^3}{3!} + \frac{x^4}{4!} + \frac{x^5}{5!} + \ldots$$

We can find the derivative of the sum of the powers as

$$\frac{dy}{dx} = 0 + 1 + \frac{x}{1!} + \frac{x^2}{2!} + \frac{x^3}{3!} + \frac{x^4}{4!} + \frac{x^5}{5!} + \ldots$$

$$= 1 + \frac{x}{1!} + \frac{x^2}{2!} + \frac{x^3}{3!} + \frac{x^4}{4!} + \frac{x^5}{5!} + \ldots$$

After simplifying the expression, we will notice that we obtained the original expression. We can therefore say that

$$\frac{dy}{dx} = e^x$$

**Method 2**

$$y = e^x$$

Similarly

$$y + \delta y = f(x + \delta x)$$
$$= e^{x+\delta x} = e^x \times e^{\delta x}$$

Therefore,

$$\delta y = [e^x \times e^{\delta x}] - [e^x]$$
$$= e^x (e^{\delta x} - 1)$$

Now we can evaluate the slope as

$$\text{Slope} \approx \frac{\delta y}{\delta x} \approx \frac{e^x (e^{\delta x} - 1)}{\delta x}$$

Thus, the derivative is

$$\frac{dy}{dx} = \lim_{\delta x \to 0}\left(\frac{\delta y}{\delta x}\right) = \lim_{\delta x \to 0}\left(\frac{e^x\left(e^{\delta x}-1\right)}{\delta x}\right)$$

$$= e^x \cdot \lim_{\delta x \to 0}\left(\frac{\left(e^{\delta x}-1\right)}{\delta x}\right)$$

It can be proved numerically that

$$\lim_{\delta x \to 0}\left(\frac{\left(e^{\delta x}-1\right)}{\delta x}\right) = 1$$

where $e = 2.718\ldots$ (3 d.p.)

$$\therefore \frac{d}{dx}[e^x] = e^x$$

## (B) PROOF OF THE DERIVATIVE OF sin $x$

We will use two known facts for angles measured in radians:

$$\sin \delta x \approx \delta x \text{ or } \sin x \approx x$$

provided the $x$ is infinitesimal. This can also be stated as

$$\lim_{x \to 0}(\sin x) = x$$

Similarly,

$$\lim_{x \to 0}(\cos x) = 1$$

Following on from the above, we have that

$$\lim_{x \to 0}\left(\frac{\sin x}{x}\right) = 1$$

---

**Method 1**

$$y = \sin x$$

Similarly,

$$y + \delta y = f(x + \delta x)$$
$$= \sin(x + \delta x)$$

Applying the addition formula, we will have

$$= \sin x \cos \delta x + \sin \delta x \cos x$$

Therefore,

$$\delta y = [\sin x \cos \delta x + \sin \delta x \cos x] - [\sin x]$$

# Appendix

Now we can evaluate the slope as

$$\text{Slope} \approx \frac{\delta y}{\delta x}$$

$$\approx \frac{\sin x \cos \delta x + \sin \delta x \cos x - \sin x}{\delta x}$$

Thus, the derivative is

$$\frac{dy}{dx} = \lim_{\delta x \to 0} \left( \frac{\delta y}{\delta x} \right)$$

$$= \lim_{\delta x \to 0} \left( \frac{\sin x \cos \delta x + \sin \delta x \cos x - \sin x}{\delta x} \right)$$

Using the facts stated above, we have

$$= \lim_{\delta x \to 0} \left( \frac{\sin x \times 1 + \delta x \times \cos x - \sin x}{\delta x} \right)$$

$$= \lim_{\delta x \to 0} \left( \frac{\delta x \times \cos x}{\delta x} \right)$$

$$\therefore \frac{dy}{dx} = \cos x$$

## Method 2

$$y = \sin x$$

Similarly,

$$y + \delta y = f(x + \delta x)$$
$$= \sin(x + \delta x)$$

Therefore,

$$\delta y = \sin(x + \delta x) - \sin x$$

Applying the product formula to the RHS, we will have

$$\delta y = 2 \sin \left[ \frac{(x + \delta x) - (x)}{2} \right] \cos \left[ \frac{(x + \delta x) + (x)}{2} \right]$$

$$= 2 \sin \left[ \frac{\delta x}{2} \right] \cos \left[ \frac{2x + \delta x}{2} \right]$$

$$= 2 \sin \left[ \frac{\delta x}{2} \right] \cos \left[ x + \frac{\delta x}{2} \right]$$

Now we can evaluate the slope as

$$\text{Slope} \approx \frac{\delta y}{\delta x} \approx \frac{2 \sin \left[ \frac{\delta x}{2} \right] \cos \left[ x + \frac{\delta x}{2} \right]}{\delta x}$$

$$\approx \left[ \frac{\sin \left( \frac{\delta x}{2} \right)}{\frac{\delta x}{2}} \right] \left[ \cos \left( x + \frac{\delta x}{2} \right) \right]$$

Using the stated fact, the derivative is

$$= \lim_{\delta x \to 0} \left[ \frac{\sin\left(\frac{\delta x}{2}\right)}{\frac{\delta x}{2}} \right] \left[ \cos\left(x + \frac{\delta x}{2}\right) \right]$$

$$= \lim_{\delta x \to 0} [1] \left[ \cos\left(x + \frac{0}{2}\right) \right] = \lim_{\delta x \to 0} [1] [\cos(x)]$$

$$\therefore \frac{dy}{dx} = \cos x$$

## (C) PROOF OF THE DERIVATIVE OF cos x

We will use two known facts for angles measured in radians:

$$\sin \delta x \approx \delta x \text{ or } \sin x \approx x$$

provided $x$ is infinitesimal. This can also be stated as

$$\lim_{x \to 0} (\sin x) = x$$

Similarly,

$$\lim_{x \to 0} (\cos x) = 1$$

**Method 1**

$$y = \cos x$$

Similarly

$$y + \delta y = f(x + \delta x)$$
$$= \cos(x + \delta x)$$

Applying the addition formula, we will have

$$= \cos x \cos \delta x - \sin x \sin \delta x$$

Therefore,

$$\delta y = [\cos x \cos \delta x - \sin x \sin \delta x] - [\cos x]$$

Now we can evaluate the slope as

$$\text{Slope} \approx \frac{\delta y}{\delta x}$$
$$\approx \frac{\cos x \cos \delta x - \sin x \sin \delta x - \cos x}{\delta x}$$

Thus, the derivative is

$$\frac{dy}{dx} = \lim_{\delta x \to 0} \left( \frac{\delta y}{\delta x} \right)$$
$$= \lim_{\delta x \to 0} \left( \frac{\cos x \cos \delta x - \sin x \sin \delta x - \cos x}{\delta x} \right)$$

Using the stated facts, we have

$$= \lim_{\delta x \to 0} \left( \frac{\cos x \times 1 - \sin x \times \delta x - \cos x}{\delta x} \right)$$

$$= \lim_{\delta x \to 0} \left( \frac{-\sin x \times \delta x}{\delta x} \right)$$

$$\therefore \frac{dy}{dx} = -\sin x$$

---

## Method 2

$$y = \cos x$$

Similarly,

$$y + \delta y = f(x + \delta x)$$
$$= \cos(x + \delta x)$$

Therefore,

$$\delta y = \cos(x + \delta x) - \cos x$$

Applying the product formula to the RHS, we will have

$$\delta y = -2 \sin\left[\frac{(x + \delta x) + (x)}{2}\right] \sin\left[\frac{(x + \delta x) - (x)}{2}\right]$$

$$= -2 \sin\left[\frac{2x + \delta x}{2}\right] \sin\left[\frac{\delta x}{2}\right]$$

$$= -2 \sin\left[x + \frac{\delta x}{2}\right] \sin\left[\frac{\delta x}{2}\right]$$

Now we can evaluate the slope as

$$\text{Slope} \approx \frac{\delta y}{\delta x} \approx \frac{-2 \sin\left[x + \frac{\delta x}{2}\right] \sin\left[\frac{\delta x}{2}\right]}{\delta x}$$

$$\approx \left[-\sin\left(x + \frac{\delta x}{2}\right)\right] \left[\frac{\sin\left(\frac{\delta x}{2}\right)}{\frac{\delta x}{2}}\right]$$

Using the stated fact, the derivative is

$$\frac{dy}{dx} = \lim_{\delta x \to 0} \left[-\sin\left(x + \frac{\delta x}{2}\right)\right] \left[\frac{\sin\left(\frac{\delta x}{2}\right)}{\frac{\delta x}{2}}\right]$$

$$= \lim_{\delta x \to 0} \left[-\sin\left(x + \frac{\delta x}{2}\right)\right] [1] = \lim_{\delta x \to 0} \left[-\sin\left(x + \frac{0}{2}\right)\right]$$

$$\therefore \frac{dy}{dx} = -\sin x$$

## (D) PROOF OF THE PRODUCT RULE FOR DIFFERENTIATION

$$y = uv$$

where

$$y = f(x), \ u = g(x), \text{ and } v = h(x)$$

$\delta x$ in $x$ will produce a corresponding $\delta y$, $\delta u$, and $\delta v$ in $y$, $u$, and $v$, respectively. Therefore,

$$y + \delta y = f(x + \delta x)$$
$$= (u + \delta u)(v + \delta v)$$

Open the bracket on the RHS, we will have

$$= uv + u\delta v + v\delta u + \delta u \delta v$$

Therefore,

$$\delta y = [uv + u\delta v + v\delta u + \delta u \delta v] - [uv]$$
$$= u\delta v + v\delta u + \delta u \delta v$$

Now we can evaluate the slope as

$$\text{Slope} \approx \frac{\delta y}{\delta x}$$
$$\approx \frac{u\delta v + v\delta u + \delta u \delta v}{\delta x}$$
$$\approx u\frac{\delta v}{\delta x} + v\frac{\delta u}{\delta x} + \delta u \frac{\delta v}{\delta x}$$

Thus, the derivative is

$$\frac{dy}{dx} = \lim_{\delta x \to 0} \left(\frac{\delta y}{\delta x}\right)$$
$$= \lim_{\delta x \to 0} \left(u\frac{\delta v}{\delta x} + v\frac{\delta u}{\delta x} + \delta u \frac{\delta v}{\delta x}\right)$$

As $\delta x \to 0$, $\frac{\delta v}{\delta x} \to \frac{dv}{dx}$, and $\frac{\delta u}{\delta x} \to \frac{du}{dx}$, hence

$$\frac{dy}{dx} = u\frac{dv}{dx} + v\frac{du}{dx} + 0 \times \frac{dv}{dx}$$

$$\therefore \frac{dy}{dx} = \frac{d}{dx}(uv) = u\frac{dv}{dx} + v\frac{du}{dx}$$

# (E) PROOF OF QUOTIENT RULE FOR DIFFERENTIATION $y = \frac{u}{v}$

$$y = \frac{u}{v}$$

where

$$y = f(x), \; u = g(x), \; \text{and} \; v = h(x)$$

$\delta x$ in $x$ will produce a corresponding $\delta y$, $\delta u$, and $\delta v$ in $y$, $u$, and $v$, respectively. Therefore,

$$y + \delta y = f(x + \delta x)$$
$$= \frac{u + \delta u}{v + \delta v}$$
$$\delta y = \left[\frac{u + \delta u}{v + \delta v}\right] - \left[\frac{u}{v}\right]$$
$$= \frac{v(u + \delta u) - u(v + \delta v)}{v(v + \delta v)}$$
$$= \frac{vu + v\delta u - uv - u\delta v}{v(v + \delta v)}$$
$$= \frac{v\delta u - u\delta v}{v(v + \delta v)}$$

Now we can evaluate the slope as

$$\text{Slope} \approx \frac{\delta y}{\delta x}$$
$$\approx \frac{\left[\frac{v\delta u - u\delta v}{v(v+\delta v)}\right]}{\delta x}$$
$$\approx \frac{v\frac{\delta u}{\delta x} - u\frac{\delta v}{\delta x}}{v(v + \delta v)}$$

Thus, the derivative is

$$\frac{dy}{dx} = \lim_{\delta x \to 0} \left(\frac{\delta y}{\delta x}\right)$$
$$= \lim_{\delta x \to 0} \left(\frac{v\frac{\delta u}{\delta x} - u\frac{\delta v}{\delta x}}{v(v + \delta v)}\right)$$

As $\delta x \to 0$, $\frac{\delta v}{\delta x} \to \frac{dv}{dx}$ and $\frac{\delta u}{\delta x} \to \frac{du}{dx}$, hence

$$\frac{dy}{dx} = \frac{v\frac{du}{dx} - u\frac{dv}{dx}}{v(v + 0)}$$

$$\therefore \frac{dy}{dx} = \frac{d}{dx}\left(\frac{u}{v}\right) = \frac{v\frac{du}{dx} - u\frac{dv}{dx}}{v^2}$$

## (F) PROOF OF THE DERIVATIVE OF tan $x$

$$y = \tan x = \frac{\sin x}{\cos x}$$

Let $u = \sin x$ and $v = \cos x$, therefore

$$u' = \cos x$$
$$v' = -\sin x$$

Using the quotient rule, we have that

$$y' = \frac{vu' - uv'}{v^2}$$
$$= \frac{(\cos x).(\cos x) - (\sin x).(-\sin x)}{(\cos x)^2}$$
$$= \frac{\cos^2 x + \sin^2 x}{\cos^2 x}$$
$$= \frac{1}{\cos^2 x} = \left(\frac{1}{\cos x}\right)^2 = (\sec x)^2$$

$$\therefore \frac{d}{dx}(\tan x) = \sec^2 x$$

## (G) PROOF OF THE DERIVATIVE OF sec $x$

**Method 1** Quotient Rule

$$y = \sec x = \frac{1}{\cos x}$$

Let $u = 1$ and $v = \cos x$, therefore

$$u' = 0$$
$$v' = -\sin x$$

Using the Quotient rule, we have that

$$y' = \frac{vu' - uv'}{v^2} = \frac{-uv'}{v^2}$$
$$= \frac{-(1).(-\sin x)}{(\cos x)^2}$$
$$= \frac{\sin x}{\cos^2 x} = \frac{\sin x}{\cos x . \cos x}$$
$$= \left(\frac{1}{\cos x}\right)\left(\frac{\sin x}{\cos x}\right) = (\sec x)(\tan x)$$

$$\therefore \frac{d}{dx}(\sec x) = \sec x . \tan x$$

# Appendix

**Method 2** Quotient Rule

$$y = \sec x = \frac{1}{\cos x} = (\cos x)^{-1}$$

If $u = \cos x$ then $y = u^{-1}$, therefore

$$\frac{du}{dx} = -\sin x$$
$$\frac{dy}{du} = -u^{-2} = -(\cos x)^{-2}$$

Using the chain rule, we have that

$$\frac{dy}{dx} = \frac{dy}{du} \times \frac{du}{dx}$$
$$= -(\cos x)^{-2} \times (-\sin x)$$
$$= \frac{\sin x}{\cos^2 x} = \frac{\sin x}{\cos x . \cos x}$$
$$= \left(\frac{1}{\cos x}\right)\left(\frac{\sin x}{\cos x}\right) = (\sec x)(\tan x)$$
$$\therefore \frac{d}{dx}(\sec x) = \sec x . \tan x$$

## (H) PROOF OF THE DERIVATIVE OF cosec x

**Method 1** Quotient Rule

$$y = \operatorname{cosec} x = \frac{1}{\sin x}$$

Let $u = 1$ and $v = \sin x$, therefore

$$u' = 0$$
$$v' = \cos x$$

Using the quotient rule, we have that

$$y' = \frac{vu' - uv'}{v^2} = \frac{-uv'}{v^2}$$
$$= \frac{-(1).(\cos x)}{(\sin x)^2}$$
$$= -\frac{\cos x}{\sin^2 x} = -\frac{\cos x}{\sin x . \sin x}$$
$$= -\left(\frac{1}{\sin x}\right)\left(\frac{\cos x}{\sin x}\right) = -(\operatorname{cosec} x)(\cot x)$$
$$\therefore \frac{d}{dx}(\operatorname{cosec} x) = -\operatorname{cosec} x . \cot x$$

## Method 2

$$y = \operatorname{cosec} x = \frac{1}{\sin x} = (\sin x)^{-1}$$

If $u = \sin x$, then $y = u^{-1}$, therefore

$$\frac{du}{dx} = \cos x$$

$$\frac{dy}{du} = -u^{-2} = -(\sin x)^{-2}$$

Using the chain rule, we have that

$$\frac{dy}{dx} = \frac{dy}{du} \times \frac{du}{dx}$$

$$= -(\sin x)^{-2} \times (\cos x)$$

$$= -\frac{\cos x}{\sin^2 x} = -\frac{\cos x}{\sin x . \sin x}$$

$$= -\left(\frac{1}{\sin x}\right)\left(\frac{\cos x}{\sin x}\right) = -(\operatorname{cosec} x)(\cot x)$$

$$\therefore \frac{d}{dx}(\operatorname{cosec} x) = -\operatorname{cosec} x . \cot x$$

## (l) PROOF OF THE DERIVATIVE OF cot $x$

$$y = \cot x = \frac{\cos x}{\sin x}$$

Let $u = \cos x$ and $v = \sin x$, therefore

$$u' = -\sin x$$
$$v' = \cos x$$

Using the quotient rule, we have that

$$y' = \frac{vu' - uv'}{v^2}$$

$$= \frac{(\sin x).(-\sin x) - (\cos x).(\cos x)}{(\sin x)^2}$$

$$= \frac{-\sin^2 x - \cos^2 x}{\sin^2 x} = \frac{-(\sin^2 x + \cos^2 x)}{\sin^2 x}$$

$$= \frac{-1}{\sin^2 x} = -\left(\frac{1}{\sin x}\right)^2 = -(\operatorname{cosec} x)^2$$

$$\therefore \frac{d}{dx}(\cot x) = -\operatorname{cosec}^2 x$$

# Appendix

## (J) PROOF OF THE DERIVATIVE OF ln $x$

$$y = \ln x$$
$$\therefore x = e^y$$

Differentiate both sides wrt $x$, we have

$$\frac{d}{dx}x = \frac{d}{dx}e^y$$

Using implicit differentiation, we have

$$1 = e^y \frac{dy}{dy}$$

Re-arrange this

$$\frac{dy}{dy} = \frac{1}{e^y}$$

By substituting $x = e^y$

$$\therefore \frac{dy}{dx} = \frac{1}{x}$$

## (K) PROOF OF THE DERIVATIVE OF $a^x$

**Method 1**

$$y = a^x$$

Let $a = e^k$, therefore

$$\therefore y = \left(e^k\right)^x = e^{kx}$$

Using the chain rule, we have

$$\frac{dy}{dx} = k.e^{kx} = k.\left(e^k\right)^x$$

But given that $a = e^k$ implies $k = \ln a$. We can substitute this to obtain

$$\frac{dy}{dx} = \ln a . a^x$$

$$\therefore \frac{dy}{dx} = a^x . \ln a$$

## Method 2

$$y = a^x$$

Taking the natural log of both sides, we have

$$\ln y = \ln a^x$$

Using the power law of logarithm

$$\ln y = x \ln a$$

Differentiating wrt $x$ and applying implicit differentiation, we have

$$\frac{1}{y} \cdot \frac{dy}{dx} = \ln a$$

$$\frac{dy}{dx} = y \ln a$$

$$= a^x \ln a$$

$$\therefore \frac{dy}{dx} = a^x \cdot \ln a$$

## (L) PROOF OF THE INVERSE DERIVATIVE

### Method 1

Given a function

$$y = f(x)$$

The inverse of this is

$$x = f^{-1}(y)$$

Let $f^{-1}(y) = g(y)$, thus

$$x = g(y)$$

Differentiating this wrt $x$

$$\frac{d}{dx}[x] = \frac{d}{dx}[g(y)]$$

Let's modify the RHS a bit as we cannot differentiate wrt $x$ as

$$\frac{d}{dx}[x] = \frac{d}{dx}[g(y)] \frac{dy}{dy}$$

Differentiate wrt $y$, we have

$$1 = \frac{d}{dy}[g(y)] \frac{dy}{dx}$$

Appendix

$$1 = \frac{d}{dy}[x]\frac{dy}{dx}$$

$$\therefore \frac{dy}{dx} = \frac{1}{\left(\frac{dx}{dy}\right)}$$

We can therefore say that

$$\frac{dy}{dx} = \frac{1}{\left(\frac{dx}{dy}\right)}$$

---

**Method 2**

Let

$$y = x$$

Differentiate wrt $x$, we have

$$\frac{dy}{dx} = 1$$

Differentiate wrt $y$, we have

$$\frac{dx}{dy} = 1$$

$$1 = \frac{1}{\left(\frac{dx}{dy}\right)}$$

We can therefore say

$$\therefore \frac{dy}{dx} = \frac{1}{\left(\frac{dx}{dy}\right)}$$

## (M) PROOF OF INTEGRATION BY PARTS FORMULA

$$y = uv$$

where

$$y = f(x), \ u = g(x), \ \text{and} \ v = h(x)$$

We have

$$\frac{dy}{dx} = u\frac{dv}{dx} + v\frac{du}{dx}$$

Let's take the integral of both sides wrt $x$

$$\int \frac{dy}{dx}dx = \int u\frac{dv}{dx}dx + \int v\frac{du}{dx}dx$$

$$y = \int u\frac{dv}{dx}dx + \int v\frac{du}{dx}dx$$

$$uv = \int u\frac{dv}{dx}dx + \int v\frac{du}{dx}dx$$

$$uv - \int v\frac{du}{dx}dx = \int u\frac{dv}{dx}dx$$

Therefore,

$$\int u\frac{dv}{dx}dx = uv - \int v\frac{du}{dx}dx$$

# Glossary

| S/N | Term | Explanation |
|---|---|---|
| 1. | Absolute value | The magnitude (or size) of a quantity without direction. In other words, its value without a positive or negative sign attached. It is indicated using two vertical straight lines. In this case, the absolute value of both **−5** and **+5** is **5**. Absolute value is also referred to as modulus. |
| 2. | Acceleration | The time rate of change of velocity; it is a vector quantity. If $v = f(t)$, where $v$ and $t$ are velocity and time, respectively, acceleration ($a$) is the differential coefficient of $v$ wrt $t$. |
| 3. | Acute angle | An angle that is less than **90°** but more than **0** such that $0 < \theta < 90°$. |
| 4. | Acute-angled triangle | A triangle in which all the three angles are acute. |
| 5. | Adjacent (adj) | Linguistically, it means 'next to' or 'close to'. In a triangle, it refers to the side that is close to the angle being referred to. |
| 6. | Algebraic expression | A mathematical statement expressed using a letter(s) and number(s), and it is made up of one or more terms. |
| 7. | Amplitude | The maximum displacement (or vertical distance) from the centre (or equilibrium position), and is used for varying quantities, e.g., alternating current. |
| 8. | Angle | The amount of relative turn, commonly measured in degrees and radians. |
| 9. | Anti-clockwise (ACW) | A turn in the direction opposite to the clock hands. It is also known as counter-clockwise (CCW). |
| 10. | Arccos | The inverse of the cosine function, and is denoted as **arccos**$(x)$ or $\cos^{-1}x$, where $-1 \leq x \leq 1$. The command **acos** is used to compute arccos in MATLAB and Excel and the angle is given in radians. |
| 11. | Arcsin | The inverse of the sine function and is denoted as **arcsin**$(x)$ or $\sin^{-1}x$, where $-1 \leq x \leq 1$. The command **asin** is used to compute arcsin in MATLAB and Excel and the angle is given in radians. |

| | | |
|---|---|---|
| 12. | Arctan | The inverse of the tangent function, and is denoted as **arctan**$(x)$ or $\tan^{-1}x$, where $-\infty <= x <= \infty$. The command **atan** is used to compute arctan in MATLAB and Excel and the angle is given in radians. |
| 13. | Argand diagram | A geometrical plot of complex numbers on a $x-y$ cartesian plane, also known as the complex plane. The $x$-axis (horizontal axis) represents the real part, and the $y$-axis (vertical axis) represents the imaginary part of complex numbers; they are called real axis and imaginary axis, respectively. |
| 14. | Argument | This is the angle between a phasor and the positive $x$-axis. |
| 15. | Asymptotes | A situation where a line (or curve) approaches another line (or curve) but never touches it. In this context, the line (or curve) is said to be asymptotic to the line (or curve) it cannot touch. |
| 16. | Average | Strictly speaking, it is one of the parameters used to measure the location of a data set. When used in language, it is synonymous with arithmetic mean or simply mean. Refer to mean. |
| 17. | Binomial expression | 'Bi' means two, and 'binomial expression' refers to an algebraic expression with two terms. |
| 18. | Bisector | A line that divides another line or angle into two equal parts. |
| 19. | Calculus | A branch of mathematics with two distinctive parts, namely differentiation and integration. |
| 20. | Cartesian coordinate system | This is a system that is used to describe the position of a point on a plane or in a space, and it is measured from a reference point or axis. The reference point is given as $(0, 0)$ for a plane or $(0, 0, 0)$ for a space. The axes are $x$-axis, $y$-axis, and $z$-axis, and they meet at $90°$ at the reference point. |
| 21. | Cartesian coordinates | The coordinates $(x, y)$ and $(x, y, z)$ describe the position of a point in terms of its distance from the origin $(0, 0)$ and $(0, 0, 0)$, respectively. The order of stating the coordinates is to write $x$-coordinate, then $y$-coordinate, and finally the $z$-coordinate (if it's a point in space). |
| 22. | CAST diagram | A diagram which divides the $x - y$ plane into four equal parts, called quadrants. |
| 23. | Chain rule (calculus) | It is used in differentiation to solve multiple or cascaded processes. It is also called 'function of a function' rule. |
| 24. | Clockwise | Turning in the direction of the clock hands. |

| 25. | Collinear | Points are said to be collinear if they all lie on the same straight line. |
|---|---|---|
| 26. | Column matrix | This is a matrix with a single (or one) column. The general dimension is $n \times 1$ where $n$ is the number of rows. |
| 27. | Combination | Any arrangement of items where the order is irrelevant. |
| 28. | Complementary angles | These are two angles that add up to make $90°$, and each of the angles is said to be the complement of the other. |
| 29. | Complex conjugate | Given a complex number $z = x + yj$, its complex conjugate is denoted as $\bar{z}$ such that $\bar{z} = x - yj$. Note that a complex conjugate pair only differs in the sign between the real and imaginary parts. |
| 30. | Complex numbers | A number that is made up of two parts, the real and the imaginary. This is expressed as $z = x + jy$, where $x$ is the real part (or component) of the complex number ($z$), abbreviated as $Re\ (z)$, and the imaginary part is $y$, shortened as $Im\ (z)$. |
| 31. | Coplanar | Points are said to be coplanar if they all lie on the same plane. |
| 32. | De Moivre's theorem | It is used to find the roots and powers of complex numbers. |
| 33. | Decreasing function | A function $y = g(x)$ is said to be decreasing, often in a particular interval, if $y$ increases as $x$ decreases and vice versa. |
| 34. | Definite integration | It is used to determine the area under a curve between two specified limits and the area bounded by two curves, linear or non-linear. |
| 35. | Diagonal matrix | This is a type of square matrix where all elements are zeros except those on the diagonal line, generally called the leading diagonal. The diagonal consists of all elements with $a_{ii}$, starting with $a_{11}$ on the top-left to $a_{ii}$ on the bottom right. |
| 36. | Differentiation | A measure of 'rate of change'. In other words, it is a technique that is used to determine the size of a change in one physical quantity (e.g., volume, distance, mass) when compared to an equivalent change in another physical quantity (e.g., time). |

| | | |
|---|---|---|
| 37. | Dividend | The number (or expression) being divided by another number (or expression) is called the dividend. It is the number (or expression) above the division line or to the left of the division sign ('\'). In a fraction, it corresponds to the numerator. |
| 38. | Divisible | A number $X$ is said to be divisible by another number $Y$ if the division process results in a whole number (without a remainder), with a pre-condition that $X > Y$. |
| 39. | Divisor | The number (or expression) that divides another number is called the divisor. It is the number (or expression) below the division line or to the right of the division sign ('\'). In a fraction, it corresponds to the denominator. |
| 40. | Element | It is used in set theory to refer to the members of a set or in a matrix to denote the members of the matrix. |
| 41. | Empty (or null) set | A set (or collection of something) without a single member. |
| 42. | Exponent | See power. |
| 43. | Factorial | It is denoted by an exclamation mark (!). It technically implies the product of all the positive consecutive integers between a particular given integer and **1**. Factorials are generally computed for (i) positive numbers and (ii) integers. |
| 44. | General solution (calculus) | An expression obtained when a function is integrated with a constant of integration included. |
| 45. | Gradient | See 'slope'. |
| 46. | Higher derivatives | These imply finding the derivative of a derivative and of another in a consecutive manner. The higher you go, the higher the number associated with the derivative. |
| 47. | Implicit differentiation | The process of finding the derivatives of an implicit function. |
| 48. | Increasing function | A function $y = f(x)$ is said to be increasing, often in a particular interval, if $y$ increases as $x$ increases and vice versa. |
| 49. | Indefinite integration | It is represented by $\int y dx$ and is read as 'the integral of $y$ with respect to $x$'. The '$dx$' must be included otherwise the integration is meaningless. In this case, $x$ is the independent variable. If the independent variable is time ($t$), we will have $dt$ and if it is volume ($V$), we will have $dV$ and so on. |

| | | |
|---|---|---|
| 50. | Infinitesimal | Very small and insignificant that can be assumed to be equal to zero (**0**). |
| 51. | Integration | A direct inverse of differentiation (or its anti-derivative). |
| 52. | Jerk | It is also known as Jolt and is the time rate of change of acceleration. If $a = f(t)$, where $a$ and $t$ are acceleration and time, respectively, jerk ($j$) is the differential coefficient of $a$ wrt $t$. |
| 53. | Limit | This refers to the boundary of a function. |
| 54. | Line of symmetry | It is also called a 'mirror line'. It is a line that divides (or can be used to divide) a plane into two equal and identical parts. There could be more than one line of symmetry for a plane or 2D object. For example, a square has four lines of symmetry, and a parallelogram has none. |
| 55. | Line segment | This is the part of a line between two given points. |
| 56. | MATLAB | It stands for MATrix LABoratory. It is a programming software that is widely used by scientists and engineers to design and analyse systems. |
| 57. | Matrix | A matrix (plural: matrices) is defined as an array of numbers (or elements), which consists of rows and columns enclosed in square brackets [ ] or parentheses ( ). |
| 58. | Mean | In statistics, it is one of the measures of location; others include median and mode. It is computed by adding all the data and dividing the result by the number of the data elements. Mathematically represented as $\frac{\sum x}{n} = \frac{x_1 + x_2 + x_3 + \cdots + x_n}{n}$. |
| 59. | Mixed number | A number that is made up of a whole number and fraction, e.g., $2\frac{1}{3}$. |
| 60. | Modulus | See 'absolute'. |
| 61. | Order of rotational symmetry | It refers to the number of times a shape fits exactly into its original form when it is rotated through a revolution or 360 degrees. |
| 62. | Particular solution | In integration, it refers to an expression obtained, having replaced $C$ with a numerical value. |
| 63. | Pascal's triangle | A pattern formed using the coefficients of the terms obtained when binomial expressions of different orders are simplified. |

| | | |
|---|---|---|
| 64. | Peak | It is the highest or maximum point on a curve. It is also called crest. |
| 65. | Periodicity | The property of an object or system being periodic, i.e., repeating in its pattern at a fixed interval or rate. |
| 66. | Permutation | The number of ways objects can be arranged (or selected) such that the order is paramount. |
| 67. | Phasor diagram | A graphical representation of voltages and currents in an AC circuit, it is similar to an Argand diagram. |
| 68. | Point of inflexion | A point on a curve where there is a change in the magnitude of the gradient without a change in its sign, i.e., the curve changes from being concave to being convex to a particular direction or vice versa. |
| 69. | Positive angle | The angle that is measured in an anti-clockwise direction. |
| 70. | Power | Power (or index or exponent) is the superscript of a number or term called the base. Power is a shorthand way of saying that a number is being multiplied by itself a number of times dictated by the value of the power. |
| 71. | Product rule (calculus) | It is used in differentiation when one function is multiplied by another function wrt the same variable. |
| 72. | Quadrant | A name given to each of the four sections formed in a plane by the $x$-axis and the $y$-axis. |
| 73. | Quotient | The result obtained when a number (or expression) divides another number (or expression). |
| 74. | Quotient rule (calculus) | It is used in differentiation when one function is divided by another function, where it is either impossible to simplify it into a single function or practically difficult to do so. |
| 75. | Remainder theorem | It states that if a polynomial $f(x)$ is divided by $x - a$, the remainder is $f(a)$. |
| 76. | Row matrix | This is a matrix with a single (or one) row. It is also called a line matrix and is the inverse of the column matrix. The general dimension is $1 \times n$ where $n$ is the number of columns. |
| 77. | Scalar (or scalar quantity) | A quantity that is completely identified or described by its magnitude (amount or size) and associated units only. Examples include time, speed, height, temperature, pressure, and power. Scalar (without quantity) refers to a real number. |

| | | |
|---|---|---|
| 78. | Scalene triangle | A triangle with no two equal sides. In other words, each of the sides is distinct in length and their angles are different. A scalene triangle has no line of symmetry. |
| 79. | Set | A collection of items or members. |
| 80. | Single (element) matrix | This is a matrix of a dimension $1 \times 1$, i.e., it has one element, one row and one column. |
| 81. | Singular matrix | A matrix whose determinant is zero. |
| 82. | Skew lines | In space, it is when lines are neither parallel nor do they intersect. |
| 83. | Skew-symmetric (matrix) | A square matrix that is equal to its transpose multiplied by $k = -1$. |
| 84. | Slope | It is also called gradient, denoted by $m$, and is a measure of the steepness of a line. The more the value of $m$, the greater the steepness or tendency to fall and thus the slope. Slope can be positive or negative. A positive slope ramps up from left to right, and a negative slope ramps down from left to right. |
| 85. | Square matrix | This is a matrix where the number of rows is the same as that of the columns. |
| 86. | Symmetry | A property which means that an object can be divided along a line or plane into two identical parts (reflection symmetry or reflective symmetry) or turned about a fixed point or line and remains unchanged (rotation symmetry or rotational symmetry). A shape that does not possess this property is said to be asymmetric. |
| 87. | Tangent | A straight line $l$ drawn tangential to the curve at a given point $A$. It is also the ratio of the sine of a given angle to the cosine of the same angle. |
| 88. | Term (or algebraic term) | The simplest algebraic unit, like cells in biology, atoms in chemistry, bits in computing, etc. It may consist of a constant (number) or constants, a variable or variables, or a combination of constant(s) and variable(s). Terms are usually separated by plus ($+$) or minus ($-$) sign. |
| 89. | Transformation | The change in position, orientation, size, or shape of an object. |
| 90. | Transpose matrix | A matrix formed when the rows have been converted to respective columns and vice versa (i.e., rows and columns are interchanged). |

| | | |
|---|---|---|
| 91. | Trapezium | Also called trapezoid, it is a quadrilateral with a pair of parallel sides and one or two sloping sides. If the sloping sides are the same in length, the resulting trapezium is called an isosceles trapezium. |
| 92. | Trigonometric equation | An equation (linear or non-linear) involving trigonometric ratios. |
| 93. | Trigonometric ratios | It refers to the sine, cosine, and tangent of an angle. They are usually shortened as **sin** $\theta$, **cos** $\theta$, and **tan** $\theta$, respectively, where $\theta$ represents the angle. |
| 94. | Trigonometry | A branch of mathematics concerned with the study of angles and their relationship in two-dimensional (plane) and three-dimensional (solid) objects. |
| 95. | Trough | The exact opposite of a crest; it represents the minimum point on a curve. |
| 96. | Unit (or identity) matrix | A diagonal matrix in which the elements along the diagonal line are ones (**1**). |
| 97. | Unit vector | A vector whose magnitude is unity. In general, a unit vector in the direction of $a$ is given as $\hat{a} = \frac{a}{|a|}$. |
| 98. | Variable | A quantity, represented by a letter or a symbol, that we cannot assign a fixed numerical value to and generally represent quantities (time, amount, force, height, etc.) that change in values. There are two types of variables: dependent and independent variables. It is also called the 'unknown'. |
| 99. | Vector quantity | (Or simply a vector) is a quantity which is completely identified by both its magnitude and direction (and associated units). In other words, it is a scalar quantity that has a direction. Examples include velocity, displacement, force, electric current, stress, and momentum. |
| 100. | Velocity | The time rate of change of position; it is a vector quantity. If $s = f(t)$, where $s$ and $t$ are position and time, respectively, velocity ($v$) is the differential coefficient of $s$ wrt $t$. |
| 101. | $x$-coordinate | The distance of the point from the origin along the $x$-axis, which is the first number given in a coordinate system notation. |
| 102. | $y$-coordinate | The distance of the point from the origin along the $y$-axis, which is the second number given in a coordinate system notation. |

# Glossary

| | | |
|---|---|---|
| 103. | *z*-coordinate | The distance of the point from the origin along the *z*-axis, which is the third number given in a coordinate system notation. |
| 104. | Zero or null matrix | This is a matrix in which all the elements are zeros; it is denoted by **0**. |
| 105. | Zero vector | It is also called a null vector, and it is a vector whose magnitude is zero. |

# Index

## A

Absolute, 751, 766 *see also* Magnitude
Acceleration, 505, 565–6, 582, 648, 679
Accuracy, 466, 751, 757
Addition, 86, 96, 117, 128, 178, 281, 288, 357, 406, 421, 423
Adj *see* Adjacent
Adjacent, 2, 205, 385
Advanced calculus, 511, 681, 717, 751
    approximate, 461, 751, 766
Angle, 1, 7, 26
    acute, 87, 129, 430, 454
    angular frequency, 595, 602
    obtuse, 87, 430, 454
Angular frequency, 319, 595, 602
Application, 1, 203, 239, 341, 408, 461
Approximation, 234, 751
Arg *see* Complex numbers, argument
ASTAC (All–sine–tangent–cosine), 23
Asymptotes, 40, 46, 603
Author's method, 138, 262, 274

## B

Behaviour, 1, 37, 47, 602, 610
Binomial, 203–4, 209, 211, 215–17, 220, 222–3, 227–9, 234, 236–7, 239, 336, 473, 477, 516, 683
    approximate, 203, 234
    array, 205, 207
    combination, 214–15
    expansion, 203, 215–16
    permutation, 215
Brackets, 351, 361, 405, 514
    curvy, 352
Branch, 1, 410, 461

## C

CAH (cosine–adjacent–hypotenuse), 3, 44
Calculus, 117, 461–2, 613, 648, 751
    cascade, 512–13, 522
    definite, 613, 647, 652, 681
    exponential, 461, 493, 634, 722
    gradient, 461, 463, 465, 566, 579, 583, 597
    higher-order, 378, 384, 461, 503
    indefinite, 613, 681, 717, 735
    nature, 242, 598
    parameter, 554, 717

Capacitance, 318–19
    capacitive, 318–19, 326
    capacitors, 319
Cartesian (representation, form, equation), 297, 304, 309, 311, 347, 413, 418, 751
Cast, 23, 152, 155
CCW (counter-clockwise) *see* Counter-clockwise
Circuits, 128, 280, 318, 410
Clockwise, 18, 20, 29, 153
Commutative, 218, 357, 365, 406–7, 422
Complex numbers
    alternating, 280, 318
    argand, 304, 311, 318
    argument, 298, 310
    associative, 291
    conjugates, 280, 290, 292, 309
    conversion, 304–5
    coplanar, 343
    equidistant, 331
    exponential, 297, 303
    imaginary, 287, 291–2, 313
    locus, 344–5
    logarithms, 311
    loss, 322, 327
    Moivre's, 311, 336, 338, 348
    quadrature, 288
Complimentary, 7, 9, 54
Components, 422, 443
Concave, 597, 602, 754
Consistency, 390, 449
Convex, 597, 602, 754
Counterbalance, 422
Counter-clockwise, 17, 153, 281, 309, 344
Crest, 37, 586
Currents, 318, 348, 410

## D

Decelerate, 581, 648
Dependent variable, 35, 512, 543
Derivative, 461, 471, 514, 548, 565–6, 614
Determinant, 351, 353, 356, 372, 376, 378–9, 384
Differential *see* Differentiation
Differentiation, 461, 487, 503, 543, 565, 611, 613
    analytical method, 470, 584
    boundary, 462, 508
    chain rule, 511–12, 516
    dee, 461, 472, 503, 508
    delta, 322, 463, 508
    explicit, 543–4

implicit, 511, 543, 546
increment, 471
linearity, 487, 623
lowercase, 463
maxima, 584–5
minima, 584–5
normal, 565–6
Dividend, 272, 536, 690
Divisor, 272, 690

# E

Elimination method, 315, 351, 389
Experience-based approach, 137

# F

Factorials, 203, 212
Force, 343, 408, 411
Foundation (mathematics), 272, 589
Fundamentals, 80, 173, 288, 498, 565

# G

Geometry, 347, 408
Global, 584

# H

Hyp *see* Hypotenuse
Hypotenuse, 2–3, 20

# I

Identity, 3, 101, 117, 242, 355
Impedance, 318, 326
Inductance, 318–19, 348
    inductive, 318–20
    inductors, 319
Infinite, 221–2, 462, 465
Infinitesimal, 463, 471
Infinity, 48, 647, 677
Inflection *see* Integration, inflexion
Integration, 613–15
    arbitrary, 614, 616
    area, 614, 647, 649, 652, 665, 670
    boundary, 677
    bounded, 647, 658, 670, 673–4
    decomposition, 661, 667
    displacement, 648
    error, 751, 766, 772
    estimate, 751, 766
    inflexion, 584, 588, 597–9

    integral, 613–14, 616–17
    integrand, 613
    ordinates, 652, 754
    width, 661, 752
Intercepts, 59, 602
Intersection, 408, 449, 670
Inverse, 1, 10, 47, 56, 351, 353, 381, 384, 511, 554, 613

# J

Jerk, 505, 566, 648–9, 679
Jolt *see* Jerk

# L

Lag, 46, 62, 70, 128, 318
Limits, 462–3, 647
Local, 584
Logarithm, 340, 461, 496, 511, 637, 722

# M

Magnitude, 281, 298, 331, 408, 410, 429, 565, 597
Many-to-one, 56
MATLAB, 11
Matrices, 351
    adjoint, 351, 385, 389
    array, 351, 405
    associative, 365, 406
    cofactor, 385
    columns, 353, 356, 380, 405
    Cramer's rule, 390–1
    diagonal, 354, 373, 406
    diagrams, 311
    dimension, 352–3, 356, 365, 406
    element, 351, 353–4
    non-singular, 381
    non-square, 381
    null, 355, 458
Maximum, 37, 39–40, 48, 50, 58, 62, 117, 128, 150, 196, 318, 584
Mechanics, 648–9
Minimum, 37, 39–40, 48, 50, 58, 62, 117, 128, 150, 196, 318, 584
Minuend, 670–1, 766
Mirror, 56, 68–9, 153, 602, 752
Mnemonic, 3, 23
Mod *see* Modulus
Modulus, 138, 298, 303, 331, 413–14, 632, 766
Multiple (angles, functions), 30, 40, 96, 179, 311, 336, 429, 449, 511
Multiplication, 280, 289, 351, 356, 364, 429, 511

## N

Non-linear, 242, 256, 269, 465, 647
Notation, 214–15, 351, 405, 410, 418, 461, 613

## O

Ohm's, 326
One-to-one (function), 56–7, 78
Operations, 288, 351, 357, 365, 406, 408
Operator, 215, 280, 318, 472
Opp, 3–10, 15–17, 20–2, 24–5, 47, 54, 56, 79
Outputs, 35, 512
Overestimation, 753–5, 757

## P

Partial fractions, 222, 239, 242
    algebraic, 239, 241
    factors, 242, 249, 256, 269
    non-repeated, 149, 242, 256
Pascal's triangle, 203–4
    ascending, 204, 206
    descending, 204
    limitations, 203, 214
Per *see* Percentage
Percent *see* Percentage
Percentage, 381, 492, 543, 565–6, 615, 627, 691, 751, 766–8, 772
Periodicity, 37, 40, 48, 50–1, 70–1, 602
Phase, 39, 62, 128, 138, 196, 318
Phasor, 318
Polar form, 297–8, 303–4, 318
Polynomials, 149, 256, 272, 623
Power, 311, 329, 336, 381, 461, 478, 613, 616, 634, 698, 738
Pythagoras' theorem, 2, 7, 414

## Q

Quadrants, 1, 18, 22–3
Quadruple, 503, 666
Quotient, 511, 533

## R

Rad *see* Radian
Radian, 303–4, 464
Reactance, 318–19
Reciprocal, 47–8, 50, 357
Reflection, 68–9, 205
Remainder theorem, 272, 282
Representation, 318, 410, 413
Resistance, 318
Resistors, 138, 318
Resultant, 411, 422–3
Revolution, 17, 23, 35, 751
Root, 287, 331, 414, 602, 751

## S

Scalar, 412, 414, 429–30, 443, 449
Simpson's rule, 751, 759
Sinusoidal waveform, 1, 595
Skew, 449, 454
Slope, 463, 465–6, 470–1
Standards, 613, 616, 632, 639, 648
Stationary, 565, 583, 585–6, 588, 597
Substitution, 243, 704, 717
Subtraction, 86–7, 117, 178, 280, 288, 304, 351, 421, 423
Subtrahend, 413, 670–1, 766
Surd, 290, 292
Symmetrical, 7, 39, 46, 175, 205, 602–3
Symmetry, 37, 46, 48, 150, 161, 176–7, 184, 188, 195, 199, 205, 602–3, 745, 753
    symmetric, 356, 406, 653
    symmetrically, 331, 349

## T

Tan *see* Tangent
Tangent, 1–4, 23, 40–1, 47, 49, 80, 565–6, 568–9, 584, 603, 637–8
Theorem, 203, 272, 311, 329, 336, 414
Transformation, 47, 57, 62, 159
Translation, 62, 165, 167, 196
Trapezia *see* Trapezium
Trapezium, 423, 613, 751–4, 759
Trapeziums *see* Trapezium
Trapezoidal *see* Trapezium
Trigonometry, 1, 10, 80
    amplitude, 37, 46, 128
    anticlockwise, 18, 23, 281
    discontinuities, 40, 46, 584
    double-angle, 96, 101
    equilateral, 7
    Euler-trigonometric, 337
    factor, 117
    half-angle, 80, 103
    peaks, 37
    phase difference, 39–40, 69
    right-angled, 1–2, 7, 23
    triple-angle, 80, 96, 101, 116, 180, 188

## U

Underestimation, 753, 766, 772
Unity (matrix, vector), 400, 413

# Index

## V

Validity, 121, 203, 223, 227, 637
    valid, 58, 84, 87, 115, 129, 137, 175, 183, 189, 204, 221, 223–7, 229–31, 233, 237, 304, 310, 329, 339, 348, 352, 355, 365, 381, 390, 411, 478, 509, 600, 632, 634, 637, 719, 722
    validate, 451, 454
Vector, 280–1, 408–9, 411
    analytically, 421
    anti-clockwise, 422
    anti-phase, 410, 458
    collinear, 410, 458
    displacement, 408, 565, 648
    distance, 408, 422
    dot, 408, 430, 459
    equality, 411
    hat, 413
    lowercase, 410
    momentum, 408
    mutually, 413
    negation, 410
    non-zero, 411, 429
    orthogonal, 431
    overbar, 410
    parallelogram, 343, 422
    polygon, 422
    wavy, 410
Velocity, 408, 457, 461, 565, 648
Velocity–time, 648
Vertex, 2, 7, 605
Voltage, 318, 322, 324, 595
Volume, 408, 461, 613, 751

## W

Waveform, 1, 128

## X

$x$-axis
    $x$-coordinate, 63, 588, 597, 670
    $x$-value, 602, 754
$x$–$y$ plane, 62, 201, 311, 414, 554

## Y

$y$-axis
    $y$-coordinate, 63, 463
    $y$-intercept, 608
    $y$-prime, 472

## Z

$z$-axis, 413–14
$z$-directions, 408
Zero, 354–5, 410, 458, 597